Progress in Polymer Research for Biomedical, Energy and Specialty Applications

With the rapid advancements in polymer research, polymers are finding newer applications such as scaffolds for tissue engineering, wound healing, flexible displays, and energy devices. In the same spirit, this book covers the key features of recent advancements in polymeric materials and their specialty applications. Divided into two sections – Polymeric Biomaterials and Polymers from Sustainable Resources, and Polymers for Energy and Specialty Applications – this book covers biopolymers, polymer-based biomaterials, polymer-based nanohybrids, polymer nanocomposites, polymer-supported regenerative medicines, and advanced polymer device fabrication techniques.

FEATURES

- Provides a comprehensive review of all different polymers for applications in tissue engineering, biomedical implants, energy storage or conversion, and so forth
- Discusses advanced strategies in development of scaffolds for tissue engineering
- Elaborates various advanced fabrication techniques for polymeric devices
- Explores the nuances in polymer-based batteries and energy harvesting
- Reviews advanced polymeric membranes for fuel cells and polymers for printed electronics applications
- Throws light on some new polymers and polymer nanocomposites for optoelectronics, next generation tires, smart sensors and stealth technology applications

This book is aimed at academic researchers, industry personnel, and graduate students in the interdisciplinary fields of polymer and materials technology, composite engineering, biomedical engineering, applied chemistry, chemical engineering, and advanced polymer manufacturing.

Progress in Polymer Research for Biomedical, Energy and Specialty Applications

Edited by
Anandhan Srinivasan
Selvakumar Murugesan
Arunjunai Raj Mahendran

CRC Press
Taylor & Francis Group
Boca Raton London New York

CRC Press is an imprint of the
Taylor & Francis Group, an **informa** business

First edition published 2023
by CRC Press
6000 Broken Sound Parkway NW, Suite 300, Boca Raton, FL 33487-2742

and by CRC Press
4 Park Square, Milton Park, Abingdon, Oxon, OX14 4RN

CRC Press is an imprint of Taylor & Francis Group, LLC

© 2023 Anandhan Srinivasan, Selvakumar Murugesan and Arunjunai Raj Mahendran

ISBN: 978-1-032-06100-9 (hbk)
ISBN: 978-1-032-06102-3 (pbk)
ISBN: 978-1-003-20071-0 (ebk)

DOI: 10.1201/9781003200710

Typeset in Times
by MPS Limited, Dehradun

Contents

Section I Polymeric Biomaterials and Polymers from Sustainable Resources

Section II Polymers for Energy and Specialty Applications

Preface

Polymers have played a pivotal role during the last century with a huge impact on human civilization. The contribution of polymers has been tremendous in enhancing the standard of living of the citizens worldwide. Polymers have been used for manufacturing a diverse range of products starting from clothes to medical implants such as heart valves. This is due to the fact that, polymers, in general, exhibit a unique combination of properties such as low temperature flexibility, dynamic mechanical strength, viscoelasticity, ease of processing, and biocompatibility. The properties of polymers can be engineered by using them in the form of either composites or blends. The idea of dispersing nanoparticles in polymer matrices to strengthen their physicochemical properties has become the state-of-the-art for various functional applications. Exemplarily, recent research has led to the use of polymers in stretchable electronics, especially in energy harvesting and semiconductor devices.

I congratulate the editors of this book for choosing a hot topic theme in the current research scenario. This book highlights some of the recent accomplishments of polymeric materials research in the field of energy and biomedical applications. **Section I** covers polymeric biomaterials and polymers from sustainable resources discussing recent advancements in the field of 3D bioprinting for tissue regeneration, bionanocomposites, recyclable thermosets from biomass-derived building blocks, tree/plant-based gum biopolymers, dissolution characteristics of cellulose biopolymer in ionic liquids, and structure-property relationship of hydrogels. **Section II** focuses on polymers for energy and specialty applications showcasing the latest developments in the areas of polymer electrolytes for rechargeable sodium-ion batteries, aromatic polymer-based proton exchange membranes, anion exchange membranes for fuel cells, polymers for printable electronic devices, smart polymer-based wearable sensors, optoelectronic polymers based on thiophenes, macromolecular gels for environmental remediation, next-generation polymers for tires, and polymer composites for stealth technology.

In my opinion, this book will be very useful for interdisciplinary researchers, industry personnel, as well as senior undergraduate students.

Prof. Dr. Thomas Scheibel
Lehrstuhl Biomaterialien, Universität Bayreuth, Germany

Acknowledgements

First and foremost, we would like to thank the Almighty God for keeping us safe and sound during the pandemic time and for providing us the strength to pursue this scholarly piece of work.

We sincerely thank Dr. Gagandeep Singh, Senior Publisher (Engineering), CRC Press, Taylor & Francis Books India Pvt Ltd., for inviting us to edit this book and making its publication a reality.

We express our deep sense of gratitude to the authors of the 16 chapters for their invaluable time and admirable efforts. We are very fortunate to have our authors from industry and academia from around the globe, who are well-experienced technologists and renowned scientists in their respective fields.

The editors acknowledge the assistance of Mr. Sawan Shetty, a research scholar working with Prof. Anandhan, for his assistance in the crucial stages of preparation of the manuscript. Our thanks are also due to the publishers of various journals for permitting us to reproduce copyrighted materials in this book.

Most importantly, we would like to thank our families for their unconditional love, patience, and support; we would like to dedicate this work to our families.

Anandhan Srinivasan and Selvakumar Murugesan
National Institute of Technology Karnataka, India
Arunjunai Raj Mahendran
Kompetenzzentrum Holz GmbH, Austria

Editors

Anandhan Srinivasan, PhD, FRSC, holds a BSc in chemistry (Madura College, Madurai Kamaraj University, India, 1995), an MSc in applied chemistry (Thiagarajar College of Engineering, Madurai Kamaraj University, India, 1997), an MTech in polymer technology (Cochin University of Science and Technology, India, 1999), and a PhD in polymer science and technology (Indian Institute of Technology Kharagpur, 2004). He was a gold medallist in both BSc (first rank among 240 students) and MSc.

He was invited as a visiting research associate by the School of Materials Science and Engineering, University of New South Wales, Australia in 2002.

He was a Postdoctoral Fellow and Lecturer at Inha University, Korea (2004–2005) in the erstwhile Textile Engineering Department (currently Chemical Engineering). Subsequently, he served as an assistant professor in the Department of Materials Science at AIMST University, Malaysia from 2005 to 2008. He joined NITK-Surathkal in 2009 as an assistant professor and rose in ranks as associate professor and professor in 2012 and 2018, respectively. He has been teaching courses in the areas of polymers, nanomaterials, materials characterization, and nanocomposites.

Under his mentorship, eight students have earned their doctoral degrees so far and four more have been working towards their PhD. He has also guided 36 master's research projects in various areas related to polymers and nanotechnology. He has published 85 full length articles in peer-reviewed international journals and his publications have so far received about 1,600 citations. He has also published a patent, three books, and 14 book chapters. His book titled *Advances in Polymer Materials and Technology* was published in 2016 by Taylor & Francis, CRC Press. He was the principal investigator of two sponsored research projects. He is currently a member of the teams that have secured highly competitive research grants from the Department of Science and Technology (DST), India and National Science Foundation, USA. His strenuous efforts have resulted in the establishment of various research facilities in his institute, such as the advanced nanofiber lab, piezoelectric materials lab, and polymers and nanomaterials lab. He has been a referee for more than 50 international journals of repute. He has refereed about 30 doctoral dissertations from India and abroad.

Dr. Srinivasan was admitted as a Fellow of the Royal Society of Chemistry, UK in September 2021 for his outstanding contribution to the advancement of the chemical sciences through his research on advanced functional nanofibers, energy materials, and polymer nanocomposites. He has won a number of awards and honors, including the Doctoral Research Fellowship from DST-India, Australia-India Council Fellowship, Postdoctoral Fellowship from Inha University-Korea, ARCNN-ARNAM Fellowship from UNSW-Australia, and the Fast Track Award for young scientists from DST-India. He was elected a Fellow of the Institution of Engineers (India) and Indian Chemical Society in 2017 and 2019, respectively.

Selvakumar Murugesan, PhD, is an assistant professor at the National Institute of Technology Karnataka (NITK) in the Department of Metallurgical and Materials Engineering. Prior to joining NITK, he was a Postdoctoral Fellow at the Chair of Biomaterials, University of Bayreuth, Germany, through the prestigious Alexander von Humboldt (AvH) Fellowship. Earlier, he had a couple of stints as a Postdoctoral Fellow at École Polytechnique de Montréal, Canada and Pohang University of Science and Technology, South Korea. He earned his doctoral degree from the Indian Institute of Technology Kharagpur in 2016. He has a master's degree in materials engineering from NITK and a bachelor's degree in polymer technology from Anna University. Dr. Murugesan won the Gandhian Young Technology Innovation award in 2016 for his doctoral research. He is an interdisciplinary researcher who focuses on functional polymers for biomedical applications. He has published 30 full-length articles in refereed journals and four book chapters. His research aims at understanding the mechanical and biological behaviour of biopolymers, copolymers, spider silk protein, nanostructured polymers, and their hybrids for regenerative medicine including 3D-printing approaches. Currently, he has been co-guiding two PhD students, and three master's students have completed their research projects under his supervision.

Arunjunai Raj Mahendran, PhD, received a bachelor's degree in polymer technology with university rank and gold medal from Madurai Kamaraj University, India. He earned a master's and PhD in polymer engineering and science at the Montanuniversität Leoben, Austria. He has been a project manager and key researcher at Kompetenzzentrum Holz GmbH, Austria since 2007. He works with researchers and scholars from different disciplines on a research team called Green Composites, which deals with renewable materials and composites. He has carried out several research and development projects with small- and medium-size industries funded by the Austrian Research Promoting Agency and also received the JEC Innovation Award, Innovation and Research Award from Carinthian Province (Austria). The notable projects he has worked on along with industry partners are small-wind turbine blades from natural fiber-reinforced composites and the development of bio-based thermoset matrix resins for automotive parts.

Dr. Mahendran has an external university teaching assignment at the University of Natural Resources and Life Sciences, Vienna, Austria. He has guided six master's research projects in various areas related to natural fibers and polymers. He mentored three doctoral students and his research interests include bio-based thermoset resins, biodegradable thermoplastics, cure kinetics of thermoset resins, isocyanate free polyurethane, renewable coatings, high temperature thermoset resins, polymer nanocomposites, and natural fiber-reinforced composites. He is the author of 60 full-length research papers in peer-reviewed international journals and his publications have so far received about 480 citations. He has authored five book chapters and presented his research works in 65 conference papers. He has been a referee for international journals and doctoral dissertation works.

Contributors

Venimadhav Adyam
Indian Institute of Technology Kharagpur
Kharagpur, West Bengal, India

Sri Bandyopadhyay
University of New South Wales
Sydney, Australia

Susanta Banerjee
Indian Institute of Technology Kharagpur
Kharagpur, India

P. M. Sabura Begum
Cochin University of Science and Technology
  (CUSAT)
Kerala, India

Sayantani Bhattacharya
Indian Institute of Science Education and
  Research Kolkata
Mohanpur, Nadia, West Bengal, India

Miroslav Černík
Technical University of Liberec (TUL)
Liberec, Czech Republic

Udaya Kumar D.
Department of Chemistry
National Institute of Technology Karnataka
  (NITK)
Surathkal, Mangaluru, India

Deepthi Anna David
Cochin University of Science and Technology
  (CUSAT)
Kerala, India

Shreyas Devanathan
University of New South Wales
Sydney, Australia

Sheila Devasahayam
Monash University
Victoria, Australia

Pritam V. Dhawale
Institute of Chemical Technology
Matunga, Mumbai, India

M. P. Drupitha
Aliaxis Research & Technology Center Asia
  Bangalore, India
and
University of Guelph
Guelph, Ontario, Canada

Jabeen M. J. Fatima
Cochin University of Science and Technology
  (CUSAT)
Kerala, India

Kausar Ahmad Fuad
Universiti Teknologi PETRONAS
Perak, Malaysia

Thomas George
Cochin University of Science and Technology
  (CUSAT)
  Kerala, India

Arijit Ghorai
Indian Institute of Technology Kharagpur
Kharagpur, India

C. Harimohan
Yokohama Off-Highway Tires
Research and Development
Tirunelveli, India

Joshua Raj Jaganathan
Universiti Teknologi PETRONAS
Perak, Malaysia

S. Janakiraman
Indian Institute of Technology Indore
Indore, Madhya Pradesh, India

Viprabha K.
National Institute of Technology Karnataka
  (NITK)
Surathkal, Mangaluru, India

Mohammed Khalifa
Competence Center for Wood Composites and
    Wood Chemistry
Linz, Austria
and
Komepetenzzentrum Holz GmbH
Wood K-plus, Linz, Austria

Akshay Kumar K. P.
Brno University of Technology
Brno, Czech Republic
and
Cochin University of Science and Technology
Kochi, India

Herfried Lammer
Competence Center for Wood Composites and
    Wood Chemistry
Linz, Austria

Arunjunai Raj Mahendran
Competence Center for Wood Composites and
    Wood Chemistry
Linz, Austria

Partheban Manoharan
Yokohama Off-Highway Tires
Research and Development
Tirunelveli, India

Suja Mathai
Mar Ivanios College
Thiruvananthapuram, Kerala, India

Manjusri Misra
University of Guelph, Guelph
Ontario, Canada

Amar Kumar Mohanty
University of Guelph, Guelph
Ontario, Canada

Titash Mondal
Indian Institute of Technology
Kharagpur, West Bengal, India

Selvakumar Murugesan
Universität Bayreuth
Bayreuth, Germany
and
National Institute of Technology Karnataka
Surathkal, Mangalore, Karnataka, India

Rajendran Muthuraj
University of Guelph, Guelph
Ontario, Canada
and
Worn Again Technologies Ltd
Nottingham, Nottinghamshire, United Kingdom

Vidhukrishnan Naiker
Institute of Chemical Technology
Matunga, Mumbai, India

Sang Yong Nam
Gyeongsang National University
Jinju, Republic of Korea

Vinod V. T. Padil
Technical University of Liberec (TUL)
Liberec, Czech Republic

Eswaran Padmanabhan
Universiti Teknologi PETRONAS
Perak, Malaysia

Prasanth Raghavan
Cochin University of Science and Technology
    (CUSAT)
Kerala, India
and
Institute of Chemical Technology
Matunga, Mumbai, India
Gyeongsang National University
Jinju, Republic of Korea

Thomas Scheibel
Universität Bayreuth
Bayreuth, Germany

C. Shamitha
Manipal Institute of Technology Bengaluru
Manipal Academy of Higher Education
    (MAHE)
Bengaluru, India

Simran Sharma
Indian Institute of Technology
Kharagpur, West Bengal, India

Sawan Shetty
Manipal Institute of Technology
Manipal Academy of Higher Education
  (MAHE)
Manipal, India

Raja Shunmugam
Indian Institute of Science Education and
  Research Kolkata
Mohanpur, Nadia, West Bengal, India

Anandhan Srinivasan
National Institute of Technology Karnataka
Surathkal, Mangalore
Karnataka, India

Mrudul Vijay Supekar
Cochin University of Science and Technology
  (CUSAT)
Kerala, India

Vijay Kumar Thakur
Biorefining and Advanced Materials Research
  Centre
Dumfries, Edinburgh, United Kingdom

Vinoy Thomas
University of Alabama at Birmingham
Birmingham, Alabama, United States

Vijayalekshmi Vijayakumar
Gyeongsang National University
Jinju, Republic of Korea

Vineeth M. Vijayan
University of Alabama at Birmingham,
  Birmingham
Alabama, United States

Günter Wuzella
Competence Center for Wood Composites and
  Wood Chemistry
Linz, Austria

Section I

Polymeric Biomaterials and Polymers
from Sustainable Resources

1 3D Bioprinting for Tissue Regeneration

Selvakumar Murugesan
Universität Bayreuth, Bayreuth, Germany

National Institute of Technology Karnataka, Surathkal, Mangalore, Karnataka, India

Thomas Scheibel
Universität Bayreuth, Bayreuth, Germany

CONTENTS

1.1 INTRODUCTION

Finding biomaterials to support regeneration of damaged tissue is one of the important issues to be addressed in the field of tissue engineering [1]. Tissue or even organ reproduction is one of the most pressured tasks in medicine, having on mind that the waiting lists of organ transplantations are very long everywhere. According to Department of Health and Human Services, USA, as of February 2021, about 107,000 people are on waiting lists in the USA, in contrast to 39,000 transplants implanted in 2020 [2]. About seven people die each day in the USA due to unavailable organs, and waiting lists are increasing exponentially as per statistics. Scientists and clinical doctors have been working to develop alternative ways to circumvent organ donor shortage. Tissue engineering could be one of the ways to save the lives of patients with the help of various functional materials combined with fabrication techniques [3]. 3D constructs or scaffolds have been used to regenerate tissue by enhancing cell interactions and providing cellular microenvironments [4]. There are many techniques, including microfabrication using gas foaming, photolithography, salt-leaching, electrospinning and 3D printing to fabricate scaffolds mimicking the extracellular matrix (ECM) of native tissue or organs. However, apart from 3D printing, no other technique allows to fabricate 3-dimensional tissues/organs [5]. 3D bioprinting even offers the ability to print a combination of living cells and biomaterials simultaneously, which is termed biofabrication [6,7]. The resolution of 3D constructs produced using bioprinting is 10–10,000 µm. This fabrication technique can even mimic the complexity of organ structures, which cannot be achieved by other techniques. Historically, 3D printing in general was first reported by Charles Hull in the early 1980s using computer-aided design (CAD) [8,9]. In the 1990s, 3D printing was increasingly used in biomedical applications for

DOI: 10.1201/9781003200710-2

fabricating dental implants and bladders [10]. Today, the demand for 3D bioprinting is constantly increasing. The combination of a biomaterial and cells, printed simultaneously, is named "bioinks". The properties of bioinks are important for successful biofabrication of 3D constructs including shape fidelity, rapid crosslinking ability and long-time mechanical stability [11]. Hence, design and preparation of bioinks is playing a pivotal role in bioprinting [12]. 3D bioprinting has been used for various purposes to fabricate tissues, organ modules and tissue/organ-on-a-chip applications due to its possibility to reconstitute ECM morphology, adaptability and microporosity [13]. In a typical 3D printing process, bioinks are pushed towards the substrate via a layer-by-layer method according to the desired model provided by CAD [14]. Subsequently, the printed constructs are either crosslinked at controlled temperature or using support baths with explicit ions contents. Recently, incorporation of nanoparticles into bioinks, which are then called nanocomposite bioinks, have been established enhancing various properties including shear thinning characteristics, mechanical stability and biological cues [15]. Nanocomposite bioinks provide additional crosslinking densities to the 3D printed constructs and help to enhance the shape fidelity of the constructs [16]. Additionally, nanoparticles are being added to mimic the composition of hard tissue. For example, hydroxyapatite nanoparticle-based bioinks have been used to mimic bone tissue [17]. A wide range of tissues and organs have been fabricated using 3D bioprinting techniques including, but not limited to, bone, cartilage, skin, liver and heart, etc [18,19]. 3D bioprinting has nowadays been commercialized by various companies for various biomedical applications including cancer therapy, organ transplantation and drug testing, etc [20]. In the present chapter, various bioprinting methodologies and the role of different bioinks for tissue regeneration applications are discussed in detail.

1.2   TYPES OF 3D BIOPRINTING TECHNIQUES

3D bioprinting techniques offer three modalities such as 1) droplet, 2) extrusion and 3) laser-induced forward transfer bioprinting. Droplet-based bioprinting (DBB) was first introduced in the early 2000s and it uses either thermal, piezo or acoustic forces to deposit droplets of bioinks or cell suspensions in a high-throughput mode. The advantage of DBB is that even low viscosity bioinks (3.5–12 mPa/s) can be printed at high velocities (1–10,000 droplets/s) without reducing the resolution of the achieved 3D constructs [22]. Extrusion bioprinting (EBB) was first introduced in the early 2000s and it usually applies mechanical (screw or piston-based) or pneumatic dispensing forces to deposit bioinks or cell suspensions in the form of filaments [23]. EBB is the easiest as well as most economical version to print tissues/organs including core/shell, multihead printing, etc. In EBB, the viscosities of bioinks are slightly higher than in DBB in the range of 30 mPa/s to $>6 \times 10^7$ mPa/s, which enables, for example, printing spheroids out of scaffold-free bioinks [24]. However, cell death is a common problem in EBB, as the bioinks' higher viscosities induce stress on cells by shear forces [25]. Laser induced forward transfer (LFT) was established in 1999 and uses laser energy to deposit droplets of bioinks or cell suspension from a donor slide to a collector slide without any nozzles like in DBB and EBB [25]. It allows the viscosities of bioinks in the range of 1–300 mPa/s. It also has capabilities of printing high cell densities of 10^8 cells/ml with the resolution of one cell per droplet [21,26]. The technique induces very little cell death, since it is nozzle-free. The various types of 3D bioprinting are shown in Figure 1.1 and key findings are summarised in Table 1.1.

1.3   3D BIOPRINTING FOR TISSUE ENGINEERING

3D bioprinting techniques are in use for various biomedical applications. This chapter focuses only on applications in bone engineering, wound dressings and in drug delivery.

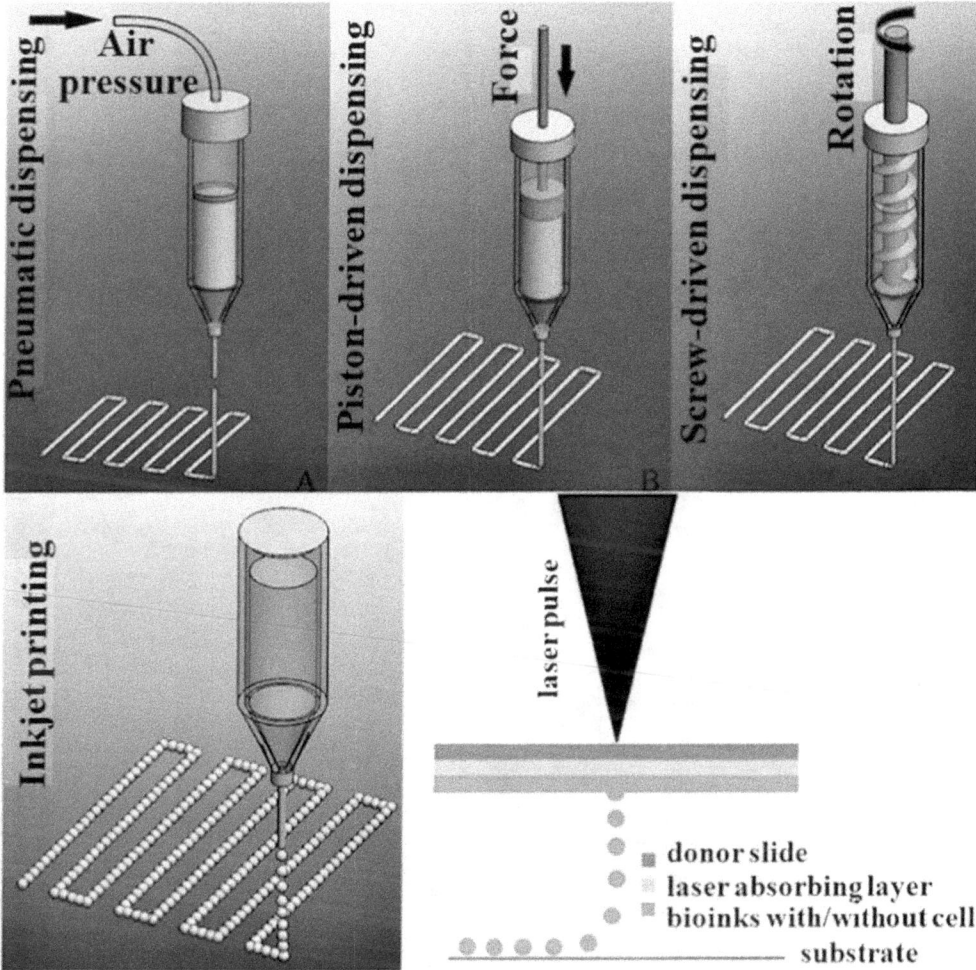

FIGURE 1.1 Top panel shows the extrusion-based bioprinting variants (pneumatic, piston and screw-driven dispensing mode) and bottom panel shows inkjet printing and laser-induced forward transfer schemes. Reproduced with permission from Soroosh Derakhshanfar *et al.* [21].

1.3.1 3D BIOPRINTING FOR BONE ENGINEERING

Bone is to some extent a calcified form of connective tissues and it shows five main functions in our body: mechanically supporting the body, protecting internal organs, aiding body movements, producing blood cells and storing fat and minerals. For bone tissue engineering, many factors have to be considered such as bone construction, mechanics of bone and tissue configuration/creation and regeneration. At present, a large number of patients require bone tissue transplantation, but the number of donors is very limited. Therefore, biofabrication could be one alternative route towards bone regeneration. In comparison to other fabrication techniques, 3D bioprinted constructs can facilitate robust cell migration, proliferation and stem cell differentiation along with vascularisation, as this fabrication technique allows mimicking the exact native bone architecture as well as shows the ability to print multiple cell lines. There are many works carried out using 3D bioprinting for various types of bone regeneration [30–41]. Falguni Pati *et al.* [42] fabricated 3D constructs using mesenchymal stromal cell (hTMSC)-laden polycaprolactone (PCL), poly (lactic-co-glycolic acid) (PLGA) and β-tricalcium phosphate (β-TCP) and mineralized ECM as bioinks for bone regeneration within a

TABLE 1.1

Various 3D Printer Types, Their Silent Features as Well as Printed Tissue/Organ Model Structures

Printer Types	Silent Features	Model Tissue/Organ Construct
Droplet-based bioprinting (DBB)	Bioinks with very low viscosity can be printed at high velocities without reducing the resolution of 3D constructs	Bioprinted tubes composed of porcine aortic smooth muscle cells in a perfusion bioreactor. Reproduced with permission from Ajay Vikram Singh et al. [27]
Extrusion-based bioprinting (EBB)	Facile and capable of printing various biomaterials as well as able to print high cell densities and high viscosity bioinks	Bioprinted human ear and blood vessel composed of Kappa-carrageenan (κCA) modified laponite-based nanosilicates reinforced GelMA. Reproduced with permission from David Chimene et al. [28]

| Laser-induced forward transfer (LFT) | High-resolution printing and deposition of biomaterials from solid as well as liquid phases |

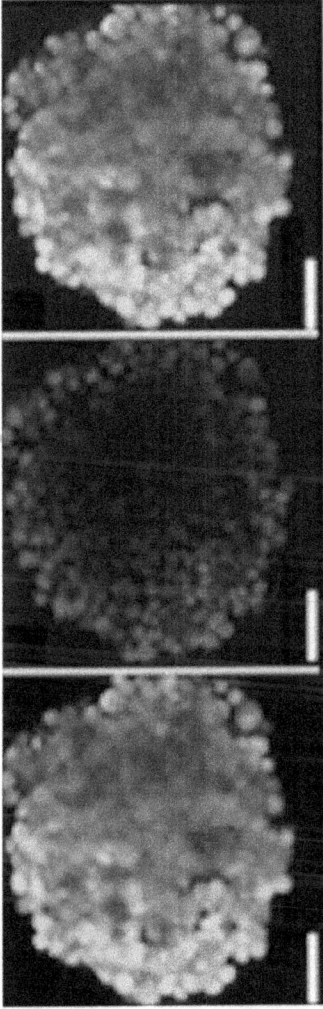

Bioprinted alginate microbeads-laden MDA-MB-231 cells (epithelial, human breast cancer). Reproduced with permission from David M.Kingsley et al. [29]

layer-by-layer approach and extrusion-based printing. F-actin expression of hTMSCs is shown in Figure 1.2, depicting that ECM ornamented composite 3D scaffolds exhibit higher expression levels in comparison to that of bare 3D polymeric scaffolds. The ECM-ornamented hTMSC-laden 3D printed scaffolds encouraged osteoblastic differentiation by up-regulating osteoblastic genes e.g., with a fourfold increment of RUNX2, threefold increment of ALP, fourfold increment of osteocalcin and fourfold increment of osteopontin as well as and increasing calcium deposition. *In-vivo* experiments also showed good levels of bone formation indicated by a better filling of a calvarial defect [42].

Minqi Wang *et al.* [43] fabricated a 3D scaffold using polycaprolactone/mesoporous bioactive glass/DOX together with an engineered progenitor cell line C3H10T1/2 as a bioink. The bioprinted

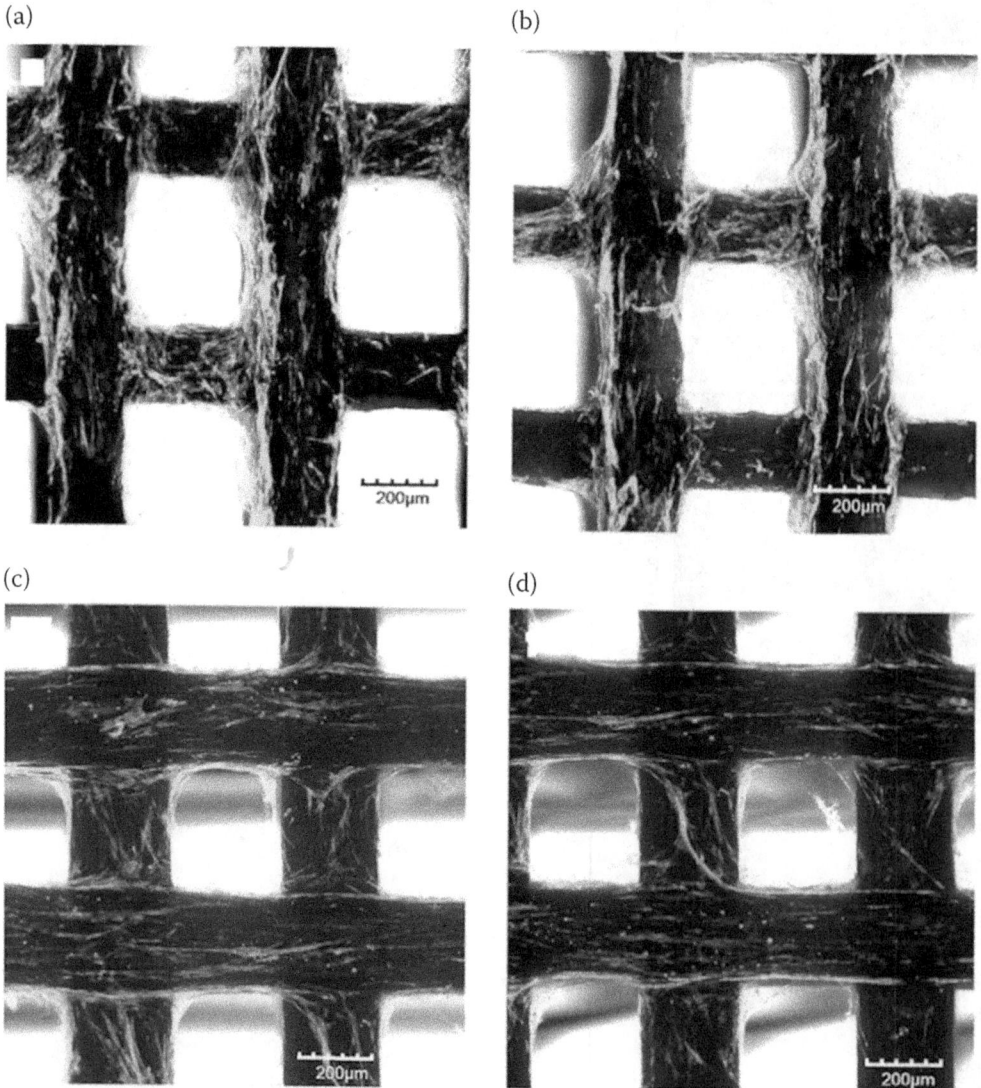

FIGURE 1.2 F-actin expression of hTMSCs-laden bioprinted scaffolds made of different materials after 24 hrs of culturing: (a) ECM/PCL/PLGA/TCP, (b) PCL/PLGA/TCP, (c) ECM/PCL/PLGA and (d) PCL/PLGA scaffolds. PCL, PLGA, TCP and ECM denotes polycaprolactone, poly (lactic-*co*-glycolic acid) and β-tricalcium phosphate and mineralized extracellular matrix. ECM ornamented composite 3D scaffolds (a&c) exhibit higher F-actin expression in comparision to bare 3D polymeric scaffolds (b&d). Reproduced with permission from Falguni Pati *et al.* [42].

3D scaffold enhanced the BMP2 growth factor secretion, antibacterial activity, promoted osteo-blast differentiation and induced ectopic bone formations. Bin Liu *et al.* [44] fabricated tissue engineered bone constructs using biomimetic nanocomposite bioinks made of rat bone marrow mesenchymal stem cell-laden nanosilicate (nSi) reinforced blends of gelatin and alginate hydro-gels. Apparently, nSi reinforcement helped to enhance the printability and shear thinning beha-viour of bioinks and mechanical stability of fabricated tissues or organ structures (up to 15 mm height). The scaffold showed excellent osteogenic differentiation. Further, *in-vivo* results de-monstrated that the nanocomposite bioinks printed 3D constructs showed excellent bone repairing characteristics.

1.3.2 3D Bioprinting for Wound Dressing Applications

In general, wounds can be classified with two major categories based on their rate of healing such as 1) an acute wound that heals faster and 2) a chronic wound taking more time to heal due to infection and larger discontinuity. There are several morphologies used including hydrogels, fiber meshes and thin films to regenerate the damaged skin or wound. However, 3D bioprinted patches so far outperform other morphologies [45].

Zhen Wu and Youliang Hong [46] have developed an antibacterial superporous wound dressing patch based on nanocomposite hydrogels made of silver nanoparticles (AgNPs) embedded poly-acrylamide (PAM)/hydroxypropyl methylcellulose (HPMC). The introduction of AgNPs in bioinks enhanced the physical crosslinking density and improved the mechanical stability as well as shape fidelity of the printed patches, *In-vivo* results of printed 3D patches showed faster wound healing rates as well as smoother surfaces upon wound healing with the smallest scarring morphology. Si Xiong *et al.* [45] fabricated a 3D printed wound dressing patch including fibroblast growth factor 2 (FGF-2)-laden gelatin-sulfonated *Bombyx mori* silk blends. Silk incorporation in 3D printed pat-ches promoted granulation. *In-vivo* rat model results demonstrated that FGF-2-laden 3D patches showed excellent full thickness wound regeneration characteristics by stimulating epidermal growth and dermal neovascularisation.

Chunlin Xu *et al.* [47] have developed nanocellulose-based 3D printed scaffolds for wound healing applications. The dual crosslinked 3D scaffolds showed roughly doubled mechanical strength, excellent fibroblast cell viability of 98%, and rapid cell proliferation. Elif Ilhan *et al.* [48] fabricated 3D printed patches from Satureja Cuneifolia (drug) embedded sodium alginate/poly-ethylene glycol bioinks for diabetic wound healing. The 3D printed drug embedded patches showed a drug loading efficiency of 93%, antimicrobial effects against gram-positive as well as negative strains, and the drug did not affect fibroblasts attachment and proliferation. Wenyang Xu *et al.* [49] fabricated 3D scaffolds from Ca^{2+} crosslinked cellulose nanofibrils (CNFs) reinforced gelatin me-thacrylate (GelMA) bioinks for wound healing applications. The addition of CNFs enhanced the 3T3 fibroblasts cell adhesion, improved compression modulus and promoted the rate of proliferation.

Hao Zhao *et al.* [50] have fabricated a 3D artificial skin patch using the combination of a poly (phenylene vinylene) derivative (PPV) and gelatin/alginate/hyaluronic acid blend bioink for augmenting wound repair. PPV acted as a photosensitizer and resulted in excellent anti-infection properties (almost 90% bacteria were killed) for the artificial skin patches through photodynamic therapy. The printed 3D patches also showed excellent biocompatibility, cell migration, vascu-larisation and rapid wound closure within 20 days upon implantation (in-vivo). Kristin Schacht *et al.* [13] used spider silk based hydrogels for biofabricating layer-by-layer 3D constructs. The recombinant protein bioinks are capable to physically crosslink without external crosslinkers. The printed 3D constructs were capable to retain their shape fidelity, and it was able to print up to 16 layers with a height of 3 mm (Figure 1.3a&b). In this first study, 70% cells were viable, but in a follow-up study with an improved bioink recipe RGD peptide tagged spider silk proteins as well as a blend with gelatin bioinks the fibroblasts showed almost 100% cell viability [6]. Figure 1.3c&d shows the live/dead cell photomicrographs of biofabricated RGD-spider silk scaffolds. These

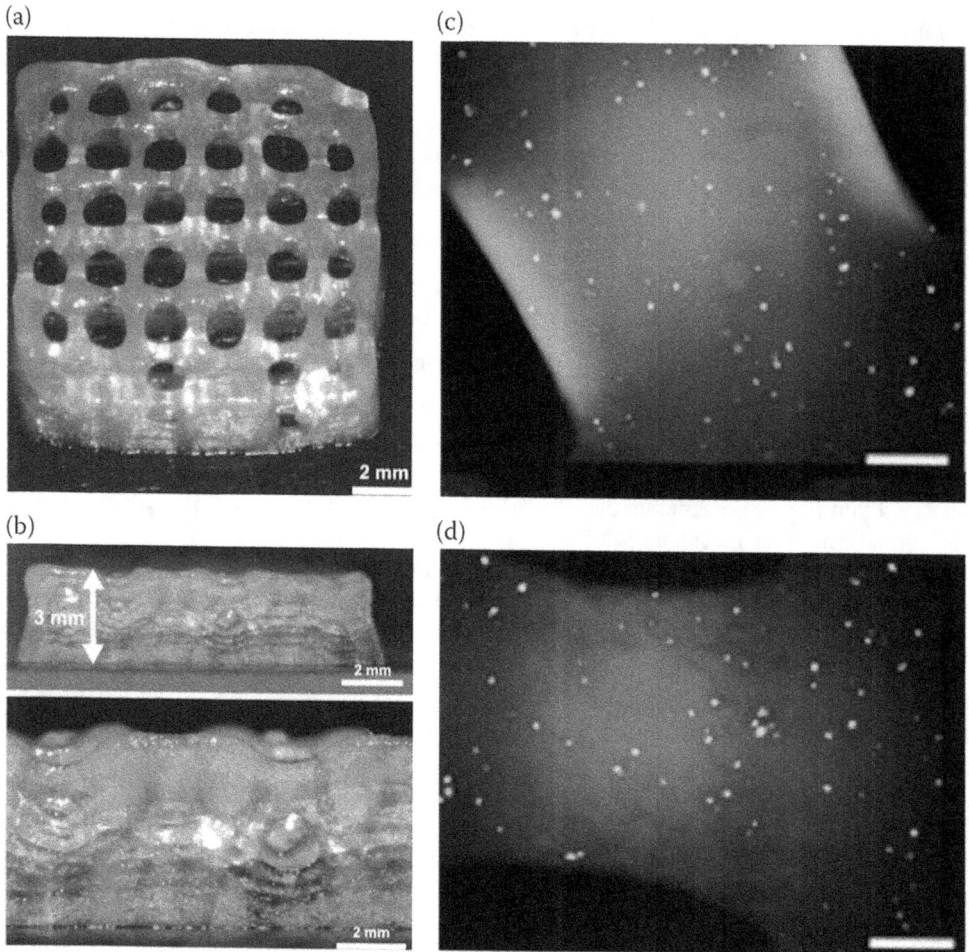

FIGURE 1.3 Biofabrication of recombinant spider silk protein-based hydrogels: (a) Top view of a 16 layer scaffold, (b) cross-sectional view of the 16-layer scaffold and (c–d) live-dead cell assay of a bioprinted construct with mouse fibroblasts BALB/3T3 encapsulated in spider silk hydrogels or in blended hydrogels with gelatine after 18 h of culturing. Importantly, the red background is based on the unspecific interaction between the red dye and the spider silk hydrogel. Reproduced with permission from Kristin Schacht *et al.* and Elise DeSimone *et al.* [6,13].

results open a new avenue of using hydrogels made of recombinant spider silk proteins for 3D bioprinting with the potential for tissue engineering applications.

1.3.3 3D BIOPRINTING FOR DRUG DELIVERY

3D bioprinting techniques also have been used for drug delivery applications to treat various diseases including cancer, retinal vascular diseases and diabetic wounds. Coaxial printing techniques are able to encapsulate drugs and, thereby, to regulate the drug release more efficiently than that of printed monolithic strands. Subsequently, sustained drug release can be easily controlled and burst drug release avoided. Various reports demonstrated that 3D bioprinting enables to manufacture multilayered tablets with different drugs in different layers for different actions against a particular disease. Further, it is possible to dispense very low volumes of precursor materials with high accuracy [51]. For instance, a bilayered tablet containing a robust release layer and a sustained release layer of guaifenesin embedded in a polymeric blend of hypromellose and polyacrylic acid achieved the desired drug release profile [52].

Jae Yon Won *et al.* [53] have used the extrusion-based coaxial printing technique to fabricate core/shell constructs to deliver two types of drugs (bevacizumab (BEV) and dexamethasone (DEX)) for retinal vascular diseases. They have used PCL (shell) and alginate (core) as precursors for printing. The dual drug loaded core/shell constructs showed excellent compatibility with human umbilical vein endothelial cells (HUVECs) and allowed controlled drug release for up to 50 days. *In-vivo* implantation of core-shell tubes in eyes of rabbits showed less invasion, enhanced vascularisation and time controlled release of BEV and DEX compared to that of commercially available intravitreal drug injection methods.

Hee-Gyeong Yi *et al.* [54] fabricated a 3D printed patch of 5-fluorouracil anticancer drug encapsulated using poly(lactide-*co*-glycolide) and polycaprolactone blend bioinks and extrusion-based printing techniques for pancreatic cancer growth suppression. *In-vitro* results showed that only 75% of MIA-PaCa-2 cells were killed at high drug loading content of 55.90 µg/ml and a low amount of drug release (20%) over 4 weeks. The *in-vivo* results displayed an excellent reduction in tumour size upon subcutaneous implantation into pancreatic cancer xenografts in a rat model. It took 4 months to degrade the 3D printed patches, and they showed no acute toxicity to any organs upon implantation. In another study, Vanessa J. Neubauer *et al.* [55] fabricated 3D scaffolds based on recombinant spider silk protein hydrogels for drug delivery applications. The prepared recombinant spider silk protein hydrogels using aqueous–organic co-solvents allow formulation of water-insoluble or hydrophobic drugs e.g., the anticancer drug 6-mercaptopurine along with adjustable organic content to yield biocompatible, biodegradable, non-toxic, non-inflammatory, transdermal, injectable and even 3D printable drug depots.

Iria Louzao *et al.* [56] fabricated a 3D printed scaffold with an antidepressive drug (paroxetine) encapsulated in poly (β-amino ester)-based photocurable bioinks using inkjet 3D printing. The printed patches exhibited a high drug release rate of above 95% and showed very good mechanical stability. Mengsuo Cui *et al.* [57] fabricated a 3D printed tablet using levetiracetam encapsulated in hydroxypropyl cellulose (HPC-M) and semi-solid extrusion 3D printing to treat epilepsy. They printed three different geometrical shapes (cylinder, torus and oval) of tablets with diameters of 12 mm.

1.4 CONCLUSIONS AND FUTURE PERSPECTIVES

Bioprinting is a fast-growing technology to be used for several biomedical applications, especially for replacement of damaged tissue. In this chapter, we have discussed some bioprinting techniques using different bioinks for bone engineering, wound dressing applications and drug delivery. Among all, the preparation of bioinks as well as their nature of crosslinking is important, as they determine the mechanical stability of the 3D constructs. Bioprinted tissue constructs show for instance native-like tissue architecture, which is one of several advantages of bioprinting. However, they also face few limitations such as 3D construct resolution and lack of vascularisation and perfusability. Such issues need to be resolved by developing novel multifunctional hydrogel systems. If these problems can be solved, future developments will enable bioprinting techniques to expand quickly from tissue engineering and pharmaceutical research to clinical uses.

1.5 ACKNOWLEDGEMENTS

The authors gratefully acknowledge the support of Alexander von Humboldt (AvH) Foundation for a postdoctoral fellowship to SM.

REFERENCES

[1] Murugesan S, Scheibel T. Copolymer/Clay Nanocomposites for Biomedical Applications. *Advanced Functional Materials*. 2020;30:1908101.

[2] Department of Health and Human Services, USA. 2021. https://www.usa.gov/federal-agencies/ u-s-department-of-health-and-human-services (accessed on 30th May 2022).

[3] Salehi S, Koeck K, Scheibel T. Spider Silk for Tissue Engineering Applications. *Molecules.* 2020;25: 737.

[4] Murugesan S, Scheibel T. Chitosan-Based Nanocomposites for Medical Applications. *Journal of Polymer Science.* 2021;59:1610–1642.

[5] DeSimone E, Schacht K, Jungst T, Groll J, Scheibel T. Biofabrication of 3D Constructs: Fabrication Technologies and Spider Silk Proteins as Bioinks. *Pure and Applied Chemistry.* 2015;87:737–749.

[6] DeSimone E, Schacht K, Pellert A, Scheibel T. Recombinant Spider Silk-Based Bioinks. *Biofabrication.* 2017;9:044104.

[7] Jungst T, Smolan W, Schacht K, Scheibel T, Groll J. Strategies and Molecular Design Criteria for 3D Printable Hydrogels. *Chemical Reviews.* 2016;116:1496–1539.

[8] Gao W, Zhang Y, Ramanujan D, Ramani K, Chen Y, Williams CB, et al. The Status, Challenges, and Future of Additive Manufacturing in Engineering. *Computer-Aided Design.* 2015;69:65–89.

[9] Melchels FP. *Celebrating three decades of stereolithography.* Taylor & Francis; 2012.

[10] Cheng L, Suresh KS, He H, Rajput RS, Feng Q, Ramesh S, et al. 3D Printing of Micro- and Nanoscale Bone Substitutes: A Review on Technical and Translational Perspectives. *International Journal of Nanomedicine.* 2021;16:4289–4319.

[11] Xu W, Jambhulkar S, Ravichandran D, Zhu Y, Kakarla M, Nian Q, et al. 3D Printing-Enabled Nanoparticle Alignment: A Review of Mechanisms and Applications. *Small.* 2021;17(45):2100817.

[12] Carvalho V, Goncalves I, Lage T, Rodrigues RO, Minas G, Teixeira S, et al. 3D Printing Techniques and Their Applications to Organ-on-a-Chip Platforms: A Systematic Review. *Sensors.* 2021;21(9):3304.

[13] Schacht K, Jungst T, Schweinlin M, Ewald A, Groll J, Scheibel T. Biofabrication of Cell-Loaded 3D Spider Silk Constructs. *Angewandte Chemie.* 2015;54:2816–2820.

[14] Thamm C, DeSimone E, Scheibel T. Characterization of Hydrogels Made of a Novel Spider Silk Protein Emasp1s and Evaluation for 3D Printing. *Macromolecular Bioscience.* 2017;17:1700141.

[15] Arefin AME, Khatri NR, Kulkarni N, Egan PF. Polymer 3D Printing Review: Materials, Process, and Design Strategies for Medical Applications. *Polymers.* 2021;13(9):1499.

[16] Rastin H, Zhang B, Mazinani A, Hassan K, Bi J, Tung TT, et al. 3D Bioprinting of Cell-Laden Electroconductive MXene Nanocomposite Bioinks. *Nanoscale.* 2020;12:16069–16080.

[17] Nadernezhad A, Caliskan OS, Topuz F, Afghah F, Erman B, Koc B. Nanocomposite Bioinks Based on Agarose and 2D Nanosilicates With Tunable Flow Properties and Bioactivity for 3D Bioprinting. *ACS Applied Bio Materials.* 2019;2:796–806.

[18] Wilson SA, Cross LM, Peak CW, Gaharwar AK. Shear-Thinning and Thermo-Reversible Nanoengineered Inks for 3D Bioprinting. *ACS Applied Materials & Interfaces.* 2017;9:43449–43458.

[19] Chimene D, Lennox KK, Kaunas RR, Gaharwar AK. Advanced Bioinks for 3D Printing: A Materials Science Perspective. *Annals of Biomedical Engineering.* 2016;44:2090–2102.

[20] Mihaila SM, Gaharwar AK, Reis RL, Marques AP, Gomes ME, Khademhosseini A. Photocrosslinkable Kappa-Carrageenan Hydrogels for Tissue Engineering Applications. *Advanced Healthcare Materials.* 2013;2:895–907.

[21] Derakhshanfar S, Mbeleck R, Xu K, Zhang X, Zhong W, Xing M. 3D Bioprinting for Biomedical Devices and Tissue Engineering: A Review of Recent Rrends and Advances. *Bioactive Materials.* 2018;3:144–156.

[22] Gudapati H, Dey M, Ozbolat I. A Comprehensive Review on Droplet-Based Bioprinting: Past, Present and Future. *Biomaterials.* 2016;102:20–42.

[23] Askari M, Naniz MA, Kouhi M, Saberi A, Zolfagharian A, Bodaghi M. Recent Progress in Extrusion 3D Bioprinting of Hydrogel Biomaterials for Tissue Regeneration: A Comprehensive Review with Focus on Advanced Fabrication Techniques. *BiomaterialsScience.* 2021;9:535–573.

[24] Ning L, Chen X. A Brief Review Of Extrusion-Based Tissue Scaffold Bio-Printing. *Biotechnology Journal.* 2017;12:1600671.

[25] Cui X, Li J, Hartanto Y, Durham M, Tang J, Zhang H, et al. Advances in Extrusion 3D Bioprinting: A Focus on Multicomponent Hydrogel-Based Bioinks. *Advanced Healthcare Materials.* 2020;9:1901648.

[26] Kačarević ŽP, Rider PM, Alkildani S, Retnasingh S, Smeets R, Jung O, et al. An Introduction to 3D Bioprinting: Possibilities, Challenges and Future Aspects. *Materials.* 2018;11:2199.

[27] Vikram Singh A, Hasan Dad Ansari M, Wang S, Laux P, Luch A, Kumar A, et al. The Adoption of Three-Dimensional Additive Manufacturing From Biomedical Material Design to 3D Organ Printing. *Applied Sciences.* 2019;9:811.

[28] Chimene D, Peak CW, Gentry JL, Carrow JK, Cross LM, Mondragon E, et al. Nanoengineered Ionic–Covalent Entanglement (NICE) Bioinks for 3D Bioprinting. *ACS Applied Materials & Interfaces*. 2018;10:9957–9968.

[29] Kingsley DM, Roberge CL, Rudkouskaya A, Faulkner DE, Barroso M, Intes X, et al. Laser-Based 3d Bioprinting for Spatial and Size Control of Tumor Spheroids and Embryoid Bodies. *Acta Biomaterialia*. 2019;95:357–370.

[30] Li K, Zhang F, Wang D, Qiu Q, Liu M, Yu A, et al. Silkworm-Inspired Electrohydrodynamic Jet 3D Printing of Composite Scaffold with Ordered Cell Scale Fibers for Bone Tissue Engineering. *International Journal of Biological Macromolecules*. 2021;172:124–132.

[31] Hann SY, Cui H, Esworthy T, Zhou X, Lee SJ, Plesniak MW, et al. Dual 3D Printing for Vascularized Bone Tissue Regeneration. *Acta Biomaterialia*. 2021;123:263–274.

[32] Dong L, Bu Z, Xiong Y, Zhang H, Fang J, Hu H, et al. FAcile Extrusion 3D Printing of Gelatine Methacrylate/Laponite Nanocomposite Hydrogel with High Concentration Nanoclay for Bone Tissue Regeneration. *International Journal of Biological Macromolecules*. 2021;188:72–81.

[33] Zhang X, Zhou J, Xu Y. Experimental Study on Preparation of Coaxial Drug-Loaded Tissue-Engineered Bone Scaffold by 3D Printing Technology. *Proceedings of the Institution of Mechanical Engineers Part H, Journal of Engineering in Medicine*. 2020;234:309–322.

[34] Wibowo A, Vyas C, Cooper G, Qulub F, Suratman R, Mahyuddin AI, et al. 3D Printing of Polycaprolactone-Polyaniline Electroactive Scaffolds for Bone Tissue Engineering. *Materials*. 2020;13(3):512.

[35] Wang C, Huang W, Zhou Y, He L, He Z, Chen Z, et al. 3D Printing of Bone Tissue Engineering Scaffolds. *Bioactive Materials*. 2020;5:82–91.

[36] Sohling N, Neijhoft J, Nienhaus V, Acker V, Harbig J, Menz F, et al. 3D-Printing of Hierarchically Designed and Osteoconductive Bone Tissue Engineering Scaffolds. *Materials*. 2020;13(8):1836.

[37] Polley C, Distler T, Detsch R, Lund H, Springer A, Boccaccini AR, et al. 3D Printing of Piezoelectric Barium Titanate-Hydroxyapatite Scaffolds with Interconnected Porosity for Bone Tissue Engineering. *Materials*. 2020;13(7):1773.

[38] Jeong JE, Park SY, Shin JY, Seok JM, Byun JH, Oh SH, et al. 3D Printing of Bone-Mimetic Scaffold Composed of Gelatin/beta-Tri-Calcium Phosphate for Bone Tissue Engineering. *Macromolecular Bioscience*. 2020;20:e2000256.

[39] Chae S, Sun Y, Choi YJ, Ha DH, Jeon IH, Cho DW. 3D Cell-Printing of Tendon-Bone Interface Using Tissue-Derived Extracellular Matrix Bioinks for Chronic Rotator Cuff Repair. *Biofabrication*. 2020;13(3):035005.

[40] Wei L, Wu S, Kuss M, Jiang X, Sun R, Reid P, et al. 3D Printing of Silk Fibroin-Based Hybrid Scaffold Treated with Platelet Rich Plasma for Bone Tissue Engineering. *Bioactive Materials*. 2019;4:256–260.

[41] Trombetta R, Inzana JA, Schwarz EM, Kates SL, Awad HA. 3D Printing of Calcium Phosphate Ceramics for Bone Tissue Engineering and Drug Delivery. *Annals of Biomedical Engineering*. 2017;45:23–44.

[42] Pati F, Song TH, Rijal G, Jang J, Kim SW, Cho DW. Ornamenting 3D Printed Scaffolds with Cell-Laid Extracellular Matrix for Bone Tissue Regeneration. *Biomaterials*. 2015;37:230–241.

[43] Wang M, Li H, Yang Y, Yuan K, Zhou F, Liu H, et al. A 3D-Bioprinted Scaffold with Doxycycline-Controlled BMP2-Expressing Cells for Inducing Bone Regeneration and Inhibiting Bacterial Infection. *Bioactive Materials*. 2021;6:1318–1329.

[44] Liu B, Li J, Lei X, Cheng P, Song Y, Gao Y, et al. 3D-Bioprinted Functional and Biomimetic Hydrogel Scaffolds Incorporated with Nanosilicates To Promote Bone Healing In Rat Calvarial Defect Model. Materials Science & Engineering C. *Materials for Biological Applications*. 2020;112:110905.

[45] Xiong S, Zhang X, Lu P, Wu Y, Wang Q, Sun H, et al. A Gelatin-sulfonated Silk Composite Scaffold Based on 3D Printing Technology Enhances Skin Regeneration by Stimulating Epidermal Growth and Dermal Neovascularization. *Scientific Reports*. 2017;7:4288.

[46] Wu Z, Hong Y. Combination of the Silver-Ethylene Interaction and 3D Printing To Develop Antibacterial Superporous Hydrogels for Wound Management. *ACS Applied Materials & Interfaces*. 2019;11:33734–33747.

[47] Xu C, Molino BZ, Wang X, Cheng F, Xu W, Molino P, et al. 3D Printing of Nanocellulose Hydrogel Scaffolds With Tunable Mechanical Strength Towards Wound Healing Application. *Journal of Materials Chemistry*. 2018; 6(43): 7066–7075.

[48] Ilhan E, Cesur S, Guler E, Topal F, Albayrak D, Guncu MM, et al. Development of Satureja Cuneifolia-Loaded Sodium Alginate/polyethylene Glycol Scaffolds Produced by 3D-Printing Technology as a Diabetic Wound Dressing Material. *International Journal of Biological Macromolecules*. 2020;161:1040–1054.

[49] Xu W, Molino BZ, Cheng F, Molino PJ, Yue Z, Su D, et al. On Low-Concentration Inks Formulated by Nanocellulose Assisted with Gelatin Methacrylate (GelMA) for 3D Printing toward Wound Healing Application. *ACS Applied Materials & Interfaces*. 2019;11:8838–8848.

[50] Zhao H, Xu J, Yuan H , Zhang E, Dai N, Gao Z, et al. 3D Printing of Artificial Skin Patches with Bioactive and Optically Active Polymer Materials for Anti-Infection and Augmenting Wound Repair. *Materials Horizons*. 2022; 9:342–349.

[51] Beg S, Almalki WH, Malik A, Farhan M, Aatif M, Rahman Z, et al. 3D Printing for Drug Delivery and Biomedical Applications. *Drug Discovery Today*. 2020;25:1668–1681.

[52] Khaled SA, Burley JC, Alexander MR, Roberts CJ. Desktop 3D Printing of Controlled Release Pharmaceutical Bilayer Tablets. *International Journal of Pharmaceutics*. 2014;461:105–111.

[53] Won JY, Kim J, Gao G, Kim J, Jang J, Park YH, et al. 3D Printing Of Drug-Loaded Multi-Shell Rods for Local Delivery of Bevacizumab and Dexamethasone: A Synergetic Therapy for Retinal Vascular Diseases. *Acta Biomaterialia*. 2020;116:174–185.

[54] Yi HG, Choi YJ, Kang KS, Hong JM, Pati RG, Park MN, et al. A 3D-Printed Local Drug Delivery Patch for Pancreatic Cancer Growth Suppression. *Journal of Controlled Release: Official Journal of the Controlled Release Society*. 2016;238:231–241.

[55] Neubauer VJ, Trossmann VT, Jacobi S, Döbl A, Scheibel T. Recombinant Spider Silk Gels Derived from Aqueous–Organic Solvents as Depots for Drugs. *Angewandte Chemie International Edition*. 2021;60:11847–11851.

[56] Louzao I, Koch B, Taresco V, Ruiz-Cantu L, Irvine DJ, Roberts CJ, et al. Identification of Novel "Inks" for 3D Printing Using High-Throughput Screening: Bioresorbable Photocurable Polymers for Controlled Drug Delivery. *ACS Applied Materials & Interfaces*. 2018;10:6841–6848.

[57] Cui M, Pan H, Fang D, Qiao S, Wang S, Pan W. Fabrication of High Drug Loading Levetiracetam Tablets Using Semi-Solid Extrusion 3D Printing. *Journal of Drug Delivery Science and Technology*. 2020;57:101683.

2 Polymeric Biomaterials and Current Trends for Advanced Applications

Vineeth M. Vijayan
University of Alabama at Birmingham, Birmingham, Alabama,
United States

Suja Mathai
Mar Ivanios College, Thiruvananthapuram, Kerala, India

Vinoy Thomas
University of Alabama at Birmingham, Birmingham, Alabama,
United States

CONTENTS

2.1 INTRODUCTION TO POLYMERIC BIOMATERIALS

The exceptional great manner of life that we have today is the quickest outcome of huge progress in the field of chemistry over the previous century. Biomaterials as we know them now did not occur many years ago. Biomaterials are widely employed in medicine, dentistry, and biotechnology to increase the quality and duration of human life at the turn of the third decade of the twenty-first century.[1] Drug delivery systems to replacing or repairing tissues, organs, or physiological processes are only a few of the roles. It wasn't until a few million years ago that the term "biomaterial" was coined. There was no production of designed engineered devices, no organized

DOI: 10.1201/9781003200710-3

approval procedure, no understanding of biocompatibility, and no academic courses on biomaterials throughout history.

A biomaterial is a component that is used in prosthetics or medical products and is intended to come into contact with the live body in a certain way and for a specific period of time. One factor that must be highlighted is that a biomaterial must come into touch with living things or bodily fluids, resulting in a living-nonliving interface.[2] The advancements resulted in a significant rise in the variety of applications and efficiency of biomaterials, allowing devices like cardiovascular stents, dental nourishments, prosthetic hips, and contact lenses to save or enhance millions of lives. Biomaterials were described as any systemically, pharmacologically inactive substance or mixture of substances used to implant inside or interface with a living system to complement or replace activities of live tissues or organs based on their application.[3]

2.1.1 HISTORY AND EVOLUTION OF BIOMATERIALS

The fascinating topic of biomaterials is undergoing a dramatic shift, with biological sciences taking equal precedence over materials science and engineering as the discipline's basis and evolution. Nanotechnology, for example, has substantially increased the sophistication with which biomaterials are constructed and has enabled simultaneous manufacturing of materials with more complicated functionalities. Self-assembly, a bottom-up technique of manufacturing that allows for the production of complex structures in a repeatable way with minimum material waste is used to build natural biomaterials on these smaller sizes.[4]

The renewed interest in polymeric biomaterials may be termed a resurgence of sorts. Sutures constructed from animal sinew have been dated back to ancient Egypt, with some claiming they were used far earlier.[5] Biomaterials were initially used by the ancient Egyptians, who used coconut shells to heal broken skulls and wood and ivory as fake teeth, dating back to 3000 BC.[6] Some of the first biomaterial uses date back to ancient Phoenicia, when gold wires were used to bind loose teeth together, attaching prosthetic teeth to surrounding teeth.[7] Some of these doctors in the early first century at India and Greece are said to have successfully treated disemboweled soldiers.[5] The first hip replacement operation was performed in Germany in 1891 AD, and ivory was utilized in this case. This material is preferred because of its low cost, but it also has important biomechanical bonding properties, making it exactly suited for working with human body tissue.[6,8]

Bone plates were successfully used in the early 1900s to support bone fractures and expedite their recovery, and by the 1950s and 1960s, hip joints and artificial heart valves, as well as vascular grafts development, were in clinical trials.[9] It is thought that understanding the history of biomaterial development can help us get a get a deeper understanding of how far the state of the art has progressed. Biomaterials' main issue is how they will perform better than existing materials. In today's fragmented and specialized culture, it is all too simple to do research to form a larger picture. In previous years, biomaterials research was fueled by clever and imaginative clinicians who tried anything they felt would help to solve an issue, sometimes with great outcomes but always with a significant risk. Researchers now are far more focused on steadily developing technology with as little harm to patients as feasible. Nature has provided us with a plethora of excellent cures that have been crafted through thousands of years of evolution.[10]

The need for biocompatible and biodegradable materials has exploded in the recent decade. Biomaterials that are biocompatible, non-immunogenic, and bioresorbable that can be functionalized with bioactive proteins and chemicals are ideal. One of the most significant properties of biomaterials is their biodegradability.[11] Biocompatibility, biofunctionality, and availability, in a smaller extent, are the most important variables in biomaterial use. Biomaterials are widely employed in medical applications, and several polymers with comparable properties are employed across disciplines. This item is about polymers in medicine, especially a range of innovative approaches for tissue engineering, artificial vessels, and cell and biomolecule transport.

One of the most significant disciplines of modern science is polymer research for the construction of biomaterials. Sir Nicholas Harold, a British ophthalmologist, created the first medical-grade synthetic polymer, poly (methyl methacrylate) (PMMA), in 1949, for producing intraocular lenses.[12] The two most significant factors to consider when choosing a polymer for biomedical usage are biostability and biodegradability. The significant hydrolytically degradable biomedical polymers include polyanhydrides, polyesters, polyamides, poly (ortho esters), poly (amido amines), and poly (amido amines) homo- and copolymers (-amino esters).[13] They're also known as biopolymers and smart polymers, and they're mostly employed in the medical area. The goal of this book is to provide an overview of polymeric biomaterials that are utilized to replace or augment tissue and organ functions. The chemical and mechanical qualities of the materials used in the medical implant of the medical devices are directly linked to the success of polymeric devices.

2.1.2 POLYMERIC BIOMATERIALS

One of the foundations of material science is polymeric biomaterials. Approaches to manufacture drug and gene delivery functions, micro patterning, microfluidics, and other technologies have become more sophisticated in recent years. The creation of polymeric biomaterials for tissue engineering to produce precise biologic effects on cells is a fantastic new invention. One of the most significant disciplines of medicine in the contemporary era is polymeric biomaterials research. Organ implants, wound healing, medicine delivery, and other applications employ biomaterials. It has been proposed that natural polymeric biomaterials are the most desired for a variety of reasons, including the fact that they are biodegradable, biocompatible, and nontoxic.[14] Biocompatible, bifunctional, bioactive, bioinert, and sterilizable polymeric biomaterials are widely employed in both uses in healthcare and pharmaceuticals; however, they must be biocompatible, biofunctional, bioactive, bioinert, and sterilizable (Figure 2.1). Sutures, intraocular lenses, artificial hearts, vascular grafts, joints, and breast prosthesis are just a few of the implants or other supporting materials used in these applications. The most fascinating applications of polymeric biomaterials include therapeutic use of extracorporeal fluid and other supportive devices, such as hemodialysis, hem perfusion, bags, IV lines, needle catheters, and blood

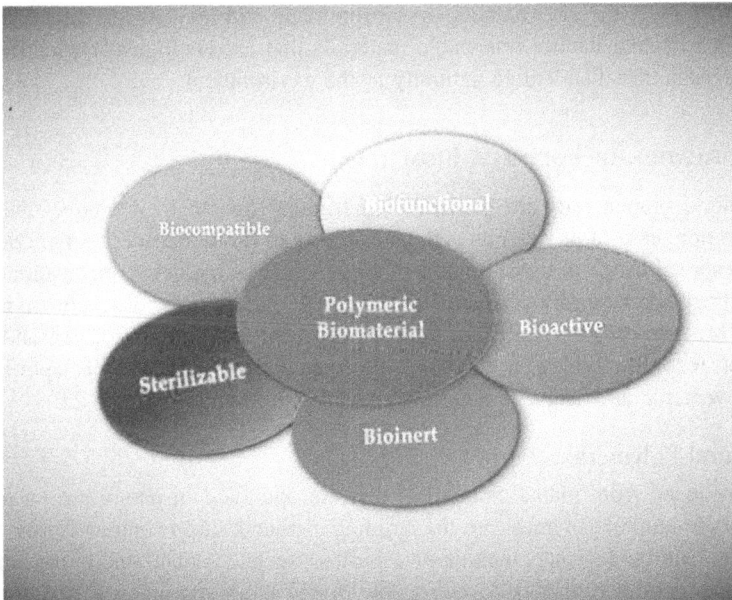

FIGURE 2.1 Polymers to be biomaterials.

oxygenation. Some of the applications include controlled release systems such as microcapsules, microspheres, transdermal drug delivery patches, and drug delivery vehicles for several routes of administration, mainly as carriers and supporting materials.[15]

Intelligent materials and structures are emerging as multidisciplinary technologies. The technology advancements of the first decade of the twenty-first century are gradually transforming the way we live by managing and regulating our daily lives with intelligent devices. This advanced innovation is being integrated into current engineering and design with the objective of allowing materials to have "intelligent" properties. The best example of combining material science and tissue engineering knowledge is the development of polymers as biomaterials with new improved functionalities.[16] The knowledge that has been obtained is responsible for the development and creation of a generation of "intelligent materials". By changing the surface qualities of biomaterials to meet the current needs of the polymer industry, existing and future end-user expectations can be met. In this regard, the focus of this book chapter will be on a unique approach for functional finishing biomaterials utilizing stimuli-responsive polymers.[17] Polymers that respond to tiny variations in chemical or physical circumstances approaching a critical state with sharp and substantial property changes are known as stimulus-responsive or intelligent polymers. In addition, they are "smart", "stimulus-responsive," and "environmentally friendly" polymers. The biological applications of smart polymers have been the subject of several reviews published in the last 20–25 years.[18]

Polymeric biomaterials are used in a variety of medical applications, including cardiology, orthopedics dentistry, sutures, ophthalmology, drug delivery, plastic and reconstructive surgery, extracorporeal devices, encapsulates, and tissue engineering, and their properties are suited to each application. Traditional polymer materials, particularly plastics, are abundant today as a result of decades of evolution. Their manufacturing is incredible in terms of raw material and energy usage, as well as waste discharge. The materials created have a variety of remarkable characteristics, including impermeability to water and microbes, high mechanical strength, low density, and low cost.[19] In recent years, significant progress has been achieved in degradable polymeric biomaterials for use in life applications. Because these materials have special physical, chemical, biological, biomechanical, and degrading features, degradable polymeric biomaterials are preferred. The major goal of this book chapter is to provide an overview of the importance of biomaterials, their classification, and applications, whether they have been created or destroyed naturally.[20] Recently, strong efforts to protect the environment have been made and are fast increasing as a life application, not only by using natural renewable materials that are ecologically friendly, but also by employing materials that disintegrate naturally in the environment.

2.1.3 Classification of Polymeric Biomaterials

Organic polymeric biomaterials are particularly intriguing due to their unique qualities of controllable preparation, ease of processing, and adaptability. In recent decades, polymeric biomaterials have seen widespread usage in tissue engineering, regenerative medicine, medication delivery, and gene therapy. They can be easily produced into a variety of dimensional structures, including one-dimensional (1D), two-dimensional (2D), three-dimensional (3D), microspheres, and porous scaffolds, to support a specific biological function.[21] Polymeric biomaterials can be split into two types based on their versatile architectures: natural and synthetic polymers (Figure 2.2).

2.1.3.1 Natural Polymers

Compounds produced from plants or animals that are abundant in nature are known as natural polymers. Because natural polymers are the building elements of our human forms, they are vital in everyday life. Natural polymers include proteins and nucleic acids found in the human body, as well as cellulose, natural rubber, silk, and wool. Honey is an example of a naturally occurring polymer that is frequently used in everyday life.[22] Natural polymers are generated during all species' growth cycles and are widely available in large amounts. Natural polymers were first used

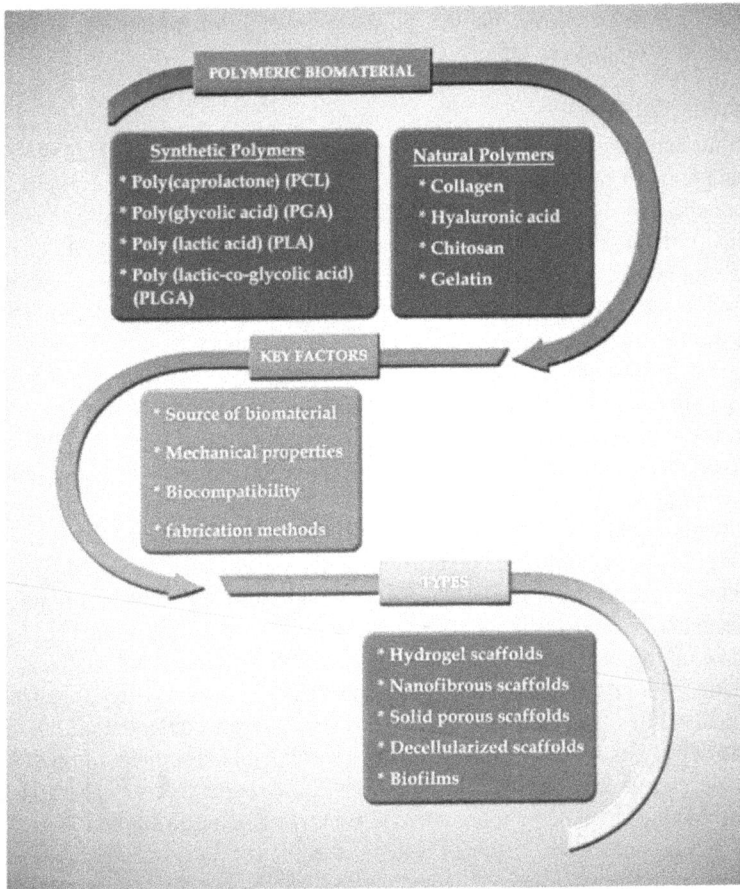

FIGURE 2.2 Classification of polymeric biomaterials.

in biomedical applications many years ago. Collagen, hyaluronic acid, chitosan, and gelatin are among the members of this group. The inherent bioactivities of natural polymers are improving, allowing them to stimulate cell adhesion, proliferation, and tissue recovery. They've been widely used for skin repair and reconstruction, and there are still great obstacles to overcome. Natural polymers can be shaped into bioactive materials for cell, medication, or gene delivery, allowing for wound healing. The most widely used natural polymers, such as alginate and hyaluronic acid, are negatively charged due to their carboxyl or sulphate groups.[23]

Natural polymers are used for drug delivey, tissue regeneration, and bioimaging. In wound treatment, they are used as bandages for chronic and acute wounds. Natural polymers can be found in plants, animals, and microorganisms, among other places. Because of their extracellular matrix mimicking capability, excellent biocompatibility mechanical tunability, and high water-holding capacity, natural polymer scaffolds are widely used for skin repair and regeneration.[24] Researchers must choose and modify the correct material for a polymeric scaffold material that has specific properties for the kind of wound being treated, as well as redesign natural polymer-based biomaterials that stimulate skin regeneration while allowing for scar-free wound closure. The most promising naturally produced polymer-based biomaterials have outstanding biocompatibility, biodegradability, and reconstructing properties, making them promising for tissue and organ repair and replacement as well as implantation. The four deoxy nucleoside monomers comprised of nucleobases, namely, guanine (G), cytosine (C), adenine (A), and thymine (T), make up DNA, which is a polymeric biomaterial chain. DNA is an excellent step in developing novel natural polymer-based biomaterials for diagnostics, medication delivery, and protein synthesis.[21]

2.1.3.2 Synthetic Polymers

Human-made polymers are known as synthetic polymers. After serving their purpose, synthetic polymers degrade and are removed from the body. They should, in an ideal world, create natural intrinsic by-products. These polymers can be synthetic, such as ethers, amides, and esters or natural polymers such as proteins and polysaccharides, and they can be non-degradable or biodegradable. Non-degradable and biodegradable polymers are used to create biomaterials. A synthetic polymer-based biomaterial is defined as "any natural or synthetic substance developed to interact with biological systems to influence medical therapy".[25] Biopolymers are attractive biomaterials because they may be broken down, excreted, or absorbed without requiring surgery or removal. Biodegradable polymers are today's "materials". These polymers are now employed in a variety of biomedical applications, including drug administration, implant placement, and a few medical devices. Moreover, biodegradable polymers are increasingly being used in human life applications since they are ecologically beneficial.

Synthetic polymers are extremely important in the biomedical industry because their qualities, including as porosity, degradation time, and mechanical qualities, may be modified to meet specific needs. Synthetic polymer-based biomaterials are frequently less expensive than biologic scaffolds, and they can be manufactured in large, consistent batches with a long shelf life. Synthetic polymers, which make up the biggest category of biodegradable polymers and can be generated under controlled conditions, have physicochemical and mechanical characteristics that are equivalent to those of biological tissues. Elastic modulus, tensile strength, and degradation are all mechanical and physical characteristics that are predictable and reproducible. The most often utilized synthetic polymers in tissue engineering are PLA, PGA, and PLGA copolymers.[26] PHA is a kind of microbial polyester that is increasingly being considered for use in tissue engineering.[27] To enable the growth of new tissues when the immune system is activated in the body, synthetic polymeric biomaterials must have the necessary mechanical characteristics. Traditional polyolefins such as polyethylene (PE), polypropylene (PP), and polymethylmethacrylate (PMMA) have been effectively produced in this sector, as have degradable polymers such as polylactic acid (PLA), polyglycolide, polypeptides, polyurethanes (PU), polycarbonates (PC), and so on.[28] Blending, strengthening, or combining synthetic polymers with natural polymers or functional biomolecules might increase their bioactivity.

2.1.4 Desired Properties of Polymeric Biomaterials

When utilized as medical implants and organ or tissue replacements, the various characteristics of polymeric biomaterials have been proven to have a significant impact on their dynamic interactions with the biological environment.[29] The characteristics and biocompatibility of the polymeric biomaterial utilized to manufacture these implants determine their performance and success. As a result, while developing any medical device or implant, it's critical to consider the material's qualities in the context of the product's future biological application. In terms of the bulk and surface of the material, the most important qualities connected with biomaterials are categorized as chemical, physical, mechanical, and biological (Figure 2.3).

In the case of polymeric biomaterials, the body contains a variety of tissues such as skin, bone, and cartilage, each of which has its own mechanical and biochemical properties. Tissues are made up of cells that are embedded in a cell membrane (Cm) that contains proteins, polysaccharides, and other bioactive substances like growth factors. Langer and colleagues from MIT33 pioneered the discipline of tissue engineering, which entails constructing new tissues using a mixture of engineering, cell biology, and materials.[30] Researchers in the field of biomaterials are attempting to create biocompatible polymer scaffolds that allow cells to adhere, proliferate, differentiate, and

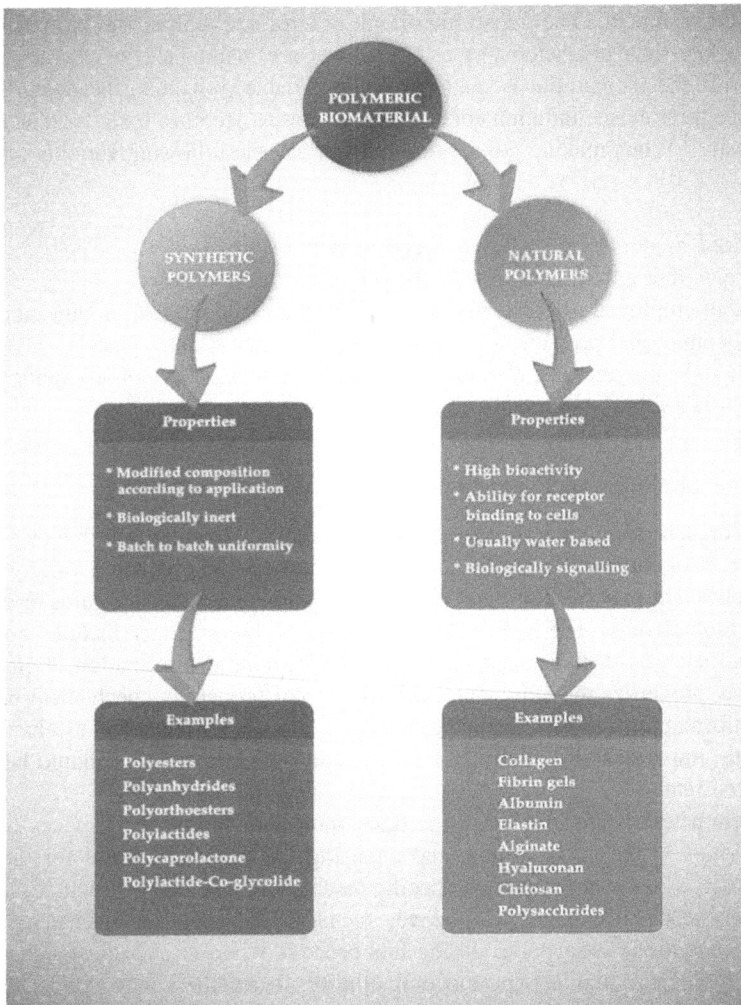

FIGURE 2.3 Desired properties of polymeric biomaterial.

construct their own matrix to repair tissue. Many cells require cell adhesion for life; thus, molecules that enhance cell adhesion have already been included into the design of polymer-based biomaterials.[30] Some characteristics, such as mechanical characteristics of biomaterials and growth factor delivery, have lately been considered. Scientists, on the other extreme, have long investigated the mechanism of cell attachment as well as cell mechanical properties.[31]

Polymers come in a wide range of compositions and qualities, both natural and manmade. Surface qualities are one of the most significant factors to consider in biological applications since they may be changed physically, chemically, or biochemically. Surface properties such as surface charge, surface energetics, type of sorbed water, surface molecule mobility, hydrophobicity, and the distribution of reactive chemical groups are all desired for polymeric biomaterials.[32] Biomechanical qualities such as tension, compression, and shear should be present in polymer-based biomaterials for specific applications. Depending on the application, they should have desirable physical structures like equilibrium swelling, entanglement and crystallinity, as well as other expected properties such as elasticity, electrical characteristics, and permeability.[33] Time, temperature, and environmental factors affect all of the characteristics of polymer-based biomaterials. The characteristics of polymeric biomaterials may alter after their application in a physiological environment. The most important criteria are polymeric biomaterials be bio-compatible and that their characteristics and functions are maintained in vivo for the necessary period of time. If the biomaterial is made of biodegradable polymers, the degradable products should not cause thrombosis, inflammatory responses, tissue necrosis, toxicity, allergic responses, or carcinogenesis.[34] When making polymeric biomaterials, the following variable qualities can be controlled:

- Polymerized monomers or copolymerized monomers.
- The reagents that are utilized to start the polymerization reaction.
- The reagent employed to crosslink the polymer chains' identification and quantity.
- The temperature and pressure under which polymerization takes place.
- The method the polymer is gathered, which might result in a more or less random polymer alignment or a fabric with chains aligned in a single orientation.[35]

2.1.5 Characterization of Polymeric Biomaterials

The analytical branch of polymer science is polymer characterization. The major purpose of the characterization is to enhance the performance of biomaterials. There are a variety of character-ization approaches that may be employed to get information about the structures or characteristics of polymeric biomaterials. The properties of polymeric biomaterials include both molecular characterization, such as degree of crystallinity, microstructural information, molecular weight, and macroscopic properties measurement, such as thermal properties, mechanical properties, mi-crostructural information, and time dependence of properties.[35] Furthermore, mechanical attributes such as strength, impermeability, thermal stability, and optical properties should be closely con-nected to desired features.[36]

Polymeric biomaterials are utilized in the development of biomedical devices that perform a variety of activities in the human body, and choosing the best biomaterial for these tasks is a complicated process that is dependent on specific mechanical, chemical, structural, and biological criteria. The physical and engineering sciences focus on the physical, mechanical, and surface qualities of the polymeric biomaterial during this process, whereas biological sciences and med-icine focus on biological qualities and biocompatibility. Researchers may extensively investigate the characteristics of polymeric biomaterials by combining diverse methodologies based on these material sciences. The following categories of surface qualities are responsible for the thorough characterization of polymeric biomaterials: Chemical attributes include elemental and molecular composition, as well as physical qualities such as specific surface area, porosity, roughness, and domain structure. Physiochemical qualities, such as the surface's capacity to interact with well-defined test medium, such as biomedically relevant compounds, and biological qualities, such as the pattern of interaction with complex biological systems.[37] The distinction between these groups is important. Physical and chemical qualities determine physicochemical and biological qualities, albeit this cannot always be defined.

Surface spectroscopies, thermal and electrochemical experiments, and microscopies are some of the characterization techniques used to collect data on polymeric biomaterials. Numerous char-acterization investigations have revealed that the primary polymer component does not alone control the interfacial characteristics of biomaterials. Processing conditions, the presence of che-micals and stabilizers, as well as sterilizing processes, have a significant impact on them.[38] Surface analysis techniques, which are often used to characterize polymeric biomaterials, may also be useful in elucidating interfacial interactions between materials and biological organisms. In-vitro model systems were shown to be beneficial for investigating protein adsorption and cell adhesion using these methodologies. Molecular mass, molecular structure, shape, thermal characteristics, and mechanical characteristics are often determined using characterization procedures. A number of experimental methodologies can be used to characterize polymer methods. Some of the most often used spectroscopic methods are nuclear magnetic resonance (NMR), infrared spectroscopy (IR), raman spectroscopy, UV-visible absorption spectroscopy, electron spin resonance (ESR), and mass spectroscopy (MS).[39] These methods are most commonly used to learn about the chemical structure of polymeric biomaterials, and they include obtaining a macroscopic polymer biomaterial specimen, usually in the form of a solid, and running tests to learn about its properties. The most relevant characterization of polymeric biomaterials is based on a selection of all imaginable molecular and physical characteristics.[40]

2.2 APPLICATION OF POLYMERIC BIOMATERIALS

Polymers have a lot of uses in the biomedical sector, and they're quite useful. These are less harmful, safer, and easier to use. Biomaterials made of smart polymers are utilized in a range of biological applications, such as blood oxygenation and cardiac surgery. Many applications benefit from the stimulus and response nature and sensitivity to physical factors. Different medical devices such as tissue adhesives, vascular grafts, dental composites, contact and intraocular lenses, suture materials and other biomaterials, all employ polymers.[41] Polymeric biomaterials are used in the production of biological screening devices, basic scientific research, and a variety of non-medical uses. These biomaterials are designed to show and organize arrays of molecules and cells on diverse length scales for mechanistic research and drug screening.[42] Novel approaches to biomaterials production, which frequently include both physical and chemical manufacturing procedures, have cleared the door for new diagnostic procedures.[43]

The capacity to develop information-rich materials that test numerous targets and enable different outputs is critical in our effort, just as it is in the creation of biomimetic medical devices.[44] The capacity to collect uncommon cell populations is a key aspect of these methods. Traditional probing tools, such as fluorescence in diagnostics,[45] and fundamental research into cell–matrix interactions have also been replaced by materials that change their properties in response to stimuli.[46]

Because of the relevance of these biomaterials, they have a wide range of applications that overlap in our daily lives. Polymeric materials are used in the creation of gadgets and prostheses for a variety of reasons due to their adaptability. Based on whether they are interacting directly with the human body, polymer applications in biomedicine may be divided into three categories.[47]

A. Utilization inside the human body.
   Typical examples are: polymethylmethacrylate (PMMA), silicones, polyurethanes, polyethylene (PE), polyamides (nylon), polyimides, hydrogels (e.g., poly-hydroxyethylmethacrylate (PHEMA) and polyesters (e.g., polyethylene terephthalate (PET), biodegradable polyesters (PGA, PLA).
B. When it comes into contact with the human body.
   Typical examples are: polyacrylonitrile (PAN), polysulfones, polyvinylchloride (PVC), and low-density polyethylene (LDPE).
C. When it does not come into contact with the human body (biomedical devices).

Typical examples are polystyrene (PS, PST), copolymers (SAN, ABS), and polycarbonate (PC). Because many polymeric biomaterials have drawn significant attention in the fields of tissue engineering scaffolds, medicine delivery, dental supplication, biosensors, gene delivery, and bioimaging, their application in the human body will be given greater consideration (Figure 2.4).

For life applications, the most important step is to choose a polymeric biomaterial. They should aid in the regeneration of new tissue while preventing inflammation. In the creation of biomaterials, polymers provide a wide range of benefits. Imaging, sensing, and treatment with extremely magnified applications have been made possible because to brilliant design and production of polymeric biomaterials. Polymeric biomaterials for bone-related applications include nondegradable and

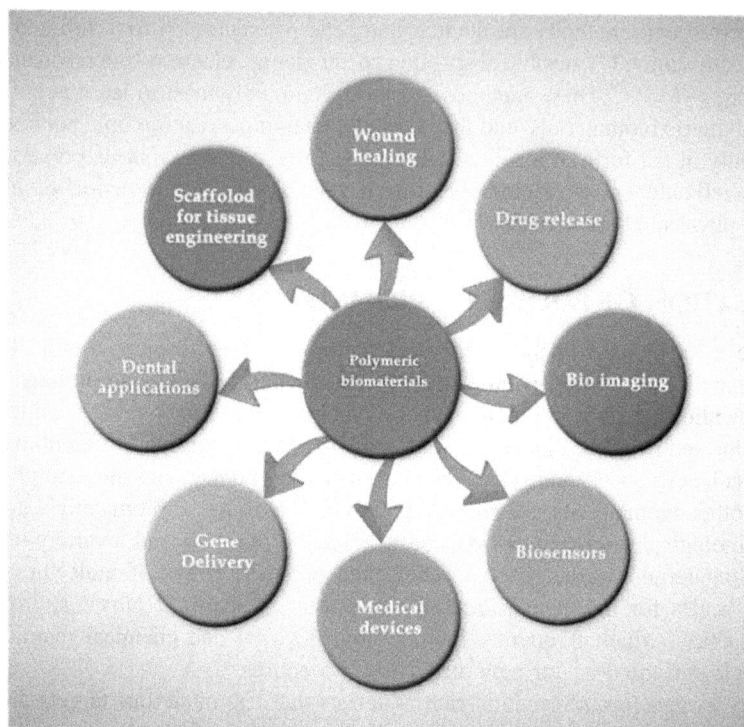

FIGURE 2.4 Applications of polymeric biomaterials in different fields.

TABLE 2.1

Some Special Applications of Polymers as Biomaterials

Polymers	Properties	Applications
PMMA	Stability and resistance to corrosion	Dental
(Polymethylmethacrylate)	Resistance to tiredness and strength	
Polyamides	Excellent adherence	
	Low allergenicity	
Polyacrylamide	Ability to create a gel or a film	Ophthalmic
PHEMA	Permeability of oxygen	
(Polyhydroxyethylmethacrylate)		
Polyethylene	Resistance to fatigue and strength	Orthopedic
Polypropylene	Bones and muscles are well integrated	
PMMA		
Silicones	Fatigue resistance	Cardiovascular
Teflon	Lubricity	
Polyurethanes	Sterilizability	
Silicones	Compatibility with drugs	Drug delivery
HEMA	Biodegradability	
Polypropylene	Good fracture toughness	Sutures
Nylon	Flexibility	
	Retention of knots	

biodegradable polymeric biomaterials for constructing prosthetic devices. Three-dimensional (3D) scaffolds are primarily used to assist tissue regeneration[48] (Table 2.1).

2.2.1 Scope of Polymeric Biomaterials

The fascinating area of polymeric biomaterials is undergoing a dramatic shift, with biological sciences taking equal precedence over materials science and engineering as the subject's basis. Advances in engineering, such as nanotechnology, have substantially increased the sophistication with which polymer-based biomaterials are developed, allowing production of more sophisticated functionalities. Such advanced polymeric biomaterials are frequently developed to imitate a subs-cluster of natural materials' physicochemical features. Nature is increasingly inspiring not just the materials themselves, but also the methods by which they are created. Self-assembly, a bottom-up approach of manufacturing that allows the configuration of natural materials on these lower sizes, is used to create natural materials on these smaller scales. Synthetic materials are generally created on the size of millimeters or greater, then machined to have micrometer-scale or nanometer-scale features.[49] Biomaterials that interact with the biology of the host are being developed utilizing information obtained through systematic study and manufacturing techniques such as self-assembly. To govern the maintenance, regeneration, or even elimination of particular tissues in the body, this is usually achieved via binding interactions with cell surface receptors.[50,51]

Polymeric biomaterials offer a lot of potential in the biomedical area. They now have a 50-year track record of performance, with applications in a variety of sectors (Figure 2.5). Successful applications include enzyme immobilization, wound dressing, tissue regeneration, surgical sutures, controlled drug delivery and gene delivery, nanotechnology, prosthetics, cosmetics, sanitation products, coatings, and many more.[52] Long-term exposure and related toxicity are emerging

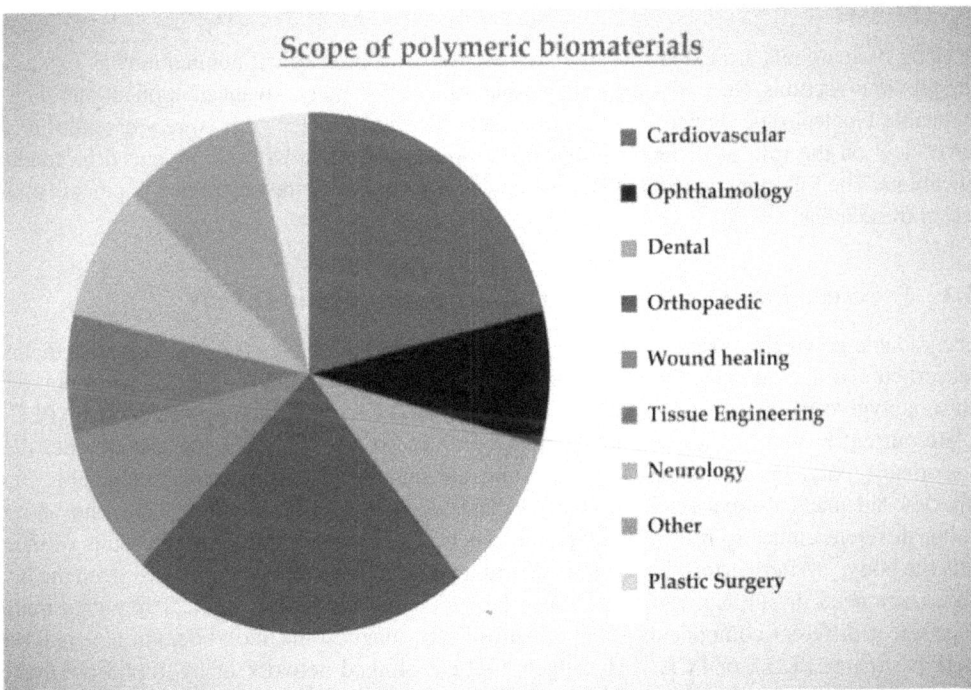

FIGURE 2.5 Application of polymeric biomaterials towards different biomedical applications.

problems with emerging uses. Long-term toxicity of tissue fillers in rats was recently found in our research, for example. In the last 20 years, biodegradable polymeric biomaterials have advanced significantly. The use of these polymers as biomaterials is growing, with novel polymers being reported on a regular basis. As a result, they are the "green biomaterials" that are most likely to replace today's conventional polymers. Biodegradable polymer-based biomaterials have a lot of potential in the field of biomedicine.[53]

Mechanical robustness was a major drawback for polymeric biomaterials based on biodegradable polymers, but they have since addressed many of these drawbacks. Newer synthetic biodegradable polymers and composites have aided in the solution of the problem. Another key trend in biodegradable polymeric biomaterial research in recent years has been tailor-made tunable characteristics. It's possible to fine-tune the chemistry to create spontaneously responsive and shape memory polymers. Biodegradable polymeric biomaterials may now be fine-tuned to disintegrate over a period of two years. They can be utilized in medical implants and devices, depots, tissue augmentation, and other applications as a result of this. However, when new applications emerge, there are increasing concerns regarding long-term exposure and toxicity.[54] The clinical laboratory has discovered long-term toxicity of tissue fillers in rats. Despite the fact that no clinical damage has been documented, there is still reason to be cautious.

In the realm of tissue engineering, composite scaffolds are currently being investigated and developed for favoring the cell adhesion, proliferation and differentiation to form new tissue. However, there is increasing interest in developing three-dimensional scaffolds that can promote tissue regeneration while simultaneously acting as a biomatrix for cues and signals to improve functional tissue connection.[55] Despite major technological, scientific, regulatory, and even ethical challenges, there has been a huge increase in study and interest in bringing 3D bioprinting to reality.[56,57]

2.3 POLYMERIC BIOMATERIALS FOR ADVANCED APPLICATIONS: CURRENT TRENDS

Polymeric biomaterials are extensively utilized for different biomedical applications as explained in the previous sections. Recently, they are getting utilized for many advanced applications such as implantable bioelectronic devices, soft robotics, and bioprinting. Hence, it is very essential to get an overview on the role of different polymeric biomaterials towards these advanced biomedical applications. The following section attempts to give an overview of the recent developments taking place in these areas.

2.3.1 POLYMERIC BIOMATERIALS FOR IMPLANTABLE BIOELECTRONIC DEVICES

Biodegradable polymers represent the major class of bioresorbable material i.e., materials that can be absorbed and degraded by the body into benign products.[58] Mainly polyester based polymers such as polyglycerol sebacate (PGS), polycaprolactone (PCL), polylactic-glycolic acid (PLGA) etc. are currently used as the common bioresorbable materials.[59] During the last decade, these bioresorbable polymers are widely used as substrate material for fabricating implantable bioelectronics and medical devices. These bioresorbable devices can be implanted into the body to monitor different clinically relevant parameters, such as pH, temperature, and pressure variation, inside the body.[60] After performing its intended function, the device gets eliminated from the body without any need of surgical removal. These bioresorbable electronic medical implants mainly comprises of different components such as nanosilicon, magnesium, molybdenum and polyester materials such as PLGA or PCL.[60] Hydrogels (3D crosslinked network of hydrophilic polymers with very high water absorbing capability) play a major role as implantable bioelectronic devices by providing a good bio-interface between the electronics components and tissues.[61] The high water content and good tissue like properties of the hydrogel make it suitable substrate material for bioelectronic devices. Thus, it was clear that different synthetic polymeric biomaterials are important in the development of new generation of implantable electronic devices. They are mainly utilized as a substrate material to embed the different degradable electronic components such as nanosilicon, magnesium, molybdenum. The polymer matrix acts as a protective layer which undergoes gradual degradation with respect to time and ensures for the clearance of the electronic device from the body. In general the operational lifetime of any bioresorbable implantable bioelectronic medical device greatly depends on the dissolution or degradation of the encapsulated polymer layer and permeation of fluids in to the electronic components inside.[59] Thus, the degradation characteristics of the bioresorbable polymeric substrate can be directly correlated with the device function and lifetime. In this regard, recently there are reports on polyanhydride and silk protein encapsulated bioelectronic devices.[62,63] The performance and lifetime of these implantable devices were correlated with the *in vitro* dissolution or degradation rate of the polyanhydride or silk protein layer. The in vitro degradation kinetics of the device gives a good idea about the in vivo degradation of these implantable bioelectronic devices. There are many recent reports regarding the development of polymer-based degradable bioelectronic devices.

Recently, Zhong *et al.* have reported a transient electronic device which was encapsulated with a UV light responsive polymer-fullerene encapsulating layer (Figure 2.6).[64] More specifically, the encapsulating layer was comprised of fullerene derivatives and a polymer namely [6,6]-phenyl-C61-butyric acid methyl esters (PCBM). This polymer-fullerene encapsulating layer has provided a smart UV light triggered degradation characteristics to the transient electronic device. The UV light induces light-assisted degradation of the PCBM, which causes great amounts of water

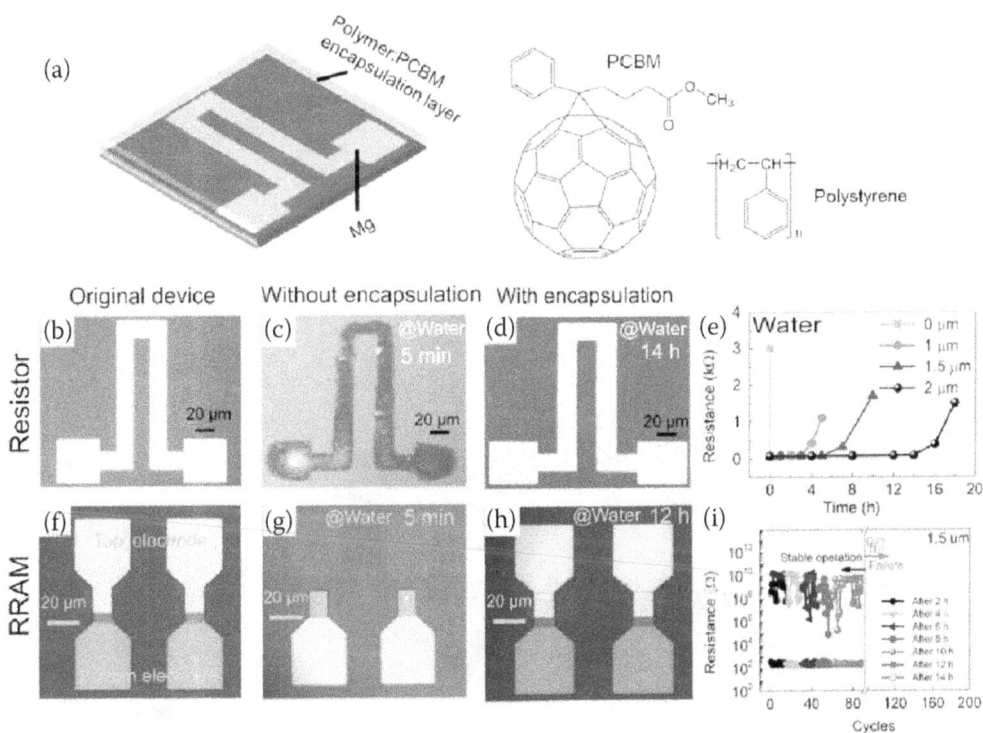

FIGURE 2.6 (a) Schematic depicting transient devices. Optical images of (b) an original Mg resistor, (c) hydrolyzed Mg resistor, (d) intact Mg resistor, (e) resistance change in Mg resistor. Optical images of (f) RRAM device, (g) device without encapsulation after 5 min water immersion, and (h) device with 1.5 μm PS:PCBM encapsulation layer after 12 h of water immersion. (i) Lifetime investigation of transient Mg/MgO/W RRAM device. "Adapted with permission from.[64] Copyright (2019) American Chemical Society".

permeation in to the electronic device, which leads to enhanced hydrolysis of magnesium and causes the device degradation. Such new light responsive polymeric layer provides new material design pathway in bioresorbable-polymeric-material-based implantable devices.

In an interesting work, Choi *et al.* have reported the usage of bioresorbable dynamic covalent polyurethane (b-DCPU) material as a substrate material for a wireless bioresorbable electrical stimulation device which can enhance the neuromuscular reinnervation after the process of denervation injury (Figure 2.7).[65] The electrical device component was mainly comprised of a

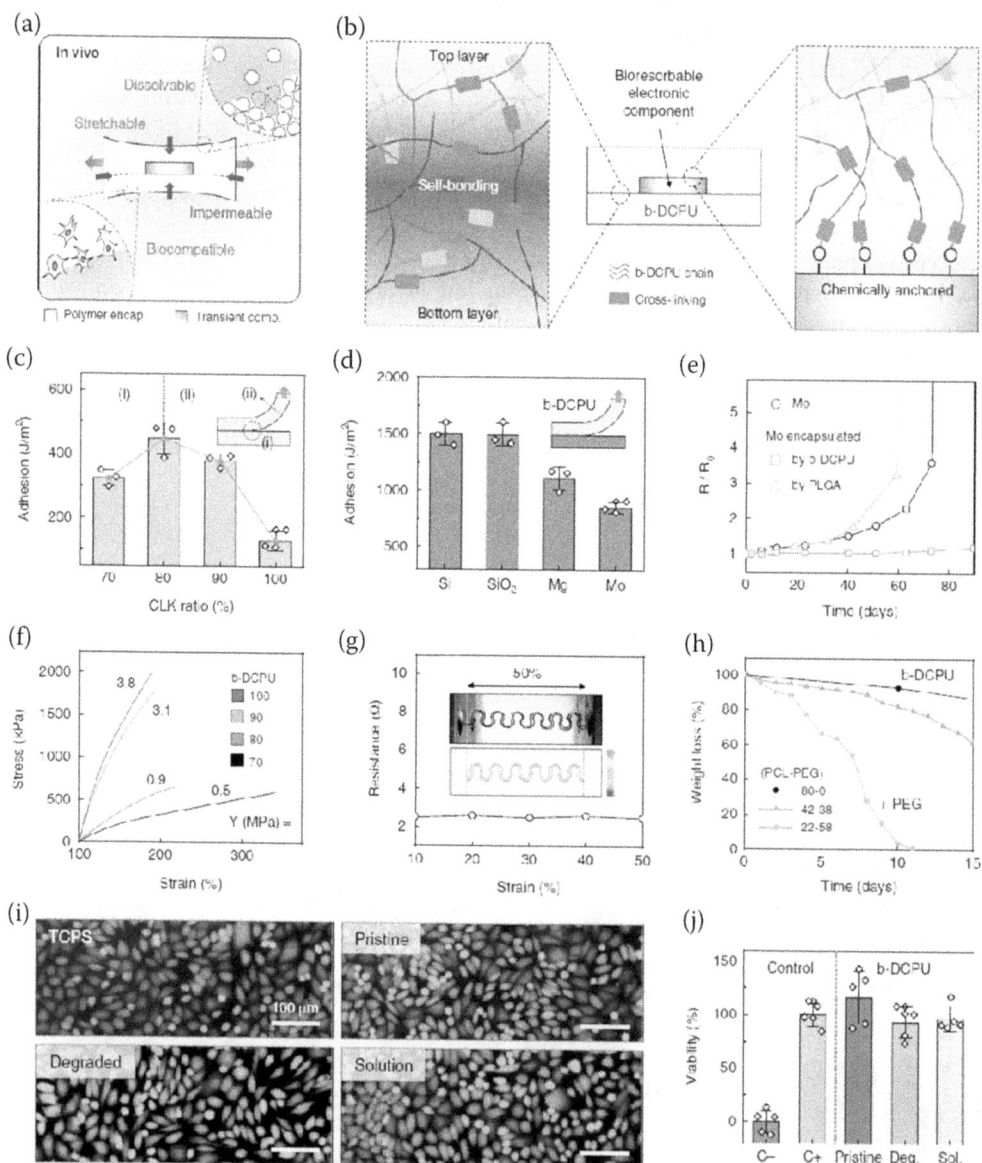

FIGURE 2.7 (a) Schematic illustration of the key requirements for the materials. (b) Schematic illustration of the adhesion mechanism. (c) Adhesion energy between bonded pieces. (d) Adhesion energy between b-DCPU 80 and other inorganic bioresorbable materials. (e) Changes in resistance. (f) Stress–strain curves for b-DCPU materials with different crosslinking ratios. (g) Changes in resistance of serpentine Mo electrodes. (h) Measured changes in weight as a function of the immersion time. (i) Biocompatibility of various b-DCPU samples. j Normalized in vitro viability assay data. *Copyright © 2020, Choi et al,*[65] *Nature Communications (2020) 11:5990.*

biodegradable harvester which is a molybdenum (Mo) loop antenna in a bilayer. This electrical component was encapsulated with stretchable polyurethane as a dielectric interlayer and a doped membrane of monocrystalline silicon. The *in vitro* and *in vivo* studies have clearly demonstrated the capability of the dynamic and flexible polyurethane polymer to act as an ideal substrate/encapsulating material which can serve for multiple purposes such as providing electrical reliability, mechanical stability, bioresorbability, and biocompatibility.

Development of such bioresorbable polymer supported electronic devices shows the potential towards the development of long-term electrical stimulation strategy for maintaining muscle receptivity and enhancing the functional performance of the device.

Optogenetics is another recently emerging area in implantable bioelectronic devices.[66] More specifically, optogenetics focuses on the target-specific spatiotemporal manipulation of neuronal activity. Polymeric materials do have an important role to play in the design of optogenetics also as a soft encapsulating or supporting layer for the device.[67] In this regard, very recently Kim *et al.* have reported a soft subdermal implant material which is capable for wireless charging and provides a programmable control for optogenetics (Figure 2.8).[68] More specifically, the designed device comprised of different components such as inorganic light emitting diodes, power management circuit, radiofrequency coil antennas, battery, and bluetooth low-energy system. The polymer encapsulation layer was composed of multiple polymers such as polydimethylsiloxane and Parylene C. This encapsulating polymer layer serves for multiple purposes such as providing device protection from biofluids and external shock, this endows conformal biointegration for more stable and reliable operation inside the body system. In the case of conventional implantable devices which work on battery, the removal of battery after its lifetime through surgery is a big stress. The wireless charging capability of this device completely eliminates this surgical removal stress associated with the implant material. The smartphone-assisted control of this device also facilitates future therapeutic inventions with the brain.

All of these recent research works clearly show the very high importance of polymeric materials as a soft supporting layer of many different bioelectronic devices. The polymer layer serves for different purposes such as providing device protection from biofluids and external shock, biocompatibility, and conformal biointegration for adaptive and reliable operation inside the body.

2.3.2 POLYMERIC BIOMATERIALS FOR BIOPRINTING

The printing process has shown advancement from two-dimensional printing in to more sophisticated additive manufacturing which is also known as 3D printing.[69,70] In 3D printing, the printing process is accomplished with the help of computer aided designs. Complex 3D shapes are achieved by the successive printing of different layers on the top of each other which finally forms the desired 3D shape designed by the computer aided software.[71,72] The term 3D printing was first coined by in 1986 by Charles W. Hull; he has shown the ultraviolet light assisted curing of thin layers of a material which can be printed in layers to form a solid 3D structure.[73] Afterwards, the usage of aqueous-based systems for the 3D printing has enabled the incorporation of biological moieties like cells inside the 3D-printed construct which has laid the foundation for 3D bioprinting.[74–76] Subsequently, the area of 3D bioprinting has rapidly evolved and immensely contributed in the cell biology and material science. Currently, researchers are exploring this technology for the fabrication of functional 3D biological constructs which has sufficient biological and mechanical properties suitable for clinical restoration of tissue and organ function.[77–79] One of the major challenges currently associated with 3D bioprinting is to mimic the complex extracellular matrix architecture of cells in sufficient resolution and mechanical strength.[80,81]

Different polymeric materials such as natural and synthetic polymeric materials are widely utilized for bioprinting process.[82,83] Natural polymers are more efficient in mimicking the extracellular matrix of cells where as synthetic polymers can be easily tailored for more efficient printing.[84] Due to these reasons, both synthetic and natural polymeric materials are widely utilized

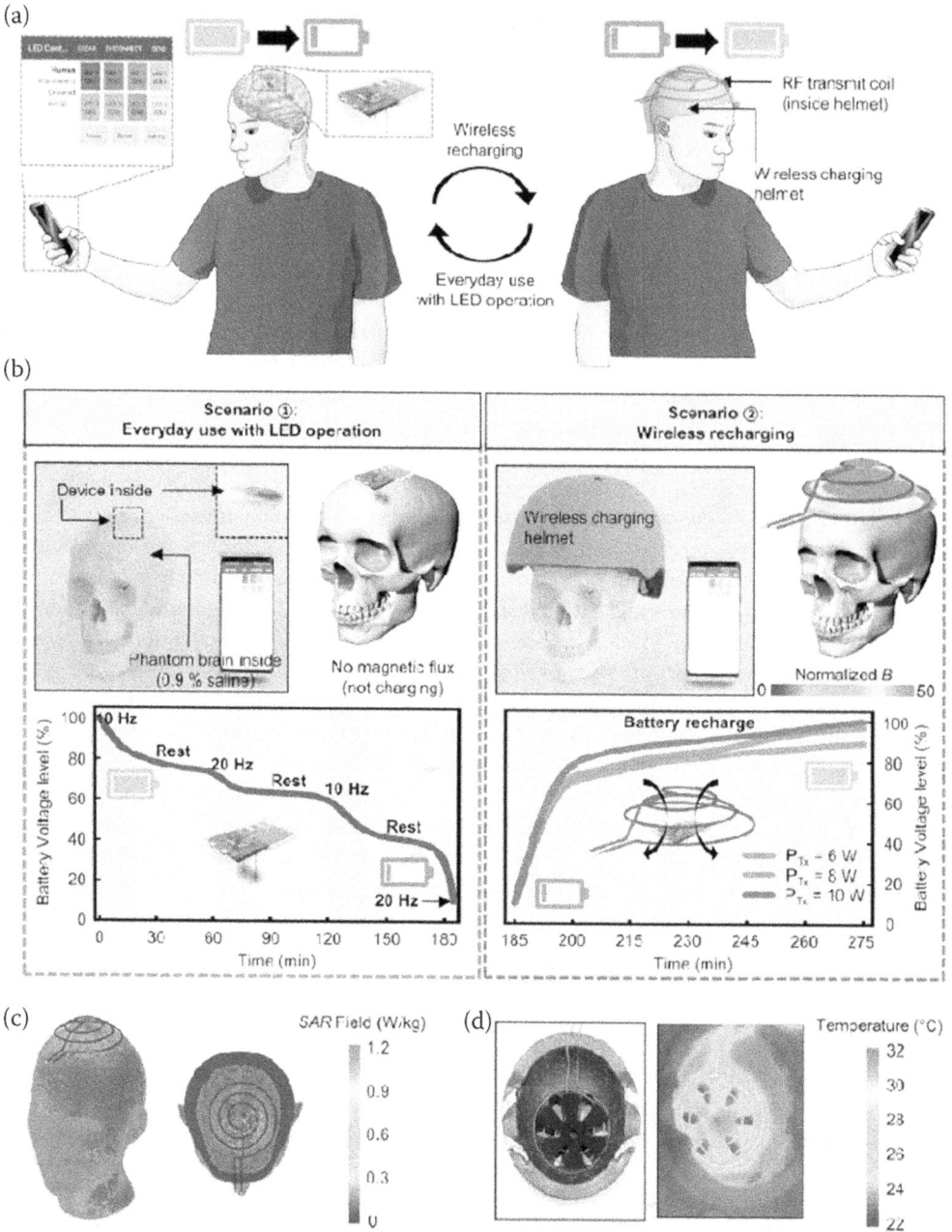

FIGURE 2.8 (a) A man with the wireless system implanted in his/her brain. (b) Optical images and electrical characteristics of the system, (1) everyday use with LED operation and (2) wireless recharging. (c, d) Simulated illustration of the specific absorption rate (SAR) over the human head. *Copyright © 2021, Kim et al.,*[68]*Nature Communications (2021) 12:535.*

for bioprinting. The synthetic polymers currently utilized for the bioprinting process are poly (ethylene glycol) (PEG), poly(lactide-*co*-glycolide) (PLGA), poly(ε-caprolactone) (PCL), and poly (l-lactic acid) (PLLA).[85] They offer tailorable mechanical properties and degradation characteristics with good printing resolution, however the hydrophobic surface characteristics of them makes it difficult for the cell attachment. The currently employed natural polymers for 3D biprinting are collagen/gelatin, alginate, fibrin, hyaluronan, and dextran.[86–90] They all have good extracellular mimicking environment with good cell adhesion characteristics, however some of the other properties such as mechanical properties and degradation rates are not tailorable. Both synthetic and natural polymers are utilized as a bioink for 3D bioprinting. A bioink is defined as a matrix comprising of cells and bioactive signals which can be printed in to desired shape and resolution without compromising the activity of the cells/bioactive signals.[91] Polymeric materials serve as an ideal matrix to form a bioink due to some of its advantageous properties such as extracellular mimicking characteristics, enhanced cell adhesion, and good mechanical properties.[92] A bioink material should retain its structural integrity after the printing process; hence, many different properties of polymeric material will affect the overall performance of the bioink. The viscoelastic property of the polymer material is one of the major properties of the polymeric bioink which affects printed structure outcome, or polymer behavior in solution which will govern the printing resolution.[93] Apart from viscoelasticity, there are different other properties controls the success of polymeric bioink such as hydration, degradation, mechanical stability, cell adhesion, and cross-linking mechanism.[93] Hence, it was clear that careful optimization strategies are required for the design of an efficient polymeric bioink material. As natural and synthetic polymeric materials have their own advantageous and disadvantageous, it is not easy to use either one of them exclusively for the design of bioink. This has led towards the development of hybrid polymeric bioink where both natural and synthetic polymers are combined together to form the bioink.[94] The hybrid polymeric bioink can combine the advantageous properties of both natural and synthetic polymer bioinks such as enhanced cell adhesion, optimal mechanical properties and good printing resolution. One such example for a hybrid polymeric bioink system is hyaluronic acid (HA)-PEG hydrogel bioink where the HA component enhances the cellular adhesion and PEG provides the mechanical properties and printing resolution.[95]

Recently, many important developments are taking place in this area; Kim et al. have reported a biopolymer silk fibroin (SF) based bioink for digital light processed 3D printing (Figure 2.9).[96] For imparting the bioprintability to this silk-based bioink, SF was metharcylated after reacting with glycidyl methacrylate (GMA). The mechanical and rheological characteristics of the SF were tailored by varying the methacrylation degree of the silk fibroin. This silk based bioink was printed in to different complex biological organs such as heart, vessel, brain, trachea, and ear. These printed organ constructs have exhibited structural integrity and good mechanical property. Development of such biocompatible bioink may find wide utility towards tissue engineering applications.

In another interesting work, Yuk et al. have reported a conducting polymer-based bioink for 3D printing (Figure 2.10).[97] More specifically, in this work, authors have reported a poly (3,4-ethylenedioxythiophene):polystyrene sulfonate (PEDOT:PSS) conducting polymer-based bioink. This conducting polymer-based bioink was used to print high aspect ratio microstructures with good printing resolution. These printed microstructures were also integrated with insulating elastomers via multi-material 3D printing. The biomedical utility of this 3D-printed conductive bioink constructs were demonstrated as a neural probe capable of recording in vivo signals.

Recently, four-dimensional (4D) bioprinting has emerged as a new discipline where time is integrated with the 3D bioprinting which acts as the fourth dimension.[98,99] In 4D printing, the printed constructs change some of their properties such as shape and functionalities with time in response to external stimuli. The different stimulus includes temperature, water, and magnetic field.[100,101] Whenever these different stimuli are coming in contact with the bioprinted scaffold materials, they can reshape and transform in to different structures. These time-dependent stimuli responsive characteristics of the bioprinted scaffold material greatly help to mimic the complex architecture of native tissues such as heart and liver.[98] This makes 4D bioprinting highly important in the biomedical device industry. Polymer materials have different functional groups which can endow it with stimuli responsive nature. This makes polymeric materials an important candidate for 4D bioprinting applications. One such example is poly (N-isopropyl acrylamide) (PNIPAAm).[102] Which forms

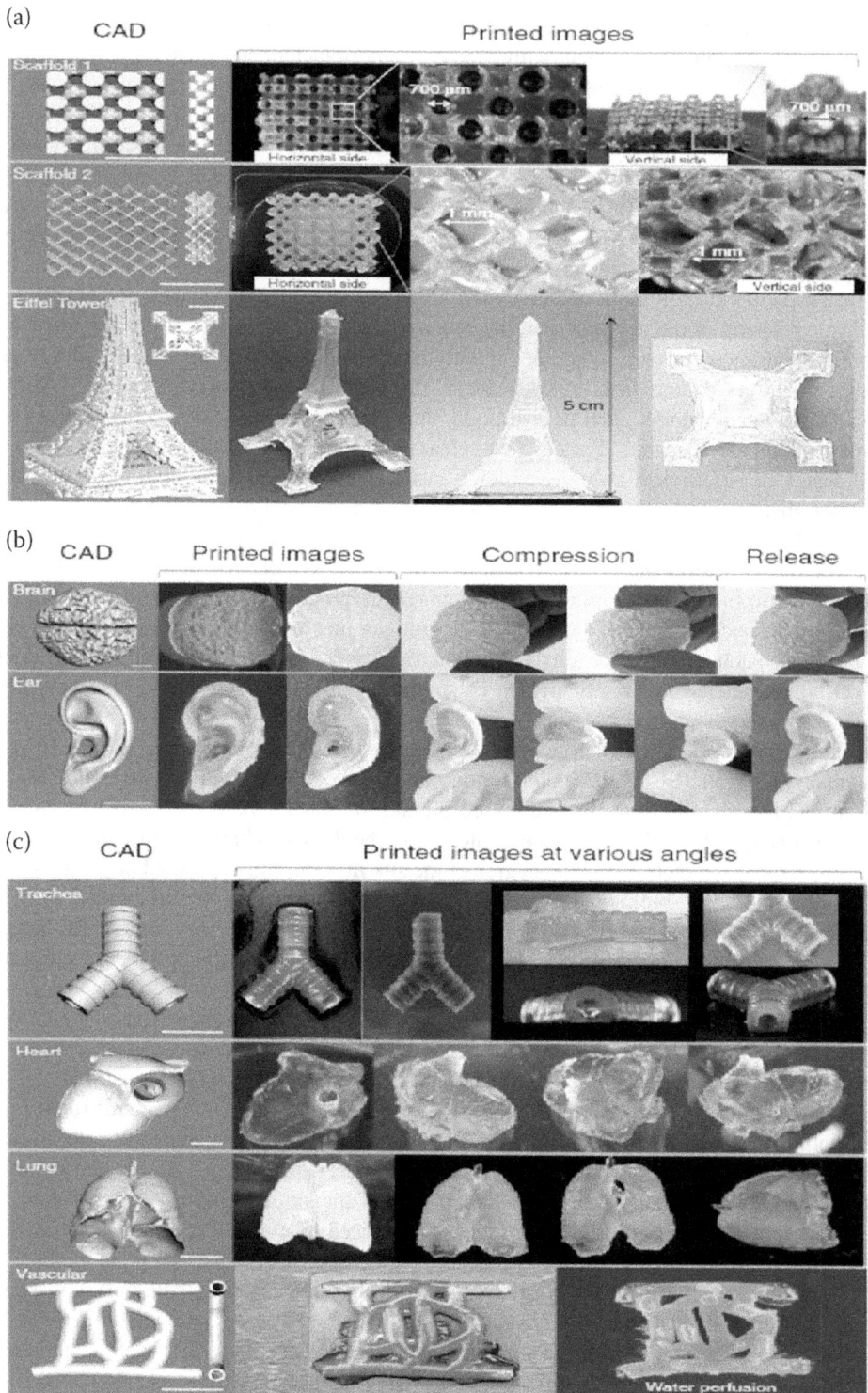

FIGURE 2.9 (a) Porous scaffold and Eiffel Tower imitation; (l) CAD images depicting scaffolds and Eiffel Tower and printed images. (b) Ear and brain mimicked shape; (l) CAD images depicting the ear and brain and printed images. (c) Trachea, heart, lung, and vessel mimicked shape. *Copyright © 2018, Kim et al.[96] Nature Communications (2018) 9:1620.*

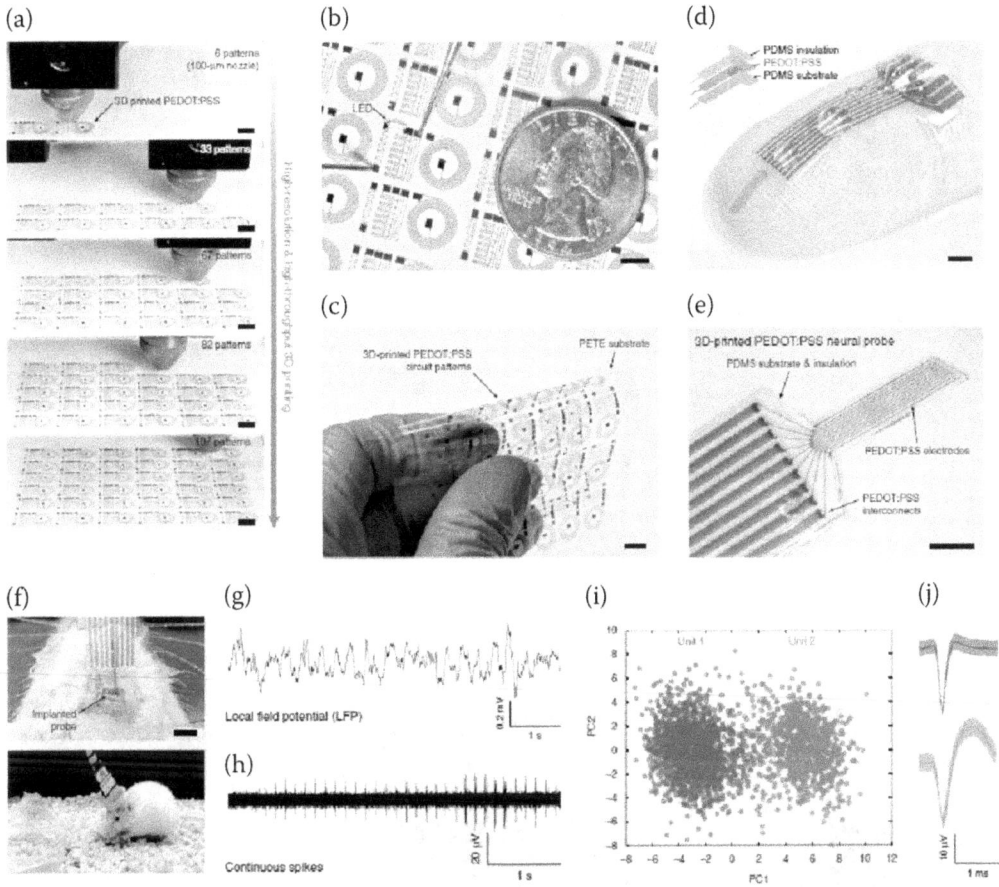

FIGURE 2.10 (a) Sequential snapshots for 3D printing of high-density flexible electronic circuit. (b) Lighting up of LED on the 3D-printed conducting polymer circuit. (c) Bending of the 3Dprinted conducting polymer circuit without failure. (d) Image of the 3D-printed soft neural probe. (e) Image of the 3D-printed soft neural probe in magnified view. (f) Images of the implanted 3D-printed soft neural probe. (g, h) Representative electrophysiological recordings in the mouse. Local field potential under freely moving conditions (g). AP) traces (300 to 40 kHz) recorded under freely moving conditions (h). (i) Principal component analysis of the recorded single-unit potentials from (h). j Average two units spike waveforms recorded over time (i). *Copyright © 2020, Yuk et al.[97]Nature Communications (2020) 11:1604.*

bilayered scaffold with water-insoluble polymer poly (e-caprolactone) (PCL) which can self-fold or self-unfold in response to temperature stimuli, this response could be utilized for yeast cell encapsulation and release.[103]

Shape memory polymers (which have the capability to return from a deformed shape to their original state under the assistance of an external stimulus like temperature) are widely utilized for 4D bioprinting applications as they have the temperature-controlled shape changes.[104,105] Currently most of the shape memory polymers are having high shape transition temperature which makes it more inconvenient for biomedical implantations. This has led towards the quest of new shape memory polymers whose transition temperature lies in the room temperature range. In this regard, Zhang *et al.* have reported a new shape memory polymer named poly (glycerol dodecanoate) acrylate (PGDA) (Figure 2.11).[106] This new polymer was synthesized by the polycondensation of glycerol and dodecanedioic acid. This prepared PGDA polymer has shape transition temperature around 20–37°C, which lies in the room temperature range. This

(a)

(b)

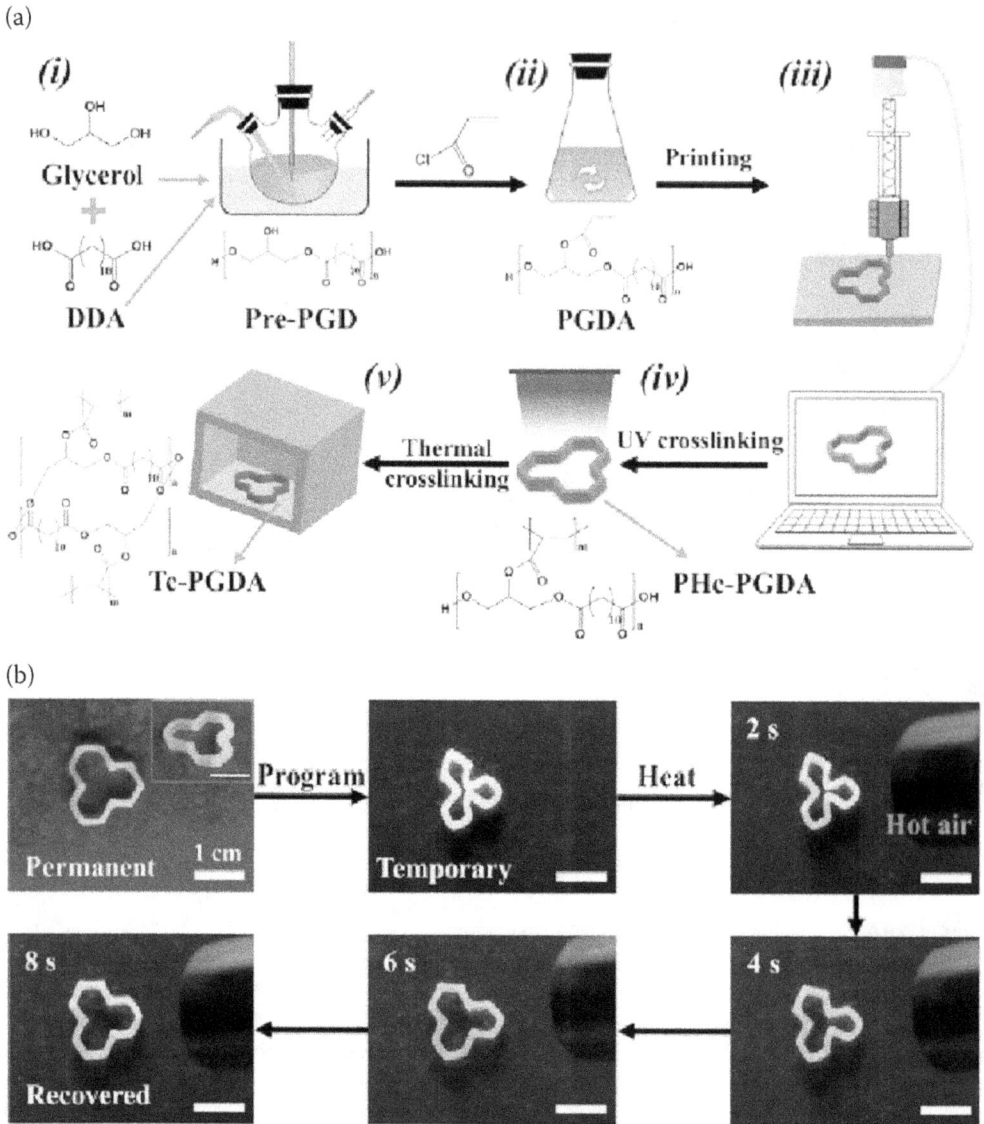

FIGURE 2.11 (a) Schematic of the synthesis and 3D printing process used for PGDA. (b) Photographs showing the shape memory behavior of a printed "triangular star" structure. "Adapted with permission from (109). Copyright (2021) Elsevier Publishers".

low shape transition temperature range makes it ideal for biological applications. This new polymer was utilized to fabricate different structures such as triangular star, six-petal flower, and honeycomb. These 3D structures have shown high shape fixity and recovery ratios. The biomedical use of this shape memory polymer was demonstrated by showing its utility as a stent and vascular graft.

2.3.3   POLYMERIC BIOMATERIALS FOR SOFT ROBOTICS

Robotics was first developed to perform human or animal like functions to be performed by machines. The basic principle behind robotics was to mimic the body plan, skeleton, muscles, and nerves of human beings/animals for useful functions with greater strength, speed, and accuracy.[107]The components for robotics such as actuators and other structural building materials are made from rigid materials and hence they are commonly referred to as hard robotics.[108] For example, rigid materials like electric motors and pneumatic pistons are used as the actuators in hard robotics. These robotic systems are used to replicate some of the functions of humans/animals such as walking or gripping (currently talking and maybe in future understanding).[107] Human mimetic hard robotic systems are usually very complex in design and hard to manage. This has led towards the development of new robotic systems which are less complex than human mimicking robots known as soft robotics.[108] Soft robotics focuses on mimicking organisms that are anatomically and functionally less complicated than humans such as starfish, worms, and other simple animals.[109] These soft robotics are useful for very complex motions and functions. In simple terms, soft robotics are mimicking simple biological models (which are less complex than human beings anatomy and other structural features) that have the potential to impart useful new functions for robotics.[109] Mimicking the structure and function of these simple biological models can be easily accomplished in comparison with higher organisms like human beings. Soft robotics is very simple and low cost compared to the hard robotics. These soft actuators have demonstrated no frictional surfaces and wear with hard robots.[107]

In the early phase of soft robotic system development, the focus was to mimic the features of organisms without internal skeletons. This may seem easy but in actual practice they are not as much easy as they appear. For example, the mimicking of the sophisticated movement of a worm is still not easy to replicate. Starfish is one of the commonly employed organisms which are mimicked to form grippers.[110] The squid and octopods are other biological organisms whose functions and movements are attempted to replicate.[107] The first example of a useful soft robotic device was a pentagonal gripper.[109] This design was inspired from a starfish structure; this soft robotic device was found to be capable of gripping an uncooked egg. Soft robotics major application lies in medicine where they can be utilized as assistive devices for healthcare workers, elderly care devices, treatment and prevention of pressure ulcers etc.[107] Polymeric material especially elastomeric poly (dimethyl siloxane) (PDMS) is widely used as a structural material to design soft robotics.[111] This was clearly showing the importance of elastomeric polymers in this recently developing field of soft robotics. PDMS possess different useful properties such as transparency, ease of sterilization, and biocompatibility which makes them an ideal structural material for soft robotics.[111] There are some recent approaches of soft robotics designs based on PDMS which are worth of reviewing.

Lu et al. have reported a bioinspired multi-legged soft robot which can move very rapidly in both dry and wet conditions (Figure 2.12).[112] This particular soft robot design was inspired from nature where different legged animals which uses their legs for effective body supporting, higher motion agility and better obstacle over cross ability. This has inspired the authors to design a soft millimeter size robot with multi-leg architecture. This soft robotic structure was built by using polydimethylsiloxane (PDMS), hexane, and magnetic nanoparticles. The designed multilegged robot was found to be capable of moving in different harsh environments with very high locomotion speed (>40 limb length/s), ultra-strong carrying capacity (>100 own weight), and excellent capability to cross obstacles. Magnetic torque and pulling force were utilized to drive the robot and the flexible legs of this device could store and release elastic energy during the locomotion process. Interestingly, it was also found that the locomotion in one gait cycle exhibits very high similarity to that of natural human walking process. The biomedical application of this soft robot was demonstrated by carrying out locomotion of the robot on a structure that was mimicking human stomach and a medical tablet was adhered on the robot body.

There are organisms in nature like African trypanosomes that can respond and change their motion and morphologies to adapt to changes with their microenvironment. Inspired from such natural small organisms and building micro-machines that can change the morphology and architecture of their bodily parts will be highly useful for a wide range of applications. In this regard,

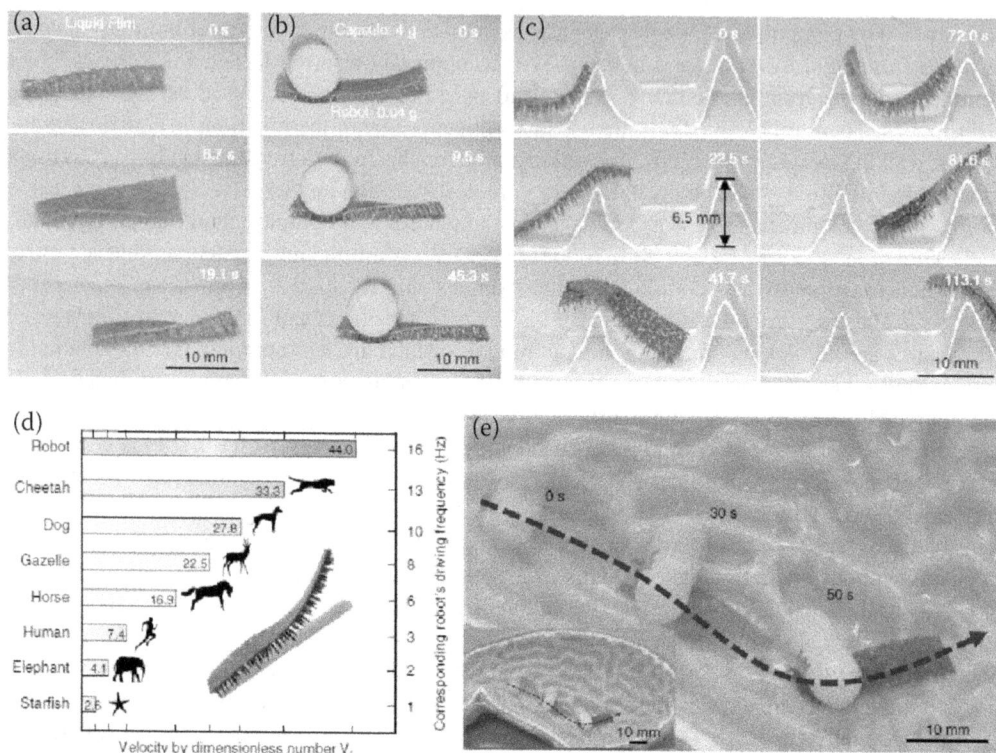

FIGURE 2.12 (a) The robot can move with an average speed of 0.5 mm/s on wet surface (b) Robot locomotion with a loading 100 times of its own weight. (c) Cross a steep obstacle with height ~10 times higher of its own leg. (d) Comparison of the normalized speed between the soft robot and other animals. (e) Demonstration of drug transport in a stomach model under wet environment. *Copyright © 2018, Lu et al.,*[112]*Nature Communications (2018) 9:3944.*

Huang *et al.* have reported an origami-inspired rapid prototyping process for preparing a self-folding and magnetically powered micro-machine (Figure 2.13).[113] This micro-machine was found to have complex body plans, reconfigurable shape, and controllable motility. It was found that the tail and body morphologies together determine swimming efficiency of the micro-machine. Lithographic patterning of hydrogel sheets formed from the ultraviolet-assisted polymerization of N-isopropylacrylamide (NIPAAm) and poly (ethylene glycol) diacrylate (PEGDA) was utilized to construct this micro-machine. Motion capability was conferred to this device by the incorporation of magnetic nanoparticles in the hydrogel layers and controlled by magnetically aligning the particles during polymerization. The magnetic nanoparticles component present in this machine also confer it with near-infrared (NIR) responsive characteristics, more specifically it allows the spontaneous folding and controllable structural characteristics without the need for further assembly or reassembly steps.

2.3.4 POLYMERIC BIOMATERIALS IN THE FUTURE

The creation of "smart", multifunctional nanoparticles or implants for usage in live organisms will most likely be a priority of polymeric biomaterials research in the new generation of bioinspired materials. Recognition of biological polymers and animals as models or templates for multifunctional, dynamic devices is a crucial intellectual step in biomaterials design.[114–116] It's hard to say which polymer is the best since there are so many natural and synthetic

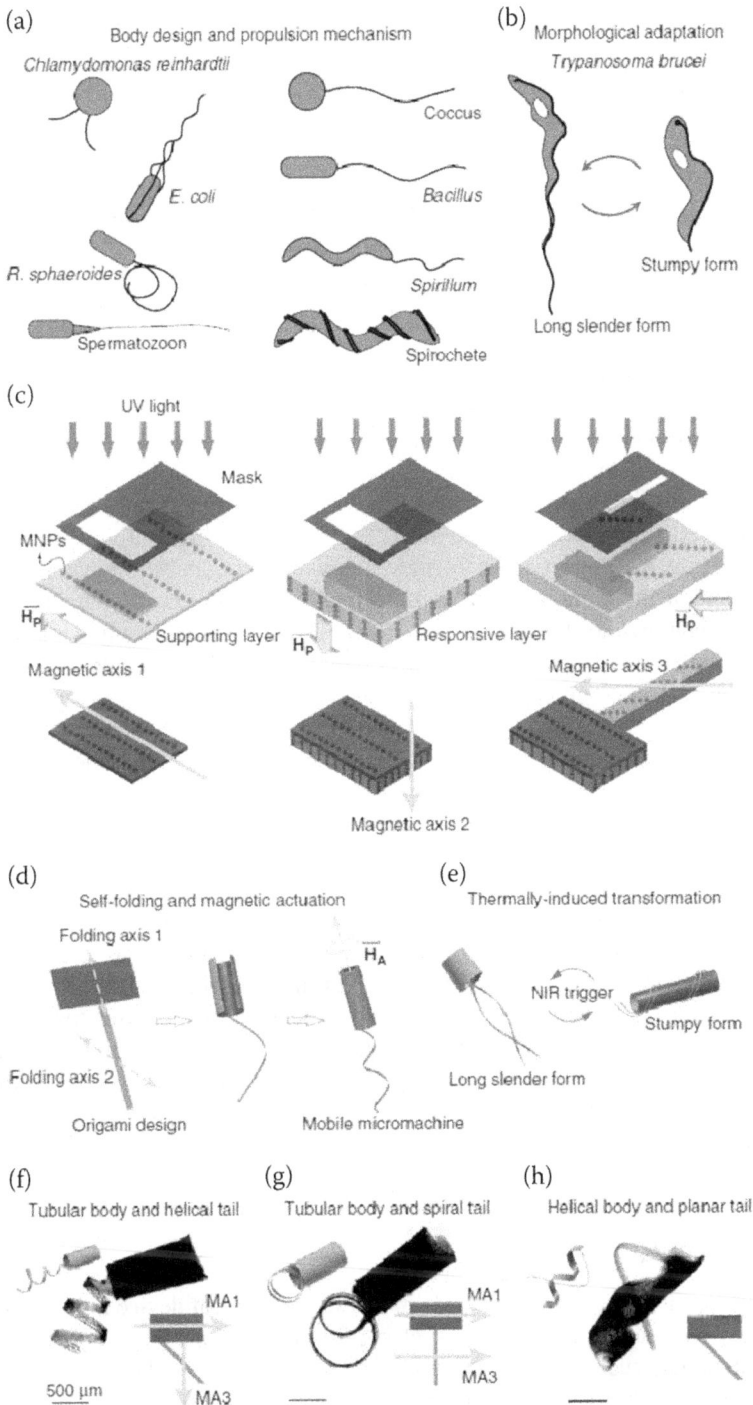

FIGURE 2.13 (a) a variety of body design and propeller mechanisms and (b) materials and architectures that can be utilized for morphological adaptation. (c) Schematic of the batch fabrication of biomimetic soft micro machines. (d) Anisotropic swelling behavior (e) The soft micro-machine different propulsion mechanism when exposed to external NIR heating. (f–h) Optical images of flagellated soft micro machines. *Copyright © 2016, Huang et al,*[113]*Nature Communications (2018) 7:12263.*

polymeric biomaterials, the formulation quality varies, and there aren't enough comparative studies of particular biomedical uses. Natural polymeric biomaterials have higher intrinsic biocompatibility than synthetic biomaterials, but they are also technically inferior since their mechanical, structural, and chemical characteristics cannot be changed as easily as synthetic biomaterials. With techniques like electrospinning to create nanotubes, nanofibers, and nanospheres, as well as the nearly limitless potential of generating numerous different combinations of natural and synthetic biomaterials, there's a good chance that more effective biomaterial-based materials can be fabricated with the proper biocoordination.[117,118] Many intriguing difficulties still exist today across most or all polymeric biomaterials, such as mass manufacturing feasibility at a cheap cost and overcoming particular biomaterial physiochemical limits. However, with the development of cloning and genetic engineering techniques for expressing native and synthetic biomaterials in a variety of host systems, polymeric biomaterials in medicine have a promising future.

Machine learning and artificial intelligence are two significant future directions in the development of polymeric biomaterials. In the future, it is likely to play a significant role. Machine learning and artificial intelligence will mostly be used to identify and create next-generation bioinspired polymeric materials. With the most efficient digital, high-throughput techniques, it is projected to increase the design space of new interesting polymeric biomaterials.[119] The machine learning and artificial intelligence assisted material development contains a virtual space which is mainly comprised of materials informatics, artificial intelligence, and computational modeling and simulations.[120] The valuable information's derived from this virtual space will be applied in to the real experimental material development space which comprises of different elements such as experimental synthesis, advanced manufacturing, and inspiration from nature. This synergistic combinatory approach of virtual space (offered by the machine learning and artificial intelligence) and real experimental space can greatly enhance the development of new exciting polymeric biomaterials.[121] There are different computational modeling of soft biomaterials such as density functional theory, fully atomistic molecular dynamics methods, coarse-grained molecular dynamics methods, and macroscale modeling and simulations are currently employed for this purpose.[119] These different approaches may revolutionize the development of soft polymeric biomaterials in the future. Even though it has enormous potential to accelerate the polymeric material synthesis space, certain challenges must be addressed, such as (i) taking into account the increased accuracy and efficiency of DFT functionals and MD forcefields, (ii) developing a common framework for easily accessible soft materials databases, and (iii) continuing to develop emerging techniques in artificial intelligence.[119] The coming decade may witness many approaches that are capable of addressing these different challenges in machine learning and artificial intelligence towards the development of different polymeric biomaterials.

2.4 CONCLUSION

Polymeric biomaterials have immensely contributed in the medical device industry in the last few decades and they represent one of the largest classes of biomaterials. Some of the properties of the polymeric biomaterials such as tunable degradation time, biocompatibility, and easy fabrication into different forms such as electrospun fibers, hydrogels, and 3D-printed scaffolds make them highly important in medicine. It is recently being employed in many emerging fields of biomedical research such as implantable bioelectronics devices, bioprinting, and soft robotics. A multidisciplinary approach comprising scientists, engineers, and physicians is required for the successful clinical translation of polymeric materials in these emerging fields into clinics. The upcoming decades may witness the clinical translation of many polymer-based medical devices in these emerging areas.

REFERENCES

[1] Ratner BD, Zhang G. History of biomaterials. *Biomat Sci* 2020;4:21–34.

[2] Ratner BD, Bryant SJ. Biomaterials: where we have been and where we are going. *Annual Review Biomed Eng* 2004;6:41–75.

[3] Rodriguez JM. Foreign body reaction to biomaterials. *Seminars Immunol* 2008;20:86–100.

[4] Huebsch N, Mooney DJ. Inspiration and application in the evolution of biomaterials. *Nature* 2009;462:426–432.

[5] Zimmerman LM, Veith I. *Great Ideas in the History of Surgery*, Norman Publishers, New York, NY, USA, 1st edition, 1993, 10.1002/bjs.18004921749.

[6] DSM Corporate Communications, A Brief History of Biomedical Materials 2009, http://www.dsm.com/en_US/cworld/public/medical downloads/.

[7] Heness G, Ben-Nissan B. Innovative bioceramics. *Mater Forum*. 2004;27:104–114.

[8] Doremus RH. Review- bioceramics. *J Mater Sci* 1992;27:285–297.

[9] Heness G, Ben Nissan B. Innovative bioceramics. *Materials Forum* 2004;27:104–114.

[10] Coburn JC, Pandit A. Development of naturally- derived biomaterials and optimization of their biomechanical properties. *Topics Tissue Eng. Supplement* 2007;13:1–23.

[11] Cao Y, Wang B. Biodegradation of silk biomaterials. *Int J of Mol Sci* 2009;10:1514–1524.

[12] Damodaran VB, Joslin J, Reynolds MM. Preparing biocompatible materials for non- permanent medical devices. *Euro Pharma Rev* 2012;17:71–77.

[13] Carmali S, Brocchini S. Polyacetals. *Natural Synthetic Biomed Polym* 2014;219–233.

[14] Kohana DS, Langer R. Polymeric biomaterials in tissue engineering. *Pedia Res* 2008; 63:487–491.

[15] Sharma K, Singh V, Arora A. Natural biodegradable polymers as matrices in transdermal drug delivery. *Int J Drug Delivery Res* 2011;3:85–103.

[16] Thomson S, Grohens Y, Ninan N. Biomaterials: Design, development and biomedical applications. *Nanotechnol Appl Tissue Eng Elsvier Inc*, 2015, 10.1016/B978-0-323-32889-0.00002-9.

[17] Wang D. Design, synthesis, characterization, and applications. *Stimuli Responsive Polym*, 2016;2016:Article ID: 6480259.10.1155/6480259.

[18] Peponi L. *Smart Polymers, Modification of Polymer Properties*, Elsevier Inc, 2017, 10.1016/B978-0-323-44353-1.00006-3.

[19] Love B. Polymeric biomaterials. *Biomat* 205–238, 2017, 10.1016/B978-0-12-809478-5.00009-2.

[20] Stratton S, Kumbar SG. Bioactive polymeric scaffolds for tissue engineering. *Bioactive Materials* 2016;1:93–108.

[21] Langer R, Tirrell DA. Designing materials for biology and medicine. *Nature* 2004;428:487–492.

[22] Nair LS, Laurencin CT. Biodegradable polymers as biomaterials. *Progress Polym Sci* 2007; 32:762–798.

[23] Gasperini L, Mano JF, Reis RL. Natural polymers for the microencapsulation of cells. *J Royal Soc Interface* 2014; 11:20140817.doi:10.1098/rsif.2014.0817.

[24] Grover LM. *Natural polymers, Biomaterials for Skin Repair and Regeneration*, Woodhead Publishing 2019:151–192.

[25] Ulery BD. Biomedical applications of biodegradable polymers. *J Poly Sci Part B, Poly Phys* 2011;49(12):832–864.

[26] Piskin E. Biodegradable polymers as biomaterials. *J Biomat Sci Polym Ed* 1994;6(9):775–795.

[27] Navaei T. Nanoengineered biomaterials for regenerative medicine, biodegradable synthetic polymer. *Non- Thermal Plasma Technology for Polymeric Materials* 2019:345–362.

[28] Simionescu, BC, Ivanov, D. Natural and Synthetic Polymers for Designing Composite Materials. In *Handbook of Bioceramics and Biocomposites*; Antoniac, IV, Ed.; Springer International Publishing: Cham, Switzerland, 2014;1–54, ISBN 978-3-319-09230-0.

[29] Mayer JE. Tissue engineering of cardiovascular structures. *Current Opinion in Cardiology*, 1997;12(6):528–532.

[30] Gribova V. A material's point of view on recent developments of polymeric biomaterials: control of mechanical and biochemical properties. *J Mat Chem* 2011;21(38):14354–14366.

[31] Wang L. Influence of the mechanical properties of biomaterials on degradability, cell behaviours and signalling pathways: current progress and challenges. *Biomater Sci* 2020;8:2714–2733.

[32] Paterlini TT. The role played by modified bioinspired surfaces in interfacial properties of biomaterials. *Biophys Rev* 2017; 9(5):683–698.

[33] Adrian JT. Polymeric biomaterials for medical implants and devices. *ACS Biomat Sci and Eng* 2016;2(4):454–472.

[34] Saad B, Suter UW. Biodegradable polymeric materials. In *Encyclopedia of Materials: Science and Technology*; Buschow, KHJ, Cahn, RW, Flemings, MC et al Eds.; Elsevier: Oxford 2001;551–555.

[35] Rehfeldt F. Cell responses to the mechanochemical microenvironment- Implications for regenerative medicine and drug delivery. *Adv Drug Deliv Rev* 2007;59:1329–1339.

[36] Kundu B. Silk fibroin biomaterials for tissue regenerations. *Adv Drug Deliv Rev* 2013 65(4):457–470.

[37] Shao W. Coaxial electrospun aligned tussah silk fibroin nanostructured fiber scaffolds embedded with hydroxyapatite-tussah silk fibroin nanoparticles for bone tissue engineering. *Mat Sci and Eng. C: Mater Biological Applications* 2016;58:342–351.

[38] Tian L. Coaxial electrospun poly(lactic acid)/silk fibroin nanofibers incorporated with nerve growth factor support the differentiation of neuronal stem cells. *RSC Adv* 2015;5(62):49838–49848.

[39] Bhardwaj N. Potential of silk fibroin/chondrocyte constructs of muga silkworm Antheraea assamensis for cartilage tissue engineering. *J Mater Chem B* 2016;4(21):3670–3684.

[40] Bhardwaj N. Silk fibroin-keratin based 3D scaffolds as a dermal substitute for skin tissue engineering. *Integrative Biol* 2015;7(1):53–63.

[41] Liu H, Slamovich EB, Webster TJ. Less harmful acidic degradation of poly (lacticco-glycolic acid) bone tissue engineering scaffolds through titania nanoparticle addition. *Inter J of Nanomed* 2006;1(4):541–545.

[42] Mayer JE, Shin'oka T, Shum-Tim D. Tissue engineering of cardiovascular structures. *Curr Opin Cardiol* 1997;12:6528–6532.

[43] Dobrowsky TM, Panorchan P, Konstantopoulos K and Wirtz D. Live-Cell Single-Molecule Force Spectroscopy. In *Biophysical Tools for Biologists, In Vivo Techniques*, Elsevier Academic Press Inc: San Diego 2018; 2(89):411.

[44] Nemir S, West JL. Synthetic materials in the study of cell response to substrate rigidity. *Annals of Biomed Eng* 2010;38:2–20.

[45] Shin H. Attachment, proliferation, and migration of marrow stromal osteoblasts cultured on biomimetic hydrogels modified with an osteopontin-derived peptide. *Biomaterials* 2004;25:895–906.

[46] Bandoh S. The development of composite stem for hip joint, an application of composite materials for medical implant device. Proceedings of the 16th International Conference on Composite Materials, Kyoto, Japan 2006;8–13.

[47] Middleton JC, Tipton AJ. Synthetic biodegradable polymers as orthopedic devices. *Biomaterials* 2000;21(23):2335–2346.

[48] Nikolova MP, Chavali MS. Recent advances in biomaterials for 3D scaffolds: A review. *Bioact Mater* 2019;4:271–292.

[49] Thiruvengadathan. Nanomaterial processing using self- assembly- bottom- up chemical and biological approaches. *Reports Progr Physics* 2013;76(6):066501.

[50] Yu I. Polymers, biotechnology and medical application. *Encyclopedia Smart Materials* 2002, 10.1 002/0471216275.esm065.

[51] Patil NV. Smart polymers are in the biotech future. *Bio Process Int* 2006;4(8):42–46.

[52] Liechty WB. Polymers for drug delivery systems. *Annual Review Chemical Biomolecular Eng* 2010;1:149–173.

[53] Galaev I, Mattiasson B. Smart polymers and what they could do in biotechnology and medicine. *Trends Biotech* 1999;17(8):335–340.

[54] Kim YH. Macromolecules. The shifting research frontiers. *Polymer Science Eng* 1994;28(4):939–944.

[55] Lahann J, Mitragotri S, Tran TN, Kaido H, Sundaram J, Choi IS, Hoffer S, Somorjai GA, Langer R. A reversibly switching surface. *Science* 2003;299:371–374.

[56] Martinez AW. Three-dimensional microfluidic devices fabricated in layered paper and tape. *Proc Natl Acad Sci, USA* 2008;105(52):19606–19611.

[57] Chawla S. Silk-Based Bioinks for 3D Bioprinting. *Adv Healthcare Mater* 2018;7(8):1701204.

[58] Feig VR, Tran H, Bao Z. Biodegradable polymeric materials in degradable electronic devices. *ACS Cent Sci.* 2018;4:337–348.

[59] La Mattina AA, Mariani S, Barillaro G. Bioresorbable materials on the rise: From electronic components and physical sensors to In Vivo monitoring systems. *Adv Sci.* 2020;7:1902872.

[60] Li R, Wang L, Kong D, Yin L. Recent progress on biodegradable materials and transient electronics. *Bioact Mater.* 2018;3:322–333.

[61] Yuk H, Lu B, Zhao X. Hydrogel bioelectronics. *Chem Soc Rev.* 2019;48:1642–1667.

[62] Kang S-K, Murphy RKJ, Hwang S-W, Lee SM, Harburg DV, Krueger NA, Shin J, Gamble P, Cheng H, Yu S, Liu Z, McCall JG, Stephen M, Ying H, Kim J, Park G, Webb RC, Lee CH, Chung S, Wie DS, Gujar AD, Vemulapalli B, Kim AH, Lee K-M, Cheng J, Huang Y, Lee SH, Braun PV, Ray WZ, Rogers JA. Bioresorbable silicon electronic sensors for the brain. *Nature.* 2016;530:71–76.

[63] Tao H, Hwang S-W, Marelli B, An B, Moreau JE, Yang M, Brenckle MA, Kim S, Kaplan DL, Rogers JA, Omenetto FG. Silk-based resorbable electronic devices for remotely controlled therapy and in vivo infection abatement. *Proc Natl Acad Sci.* 2014;111:17385–17389.

[64] Zhong S, Wong HC, Low HY, Zhao R. Phototriggerable Transient Electronics via Fullerene-Mediated Degradation of Polymer:Fullerene Encapsulation Layer. *ACS Appl Mater Interfaces.* 2021;13:904–911.

[65] Choi YS, Hsueh Y-Y, Koo J, Yang Q, Avila R, Hu B, Xie Z, Lee G, Ning Z, Liu C, Xu Y, Lee YJ, Zhao W, Fang J, Deng Y, Lee SM, Vázquez-Guardado A, Stepien I, Yan Y, Song JW, Haney C, Oh YS, Liu W, Yoon H-J, Banks A, MacEwan MR, Ameer GA, Ray WZ, Huang Y, Xie T, Franz CK, Li S, Rogers JA. Stretchable, dynamic covalent polymers for soft, long-lived bioresorbable electronic stimulators designed to facilitate neuromuscular regeneration. *Nat Commun.* 2020;11:1–14.

[66] Joshi J, Rubart M, Zhu W. Optogenetics: Background, methodological advances and potential applications for cardiovascular research and medicine. *Front Bioeng Biotechnol [Internet].* 2020 [cited 2021 Mar 27];7. Available from: https://www.frontiersin.org/articles/10.3389/fbioe.2019.00466/full

[67] Iseri E, Kuzum D. Implantable optoelectronic probes forin vivooptogenetics. *J Neural Eng.* 2017;14:031001.

[68] Kim CY, Ku MJ, Qazi R, Nam HJ, Park JW, Nam KS, Oh S, Kang I, Jang J-H, Kim WY, Kim J-H, Jeong J-W. Soft subdermal implant capable of wireless battery charging and programmable controls for applications in optogenetics. *Nat Commun.* 2021;12:1–13.

[69] Vayre B, Vignat F, Villeneuve F. Designing for additive manufacturing. *Procedia CIRP.* 2012;3:632–637.

[70] Ngo TD, Kashani A, Imbalzano G, Nguyen KTQ, Hui D. Additive manufacturing (3D printing): A review of materials, methods, applications and challenges. *Compos Part B Eng.* 2018;143:172–196.

[71] Li Q, Kucukkoc I, Zhang DZ. Production planning in additive manufacturing and 3D printing. *Comput Oper Res.* 2017;83:157–172.

[72] Bhushan B, Caspers M. An overview of additive manufacturing (3D printing) for microfabrication. *Microsyst Technol.* 2017;23:1117–1124.

[73] Carrow JK, Kerativitayanan P, Jaiswal MK, Lokhande G, Gaharwar AK. Chapter 13 - Polymers for Bioprinting. In *Essent 3D Biofabrication Transl*; Atala, A, Yoo, JJ, Eds.; Academic Press: Boston, 2015. Available from: https://www.sciencedirect.com/science/article/pii/B978012800972700013X

[74] Murphy SV, Atala A. 3D bioprinting of tissues and organs. *Nat Biotechnol.* 2014;32:773–785.

[75] Vijayavenkataraman S, Yan W-C, Lu WF, Wang C-H, Fuh JYH. 3D bioprinting of tissues and organs for regenerative medicine. *Adv Drug Deliv Rev.* 2018;132:296–332.

[76] Leberfinger AN, Dinda S, Wu Y, Koduru SV, Ozbolat V, Ravnic DJ, Ozbolat IT. Bioprinting functional tissues. *Acta Biomater.* 2019;95:32–49.

[77] Ma X, Liu J, Zhu W, Tang M, Lawrence N, Yu C, Gou M, Chen S. 3D bioprinting of functional tissue models for personalized drug screening and in vitro disease modeling. *Adv Drug Deliv Rev.* 2018;132:235–251.

[78] Sorkio A, Koch L, Koivusalo L, Deiwick A, Miettinen S, Chichkov B, Skottman H. Human stem cell based corneal tissue mimicking structures using laser-assisted 3D bioprinting and functional bioinks. *Biomaterials.* 2018;171:57–71.

[79] Kim W, Kim G. 3D bioprinting of functional cell-laden bioinks and its application for cell-alignment and maturation. *Appl Mater Today.* 2020;19:100588.

[80] Kačarević ŽP, Rider PM, Alkildani S, Retnasingh S, Smeets R, Jung O, Ivanišević Z, Barbeck M. An Introduction to 3D Bioprinting: Possibilities, Challenges and Future Aspects. *Materials.* 2018;11:2199.

[81] Zhang B, Luo Y, Ma L, Gao L, Li Y, Xue Q, Yang H, Cui Z. 3D bioprinting: an emerging technology full of opportunities and challenges. *Bio-Des Manuf.* 2018;1:2–13.

[82] Donderwinkel I, van Hest JCM, Cameron NR. Bio-inks for 3D bioprinting: recent advances and future prospects. *Polym Chem.* 2017;8:4451–4471.

[83] Stanton MM, Samitier J, Sánchez S. Bioprinting of 3D hydrogels. *Lab Chip.* 2015;15:3111–3115.

[84] Liu F, Chen Q, Liu C, Ao Q, Tian X, Fan J, Tong H, Wang X. Natural Polymers for Organ 3D Bioprinting. *Polymers.* 2018;10:1278.

[85] Gungor-Ozkerim PS, Inci I, Zhang YS, Khademhosseini A, Dokmeci MR. Bioinks for 3D bioprinting: an overview. *Biomater Sci.* 2018;6:915–946.

[86] Yin J, Yan M, Wang Y, Fu J, Suo H. 3D bioprinting of low-concentration Cell-Laden Gelatin Methacrylate (GelMA) Bioinks with a two-step cross-linking strategy. *ACS Appl Mater Interfaces.* 2018;10:6849–6857.

[87] Axpe E, Oyen ML. Applications of Alginate-Based Bioinks in 3D Bioprinting. *Int J Mol Sci.* 2016;17:1976.

[88] De Melo BAG, Jodat YA, Cruz EM, Benincasa JC, Shin SR, Porcionatto MA. Strategies to use fibrinogen as bioink for 3D bioprinting fibrin-based soft and hard tissues. *Acta Biomater.* 2020;117:60–76.

[89] Poldervaart MT, Goversen B, de Ruijter M, Abbadessa A, Melchels FPW, Öner FC, Dhert WJA, Vermonden T, Alblas J. 3D bioprinting of methacrylated hyaluronic acid (MeHA) hydrogel with intrinsic osteogenicity. *Plos One.* 2017;12:e0177628.

[90] Du Z, Li N, Hua Y, Shi Y, Bao C, Zhang H, Yang Y, Lin Q, Zhu L. Physiological pH-dependent gelation for 3D printing based on the phase separation of gelatin and oxidized dextran. *Chem Commun.* 2017;53:13023–13026.

[91] Chung JHY, Naficy S, Yue Z, Kapsa R, Quigley A, Moulton SE, Wallace GG. Bio-ink properties and printability for extrusion printing living cells. *Biomater Sci.* 2013;1:763–773.

[92] GhavamiNejad A, Ashammakhi N, Wu XY, Khademhosseini A. Crosslinking strategies for 3D bioprinting of polymeric hydrogels. *Small.* 2020;16:2002931.

[93] Chimene D, Kaunas R, Gaharwar AK. Hydrogel bioink reinforcement for additive manufacturing: A focused review of emerging strategies. *Adv Mater.* 2020;32:1902026.

[94] Habib A, Khoda B. Development of clay based novel hybrid bio-ink for 3D bio-printing process. *J Manuf Process.* 2019;38:76–87.

[95] Skardal A, Zhang J, Prestwich GD. Bioprinting vessel-like constructs using hyaluronan hydrogels crosslinked with tetrahedral polyethylene glycol tetracrylates. *Biomaterials.* 2010;31:6173–6181.

[96] Kim SH, Yeon YK, Lee JM, Chao JR, Lee YJ, Seo YB, Sultan MT, Lee OJ, Lee JS, Yoon S, Hong I-S, Khang G, Lee SJ, Yoo JJ, Park CH. Precisely printable and biocompatible silk fibroin bioink for digital light processing 3D printing. *Nat Commun.* 2018;9:1–14.

[97] Yuk H, Lu B, Lin S, Qu K, Xu J, Luo J, Zhao X. 3D printing of conducting polymers. *Nat Commun.* 2020;11:1–8.

[98] Gao B, Yang Q, Zhao X, Jin G, Ma Y, Xu F. 4D Bioprinting for Biomedical Applications. *Trends Biotechnol.* 2016;34:746–756.

[99] Ashammakhi N, Ahadian S, Zengjie F, Suthiwanich K, Lorestani F, Orive G, Ostrovidov S, Khademhosseini A. Advances and future perspectives in 4D bioprinting. *Biotechnol J.* 2018;13:1800148.

[100] Li Y-C, Zhang YS, Akpek A, Shin SR, Khademhosseini A. 4D bioprinting: the next-generation technology for biofabrication enabled by stimuli-responsive materials. *Biofabrication.* 2016;9:012001.

[101] Miao S, Cui H, Nowicki M, Xia L, Zhou X, Lee S-J, Zhu W, Sarkar K, Zhang Z, Zhang LG. Stereolithographic 4D bioprinting of multiresponsive architectures for neural engineering. *Adv Biosyst.* 2018;2:1800101.

[102] Kuang X, Roach DJ, Wu J, Hamel CM, Ding Z, Wang T, Dunn ML, Qi HJ. Advances in 4D printing: Materials and applications. *Adv Funct Mater.* 2019;29:1805290.

[103] Stoychev G, Puretskiy N, Ionov L. Self-folding all-polymer thermoresponsive microcapsules. *Soft Matter.* 2011;7:3277–3279.

[104] Zhang Y, Huang L, Song H, Ni C, Wu J, Zhao Q, Xie T. 4D Printing of a digital shape memory polymer with tunable high performance. *ACS Appl Mater Interfaces.* 2019;11:32408–32413.

[105] Choong YYC, Maleksaeedi S, Eng H, Wei J, Su P-C. 4D printing of high performance shape memory polymer using stereolithography. *Mater Des.* 2017;126:219–225.

[106] Zhang C, Cai D, Liao P, Su J-W, Deng H, Vardhanabhuti B, Ulery BD, Chen S-Y, Lin J. 4D Printing of shape-memory polymeric scaffolds for adaptive biomedical implantation. *Acta Biomater.* 2021;122:101–110.

[107] Whitesides GM. Soft robotics. *Angew Chem Int Ed.* 2018;57:4258–4273.

[108] Ilievski F, Mazzeo AD, Shepherd RF, Chen X, Whitesides GM. Soft robotics for chemists. *Angew Chem.* 2011;123:1930–1935.

[109] Kim S, Laschi C, Trimmer B. Soft robotics: a bioinspired evolution in robotics. *Trends Biotechnol.* 2013;31:287–294.

[110] Jin H, Dong E, Alici G, Mao S, Min X, Liu C, Low KH, Yang J. A starfish robot based on soft and smart modular structure (SMS) actuated by SMA wires. *Bioinspir Biomim.* 2016;11:056012.

[111] Majidi C. Soft Robotics: A Perspective—Current Trends and Prospects for the Future. *Soft Robot.* 2013;1:5–11.

[112] Lu H, Zhang M, Yang Y, Huang Q, Fukuda T, Wang Z, Shen Y. A bioinspired multilegged soft millirobot that functions in both dry and wet conditions. *Nat Commun.* 2018;9:1–7.

[113] Huang H-W, Sakar MS, Petruska AJ, Pané S, Nelson BJ. Soft micromachines with programmable motility and morphology. *Nat Commun.* 2016;7:1–10.

[114] Ulery BD, et al. Biomedical applications of biodegradable polymers. *J. of Poly. Science Part B: Polymer Physics* 2011;49(12):832–864.

[115] Xuehui Yan et al. Bottom- up self- assembly based on DNA nanotechnology. *Nanomaterials* 2020;10(10):2047, 10.3390/nano10102047.

[116] Dingying Shan et al. Polymeric biomaterials for biophotonic applications. *Bioactive Materials* 2018;3(4):434–445.

[117] Nair LS, Laurencin CT. Polymers as biomaterials for tissue engineering and controlled drug delivery. *Advances in Biochem. Eng./Biotechnology* 2006;102: 47–90, 10.1007/b137240.

[118] Liao S, et al. Biomimetic electrospun nanofibrous for tissue engineering. *Biomed. Mat.* 2006;1(3): 45–53, 1088/1748-6041/1/3/R01.

[119] Zhai C, Li T, Shi H, Yeo J. Discovery and design of soft polymeric bio-inspired materials with multiscale simulations and artificial intelligence. *J Mater Chem B.* 2020;8:6562–6587.

[120] Costache AD, Ghosh J, Knight DD, Kohn J. Computational Methods for the Development of Polymeric Biomaterials. *Adv Eng Mater.* 2010;12:B3–B17.

[121] Singh AV, Rosenkranz D, Ansari MHD, Singh R, Kanase A, Singh SP, Johnston B, Tentschert J, Laux P, Luch A. Artificial Intelligence and Machine Learning Empower Advanced Biomedical Material Design to Toxicity Prediction. *Adv Intell Syst.* 2020;2:2000084.

3 Sustainable Biofillers and Their Biocomposites
Opportunities and Challenges

M. P. Drupitha
Aliaxis Research & Technology Center Asia Bangalore, India

University of Guelph, Guelph, Ontario, Canada

Rajendran Muthuraj
University of Guelph, Guelph, Ontario, Canada

Worn Again Technologies Ltd, Nottingham, Nottinghamshire, United Kingdom

Manjusri Misra
University of Guelph, Guelph, Ontario, Canada

Amar Kumar Mohanty
University of Guelph, Guelph, Ontario, Canada

CONTENTS

3.1 INTRODUCTION

Renewable materials development can reduce the dependence on fossil fuels and environmental problems associated with unsustainable materials [1]. Developing sustainable materials can keep the resources in constant circulation while providing economic, environmental, and societal benefits. Therefore, the interest in biobased composites is continuously increasing. Biobased and/or biodegradable polymeric materials have been well recognized as sustainable materials for different applications [2,3]. Lignocellulosic materials including traditional plant fibers and agro and forestry residues can be incorporated into fossil fuel-based biodegradable polymers to increase the renewable content/carbon in the developed polymeric materials while reducing the carbon footprint [4]. Developing composites from lignocellulosic materials and biopolymers have great potential for packaging, automotive, and consumer goods applications. The incorporation of lignocellulosic materials into biodegradable polymer matrices can provide fully biodegradable composites. Such fully biodegradable polymer composites can be used in packaging applications where biodegradability is essential after their end of life [5] depends on the reinforcement, matrix, and application requirements; the biocomposites can be developed by extrusion, injection, hand-lay up, pultrusion, resin transfer, and compression molding.

Despite good specific mechanical properties, renewability, and lightweight, lignocellulosic fibers/fillers often do not fulfill the durable composites application requirements such as consistency, durability, thermal stability, flammability, and resistance to water absorption for durable composites development [6]. A by-product of the pyrolyzed lignocellulosic biomass (biochar) could be used to develop biocomposites with reduced flammability and water absorption while increasing thermal stability [7–9]. In some cases, the biocomposites developed from single reinforcements and/or matrix polymer showed insufficient cost and performance for durable applications. Therefore, biocomposites are produced by hybridization of two or more reinforcements and/or matrices to achieve balanced performances [10]. Biobased durable plastics with synthetic fibers and lignocellulosic fibers can be used to develop sustainable biocomposites for durable applications [11,12]. However, most of the biocomposites showed poor interaction between the fillers and matrix because of their polarity differences [13–18]. The poor compatibility in the composites can be effectively addressed by different compatibilization strategies including fiber modification and matrix modifications [4,16]. Therefore, this chapter aims to summarize recent developments on biopolymer and lignocellulosic fibers-based composites and their applications and challenges.

3.2 POLYMER MATRICES FOR BIOCOMPOSITES DEVELOPMENT

Based on the polymer origin and biodegradability nature, polymers can be classified into four different groups, as shown in Figure 3.1. Petroleum-based and non-biodegradable polymers such as polypropylene (PP), polyethylene (PE), epoxy, phenol-formaldehyde, and unsaturated polyesters are the most studied polymer matrix for composites development because of their availability, good physical, and mechanical properties. The increasing interest in developing sustainable polymers leads to the use of bioplastics as a matrix for composites development. Bioplastics can be biobased and biodegradable (PLA, PHAs, PPC, and Bio-PBS), petroleum-based and biodegradable (PCL, PBAT and PBS), and biobased and non-biodegradable (Bio-PP, Bio-PE, Bio-PTT, Bio-PA, and Bio-PET) [4]. Cellulose acetate, starch, and proteinaceous materials (e.g., soy protein, corn zein, keratin, etc.) are also modified/plasticized to use as a matrix for composites development because of the renewable and biodegradable nature [19–21]. Similar to biodegradable polymers, biobased non-biodegradable polymers (e.g., Bio-PE, Bio-PP Bio-PA, Bio-PTT, and Bio-PET) are commercially available in the market.

Biobased polypropylene (Bio-PP) was introduced in the market in 2019 and its production is expected to increase fourfold by 2025 due to its widespread applications [22]. A significant research effort has also been made to produce biobased epoxy from vegetable oils and phenol-formaldehyde resin cashew nutshell liquid-based monomers [23]. Similarly, poly(furfuryl alcohol) is another biobased thermoset resin that is produced from furanic derivatives [24]. According to the European bioplastics-nova-Institute, the global bioplastics production was around 2.11 Mt in 2020, and it is forecasted to increase to around 2.87 Mt by 2025. Currently, the biodegradable bioplastics market is dominating compared to biobased non-biodegradable bioplastics [22]. The main limitation in bioplastics is the cost, which can be overcome by large-scale production. The incorporation of inexpensive sustainable fillers into bioplastics is another way to offset the price of the bioplastics while maintaining or enhancing the performances.

3.3 LIGNOCELLULOSIC FIBERS

Lignocellulosic fibers are derived from both plant and animal origin with different characteristics, as shown in Figure 3.2. In addition to traditional natural fibers, biofibers can be obtained from agro and forestry residues. These biofibers are an attractive material for composite applications because of

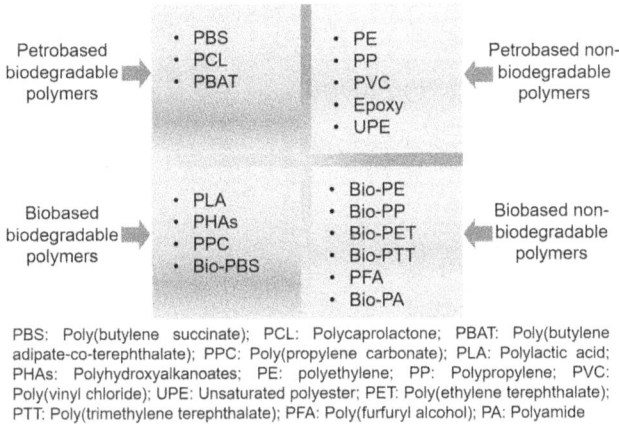

PBS: Poly(butylene succinate); PCL: Polycaprolactone; PBAT: Poly(butylene adipate-co-terephthalate); PPC: Poly(propylene carbonate); PLA: Polylactic acid; PHAs: Polyhydroxyalkanoates; PE: polyethylene; PP: Polypropylene; PVC: Poly(vinyl chloride); UPE: Unsaturated polyester; PET: Poly(ethylene terephthalate); PTT: Poly(trimethylene terephthalate); PFA: Poly(furfuryl alcohol); PA: Polyamide

FIGURE 3.1 Matrix polymers for biocomposites.

their availability, relatively inexpensive, and sustainability. Lignocellulosic fibers are obtained from different parts of the plants including leaf (e.g., sisal, abaca, pineapple, etc.), seed/fruit (e.g., cotton, kapok, coir, etc.), core (e.g., jute, kenaf, hemp, etc.), bast (e.g., jute, flax, hemp, ramie, kenaf, etc.), reed/grass (e.g., miscanthus, switchgrass, bamboo, etc.), and straw (e.g., rice, wheat, soy, corn, etc.) [25]. Similarly, wood-based fillers are also considered as sustainable fillers for composite applications. The physical, thermal, and mechanical properties of the lignocellulosic filers are mainly influenced by chemical composition, microfibril angle, and lumen size (Table 3.1). Recently, the undervalued lignocellulosic materials are used to produce biocarbon/biochar by pyrolysis. The pyrolysis process conditions and nature of feedstock can control the resulting biocarbon characteristics such as carbon content, morphology, and crystallinity. Such biocarbon can be used as a sustainable alternative filler for composite development because of its high surface area, light weight, hydrophobic nature, and high thermal stability [26,27]. Reinforcement hybridization strategy has been extensively studied to exploit the advantages of the different fillers to make sustainable composites with enhanced performances [28–30].

3.4 BIOCOMPOSITE PREPARATION METHODS

Biocomposites are produced either from biobased fillers or biobased matrices or a combination of both biobased fillers and matrices (Figure 3.3). Based on the chemical characteristics of matrix and reinforcement, biocomposites can be biodegradable or non-biodegradable. Biocomposites can be produced by different manufacturing processes including extrusion, injection, compression, pultrusion, resin transfer molding, 3D printing, and 4D printing. Depends on the matrix, filler, and application, the biocomposites manufacturing process can vary. Fabrication of natural fiber composites with thermoset resins, techniques such as hand lay-up, compression molding, resin

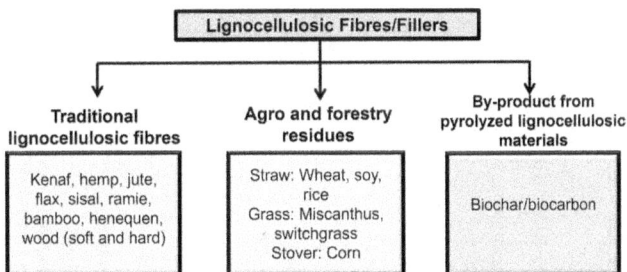

FIGURE 3.2 Lignocellulosic fibers/fillers classifications.

TABLE 3.1
Tensile Modulus, Tensile Strength, Chemical Composition, and Microfibrillar Angle (MFA) of Different Lignocellulosic Fibers/Fillers

Lignocellulosic Fibers/Fillers	Tensile Modulus (GPa)	Tensile Strength (MPa)	Cellulose [%]	Hemicelluloses [%]	Lignin [%]	Pectins [%]	MFA (°)	Ref.
Flax	46–85	600–2000	64–85	11–17	2–3	1.8–2.0	10	[31]
Jute	24.7–26.5	393–773	61–75	13.6–20.4	12–13	0.6 ± 0.6	7–12	
Kenaf	11–60	223–930	45–57	21.5	8–13	3–5	7–12	
Coir	3.44–4.16	120–304	32–46	0.15–0.3	40–45	3–4	30–49	
Wood	15.4–27.5	553–1500	38–45	19–39	22–34	0.4–5	5–45	
Bamboo	10–40	340–510	34.5–50	20.5	26	<1	2–10	
Harakeke	14–33	440–990						
Hemp	14.4–44.5	270–889	55–90	4–16	2–5	0.8–8	6.2–11.2	[32,33]
Switch grass	6.4	37.7	36	28	10	–	–	[32–34]
Miscanthus	–	6.7	50	20	15	–	–	[32]
Soy stalk	–	–	45	11	14	–	–	[32]
Corn stover	–	–	45	23	11	–	–	[32]
Wheat straw	–	–	48	24	11	–	–	
Biochar/biocarbon	5–11							[35,36]

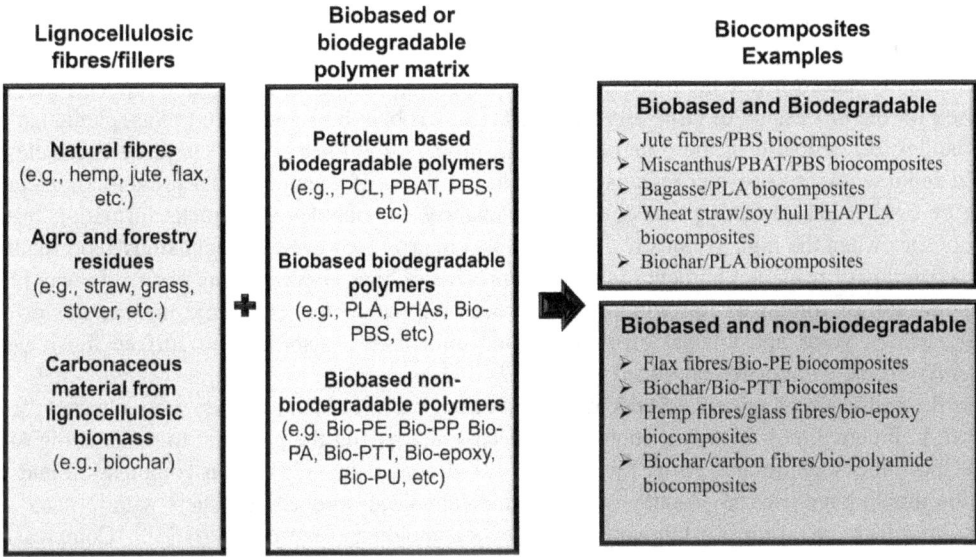

Lignocellulosic fibres/fillers

Natural fibres
(e.g., hemp, jute, flax, etc.)

Agro and forestry residues
(e.g., straw, grass, stover, etc.)

Carbonaceous material from lignocellulosic biomass
(e.g., biochar)

Biobased or biodegradable polymer matrix

Petroleum based biodegradable polymers
(e.g., PCL, PBAT, PBS, etc)

Biobased biodegradable polymers
(e.g., PLA, PHAs, Bio-PBS, etc)

Biobased non-biodegradable polymers
(e.g. Bio-PE, Bio-PP, Bio-PA, Bio-PTT, Bio-epoxy, Bio-PU, etc)

Biocomposites Examples

Biobased and Biodegradable
➢ Jute fibres/PBS biocomposites
➢ Miscanthus/PBAT/PBS biocomposites
➢ Bagasse/PLA biocomposites
➢ Wheat straw/soy hull PHA/PLA biocomposites
➢ Biochar/PLA biocomposites

Biobased and non-biodegradable
➢ Flax fibres/Bio-PE biocomposites
➢ Biochar/Bio-PTT biocomposites
➢ Hemp fibres/glass fibres/bio-epoxy biocomposites
➢ Biochar/carbon fibres/bio-polyamide biocomposites

FIGURE 3.3   Examples of biocomposites from different lignocellulosic fibers/fillers and polymer matrices.

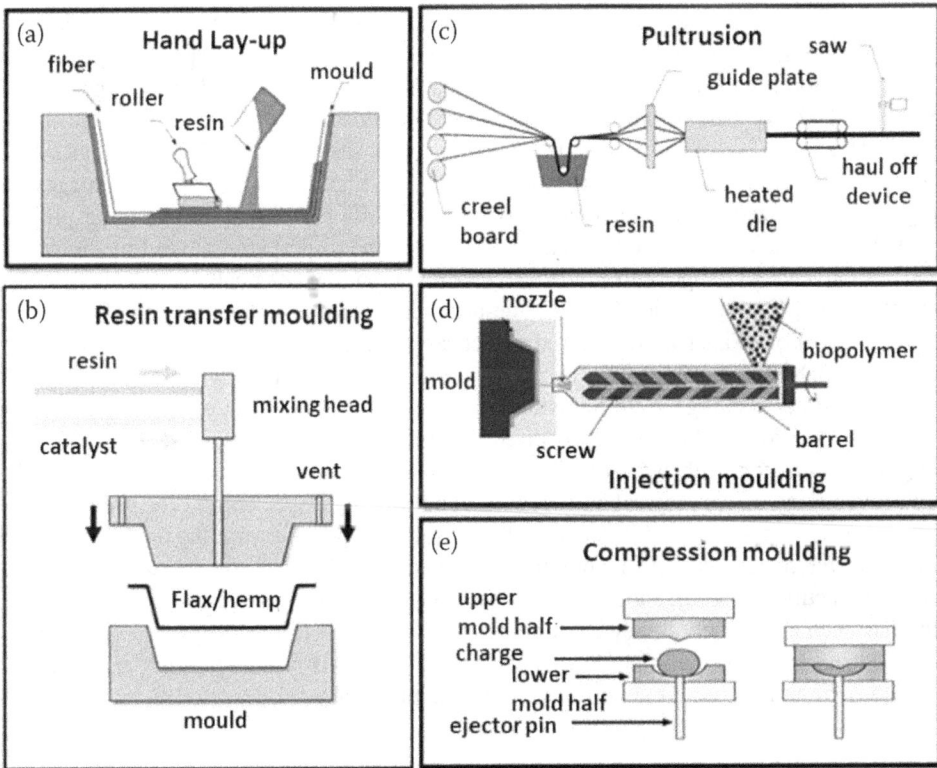

FIGURE 3.4   Different techniques for biocomposite fabrication. Modified and reproduced with permission from ref. [39] Copyright 2019 Elsevier.

transfer molding (RTM), and pultrusion are widely used [37] (Figure 3.4). Hand lay-up is the simple open molding technique (Figure 3.4(a)), in which the reinforcements (mat or woven fibers) are manually placed inside the mold followed by applying thermoset resins [38–40]. This process is used for the processing of large-sized products such as boat hulls, portable toilets, picnic tables, car bodies, etc. [38,40]. In the case of the RTM process, the fiber preform is placed in a closed mold and low viscous polymer resin is injected to wet the reinforcement under pressure or vacuum (Figure 3.4(b)) [41]. In this process, there are chances for bubbles and cracks formation in the composites when the high viscous resin is used as a matrix. However, a recent study reported that the separation of resin and hardener allowed to process of high viscous resins in a facile way [42]. Optimization of composite formulations and processing parameters can avoid the chances of defects such as voids and cracks. Moreover, good dimensional stability and surface finish were observed in the composites produced using RTM [43].

In the pultrusion process, the fibers are pulled through a resin bath and then onto a heated die as shown in Figure 3.4(c). This is a continuous process and having the potential to make composites with high fiber loading as well as hybrid filler-based composites [25]. Both compression and injection molding are used to produce small and medium-sized products. Products with intricate designs and high precision are fabricated by injection molding (Figure 3.4(d)) [38]. Compression molding is one of the most used techniques for thermoplastics, thermosetting, and elastomers-based products. In compression molding, different types of fillers including particulate, short fibers, and long fibers can be incorporated into polymer matrices to produce composites (Figure 3.4(e)).

Three-dimensional (3D) printing is a process that involves the principle of extrusion controlled by computer-aided design models for the fabrication of complex structured materials [44]. This technique comprises the layer-by-layer deposition of a molten polymer passed through a heated nozzle deposited on a printing platform [44]. Both particulate and fibrous fillers are incorporated into polymer resin to manufacture biocomposites by 3D printing [44–46]. Similar to other polymer processing methods, optimization of 3D-printing parameters is important to produce products with satisfactory performances and it easily allows the fabrication of small and large-sized biocomposites [44,47,48]. Due to the less waste generation, cost efficiency, and precision in manufacturing, 3D printing is extensively used to develop composites for several applications including automotive, biomedical, and consumer goods [31,49–51]. Figure 3.5(a) shows the images of 3D printing of a large-sized PLA/ poplar biocomposite [48]. 4D printing is an emerging processing technique in which a 3D-printed object can change its shape or properties over time according to external stimuli such as temperature, moisture, light, pH, and mechanical, electric, magnetic, and biological factors [52]. By considering the time-dependent change in functionality of a 3D-printed object, the fabrication of a smart 4D-printed object can be predicted [53]. Grigsby et.al reported the fabrication of 3D-printed composites based on keratin and lignin showing fourth-dimensional response after water soaking. Water/moisture stimulated 4D printability of cellulosic material is well explored [54,55]. Gladman et al. reported the cellulose fibrils reinforced acrylamide matrix showing 4D printability with plant-inspired shape-changing properties [54]. Water-induced (hydration/dehydration) morphological changes of cellulose- hydrogel composite prepared by 3D printing are given in Figure 3.5(b-f) [55]. This stimuli-responsive printing technology can be utilized in different applications including biomedical, textile, soft robotics, packaging, automotive, construction, clothing, and aerospace [56]. Biocomposites fabrication by both 3D and 4D printing has been extensively reviewed elsewhere [57–59]. Due to the unique features in the 4D-printed products, it is expected that the global market of the 4D-printed product applications can reach USD ~162 million by 2022 with a growth rate of ~39% [31].

3.5 BIOCOMPOSITE PERFORMANCES

The performances of the biocomposites are dependent on the filler types, the chemical composition of the fillers, properties of the matrix, fiber orientation, volume fraction of the fibers, fiber

FIGURE 3.5 (a) Large-scale 3D printing process using poplar/PLA composite and a 3D-printed product. Reproduced with permission from ref. [60]. Copyright 2019 ACS (b) 3D printing pathway of cellulose-hydrogel composite. (c) 3D printed cellulose-hydrogel composite. (d) Drying at room temperature starts the morphing. (e) Crosslinking allows to maximize and fix the 3D shape. (f) Organized to flat configuration upon hydration. (g) Drying (dehydration) recovers the petal shape showing 4D printability. Reproduced with permission from ref. [55]. Copyright 2018 Elsevier.

geometry, interfacial adhesion between the reinforcement and polymer, and manufacturing methods [33]. A large number of studies have been investigated the performances of the biocomposites developed from different lignocellulosic fibers [35,61–67]. Table 3.2 summarizes the performance of the biocomposites developed by different manufacturing methods. Regardless of the polymer matrix, unidirectionally oriented fiber composites showed superior mechanical performances compared to randomly oriented (bio)fiber composites because of the uniform load transfer axially along the fibers [68,69]. For example, the tensile strength of the unidirectional flax fiber/PLA (39/61 wt.%) and random flax fiber/PLA composites showed 150 and 83 MPa, respectively [68]. On the other hand, unidirectional flax fiber/PLA composites had less biodegradability than the randomly oriented flax fiber/PLA composites. Such a reduced biodegradation rate in the aligned flax/PLA composites could be attributed to the less water swelling behavior.

Among the different factors, the aspect ratio and critical length of the fillers can significantly influence the mechanical performances of the resulting composites. For example, composites prepared with a filler having an aspect ratio of >10 can provide better mechanical performances because of the effective reinforceability [70,71]. It has been observed that recycling of natural fibers composites can reduce the fiber aspect ratio and thus better fiber dispersion can be achieved in the matrix [72,73]. Figure 3.6 shows the tensile strength and tensile stiffness of the 3D printed, injection, and compression-molded composites. Due to the differences in the composite formulations (e.g., fiber volume fraction, fiber type, polymer type, etc.) and manufacturing parameters, the tensile strength,

TABLE 3.2

Biopolymers and Lignocellulosic Fiber-Based Biocomposites and Their Characteristics

Matrix Polymer	Reinforcement and Compatibilizer	Manufacturing Method	Performances	Reference
Biobased epoxy or polyurethane	Flax fibers	Resin transfer molding (RTM)	The composites had different failure modes because of the water plasticizing effect after being exposed to 90% RH at 30°C. The plasticizing effect occurs very slow in the polyurethane composites compared to epoxy composites. Similarly, polyurethane/flax fiber composites formed good compatibility by reacting the isocyanate groups in the polyurethane and the hydroxyl groups in the fibers to improve the mechanical performances as compared to the flax/bio-epoxy composites.	[119]
PBS	Miscanthus fiber and MA-g-PBS	Injection molding	Compatibilized composites showed an increase in tensile strength (22%), flexural strength (47%), and impact strength (139%) compared to matrix.	[17]
Bio-PBS	Apple pomace and MA-g-bio-PBS		Modulus, flexural strength, and impact strength were significantly enhanced compared to matrix polymer. Optimal performances were achieved with 3 wt.% compatibilizer and 40 wt.% apple pomaces.	[120]
PHB	Agave fiber and organic peroxide		Organic peroxide compatibilized composites resulted in a 45% increase in flexural and impact strength.	[121]
PHBV	Miscanthus biocarbon		The heat deflection temperature of the PHBV was increased from 142 to 155°C when 30% biocarbon was incorporated while the coefficient of linear thermal expansion decreased from 100.6 to 75.6 μm/(m·°C).	[122]
PLA	Bamboo charcoal		Due to good dispersion of the filler in the matrix, mechanical properties of the composite were improved up to 7.5 wt.% of biochar addition.	[123]
PLA	Wheat straw and PLA-g-MA		Compatibilized composites showed 20% improvement in the tensile strength and 14% improvement in flexural strength compared to matrix.	[124]
PLA or PBAT	Date seed powder cleaned with sulphuric acid		Modulus of the PBAT was increased by 300% after incorporation of 40% filler. Similarly, a 20% modulus increase was reported for the PLA in the presence of 40% filler. Soil biodegradation of the developed composites with 40% filler was significantly higher compared to neat polymers.	[125]

Matrix	Filler	Processing method	Remarks	Reference
Thermoplastic starch	Almond shell powder and epoxidized linseed oil		The developed composites can be recycled at least three times without affecting the mechanical properties. However, the mechanical properties are reduced gradually with increasing the number of reprocessing of the biocomposites.	[126]
PCL	Palm fruit bunch fiber	Compression molding	Lower thermal stability of the observed due to the inclusion of less thermally resistant fibers. It was observed that the filler-matrix interaction was mainly physical.	[127]
PCL	Palm fruit bunch grafted with methyl methacrylate (MMA)		MMA modified fibers showed good interaction with matrix to enhance the tensile, flexural, and impact strength compared to corresponding composites prepared with unmodified fibers.	[128]
Poly(furfuryl alcohol)	Kenaf fiber		Composites with 20 wt.% fiber loading showed a 123% increase in storage modulus while a 48% improvement in flexural strength and a 310% improvement in tensile strength.	[129]
Biobased phenol-formaldehyde resin (Novolac resin)	Maple wood particles		Low water absorption and good dimensional stability under water immersion were observed for the developed composites, suggesting great potential for developing biobased wood composites by compression molding.	[130]
PLA	Coconut shell biochar	3D printing	Mechanical and thermal properties were increased.	[131,132]
PLA	Continuous flax fibers		Composites produced with a layer thickness of 0.6 to 0.2 mm resulted in less porosity, and a 210% increase in tensile properties. On the other hand, 50% tensile modulus and 73% tensile strength increase were observed when the composites were produced with a layer thickness of 1 to 10 mm.	[133]
PTT	Miscanthus biocarbon		The addition of 5 wt.% biocarbon resulted in optimum performances in terms of dimensional stability, ease of printing, and surface finish.	[134]
Bio-PE	Spruce Thermomechanical pulp and MA-g-PE		Fully biobased composites were developed with low water uptake in the compatibilized composites.	[135]
Carboxymethyl cellulose	Cellulosic fiber	4D Printing	The shape memory effect of the 3D-printed composite was demonstrated based on the hydration/dehydration mechanism.	[55]

(a)

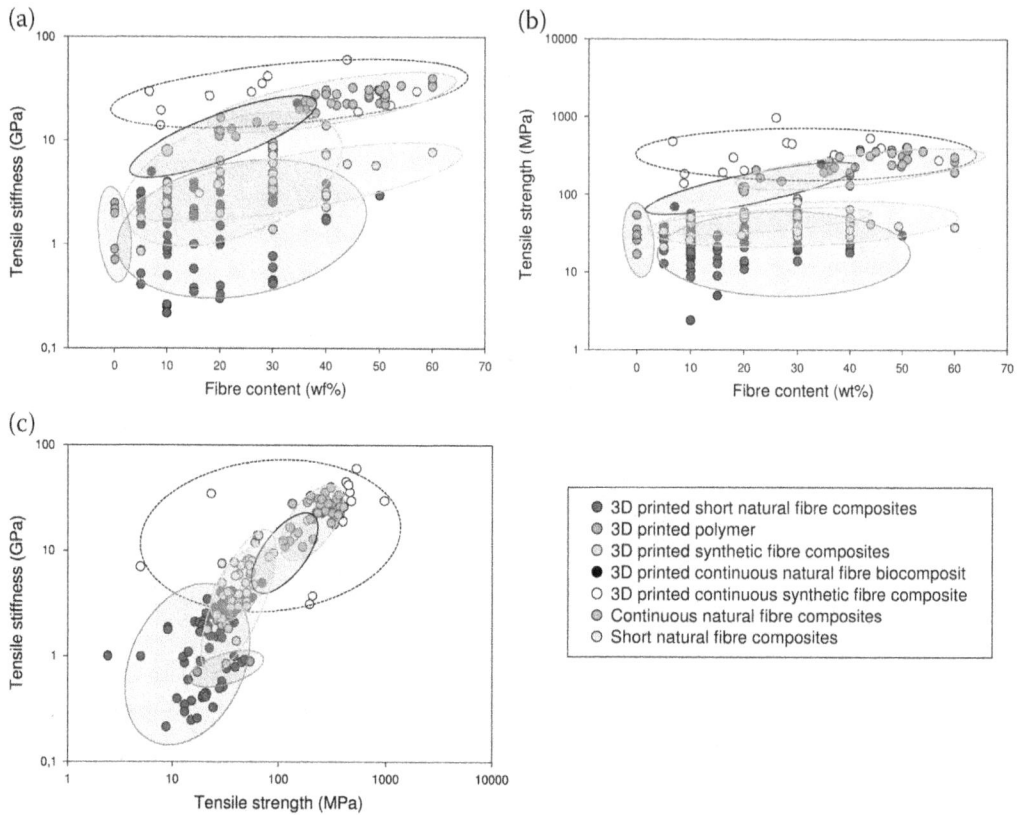

(b)

(c)

FIGURE 3.6 The tensile properties of natural fiber-based biocomposites fabricated by 3D printing. (a) Tensile stiffness and (b) strength versus fiber content; (c) tensile stiffness versus the strength of 3D printed pristine polymer and composites. Reproduced with permission from ref. [31]. Copyright 2020 Elsevier.

and stiffness of the composites can widely vary as shown in Figure 3.6(a and b) [31]. Furthermore, the lack of standards for 3D-printed mechanical test samples preparation has influenced the performances of the 3D-printed samples [74,75]. Regardless of the manufacturing methods, the continuous fiber-reinforced composites showed better tensile strength and tensile stiffness compared to short fiber-reinforced composites because of the high aspect ratio [31]. Furthermore, the fiber critical length is an important parameter that can significantly influence the stress transfer between the matrix and fiber [76]. Therefore, the composites developed with fibers above critical length can provide better mechanical performances/reinforcement compared to their corresponding composites developed with fibers below critical fiber length [77]. Similar to fibrous fillers, particulate fillers with an optimal size can provide superior performances in the resulting composites [78,79].

Composites with strong interfacial interaction can provide better mechanical performances because of their effective stress transfer from one phase to another [33]. However, most of the biocomposites are not able to form strong interfacial interaction because of the polarity difference between the fillers and polymer matrix [80]. The weak interfacial interaction between the fillers and matrix can lead to forming insufficient performances in the resulting composites (Figure 3.7(a)). Strategies such as fiber modification, matrix modification, and/or addition of compatibilizer are employed to overcome the weak interfacial interaction between the fibers and matrix (Figure 3.7(b)). Generally, the compatibility between the biofibers, and polymer matrix can form by electrostatic bonding, molecular inter diffusion, mechanical interlocking, and chemical bonding [81]. Natural fiber treatments such as alkali treatment, grafting, cyanoethylation,

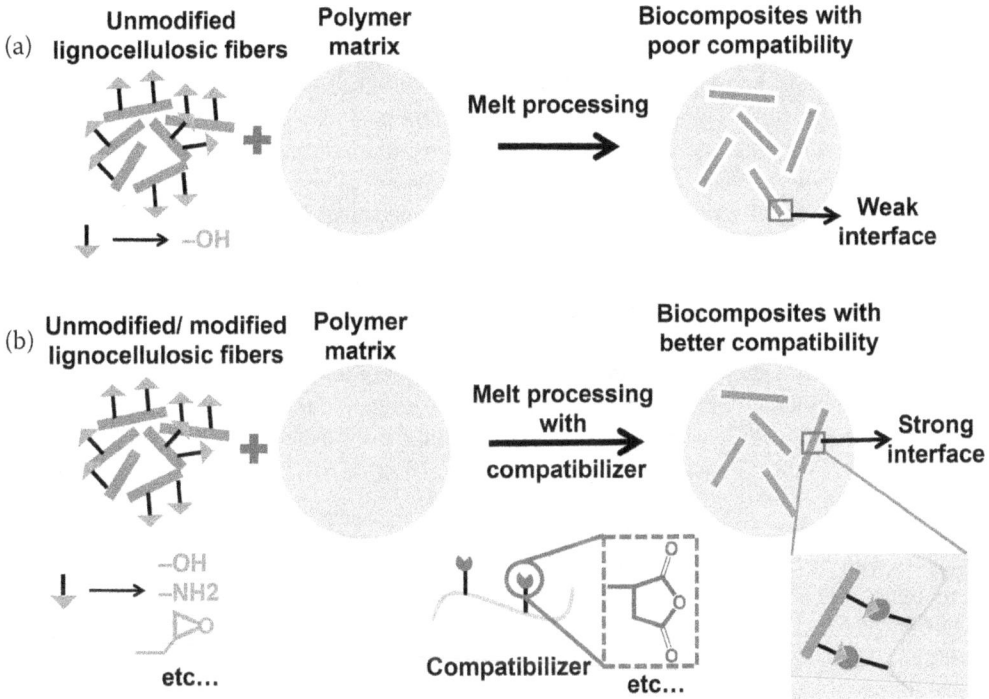

FIGURE 3.7 (a) Uncompatibilized and (b) compatibilized biocomposites preparation strategies [4].

acetylation, peroxide treatment, sizing with polymeric isocyanates, treatment with silane, and other coupling agents have been successfully used to enhance the fiber-matrix adhesion as well as the performances of the resulting composites [64]. For example, composites produce from biobased epoxy and chemically modified (alkali-treated and poly(hexafluorobutyl acrylate) coating) ground coconut waste particles showed improved compatibility [82]. Especially, poly(hexafluorobutyl acrylate) modified ground coconut waste particles offered better compatibility with epoxy resin compared to the alkali treat coconut waste particles. The alkali-treated jute fibers, kenaf fibers, hemp fibers, and coconut coir fibers had good compatibility with biobased phenol formaldehyde [23]. The compatibility of the chemically modified rice husk [83], wood fibers/flour [84,85], and cellulose fibers [86,87] was improved with different polymer matrices compared to unmodified fillers. Polymers grafted with maleic anhydride or epoxy functionality are the most widely used as a compatibilizer to improve the interfacial interaction between the blended components [14,17,88]. The maleic anhydride grafted polymers can react with the hydroxyl groups in the lignocellulosic fibers while the polymer backbone can be miscible in the bulk matrix, thereby improving the interfacial adhesion between the fillers and matrix. Similarly, peroxide initiators have also been utilized to improve the compatibility between the polymer and lignocellulosic fillers by free radical graft copolymer formation between the fillers and matrix during melt processing [18,89,90]. For instance, a significant improvement in mechanical properties was observed in the peroxide compatibilized miscanthus-based composites [18]. Polymeric methylene diphenyl diisocyanate (pMDI) is another reactive compatibilizer that has been widely used to improve the compatibility between the fiber and matrix [91,92]. The combination of pMDI and poly(2-ethyl-2-oxazoline) has also shown very effective to improve the interfacial interaction between the fiber and matrix [93]. In a recent study, the compatibilization of the biocomposites with hybrid compatibilizers (octadecyl organically modified montmorillonite and maleic anhydride grafted polyethylene octene elastomer) showed enhanced fiber/matrix interaction as well as mechanical properties [94]. It has been established that the compatibility between lignocellulosic fibers and polymer matrix can be improved by different methods. To achieve optimal performances in the compatibilized composites, the composites should be produced with a proper compatibilization strategy.

Melt processing of lignocellulosic fibers with a high melting temperature polymer (>225°C) for a long time is challenging due to the thermal decomposition nature of their constituents [35,95]. For example, PLA/wood fibers composites produced at 230°C showed more than a 5% reduction in their tensile and flexural properties compared to PLA/wood fiber composites produced at 200°C [96]. In addition to deteriorating the performances of the biocomposites, the thermal degradation of the lignocellulosic materials can produce an unpleasant odor and surface defects. Enzymatic treatment has also been successfully used to remove the low thermal stability components (e.g., fatty acids, glycerol, dextrins, and amino acids) from the lignocellulosic materials (rice husks, and rye brans) before using them as a reinforcement in the polymer matrix [97]. Despite the low carbon footprint, the enzymatic treatment of the lignocellulosic materials to enhance the thermal stability would make the lignocellulosic materials expensive for the composite reinforcement. High melting point polymers (>200°C) such as bio-polyamides 6, 10 (PA6,10), poly(ethylene terephthalate) (PET)) and poly(trimethylene terephthalate) (PTT) make it difficult to blend with lignocellulosic fibers [35]. In the presence of lithium chloride, the melting point of PA6 can be reduced to make composites with lignocellulosic fibers [64]. The incorporation of carbonaceous filler (biochar) into high melting point biopolymers (e.g., PA6,10, PTT, and PET) is also a promising approach for the preparation of biobased composites [98]. Biocomposites developed from biobased PA6,10 and 20 wt.% biochar (particle size: <500 microns) showed a 50% increase in flexural modulus and 83% increase in tensile modulus compared to neat PA 6,10 [99]. The tensile strength of the PA6,10/ biochar composite was unchanged while a 61% increase in the flexural strength was observed. A significant increase (200%) in the impact strength was observed when the bio-PA6,10 composites were prepared from biochar particles sized at <63 μm. The observed enhanced performances in the bio-PA6,10/biochar composites were attributed to the high surface area, good dispersion, and wetting of biochar in the PA6,10 matrix [99]. The tribological properties of the 3D printed PLA/ wheat straw-based biochar (30 wt.%) were significantly improved because of the strong reinforcing effect of the biochar in the PLA matrix [100]. A recent study developed injection-molded biocomposites from miscanthus grass-derived carbonaceous filler (biochar/biocarbon) and poly-phthalamide (PPA) to compare the performance of the resulting composites with the same 20 wt.% talc-filled PPA composites [101]. The biocarbon/PPA composites showed performances (except impact strength and heat deflection temperature) comparable to talc/PPA composites. Therefore, the biocarbon could be a potential substitute filler for talc in the automotive composite application while reducing the density of the composites [101].

The majority of the lignocellulosic fibers are vulnerable to moisture because they are composed of hydrophilic components such as cellulose. The performances of such fibers incorporated composite can be significantly affected when exposed to moisture/water [102]. Especially, the water absorption rate of the biocomposites is very high at elevated temperatures because the water molecules can easily diffuse and migrate into voids and fiber-matrix interface to form filler-matrix debonding [103]. Surface modification of the lignocellulosic fibers can reduce the hydrophilic functional groups on the surface, thereby increasing the resistance to water absorption [104,105]. Furthermore, the water resistance of the lignocellulosic fibers-based composites can be improved by the incorporation of a reactive compatibilizer which tends to react with hydrophilic functional groups [106]. Outdoor durability and fire resistance of the biocomposites are not sufficient when the composites are developed without UV stabilizers and flame-retardant additives [103]. Incorporation of these additives with an optimal concentration can provide balanced durability and strength for the resulting biocomposites.

Most of the lignocellulosic fibers are sensitive to fire because of their low thermal stability. The flammability of the lignocellulosic fibers depends on the chemical composition of the lignocellulosic fibers. For instance, the lignocellulosic fibers with a high lignin content can be more resistant to fire because they can form char to limit flame propagation [107]. Unlike lignocellulosic filler reinforced composites, the biocomposites developed from wools had a high flame retardancy with a low thermal conductivity because of the high char forming ability of the wool [108]. Such composites can be used in building interior applications. The flammability of the lignocellulosic fibers can be reduced by different types of surface treatment including silane and phosphorus compounds [109,110]. The inherent nature of the flame retardant properties of the proteinaceous materials, the composites produced with proteinaceous materials showed better char forming ability and less flammability compared to polymer matrix [111]. Halogenated flame retardants (e.g., polybrominated diphenyl ether, hexabromocyclododecane, etc.) have been incorporated to reduce the flammability of the biocomposites [112]. However, halogenated flame retardants are harmful to the environment. Therefore, non-halogenated flame retardants (e.g., magnesium hydroxide ($Mg(OH)_2$), aluminum hydroxide ($Al(OH)_3$), modified and unmodified carbonaceous materials, alumina trihydrate (ATH), organoclay (OMMT), and zinc borate, aluminum phosphonate, ionic liquid-based metal-organic hybrid and oligomeric phosphorus-nitrogen containing intumescent flame retardant) have been used to reduce the flammability of the polymers [113]. Recently, next-generation flame retardants are produced from biosources including lignin [114], phytic acid [115], and tartaric acid [116] have been used to improve the fire resistance of the biocomposites [107]. Due to the high thermal stability of the carbonaceous materials (e.g., fullerene, carbon nanotubes, graphene, graphite, biochar, etc.) and effectively form char to reduce the oxygen transfer into polymers, the carbonaceous materials incorporated composites are less flammable compared to neat polymers [36,117]. Among carbonaceous materials, biochar is one of the attractive and inexpensive materials with flame retardant properties. It has been shown that the polypropylene composites produced with biochar can reduce flammability while increasing rigidity as well as flexural strength [118]. The biochar could be combined with other flame retardant additives to achieve adequate flame retardant properties.

3.5.1 HYBRID BIOCOMPOSITE PERFORMANCES

Hybrid composites are fabricated to achieve a high-performance material by taking advantage of the individual components. The hybrid composites can be developed from fully biodegradable components, or a combination of biodegradable and non-biodegradable components as summarized in Figure 3.8 and Table 3.3. In hybrid composites, the matrix can be a single polymer or a blend of two or more polymers based on the requirement. Similarly, a combination of different fibers/fillers can be used to develop hybrid composites. In addition to processing parameters optimization, optimal functionalization/ treatments of the matrix or fillers or both fillers and matrix may require to achieve synergistic effects in the resulting composites [136]. Matrix polymer is the major component in a composite that imparts overall strength, durability, and dimensional stability. Therefore, hybridizing matrix polymers could provide a cost-performance balance. The toughening of the brittle bioplastics (e.g., PHBV and PLA) can be upgraded by the addition of elastic bioplastics such as PBAT and PCL [137,138]. The cost of such blends is still relatively high compared to conventional plastics [32]. Therefore, biocomposites are developed using inexpensive fillers as reinforcements to improve the cost effectiveness of the product while maintaining performance [139,140]. It is established that the high-performance composites are obtained when the composites form a strong interface between matrix-filler with the help of optimal compatibility. Zhang et al. reported ternary blends of PHBV, PBAT, and epoxidized natural rubber for the incorporation of miscanthus fiber in the presence of an organic peroxide, which leads to a composite having a very good toughness-stiffness balance [89]. A similar approach has been carried in binary biodegradable and nonbiodegradable matrix-based biocomposites to achieve a cost-performance balance [14,141].

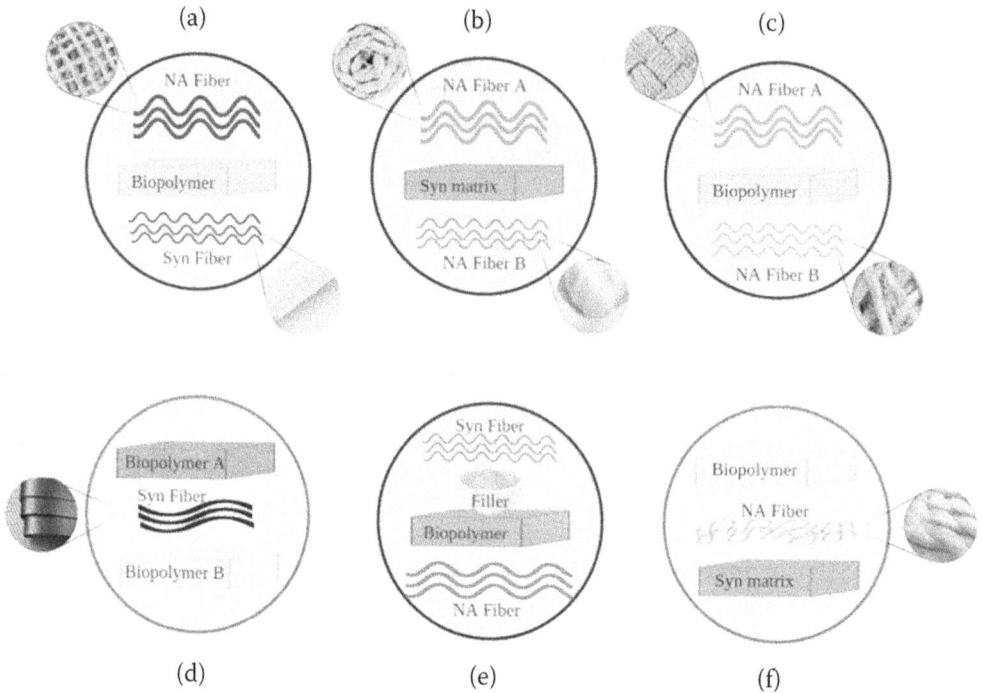

FIGURE 3.8 Schematic diagram showing different types of hybrid biocomposites based on natural and synthetic components as matrices and reinforcements. Reproduced with permission from ref. [142]. Copyright 2020 under the Creative Commons Attribution License.

Hybrid fillers can be two different types of fibers or particulate fillers or a combination of both as shown in Figure 3.8. Hybridization of natural fibers is used when the two reinforcements possess distinct characteristics such as different chemical composition, toughness, stiffness, density, porosity, cost, etc. A higher cellulose content promotes the stiffness of the fiber in composites [41] and similarly, a combination of bast and leaf fiber results in excellent stiffness/toughness balance [136]. Unidirectional flax-paper hybrid reinforced epoxy composites resulted in better specific stiffness compared to unidirectional E-glass/epoxy composites [143]. Such hybrid composites exhibit high specific strength and stiffness compared to composites developed with the same volume fraction of only flax fibers. In another study, the specific tensile properties of the unidirectional flax-paper hybrid reinforced PLA composites (217 MPa·cm^3·g^{-1}) were similar to epoxy/woven glass fabric composites (227–278 MPa·cm^3·g^{-1}) [144]. By considering the cost and performance-related characteristics, a hybrid of synthetic and natural fibers can be used as reinforcement for biocomposite development. Wool and biochar-based hybrid fillers incorporated biocomposites showed enhanced flame and thermal resistant properties [36]. The improved thermal resistance was attributed to the inertness of biochar against fire while wool prevents fire propagation through char formation. Wool and hemp fiber hybrid-based composites possess adequate properties for applications such as construction boards, insulation boards, and bulletproof materials [145,146]. For better mechanical properties, the reinforcement can be hybridized with synthetic fibers at a lower percentage [147,148]. The effect of the hybridization of two fibers in vinyl ester matrix forming a strong interface can be observed in Figure 3.9. In the case of a woven flax fiber-based vinyl ester composite, microstructural defects; shear kink bands and nodes caused failure at matrix fiber interface. The fractured surfaces of hybrid composites of unstitched flax and basalt fibers showed delamination of fibers and stitched fibers resulted in a strong fiber-matrix interface, as shown in Figure 3.9(d).

TABLE 3.3

Different Hybrid Composites and Their Performances

Matrix Polymer	Reinforcement	Manufacturing Method	Performances	Reference
PLA	Hemp fibers (hybrid yarn)	Compression molding	The composites prepared with 45 wt.% fiber showed around a 3.3-fold increase in flexural strength while both tensile and impact strengths showed around a twofold increase compared to pristine PLA.	[150]
PLA	Unidirectional flax and paper		The tensile properties of the flax-paper/PLA (217 MPa·cm^3·g^{-1}) composites are similar to epoxy/woven glass fabrics composites (227–278 MPa·cm^3·g^{-1}).	[144]
PLA/PBAT blend	Rice husk, wheat husk, and wood fibers		All the developed composites exhibit adequate density (378–488 kg/m^3), flexural strength (0.80–2.25 MPa), and thermal insulation (0.08–0.14 W/m.K) properties for indoor building insulation applications.	[151]
Bio-epoxy	Kenaf and sisal		Composites were thermally stable up to 295°C. The reinforcing effect of the fibers is not affected after weathering test, suggesting suitability for semi-structural applications.	[152]
PLA and PTT	Miscanthus biochar	Injection molding	Toughness and overall mechanical and thermal properties were enhanced by the hybridization.	[79]
PBS/PBAT	Miscanthus fibers and maleic anhydride functionalized PBS/PBAT blend		The mechanical properties of the compatibilized composites were significantly improved compared to their uncompatibilized counterparts. Furthermore, the strong reinforcing effect of the miscanthus fiber leads to an increase in the modulus of the composites with fiber percentage.	[14]
PHBV/PBAT	Soy stalk, wheat straw, and corn stover		Among the developed composites, the composites developed from miscanthus fibers showed slightly higher tensile strength, Young's modulus, and heat deflection temperature.	[32]
Biopolyamide (Bio-PA), PP	Carbon and wood fiber		The addition of 30 wt.% of wood fiber into bio-PA/PP blend enhanced the heat deflection temperature from 45°C to 100°C. Such composites could be suitable for automotive applications.	[153]

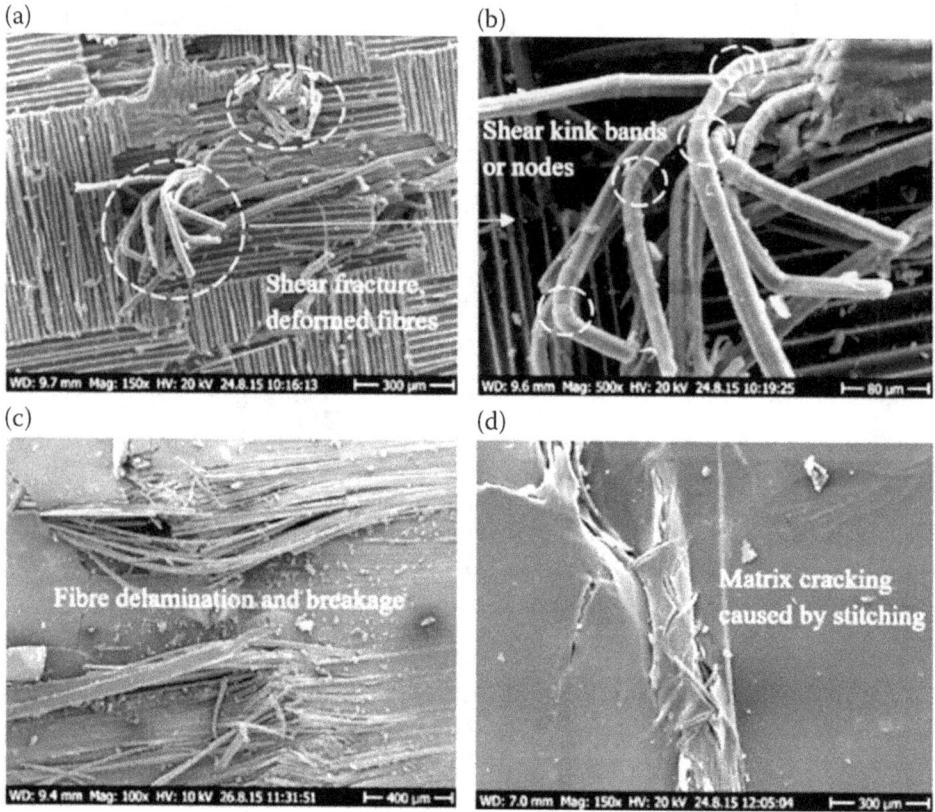

FIGURE 3.9 SEM images of fractured surfaces of (a) shear fracture which deformed fibers for flax fiber reinforced vinyl ester (b) failure of flax fibers due to shear (c) failure of the unstitched reinforcement due to the delamination from vinyl ester matrix (d) matrix cracking caused by stitching of flax fiber reinforced vinyl ester. Reproduced with permission from ref. [149]. Copyright 2018 Elsevier.

3.6 APPLICATION OF BIOPOLYMER-BASED COMPOSITES

The growing interest and government regulations are accelerating the sustainable material design and development for various applications including packaging, automotive, aerospace, and other high-performance applications [154]. The packaging industries consume the highest percentage of world polymer production because of their attractive properties including lightweight and low cost [5]. The environmental and waste management issues associated with fully petroleum-based and non-biodegradable plastic packaging trigged the utilization of sustainable alternatives. Effective recycling can be a solution to tackle the end-of-life plastic waste problems. However, there are several challenges (e.g., contamination removal and multi-layer packaging) that limit the end-of-life plastic packaging waste recycling [155]. Biodegradable biocomposites have attracted packaging applications because they can provide both economic and environmental benefits. Especially, biodegradable composites can be used for short-life span packaging applications such ass flexible films and cutleries [5]. Other than packaging application, net clip for vineyard has been produced from biodegradable aliphatic polyester and grape pomace [156]. PLA/Kiwi fruit residue-based biodegradable spoon-knife was developed and commercialized by ZESPRI® [156]. Another biocomposite produced from biodegradable hybrid polymer matrix and coffee chaff has been used for coffee pod application [4]. The fully biodegradable composites with micro-fillers for gas barrier packaging application are still limited because of the insufficient barrier properties against oxygen, water vapor, and CO_2. The incorporation of biobased nanofillers (e.g., nano-cellulose) into biodegradable polymers can create a tortures path to minimize the gas and vapor diffusion rate [157]. Such composites showed better gas barrier properties when the nanofiller is uniformly dispersed in the matrix. Barrier properties of the biodegradable composites can be enhanced by developing multilayer films [158]. In addition to the barrier and mechanical properties, transparency is one of the desired properties for food packaging application. Therefore, the addition of a small quantity of nanostructure filler with uniform dispersion could help to maintain the matrix transparency while improving the performances [159]. Antimicrobial properties are essential for food packaging application to enhance the shelf-life of the packed food [160].

Henry Ford had explored the significance of biocomposites in automobile applications in the early 1940s. Since then, the use of natural fibers in composite fabrication for automotive applications has been growing because of their interesting properties [161]. Biocomposites developed from natural fibers (e.g., flax, coir, bamboo, sisal, cork, cellulose fibers, coconut shell powder, etc.) with different polymer matrices (e.g., Bio-PBS, Bio-PA, Bio-PET, epoxy, etc.) have shown great potential in automotive applications [162]. Therefore, automotive companies such as Ford, Hyundai, Mercedes, Toyota, and Volkswagen have demonstrated the biocomposites for several applications including console armrests, veneer, radiator end tank, and interior components [4]. Akampumuza et al. have comprehensively reviewed the biocomposites application progressive in the automotive industries [63]. To achieve desired properties in the biocomposites, the natural fibers are hybridized at different loading [147]. For instance, sisal and flax fiber mats were used as hybrid reinforcement in the epoxy matrix to produce interior door panels by Mercedes Benz [63]. Similarly, biocomposites from polyurethane and sisal/flax hybrid were developed for door trim panels by Volkswagen [63].

The sustainable composites used for aerospace applications mainly include the composites with petroleum-based matrices with sustainable reinforcement. Biodegradable polymers are less used in composite fabrication in aircraft components due to the lower thermal stability and degradation behavior. The aerospace field utilizes a variety of advanced fibrous and particulate reinforcements such as glass fibers, Kevlar, carbon fiber, carbon nanotubes, graphene, etc. [163–165]. The most widely used sustainable material in aerospace applications is natural fibers owing to their higher strength and lower density and most importantly the sustainability [166]. Furthermore, a methodological guide was developed to promote the use of sustainable materials in aircraft interiors as the natural fiber composites showed promising characteristics to substitute the current aircraft panels [167]. A study by Boegler et al. showed that the epoxy composites with hemp, flax, sisal, and ramie can be used to design aircraft wings because of their good mechanical properties as well as their lightweight characteristics [168].

Recently, PLA/lignin hybrid composites with flax, hemp, and waxes have been developed for extrusion, injection molding, and thermoforming application by a German compounding company, TECNARO [169]. The trademark of this formulation is called ARBOFORM® as well as it is called "liquid wood". In addition to construction, electronics, and furniture applications, the "liquid wood" biocomposite formulation was used to develop earcups in the commercially available headphones [169]. Agro and forestry residues based on biocomposites showed great potential for interior building thermal insulation applications because of their attractive thermal insulation, density, and mechanical properties [151,170,171]. Similarly, highly filled wood fiber composites have been proposed for crates and thermo-formable sheets [172]. Biocomposites developed from PHA-based formulations showed good mechanical properties as well as harmless biodegradation in soil [173]. As a result, the PHA/lignocellulosic fibers composites could be used to develop biodegradable flowerpots as a sustainable alternative to petroleum-based non-biodegradable polypropylene flowerpots.

3.7 CHALLENGES AND OPPORTUNITIES

The utilization of lignocellulosic fillers in composites development has been widely accepted because of their sustainability, lower density, easier availability, and cost-effectiveness. The search

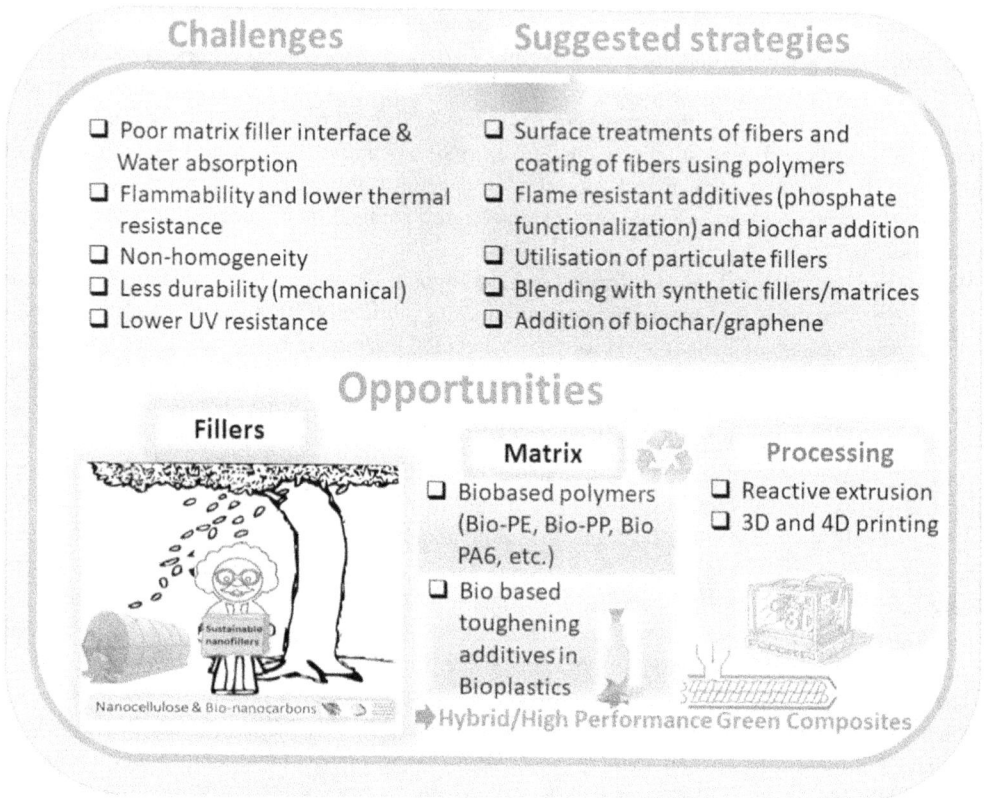

FIGURE 3.10 Challenges, suggested strategies and opportunities of biocomposites.

for bioplastics was not pronounced until the issues associated with single-use plastic waste become critical. The eco-friendliness of bioplastics in terms of their renewability and/or biodegradability created a platform to work with biofillers for the development of sustainable biocomposites. The research conducted in the field of biocomposite mainly focussed on specific property improvement, biodegradability, and cost reduction. The concerns about the poor matrix filler interface, water absorption, flammability, lower thermal stability, non-homogeneity, durability, and UV resistance are still existing to some extent in the biocomposites reported so far [64] (Figure 3.10). The poor matrix-filler interface is one of the major issues observed in biocomposites, which can be solved by appropriate compatibilization strategies [162].

Hybridization of biofillers and nonbiodegradable polymers with compatibilizers can be used as a general sustainable strategy for expanding the usage of biocomposites. Apart from the idea of hybridization of different components in composites, the development of bionanofillers is a promising field that is expected to open enticing opportunities for high-performance biocomposites development. Composites based on nanocellulose possess significant use in various applications including packaging [174,175]. However, its lower thermal stability limits its usage for high-temperature processing and hence the nanofillers based on carbonaceous material are essential for achieving potential growth of sustainable composites for advanced applications. Carbonaceous nanofillers include nanobiochar [176], Bio-CNT [177], graphene-like carbons derived from biomass [178,179]). Biochar can be utilized as a substitute for mineral fillers too. The layered structure of graphene-like carbons derived from biomass can improve the barrier properties of biopolymer matrices such as PLA, PHAs, etc. The enhanced properties of the aforementioned lightweight fillers are attractive for automobile applications. Recently introduced biobased nonbiodegradable polymers (e.g., Bio-PE, Bio-PP, Bio-PA6, etc.) are also offers a huge scope in finding biobased alternatives for petroleum-based polymers. Such bioplastics would gradually help in replacing petroleum-based general purpose plastics in packaging and automotive applications [153]. Similar to the material properties, selecting an appropriate biocomposite manufacturing utilizing green chemistry approaches can minimize carbon footprint, energy consumption, and cost. Furthermore, recycling and upcycling of biocomposites are also supporting sustainability through environmental and economical aspects.

3.8 CONCLUSIONS

The preparation of sustainable materials from renewable sources or via a green pathway is essential for a sustainable future. According to the sustainability definition, cleaner, greener, and inexpensive materials are needed in all sectors to achieve sustainable development [180]. Balance in cost, performance, and sustainability are vital to developing products that can be used in real applications. In this chapter, lignocellulosic fibers/fillers based on sustainable composite developments and their performances have been summarized. The major advantages of the lignocellulosic fibers/fillers-based composites are the lightweight, cost, good specific strength, and reduced carbon footprint compared to synthetic fillers. However, the biocomposites showed inferior performances when the interfacial interaction between the filler and matrix is not sufficient in the resulting composites. Therefore, compatibilization strategy plays a key role in successful biocomposites development. Commonly used compatibilization strategies are surface modification of the lignocellulosic fibers/fillers, modification of the matrix, and incorporation of reactive additives such as compatibilizer and peroxides. Biocomposites are often not satisfying the required durability, water absorption, and flammability. Lignocellulosic fibers/fillers can be hybridized with synthetic fillers to overcome some of the drawbacks while partially replacing petroleum-based fillers in the composites. Especially, composites prepared by hybridizing natural fiber or biochar with synthetic fibers/fillers showed performances similar to the fully synthetic filler-based counterparts. Such a hybridization strategy could reduce the overall cost of the composites besides minimizing the dependence on petroleum resources.

ACKNOWLEDGMENTS

The authors would like to thank the financial support of (i) the Ontario Research Fund, Research Excellence Program; Round 9 (ORF-RE09) Ontario Ministry of Economic Development, Job Creation and Trade (Project Nos. 053970 and 054345); ((ii) the Ontario Ministry of Agriculture, Food and Rural Affairs (OMAFRA)/University of Guelph – Bioeconomy for Industrial Uses Research Program (Project Nos. 030251, 030332, 030486 and 030578); (iii) the Agriculture and Agri-Food Canada (AAFC), Maple Leaf Food, Canada and Bank of Montreal (BMO), Canada through Bioindustrial Innovation Canada (BIC) Bioproducts AgSci Cluster Program (Project Nos 054015, 054449 and 800148); and (iv) the Natural Sciences and Engineering Research Council of Canada (NSERC), Canada Research Chair (CRC) program Project No. 460788 to carry out this study. This research was also benefited from the facility funding to the Bioproducts Discovery and Development Centre (BDDC) lab by FedDev Ontario; Ontario Ministry of Agriculture, Food, and Rural affairs (OMAFRA); Canada Foundation for Innovation (CFI); Federal Post-Secondary Institutions Strategic Investment Fund (SIF); and matching funds from the province of Ontario and numerous University of Guelph's Alumni.

REFERENCES

[1] J.N. Hahladakis, E. Iacovidou, S. Gerassimidou, *Plastic waste in a circular economy*, Elsevier Inc., 2020. doi:10.1016/b978-0-12-817880-5.00019-0.

[2] D. Andrews, The circular economy, design thinking and education for sustainability, *Local Econ.* 30 (2015) 305–315. doi:10.1177/0269094215578226.

[3] O.D. Elisha, moving beyond take-make-dispose to take-make- use for sustainable economy, *Int. J. Sci. Res. Educ.* 13 (2020) 497–516.

[4] A.K. Mohanty, S. Vivekanandhan, J.M. Pin, M. Misra, composites from renewable and sustainable resources: Challenges and innovations, *Science.* 362 (2018) 536–542. doi:10.1126/science.aat9072.

[5] R. Muthuraj, M. Misra, A.K. Mohanty, Biodegradable compatibilized polymer blends for packaging applications: A literature review, *J. Appl. Polym. Sci.* 135 (2018) 1–37. doi:10.1002/app.45726.

[6] T. Väisänen, O. Das, L. Tomppo, A review on new bio-based constituents for natural fiber-polymer composites, *J. Clean. Prod.* 149 (2017) 582–596.

[7] O. Das, D. Bhattacharyya, D. Hui, K.-T. Lau, Mechanical and flammability characterisations of biochar/polypropylene biocomposites, *Compos. Part B Eng.* 106 (2016) 120–128.

[8] O. Das, D. Bhattacharyya, A.K. Sarmah, Sustainable eco–composites obtained from waste derived biochar: A consideration in performance properties, production costs, and environmental impact, *J. Clean. Prod.* 129 (2016) 159–168.

[9] O. Das, M.S. Hedenqvist, E. Johansson, R.T. Olsson, T.A. Loho, A.J. Capezza, R.K.S. Raman, S. Holder, An all-gluten biocomposite: comparisons with carbon black and pine char composites, *Compos. Part A Appl. Sci. Manuf.* 120 (2019) 42–48.

[10] V. Guna, M. Ilangovan, M.G. Ananthaprasad, N. Reddy, Hybrid biocomposites, *Polym. Compos.* 39 (2018) E30–E54.

[11] S. Vivekanandhan, M. Misra, A.K. Mohanty, Thermal, mechanical, and morphological investigation of injection molded poly (trimethylene terephthalate)/carbon fiber composites, *Polym. Compos.* 33 (2012) 1933–1940.

[12] X.-F. Wei, K.J. Kallio, S. Bruder, M. Bellander, R.T. Olsson, M.S. Hedenqvist, High-performance glass-fibre reinforced biobased aromatic polyamide in automotive biofuel supply systems, *J. Clean. Prod.* 263 (2020) 1–9.

[13] R. Muthuraj, M. Misra, A.K. Mohanty, Biodegradable biocomposites from poly(butylene adipate-co-terephthalate) and miscanthus: Preparation, compatibilization, and performance evaluation, *J. Appl. Polym. Sci.* (2017) 1–9. doi:10.1002/app.45448.

[14] R. Muthuraj, M. Misra, A.K. Mohanty, Biocomposite consisting of miscanthus fiber and biodegradable binary blend matrix: compatibilization and performance evaluation, *RSC Adv.* 7 (2017) 27538–27548. doi:10.1039/C6RA27987B.

[15] R. Muthuraj, M. Misra, F. Defersha, A.K. Mohanty, Influence of processing parameters on the impact strength of biocomposites: A statistical approach, *Compos. Part A Appl. Sci. Manuf.* 83 (2016) 120–129. doi:10.1016/j.compositesa.2015.09.003.

[16] R. Muthuraj, T. Mekonnen, Carbon dioxide–derived poly (propylene carbonate) as a matrix for composites and nanocomposites: performances and applications, *Macromol. Mater. Eng.* (2018) 1–19.

[17] R. Muthuraj, M. Misra, A.K. Mohanty, Injection molded sustainable biocomposites from poly(butylene succinate) bioplastic and perennial grass, *ACS sustain. Chem. Eng.* 3 (2015) 2767–2776. 10.1021/acssuschemeng.5b00646.

[18] R. Muthuraj, M. Misra, A.K. Mohanty, Reactive compatibilization and performance evaluation of miscanthus biofiber reinforced poly(hydroxybutyrate-co-hydroxyvalerate) biocomposites, *J. Appl. Polym. Sci.* 134 (2017) 1–10. doi:10.1002/app.44860.

[19] N. Vahedikia, F. Garavand, B. Tajeddin, I. Cacciotti, S.M. Jafari, T. Omidi, Z. Zahedi, Biodegradable zein film composites reinforced with chitosan nanoparticles and cinnamon essential oil: Physical, mechanical, structural and antimicrobial attributes, *Colloids Surfaces B Biointerfaces.* 177 (2019) 25–32.

[20] A. Colombo, A.E. León, P.D. Ribotta, Rheological and calorimetric properties of corn-, wheat-, and cassava-starches and soybean protein concentrate composites, *Starch-Stärke.* 63 (2011) 83–95.

[21] A.K. Mohanty, A. Wibowo, M. Misra, L.T. Drzal, Effect of process engineering on the performance of natural fiber reinforced cellulose acetate biocomposites, *Compos. Part A Appl. Sci. Manuf.* 35 (2004) 363–370.

[22] https://www.european-bioplastics.org/market/, (n.d.).

[23] R.L. Quirino, T.F. Garrison, M.R. Kessler, Matrices from vegetable oils, cashew nut shell liquid, and other relevant systems for biocomposite applications, *Green Chem.* 16 (2014) 1700–1715.

[24] H. Wang, J. Yao, Use of poly (furfuryl alcohol) in the fabrication of nanostructured carbons and nanocomposites, *Ind. Eng. Chem. Res.* 45 (2006) 6393–6404.

[25] O. Faruk, A.K. Bledzki, H.-P. Fink, M. Sain, Biocomposites reinforced with natural fibers: 2000–2010, *Prog. Polym. Sci.* 37 (2012) 1552–1596. doi:10.1016/j.progpolymsci.2012.04.003.

[26] D. Tadele, P. Roy, F. Defersha, M. Misra, A.K. Mohanty, A comparative life-cycle assessment of talc-and biochar-reinforced composites for lightweight automotive parts, *Clean Technol. Environ. Policy.* 22 (2020) 639–649.

[27] O. Das, A.K. Sarmah, D. Bhattacharyya, A sustainable and resilient approach through biochar addition in wood polymer composites, *Sci. Total Environ.* 512 (2015) 326–336.

[28] S. Sajith, V. Arumugam, H.N. Dhakal, Comparison on mechanical properties of lignocellulosic flour epoxy composites prepared by using coconut shell, rice husk and teakwood as fillers, *Polym. Test.* 58 (2017) 60–69.

[29] M.K. Gupta, R.K. Srivastava, Mechanical properties of hybrid fibers-reinforced polymer composite: A review, *Polym. Plast. Technol. Eng.* 55 (2016) 626–642.

[30] R. Várdai, T. Lummerstorfer, C. Pretschuh, M. Jerabek, M. Gahleitner, B. Pukánszky, K. Renner, Impact modification of PP/wood composites: a new approach using hybrid fibers, *Express Polym. Lett.* 13 (2019) 223–234.

[31] A. Le Duigou, D. Correa, M. Ueda, R. Matsuzaki, M. Castro, A review of 3D and 4D printing of natural fibre biocomposites, *Mater. Des.* 194 (2020) 1–26. doi:10.1016/j.matdes.2020.108911.

[32] V. Nagarajan, A.K. Mohanty, M. Misra, Sustainable green composites: Value addition to agricultural residues and perennial grasses, ACS sustain. *Chem. Eng.* 1 (2013) 325–333. 10.1021/sc300084z.

[33] R. Muthuraj, M. Misra, A.K. Mohanty, Studies on mechanical, thermal, and morphological characteristics of biocomposites from biodegradable polymer blends and natural fibers, *In Biocomposites* (pp. 93–140) Woodhead Publishing, 2015. doi: 10.1016/B978-1-78242-373-7.00014-7.

[34] K. Van der Cruijsen, M. Al Hassan, G. van Erven, O. Dolstra, L.M. Trindade, breeding targets to improve biomass quality in Miscanthus, *Molecules.* 26 (2021) 1–28.

[35] E.O. Ogunsona, A. Codou, M. Misra, A.K. Mohanty, A critical review on the fabrication processes and performance of polyamide biocomposites from a biofiller perspective, *Mater. Today Sustain.* 5 (2019) 1–18.

[36] O. Das, N.K. Kim, A.K. Sarmah, D. Bhattacharyya, Development of waste based biochar/wool hybrid biocomposites: Flammability characteristics and mechanical properties, *J. Clean. Prod.* 144 (2017) 79–89. doi:10.1016/j.jclepro.2016.12.155.

[37] N. Saba, P.M. Tahir, M. Jawaid, A review on potentiality of nano filler/natural fiber filled polymer hybrid composites, *Polymers (Basel).* 6 (2014) 2247–2273. doi:10.3390/polym6082247.

[38] P.K. Rakesh, I. Singh (Eds.).*Processing of Green Composites*, Singapore, Springer, 2019.

[39] A. Sharma, M. Thakur, M. Bhattacharya, T. Mandal, S. Goswami, Commercial application of cellulose nano-composites – A review, *Biotechnol. Reports.* 21 (2019) 1–18, e00316. doi:10.1016/j.btre.2019.e00316.

[40] S.-A. Riyajan, Green natural fibre reinforced natural rubber composites, *Natural Rubber Materials: Volume 2: Composites and Nanocomposites*, 8 (2013) 353. doi:10.1039/9781849737654-00353.

[41] R. Yahaya, S.M. Sapuan, M. Jawaid, Z. Leman, E.S. Zainudin, Review of kenaf reinforced hybrid biocomposites: Potential in defence applications, *Curr. Anal. Chem.* 13 (2017). doi:10.2174/15 73411013666171113150225.

[42] S. Yu, X. Zhang, C. Liu, C. Rudd, X. Yi, A conceptional approach of resin-transfer-molding to rosin-sourced epoxy matrix green composites†, *Aerospace.* 8 (2021) 1–12. doi:10.3390/ aerospace8010005.

[43] Z. Sun, J. Xiao, L. Tao, Y. Wei, S. Wang, H. Zhang, S. Zhu, M. Yu, Preparation of high-performance carbon fiber-reinforced epoxy composites by compression resin transfer molding, *Materials (Basel).* 12 (2018) 1–13. doi:10.3390/ma12010013.

[44] A. Ji, S. Zhang, S. Bhagia, C.G. Yoo, A.J. Ragauskas, 3D printing of biomass-derived composites: Application and characterization approaches, *RSC Adv.* 10 (2020) 21698–21723. doi:10.1039/d0ra03620j.

[45] S.H. Hong, J.H. Park, O.Y. Kim, S.H. Hwang, Preparation of chemically modified lignin-reinforced pla biocomposites and their 3D printing performance, *Polymers (Basel).* 13 (2021) 1–10. doi:10.33 90/polym13040667.

[46] M. Tanase-Opedal, E. Espinosa, A. Rodríguez, G. Chinga-Carrasco, Lignin: A biopolymer from forestry biomass for biocomposites and 3D printing, *Materials (Basel).* 12 (2019) 1–15. doi:10.3390/ ma12183006.

[47] T. Yao, Z. Deng, K. Zhang, S. Li, A method to predict the ultimate tensile strength of 3D printing polylactic acid (PLA) materials with different printing orientations, *Compos. Part B Eng.* 163 (2019) 393–402.

[48] N.D. Sanandiya, Y. Vijay, M. Dimopoulou, S. Dritsas, J.G. Fernandez, Large-scale additive manufacturing with bioinspired cellulosic materials, *Sci. Rep.* 8 (2018) 1–8. doi:10.1038/s41598-018-26985-2.

[49] R.A. Ilyas, S.M. Sapuan, M.M. Harussani, M.Y.A.Y. Hakimi, M.Z.M. Haziq, M.S.N. Atikah, Polylactic acid (PLA) biocomposite: Processing, *Additive Manufact Advanced Appl* 13 (2021) 1–34.

[50] A. Le Duigou, A. Barbé, E. Guillou, M. Castro, 3D printing of continuous flax fibre reinforced biocomposites for structural applications, *Mater. Des.* 180 (2019) 1–8. doi:10.1016/j.matdes.2019.107884.

[51] A. Le Duigou, M. Castro, R. Bevan, N. Martin, 3D printing of wood fibre biocomposites: From mechanical to actuation functionality, *Mater. Des.* 96 (2016) 106–114. doi:10.1016/j.matdes.2016.02.018.

[52] manufacturing, *Met. Mater. Int.* 26 (2020) 564–585. doi:10.1007/s12540-019-00441-w.

[53] S. Miao, N. Castro, M. Nowicki, L. Xia, H. Cui, X. Zhou, W. Zhu, S. Jun Lee, K. Sarkar, G. Vozzi, Y. Tabata, J. Fisher, L.G. Zhang, 4D printing of polymeric materials for tissue and organ regeneration, *Mater. Today.* 20 (2017) 577–591. doi:10.1016/j.mattod.2017.06.005.

[54] A. Sydney Gladman, E.A. Matsumoto, R.G. Nuzzo, L. Mahadevan, J.A. Lewis, Biomimetic 4D printing, *Nat. Mater.* 15 (2016) 413–418. doi:10.1038/nmat4544.

[55] M.C. Mulakkal, R.S. Trask, V.P. Ting, A.M. Seddon, Responsive cellulose-hydrogel composite ink for 4D printing, *Mater. Des.* 160 (2018) 108–118. doi:10.1016/j.matdes.2018.09.009.

[56] M.C. Biswas, S. Chakraborty, A. Bhattacharjee, Z. Mohammed, 4D printing of shape memory materials for textiles: Mechanism, mathematical modeling, and challenges, *Adv. Funct. Mater.* 31 (2021) 1–25. doi:10.1002/adfm.202100257.

[57] L.R. Hart, Y. He, L. Ruiz-Cantu, Z. Zhou, D. Irvine, R. Wildman, W. Hayes, 3D and 4D printing of biomaterials and biocomposites, bioinspired composites, and related transformers *3D 4D Printing of Polymer Nanocomposite Materials*, Elsevier, 2020: pp. 467–504.

[58] C. Zarna, M.T. Opedal, A.T. Echtermeyer, G. Chinga-Carrasco, Reinforcement ability of lignocellulosic components in biocomposites and their 3D printed applications–A review, *Compos. Part C Open Access* 6(2021) 1–17.

[59] F. Momeni, X. Liu, J. Ni, A review of 4D printing, *Mater. Des.* 122 (2017) 42–79.

[60] X. Zhao, H. Tekinalp, X. Meng, D. Ker, B. Benson, Y. Pu, A.J. Ragauskas, Y. Wang, K. Li, E. Webb, D.J. Gardner, J. Anderson, S. Ozcan, poplar as biofiber reinforcement in composites for large-scale 3D printing, *ACS Appl. Bio Mater.* 2 (2019) 4557–4570. doi:10.1021/acsabm.9b00675.

[61] H.P.S.A. Khalil, I.U.H. Bhat, M. Jawaid, A. Zaidon, D. Hermawan, Y.S. Hadi, Bamboo fibre reinforced biocomposites: A review, *Mater. Des.* 42 (2012) 353–368.

[62] M.E. Mngomezulu, M.J. John, V. Jacobs, A.S. Luyt, Review on flammability of biofibres and biocomposites, *Carbohydr. Polym.* 111 (2014) 149–182.

[63] O. Akampumuza, P.M. Wambua, A. Ahmed, W. Li, X. Qin, Review of the applications of biocomposites in the automotive industry, *Polym. Compos.* 38 (2017) 2553–2569.

[64] B.P. Chang, A.K. Mohanty, M. Misra, Studies on durability of sustainable biobased composites: A review, *RSC Adv.* 10 (2020) 17955–17999. doi:10.1039/c9ra09554c.

[65] M.L. Sanyang, S.M. Sapuan, M. Jawaid, M.R. Ishak, J. Sahari, Recent developments in sugar palm (Arenga pinnata) based biocomposites and their potential industrial applications: A review, *Renew. Sustain. Energy Rev.* 54 (2016) 533–549.

[66] M. Ramesh, K. Palanikumar, K.H. Reddy, Plant fibre based bio-composites: Sustainable and renewable green materials, *Renew. Sustain. Energy Rev.* 79 (2017) 558–584.

[67] M. Berthet, H. Angellier-Coussy, V. Guillard, N. Gontard, Vegetal fiber-based biocomposites: Which stakes for food packaging applications?, *J. Appl. Polym. Sci.* 133 (2016) 1–18.

[68] M. Akonda, S. Alimuzzaman, D.U. Shah, A.N.M. Rahman, Physico-mechanical, thermal and biodegradation performance of random flax/polylactic acid and unidirectional flax/polylactic acid biocomposites, *Fibers.* 6 (2018) 1–19.

[69] I.M. De Rosa, C. Santulli, F. Sarasini, Mechanical and thermal characterization of epoxy composites reinforced with random and quasi-unidirectional untreated Phormium tenax leaf fibers, *Mater. Des.* 31 (2010) 2397–2405.

[70] D. Kraiem, S. Pimbert, A. Ayadi, C. Bradai, Effect of low content reed (Phragmite australis) fibers on the mechanical properties of recycled HDPE composites, *Compos. Part B Eng.* 44 (2013) 368–374.

[71] P.J. Herrera-Franco, A. Valadez-Gonzalez, A study of the mechanical properties of short natural-fiber reinforced composites, *Compos. Part B Eng.* 36 (2005) 597–608.

[72] A. Bourmaud, D. Åkesson, J. Beaugrand, A. Le Duigou, M. Skrifvars, C. Baley, Recycling of L-Poly-(lactide)-Poly-(butylene-succinate)-flax biocomposite, *Polym. Degrad. Stab.* 128 (2016) 77–88.

[73] A. Bourmaud, C. Baley, Investigations on the recycling of hemp and sisal fibre reinforced poly-propylene composites, *Polym. Degrad. Stab.* 92 (2007) 1034–1045.

[74] A.R. Torrado, D.A. Roberson, Failure analysis and anisotropy evaluation of 3D-printed tensile test specimens of different geometries and print raster patterns, *J. Fail. Anal. Prev.* 16 (2016) 154–164.

[75] L. Pyl, K.-A. Kalteremidou, D. Van Hemelrijck, Exploration of specimen geometry and tab con-figuration for tensile testing exploiting the potential of 3D printing freeform shape continuous carbon fibre-reinforced nylon matrix composites, *Polym. Test.* 71 (2018) 318–328.

[76] A. Kelly, and W.R. Tyson, Tensile properties of fibre-reinforced metals: copper/tungsten and copper/molybdenum, *J. Mech. Phys. Solids.* 13 (1965) 329–350.

[77] C. Badouard, F. Traon, C. Denoual, C. Mayer-Laigle, G. Paës, A. Bourmaud, Exploring mechanical properties of fully compostable flax reinforced composite filaments for 3D printing applications, *Ind. Crops Prod.* 135 (2019) 246–250.

[78] A. Codou, M. Misra, A.K. Mohanty, Sustainable biocarbon reinforced nylon 6/polypropylene compatibilized blends: Effect of particle size and morphology on performance of the biocomposites, *Compos. Part A Appl. Sci. Manuf.* 112 (2018) 1–10. doi:10.1016/j.compositesa.2018.05.018.

[79] V. Nagarajan, A.K. Mohanty, M. Misra, Biocomposites with size-fractionated biocarbon: Influence of the microstructure on Macroscopic Properties, *ACS Omega* 1 (2016) 636–647. doi:10.1021/acsomega.6b00175.

[80] M.M.A. Nassar, K.I. Alzebdeh, T. Pervez, N. Al-Hinai, A. Munam, Progress and challenges in sustainability, compatibility, and production of eco-composites: A state-of-art review, *J. Appl. Polym. Sci.* 138 (2021) 1–31.

[81] R. Latif, S. Wakeel, N. Zaman Khan, A. Noor Siddiquee, S. Lal Verma, Z. Akhtar Khan, Surface treatments of plant fibers and their effects on mechanical properties of fiber-reinforced composites: A review, *J. Reinf. Plast. Compos.* 38 (2019) 15–30.

[82] S. Kocaman, M. Karaman, M. Gursoy, G. Ahmetli, Chemical and plasma surface modification of lignocellulose coconut waste for the preparation of advanced biobased composite materials, *Carbohydr. Polym.* 159 (2017) 48–57.

[83] J. Sun, Y. Pang, Y. Yang, J. Zhao, R. Xia, Y. Li, Y. Liu, H. Guo, Improvement of rice husk/HDPE bio-composites interfacial properties by silane coupling agent and compatibilizer complementary modification, *Polymers (Basel).* 11 (2019) 1–13.

[84] L. Dányádi, J. Móczó, B. Pukánszky, Effect of various surface modifications of wood flour on the properties of PP/wood composites, *Compos. Part A Appl. Sci. Manuf.* 41 (2010) 199–206.

[85] Z. Dominkovics, L. Dányádi, B. Pukanszky, Surface modification of wood flour and its effect on the properties of PP/wood composites, *Compos. Part A Appl. Sci. Manuf.* 38 (2007) 1893–1901.

[86] E. Fortunati, M. Peltzer, I. Armentano, L. Torre, A. Jiménez, J.M. Kenny, Effects of modified cellulose nanocrystals on the barrier and migration properties of PLA nano-biocomposites, *Carbohydr. Polym.* 90 (2012) 948–956.

[87] S.K. Ramamoorthy, F. Bakare, R. Herrmann, M. Skrifvars, Performance of biocomposites from surface modified regenerated cellulose fibers and lactic acid thermoset bioresin, *Cellulose.* 22 (2015) 2507–2528.

[88] R. Muthuraj, M. Misra, A.K. Mohanty, Biodegradable compatibilized polymer blends for packaging applications: A literature review, *J. Appl. Polym. Sci.* 135 (2018) 1–35. doi:10.1002/app.45726.

[89] K. Zhang, M. Misra, A.K. Mohanty, Toughened sustainable green composites from Poly(3-hydroxybutyrate-co-3-hydroxyvalerate) based ternary blends and miscanthus biofiber, *ACS Sustain. Chem. Eng.* 2 (2014) 2345–2354. 10.1021/sc500353v.

[90] F. Wu, M. Misra, A.K. Mohanty, Sustainable green composites from biodegradable plastics blend and natural fibre with balanced performance: Synergy of nano-structured blend and reactive ex-trusion, *Compos. Sci. Technol.* 200 (2020) 1–8.

[91] N. Zarrinbakhsh, A.K. Mohanty, M. Misra, Improving the interfacial adhesion in a new renewable resource-based biocomposites from biofuel coproduct and biodegradable plastic, *J. Mater. Sci.* 48 (2013) 6025–6038.

[92] S. Sahoo, M. Misra, A.K. Mohanty, Effect of compatibilizer and fillers on the properties of injection molded lignin-based hybrid green composites, *J. Appl. Polym. Sci.* 127 (2013) 4110–4121.

[93] B. Liu, L. Jiang, H. Liu, J. Zhang, Synergetic effect of dual compatibilizers on in situ formed poly (lactic acid)/soy protein composites, *Ind. Eng. Chem. Res.* 49 (2010) 6399–6406. doi:10.1021/ie1 00218t.

[94] M.A.M. Elamin, Z.A. Osman, T.A. Otitoju, Preparation and characterization of wood-plastic composite by utilizing a hybrid compatibilizer system, *Ind. Crops Prod.* 154 (2020) 1–8.

[95] A. Kandemir, T.R. Pozegic, I. Hamerton, S.J. Eichhorn, M.L. Longana, Characterisation of natural fibres for sustainable discontinuous fibre composite materials, *Materials (Basel).* 13 (2020) 1–17.

[96] T.-C. Yang, Effect of extrusion temperature on the physico-mechanical properties of unidirectional wood fiber-reinforced polylactic acid composite (WFRPC) components using fused deposition modeling, *Polymers (Basel).* 10 (2018)1–11.

[97] A.K. Bledzki, P. Franciszczak, A. Mamun, The utilization of biochemically modified microfibers from grain by-products as reinforcement for polypropylene biocomposite, *Express Polym. Lett.* 8 (2014) 767–778.

[98] A. Rodriguez-Uribe, M.R. Snowdon, M.A. Abdelwahab, A. Codou, M. Misra, A.K. Mohanty, Impact of renewable carbon on the properties of composites made by using three types of polymers having different polarity, *J. Appl. Polym. Sci.* (2020) 1–15. doi:10.1002/app.49948.

[99] E.O. Ogunsona, M. Misra, A.K. Mohanty, Sustainable biocomposites from biobased polyamide 6, 10 and biocarbon from pyrolyzed miscanthus fibers, *J. Appl. Polym. Sci.* 134 (2017) 1–11.

[100] T. Welzel, E.G. Ertane, A. Dorner-Reisel, V. Matner, O. Baran, S. Svoboda, processing and wear behaviour of 3D printed PLA reinforced with biogenic carbon, *Adv. Tribol.* 2018 (2018) 1–11. doi:10.1155/2018/1763182.

[101] M. Gonzalez De Gortari, A. Rodriguez-Uribe, M. Misra, A.K. Mohanty, Insights on the structure-performance relationship of polyphthalamide (PPA) composites reinforced with high-temperature produced biocarbon, *RSC Adv.* 10 (2020) 26917–26927. doi:10.1039/d0ra03629c.

[102] H.N. Dhakal, Z.Y. Zhang, M.O.W. Richardson, Effect of water absorption on the mechanical properties of hemp fibre reinforced unsaturated polyester composites, *Compos. Sci. Technol.* 67 (2007) 1674–1683. doi:10.1016/j.compscitech.2006.06.019.

[103] Z.N. Azwa, B.F. Yousif, A.C. Manalo, W. Karunasena, A review on the degradability of polymeric composites based on natural fibres, *Mater. Des.* 47 (2013) 424–442.

[104] H.M. Akil, L.W. Cheng, Z.A. Mohd Ishak, A. Abu Bakar, M.A. Abd Rahman, Water absorption study on pultruded jute fibre reinforced unsaturated polyester composites, *Compos. Sci. Technol.* 69 (2009) 1942–1948. doi:10.1016/j.compscitech.2009.04.014.

[105] M. Gallardo-Cervantes, Y. González-García, A.A. Pérez-Fonseca, M.E. González-López, R. Manríquez-González, D. Rodrigue, J.R. Robledo-Ortíz, Biodegradability and improved mechanical performance of polyhydroxyalkanoates/agave fiber biocomposites compatibilized by different strategies, *J. Appl. Polym. Sci.* 138 (2021) 1–14.

[106] S. Mishra, J. Verma, Effect of compatibilizers on water absorption kinetics of polypropylene/wood flour foamed composites, *J. Appl. Polym. Sci.* 101 (2006) 2530–2537.

[107] L. Costes, F. Laoutid, S. Brohez, P. Dubois, Bio-based flame retardants: When nature meets fire protection, *Mater. Sci. Eng. R Reports.* 117 (2017) 1–25.

[108] V. Guna, M. Ilangovan, H.R. Vighnesh, B.R. Sreehari, S. Abhijith, H.E. Sachin, C.B. Mohan, N. Reddy, Engineering sustainable waste wool biocomposites with high flame resistance and noise insulation for green building and automotive applications, *J. Nat. Fibers.* 18 (2019) 1–11.

[109] F. Samyn, M. Vandewalle, S. Bellayer, S. Duquesne, Sol–gel treatments to flame retard PA11/flax composites, *Fibers.* 7 (2019) 1–15.

[110] R. Hajj, R. El Hage, R. Sonnier, B. Otazaghine, S. Rouif, M. Nakhl, J.-M. Lopez-Cuesta, Influence of lignocellulosic substrate and phosphorus flame retardant type on grafting yield and flame retardancy, *React. Funct. Polym.* 153 (2020) 1–13.

[111] N.K. Kim, S. Dutta, D. Bhattacharyya, A review of flammability of natural fibre reinforced polymeric composites, *Compos. Sci. Technol.* 162 (2018) 64–78.

[112] J. Zhang, Q. Wu, G. Li, M.-C. Li, X. Sun, D. Ring, Synergistic influence of halogenated flame retardants and nanoclay on flame performance of high density polyethylene and wood flour composites, *RSC Adv.* 7 (2017) 24895–24902.

[113] R.A. Ilyas, S.M. Sapuan, M.R.M. Asyraf, D. Dayana, J.J.N. Amelia, M.S.A. Rani, M.N.F. Norrrahim, N.M. Nurazzi, H.A. Aisyah, S. Sharma, Polymer composites filled with metal derivatives: A review of flame retardants, *Polymers (Basel).* 13 (2021) 1–21.

[114] H. Yang, B. Yu, X. Xu, S. Bourbigot, H. Wang, P. Song, Lignin-derived bio-based flame retardants toward high-performance sustainable polymeric materials, *Green Chem.* 22 (2020) 2129–2161.

[115] Y.-Y. Gao, C. Deng, Y.-Y. Du, S.-C. Huang, Y.-Z. Wang, A novel bio-based flame retardant for polypropylene from phytic acid, *Polym. Degrad. Stab.* 161 (2019) 298–308.

[116] B.A. Howell, W. Sun, Biobased flame retardants from tartaric acid and derivatives, *Polym. Degrad. Stab.* 157 (2018) 199–211.

[117] Q. Zhang, D. Zhang, H. Xu, W. Lu, X. Ren, H. Cai, H. Lei, E. Huo, Y. Zhao, M. Qian, Biochar filled high-density polyethylene composites with excellent properties: Towards maximizing the utilization of agricultural wastes, *Ind. Crops Prod.* 146 (2020) 1–9.

[118] O. Das, N.K. Kim, A.L. Kalamkarov, A.K. Sarmah, D. Bhattacharyya, Biochar to the rescue: Balancing the fire performance and mechanical properties of polypropylene composites, *Polym. Degrad. Stab.* 144 (2017) 485–496.

[119] N. Cuinat-Guerraz, M.-J. Dumont, P. Hubert, Environmental resistance of flax/bio-based epoxy and flax/polyurethane composites manufactured by resin transfer moulding, *Compos. Part A Appl. Sci. Manuf.* 88 (2016) 140–147.

[120] M.C. Picard, A. Rodriguez-Uribe, M. Thimmanagari, M. Misra, A.K. Mohanty, Sustainable bio-composites from poly (butylene succinate) and apple pomace: A study on compatibilization performance, *Waste and Biomass Valorization.* 11 (2020) 3775–3787.

[121] M.K.M. Smith, D.M. Paleri, M. Abdelwahab, D.F. Mielewski, M. Misra, A.K. Mohanty, Sustainable composites from poly(3-hydroxybutyrate) (PHB) bioplastic and agave natural fibre, *Green Chem.* 22 (2020) 3906–3916. doi:10.1039/d0gc00365d.

[122] Z. Li, C. Reimer, T. Wang, A.K. Mohanty, M. Misra, Thermal and mechanical properties of the biocomposites of miscanthus biocarbon and poly (3-Hydroxybutyrate-co-3-Hydroxyvalerate) (PHBV), *Polymers (Basel).* 12 (2020) 1–13.

[123] M.P. Ho, K.T. Lau, H. Wang, D. Hui, Improvement on the properties of polylactic acid (PLA) using bamboo charcoal particles, *Compos. Part B Eng.* 81 (2015) 14–25. doi:10.1016/j.compositesb.2015.05.048.

[124] C. Nyambo, A.K. Mohanty, M. Misra, Effect of maleated compatibilizer on performance of PLA/wheat straw-based green composites, *Macromol. Mater. Eng.* 296 (2011) 710–718. doi:10.1002/mame.201000403.

[125] V. Mittal, G.E. Luckachan, B. Chernev, N.B. Matsko, Bio-polyester–date seed powder composites: Morphology and component migration, *Polym. Eng. Sci.* 55 (2015) 877–888.

[126] A. Ibáñez-García, A. Martínez-García, S. Ferrándiz-Bou, Recyclability analysis of starch thermoplastic/almond shell biocomposite, *Polymers (Basel).* 13 (2021) 1–16.

[127] A.F. Ahmad, Z. Abbas, S.J. Obaiys, M.F. Zainuddin, Effect of untreated fiber loading on the thermal, mechanical, dielectric, and microwave absorption properties of polycaprolactone reinforced with oil palm empty fruit bunch biocomposites, *Polym. Compos.* 39 (2018) E1778–E1787.

[128] M.Z.A. Hamid, N.A. Ibrahim, W.M.Z.W. Yunus, K. Zaman, M. Dahlan, Effect of grafting on properties of oil palm empty fruit bunch fiber reinforced polycaprolactone biocomposites, *J. Reinf. Plast. Compos.* 29 (2010) 2723–2731.

[129] H. Deka, M. Misra, A. Mohanty, Renewable resource based "all green composites" from kenaf biofiber and poly (furfuryl alcohol) bioresin, *Ind. Crops Prod.* 41 (2013) 94–101.

[130] N. Yan, B. Zhang, Y. Zhao, R.R. Farnood, J. Shi, Application of biobased phenol formaldehyde novolac resin derived from beetle infested lodgepole pine barks for thermal molding of wood composites, *Ind. Eng. Chem. Res.* 56 (2017) 6369–6377.

[131] A. Lotfi, H. Li, D.V. Dao, G. Prusty, Natural fiber–reinforced composites: A review on material, manufacturing, and machinability, *J. Thermoplast. Compos. Mater.* 34 (2021) 238–284. doi:10.1177/0892705719844546.

[132] C.O. Umerah, D. Kodali, S. Head, S. Jeelani, V.K. Rangari, Synthesis of carbon from waste co-conutshell and their application as filler in bioplast polymer filaments for 3D printing, *Compos. Part B Eng.* 202 (2020) 1–9. doi:10.1016/j.compositesb.2020.108428.

[133] A. Le Duigou, G. Chabaud, R. Matsuzaki, M. Castro, Tailoring the mechanical properties of 3D-printed continuous flax/PLA biocomposites by controlling the slicing parameters, *Compos. Part B Eng.* 203 (2020) 1–11.

[134] E. Diederichs, M. Picard, B.P. Chang, M. Misra, A. Mohanty, Extrusion based 3D printing of sustainable biocomposites from biocarbon and poly(trimethylene terephthalate), *Molecules,* 26 (2021) 1–14.

[135] D. Filgueira, S. Holmen, J.K. Melbø, D. Moldes, A.T. Echtermeyer, G. Chinga-Carrasco, 3D printable filaments made of biobased polyethylene biocomposites, *Polymers (Basel)*. 10 (2018). 1–15 doi:10.3390/polym10030314.

[136] M.J. John, S. Thomas, biofibres and biocomposites, *Carbohydr. Polym.* 71 (2008) 343–364. doi:10. 1016/j.carbpol.2007.05.040.

[137] Y. Parulekar, A.K. Mohanty, Biodegradable toughened polymers from renewable resources: Blends of polyhydroxybutyrate with epoxidized natural rubber and maleated polybutadiene, *Green Chem.* 8 (2006) 206–213. doi:10.1039/b508213g.

[138] N.K. Kalita, S.M. Bhasney, C. Mudenur, A. Kalamdhad, V. Katiyar, End-of-life evaluation and biodegradation of Poly (lactic acid)(PLA)/Polycaprolactone (PCL)/Microcrystalline cellulose (MCC) polyblends under composting conditions, *Chemosphere*. 247 (2020) 1–8.

[139] V. Nagarajan, M. Misra, A.K. Mohanty, New engineered biocomposites from poly(3-hydroxybutyrate-co-3-hydroxyvalerate) (PHBV)/poly(butylene adipate-co-terephthalate) (PBAT) blends and switchgrass: Fabrication and performance evaluation, *Ind. Crops Prod.* 42 (2013) 461–468. doi:10.1016/j.indcrop.2012.05.042.

[140] K. Zhang, V. Nagarajan, N. Zarrinbakhsh, A.K. Mohanty, M. Misra, Co-injection molded new green composites from biodegradable polyesters and miscanthus fibers, *Macromol. Mater. Eng.* 299 (2014) 436–446. 10.1002/mame.201300189.

[141] A. Codou, M. Misra, A.K. Mohanty, Sustainable biocomposites from Nylon 6 and polypropylene blends and biocarbon–Studies on tailored morphologies and complex composite structures, *Compos. Part A Appl. Sci. Manuf.* 129 (2020) 1–13.

[142] M. Bahrami, J. Abenojar, M.Á. Martínez, Recent progress in hybrid biocomposites: Mechanical properties, water absorption, and flame retardancy, *Materials*. 13 (2020) 1–46. doi:10.3390/ma13225145.

[143] E. Ameri, L. Laperrière, G. Lebrun, Mechanical characterization and optimization of a new unidirectional flax/paper/epoxy composite, *Compos. Part B Eng.* 97 (2016) 282–291.

[144] A. Couture, G. Lebrun, L. Laperrière, Mechanical properties of polylactic acid (PLA) composites reinforced with unidirectional flax and flax-paper layers, *Compos. Struct.* 154 (2016) 286–295.

[145] X. Peng, M. Fan, J. Hartley, M. Al-Zubaidy, Properties of natural fiber composites made by pultrusion process, *J. Compos. Mater.* 46 (2012) 237–246. doi:10.1177/0021998311410474.

[146] R. Menezes Bezerra, Modelling and Simulation of the Closed Injection Pultrusion Process (Doctoral dissertation, Dissertation, Karlsruhe, Karlsruher Institut für Technologie, (2017) 151.

[147] J. Agarwal, S. Sahoo, S. Mohanty, S.K. Nayak, Progress of novel techniques for lightweight automobile applications through innovative eco-friendly composite materials: A review, *J. Thermoplast. Compos. Mater.* 33 (2020) 978–1013. doi:10.1177/0892705718815530.

[148] Y. Yuryev, A.K. Mohanty, M. Misra, Novel biocomposites from biobased PC/PLA blend matrix system for durable applications, *Compos. Part B Eng.* 130 (2017) 158–166.

[149] F.A. Almansour, H.N. Dhakal, Z.Y. Zhang, Investigation into Mode II interlaminar fracture toughness characteristics of flax/basalt reinforced vinyl ester hybrid composites, *Compos. Sci. Technol.* 154 (2018) 117–127. doi:10.1016/j.compscitech.2017.11.016.

[150] B. Baghaei, M. Skrifvars, L. Berglin, Manufacture and characterisation of thermoplastic composites made from PLA/hemp co-wrapped hybrid yarn prepregs, *Compos. Part A Appl. Sci. Manuf.* 50 (2013) 93–101.

[151] R. Muthuraj, C. Lacoste, P. Lacroix, A. Bergeret, Sustainable thermal insulation biocomposites from rice husk, wheat husk, wood fibers and textile waste fibers: Elaboration and performances evaluation, *Ind. Crops Prod.* 135 (2019) 238–245.

[152] K. Yorseng, S.M. Rangappa, H. Pulikkalparambil, S. Siengchin, J. Parameswaranpillai, Accelerated weathering studies of kenaf/sisal fiber fabric reinforced fully biobased hybrid bioepoxy composites for semi-structural applications: Morphology, thermo-mechanical, water absorption behavior and surface hydrophobicity, *Constr. Build. Mater.* 235 (2020) 1–14. doi:10.1016/j.conbuildmat.2019.117464.

[153] S. Armioun, S. Panthapulakkal, J. Scheel, J. Tjong, M. Sain, Sustainable and lightweight bio-polyamide hybrid composites for greener auto parts, *Can. J. Chem. Eng.* 94 (2016) 2052–2060. doi:10.1002/cjce.22609.

[154] F. Dilucia, V. Lacivita, A. Conte, M.A. Del Nobile, Sustainable use of fruit and vegetable by-products to enhance food packaging performance, *Foods*. 9 (2020) 1–19.

[155] O. Valerio, R. Muthuraj, A. Codou, Strategies for polymer to polymer recycling from waste: Current trends and opportunities for improving the circular economy of polymers in South America, *Curr. Opin. Green Sustain. Chem.* 25 (2020) 1–8.

[156] F.H.M. Graichen, W.J. Grigsby, S.J. Hill, L.G. Raymond, M. Sanglard, D.A. Smith, G.J. Thorlby, K.M. Torr, J.M. Warnes, Yes, we can make money out of lignin and other bio-based resources, *Ind. Crops Prod.* 106 (2017) 74–85.

[157] G. Jiang, M. Zhang, J. Feng, S. Zhang, X. Wang, High oxygen barrier property of poly(propylene carbonate)/polyethylene glycol nanocomposites with low loading of cellulose nanocrytals, ACS Sustain. *Chem. Eng.* 5 (2017) 11246–11254. doi:10.1021/acssuschemeng.7b01674.

[158] C.C. Chang, B.M. Trinh, T.H. Mekonnen, Robust multiphase and multilayer starch/polymer (TPS/PBAT) film with simultaneous oxygen/moisture barrier properties, *J. Colloid Interface Sci.* 593 (2021) 290–303.

[159] E. Ojogbo, J. Jardin, T.H. Mekonnen, Robust and sustainable starch ester nanocomposite films for packaging applications, *Ind. Crops Prod.* 160 (2021) 1–10.

[160] E.O. Ogunsona, R. Muthuraj, E. Ojogbo, O. Valerio, T.H. Mekonnen, Engineered nanomaterials for antimicrobial applications: A review, *Appl. Mater. Today.* 18 (2020) 1–32. doi:10.1016/j.apmt.2019.100473.

[161] F.M. AL-Oqla, S.M. Sapuan, Natural fiber reinforced polymer composites in industrial applications: Feasibility of date palm fibers for sustainable automotive industry, *J. Clean. Prod.* 66 (2014) 347–354. doi:10.1016/j.jclepro.2013.10.050.

[162] L. Mohammed, M.N.M. Ansari, G. Pua, M. Jawaid, M.S. Islam, A Review on natural fiber reinforced polymer composite and its applications, *Int. J. Polym. Sci.* 2015 (2015). doi:10.1155/2015/243947.

[163] M Jawaid, M Thariq, *Sustainable composites for aerospace applications*, Woodhead Publishing, 2018.

[164] V.D. Punetha, S. Rana, H.J. Yoo, A. Chaurasia, J.T. McLeskey, M.S. Ramasamy, N.G. Sahoo, J.W. Cho, Functionalization of carbon nanomaterials for advanced polymer nanocomposites: A comparison study between CNT and graphene, *Prog. Polym. Sci.* 67 (2017) 1–47. doi:10.1016/j.progpolymsci.2016.12.010.

[165] R. Kumar, R.S. Tiwari, O.N. Srivastava, Scalable synthesis of aligned carbon nanotubes bundles using green natural precursor: Neem oil, *Nanoscale Res. Lett.* 6 (2011) 2–7. doi:10.1186/1556-2 76X-6-92.

[166] K. Senthilkumar, I. Siva, N. Rajini, J.T.W. Jappes, S. Siengchin, Mechanical characteristics of tri-layer eco-friendly polymer composites for interior parts of aerospace application. In*sustainable composites for aerospace applications* (pp. 35–53), Woodhead Publishing, 2018. doi:10.1016/B978-0-08-102131-6.00003-7

[167] C.V. Dos Santos, D.R. Leiva, F.R. Costa, J.A.R. Gregolin, Materials selection for sustainable executive aircraft interiors, *Mater. Res.* 19 (2016) 339–352. doi:10.1590/1980-5373-MR-2015-0290.

[168] O. Boegler, U. Kling, A. Empl, A.T. Isikveren, potential of sustainable materials in wing structural design, *Dtsch. Luft- Und Raumfahrtkongress.* 327 (2014) 1–6.

[169] D. Kun, B. Pukánszky, Polymer/lignin blends: Interactions, properties, applications, *Eur. Polym. J.* 93 (2017) 618–641.

[170] Y. Khalaf, P. El Hage, J.D. Mihajlova, A. Bergeret, P. Lacroix, R. El Hage, Influence of agricultural fibers size on mechanical and insulating properties of innovative chitosan-based insulators, *Constr. Build. Mater.* 287 (2021) 1–10.

[171] C. Lacoste, R. El Hage, A. Bergeret, S. Corn, P. Lacroix, Sodium alginate adhesives as binders in wood fibers/textile waste fibers biocomposites for building insulation, *Carbohydr. Polym.* 184 (2018) 1–8. doi:10.1016/j.carbpol.2017.12.019.

[172] B.M. Trinh, E.O. Ogunsona, T.H. Mekonnen, Thin-structured and compostable wood fiber-polymer biocomposites: Fabrication and performance evaluation, *Compos. Part A Appl. Sci. Manuf.* 140 (2021) 1–12.

[173] S.A. Madbouly, J.A. Schrader, G. Srinivasan, K. Liu, K.G. McCabe, D. Grewell, W.R. Graves, M.R. Kessler, Biodegradation behavior of bacterial-based polyhydroxyalkanoate (PHA) and DDGS composites, *Green Chem.* 16 (2014) 1911–1920. doi:10.1039/C3GC41503A.

[174] B. Thomas, M.C. Raj, B.K. Athira, H.M. Rubiyah, J. Joy, A. Moores, G.L. Drisko, C. Sanchez, Nanocellulose, a versatile green platform: From biosources to materials and their applications, *Chem. Rev.* 118 (2018) 11575–11625. doi:10.1021/acs.chemrev.7b00627.

[175] D. Mohan, Z.K. Teong, A.N. Bakir, M.S. Sajab, H. Kaco, Extending cellulose-based polymers application in additive manufacturing technology: A review of recent approaches, *Polymers (Basel).* 12 (2020) 1–31. doi:10.3390/POLYM12091876.

[176] P. Toth, T. Vikström, R. Molinder, H. Wiinikka, Structure of carbon black continuously produced from biomass pyrolysis oil, *Green Chem.* 20 (2018) 3981–3992. doi:10.1039/c8gc01539b.

[177] S. Vivekanandhan, M. Schreiber, S. Muthuramkumar, M. Misra, A.K. Mohanty, Carbon nanotubes from renewable feedstocks: A move toward sustainable nanofabrication, *J. Appl. Polym. Sci.* 134 (2017) 1–15. doi:10.1002/app.44255.

[178] S.Y. Lu, M. Jin, Y. Zhang, Y.B. Niu, J.C. Gao, C.M. Li, chemically exfoliating biomass into a graphene-like porous active carbon with rational pore structure, good conductivity, and large surface area for high-performance supercapacitors, *Adv. Energy Mater.* 8 (2018) 1–9. doi:10.1002/aenm.201702545.

[179] V.K. Das, Z.B. Shifrina, L.M. Bronstein, Graphene and graphene-like materials in biomass conversion: Paving the way to the future, *J. Mater. Chem. A.* 5 (2017) 25131–25143. doi:10.1039/c7ta09418c.

[180] V.P. Sharma, V. Agarwal, S. Umar, A.K. Singh, Polymer composites sustainability: Environmental perspective, *Future Trends Minimization Health Risk*, 4 (2011) 259–261.

4 Renewable Vitrimer—A Novel Route Towards Reprocessable and Recyclable Thermosets from Biomass-Derived Building Block

Arunjunai Raj Mahendran, Mohammed Khalifa, Günter Wuzella, and Herfried Lammer

Competence Center for Wood Composites and Wood Chemistry, Linz, Austria

CONTENTS

4.1 INTRODUCTION

The thermosetting materials are cross-linked networks formed by the reaction of several comonomers and they are used in a wide variety of applications that includes construction, furniture, coatings, automobiles, electrical appliances and insulators, composite industries, agricultural products, aerospace, etc. Some examples of thermosetting resins are phenol-formaldehyde, epoxy resin, melamine and urea-formaldehyde resin, unsaturated polyester resin, vinyl ester resin, polyurethane, bismaleimides, and cyanoacrylate. The advantages of thermoset resins are resistance to heat, high strength-to-weight ratio, excellent dielectric strength, low thermal conductivity, excellent chemical resistance, good dimensional stability at high temperature, and high fatigue strength and toughness. The thermoset resins can be processed into reinforced composites using different manufacturing techniques and diverse fiber reinforcements, which is a cost-effective, lightweight, and high strength-to-weight ratio. The thermoset resin major disadvantage is that if they form a highly cross-linked 3D structure, it cannot be reshaped just by applying heat or other external energy. Besides, the composite structures made using thermoset resins cannot be easily recycled or reprocessed. Consequently, several efforts are made to facilitate remolding or restoring the damage in the thermoset resin by introducing exchangeable covalent bonds in the polymer network or non-covalent bonds. Several investigations have been carried out in introducing

DOI: 10.1201/9781003200710-5

73

non-covalent bonds like hydrogen bonds (Campanella, Döhler, & Binder, 2018), π-π stacking (Thakur & Kessler, 2015) and metal-ligand bonds (Fischer, 2010) in the thermoset network. However, this non-covalent bond leads to low mechanical strength compared to covalent bonds due to weak bonding energies.

In the case of an exchangeable covalent network, the covalent bonds can undergo bond cleavage or reconstruction under the external thermal or other energy sources, which enables reprocessability and reparability. The advantage of these exchangeable chemical bonds is that they can form dynamic cross-linking sites based on intermolecular interactions (Cordier, Tournilhac, Soulié-Ziakovic, & Leibler, 2008; Seiffert & Sprakel, 2012) and reversible covalent or dynamic covalent bonds (Bowman & Kloxin, 2012; Kloxin, Scott, Adzima, & Bowman, 2010)(Kloxin & Bowman, 2013). The exchangeable chemical bonds are otherwise called covalent adaptable network (CAN). Jourdain et al. (Jourdain et al., 2020) summarized the major covalent exchange reactions that can be used in CANs and they are classified into two categories (a) associative and (b) dissociative CAN (see Figure 4.1). The dissociative CAN happens via reversible covalent bond formation between two functional groups. A well-known example of the dissociative CAN is the reversible Diels-Alder (DA) reaction. For example, DA adduct between furans and maleimides is stable at room temperature, but the reversible bond exchange reaction dominates at a high temperature. The bond dissociation due to external stimuli like temperature can cause a decrease in connectivity and also a drop in viscosity and dimensional stability. The system undergoes sol-gel transition at this stage and again after cooling to ambient temperature, network density can be restored. The limitations of DA reactions are a drop in the cross-linking density, sol-gel transition, and sensitivity to the presence of solvents at elevated temperature (Jin, Lei, Taynton, Huang, & Zhang, 2019).

The most effective is the associative CAN and in this reaction mechanism there is no loss in connectivity. The reason is that a cross-link between chains is broken down beforehand a bond with another polymer chain has been formed. Therefore, it shows more fixed cross-link density during bond exchange. The associative CAN exchange reaction rate increases with temperature and shows an Arrhenius-like viscosity dependence (Denissen et al., 2015). In 2005, Bowman reported the behavior of photoinduced cross-linking in a polymer system and they found that a cross-linked polymer network exposed to light has achieved a stress/strain relaxation which doesn't changed the properties (Scott, Schneider, Cook, & Bowman, 2005). In addition, they reported that CANs yield the properties such as recyclability, healability, tenability, and change in shape of the cross-linked network (Kloxin et al., 2010). Covalent adaptable networks are smart materials that are capable of responding to a stimulus accompanied by change in the network structure and shape. At first, Professor Leibler from ESPCI Paris Tech designed a thermally triggered associative CAN in thermosetting polymers and reported their work on epoxy polymer networks (Montarnal, Capelot, Tournilhac, & Leibler, 2011). They studied the unique properties of the thermoset materials made using epoxy resin cured with di- and tri-carboxylic acid in the presence of mild Lewis acid catalyst zinc acetate and due to transesterification reaction, it showed vitreous silica like behavior (Montarnal et al., 2011). The mechanical properties of the cross-linked network are similar to epoxy resin, but the cross-linked network can be broken at elevated temperatures and reprocessed again. The term "vitrimer" was designated for this thermally triggered associative CAN and this kind of network structure has also been reported for the olefin (Lu & Guan, 2012) or disulfide (Rekondo et al., 2014)(Lei, Xiang, Yuan, Rong, & Zhang, 2014) metathesis reaction. Two transition temperatures are characteristics of this vitrimer: glass transition (T_g) and topology freezing point temperatures (T_v) (see Figure 4.2). Topology freezing transition temperature is the temperature in the vitrimer system where it changes from viscoelastic solid to viscoelastic liquid (Montarnal et al., 2011). The rearrangement process of vitrimers could be controlled between the previously mentioned two temperatures. If the vitrimers are heated above the glass transition region, then segmental motion starts, at the same time exchange reactions are already fast enough to form molecular rearrnagement. So initially, due to network rearrangement, it is a diffusion-controlled process and then network topology rearrangement occurs attributed to

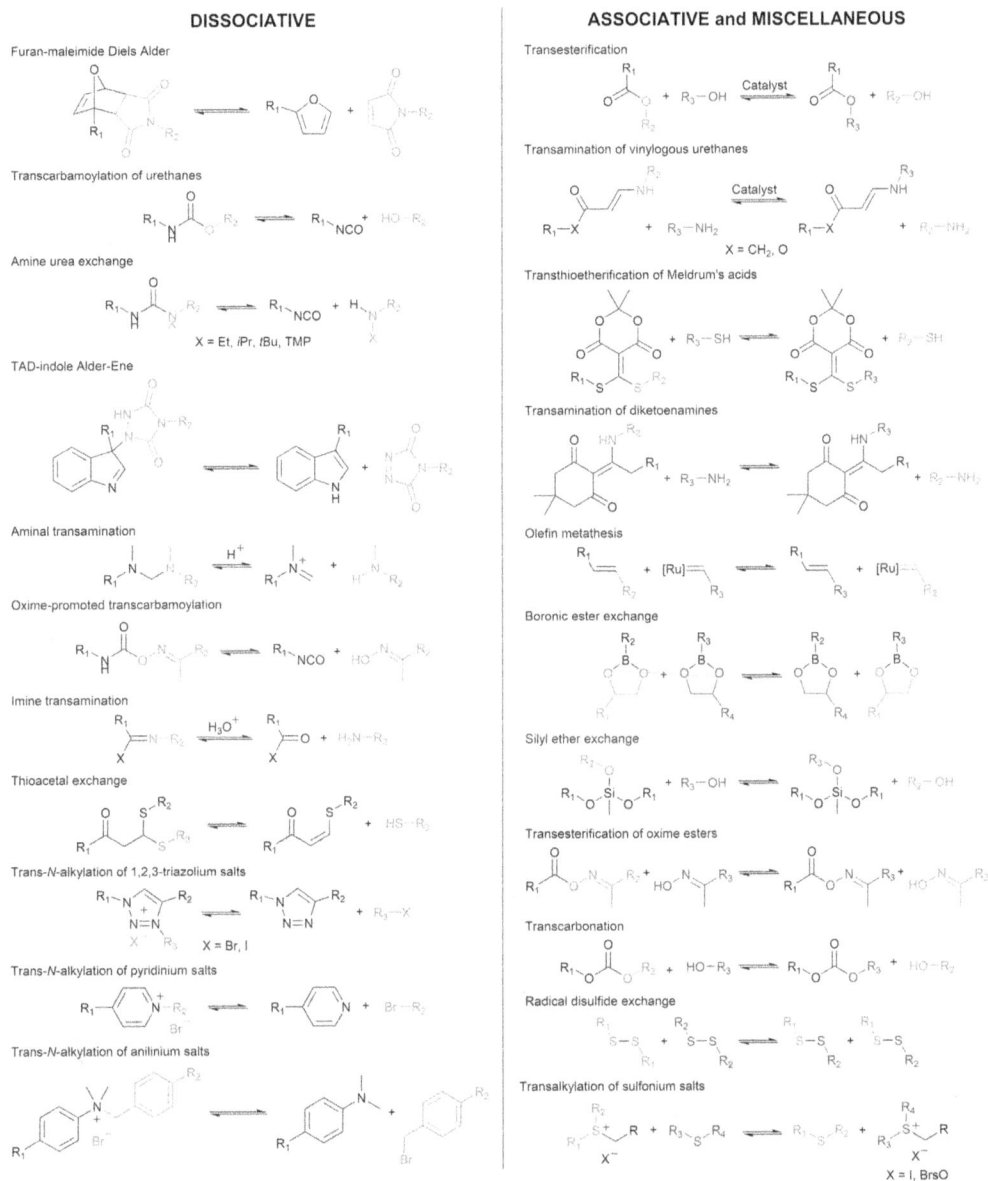

DISSOCIATIVE

Furan-maleimide Diels Alder

Transcarbamoylation of urethanes

Amine urea exchange

X = Et, iPr, tBu, TMP

TAD-indole Alder-Ene

Aminal transamination

Oxime-promoted transcarbamoylation

Imine transamination

Thioacetal exchange

Trans-N-alkylation of 1,2,3-triazolium salts

X = Br, I

Trans-N-alkylation of pyridinium salts

Trans-N-alkylation of anilinium salts

ASSOCIATIVE and MISCELLANEOUS

Transesterification

Transamination of vinylogous urethanes

X = CH$_2$, O

Transthioetherification of Meldrum's acids

Transamination of diketoenamines

Olefin metathesis

Boronic ester exchange

Silyl ether exchange

Transesterification of oxime esters

Transcarbonation

Radical disulfide exchange

Transalkylation of sulfonium salts

X = I, BrsO

FIGURE 4.1 Major covalent exchange reactions that enable malleability in CANs Reproduced with permissions ©2020 American chemical society (Jourdain et al., 2020).

segmental motions. Nevertheless, after increasing temperature, it follows the Arrhenius law, which means it changes from a diffusion to an exchange reaction controlled process (Denissen, Winne, & Du Prez, 2016).

The chemistry behind associative CAN is recently reviewed by several authors (Jin 2019; Krishnakumar et al., 2020; Guerre, Taplan, Winne, & Du Prez, 2020; Zou, Dong, Luo, Zhao, & Xie, 2017). Vitrimers can be synthesized via many different chemistries and the major associative covalent adaptable networks are based on carboxylic acid chemistry (transesterification), imine amine exchange chemistry, olefin metathesis, disulfide exchange chemistry, transamination of vinylogous urethane, and boronic acid chemistry. As mentioned earlier, most of the vitrimer networks are from fossil fuel–derived chemicals and currently due to environmental concern, bio-based

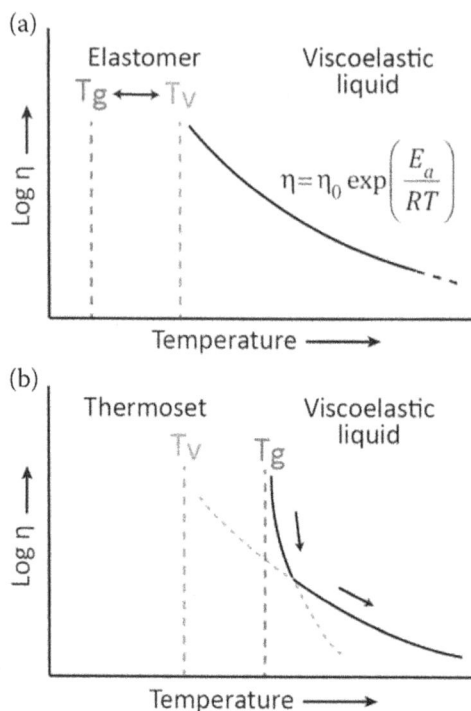

(a)

Elastomer   Viscoelastic
                        liquid

$T_g \longleftrightarrow T_v$

$\eta = \eta_0 \exp\left(\dfrac{E_a}{RT}\right)$

Log η

Temperature ⟶

(b)

Thermoset   Viscoelastic
                        liquid

T_v T_g

Log η

Temperature ⟶

FIGURE 4.2 (a) The glass transition temperature is higher than the topology freezing point which means the viscoelastic behavior of the vitrimer follows the Arrhenius law, (b) the topological freezing point is well below the glass transition temperature therefore upon heating visoelastic behavior of vitrimer follows the Williams-Landel-Ferry (WLF) and then Arhenius law via exchange kinetics [Reproduced with permissions ©2016 Royal society of chemistry] (Denissen et al., 2016).

polymer networks are acquiring attention among industries and consumers. Renewable resources like vegetable oils, lignin, tannins, isosorbide, furan, etc., are the suitable raw material for replacing fossil fuel–derived chemicals either partially or fully in bio-based thermosets. Therefore, this review focuses on preparing various renewable vitrimers from biomass-derived building blocks and their properties. Initially, the chemistry behind the different exchangeable covalent networks is explained and then the research works done by several authors in the area of renewable vitrimers are summarized. Finally, the shortcoming and the future perspective of this vitrimer is discussed.

4.2   EPOXY VITRIMERS

Aforementioned, the transesterification reaction between epoxy-carboxylic acid (see Figure 4.3) enables their behavior as vitrimers. The transesterifications can occur without the presence of catalysts, but the reaction can be speeded up by various catalysts such as tertiary amines, triphenylphosphine, zinc salts, Lewis and Bronsted acids (Altuna, Pettarin, & Williams, 2013). The first reported epoxy vitrimers was prepared by using Lewis acid catalyst, which is reprocessable and malleable. The activation energy of the vitrimer system was 80 kJ mol^{-1} K^{-1} and topology freezing point temperature (T_v) for 10 mol% of catalyst was 53°C (Montarnal et al., 2011).

Synthesis of epoxy-based vitrimers possessing high T_g is one of the stimulating task to achieve. For practical applications, T_g of epoxy is habitually maintained above 100°C, to attain high service temperature requirements. However, several reports demonstrate relatively low T_g than the desired value, which hinders its practical use. Epoxy-based vitrimers are often not completely cured intentionally by utilizing a low stoichiometric quantity of curing agent containing anhydride or carboxylic acid. Thus, it limits the cross-link density of vitrimers. Furthermore, sufficient molecular movement at high temperatures is attained in epoxy-based vitrimers by building them with soft monomers and curing agents. As a result, the T_g of epoxy-based vitrimers is relatively low (Liu, Hao, et al., 2017; Liu, Zhang, et al., 2017; Yang & Urban, 2013). Basically, T_g of epoxy is

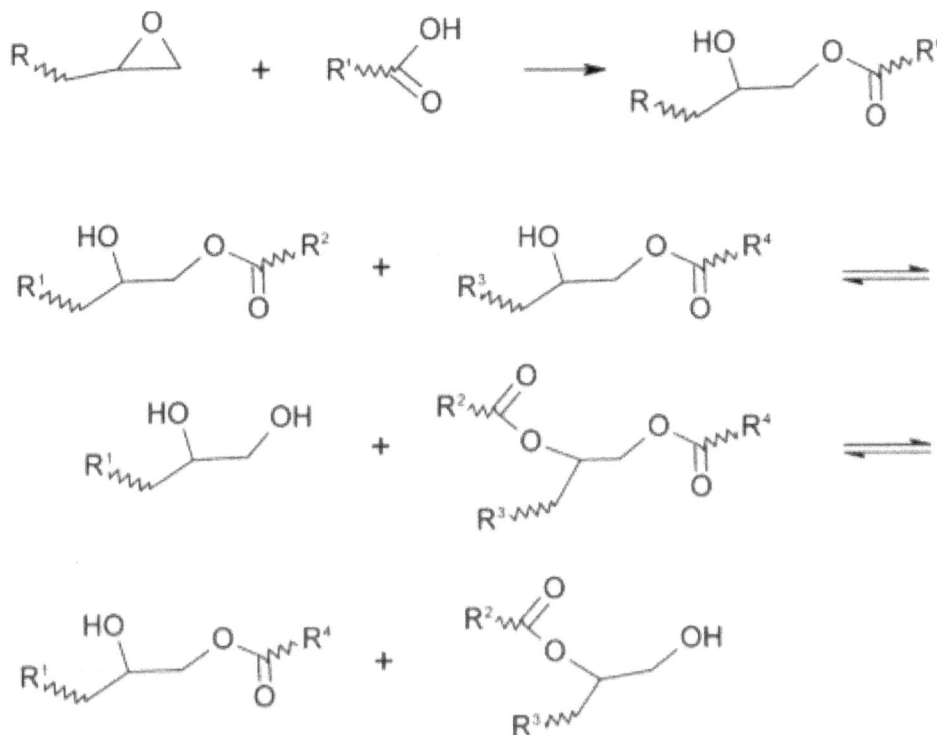

FIGURE 4.3 Transesterification reaction [Reproduced with permissions ©2019 Elsevier] (Altuna, Hoppe, & Williams, 2019).

strongly influenced by the curing agent, curing conditions, chemical structure of resin, and curing agent along with their molar ratios. Also, the stiffness of the chain and cross-linking density of the network defines the T_g of the epoxy. Diglycidyl ether of bisphenol A and N,N,N′,N′- tetraglycidyl-4,4′-diaminodiphenylmethane epoxies are often used to achieve high T_g (>150°C) (Dobáš et al., 1992; Keeratitham & Somwangthanaroj, 2016).

Giebler et al. (Giebler, Sperling, Kaiser, Duretek, & Schlögl, 2020) prepared epoxy-anhydride vitrimer, which experiences associated bond exchange reaction at higher temperature. The vitrimer was synthesized by reacting 4,4′-methylenebis(N,N-diglycidylaniline) (4-DGA) or N,N-diglycidyl-4-glycidyloxyaniline (3-DGOA) with glutaric anhydride and zinc acetate as catalyst. The highest T_g of 140°C attained through non-catalyzed 4-DGA/glutaric anhydride systems, which may perhaps relax stresses effectively. In peculiar, quick stress relaxation was achieved i.e., at 113°C (T_v) through catalysed 3-DGOA/glutaric anhydride networks.

Liu et al. (Liu et al., 2017) prepared eugenol-derived vitrimer using a bio-based epoxy, the synthesized bio-based epoxy (Eu-EP) reacted with succinic anhydride (SA) at distinct concentrations (1:0.5, 1:0.75, and 1:1), and a final cured network exhibited excellent shape memory properties. They found that all Eu-EP/SAs could be disintegrated in ethanol at 160°C and they can be transformed into novel thermosetting polymers after exposure at 190°C for 3 h.

Ji et al. (Ji, Liu, Sheng, & Yang, 2020) synthesized epoxy vitrimer composite based on exchangeable aromatic disulphide bonds. The authors found that in the epoxy/polycaprolactone (PCL) system, PCL chains are interpenetrated into the cross-linked epoxy due to the hydrogen bonding interaction between the OH groups in the epoxy network and the carbonyl groups in the PCL chains. The multi-shape memory affects the vitrimer composites characterized at different temperatures and it was achieved through the "welding" method and constructing a multilayer

(a)

(b)

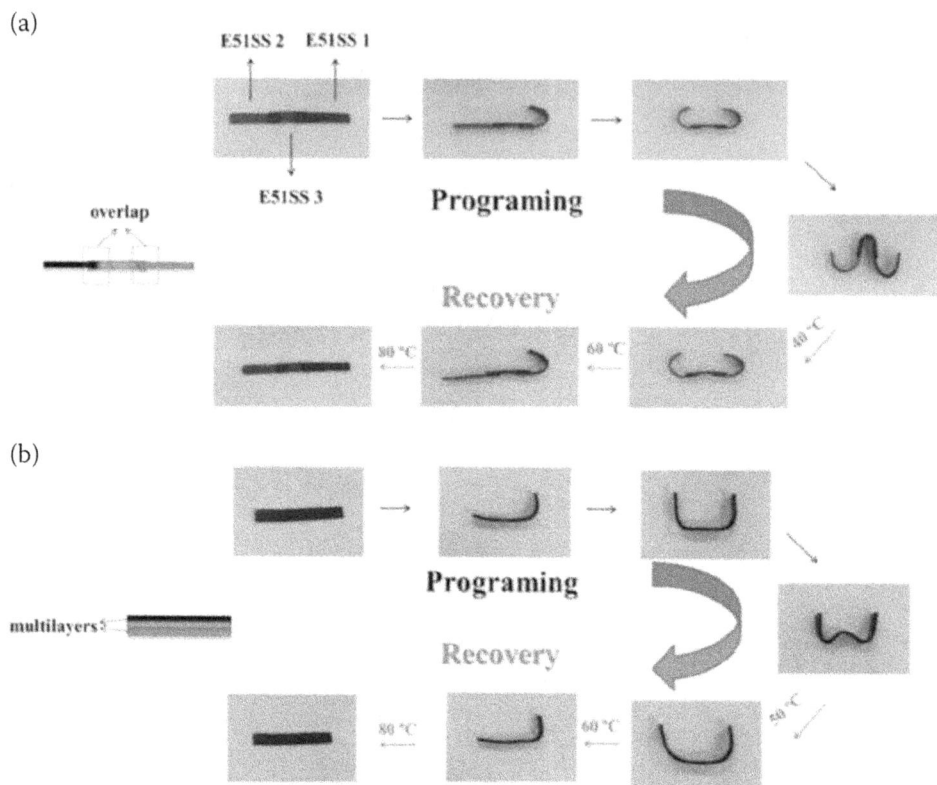

FIGURE 4.4   (a) Multi-shape memory circles of welding vitrimer composites, (b) multi-shape memory circles of multilayered vitrimer composites. [Reproduced with permissions ©2020 Elsevier] (Ji et al., 2020).

system (see Figure 4.4). The multi-shape memory effect was realized by blending the vitrimer composites of different glass transitions using a simple hot-pressing method.

Di Mauro et al. (Di Mauro, Malburet, Graillot, & Mija, 2020) prepared and studied the shape memory characteristics of the bio-based epoxidized vegetable oils. The vitrimer was prepared with dynamic disulfide covalent bonds by copolymerizing 12 epoxidized vegetable oils with 2,2′-dithiodibenzoic acid. The investigation results show that the disulfide and transesterification reaction assured complete recycling of epoxidized vegetable oils. The repairable, recyclable, and reshapable behavior of the thermosets were confirmed in the investigation, which could be interesting in the construction of renewable and sustainable materials.

Bio-based vitrimer based on isosorbide derived epoxy (IS-EPO) and aromatic diamines containing disulfide bonds was synthesized and its physico-chemical characteristics have been investigated. The epoxy was synthesized by reacting the isosorbide with epichlorohydrin in the presence of sodium hydroxide. The final product yield was >95% and the resin was transparent and viscous in nature with an equivalent weight of 303 g/eq. The resin was then cured using 4, 4′-Methylenedianiline (MDA) and 4,4′-disulfanediyldianiline (MDS) curing agents. The properties of IS-EPO varied with the curing agent. The resin cured with MDA showed higher modulus and thermal stability than resin cured with MDS. T_g of MDS-EPO was 40.6°C, while it was 37.3°C for MDA-EPO. From TGA results, the temperature for 5% weight loss was 291.4°C for MDA-EPO, which is ~22°C higher than MDS-EPO. The lower thermal stability of MDS-EPO may be attributed to labile disulfide bonds. The storage modulus at 20°C was 1,571 MPa and 1,051 MPa for MDA-EPO and MDS-EPO, respectively. On the other hand, recovery efficacy of Young's modulus, tensile strength, and elongation at break of MDS-EPO after reprocessing of vitrimers was

(a)

(b)

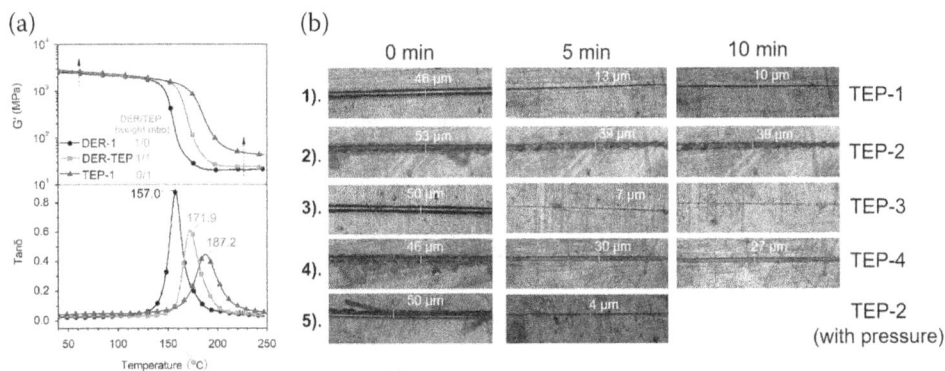

FIGURE 4.5 (a) Storage modulus and Tan δ of TEP and DER epoxy; (b) thermal repairing of cured epoxies at 220°C [Reproduced with permissions ©2018 American chemical society] (Liu et al., 2018).

103%, 82.6%, and 80%, respectively. However, the efficacy declined after three cycles, which may be due to the oxidation of sulfur radicals at high temperatures. As a result, there was scissoring of covalent bonds and damage in the disulfide linkages (Ma, Wang, Zhu, Yu, & Hu, 2017).

Lignin-based thermosets, including epoxy and phenol resins, have been synthesized in order to make the polymer renewable. For lignin-based epoxy, fractionation and chemical modifications have been implemented for the synthesis. However, preparation of these thermosets often faces difficulties due to poor reactivity and solubility. Lignin-derived monomers such as vanillin have been extensively reported for the synthesis of thermosets (Hernandez, Bassett, Sadler, La Scala, & Stanzione, 2016; Zhao & Abu-Omar, 2016, 2017). Liu et al. (Liu et al., 2018) reported bio-based triepoxy (TEP) with high T_g and it was prepared by curing TEP with anhydride (4-methylcyclohexane-1,2-dicarboxylic anhydride (MHHPA)) in the presence of a zinc catalyst. Commercial bisphenol A (DER 1) was used as a reference for the assessment of TEP properties. Figure 4.5a demonstrates the storage modulus and Tan δ TEP and DER epoxy as a function of temperature. From the Tan δ curves, one peak was observed, indicating only one T_g was existed. With an increase in TEP stoichiometric ratio, storage modulus increased gradually from 2.5 GPa to 2.9 GPa at 50°C along with the increase in Tg from 157°C to 187°C. The result indicates that the T_g and storage modulus of the TEP epoxy can be modulated by varying the TEP: MHHPA ratio. High T_g of TEP is attributed to the more number of epoxy groups than the DER 331, which facilitated the formation of a highly dense cross-linking network. Also, the rigid benzene rings and epoxy groups are distant, which eliminates the side reactions. Thus, the epoxy groups efficiently participate in the reaction to achieve a highly dense cross-linking network. For TEP, T_g up to ~220°C was achieved, which designate that below the T_g, the cross-linking network's segmental mobility is rigid and restricts the dynamic reaction. Also, the catalyst loading is an important parameter in modulating the stress relaxation rate. However, higher loading often results in poor solubility and improper mixing. Stress relaxation is an important parameter that reflects the presence of thermally activated transesterification reactions, which helps in the revamping of the material from the cracks (Demongeot, Mougnier, Okada, Soulié-Ziakovic, & Tournilhac, 2016; Giebler et al., 2020; Kuang, Liu, Dong, & Wang, 2017). TEP's reparable characteristics were studied by creating a scratch on the surface and placed between the metal plates. Test was carried out at 220°C and the microscopy images before and after the repair are shown in Figure 4.5b. The crack was mended by almost 70% in a few minutes. As the TEP ratio was increased, the reparability improved (TEP-3; R = 1:1, where "R" is the stoichiometric ratio between the epoxy group and anhydride group). Also, the width of the crack reparability improved with the increase in the catalyst loading (TEP-4; R = 1:0.5). To further improve the reparability, vitrimers can be pressed under a hot press machine, which decreased the crack width by 90%. The good reparable characterisitics of TEP could be attributed to a co-planar structure that facilitated the free movement of the polymer chain above T_g temperature. High T_g, tunable mechanical properties, and self-reparable characteristics of TEP make it a potential prospect for practical applications.

A completely bio-based polymer vitrimer was synthesized by cross-linking soybean oil with an emulsion of citric acid solution. To optimize the polymer network and its properties, the ratio of epoxidized soybean oil to citric acid was varied (R = 1, 0.8, and 0.5). In the cross-linking of epoxidized soybean oil by citric acid (CA), the detachment of citric acid in the solution produces the protons to catalyze the epoxy-acid reaction. Initially, the protonation of epoxide takes place subsequently, followed by the occurrence of carboxylate anion or by H_2O molecules. The carboxylate anion reaction results in β-hydroxy ester. On the other hand, presence of water in CA, hydrolysis of epoxy groups may also occur when the stoichiometric ratio is less than one. Furthermore, esterification reaction may also occur between residual COOH and hydroxyl groups (Ahn, Kraft, & Sun, 2012; Schuster, Rios, Weckes, & Hoelderich, 2008). The polymer network with different stoichiometric ratios was post-cured at 160°C for different periods of time. At 120°C, the samples showed cross-linking behavior with alpha relaxations at ambient temperature. Post-curing of the polymer network with R = 1 displayed a substantial increase in rubbery modulus, while the polymer network with R = 0.5, the increase was not significant. However, after the cure step at 120°C, the storage modulus of polymer network with R = 0.5 offered higher storage modulus compared to other stoichiometric ratio. On the other hand, the stress relaxation rate decreased with the increase in the R value. As the R value decreased, the cross-linking density decreased, while the residual OH concentration increases. An increase in the concentration of OH groups augments the stress relaxation and transesterification reaction rate (Capelot, Montarnal, Tournilhac, & Leibler, 2012). These result insights provide indirect proof of thermally activated transesterification reaction in epoxidized soybean oil/citric acid–based polymer network.

A fully bio-based vitrimers derived from gycyrrhizic acid and epoxidized soybean oil was synthesized, and its shape memory, weldability, and recyclable characteristics have been investigated. The raw products were used without any chemical modification to avoid using petroleum-based resources and making the vitrimers 100% bio-based. Glycyrrhizic acid is a triterpenoid glycoside complex and a weak acid with three carboxyl and five OH groups. The presence of carboxyl and OH groups play a vital role in the transesterification reaction of vitrimers. Also, the rigid structure of glycyrrhizic augments the mechanical and thermal properties of the vitrimer (Fan, 2020). The carboxyl groups of glycyrrhizic acid react with the epoxidized soybean oil to form ester bonds along with the formation of OH groups. Further, the OH groups of glycyrrhizic acid react with epoxy to form ether bonds, resulting in the formation of a cross-linked network. The cross-linked network of the vitrimers resulted in achieving good thermal stability, mechanical properties. The T_g of the vitrimers can be tuned in the range of 39 to 64°C by modulating the stoichiometric ratio of the compounds. The stress-relaxation, creep-recovery, and thermos dilatometry analysis indicated the transesterification reaction in the vitrimer at elevated temperature. The vitrimer showed effective welding, reparability, and recyclable adhesive characteristics. Furthermore, vitrimer derived from gycyrrhizic acid and epoxidized soybean oil can be effectively degraded by ethylene glycol and offer biocompatibility. Figure 4.6 shows the shape memory characteristics of vitrimer. The strip of vitrimer was heated above T_g for 10 s to deform the strip (I) followed with heating it above the malleable temperature for 10 min; because of stress relaxation, the strip retains its shape when cooled at room temperature (ii). This permanent form could be recuperated by heating the strip above T_g followed by cooling it to room temperature (iii and iv). When the external stress is applied, the chain mobility activates above T_g to achieve macroscopic deformation. The shape recovery of the vitrimer can be recovered by activation of chain movement assisted through heating and cooling.

Lignin-derived vitrimer was synthesized by introducing ozonated lignin with sebacic acid-derived epoxy and zinc catalyst. Figure 4.7a illustrates the general method and chemical structure of synthesized bio-based vitrimer. The ozonating of lignin resulted in scissoring the benzene ring, which reduced the lignin stiffness and presented additional carboxylic acid. The ozone-treated

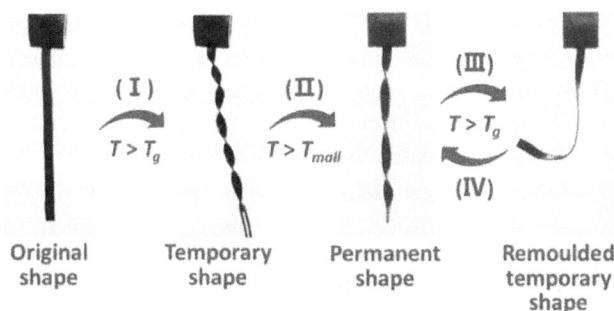

FIGURE 4.6 Shape recovery characteristics of vitrimer derived from gycyrrhizic acid and epoxidised soybean oil [Reproduced with permissions ©2020 American chemical society] (Wu et al., 2020).

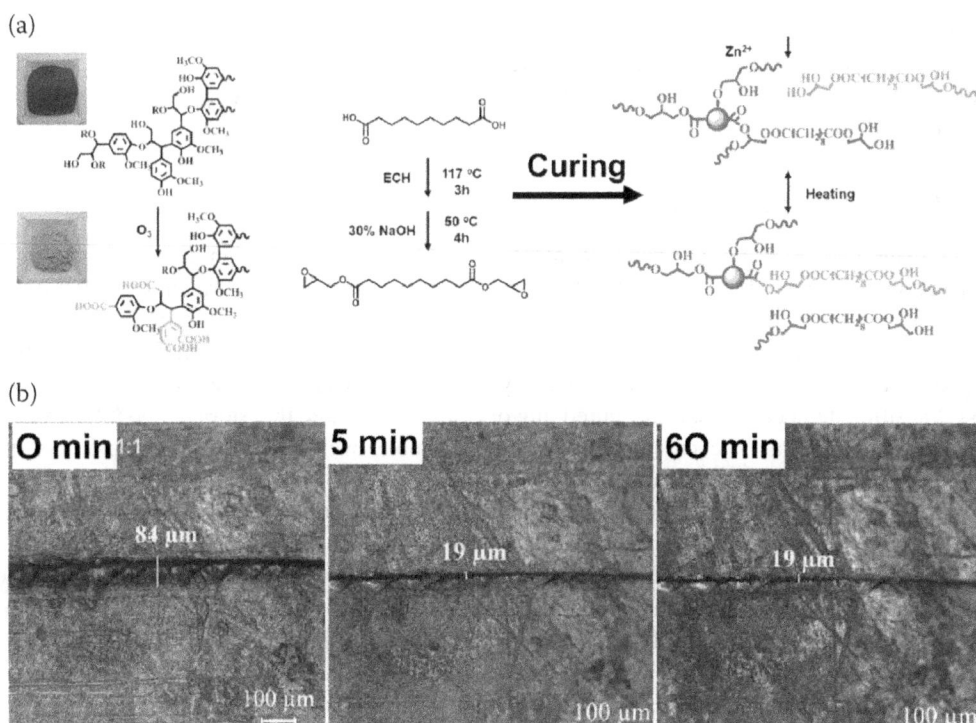

FIGURE 4.7 (a) Schematic illustration of ozone-treated Kraft lignin cured with sebacic acid epoxy, (b) thermal reparability of vitrimers as a function of temperature [Reproduced with permissions ©2018 Royal society of chemistry] (Zhang et al., 2018).

lignin was confirmed from FTIR spectroscopy results. The cleaving of the benzene rings may lead to a decline in the thermal stability of the structure. The presence of high carboxylic and OH groups in the ozonated lignin can function as a curing agent for epoxy resins. Diglycidyl ester derived from sebacic acid was reacted with ozonated lignin. The presence of carboxylic and OH groups of ozonated lignin reacted with the epoxy in the presence catalyst to form ester linkages. The epoxy reacts with carboxylic acid at lower temperature compared with the phenol groups. To ensure the completion of curing reaction and to form maximum ester linkage, the mixture must be cured above 150°C. The vitrimers stoichiometric ratio was based on the epoxy and the sum of carboxylic and phenolic OH groups. With the increase in the ratio of carboxylic and phenolic OH groups, the mechanical properties and T_g increased drastically. The increase in the mechanical properties and T_g of the vitrimers is directly related to the stiffness of the backbone and degree of the cross-linking network. Higher the lignin content, more the stiffness and cross-link density. As a result, the mechanical properties are augmented. Self-healing and thermal reparability of lignin derived was investigated by a scratch reparability test at 190°C and 0.14 MPa pressure (Figure 4.7b). The scratch width decreased as a function of time. The scratch width decreased by ~70 min in less than 5 min, which could be attributed to the transesterification reaction induced thermal-shape changing characteristics. The transesterification reactions generally take place between the ester linkages and the carboxylic and phenolic OH groups at high temperature in the presence of a catalyst. The thermal reparability, self-healing, and recoverable adhesive characteristics at elevated temperatures could be potential material as vitrimers (Zhang et al., 2018).

Yu et al. (Yu et al., 2019) presented vanillin-based epoxy vitrimer synthesized by monoglycidyl and isophorone diamine. The vitrimer exhibited good tensile properties and thermal stability comparable to the bisphenol A epoxy. The glass transition temperature of the vitrimer reached as high as 121°C, while the tensile strength up to 65 MPa was achieved. Because the low density of cross-linking and the presence of a higher amount of dynamic imine bonds, the vitrimer network facilitated quick stress relaxation characteristics. Interestingly, the mechanical properties and T_g of the vitrimer were retained even after reprocessing. Also, the vitrimer was degradable under an acid solution, which is attributed to its Schiff structure.

4.3 URETHANE VITRIMERS

The urethane networks can also form covalent adaptable networks via transamidation reactions. Denisson et al. (Denissen et al., 2015) for the first time explored the vitrimer behavior of the vinylogous amides, which can undergo associative transamination reactions at elevated temperatures without the presence of a catalyst. The authors synthesized vitrimer by bulk polymerization of cyclohexane dimethanol bisacetoacetate, m-xylylene diamine, and tris(2-aminoethyl)amine. The glass transition temperature of the cured ntwork was 87°C and the storage modulus was in the range of 2.4 GPa. The cured networks are insoluble even at elevated temperatures; the stress–relaxation and creep experiments showed a viscoelastic liquid behavior with excellent recyclability without loss of mechanical properties. Chen et al. (F. Chen, Gao, Zhong, Shen, & Lin, 2020) synthesized vinylogous urethane vitrimers from renewable cardanol and the building blocks contain acetoacetate functionalities (acetoacetate modified cardanol- AMCA). Different diamines reacted with AMCA and three rounds of remolding were performed for the vitrimers. The thermal and mechanical properties of the film are nearly unchanged and good recyclability was observed. The vitrimer like behavior was observed for the cross-linked polyurethanes containing free hydroxyl groups; it could be achieved by reacting polyfunctional cyclic carbonates and amines (see Figure 4.8). The polyhydroxy urethanes containing carbamate linkages do not contain any toxic isocyanates and at the same time it can be synthesized from renewable resources. Fortman et al. (Fortman, Brutman, Cramer, Hillmyer, & Dichtel, 2015) synthesized polyhydroxyurethane

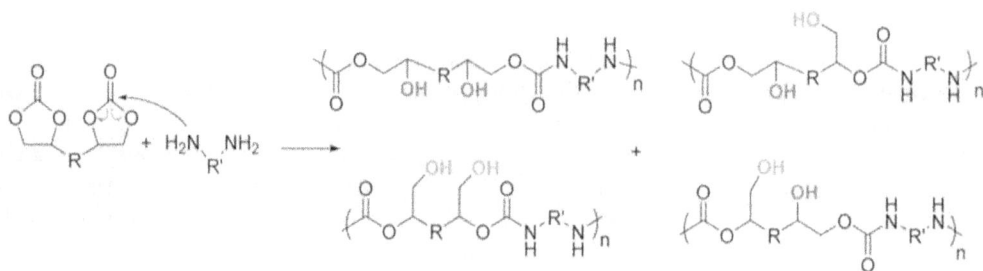

FIGURE 4.8 Polyhydroxyurethane from cyclic carbonate and diamine [Reproduced with permissions ©2015 American chemical society] (Maisonneuve, Lamarzelle, Rix, Grau, & Cramail, 2015).

vitrimers from six-membered cyclic carbonates and amines. Due to the nucleophilic addition of free hydroxyl groups to the carbamate linkages, stress relaxation was observed, which exhibited an Arrhenius activation energy of 111 ± 10 kJ/mol. The polyhydroxyurethanes (PHU) displayed vitrimer behavior in the absence of catalyst and they had mechanical properties comparable to the traditional PU thermosets apart from that they exhibited reshaping and repairable behavior. The same research group (Fortman, Snyder, Sheppard, & Dichtel, 2018) has dramatically shortened the reprocessing times of PHU networks by incorporating dynamic disulfide bonds. The use of cystamin comonomer has shown rapid stress relaxation, but the mechanical properties were similar to the other rigid cross-linked PHU and they had relaxation time as low as 30 s at 150°C. Chen et al. (X. Chen, Li, Jin, & Torkelson, 2017) studied the concurrent associative and dynamic dissociative chemistry via transcarbamoylation and reversible cyclic carbonate aminolysis for the polyhydroxyurethanes. Their investigation indicates that some reprocessable PHU networks cannot be considered as vitrimer due to the existence of both associative and dissociative mechanisms during reprocessing.

The vinylogous urethane-based vitrimer can be synthesized through a condensation reaction of ketoester and primary amines (Denissen et al., 2015). The vitrimer is readily prepared by the bulk polymerization that exhibits T_g around 80°C along with exceptional mechanical properties. Vinylogous urethane vitrimers are insoluble and offer short relaxation time at elevated temperatures (Denissen et al., 2015). Similarly, bio-based products as the raw material are modifiable to synthesize a catalyst-free urethane vitrimer and exhibit the reprocessing ability.

Zhu et al. (Zhu, Gao, Zhong, Shen, & Lin, 2020) reported castor oil and DL-limonene-derived vinylogous urethane vitrimer. Castor oil with readily available functional groups can be used as a starting material for the synthesis of polymer network. DL-limonene attached with two amine groups considered as the good candidate as network cross-linkers (Allauddin, Narayan, & Raju, 2013; M. Wang, Xia, Jiang, Li, & Li, 2016). Introduction of condensation of acetoacetylated castor oil and aminate DL-limonene ratio was maintained with 1:1 offer the required vinylogous urethane vitrimer with ~62% bio-based content. The vitrimer showed excellent stability when immersed in xylene and dimethylformamide at 120°C, while it dissolved completely when immersed in the benzylamine. The formation of a homogenous solution of vitrimer immersed in benzylamine is attributed to the exchange of external amine group, leading to the breakdown of the polymer network. The tensile properties of the vitrimer retained even after several remolding processes, which could be due to the availability of plentiful dynamic vinylogous urethanes that enables the exchange of polymer chains during remolding process, which efficiently eliminates the diverse interfaces. The stress relaxation analysis of vitrimer showed that with the increase in the temperature, the relaxation time decreased, indicating the temperature-dependent exchange reactions in the vitrimers chain network. As a result, vitrimers can be reshaped when heated above 70°C ($> T_g$).

Bismaleimide with different molar ratios and chemical structures was introduced as cross-linker to synthesize thermally responsive bio-based polyurethane comprising pendant furan rings. The diol structure with furan groups was extracted from oleic acid and represented as the furan oligomers. Different proportions of furan oligomers (10, 20, and 30%) in polyurethane cross-linked with 1,1′ (Methylenedi-4,1-phenylene)bismaleimide showed excellent mechanical properties, while the thermal recyclability declined. The introduction of phase segregation with the hard polyurethane segments resulted in superior mechanical properties, thermal recyclability, and adhesion properties (Tremblay-Parrado & Avérous, 2020).

A fusion of two vinylogous urethane vitrimers was prepared by reacting acetoacetate-modified cardanol with meta xylene diamine (aliphatic) and bis-(p-aminocyclohexyl) methane (cycloaliphatic). The melding of the vitrimer took place at molecular level with high homogeneity due to the enabling of dynamic vinylogous urethane. Introducing acetoacetate and amine groups into the polymer network is a facile route to design a high-performance vitrimer with tunable mechanical properties (F. Chen et al., 2020).

In another study, a heterogeneous polyurethane network comprising of an oxetane-subsituted chitosan precursor reacted with hexamethylene and polyethylene glycol. The polyurethane network gave necessary heterogeneity, while chitosan offer necessary cleaving of a four-member oxetane ring. The oxetane ring opens to form two reactive ends whenever there is damage in the network. Upon exposure to the ultraviolet light (UV light), the chitosan chain scissoring occurs, which cross-links with the oxetane reactive ends. As a result, the damage is healed (Weiner & Biomineralization, 2009).

4.4   POLYIMINE VITRIMERS

The imine bonds can also form dynamic covalent bonds; hence, several researchers investigated the vitrimer behavior of the polyimine network especially from renewable monomers. Dhers et al. (Dhers, Vantomme, & Avérous, 2019) investigated the properties of vitrimer behavior of the cross-linked network synthesized from monomers 2,5- furandicarboxaldehyde (FDC) and dimeric and trimeric amines bio-based mixture (DTA) (see Figure 4.9). The transamination reaction is possible only through excess of amine; it promotes the associative exchange reaction. The activation energy obtained from the stress relaxation for the vitrimer was 64 kJ/mol and any significant loss in mechanical properties was observed for the reprocessed vitrimer samples.

The onset temperature for degradation of fructose-derived vitrimers at 5% weight loss was ~300°C. Also, T_g of the vitrimer was around ~10°C and the found to be stable when heated at 120°C for 2 h. For stress relaxation analysis of fructose-derived vitrimers, the relaxation modulus followed as a function of time in the range of 40 to 80°C at 2% strain (constant) (Figure 4.10a). At room temperature, vitrimer showed a complete stress relaxation within an hour. These results indicate the thermal reversibility characteristics of imine bonds in the polymer network. The vitrimer was recycled via compression molding at 120°C and its tensile properties was investigated (Figure 4.10b). After three recycling steps of vitrimer, the tensile strength increased from 0.69 to

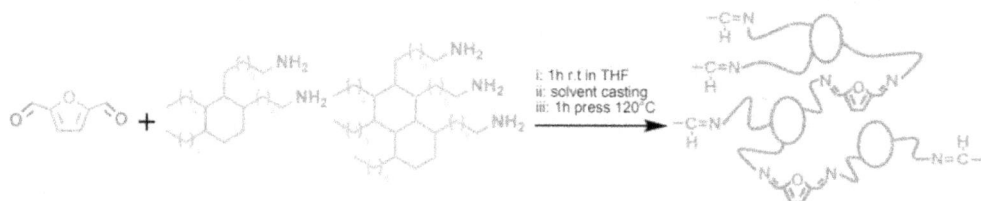

FIGURE 4.9   Synthesis of polyimine vitrimer from furandicarboxaldehyde and dimeric and trimeric amines [Reproduced with permissions ©2019 Royal society of chemistry] (Dhers et al., 2019).

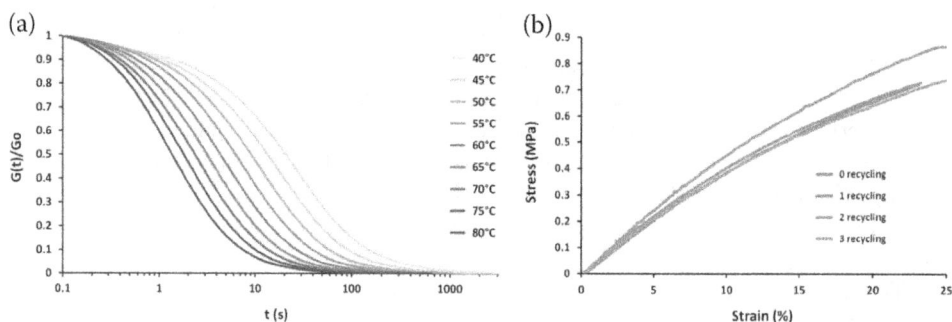

FIGURE 4.10   Fructose derived vitrimers a) Stress relaxation curves, b) stress-strain curve before and after recycling [Reproduced with permissions ©2019 Royal society of chemistry] (Dhers et al., 2019).

0.73 MPa, while the elongation at break seen a marginal decrement (24% to 21%). The retention of tensile properties after recycling of vitrimer is attributed to a high density of cross-linking network (Dhers et al., 2019).

In another study, lignin-derived vanillin-based vitrimers have been synthesized by introducing imine covalent bonds into its structure and its reprocessing ability and reusability characteristics have been investigated. The polyschiff vitrimers film was prepared by dissolving dialdehyde in dichloromethane followed by the addition of diethylenetriamine and tris(2-aminoethyl)amine. The resultant mixture was prepared in the form of film through solution cast process to obtain poly-schiff vitrimer. Tris(2-aminoethyl)amine cross-linker to diethylenetriamine ratio was tuned to achieve vitrimer with various degrees of cross-linking network. With the increase in diethylene-triamine ratio, the degree of cross-linking increased in poyimine films. As a result, T_g increased from 48°C to 64°C. Also, degradation temperature was in the range of 350 to 370°C, which is directly linked with the degree of the cross-linking network. Addition to this, tensile strength of the polyimine film was in the range of ~47 MPa to ~57 MPa.

Conventional thermosets do not have the characteristics to restructuring into its bulk structure once they are scratched or cracked. By introducing the dynamic bonds into the network. With the introduction of dynamic imine bonds, condensation, exchange, and metathesis converse reactions are desirable in achieving the dynamic characteristics of the thermosets. Figure 4.11 shows the self-healing characteristics of polyschiff vitrimers as a function of time. The scratch on the polyschiff vitrimers was exposed to heating at 180°C. The scratch width decreased within 15 min and completely healed after 2 h. The self-healing characteristics of polyschiff vitrimers is governed by the metathesis reverse imine reaction (Geng et al., 2018).

4.5 ELASTOMER AND UV CURABLE VITRIMERS

A catalyst-free and recyclable elastomer vitrimer related with exchangeable β-hydroxyl ester bonds was synthesized by blending the natural rubber with carboxylated nitrile rubber (J. Wang et al., 2020). The elastomer vitrimer displayed self-crosslinking ability without the aid of curing agents and accelerators. It is well known that the presence of dynamic ester linkages and reversible exchange reactions can rearrange the topology network under suitable conditions. Use of β-hydroxyl ester linkages, the natural rubber/nitrile rubber elastomer vitrimer, can rearrange themselves through transesterification reactions.

Figure 4.12 shows the stress-strain curves of natural rubber/nitrile rubber elastomer vitrimer with a weight ratio of 1:1. Upon recycling, the tensile strength of the elastomer vitrimer increased from 2.8 MPa to 5.1 MPa, while the elongation at break increased from 342% to 835%. These

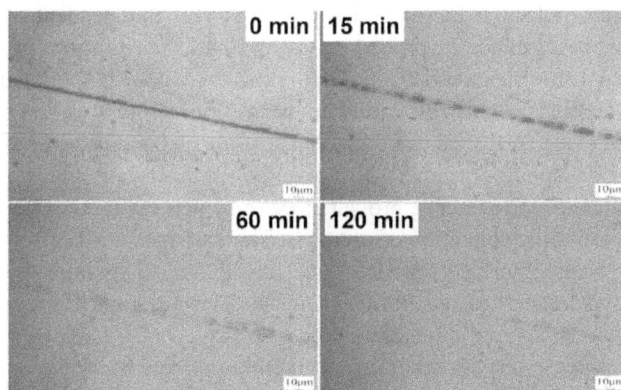

FIGURE 4.11 Optical microscopy photographs of self-healing characteristics of polyschiff vitrimer as a function of time [Reproduced with permissions © 2018 American chemical society] (Geng et al., 2018).

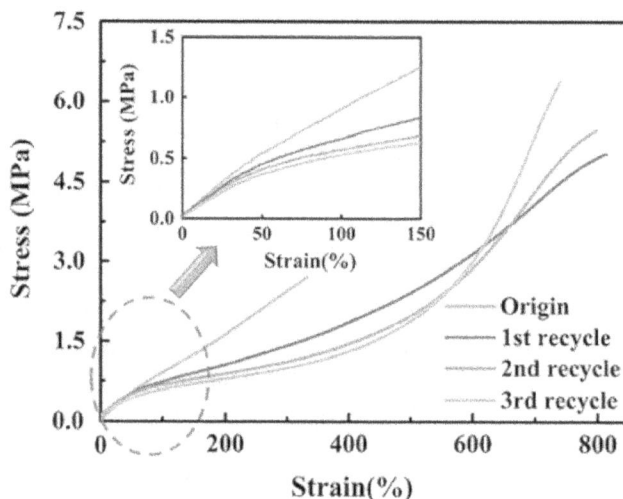

FIGURE 4.12   Stress-strain curves of natural rubber/nitrile rubber elastomer vitrimer with the weight ratio of 1:1 [Reproduced with permissions ©2020 Royal society of chemistry] (J. Wang et al., 2020).

results indicate that the recycled samples possess higher elasticity due to the easier slippage and alignment of the elastomer chain under strain conditions.

UV curable vitrimers offer facile process, excellent curing efficacy, fewer volatile compounds, and relatively less energy consumption. A combination of catalyst-free vitrimer with UV-curable resin is of significance, but often challenging. Wang et al. (S. Wang et al., 2020) synthesized the glycidyl methacrylate reacted with bio-based malic and succinic acid to prepare a UV-curable resin comprising of ester linkages. With the vicinity of carbonyl and hydroxyl groups of malic acid, six-membered ring intermolecular hydrogen bonds were created in the network. As a result, the vitrimer offered excellent tensile strength of 117 MPa and Young's modulus of 3.66 GPa.

The ^{1}H NMR analysis depicting the chemical structure of succinic acid-glycidyl methacrylate and malic acid-glycidyl methacrylate is shown in Figure 4.13a. Bands arising at 5.6-6.1 ppm relate to the protons on the double bonds of methyl allyl in succinic acid-glycidyl methacrylate and malic acid-glycidyl methacrylate. The absence of bands relating to carboxylic acid suggests that the carboxylic acid completely reacted with epoxy groups. The development of hydrogen bonds can be detected by the chemical shift in the band related to the OH bond. –OH group facilitate in hydrogen bonding interactions; thus, a downshift can be evidently seen. It is noteworthy that the chemical shift related to the intramolecular hydrogen-bonded –OH was independent of the concentration. The hydrogen bonding features can be probed by varying the concentrations of malic acid-glycidyl methacrylate (Figure 4.13b). Bands related to -OH signals have been highlighted as area 5 and 7. In area 5, the signals were unaffected with varied concentration, while area 7, a downfield shift was observed with the increase in the concentration. These results indicate the formation of strong intramolecular hydrogen bonding in malic acid-glycidyl methacrylate. Fig. R8c demonstrates the reparable characteristics of vitrimer. The broken piece (tail) was exposed to UV light and heated at 160°C for 1 h along with the application of pressure. The damaged surface was completely healed and the color of the treated surface remained unchanged due to its high thermal stability. The restored samples showed excellent regaining efficiency. The mechanical properties up to 94% of the original sample was recovered after the reprocessibility.

In another study, β-ketoesters incorporated with isobornyl methacrylate and functional amine derived from vegetable was used to prepare a catalyst-free vitrimer. Further, the vitrimer was incorporated with amine-functionalized polyhedral oligomeric silesquioxane nanoparticles (POSS-NH$_2$) at different loadings. The tensile strength of the vitrimer increased significantly when loaded

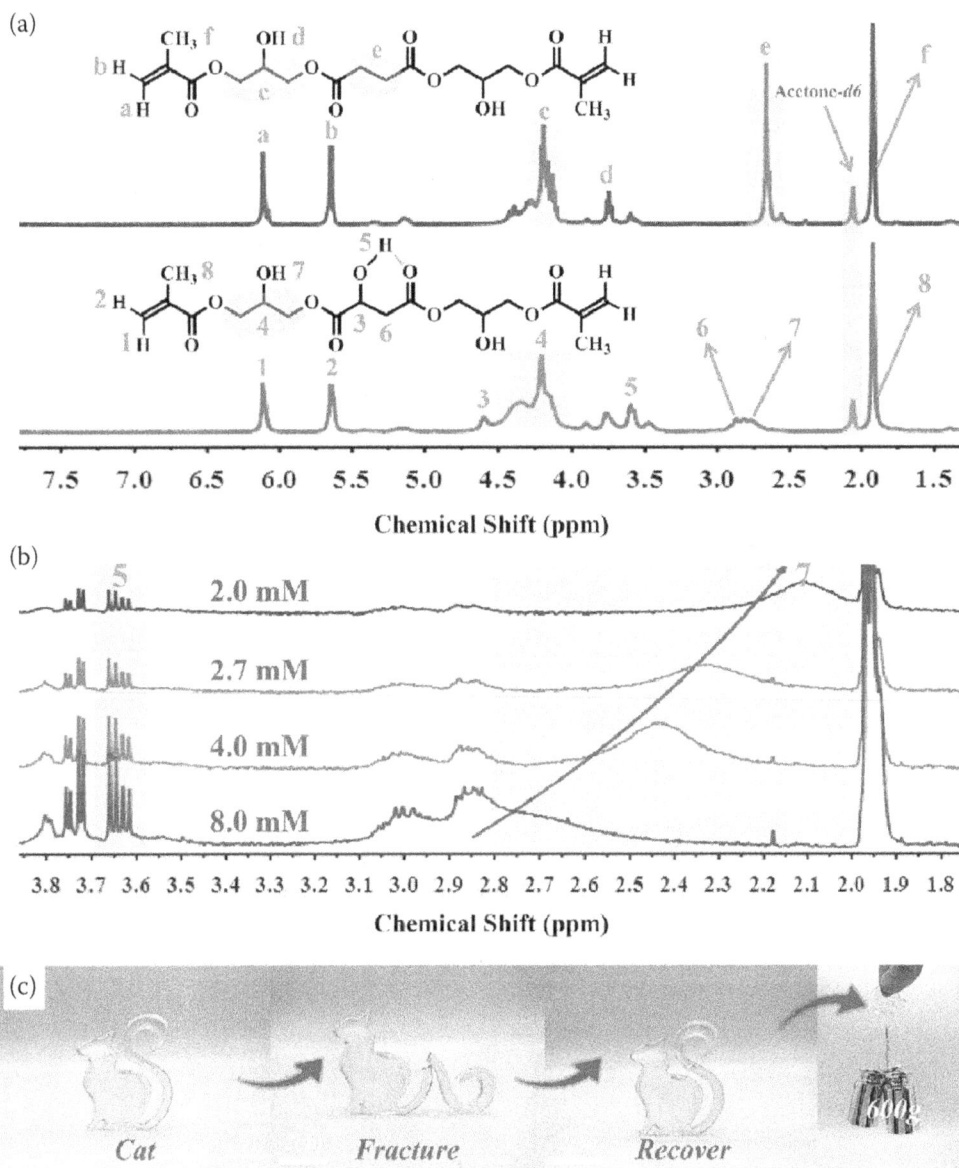

FIGURE 4.13 (a) 1 H NMR spectra of succinic acid-glycidyl methacrylate and malic acid-glycidyl methacrylate, (b) 1 H NMR spectra of malic acid-glycidyl methacrylate at different concentrations, (c) weldability and restore of damaged vitrimer [Reproduced with permissions ©2020 Elsevier] (S. Wang et al., 2020)

with 20 wt% POSS- NH_2. On the other hand, the degradation temperature increased from 225 to 255°C at 5% weight loss and the activation energy of stress relaxation increased. Furthermore, the cross-linked vinyligous urethane networks in the nanocomposite can be reversed by dissolving in excess mono-functional amine at 65°C (Hajiali, Tajbakhsh, & Marić, 2021).

4.6 NANO REINFORCEMENT IN VITRIMER STRUCTURE

The nanoscale reinforcement along with the polymer matrix can improve the mechanical, thermal, and other properties as well as the functional performance of the composites. Vitrimer incorporated with CNT significantly improve the mechanical properties and transesterification reaction. The presence of high thermal conductive nanofillers such as CNT, graphene, carbon black, borosilicate, etc., could significantly absorb heat, which triggers the transesterification reaction. One significant benefit of these nanofillers in vitrimers could possibly demonstrate the photo-thermal effect, which could easily activate the reaction process using light in any time required. Also, the stress relaxation rate, cross-link density, and T_g can be tuned effortlessly, which significantly provide flexibility in the processing of the vitrimers.

Nanoscale reinforcement for the renewable vitrimers is yet to be investigated only few are reported and here we summarized some of results carried out for the vitrimers from fossil fuel derived monomers. Legrand et al. (Legrand & Soulié-Ziakovic, 2016) investigated the vitrimer behavior of the nano silica-reinforced epoxy nanocomposites. They found that the silica loading was the predominant factor in achieving high modulus in glassy and rubbery region. To improve the dispersion of silica in the resin, silica particles were functionalized and it accelerated the relaxation process via exchangeable bonds, which contribute to the relaxation of the composite. Monodisperse nano-silica was loaded up to 40 wt% in the matrix resin and the nanoparticles were covalently linked to the network through exchangeable β-hydroxy ester bonds formed within the vitrimer matrix. Similarly, Huang et al. (Huang, Wang, Zhu, Yu, & Hu, 2018) investigated the vitrimer behavior of the cross-linked network reinforced with epoxy and thiol groups functionalized nano-silica. The modified silica particle was used as reinforcement for the disulfide-based vitrimer composites. The results showed that the thiol functionalized silica exhibited slightly inferior mechanical behavior than epoxy functionalized silica, but stress relaxation for the thiol functionalization is much faster than the other system. The reason for the quick stress relaxation is due to the faster kinetics of thiol-disulfide exchange reaction. Chen et al. (J. Chen et al., 2019) investigated the recyclable behavior of the graphene/epoxy composites and they fabricated aligned graphene nanoplate (GnP)/epoxy composites by using hot press method. Bisphenol A diglycidyl ether (DGEBA) with commercial trimeric acid (Pripol 1040) was used as matrix resin along with graphene nanoplate (Figure 4.14). The bond exchange and topological rearrangement happen during the hot press, enabling spontaneous orientation of graphene nanoplate in matrix. The scanning electron microscope results showed the orientation of nanoreinforcement as well as the increase in mechanical properties was observed. The above dynamic bond exchange reactions in the graphene/epoxy composites provided very good healability and recyclability of the epoxy vitrimer.

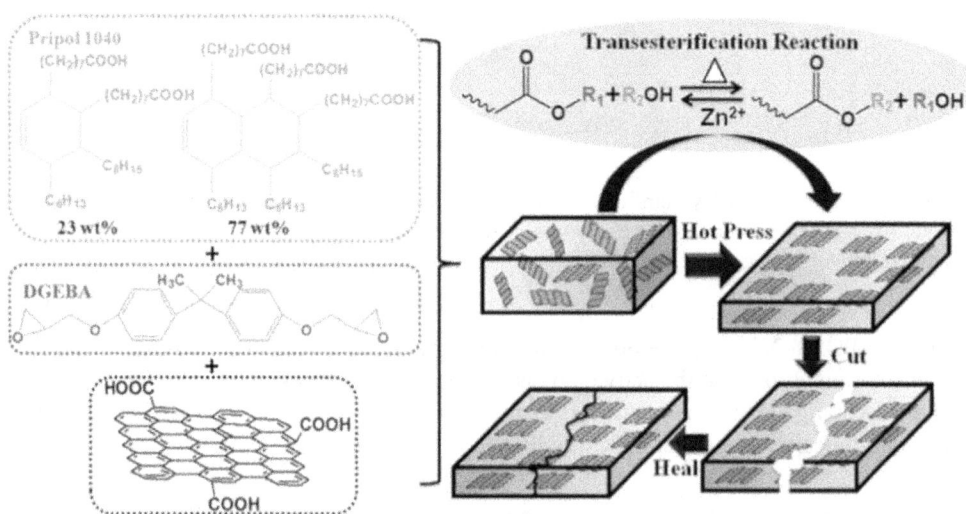

FIGURE 4.14   Schematic representation of transesterification reaction for the graphene nanoplate (GnP)/epoxy composites prepared using hot press [Reproduced with permissions ©2019 Frontiers] (J. Chen et al., 2019).

The concept of vitrimer have been introduced to the elastomer network to provide covalently cross-linked elastomers with recyclable characteristics. Elastomer vitrimer consisting of β–hydroxyl ester cross-links between the epoxidized natural rubber and carbon nanodots in the presence of zinc and 1,2-dimethylimidazole as catalyst was synthesized. Similarly, epoxidized vitrimer/carbon black composite consisting of β–hydroxyl ester linkages was synthesized, which showed covalent adaptive network, mechanical strength, and reprocessable characteristics (Qiu, Wu, Fang, Tang, & Guo, 2018; Tang, Liu, Guo, & Zhang, 2017). The reported elastomer vitrimers were basically fabricated by exchangeable cross-linkages between the rubber and nanofillers. Filler could act as a cross-linking and reinforcing agent to build a 3D network. Also, the nanofillers introduction would result in the retarding of the topology network reorganization due to hindering the chain movement. Apart from inorganic filler, cellulose and chitin nanofibrils were investigated as reinforcement for the vitrimers. The most commonly used mechanism in the nanoscale reinforcement for the vitrimer applications is transesterification chemistry because of the availability of hydroxyl functionalized fillers (Legrand & Soulié-Ziakovic, 2016).

4.7 FUTURE PROSPECTS AND APPLICATIONS

Overall, bio-based vitrimers provide enough compelling evidence and prospects for sustainable and recyclable thermoset material. However, synthesis and characterization of bio-based vitrimers are still at the initial development stage and very few publications have been reported. Nevertheless, vitrimer have confined to academic research and only a handful of companies have commercialized the vitrimers. Vitrimers can be thermally reprocessed with trailing the cross-link networks. However, poor mechanical properties, catalysts, and time-consuming processes are still major concerns for their practical implementation. Although the vitrimer network is reshapable, it is difficult to regain its original state accompanying the original mechanical properties and thermal stability. Vitrimers combine with the desirable mechanical properties and the ability to reprocess quickly above high T_g could be an exciting breakthrough.

Parameters such as T_g, stress-relaxation rate, and cross-linking network density are considered vital for achieving high performance of vitrimers. Besides, high mechanical properties and thermal stability is a primary requirement for the practical applications. Hence, researchers are keen to fabricate vitrimers in the form of reinforced composites. Various industries such as automobile, aerospace, electronics, biomedical, household appliances, construction and sports equipment, and epoxy-based composites have been profoundly ingrained in our daily lives. Recycling and reparability of characteristics of epoxy could be of high importance.

Fiber-reinforced vitrimers have received considerable attention in the composite industry due to saving energy and increasing resource efficiency. Fiber-reinforced vitrimers could be the potential and most fascinating material of choice for automobile, aerospace and construction industries. From the viewpoint of recycling and reprocessing, fiber-reinforced vitrimers with fast and robust exchange reactions present several opportunities. This enables the composites with ability to reverse the damage through heating/photo effect or recycle the material and promote efficient and sustainable management of resources along with saving the time and product cost. Especially in the composite industries, carbon fiber–reinforced epoxy composite provides the most efficient way to reduce weight and is seen as an alternative to metal/ceramic products without compromising the mechanical properties. One such example of dynamic epoxy/carbon fiber–reinforced composite is shown in Figure 4.15a-c. Nonetheless, the adoption of these composites is persistently limited due to several manufacturing challenges such as high cost and life span of the material. These challenges can be addressed by vitrimer composites by enabling recyclable, self-reparable, and reprocessing characteristics.

The low viscosity of the uncured resin permits uniform flow to integrate with reinforced fibers, indicating that the vitrimer resins can also be made in the form of prepregs. However, vitrimer prepregs are cured instantaneously after the infusion. Several studies have indicated that the recycling of vitrimer composite does not damage the fibers with excellent recovery efficiency without losing

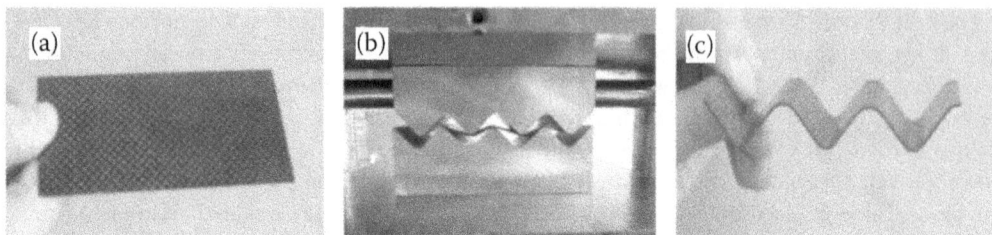

FIGURE 4.15 Digital photograph of dynamic epoxy/carbon-fiber composite laminate: (a) Compression-molded composite, (b) depiction of zig-zag shape mold, (c) illustrating the thermoformed 3D composite [Reproduced with permissions ©2016 Royal society of chemistry [65].

the mechanical properties. Hence, in applications such as sports equipment, wind energy, automobiles, pressure vessels, and coating industries, vitrimers can make a tremendous impact.

Thanks to the huge diversity in polymer chemistry, gleaning and breakthrough potential could enable the practical applications of vitrimer with its signature features and fascinating properties, including self-healing, shape recovery, recyclability, weldability, and reprocessing characteristics. Bio-based vitrimers may further expand the horizons in terms of eco-friendliness and sustainability.

4.8   SUMMARY

In this chapter, the authors provided an overview about covalent adaptable network and the new class of material called "vitrimer" which is based on the dynamic covalent bond exchange reaction chemistry. The cross-linked network can be recycled, reshaped, and welded by using thermal energy or in the presence of light-assisted resources. The recyclability and healability of the different vitrimer networks have been reported in the literature. The basic research on utilizing renewable monomer for the vitrimer is in the early stage. The earlier investigations show that nanoscale reinforcements improve the mechanical strength at the same time faster the stress relaxation behavior of the vitrimer. The research on utilizing renewable fillers as well as inorganic and carbon based fillers are ongoing and to improve dispersion and to promote bond exchange reaction introducing different functionalities on the fiber surface is needed. The bond exchange reaction can be promoted by adding catalyst or some modified nanofillers and the most simple and efficient reaction mechanism is the disulfide exchange chemistry. Additionally, polyimine-based networks are scaled and some of them are commercially available for the composite application. The application of this vitrimer for the composite and elastomer industries are encouraging but still lot of research work is needed to reduce the reprocessing temperature. In the future, vitrimers can contribute to a circular economy and recycling including waste prevention via repairability.

REFERENCES

Ahn, B. J. K., Kraft, S., & Sun, X. S. (2012). Solvent-free acid-catalyzed ring-opening of epoxidized oleochemicals using stearates/stearic acid, and its applications. *Journal of Agricultural and Food Chemistry*, *60*(9), 2179–2189. 10.1021/jf204275q

Allauddin, S., Narayan, R., & Raju, K. V. S. N. (2013). Synthesis and properties of alkoxysilane castor oil and their polyurethane/urea-silica hybrid coating films. *ACS Sustainable Chemistry and Engineering*, *1*(8), 910–918. 10.1021/sc3001756

Altuna, F. I., Hoppe, C. E., & Williams, R. J. J. (2019). Epoxy vitrimers with a covalently bonded tertiary amine as catalyst of the transesterification reaction. *European Polymer Journal*, *113*(January), 297–304. 10.1016/j.eurpolymj.2019.01.045

Altuna, F. I., Pettarin, V., & Williams, R. J. J. (2013). Self-healable polymer networks based on the cross-linking of epoxidised soybean oil by an aqueous citric acid solution. *Green Chemistry*, *15*(12), 3360–3366. 10.1039/c3gc41384e

Bowman, C. N., & Kloxin, C. J. (2012). Covalent adaptable networks: reversible bond structures incorporated in polymer networks. *Angewandte Chemie International Edition*, *51*(18), 4272–4274. 10.1002/anie.2 01200708

Campanella, A., Döhler, D., & Binder, W. H. (2018). Self-Healing in Supramolecular Polymers. *Macromolecular Rapid Communications*, *39*(17), 1700739. 10.1002/marc.201700739

Capelot, M., Montarnal, D., Tournilhac, F., & Leibler, L. (2012). Metal-catalyzed transesterification for healing and assembling of thermosets. *Journal of the American Chemical Society*, *134*(18), 7664–7667. 10.1021/ja302894k

Chen, F., Gao, F., Zhong, J., Shen, L., & Lin, Y. (2020). Fusion of biobased vinylogous urethane vitrimers with distinct mechanical properties. *Materials Chemistry Frontiers*, *4*(9), 2723–2730. 10.1039/d0qm00302f

Chen, J., Huang, H., Fan, J., Wang, Y., Yu, J., Zhu, J., & Hu, Z. (2019). Vitrimer chemistry assisted fabrication of aligned, healable, and recyclable graphene/epoxy composites. *Frontiers in Chemistry*, 7, 632. 10.3389/fchem.2019.00632

Chen, X., Li, L., Jin, K., & Torkelson, J. M. (2017). Reprocessable polyhydroxyurethane networks exhibiting full property recovery and concurrent associative and dissociative dynamic chemistry: Via transcarbamoylation and reversible cyclic carbonate aminolysis. *Polymer Chemistry*, *8*(41), 6349–6355. 10.1039/c7py01160a

Cordier, P., Tournilhac, F., Soulié-Ziakovic, C., & Leibler, L. (2008). Self-healing and thermoreversible rubber from supramolecular assembly. *Nature*, *451*(7181), 977–980. 10.1038/nature06669

Demongeot, A., Mougnier, S. J., Okada, S., Soulié-Ziakovic, C., & Tournilhac, F. (2016). Coordination and catalysis of Zn2+ in epoxy-based vitrimers. *Polymer Chemistry*, *7*(27), 4486–4493. 10.1039/c6py00752j

Denissen, W., Rivero, G., Nicolaÿ, R., Leibler, L., Winne, J. M., & Du Prez, F. E. (2015). Vinylogous urethane vitrimers. *Advanced Functional Materials*, *25*(16), 2451–2457. 10.1002/adfm.201404553

Denissen, W., Winne, J. M., & Du Prez, F. E. (2016). Vitrimers: Permanent organic networks with glass-like fluidity. *Chemical Science*, *7*(1), 30–38. 10.1039/c5sc02223a

Dhers, S., Vantomme, G., & Avérous, L. (2019). A fully bio-based polyimine vitrimer derived from fructose. *Green Chemistry*, *21*(7), 1596–1601. 10.1039/c9gc00540d

Di Mauro, C., Malburet, S., Graillot, A., & Mija, A. (2020). Recyclable, repairable, and reshapable (3R) thermoset materials with shape memory properties from bio-based epoxidized vegetable oils. *ACS Applied Bio Materials*, 3, 8094–8104. 10.1021/acsabm.0c01199

Dobáš, I., Luňák, S., Zelenka, J., Horálek, J., Špaček, V., & de Petris, S. (1992). Epoxy resins based on aromatic glycidylamines. VI. Toughening of N,N,N′,N′-tetraglycidyl-4,4′-diaminodiphenylmethane by reactive additives. *Journal of Applied Polymer Science*, *45*(5), 923–930. 10.1002/app.1992.070450520

Fan, R. (2020). Separation of glycyrrhizic acid and its derivants from hydrolyzation in subcritical water by macroporous resin. *Molecules*, *25*(18). 10.3390/molecules25184305

Fischer, H. (2010). Self-repairing material systems—a dream or a reality? *Natural Science*, *2*(8), 873–901. 10.4236/ns.2010.28110

Fortman, D. J., Brutman, J. P., Cramer, C. J., Hillmyer, M. A., & Dichtel, W. R. (2015). Mechanically activated, catalyst-free polyhydroxyurethane vitrimers. *Journal of the American Chemical Society*, *137*(44), 14019–14022. 10.1021/jacs.5b08084

Fortman, D. J., Snyder, R. L., Sheppard, D. T., & Dichtel, W. R. (2018). Rapidly reprocessable cross-linked polyhydroxyurethanes based on disulfide exchange. *ACS Macro Letters*, *7*(10), 1226–1231. 10.1021/acsmacrolett.8b00667

Geng, H., Wang, Y., Yu, Q., Gu, S., Zhou, Y., Xu, W., & … Ye, D. (2018). Vanillin-based polyschiff vitrimers: reprocessability and chemical recyclability. *ACS Sustainable Chemistry and Engineering*, *6*(11), 15463–15470. 10.1021/acssuschemeng.8b03925

Giebler, M., Sperling, C., Kaiser, S., Duretek, I., & Schlögl, S. (2020). Epoxy-anhydride vitrimers from aminoglycidyl resins with high glass transition temperature and efficient stress relaxation. *Polymers*, *12*(5), 1–14. 10.3390/POLYM12051148

Guerre, M., Taplan, C., Winne, J. M., & Du Prez, F. E. (2020). Vitrimers: directing chemical reactivity to control material properties. *Chemical Science*, *11*, 4855–4870. 10.1039/d0sc01069c

Hajiali, F., Tajbakhsh, S., & Marić, M. (2021). Thermally reprocessable bio-based polymethacrylate vitrimers and nanocomposites. *Polymer*, *212*, 123126–123138. 10.1016/j.polymer.2020.123126

Hernandez, E. D., Bassett, A. W., Sadler, J. M., La Scala, J. J., & Stanzione, J. F. (2016). Synthesis and characterization of bio-based epoxy resins derived from vanillyl alcohol. *ACS Sustainable Chemistry and Engineering*, *4*(8), 4328–4339. 10.1021/acssuschemeng.6b00835

Huang, Z., Wang, Y., Zhu, J., Yu, J., & Hu, Z. (2018). Surface engineering of nanosilica for vitrimer composites. *Composites Science and Technology*, *154*, 18–27. 10.1016/j.compscitech.2017.11.006

Ji, F., Liu, X., Sheng, D., & Yang, Y. (2020). Epoxy-vitrimer composites based on exchangeable aromatic disulfide bonds: Reprocessibility, adhesive, multi-shape memory effect. *Polymer*, *197*(February), 122514. 10.1016/j.polymer.2020.122514

Jin, Y., Lei, Z., Taynton, P., Huang, S., & Zhang, W. (2019). Malleable and recyclable thermosets: The next generation of plastics. *Matter*, *1*(6), 1456–1493. 10.1016/j.matt.2019.09.004

Jourdain, A., Asbai, R., Anaya, O., Chehimi, M. M., Drockenmuller, E., & Montarnal, D. (2020). Rheological properties of covalent adaptable networks with 1,2,3-triazolium cross-links: the missing link between vitrimers and dissociative networks. *ACS Applied Materials and Interfaces*, *53*, 1884–1900. 10.1021/acs.macromol.9b02204

Keeratitham, W., & Somwangthanaroj, A. (2016). High Tg and fast curing epoxy-based anisotropic conductive paste for electronic packaging. *AIP Conference Proceedings*, *1713*(March 2016). 10.1063/1.4942341

Kloxin, C. J., & Bowman, C. N. (2013). Covalent adaptable networks: Smart, reconfigurable and responsive network systems. *Chemical Society Reviews*, *42*(17), 7161–7173. 10.1039/c3cs60046g

Kloxin, C. J., Scott, T. F., Adzima, B. J., & Bowman, C. N. (2010). Covalent adaptable networks (CANs): A unique paradigm in cross-linked polymers. *Macromolecules*, *43*, 2643–2653. 10.1021/ma902596s

Krishnakumar, B., Sanka, R. V. S. P., Binder, W. H., Parthasarthy, V., Rana, S., & Karak, N. (2020). Vitrimers: Associative dynamic covalent adaptive networks in thermoset polymers. *Chemical Engineering Journal*, *385*, 123820. 10.1016/j.cej.2019.123820

Kuang, X., Liu, G., Dong, X., & Wang, D. (2017). Correlation between stress relaxation dynamics and thermochemistry for covalent adaptive networks polymers. *Materials Chemistry Frontiers*, *1*(1), 111–118. 10.1039/c6qm00094k

Legrand, A., & Soulié-Ziakovic, C. (2016). Silica-epoxy vitrimer nanocomposites. *Macromolecules*, *49*(16), 5893–5902. 10.1021/acs.macromol.6b00826

Lei, Z. Q., Xiang, H. P., Yuan, Y. J., Rong, M. Z., & Zhang, M. Q. (2014). Room-temperature self-healable and remoldable cross-linked polymer based on the dynamic exchange of disulfide bonds. *Chemistry of Materials*, *26*(6), 2038–2046. 10.1021/cm4040616

Liu, T., Hao, C., Wang, L., Li, Y., Liu, W., Xin, J., & Zhang, J. (2017). Eugenol-Derived Biobased Epoxy: Shape Memory, Repairing, and Recyclability. *Macromolecules*, *50*(21), 8588–8597. 10.1021/acs.macromol.7b01889

Liu, T., Hao, C., Zhang, S., Yang, X., Wang, L., Han, J., & … Zhang, J. (2018). A self-healable high glass transition temperature bioepoxy material based on Vitrimer chemistry. *Macromolecules*, *51*(15), 5577–5585. 10.1021/acs.macromol.8b01010

Liu, T., Zhang, L., Chen, R., Wang, L., Han, B., Meng, Y., & Li, X. (2017). Nitrogen-free tetrafunctional epoxy and its dds-cured high-performance matrix for aerospace applications. *Industrial and Engineering Chemistry Research*, *56*(27), 7708–7719. 10.1021/acs.iecr.7b00096

Lu, Y. X., & Guan, Z. (2012). Olefin metathesis for effective polymer healing via dynamic exchange of strong carbon-carbon double bonds. *Journal of the American Chemical Society*, *134*(34), 14226–14231. 10.1021/ja306287s

Ma, Z., Wang, Y., Zhu, J., Yu, J., & Hu, Z. (2017). Bio-based epoxy vitrimers: Reprocessibility, controllable shape memory, and degradability. *Journal of Polymer Science, Part A: Polymer Chemistry*, *55*(10), 1790–1799. 10.1002/pola.28544

Maisonneuve, L., Lamarzelle, O., Rix, E., Grau, E., & Cramail, H. (2015). Isocyanate-free routes to polyurethanes and poly(hydroxy urethane)s. *Chemical Reviews*, *115*(22), 12407–12439. 10.1021/acs.chemrev.5b00355

Montarnal, D., Capelot, M., Tournilhac, F., & Leibler, L. (2011). Silica-like malleable materials from permanent organic networks. *Science*, *334*(6058), 965–968. 10.1126/science.1212648

Qiu, M., Wu, S., Fang, S., Tang, Z., & Guo, B. (2018). Sustainable, recyclable and robust elastomers enabled by exchangeable interfacial cross-linking. *Journal of Materials Chemistry A*, *6*(28), 13607–13612. 10.1039/c8ta04173c

Rekondo, A., Martin, R., Ruiz De Luzuriaga, A., Cabañero, G., Grande, H. J., & Odriozola, I. (2014). Catalyst-free room-temperature self-healing elastomers based on aromatic disulfide metathesis. *Materials Horizons*, *1*(2), 237–240. 10.1039/c3mh00061c

Schuster, H., Rios, L. A., Weckes, P. P., & Hoelderich, W. F. (2008). Heterogeneous catalysts for the production of new lubricants with unique aproperties. *Applied Catalysis A: General*, *348*(2), 266–270. 10.1016/j.apcata.2008.07.004

Scott, T. F., Schneider, A. D., Cook, W. D., & Bowman, C. N. (2005). Chemistry: Photoinduced plasticity in cross-linked polymers. *Science*, *308*(5728), 1615–1617. 10.1126/science.1110505

Seiffert, S., & Sprakel, J. (2012). Physical chemistry of supramolecular polymer networks. *Chemical Society Reviews*, *41*(2), 909–930. 10.1039/c1cs15191f

Tang, Z., Liu, Y., Guo, B., & Zhang, L. (2017). Malleable, mechanically strong, and adaptive elastomers enabled by interfacial exchangeable bonds. *Macromolecules*, *50*(19), 7584–7592. 10.1021/acs.macromol.7b01261

Thakur, V. K., & Kessler, M. R. (2015). Self-healing polymer nanocomposite materials: A review. *Polymer*, *69*, 369–383. 10.1016/j.polymer.2015.04.086

Tremblay-Parrado, K. K., & Avérous, L. (2020). Synthesis and behavior of responsive biobased polyurethane networks cross-linked by click chemistry: Effect of the cross-linkers and backbone structures. *European Polymer Journal*, *135*, 109840. 10.1016/j.eurpolymj.2020.109840

Wang, J., Chen, S., Lin, T., Ke, J., Chen, T., Wu, X., & Lin, C. (2020). A catalyst-free and recycle-reinforcing elastomer vitrimer with exchangeable links. *RSC Advances*, *10*(64), 39271–39276. 10.1039/d0ra07728c

Wang, M., Xia, J., Jiang, J., Li, S., & Li, M. (2016). Mixed calcium and zinc salts of N-(3-amino-benzoic acid)terpene-maleamic acid: Preparation and its application as novel thermal stabilizer for poly(vinyl chloride). *RSC Advances*, *6*(99), 97036–97047. 10.1039/c6ra19523g

Wang, S., Teng, N., Dai, J., Liu, J., Cao, L., Zhao, W., & Liu, X. (2020). Taking advantages of intramolecular hydrogen bonding to prepare mechanically robust and catalyst-free vitrimer. *Polymer*, *210*(September), 123004. 10.1016/j.polymer.2020.123004

Weiner, S., & Biomineralization, O. (2009). Self-repairing oxetane-substituted chitosan polyurethane networks. *Science*, *323*(March), 1–3. 10.1126/science.1167391

Wu, J., Yu, X., Zhang, H., Guo, J., Hu, J., & Li, M.-H. (2020). Fully biobased vitrimers from glycyrrhizic acid and soybean oil for self-healing, shape memory, weldable, and recyclable materials. *ACS Sustainable Chemistry & Engineering*, *8*(16), 6479–6487. 10.1021/acssuschemeng.0c01047

Yang, Y., & Urban, M. W. (2013). Self-healing polymeric materials. *Chemical Society Reviews*, *42*(17), 7446–7467. 10.1039/c3cs60109a

Yu, Q., Peng, X., Wang, Y., Geng, H., Xu, A., Zhang, X., & … Ye, D. (2019). Vanillin-based degradable epoxy vitrimers: Reprocessability and mechanical properties study. *European Polymer Journal*, *117*(1), 55–63. 10.1016/j.eurpolymj.2019.04.053

Zhang, S., Liu, T., Hao, C., Wang, L., Han, J., Liu, H., & Zhang, J. (2018). Preparation of a lignin-based vitrimer material and its potential use for recoverable adhesives. *Green Chemistry*, *20*(13), 2995–3000. 10.1039/c8gc01299g

Zhao, S., & Abu-Omar, M. M. (2016). Renewable epoxy networks derived from lignin-based monomers: Effect of cross-linking density. *ACS Sustainable Chemistry and Engineering*, *4*(11), 6082–6089. 10.1021/acssuschemeng.6b01446

Zhao, S., & Abu-Omar, M. M. (2017). Synthesis of renewable thermoset polymers through successive lignin modification using lignin-derived phenols. *ACS Sustainable Chemistry and Engineering*, *5*(6), 5059–5066. 10.1021/acssuschemeng.7b00440

Zhu, Y., Gao, F., Zhong, J., Shen, L., & Lin, Y. (2020). Renewable castor oil and DL-limonene derived fully bio-based vinylogous urethane vitrimers. *European Polymer Journal*, *135*(June), 109865. 10.1016/j.eurpolymj.2020.109865

Zou, W., Dong, J., Luo, Y., Zhao, Q., & Xie, T. (2017). Dynamic covalent polymer networks: From old chemistry to modern day innovations. *Advanced Materials*, *29*(14), 1606100. 10.1002/adma.201606100

5 Recent Advances in Tree Gum Polymers for Food, Energy, and Environmental Domains

Akshay Kumar K. P.
Brno University of Technology, Brno, Czech Republic

Cochin University of Science and Technology, Kochi, India

Miroslav Černík
Technical University of Liberec (TUL), Liberec, Czech Republic

Vinod V. T. Padil
Technical University of Liberec (TUL), Liberec, Czech Republic

CONTENTS

5.1 INTRODUCTION

Bio-based polymers hold a prestigious position among the polymer family due to their unique properties, functionalities, and versatile applications. The researchers and polymer scientists are continually searching for novel bio-based materials, an alternative for petroleum-based polymers that can replace them, reduce global warming, and control environmental pollution. Primary sources such as plant biomass or gums, agricultural and marine wastes contributes towards a high yield bio-based polymers and their high-value products.

Tree gum polymers are versatile, compositional, physiochemical properties, functional, structural, morphological, rheological, and applicable for both food and non-food spheres. There are many natural sources for gums, including botanical, animal, seaweeds, bacterial, and plants. Even though many gum producing trees are available in nature, specific gum polymers only are exploited for their commercial applications. Tree gum polymers are extracted or exudates from the tree's bark or their seed and consist of carbohydrates, proteins, fatty acids, metal ions, and wholly or partially soluble in water (Padil et al., 2018; Barak, Mudgil, and Taneja, 2020). The present

DOI: 10.1201/9781003200710-6

chapter demonstrates the following commercial tree gums, which include: gum Arabic (GA) (*Acacia senegal*), gum karaya (GK) (*Sterculia urens*), gum ghatti (GG) (*Anogeissus latifolia*), neem gum (*Azadirachta indica*), gum kondagogu (GK) (*Cochlospermum gossypium*), gum tragacanth (GT) (*Astragalus gummifer*), cashew gum (CG) (*Anacardium occidentale L.*), and their research advances in a variety of fabrication techniques such as nanoparticles formation, nanofibers production, 3D-printing structures, sponges, films, nanocomposites and applications in food packaging, energy, environmental, and biomedical fields.

5.2 TREE GUM–BASED NANOPARTICLES (NPs) AND THEIR MULTIFACETED APPLICATIONS

Polymers, either natural or synthetic supported nanomaterials, possess uniques physicochemical, structural, morphological, and biological properties. Tree gums have been exploited to synthesize and stabilize metal, bimetallic, metal oxide, and carbon-based NPs, which provided multidimensional applications in catalysis, medicine, tissue engineering, biotechnology, energy harvesting/storage, biological labeling, environmental, water treatment, photonics, and optoelectronics (Kattumuri et al., 2007; Padil, Wacławek, and Černík, 2016; Kalaignana Selvi, Mahesh Kumar, and Sashidhar, 2017; Axiak-Bechtel et al., 2014).

5.2.1 SILVER (AG) NPS BASED ON TREE GUMS AND THEIR PROGRESS IN MULTIDIMENSIONAL AREAS

Most of the tree gum polymers, such as GA (gum arabic), GK (gum Karaya), tragacanth gum (TG), Kondagogu gum (KG), guar gum (GG), cashew gum (CG), etc., have been extensively employed for the "green" synthesis and stabilization of Ag NPs. The tree gum polymers provided a reducing agent and a capping agent for the synthesized Ag NPs, which prevented them from particle aggregation (Sanyasi et al., 2016). The size, shape, and morphology of various Ag NPs synthesized depend on many factors such as preparation and reaction environments (contact time, concentration, pH, temperature, etc.). The green synthesis of NPs based on gum polymers have shown tremendous advances in medicine and biotechnology, such as antimicrobial agents, drug carriers, wound dressings, and medical equipment (Rajeshkumar, Bharath, and Geetha, 2019).

An Ag nanocomposite was prepared using GG and gelatin as template and maltose as reducing agent and developed for antibacterial against pathogenic bacteria such as *S. aureus* (*Staphylococcus aureus*), *E.coli* (*Escherichia coli*), and *P. aeruginosa* (*Pseudomonas aeruginosa*) (Khan, Kumar, and Kumar, 2020). This synthesis protocol's significant advantage relies entirely on greener medium such as water, and no energy and pH maintenance was required in the reduction process. The particle diameter of the Ag NPs was estimated to be 6–18.6 nm range and crystalline nature by TEM and XRD analysis, respectively.

The γ-ray irradiation-induced synthesis of bimetallic (Ag-Au NPs) nanoparticles using GA as a capping agent were illustrated (El-Batal et al., 2020). The ensued Ag-Au bimetallic NPs had potential antifilm and antimicrobial efficacy against *B. subtilis, E. coli,* and *C. tropicalis,* isolated from diabetic foot infections. The Ag-Au NPs formation depended on precursor concentration, and the size of the particle and morphology were determined to be 18.6 nm, spherical, and crystalline. This research highlighted that the produced Ag-Au NPs had the highest stability at low concentrations of NPs, encompassing promising future drug and pharmaceutical research applications.

Cashew gum (CG)–based Ag nanocomposite, with palygorskite (Pal) clay as a supporting matrix, was synthesized (Araújo et al., 2020). The CG acted as a reducing and stabilizing agent in this process. Further, additional Pal enhanced the antibacterial activity (70.2% against *E. coli* and 85.3% against *S. aureus*) of nanocomposites (Ag NPs-CG/Pal) compared to the Ag NPs-CG matrix.

Gum polymers have been utilized for the development of NPs-based colorimetric sensors for intelligent food packaging. Herein, gellan gum-based Ag NPs have been developed and employed for colorimetric H_2S sensors based on silver nanoparticles (Ag NPs) as a meat spoilage indicator for intelligent packaging (Zhai et al., 2019). The developed H_2S sensor showed high selectivity, efficiency (limit of detection of 0.81 μM for detecting H_2S), practicality, low toxicity, and less cost and functions as a distinctive local surface plasmon resonance (LSPR) property of Ag NPs. Additionally, GG provided highly stability, reducing, and capping agent for the Ag NPs.

Bimetallic (Ag/Cu) NPs incorporated GG-based composite films have been developed for active food packaging (Arfat et al., 2017). The addition of NPs to the gum matrix improved their physical, chemical, mechanical, thermal, and rheological properties (Taner et al., 2011). The addition of bimetallic NPs further enhanced the antibacterial, catalytic, and optical properties of the polymeric films (Tan and Cheong, 2013). The developed GG-Ag/Cu nanocomposite films had high tensile strength, barrier properties against U.V. light, and antibacterial potency against *L. monocytogenes* and *S. typhimurium*. The GG-Ag/Cu nanocomposite films' overall properties are capable of using them as food packaging films.

5.2.2 Au NPs Based on Tree Gums

Many tree gum-based NPs, such as Ag, Au, Pt, Pd, Cu, CuO, Fe_3O_4, etc., have been recognized as "green" catalysts for many potential applications. The Au NPs are fabricated by various polymeric gums such as gellan gum, hyaluronic acid, gum Arabic, kondagogu gum, karaya gum, tragacanth guar gum, and pectin have been deliberated and expanded in environmental, catalysis, biomedical, cancer treatment, sensors, tissue engineering, and food packaging areas (Veisi et al., 2018; Dhar et al., 2011; Kemp et al., 2009; Renuga Devi et al., 2015; Pandey, Goswami, and Nanda 2013; de Almeida et al., 2020).

Sustainable catalysis based on tree gums are gaining prominence in the areas of environmental catalytic degradation of pollutants. A nanocomposite was developed using Au NP/pectin functionalized Fe_3O_4 as a catalyst to reduce nitroarenes to amines (Y. Li et al., 2020). The pectin functionalized as a reducing agent to reduce the Au salt to Au NPs and further stabilized it by ligation. The Fe_3O_4 NPs provided strong magnetic interaction, which promoted the separation of nano-catalyst efficiently and conserved the catalytic performance in several cycles. Further evaluation of this nanocomposite for their cytotoxicity against human lung cancer cell lines found their anticancer potential interrelated to antioxidant properties.

Multidimensional NPs Conjugate (FITC-SiO2@Au-DOX-GGS-FA NPs) based on fluorescent dye-loaded SiO_2 and Au NPs, GGC (guar gum succinate), folic acid, and doxorubicin could be excellent nano-platforms for dual C.T./fluorescence cellular imaging, and cancer therapy was advanced (Rajkumar and Prabaharan, 2020). The stability, direct tumor detectability, targeting, superior cellular uptake, cytotoxicity concerning the HeLa cells, and circulation in the prepared NPs conjugate in the bloodstream were enhanced.

Another research based on the expansion of dual biopolymers such as GA and chitosan conjugated with Au NPs was prepared for drug carriers and antibacterial action (Raja Namasivayam, Venkatachalam, and Arvind Bharani, 2020). The ensued composite had shown biocompatibility and anti-quorum sensing activity against a bacterium (*Pseudomonas aeruginosa*) that highlighted their advances in biomedical drug development.

The tree gum polysaccharide and protein complexes such as GA and whey protein isolate (WPI) with iron NPs were established as iron enrichment complexes, specifically in the food emulsion system, for improving the deficiency of iron in human beings (Yao et al., 2021). The formed complexes deliver ferric ions under control through pH and heat treatment and their adsorption ability on the oil-water interface with significantly higher steric hindrance and much lower interfacial energy.

Carbon-based nanoarchitecture based on tree gums has shown tremendous advances in many research fields, such as biomedicine, food packaging, and drug development. Among nanoscale-based carbonaceous materials, graphene oxide (GO) provides remarkable breakthroughs by incorporating polymers due to their outstanding properties, such as high surface area, hydrophilicity, and significant functionalities modification/interaction, and easy preparation (Joz Majidi et al., 2019; Bagheri, Jafari, and Eikani, 2019). In this context, GA and chitosan (Ch) complex NPS were formed via "ionic gelation" (IG), and the assembled nanostructures interacted with GO to construct 3D nanocomposites (Rajabi et al., 2020). The GA/Ch NPs via interaction between –COOH group of GA with –NH2 group of Ch by ionic interaction and hydrogen bonding. The ensued in forming micro/nano-size structures, depending on the factors such as pH and GA's weight ratios to Ch. The supervened nanocomposite would be a potential application in adsorption, filtration, and purification.

5.2.3 METAL OXIDE–TREE GUM NANOPARTICLES

As a chelating and stabilizing agent, GG's involvement in the synthesis of NiO nanoparticles and their nitroarenes' catalytic performance in the presence of sodium borohydride was demonstrated (Baranwal et al., 2018). The ensued NiO NPs had a particle size of 3 nm; crystalline, high surface area (30.45 m^2/g), respectively. The conversion of 4-nitrophenol to 4-aminophenol took 2 min, which followed zero-order kinetics. The catalytic activity of the NPs was shown still active during the storage of the NPs at room temperature even after eight months.

A combination of GG, Pd NPs, and cellulose nanofibers consists of stimuli-responsive hydrogel prepared and utilized for catalytic transformation of Suzuki coupling reactions between aryl bromides phenyl boronic acids (B. Wang et al., 2019). The prepared Pd NPs had a particle size of 3 nm and exhibited outstanding catalysis efficiency (over 85% yield within 2 h), and showed excellent recyclability. The whole catalytic reaction is controlled via a simple heating/cooling approach that will lead to the sol-gel transition, allowing the catalytic reactions and reuse/recycling of the catalysts.

5.3 ELECTROSPUN FIBERS BASED ON TREE GUM POLYMERS

Tree gum polymers' electrospinning is gaining prominence, and "green electrospun fibers" based on these polymers found extensive application in medical, biotechnology, and tissue engineering fields. There are many fabrication methods for nanofibers, such as side-by-side and coaxial electrospinning for composite nanofiber synthesis. Tree gum nanofibers' extraordinary performance, such as porosity, large surface area, functional fibers, mechanical attributes, and thermal stability, benefits their unique applications (Akshay Kumar et al., 2021). The diameter and surface smoothness are depending on the electric charge, gum concentration, and viscosity of the gum composites. Some active components and metallic nanoparticles are incorporated in action films, bioremediation membranes, and biomedical scaffolds (Prabhu, 2019). The bioactive molecules or drugs are even encapsulated with tree gum-based electrospun fibers before or after electrospinning could be significantly utilized to develop novel drug delivery systems (Bhattarai et al., 2018). TEM analysis, SEM (scanning electron microscopic) analysis, TGA (thermogravimetric analysis), DSC (differential calorimetric analysis) analysis, and FTIR (Fourier transform infrared analysis) are usually characterized the nanofibers.

Large numbers of tree gums to produce nanofibers which have been elaborated in recent times, are impending owing to their excellent material goods such as excellent water vapor permeability, a pronounced surface to volume ratio, extraordinary porosity, and brilliant pore-interconnectivity, water absorption rate, biocompatibility, hydrophilicity, etc. are suitable materials for tissue engineering applications.

The combination of GA, Ag NPs, with other biocompatible and biodegradable synthetic polymers (PVA and PCL) were electrospun into nanofibers (average diameter = 100–600 nm range) for the application of wound dressing (Eghbalifam et al., 2020). In these composite nanofibers, Ag could act as an antimicrobial agent. PVA and PCL function as additives for enhancing the electrospinning process and improving the ensured fibers' mechanical attributes. The developed fiber mat showed porosity (22–49%), water absorption (119–540%), and water vapor permeability (1,998–2,322 g m^{-2} Day^{-1}), respectively. Furthermore, the fiber mat explored their potential antibacterial activity towards common wound infectious bacteria such as *S. aureus, E. coli, P. aeruginosa*, and fungal strain (*C. albicans*). It demonstrated that the developed nanoparticle composite fibers would be impending wound healing.

GA's importance as a viscosity enhancer in the electrospinning of a mixture of gelatin/chitosan/PVA in an acetic acid/water solvent system was established (Tsai et al., 2015). This research's significant impacts highlighted the utilization of greener solvents, higher amount of natural polymeric content (both gelatin and chitosan) incorporation into the final nanofiber composites, and avoiding the toxic and expensive solvents (e.g., trifluoroacetic acid or hexafluoroisopropanol) for the production of gelatin-chitosan fibers. The produced nanofibers with a higher amount of natural polymeric content have precious tissue engineering benefits (Tsai et al., 2015).

Electrospun drug delivery fiber mat scaffold comprised of TG, zein, PLA, and tetracycline hydrochloride (TCH) was fabricated for antibacterial wound dressing (Ghorbani et al., 2020). The fiber mat diameter (253 ± 15.3 to 547 ± 56.4 nm) with consistent pores had influenced the blending ratio of zein/TG in the electrospinning mixture. The introduction of PLA into the scaffold mat enhanced the overall mechanical properties and biodegradability. Additionally, the cytocompatibility, cell adhesion study (against NIH-3 T3 fibroblast cell), and antibacterial against gram-positive and gram-negative bacteria (*S. aureus*, and *Pseudo aeruginosa*) exposed the developed nanofibrous mat for an efficient delivery route of TCH in wound healing application.

In another research, a combination of TG, PVA, GO (graphene oxide), and TCH nanofiber with fiber diameter (~100 nm) was constructed via electrospinning for a transdermal drug delivery system (Abdoli et al., 2020). Due to their solubility in an aqueous medium, large surface functionality, surface area, and biocompatibility endorsed the ensued fiber scaffold for the biomedical area. Further, the mechanical attributes, antibacterial, cytotoxicity (high cell viability of nanofiber against the normal human cells from umbilical vein endothelial cells), and biocompatibility studies suggest that this nanocomposite fiber could be a promising candidate for controlled drug delivery.

The Co-electrospinning technique was used to generate composite electrospun fibers with fiber diameters (131.6 ± 27.5 nm) based on TG, PCL, and PVA (Zarekhalili et al., 2017). The ensued nanofibers tested for antibacterial activity against *E. coli* (gram-negative bacteria), *S. aureus* (Gram-positive bacteria). It was revealed that *E. coli* was more susceptible to nanofibers due to the action of sugar residues such as (L-arabinose and L-fucose) present in TG (Zarekhalili et al., 2017). Further, the nanofibers exposed cell proliferation and registered non-toxicity towards the MTT assay and NIH3T3 cells.

For the well-organized nanofibers in the field of nerve tissue regeneration, an aligned and random nanofiber scaffold was constructed on blending with TG and PLLA (poly (L-lactic acid)) with a PLLA/TG ratio of 75:25. The organized nanofibers had outstanding physicochemical and mechanical properties.

Curcumin incorporated nanofibers consist of TG, PCL was fabricated via electrospinning employing acetic acid/water (90% v/v) at room temperature (Ranjbar-Mohammadi et al., 2016). Systematic release studies of curcumin wound healing potential and antibacterial effectiveness of the produced nanofibers were carried out. The TG/PCL/curcumin electrospun fibers were much effective in healing diabetic wounding created in rat models.

The electrospun scaffold of TG/PCL[(3:1; 7% (GT), 10–20% (PCL)] with excellent antibacterial attributes, mechanical strength, porosity, hydrophilicity, and degradation behavior were contrived (Ranjbar-Mohammadi et al., 2016). The developed nanofibers had an average diameter of 156 ± 25 nm and tested their cytotoxicity, and antibacterial potencies of these TG/PCL fibers demonstrated their flawless skin and wound healing scaffolds properties.

In other research, TG combined with PVA at different GT/PVA ratios such (GT/PVA: 0/100 to 100/0) and optimized the blended ratio of GT/PVA (40/60) for the smooth, beadles free, uniform diameters of nanofibers production via electrospinning (Ranjbar-Mohammadi et al., 2016).

Further, the antibacterial efficacy of GT/PVA was found to be excellent towards gram-negative bacteria (*P. aeruginosa*). The cell proliferation and biological compatibility of the nanofibers were also shown to be fabulous.

Electrospinning types of both blending and coaxial were introduced to construct conjugates composites of TG, PLGA, and TCH, tetracycline hydrochloride (as a hydrophilic model drug) (Ranjbar-Mohammadi et al., 2016). Three variants of electrospun fibers, such as pristine PLGA, PLGA/TG core-shell, and PLGA/GT-TCH core-shell, were fabricated N-2-hydroxyethyl piperazine (HEP) as a solvent. The higher amount of TG in the blended mixtures resultant in nanowires with smaller fiber diameters. The stability of nanofiber, drug-releasing behavior, cell culture, and proliferation kinetics of nanowires were systematically premeditated. The PLGA/GT-TCH core-shell nanofibers were found to be exceedingly proficient in the treatment of periodontal disease.

Electrospinning of GK blended with PVA in a greener environment has been done. The blend ratio for producing nanofibers with excellent uniformity, shape, and size was maintained by mixing PVA with GK using various weight combinations (Padil et al., 2015; Padil et al., 2016, and Padil et al., 2018). Greener technologies, such as heat and plasma treatments to improve the nanofiber performance to endorse cross-linking, water-resistance, water contact angle, BET surface area, the porosity of the fibers, all are substantiated to be successful. Furthermore, plasma-treated electrospun fibers displayed potential antibacterial influence against gram-negative and positive bacterial colonies.

KG fibers were produced via electrospinning of the blend mixtures of KG with PVA. Further evaluation of KG electrospun fibers (both untreated and plasma-treated) was also applied for antimicrobial efficiency tests, indicating that plasma-treated fibers have exhibited enhanced antibacterial potency against *E.coli* and *S. aureus* (Padil et al., 2015).

5.4   TREE GUM POLYMER REINFORCED FILMS

To develop a sustainable and "green" environment, bioplastic films based on renewable sources are the only alternative solution for replacing petroleum-based polymers in the context of growing the public demands for food and packaging industries. To develop a recyclable, improving the shelf life of food and environmentally viable food packaging system is critical and essential for this juncture. Many bio-based materials sourced from varieties of plant, animal, seaweeds, and agricultural wastes have recently come up as films or fibers as low-cost and biodegradable formulations for many potential applications. Polymeric blends, such as natural polymers with synthetic ones, are chosen compared to individual polymers to enhance the physical, chemical, barrier, mechanical, and thermal properties (Trikkaliotis et al., 2020; Kanatt and Makwana 2020; Bano et al., 2019). Edible coating and films consist of polysaccharides, proteins, lipids, or a mixture of these materials on foods have been established to improve the properties such as moisture resistivity, oxygen barrier properties, temperature, and UV resistance and also enhanced the shelf life of the food as well as improved the organoleptic, nutritional, and microbiological properties of food products (Khedri et al., 2021;Khezerlou et al., 2019; Kouhi, Prabhakaran, and Ramakrishna, 2020; Mohammadi Nafchi et al., 2017). Gum polymers such as alginate, agar, xanthan, carrageenan, guar, pectin, arabic, karaya, kondagogu, larch gum, and locust bean gum have been widely used to produce edible films (Zibaei et al., 2021). The search for new tree gums and their utilization for making edible films for food packaging is underway, and the progress of this research is being carried out. Many emerging gum polymers such as Persian Gum (*Amygdalus scoparia Spach*), Balangu (*Lallemantia royleana*) seed gum, Qodume Shirazi (*Alyssum homolocarpum*) seed gum, Sage (*Salvia macrosiphon*) seed gum, Brea (*Cercidium praecox*) gum, cashew (*Anacardium Occidentale L.*) gum, Opuntia ficus-indica and Tragacanth gum (*Astragalus gummifer Labillardiere*), kondagogu gum, and karaya gum are in this competition. The essential parameters for selecting various gums depend on their different functionality in structural and composition, functional groups, gelling properties, molecular weight and ionic charge of the molecule, rheological behavior, film formability, etc. The barrier properties such as water vapor, oxygen, and CO_2 are the critical concerns for edible films' progress. Their values depend on several factors, such as molecular structure and molecular weight of the concerned polymer, crystallinity, ratio between hydrophobicity/hydrophilicity of the films, functional groups interaction in the polymers, additives/plasticizer present in the films (Davachi et al., 2018; Davachi and Shekarabi, 2018).

Active food packaging film based on the combination of GG, chitosan, and PVA was prepared via solution casting (Bhat et al., 2021). Hydroxy citric acid (HCA) was used as the cross-linking agent, and the films were investigated their physical, chemical, mechanical, barrier, thermal, water vapor permeability, and antibacterial properties. The developed films had lesser water solubility and a reduced degree of swelling corresponding to increased cross-linking density. Further explored their property such as antibacterial efficiency against *S. aureus* and *E. coli* bacteria and barrier against UV radiation, highlighted the potentiality in the packaging of foods.

Many studies have been reported that GT can perform edible food coating agents to prevent food spoilage as a function as a semipermeable membrane (Jafari, Hojjati, and Noshad, 2018) (Tonyali, Cikrikci, and Oztop, 2018). The combination of natural polymers such as GT and chitosan blended with GO and the biocomposite film was prepared by solution casting. The ensured biocomposite film had higher mechanical and barrier properties. GT also combined with other natural sources such as soy protein/nanocellulose, potato starch, and chitosan to form biocomposite films for food packaging (Nazarzadeh Zare, Makvandi, and Tay, 2019).

A flexible food packaging film based on KG and sodium alginate has been fabricated via solution casting by mixing the natural polymers' optimum ratios (Ramakrishnan et al., 2021). Both polymer solutions were prepared in aqueous environments and used glycerol as a plasticizer to prepare the biocomposite films. The addition of sodium alginate facilitated overcoming the brittleness nature of the KG and further improved the flexibility, surface texture, and elastic nature of the composite films. Additionally, the composite films showed high mechanical attributes (tensile strength = 24 MPa), increased water contact angle (81°), barrier properties, and film transparency. The preparation and film scheme is presented in Figure 5.1. Further evaluation of the KG-sodium alginate films was found to be biodegradable in a natural environment within one month. (Table 5.1)

5.5 SPONGES BASED ON TREE GUM POLYMERS

Natural biopolymers such as KG polysaccharide and a protein (gelatin) were blended and freeze-dried to form a 3D KG-gelatin sponge (Rathore et al., 2016). The properties of the ensued sponges were carried out by SEM, FTIR, thermal study, swelling capacity, and their in-vitro drug loading/release mechanism and antibacterial efficiency were assessed. The KG-gelatin sponge exhibited 95% of drug release (ciprofloxacin) at the end of 48 h and antibacterial efficacy against *S. aureus* and *E. coli,* respectively. Further, the sponge explored their biocompatibility by testing cell adhesion and proliferation using NIH 3T3 fibroblast and Human keratinocytes cell lines. The overall result highlighted that the developed sponge would be the potential would dressing scaffold.

Many natural or synthetic-based sponges have been utilized for environmental pollution remediation, such as oil cleaning from water, absorption of toxic dyes or organic pollutants, and toxic metal removal from water. Many places in the universe have severe environmental and ecological problems created due to oil spillage or pollution. Several methods such as biodegradation, burning, mechanical collection, and chemical treatment are some of the most common techniques employed. However, the permanent and safe way of cleaning the oil or organic pollutants removal is still a highly global problem. Efficient cleaning of oil pollutants from water, mechanically robust, and environmentally friendly materials with high selectivity for oil and porous materials are required. Natural-based polymers sponges also have certain disadvantages such as low mechanical strength, hydrophilicity, brittleness, etc. To overcome these, chemical modification or surface functionalization with certain groups or molecules can enhance sponges' overall functionality.

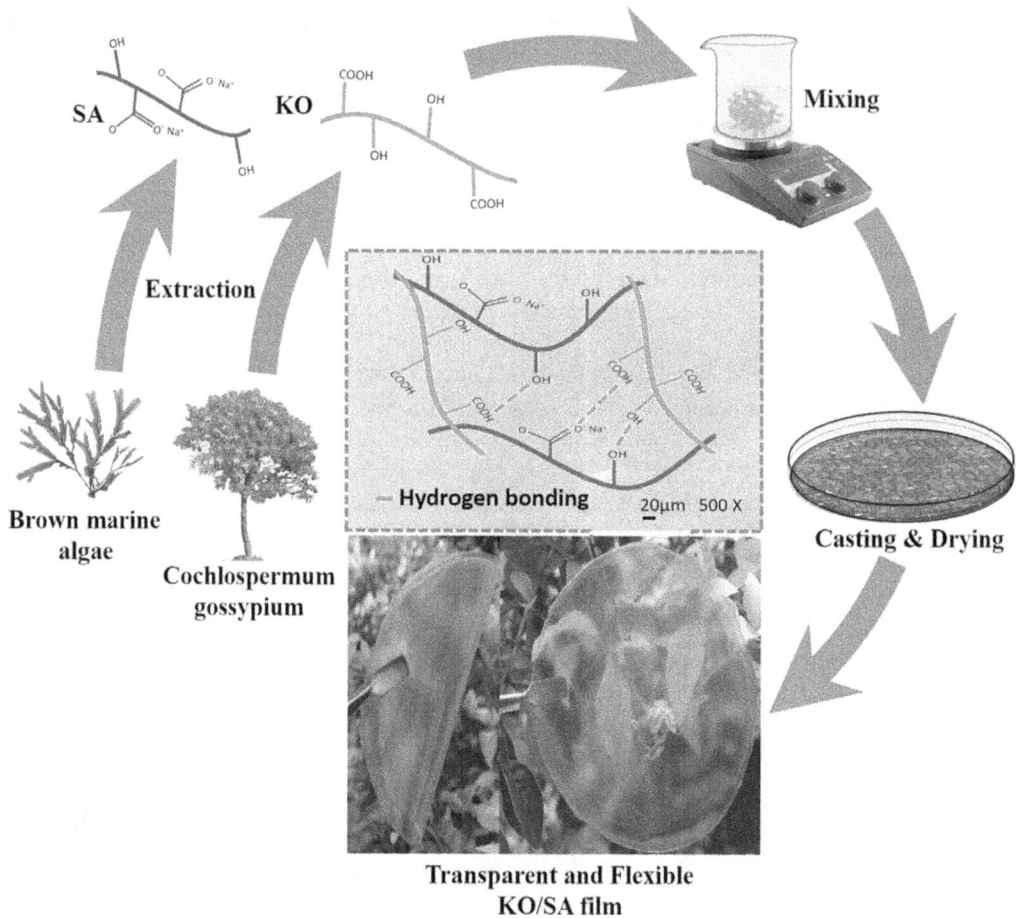

FIGURE 5.1 Schematic representation of the fabrication of films based kondagogu gum-sodium alginate bio-based polymers. Permission for reproduction from Elsevier publications (Ref. Ramakrishnan et al., International Journal of Biological Macromolecules, 177, 2021, 526–534).

Recently, a bio-based tree gum (KG) with sodium alginate (a seaweed polysaccharide) sponge was prepared. Their hydrophobic functionalization has been carried to apply selective oil/organic solvent removal (Ramakrishnan et al., 2021). The ensured silylated sponge was characterized by spectroscopic and microscopic analysis and found to be higher is exceptionally hydrophobic (water contact angle >133°), with an exceptionally low density (~18.4 mg/cm³), and high porosity (94.1%). The preparation and removal mechanism of oil/organic compounds by the KG-sodium alginate sponge is presented in Figure 5.2. Further. The ensued sponge had the potential to absorb organic solvents and oils up to 19–43 times its weight, high reusability, and could use continuously more than nine absorption-squeezing cycles. Further exploration of the bio-based sponge for their biodegradation study (determining biological oxygen demand) observed that both pristine and silylated sponge were biodegradable for over 28 days by 92%, 76%, respectively, at natural bacterial strains found in wastewater.

5.6   3D PRINTING OF BIO-BASED POLYMERS

Three-dimensional (3D) printing or additive manufacturing (AM) is a vibrant technique to fabricate 3D architectures with precise shapes and structures with versatile properties (tensile, impact,

TABLE 5.1
Tree Gum–Based Films, Preparation, Properties, and Applications

Tree Gum Polymers	Preparation Method	Properties and Applications	Reference
Cassava starch, cashew tree gum (CTG) and carnauba wax	Emulsion based casting	i. Enhanced water vapor barrier, water resistance, and opacity. ii. Food packaging	(Rodrigues et al., 2014)
Karaya gum/Schisandra chinensis oil/oleogel film	Casting	i. Reduced the water vapor permeability, improved elongation at break, and antioxidant property. ii. Edible packaging	(Yousuf, Wu, and Gao, 2021)
Cordia dichotoma gum/Salvia Mirzayani essential oil	casting	i. Enhancement of moisture content, water-solubility, ultimate tensile strength, water vapor permeability, percentage of light transmission ii. Antioxidant and antibacterial activity of the films. iii. Edible films	(Hasheminya and Dehghannya, 2021)
Gelatin/Tragacanth gum/Persian gum blended films	Solution casting	i. Enhanced tensile strength and elongation at break of blended films. ii. Edible films	(Khodaei, Oltrogge, and Hamidi-Esfahani, 2020)
methoxyl pectin/tara gum/ellagitannins.	Solution casting.	i. Improved water resistance and mechanical properties. ii. Antioxidant activity and antibacterial action against *Escherichia coli* and *Staphylococcus aureus*. iii. Edible food packaging.	(Y. Chen et al., 2020)
PVA/TG/Gallic acid blend films	Solvent casting technique	i. Higher thermal stability, tensile properties, WVTR, the water solubility of the blended films. ii. Films showed soil and biodegradation and antioxidant property. iii. Food packaging material.	(Goudar et al., 2020)
Chitosan/gum Arabic/cinnamon essential oil (CEO)/ and clove essential oil (CLO) -based polyelectrolyte complexed films	Solvent casting	i. Lower tensile strength, higher elongation, enhanced water barrier properties, and thermal stability. ii. Antibacterial film.	(Xu et al., 2019)

(Continued)

TABLE 5.1 (Continued)
Tree Gum–Based Films, Preparation, Properties, and Applications

Tree Gum Polymers	Preparation Method	Properties and Applications	Reference
GA reinforced cellulose nanocrystals film	Casting method	i. Tensile strength (2.21 MPa) and elongation at break (62.79%) of film incorporated with 4% (w/w) CNC. ii. Enhanced water vapor (8.48 to 7.58 × 10-10 g Pa$^-$ s^{-1}m^{-1}) and oxygen permeability by 10.61% and 25.30%, respectively, improved ultraviolet light barrier and film's thermal stability. iii. Eco-friendly food packaging film.	(Kang et al., 2021)
Polyvinyl alcohol (PVA), gum arabic (GA) and chitosan (CS) incorporated black pepper essential oil (BPEO) and ginger essential oil	Casting method	i. Highly flexible with improved heat stability. ii. Potential antibacterial action against *Bacillus cereus*, *Staphylococcus aureus*, *Escherichia coli*, and *Salmonella typhimurium*. iii. Applications in wound dressing and food packaging.	(Amalraj et al., 2020)
Chitosan/ OSA (Octenyl succinic Anhydride)-modified gum arabic-stabilized cinnamon essential oil films	Solution casting	i. Improved water resistance and reduced tensile strength. ii. Antimicrobial activities of the films against *E. coli* and *S. aureus*. iii. Development of edible film.	(Xu et al., 2020)
GA/wild almond protein film	Casting	i. Higher tensile strength and elongation and reduced water vapor permeability. ii. Food packaging films	(Tahsiri et al., 2019)
Gum arabic (GA) and sodium alginate (SA), and whey protein concentrate (WPC)	Hydrothermal method	i. Improved mechanical, surface hydrophobicity, and stiffness. ii. Edible food packaging applications.	(Erben et al., 2019)
Chitosan/xanthan gum,/Pullulan/gum tragacanth and Arabic gum bio-composite film	Casting technique	i. Lowered tensile strength and swelling degree, but increased the film solubility. ii. Antifungal effects on a nectarine. iii. Active packaging films	(Lian et al., 2020)

Chitosan/gum Arabic/ZnO	Casting	i. Antibacterial properties against several bacteria, including Staphylococcus aureus, Escherichia coli, and Bacillus subtilis. ii. It has improved the quality and shelf life of bananas at a temperature of 35°C and relative humidity (RH) of 54%. iii. Edible coating composite for banana preservation.	(La et al., 2021)
Zedo gum/PVA	Casting	i. Blend PVA/ZG films produced homogeneous, flexible, and transparent films. ii. Reduced moisture content, water solubility, and water vapor permeability were observed in the blended films.	(Razmjoo et al., 2021)
Persian gum/gum tragcanth/gelatin	Casting	i. Edible film for packaging of foods like fruits and vegetables as well as for pharmaceutical industry. ii. There is a high water vapor barrier, Young modulus and tensile strength, elongation at break, and water vapor permeability. iii. Blended edible films.	(Khodaei, Oltrogge, and Hamidi-Esfahani, 2020)

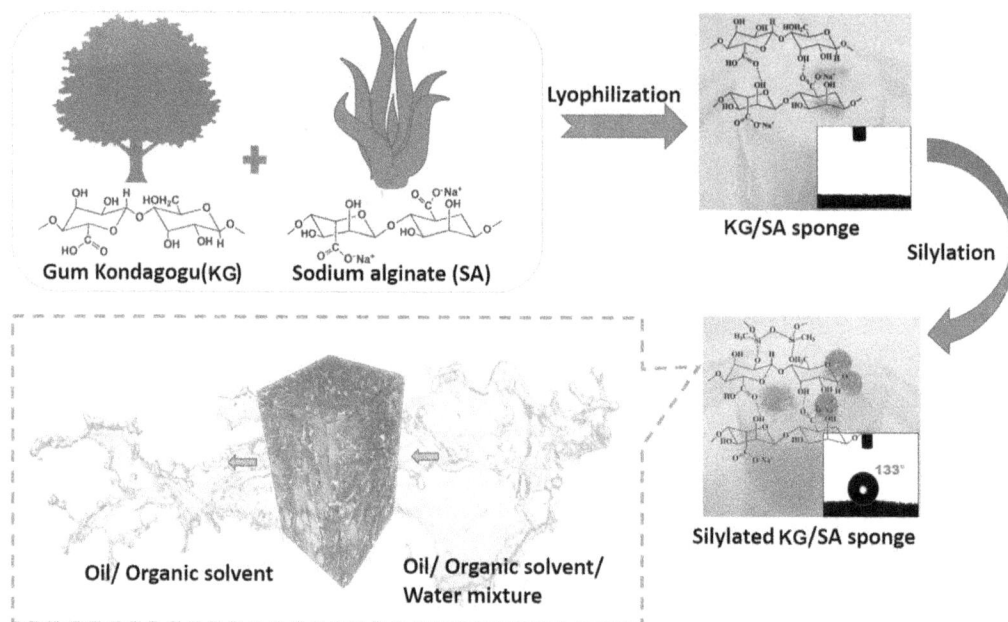

FIGURE 5.2 Fabrication of silylated sponge based on the combination of gum kondagogu and sodium alginate polymers and their application in oil/water separation. Permission from Rohith K. Ramakrishnan, Vinod V. T. Padil, Marcela Škodová, et al, Hierarchically Porous Bio-Based Sustainable Conjugate Sponge for Highly Selective Oil/Organic Solvent Absorption, Advanced Functional Materials, DOI: https://doi.org/10.1002/adfm.202100640; Publisher: John Wiley and Sons.

flexural, thermal, and morphological) (Ji et al., 2020). Recently, 3D-printed materials are widely used in aerospace, medical equipment, Li-ion battery, biosensors, automotive, biotechnology, tissue engineering, and construction (Kandambath Padinjareveetil Akshay Kumar and Pumera, 2021) (J. Li and Pumera, 2021) (Tetsuka and Shin, 2020) (Akshay Kumar et al., 2020) (Gao and Pumera, 2021) (Ligon et al., 2017). The advantages such as fewer consumables, no waste generation, cost-effectiveness, rapid fabrication, reproducibility, and automation enlighten the 3D-printing technology at a higher level in compared to any other fabrication methods (Dai et al., 2019) (Yang et al., 2020). Recent progress in the 3D printing of bio-based materials (proteins and carbohydrate polymers) in bioengineering, pharmaceutical, and food sectors have been reported (Shahbazi and Jäger, 2021) (He, Zhang, and Fang, 2020).

Formulation of printable ink based on electrochemically active 2D materials is interesting. This developing technique of solution-processed 2D-printing materials can accelerate this process by allowing additive patterning for flexible device strategy, a comprehensive method, high-speed and cost-effective engineering. Assortment of printable ink is a crucial component in 3D-printable architectures, determining the final object's structure and geometry for their respective applications in various fields.

Developing suitable ink is a critical aspect before printing. It is essential that the exfoliated 2D materials/2D material flakes have to be made into a stable colloidal dispersion. 3D-printed structures using 2D material inks from various tree gums are at their early stage. For rapid prototyping and advanced manufacturing, the capacity to transform materials into devices through printing has been considered one of the most promising and emerging solutions. Over the years, large varieties of printing strategies based on nanomaterials, including metals, semiconductors, and insulators, have been developed for a significant number of applications.

The principal tool for preparing ink and fabrication of various 3D architectures via 3D printing depends on (i) Ink preparation and their rheological characterization, (ii) formulation of printable ink with additives, fillers, binders, and solvents.

Rheological properties (viscosity and dynamic moduli), flow behavior, chemical structure, degree of chain entanglement, and crossing effect among the blended polymers or additives of the prepared ink for 3D printing are the main task for the successful printing of structures of different dimensions (Karavasili et al., 2020). The ink rheological properties rely on the deformation and flow of materials throughout the printing process.

The relation between various rheological parameters such as shear stress (τ), apparent viscosity (η), and shear rate ($\dot{\gamma}$) is given by Herschel–Bulkley model.

$$\tau = \eta\gamma = \tau_y + K\gamma^n \tag{5.1}$$

where the shear viscosity (τ), shear rate ($\dot{\gamma}$), the viscosity parameter (K), and the power-law exponent (n) determine applied shear stress (τ), yield point (τ_y) of the ink, and the flow behavior.

Generally, printable ink with distinct rheological behaviors is used; (i) low viscosity inks, (ii) shear-thinning inks, and (iii) dynamic moduli.

The significant advantage of the low viscosity inks is that they exhibit Newtonian fluid behavior (e.g., thermoplastic polyurethane (TPU)) where we find an effortless/smooth movement of ink down the nozzle and prompt curved to generate 3D structures via gelation or photo-polymerization comportments (Valentine et al., 2017; Friedrich and Begley, 2018; M'Barki, Bocquet, and Stevenson, 2017). The other type of printing inks is shear-thinning inks exhibiting non-Newtonian behavior of fluids where the viscosity decreases upon increasing shear rates, i.e., they demonstrate a constant and steady ink flow through the nozzle under pressure.

Thermoplastic polyurethane-silver composite inks and graphene oxide-based electrode inks (lithium-ion phosphate/graphene oxide and lithium titanate oxide/graphene oxide) are included in such type of 3D printable inks recently used for fabrication of 3D structures and applied for lithium-ion batteries and soft electronics (Fu et al., 2016).

Dynamic moduli play a prime role in printable ink development. It is compartmentalized to the storage modulus (G') and loss modulus (G''). For consistent 3D-printable geometries, the printable ink must have enhanced storage modulus and yield strength (Compton and Lewis, 2014). Further, the value of both elastic and viscous modulus of the printable inks (G' and G'') also determine the rate of deformation after extrusion, as with excellent shear stress, $G' > G''$, such that the printable ink are deformable under high shear stress during extrusion (Tang et al., 2018).

The composition of functional printable ink for 3D printing of 2D materials is pivotal, determining the printing procedures and final printed strategies. Table 5.2 reflects the critical ink components (fillers, binders, additives, and solvents), compositions, rheological attributes, printing techniques, and applications of printed natural bio-based polymers are summarized.

5.7 CONCLUSION AND FUTURE PROSPECTIVE

Tree gum polymers are natural hydrocolloids, which are renewable, non-toxic, and biodegradable. It can be extracted from botanical, algae, plant, or animal sources, including industrial by-products. Extensive barrier properties, flexibility, and availability of raw materials are the advantages of synthetic polymers. However, lack of biodegradability, low recycling rates, and the migration of hazardous compounds into food are the significant challenges toward employing synthetic packaging materials. Therefore, biopolymers' utilization in the packaging system's development is a promising alternative to artificial packaging systems. Therefore, the development of edible films and coatings from biopolymers is one of the promising techniques for replacing synthetic polymers. The present chapter extensively reviewed the distinctive products such as films, fibers, and sponges based on tree gum polymers for advanced applications.

TABLE 5.2

3D-Printing Techniques and Ink Formulations and Applications of Bio-Based Polymers

Bio-Based Polymers	3D Printing Ink Formulation/Mechanism/Polymer Blends/Other Additives	Printing Techniques and Properties	Ref.
Gellan gum	Ionic- and photo-cross-linking mechanism; chitosan (a polycation)/lithium phenyl-2,4,6-trimethylbenzoylphosphinate.	i. Extrusion; ii. Diameter of microstructures >1 µm and a tensile strength range of ~1 MPa.	(Robinson et al., 2020)
Glycol chitosan	Lactoferrin-loaded carboxymethyl cellulose (CMC)-glycol chitosan (GC) hydrogels; Preparation: Schiff's base reaction & ionic interaction;	i. Continuous extrusion; ii. Stable structure and high biocompatibility. iii. Application of CMC-GC hydrogel ink designed for proteins or drugs for tissue engineering applications.	(Janarthanan et al., 2020)
Carrageenan	Xanthan/starch	i. Extrusion ii. Addition of starch and xanthan gum in κ-carrageenan based inks increased inks' gelation temperature (Tgelation), viscosity (within the shear rate of 0.01–100 1/s), yield stress (G*), enhanced shear-thinning (thixotropic) behavior, and reduced time-dependence of modulus (temperature recovery). iii. Inks' gelation temperature (Tgelation) and time-dependent behavior (gelation time, gel) significantly affected their printability and shape retention performance. iv. 3D food printing.	(Liu et al., 2019)
Alginate	Soy protein isolate/gelatin	i. Extrusion ii. Shear-thinning behavior in soy protein isolate (SPI) mixtures with sodium alginate and gelatin. iii. 3D food printing.	(J. Chen et al., 2019)
Alginate	Gelatin	i. Extrusion-based 3D printing. ii. Gel strength and surface smoothness significantly improved by soaking the 3D-printed geometries in the calcium chloride solution. iii. Bio-scaffold using natural biopolymers to deliver pharmaceutical or supplemental ingredients.	(Y. Wang et al., 2017)

Hyaluronic acid	Hydroxyapatite	i. Laser bio-printing.	(Warner, Norton, and Mills, 2019)
Kappa-carrageenan	Gelatin/Cellulose powder	i. Fused deposition Modeling (FDM). ii. Concentration dependent and increased gelling temperature resultant in inadequate 3D structure. iii. The printing fidelity depended on the magnitude of the storage modulus (>23 kPa) and the rapid formation of an elastic network. iv. Food bio- printing.	(Holland et al., 2018)
Xanthan Gum	bacterial cellulose/nanoclay	i. Binder jetting. ii. Xanthan gum synergistically with cellulose to reinforce the binding mechanism. iii. Functional ingredients for food products.	(Wei et al., 2020)
Alginate	$CaCl_2$	i. Extrusion ii. Excellent printability for complexity and high shape fidelity by incorporation of nano clay. iii. Drug release, biomedical devices, and tissue engineering.	(Vancauwenberghe et al., 2018)
Pectin	Sodium alginate/sodium carboxymethyl cellulose/ xanthan gum/whey protein/taro paste	i. Coaxial and simple extrusion printing. ii. Development of 3D-printed food stimulant.	(Huang et al., 2020)
Guar gum	Pectin/drug lidocaine hydrochloride.	i. Extrusion ii. Mixtures of all the materials improved the rheological characterization and printing behavior with a smooth outlook and less dimensional variability. iii. Application in 3D food printing.	(Long et al., 2019)
Chitosan	Xanthan gum	i. Extrusion ii. High swelling ratio and water absorption of the developed 3D-printed hydrogel. iii. 3D-printed hydrogel for wound dressings.	(García-Segovia et al., 2020)
Konjac gums	Xanthan gum/gelatin/pureed carrots	i. Extrusion ii. Variables for 3D printing- printing temperature and the composition of the product analyzed using rheological and textural characterization techniques. iii. Development of personalized food with specific nutritional characteristics.	(Strother, Moss, and McSweeney 2020)

(Continued)

TABLE 5.2 (Continued)

3D-Printing Techniques and Ink Formulations and Applications of Bio-Based Polymers

Bio-Based Polymers	3D Printing Ink Formulation/Mechanism/Polymer Blends/Other Additives	Printing Techniques and Properties	Ref.
Guar gum	Gelatin/carbon nanotubes	i. Extrusion ii. Food printing.	(L. Li et al., 2020)
Alginate	Nanoclay	i. Extrusion ii. Improved mechanical attributes and printability of the 3D-printed scaffold. iii. Development of 3D-printed vascularized constructs for tissue engineering.	(Shahbazi et al., 2020)
Alginate	Gelatin/chondroitin sulfate/Graphene oxide	i. Extrusion ii. Improved thermal stability and shape of 3D structures via photo-crosslinked irradiation. iii. Adsorptive removal of toxic metals from water.	(Olate-Moya et al., 2020)
Alginate	Gelatin/TiO$_2$/ β-tricalcium phosphate	i. Extrusion ii. Improved shape fidelity and resolution of 3D-printed scaffolds. iii. Application in cartilage tissue engineering.	(Urruela-Barrios et al., 2019)
Alginate	Glycol chitosan/iron oxide nanoparticles	i. Extrusion ii. Enhanced mechanical properties(elastic modulus= 20 MPa) iii. Tissue engineering applications.	(Ko et al., 2020)
Oxidized hyaluronate	Bacterial-cellulose/copper nanoparticle	i. Extrusion ii. Improved self-healing property and shear thinning behavior. iii. (iii) Drug delivery and tissue engineering scaffolds.	(Gutierrez et al., 2019)
Alginate	Gelatin	i. Extrusion printing. ii. Stability, better printability, and antimicrobial potential against *Escherichia coli* and *Staphylococcus aureus*. iii. Tissue engineering; regenerative medicine.	(Amr et al., 2021)

Further, a substantial amount of work has been discussed to develop new solutions for 3D printing using various gums, although some challenges such as creating unique ink formations and their stability, mechanical, reinforcement with other polymers/additives/nanoparticles, and rheological properties, etc. are remaining. However, the technology for 3D-printing tree gum–based formulations has been in constant development with new outstanding properties and applications being reported to develop bio-based biodegradable, environmentally friendly materials due to the recent issues relating to micro-plastics. Such materials are expected to be biocompatible and have a low carbon footprint, recyclable and biodegradable, and less hazardous to the environment.

REFERENCES

Akshay Kumar, K.P., Ehsan Nazarzadeh Zare, Rafael Torres-Mendieta, Stanisław Wacławek, Pooyan Makvandi, Miroslav Černík, Vinod V.T. Padil, and Rajender S. Varma. 2021. "Electrospun Fibers Based on Botanical, Seaweed, Microbial, and Animal Sourced Biomacromolecules and Their Multidimensional Applications." *International Journal of Biological Macromolecules* 171: 130–149. 10.1016/j.ijbiomac.2020.12.205.

Abdoli, Mohadese, Komail Sadrjavadi, Elham Arkan, Mohammad Mahdi Zangeneh, Sajad Moradi, Akram Zangeneh, Mohsen Shahlaei, and Salar Khaledian 2020. "Polyvinyl Alcohol/Gum Tragacanth/ Graphene Oxide Composite Nanofiber for Antibiotic Delivery." *Journal of Drug Delivery Science and Technology* 60:10204410.1016/j.jddst.2020.102044.

Almeida, Débora A. de, Roberta M. Sabino, Paulo R. Souza, Elton G. Bonafé, Sandro A.S. Venter, Ketul C. Popat, Alessandro F. Martins, and Johny P. Monteiro. 2020. "Pectin-Capped Gold Nanoparticles Synthesis in-Situ for Producing Durable, Cytocompatible, and Superabsorbent Hydrogel Composites with Chitosan." *International Journal of Biological Macromolecules* 147 (March): 138–149. 10.1016/ j.ijbiomac.2020.01.058.

Amalraj, Augustine, Józef T. Haponiuk, Sabu Thomas, and Sreeraj Gopi. 2020. "Preparation, Characterization and Antimicrobial Activity of Polyvinyl Alcohol/Gum Arabic/Chitosan Composite Films Incorporated with Black Pepper Essential Oil and Ginger Essential Oil." *International Journal of Biological Macromolecules* 151: 366–375. 10.1016/j.ijbiomac.2020.02.176.

Amr, Mahmoud, India Dykes, Michele Counts, Joshua Kernan, Alia Mallah, Juana Mendenhall, Bernard Van Wie, Nehal Abu-Lail, and B. Arda Gozen. 2021. "3D Printed, Mechanically Tunable, Composite Sodium Alginate, Gelatin and Gum Arabic (SA-GEL-GA) Scaffolds." *Bioprinting* 22 (February): e00133. 10.1016/j.bprint.2021.e00133.

Araújo, Cristiany Marinho, Moisés das Virgens Santana, Antonio do Nascimento Cavalcante, Lívio César Cunha Nunes, Luiz Carlos Bertolino, Carla Adriana Rodrigues de Sousa Brito, Humberto Medeiros Barreto, and Carla Eiras. 2020. "Cashew-Gum-Based Silver Nanoparticles and Palygorskite as Green Nanocomposites for Antibacterial Applications." *Materials Science and Engineering: C* 115 (October): 110927. 10.1016/j.msec.2020.110927.

Arfat, Yasir Ali, Mohammed Ejaz, Harsha Jacob, and Jasim Ahmed. 2017. "Deciphering the Potential of Guar Gum/Ag-Cu Nanocomposite Films as an Active Food Packaging Material." *Carbohydrate Polymers* 157 (February): 65–71. 10.1016/j.carbpol.2016.09.069.

Axiak-Bechtel, Sandra, Anandhi Upendran, Jimmy Lattimer, James Kelsey, Cathy Cutler, Kim Selting, Jeffrey Bryan, et al. 2014. "Gum Arabic-Coated Radioactive Gold Nanoparticles Cause No Short-Term Local or Systemic Toxicity in the Clinically Relevant Canine Model of Prostate Cancer." *International Journal of Nanomedicine*, 5001: 5001–5011. 10.2147/IJN.S67333.

Bagheri, Mahsa, Seid Mahdi Jafari, and Mohammad H. Eikani. 2019. "Development of Ternary Nanoadsorbent Composites of Graphene Oxide, Activated Carbon, and Zero-valent Iron Nanoparticles for Food Applications." *Food Science & Nutrition* 7 (9): 2827–2835. 10.1002/fsn3.1080.

Bano, Ijaz, Muhammad Arshad, Tariq Yasin, and Muhammad Afzal Ghauri. 2019. "Preparation, Characterization and Evaluation of Glycerol Plasticized Chitosan/PVA Blends for Burn Wounds." *International Journal of Biological Macromolecules* 124 (March): 155–162. 10.1016/j.ijbiomac.201 8.11.073.

Barak, Sheweta, Deepak Mudgil, and Shelly Taneja. 2020. "Exudate Gums: Chemistry, Properties and Food Applications – a Review." *Journal of the Science of Food and Agriculture* 100 (7): 2828–2835. 10.1002/ jsfa.10302.

Baranwal, Kirti, Lalit Mohon Dwivedi, Shehala, and Vandana Singh. 2018. "Guar Gum Mediated Synthesis of NiO Nanoparticles: An Efficient Catalyst for Reduction of Nitroarenes with Sodium Borohydride." *International Journal of Biological Macromolecules* 120 (December): 2431–2441. 10.1016/j.ijbiomac.2018.09.013.

Bhat, Veena G., Shivayogi S. Narasagoudr, Saraswati P. Masti, Ravindra B. Chougale, and Yogesh Shanbhag. 2021. "Hydroxy Citric Acid Cross-Linked Chitosan/Guar Gum/Poly(Vinyl Alcohol) Active Films for Food Packaging Applications." *International Journal of Biological Macromolecules* 177 (April): 166–175. 10.1016/j.ijbiomac.2021.02.109.

Bhattarai, Rajan, Rinda Bachu, Sai Boddu, and Sarit Bhaduri. 2018. "Biomedical Applications of Electrospun Nanofibers: Drug and Nanoparticle Delivery." *Pharmaceutics* 11 (1): 5. 10.3390/pharmaceutics11010005.

Chen, Jingwang, Taihua Mu, Dorothée Goffin, Christophe Blecker, Gaëtan Richard, Aurore Richel, and Eric Haubruge. 2019. "Application of Soy Protein Isolate and Hydrocolloids Based Mixtures as Promising Food Material in 3D Food Printing." *Journal of Food Engineering* 261 (December 2018): 76–86. 10.1016/j.jfoodeng.2019.03.016.

Chen, Yue, Leilei Xu, Yajie Wang, Zhongqin Chen, Min Zhang, and Haixia Chen. 2020. "Characterization and Functional Properties of a Pectin/Tara Gum Based Edible Film with Ellagitannins from the Unripe Fruits of Rubus Chingii Hu." *Food Chemistry* 325 (May): 126964. 10.1016/j.foodchem.2020.126964.

Compton, Brett G., and Jennifer A. Lewis. 2014. "3D-Printing of Lightweight Cellular Composites." *Advanced Materials* 26 (34): 5930–5935. 10.1002/adma.201401804.

Dai, Lei, Ting Cheng, Chao Duan, Wei Zhao, Weipeng Zhang, Xuejun Zou, Joseph Aspler, and Yonghao Ni. 2019. "3D Printing Using Plant-Derived Cellulose and Its Derivatives: A Review." *Carbohydrate Polymers* 203: 71–86. 10.1016/j.carbpol.2018.09.027.

Davachi, Seyed Mohammad, Saeed Bakhtiari, Peyman Pouresmaeel-Selakjani, Jamshid Mohammadi-Rovshandeh, Babak Kaffashi, Saeed Davoodi, and Ardavan Yousefi. 2018. "Investigating the Effect of Treated Rice Straw in PLLA/Starch Composite: Mechanical, Thermal, Rheological, and Morphological Study." *Advances in Polymer Technology* 37 (1): 5–16. 10.1002/adv.21634.

Davachi, Seyed Mohammad, and Azadeh Sadat Shekarabi. 2018. "Preparation and Characterization of Antibacterial, Eco-Friendly Edible Nanocomposite Films Containing Salvia Macrosiphon and Nanoclay." *International Journal of Biological Macromolecules* 113 (July): 66–72. 10.1016/j.ijbiomac.2018.02.106.

Dhar, Sheetal, Vishal Mali, Subhash Bodhankar, Anjali Shiras, B. L. V. Prasad, and Varsha Pokharkar. 2011. "Biocompatible Gellan Gum-Reduced Gold Nanoparticles: Cellular Uptake and Subacute Oral Toxicity Studies." *Journal of Applied Toxicology* 31 (5): 411–420. 10.1002/jat.1595.

El-Batal, Ahmed I., M. Abd Elkodous, Gharieb S. El-Sayyad, Nawal E. Al-Hazmi, Mohamed Gobara, and Ahmad Baraka. 2020. "Gum Arabic Polymer-Stabilized and Gamma Rays-Assisted Synthesis of Bimetallic Silver-Gold Nanoparticles: Powerful Antimicrobial and Antibiofilm Activities against Pathogenic Microbes Isolated from Diabetic Foot Patients." *International Journal of Biological Macromolecules* 165 (December): 169–186. 10.1016/j.ijbiomac.2020.09.160.

Eghbalifam, Naeimeh, Seyed Abbas Shojaosadati, Sameereh Hashemi-Najafabadi, and Alireza Chackoshian Khorasani. 2020. "Synthesis and Characterization of Antimicrobial Wound Dressing Material Based on Silver Nanoparticles Loaded Gum Arabic Nanofibers." *International Journal of Biological Macromolecules* 155: 119–130. 10.1016/j.ijbiomac.2020.03.19.

Erben, Melina, Adrián A. Pérez, Carlos A. Osella, Vera A. Alvarez, and Liliana G. Santiago. 2019. "Impact of Gum Arabic and Sodium Alginate and Their Interactions with Whey Protein Aggregates on Bio-Based Films Characteristics." *International Journal of Biological Macromolecules* 125: 999–1007. 10.1016/j.ijbiomac.2018.12.131.

Friedrich, Leanne, and Matthew Begley. 2018. "In Situ Characterization of Low-Viscosity Direct Ink Writing: Stability, Wetting, and Rotational Flows." *Journal of Colloid and Interface Science* 529: 599–609. 10.1016/j.jcis.2018.05.110.

Fu, Kun, Yibo Wang, Chaoyi Yan, Yonggang Yao, Yanan Chen, Jiaqi Dai, Steven Lacey, et al. 2016. "Graphene Oxide-Based Electrode Inks for 3D-Printed Lithium-Ion Batteries." *Advanced Materials* 28 (13): 2587–2594. 10.1002/adma.201505391.

Gao, Wanli, and Martin Pumera. 2021. "3D Printed Nanocarbon Frameworks for Li-Ion Battery Cathodes." *Advanced Functional Materials* 31 (11): 1–10. 10.1002/adfm.202007285.

García-Segovia, P., V. García-Alcaraz, S. Balasch-Parisi, and J. Martínez-Monzó. 2020. "3D Printing of Gels Based on Xanthan/Konjac Gums." *Innovative Food Science and Emerging Technologies* 64 (March): 102343. 10.1016/j.ifset.2020.102343.

Ghorbani Marjan, Farideh Mahmoodzadeh, Leila Yavari Maroufi, and Parinaz Nezhad-Mokhtari. 2020. "Electrospun Tetracycline Hydrochloride Loaded Zein/Gum Tragacanth/Poly Lactic Acid Nanofibers for Biomedical Application." *International Journal of Biological Macromolecules* 165: 1312–132210.1016/j.ijbiomac.2020.09.225.

Goudar, Naganagouda, Vinayak N. Vanjeri, Shruti Dixit, Vishram Hiremani, Sarala Sataraddi, Tilak Gasti, Shyam Kumar Vootla, Saraswati P. Masti, and Ravindra B. Chougale. 2020. "Evaluation of Multifunctional Properties of Gallic Acid Crosslinked Poly (Vinyl Alcohol)/Tragacanth Gum Blend Films for Food Packaging Applications." *International Journal of Biological Macromolecules* 158: 139–149. 10.1016/j.ijbiomac.2020.04.223.

Gutierrez, Elena, Patricio A. Burdiles, Franck Quero, Patricia Palma, Felipe Olate-Moya, and Humberto Palza. 2019. "3D Printing of Antimicrobial Alginate/Bacterial-Cellulose Composite Hydrogels by Incorporating Copper Nanostructures." *ACS Biomaterials Science and Engineering* 5 (11): 6290–6299. 10.1021/acsbiomaterials.9b01048.

Hasheminya, Seyedeh Maryam, and Jalal Dehghannya. 2021. "Development and Characterization of Novel Edible Films Based on Cordia Dichotoma Gum Incorporated with Salvia Mirzayanii Essential Oil Nanoemulsion." *Carbohydrate Polymers* 257 (January): 117606. 10.1016/j.carbpol.2020.117606.

He, Chang, Min Zhang, and Zhongxiang Fang. 2020. "3D Printing of Food: Pretreatment and Post-Treatment of Materials." *Critical Reviews in Food Science and Nutrition* 60 (14): 2379–2392. 10.1080/10408398.2019.1641065.

Holland, Sonia, Tim Foster, William MacNaughtan, and Chris Tuck. 2018. "Design and Characterisation of Food Grade Powders and Inks for Microstructure Control Using 3D Printing." *Journal of Food Engineering* 220: 12–19. 10.1016/j.jfoodeng.2017.06.008.

Huang, Meng sha, Min Zhang, Bhesh Bhandari, and Yaping Liu. 2020. "Improving the Three-Dimensional Printability of Taro Paste by the Addition of Additives." *Journal of Food Process Engineering* 43 (5): 1–9. 10.1111/jfpe.13090.

Jafari, Saeid, Mohammad Hojjati, and Mohammad Noshad. 2018. "Influence of Soluble Soybean Polysaccharide and Tragacanth Gum Based Edible Coating to Improve the Quality of Fresh-Cut Apple Slices." *Journal of Food Processing and Preservation* 42 (6): e13638. 10.1111/jfpp.13638.

Janarthanan, Gopinathan, Hao Nguyen Tran, Eunchong Cha, Chibum Lee, Dipankar Das, and Insup Noh. 2020. "3D Printable and Injectable Lactoferrin-Loaded Carboxymethyl Cellulose-Glycol Chitosan Hydrogels for Tissue Engineering Applications." *Materials Science and Engineering C* 113 (March): 111008. 10.1016/j.msec.2020.111008.

Ji, Anqi, Shuyang Zhang, Samarthya Bhagia, Chang Geun Yoo, and Arthur J. Ragauskas. 2020. "3D Printing of Biomass-Derived Composites: Application and Characterization Approaches." *RSC Advances* 10 (37): 21698–21723. 10.1039/d0ra03620j.

Joz Majidi, Hoomaan, Amir Babaei, Zahra Arab Bafrani, Dina Shahrampour, Erfan Zabihi, and Seid Mahdi Jafari. 2019. "Investigating the Best Strategy to Diminish the Toxicity and Enhance the Antibacterial Activity of Graphene Oxide by Chitosan Addition." *Carbohydrate Polymers* 225 (December): 115220. 10.1016/j.carbpol.2019.115220.

Kalaignana Selvi, S., Jerald Mahesh Kumar, and R.B. Sashidhar. 2017. "Anti-Proliferative Activity of Gum Kondagogu (Cochlospermum Gossypium)-Gold Nanoparticle Constructs on B16F10 Melanoma Cells: An in Vitro Model." *Bioactive Carbohydrates and Dietary Fibre* 11 (July): 38–47. 10.1016/j.bcdf.2017.07.002.

Kanatt, Sweetie R., and Sweta H. Makwana. 2020. "Development of Active, Water-Resistant Carboxymethyl Cellulose-Poly Vinyl Alcohol-Aloe Vera Packaging Film." *Carbohydrate Polymers* 227 (January): 115303. 10.1016/j.carbpol.2019.115303.

Kang, Shufang, Yaqing Xiao, Xinyu Guo, Aiyun Huang, and Huaide Xu. 2021. "Development of Gum Arabic-Based Nanocomposite Films Reinforced with Cellulose Nanocrystals for Strawberry Preservation." *Food Chemistry* 350 (January): 129199. 10.1016/j.foodchem.2021.129199.

Karavasili, Christina, Konstantinos Tsongas, Ioannis I. Andreadis, Eleftherios G. Andriotis, Eleni T. Papachristou, Rigini M. Papi, Dimitrios Tzetzis, and Dimitrios G. Fatouros. 2020. "Physico-Mechanical and Finite Element Analysis Evaluation of 3D Printable Alginate-Methylcellulose Inks for Wound Healing Applications." *Carbohydrate Polymers* 247 (March): 116666. 10.1016/j.carbpol.2020.116666.

Kattumuri, Vijaya, Kavita Katti, Sharanya Bhaskaran, Evan J. Boote, Stan W. Casteel, Genevieve M. Fent, David J. Robertson, Meera Chandrasekhar, Raghuraman Kannan, and Kattesh V. Katti. 2007. "Gum Arabic as a Phytochemical Construct for the Stabilization of Gold Nanoparticles: In Vivo Pharmacokinetics and X-Ray-Contrast-Imaging Studies." *Small* 3 (2): 333–341. 10.1002/smll.200600427.

Kemp, Melissa M., Ashavani Kumar, Shaymaa Mousa, Tae-Joon Park, Pulickel Ajayan, Natsuki Kubotera, Shaker A. Mousa, and Robert J. Linhardt. 2009. "Synthesis of Gold and Silver Nanoparticles Stabilized with Glycosaminoglycans Having Distinctive Biological Activities." *Biomacromolecules* 10 (3): 589–595. 10.1021/bm801266t.

Khan, Nida, Deepak Kumar, and Pramendra Kumar. 2020. "Silver Nanoparticles Embedded Guar Gum/ Gelatin Nanocomposite: Green Synthesis, Characterization and Antibacterial Activity." *Colloid and Interface Science Communications* 35 (March): 100242. 10.1016/j.colcom.2020.100242.

Khedri, Sara, Ehsan Sadeghi, Milad Rouhi, Zohre Delshadian, Amir Mohammad Mortazavian, Jonas de Toledo Guimarães, Maryam Fallah, and Reza Mohammadi. 2021. "Bioactive Edible Films: Development and Characterization of Gelatin Edible Films Incorporated with Casein Phosphopeptides." *Lwt* 138: 110649. 10.1016/j.lwt.2020.110649.

Khezerlou, Arezou, Ali Ehsani, Mahnaz Tabibiazar, and Ehsan Moghaddas Kia. 2019. "Development and Characterization of a Persian Gum–Sodium Caseinate Biocomposite Film Accompanied by Zingiber Officinale Extract." *Journal of Applied Polymer Science* 136 (12): 1–9. 10.1002/app.47215.

Khodaei, Diako, Kristina Oltrogge, and Zohreh Hamidi-Esfahani. 2020. "Preparation and Characterization of Blended Edible Films Manufactured Using Gelatin, Tragacanth Gum and, Persian Gum." *Lwt* 117 (February 2019): 108617. 10.1016/j.lwt.2019.108617.

Ko, Eun Seok, Choonggu Kim, Youngtae Choi, and Kuen Yong Lee. 2020. "3D Printing of Self-Healing Ferrogel Prepared from Glycol Chitosan, Oxidized Hyaluronate, and Iron Oxide Nanoparticles." *Carbohydrate Polymers* 245 (February): 116496. 10.1016/j.carbpol.2020.116496.

Kouhi, Monireh, Molamma P. Prabhakaran, and Seeram Ramakrishna. 2020. "Edible Polymers: An Insight into Its Application in Food, Biomedicine and Cosmetics." *Trends in Food Science & Technology* 103 (September): 248–263. 10.1016/j.tifs.2020.05.025.

Kumar, K. P. Akshay, Kalyan Ghosh, Osamah Alduhaish, and Martin Pumera. 2020. "Metal-Plated 3D-Printed Electrode for Electrochemical Detection of Carbohydrates." *Electrochemistry Communications* 120 (August): 106827. 10.1016/j.elecom.2020.106827.

Kumar, Kandambath Padinjareveetil Akshay, and Martin Pumera. 2021. "3D-Printing to Mitigate COVID-19 Pandemic." *Advanced Functional Materials* 2100450 (91): 1–17. 10.1002/adfm.202100450.

La, Duc D., Phuong Nguyen-Tri, Khoa H. Le, Phuong T.M. Nguyen, M. Dac Binh Nguyen, Anh T.K. Vo, Minh T.H. Nguyen, et al. 2021. "Effects of Antibacterial ZnO Nanoparticles on the Performance of a Chitosan/Gum Arabic Edible Coating for Post-Harvest Banana Preservation." *Progress in Organic Coatings* 151 (October 2020): 106057. 10.1016/j.porgcoat.2020.106057.

Li, Jinhua, and Martin Pumera. 2021. "3D Printing of Functional Microrobots." *Chemical Society Reviews* 50 (4): 2794–2838. 10.1039/d0cs01062f.

Li, Liying, Shuai Qin, Jun Peng, Ang Chen, Yi Nie, Tianqing Liu, and Kedong Song. 2020. "Engineering Gelatin-Based Alginate/Carbon Nanotubes Blend Bioink for Direct 3D Printing of Vessel Constructs." *International Journal of Biological Macromolecules* 145: 262–271. 10.1016/j.ijbiomac.2019.12.174.

Li, Yun, Na Li, Wei Jiang, Guoyuan Ma, and Mohammad Mahdi Zangeneh. 2020. "In Situ Decorated Au NPs on Pectin-Modified Fe3O4 NPs as a Novel Magnetic Nanocomposite (Fe3O4/Pectin/Au) for Catalytic Reduction of Nitroarenes and Investigation of Its Anti-Human Lung Cancer Activities." *International Journal of Biological Macromolecules* 163 (November): 2162–2171. 10.1016/j.ijbiomac.2020.09.102.

Lian, Huan, Jingying Shi, Xiaoyan Zhang, and Yong Peng. 2020. "Effect of the Added Polysaccharide on the Release of Thyme Essential Oil and Structure Properties of Chitosan Based Film." *Food Packaging and Shelf Life* 23 (61): 100467. 10.1016/j.fpsl.2020.100467.

Ligon, Samuel Clark, Robert Liska, Jürgen Stampfl, Matthias Gurr, and Rolf Mülhaupt. 2017. "Polymers for 3D Printing and Customized Additive Manufacturing." *Chemical Reviews* 117 (15): 10212–10290. 10.1021/acs.chemrev.7b00074.

Liu, Zhenbin, Bhesh Bhandari, Sangeeta Prakash, Sylvester Mantihal, and Min Zhang. 2019. "Linking Rheology and Printability of a Multicomponent Gel System of Carrageenan-Xanthan-Starch in Extrusion Based Additive Manufacturing." *Food Hydrocolloids* 87 (August 2018): 413–424. 10.1016/j.foodhyd.2018.08.026.

Long, Jingjunjiao, Alaitz Etxabide Etxeberria, Ashveen V. Nand, Craig R. Bunt, Sudip Ray, and Ali Seyfoddin. 2019. "A 3D Printed Chitosan-Pectin Hydrogel Wound Dressing for Lidocaine Hydrochloride Delivery." *Materials Science and Engineering C* 104 (May): 109873. 10.1016/j.msec.2019.109873.

M'Barki, Amin, Lydéric Bocquet, and Adam Stevenson. 2017. "Linking Rheology and Printability for Dense and Strong Ceramics by Direct Ink Writing." *Scientific Reports* 7 (1): 1–10. 10.1038/s41598-017-06115-0.

Mohammadi Nafchi, Abdorreza, Ali Olfat, Mina Bagheri, Leila Nouri, A. A. Karim, and Fazilah Ariffin. 2017. "Preparation and Characterization of a Novel Edible Film Based on Alyssum Homolocarpum Seed Gum." *Journal of Food Science and Technology* 54 (6): 1703–1710. 10.1007/s13197-017-2602-z.

Nazarzadeh Zare, Ehsan, Pooyan Makvandi, and Franklin R. Tay. 2019. "Recent Progress in the Industrial and Biomedical Applications of Tragacanth Gum: A Review." *Carbohydrate Polymers* 212 (May): 450–467. 10.1016/j.carbpol.2019.02.076.

Olate-Moya, Felipe, Lukas Arens, Manfred Wilhelm, Miguel Angel Mateos-Timoneda, Elisabeth Engel, and Humberto Palza. 2020. "Chondroinductive Alginate-Based Hydrogels Having Graphene Oxide for 3D Printed Scaffold Fabrication." *ACS Applied Materials and Interfaces* 12 (4): 4343–4357. 10.1021/acsami.9b22062.

Padil, Vinod V.T., Martin Stuchlík, and Miroslav Černík. 2015. "Plasma Modified Nanofibres Based on Gum Kondagogu and Their use for Collection of Nanoparticulate Silver, Gold and Platinum."*Carbohydrate Polymers* 121: 468-476, https://doi.org/10.1016/j.carbpol.2014.11.074.

Padil, Vinod V.T., Stanisław Wacławek, Miroslav Černík, and Rajender S. Varma. 2018. "Tree Gum-Based Renewable Materials: Sustainable Applications in Nanotechnology, Biomedical and Environmental Fields." *Biotechnology Advances*. Elsevier Inc. 10.1016/j.biotechadv.2018.08.008.

Padil, Vinod Vellora Thekkae, Stanisław Wacławek, and Miroslav Černík. 2016. "Green Synthesis: Nanoparticles and Nanofibres Based on Tree Gums for Environmental Applications." *Ecological Chemistry and Engineering S* 23 (4): 533–557. 10.1515/eces-2016-0038.

Pandey, Sadanand, Gopal K. Goswami, and Karuna K. Nanda. 2013. "Green Synthesis of Polysaccharide/ Gold Nanoparticle Nanocomposite: An Efficient Ammonia Sensor." *Carbohydrate Polymers* 94 (1): 229–234. 10.1016/j.carbpol.2013.01.009.

Prabhu, Priyanka. 2019. "Nanofibers for Medical Diagnosis and Therapy." In *Handbook of Nanofibers*, 831–867. Cham: Springer International Publishing. 10.1007/978-3-319-53655-2_48.

Ranjbar-Mohammadi, Marziyeh, M. Zamani, M.P. Prabhakaran, S. Hajir Bahrami, and S. Ramakrishna. 2016."Electrospinning of PLGA/Gum Tragacanth Nanofibers Containing Tetracycline Hydrochloride for Periodontal Regeneration."*Materials Science and Engineering* 58:521–531.

Raja Namasivayam, S. Karthick, Gayathri Venkatachalam, and R.S. Arvind Bharani. 2020. "Immuno Biocompatibility and Anti-Quorum Sensing Activities of Chitosan-Gum Acacia Gold Nanocomposite (CS-GA-AuNC) against Pseudomonas Aeruginosa Drug-Resistant Pathogen." *Sustainable Chemistry and Pharmacy* 17 (September): 100300. 10.1016/j.scp.2020.100300.

Rajabi, Hamid, Seid Mahdi Jafari, Javad Feizy, Mohammad Ghorbani, and Seyed Ahmad Mohajeri. 2020. "Preparation and Characterization of 3D Graphene Oxide Nanostructures Embedded with Nanocomplexes of Chitosan- Gum Arabic Biopolymers." *International Journal of Biological Macromolecules* 162 (November): 163–174. 10.1016/j.ijbiomac.2020.06.076.

Rajeshkumar, S., L.V. Bharath, and R. Geetha. 2019. "Broad Spectrum Antibacterial Silver Nanoparticle Green Synthesis: Characterization, and Mechanism of Action." In *Green Synthesis, Characterization and Applications of Nanoparticles*, 429–444. Elsevier. 10.1016/B978-0-08-102579-6.00018-6.

Ramakrishnan, Rohith K., Vinod V.T. Padil, Marcela Škodová, Stanisław Wacławek, Miroslav Černík, and Seema Agarwal. 2021. "Hierarchically Porous Bio-Based Sustainable Conjugate Sponge for Highly Selective Oil/ Organic Solvent Absorption." *Advanced Functional Materials* 2100640: 1–9. 10.1002/adfm.202100640.

Ramakrishnan, Rohith K., Stanisław Wacławek, Miroslav Černík, and Vinod V.T. Padil. 2021. "Biomacromolecule Assembly Based on Gum Kondagogu-Sodium Alginate Composites and Their Expediency in Flexible Packaging Films." *International Journal of Biological Macromolecules* 177: 526–534. 10.1016/j.ijbiomac.2021.02.156.

Rathore, Hanumant Singh, M. Sarubala, Giriprasath Ramanathan, Sivakumar Singaravelu, M.D. Raja, Sanjeev Gupta, and Uma Tirichurapalli Sivagnanam. 2016. "Fabrication of Biomimetic Porous Novel Sponge from Gum Kondagogu for Wound Dressing." *Materials Letters* 177: 108–111. 10.1016/ j.matlet.2016.04.185.

Razmjoo, Fatemeh, Ehsan Sadeghi, Milad Rouhi, Reza Mohammadi, Razieh Noroozi, and Saeede Safajoo. 2021. "Polyvinyl Alcohol – Zedo Gum Edible Film: Physical, Mechanical and Thermal Properties." *Journal of Applied Polymer Science* 138 (8): 1–10. 10.1002/app.49875.

Renuga Devi, P., C. Senthil Kumar, P. Selvamani, N. Subramanian, and K. Ruckmani. 2015. "Synthesis and Characterization of Arabic Gum Capped Gold Nanoparticles for Tumor-Targeted Drug Delivery." *Materials Letters* 139: 241–244. 10.1016/j.matlet.2014.10.010.

Robinson, Thomas M., Sepehr Talebian, Javad Foroughi, Zhilian Yue, Cormac D. Fay, and Gordon G. Wallace. 2020. "Fabrication of Aligned Biomimetic Gellan Gum-Chitosan Microstructures through 3D Printed Microfluidic Channels and Multiple in Situ Cross-Linking Mechanisms." *ACS Biomaterials Science and Engineering* 6 (6): 3638–3648. 10.1021/acsbiomaterials.0c00260.

Rodrigues, Delane C., Carlos Alberto Caceres, Hálisson L. Ribeiro, Rosa F.A. de Abreu, Arcelina P. Cunha, and Henriette M.C. Azeredo. 2014. "Influence of Cassava Starch and Carnauba Wax on Physical Properties of Cashew Tree Gum-Based Films." *Food Hydrocolloids* 38: 147–151. 10.1016/j.foodhyd.2013.12.010.

Rajkumar, S., and M. Prabaharan 2020. "Multi-Functional FITC-Silica@gold Nanoparticles Conjugated with Guar Gum Succinate, Folic Acid and Doxorubicin for CT/Fluorescence Dual Imaging and Combined Chemo/PTT of Cancer." *Colloids and Surfaces B: Biointerfaces* 186 (February): 110701. 10.1016/j.colsurfb.2019.110701.

Sanyasi, Sridhar, Rakesh Kumar Majhi, Satish Kumar, Mitali Mishra, Arnab Ghosh, Mrutyunjay Suar, Parlapalli Venkata Satyam, Harapriya Mohapatra, Chandan Goswami, and Luna Goswami. 2016. "Polysaccharide-Capped Silver Nanoparticles Inhibit Biofilm Formation and Eliminate Multi-Drug-Resistant Bacteria by Disrupting Bacterial Cytoskeleton with Reduced Cytotoxicity towards Mammalian Cells." *Scientific Reports* 6 (1): 24929. 10.1038/srep24929.

Shahbazi, Mahdiyar, and Henry Jäger. 2021. "Current Status in the Utilization of Bio-based Polymers for 3D Printing Process: A Systematic Review of the Materials, Processes, and Challenges." *ACS Applied Bio Materials* 4 (1): 325–369. 10.1021/acsabm.0c01379.

Shahbazi, Mahdiyar, Henry Jäger, Seyed Javad Ahmadi, and Monique Lacroix. 2020. "Electron Beam Crosslinking of Alginate/Nanoclay Ink to Improve Functional Properties of 3D Printed Hydrogel for Removing Heavy Metal Ions." *Carbohydrate Polymers* 240: 116211. 10.1016/j.carbpol.2020.116211.

Strother, Heather, Rachael Moss, and Matthew B. McSweeney. 2020. "Comparison of 3D Printed and Molded Carrots Produced with Gelatin, Guar Gum and Xanthan Gum." *Journal of Texture Studies* 51 (6): 852–860. 10.1111/jtxs.12545.

Tahsiri, Zahra, Hamideh Mirzaei, Seyed Mohammad Hashem Hosseini, and Mohammadreza Khalesi. 2019. "Gum Arabic Improves the Mechanical Properties of Wild Almond Protein Film." *Carbohydrate Polymers* 222: 114994. 10.1016/j.carbpol.2019.114994.

Tan, Kim Seah, and Kuan Yew Cheong. 2013. "Advances of Ag, Cu, and Ag–Cu Alloy Nanoparticles Synthesized via Chemical Reduction Route." *Journal of Nanoparticle Research* 15 (4): 1537. 10.1007/s11051-013-1537-1.

Taner, Merve, Nilufer Sayar, Isik G. Yulug, and Sefik Suzer. 2011. "Synthesis, Characterization and Antibacterial Investigation of Silver–Copper Nanoalloys." *Journal of Materials Chemistry* 21 (35): 13150. 10.1039/c1jm11718a.

Tang, Xingwei, Han Zhou, Zuocheng Cai, Dongdong Cheng, Peisheng He, Peiwen Xie, Di Zhang, and Tongxiang Fan. 2018. "Generalized 3D Printing of Graphene-Based Mixed-Dimensional Hybrid Aerogels." *ACS Nano* 12 (4): 3502–3511. 10.1021/acsnano.8b00304.

Tetsuka, Hiroyuki, and Su Ryon Shin. 2020. "Materials and Technical Innovations in 3D Printing in Biomedical Applications." *Journal of Materials Chemistry B* 8 (15): 2930–2950. 10.1039/d0tb00034e.

Tonyali, Bade, Sevil Cikrikci, and Mecit Halil Oztop. 2018. "Physicochemical and Microstructural Characterization of Gum Tragacanth Added Whey Protein Based Films." *Food Research International* 105 (March): 1–9. 10.1016/j.foodres.2017.10.071.

Trikkaliotis, Dimitrios G., Achilleas K. Christoforidis, Athanasios C. Mitropoulos, and George Z. Kyzas. 2020. "Adsorption of Copper Ions onto Chitosan/Poly(Vinyl Alcohol) Beads Functionalized with Poly (Ethylene Glycol)." *Carbohydrate Polymers* 234 (April): 115890. 10.1016/j.carbpol.2020.115890.

Tsai, Ruei-Yi, Kuo, Ting-Yun, Hung, Shih-Chieh, Lin, Che-Min, Hsien, Tzu-Yang, Wang, Da-Ming, and Hsieh, Hsyue-Jen2015. "Use of Gum Arabic to Improve the Fabrication of Chitosan–Gelatin-Based Nanofibers for Tissue Engineering." *Carbohydrate Polymers* 115:525–53210.1016/j.carbpol.2014.08.108.

Urruela-Barrios, Rodrigo, Erick Ramírez-Cedillo, A. Díaz de León, Alejandro J. Alvarez, and Wendy Ortega-Lara. 2019. "Alginate/Gelatin Hydrogels Reinforced with TiO2 and β-TCP Fabricated by Microextrusion-Based Printing for Tissue Regeneration." *Polymers* 11 (3). 10.3390/polym11030457.

Valentine, Alexander D., Travis A. Busbee, John William Boley, Jordan R. Raney, Alex Chortos, Arda Kotikian, John Daniel Berrigan, Michael F. Durstock, and Jennifer A. Lewis. 2017. "Hybrid 3D Printing of Soft Electronics." *Advanced Materials* 29 (40): 1–8. 10.1002/adma.201703817.

Vancauwenberghe, Valérie, Pieter Verboven, Jeroen Lammertyn, and Bart Nicolaï. 2018. "Development of a Coaxial Extrusion Deposition for 3D Printing of Customizable Pectin-Based Food Simulant." *Journal of Food Engineering* 225: 42–52. 10.1016/j.jfoodeng.2018.01.008.

Veisi, Hojat, Maliheh Farokhi, Mona Hamelian, and Saba Hemmati. 2018. "Green Synthesis of Au Nanoparticles Using an Aqueous Extract of Stachys Lavandulifolia and Their Catalytic Performance for Alkyne/Aldehyde/Amine A 3 Coupling Reactions." *RSC Advances* 8 (67): 38186–38195. 10.1039/C8RA06819D.

Wang, Baobin, Lei Dai, Guihua Yang, Guida Bendrich, Yonghao Ni, and Guigan Fang. 2019. "A Highly Efficient Thermo Responsive Palladium Nanoparticles Incorporated Guar Gum Hydrogel for Effective Catalytic Reactions." *Carbohydrate Polymers* 226 (December): 115289. 10.1016/j.carbpol.2019.115289.

Wang, Ying, Shaohua Wu, Mitchell A. Kuss, Philipp N. Streubel, and Bin Duan. 2017. "Effects of Hydroxyapatite and Hypoxia on Chondrogenesis and Hypertrophy in 3D Bioprinted ADMSC Laden Constructs." *ACS Biomaterials Science and Engineering* 3 (5): 826–835. 10.1021/acsbiomaterials.7b00101.

Warner, E. L., I. T. Norton, and T. B. Mills. 2019. "Comparing the Viscoelastic Properties of Gelatin and Different Concentrations of Kappa-Carrageenan Mixtures for Additive Manufacturing Applications." *Journal of Food Engineering* 246: 58–66. 10.1016/j.jfoodeng.2018.10.033.

Wei, Jiaxin, Baoxiu Wang, Zhe Li, Zhuotong Wu, Minghao Zhang, Nan Sheng, Qianqian Liang, Huaping Wang, and Shiyan Chen. 2020. "A 3D-Printable TEMPO-Oxidized Bacterial Cellulose/Alginate Hydrogel with Enhanced Stability via Nanoclay Incorporation." *Carbohydrate Polymers* 238 (March): 116207. 10.1016/j.carbpol.2020.116207.

Xu, Tian, Cheng Cheng Gao, Xiao Feng, Meigui Huang, Yuling Yang, Xinchun Shen, and Xiaozhi Tang. 2019. "Cinnamon and Clove Essential Oils to Improve Physical, Thermal and Antimicrobial Properties of Chitosan-Gum Arabic Polyelectrolyte Complexed Films." *Carbohydrate Polymers* 217 (December 2018): 116–125. 10.1016/j.carbpol.2019.03.084.

Xu, Tian, Cheng Cheng Gao, Xiao Feng, Di Wu, Linghan Meng, Weiwei Cheng, Yan Zhang, and Xiaozhi Tang. 2020. "Characterization of Chitosan Based Polyelectrolyte Films Incorporated with OSA-Modified Gum Arabic-Stabilized Cinnamon Essential Oil Emulsions." *International Journal of Biological Macromolecules* 150: 362–370. 10.1016/j.ijbiomac.2020.02.108.

Yang, Jian, Xingye An, Liqin Liu, Shiyu Tang, Haibing Cao, Qingliang Xu, and Hongbin Liu. 2020. "Cellulose, Hemicellulose, Lignin, and Their Derivatives as Multi-Components of Bio-Based Feedstocks for 3D Printing." *Carbohydrate Polymers* 250 (29): 116881. 10.1016/j.carbpol.2020.116881.

Yao, Xiaolin, Kai Xu, Meng Shu, Ning Liu, Na Li, Xiaoyu Chen, Katsuyoshi Nishinari, Glyn O. Phillips, and Fatang Jiang. 2021. "Fabrication of Iron Loaded Whey Protein Isolate/Gum Arabic Nanoparticles and Its Adsorption Activity on Oil-Water Interface." *Food Hydrocolloids* 115: 106610. 10.1016/j.foodhyd.2021.106610.

Yousuf, Basharat, Shimin Wu, and Yuan Gao. 2021. "Characteristics of Karaya Gum Based Films: Amelioration by Inclusion of Schisandra Chinensis Oil and Its Oleogel in the Film Formulation." *Food Chemistry* 345: 128859. 10.1016/j.foodchem.2020.128859.

Zarekhalili, Zahra, S. HajirBahrami, M.Ranjbar-Mohammadi, andPeiman Brouki, Milan2017. "Fabrication and Characterization of PVA/Gum Tragacanth/PCLHybrid Nanofibrous Scaffolds for Skin Substitutes." *International Journal of Biological Macromolecules*94: 679–69010.1016/j.ijbiomac.2016.10.042.

Zhai, Xiaodong, Zhihua Li, Jiyong Shi, Xiaowei Huang, Zongbao Sun, Di Zhang, Xiaobo Zou, et al. 2019. "A Colorimetric Hydrogen Sulfide Sensor Based on Gellan Gum-Silver Nanoparticles Bionanocomposite for Monitoring of Meat Spoilage in Intelligent Packaging." *Food Chemistry* 290: 135–143. 10.1016/j.foodchem.2019.03.138.

Zibaei, Rezvan, Sara Hasanvand, Zahra Hashami, Zahra Roshandel, Milad Rouhi, Jonas de Toledo Guimarães, Amir Mohammad Mortazavian, Zahra Sarlak, and Reza Mohammadi. 2021. "Applications of Emerging Botanical Hydrocolloids for Edible Films: A Review." *Carbohydrate Polymers* 256 (March): 117554. 10.1016/j.carbpol.2020.117554.

6 Dissolution of Cellulose Biopolymer Using Alkoxy Linked Dicationic Ionic Liquids

Joshua Raj Jaganathan, Kausar Ahmad Fuad, and Eswaran Padmanabhan
Universiti Teknologi PETRONAS, Perak, Malaysia

CONTENTS

6.1 INTRODUCTION: BACKGROUND AND DRIVING FORCES

Study nature, love nature, stay close to nature. It will never fail you.

—Frank Lloyd Wright

Based on the famous quote above, it's very true that nature has been supporting mankind from the very beginning of the universe. Nature offers widespread biopolymers such as cellulose, hemicellulose and lignin under the category of biomass, which are widely used in various applications (Pettersen, 1984). Cellulose particularly is the most abundant one used in vital applications such as the source of biocompatible and biodegradable material (Potthast, Rosenau, and Kosma, 2006), biofuel/biochemical production (Brandt et al., 2013), biorefinery (FitzPatrick et al., 2010), textile recycling (Michud et al., 2016), cellulose nanocrystals (Wohlhauser et al., 2018) and many more. These applications have been drawing immense attention in recent years, making cellulose one of the widely studied biomaterials.

DOI: 10.1201/9781003200710-7

Nevertheless, cellulose also has its own drawbacks as every other material available. This intricacy arises from its complex molecular structure which consists of linear D-glucose chains linked by 1-4, β-glycosidic bonds. They are aligned parallelly as flat sheets and stacked to form microfibrils. The presence of strong inter- and intra-molecular cohesive force (hydrogen bonding) between the molecules inhibits effective dissolution of cellulose and leads to difficulty in processing. Although cellulose dissolution once began with the usage of various conventional solvents such as dimethyl sulfoxide (DMF), dimethysulfoxide (DMSO) and acetonitrile (Hammer and Turbak, 1977), a huge number drawbacks such as high cost due to high loss of solvents (high vapor pressure), high toxicity and biodegradability and poor thermal stability acted as huge barriers, preventing ways forward. Thus, the intervention of ionic liquids (ILs) over recent years has created an architectural platform as an alternative solvent for the dissolution of cellulose. ILs mainly offer various advantages, such as low vapor pressure, high thermal stability, wide electrochemical window, wide liquidus range and high solvation capability (Raj et al., 2017; Wasserscheid and Welton, 2008).

Although the use of ILs for cellulose dissolution isn't an entirely new trend anymore, the race is still alive to further tune the properties of ILs in order to create the best ILs with the suitable combination of anions and cations. These unique properties of ILs made them highly explored as potential solvents in cellulose dissolution. The first use of ILs in cellulose dissolution kicked off with first-generation ILs, 1-butyl-3-methylimidazolium chloride (C_4MIM) by Rogers group (Swatloski et al., 2002), which recorded up to 25 wt% of pulp dissolution under microwave conditions. Since then, many other ILs (mostly chloride and acetate based) emerged to take part in the race in order to find the best potential ILs for the dissolution of ILs (Fukaya et al., 2008; Gericke, Fardim, and Heinze, 2012; Li et al., 2020; Stolarska et al., 2017). On the other hand, basic ILs containing hydroxide anions (tetrabutylammonium hydroxide) also have been reported widely in the area of cellulose dissolution. However, the low thermal stability and poor regeneration of the hydroxide ILs act as a major drawback (Abe et al., 2017; Sirviö and Heiskanen, 2020). Unconventional anions such as phenolate-based ILs have been also reported with considerably good cellulose dissolution up 45 wt% at 100°C (Lethesh et al., 2020). As a summary, various types of ILs are still being employed to date in cellulose dissolution despite drawbacks such as low dissolution efficiency, high dissolution temperature, poor regeneration and reusability and low thermal stability. Table 6.1 enlists some common categories of ILs in the area of MCC cellulose dissolution.

Interestingly, it's very notable that most cellulose dissolution engages with monocationic ILs ranging from imidazolium, ammonium and pyridinium as cations (Isik, Sardon, and Mecerreyes, 2014). However, dicationic ILs have not been reported for cellulose dissolution to the best of our knowledge, which explains the novelty of this work. Symmetrical or geminal dicationic are composed of two cations and two anions that can be obtained by joining two of the same cation candidates such as imidazolium or pyrrolidinium linked by either cyclic or aliphatic chain (spacer). In the case of monocationic ILs, it contains only one cation and anion. Interestingly, the application of dicationic ILs in various area is still limited compared to monocationic ILs, despite having several advantages such as high thermal stability and volatility which allow their use as an excellent solvent for high temperature applications (Anderson et al., 2005; Fan et al., 2013), lubricants and catalyst for esterification and transesterification reactions (Wei-Li et al., 2014). Therefore, the aim of the work is to explore the potential of dicationic ILs towards cellulose dissolution. Here, we have synthesized and characterized alkoxy group (ether) linked imidazolium-based dicationic ILs using acetate and chloride anions for the dissolution of MCC cellulose.

6.2 EXPERIMENTAL

6.2.1 MATERIALS AND REAGENTS

All the starting materials were used without further purification. Microcrystalline cellulose (MCC) with a degree of polymerization (DP = ca.250) was purchased from Sigma Aldrich. The materials

TABLE 6.1

Some Common ILs Used in Cellulose Dissolution

ILs	Dissolution Temperature (°C)	Solubility (wt%)	Cellulose Type	Ref.
1-allyl-3-methylimidazolium chloride ([Amim][Cl])	100	10	MCC	(Fukaya, Sugimoto, and Ohno, 2006)
1-butyl-3-methylimidazolium benzoate ([C4mim][PhCO2])	70	12	MCC	(Xu, Wang, and Wang, 2010)
1-butyl-3-methylimidazolium formate ([C4mim][HCOO])	110	8	MCC	(Zhao et al., 2008)
1-butyl-3-methylimidazolium dicyanamide ([Bmim][N(CN)2])	110	1	MCC	(Zhao et al., 2008)
1-allyl-3-methylimidazolium formate ([Amim][HCOO])	85	22	MCC	(Fukaya, Sugimoto, and Ohno, 2006)
1-butyl-3-methylimidazolium bis [(trifluoromethyl)sulphonyl]imide ([Bmim][TFSI])	110	0.5	MCC	(Zhao et al., 2008)
N,N-dimethyl-2-methoxyethylammonium acetate ([MM(MeOEt)NH][OAc])	110	12	MCC	(Zhao et al., 2008)
3-methyl-N-butylpyridinium chloride ([MNBuPy][Cl])	105	12	MCC	(Zhao et al., 2008)
1-ethyl-3-methylimidazolium phenolate [C2mim] [OPh]	100	45	MCC	(Lethesh et al., 2020)

and reagents used for the synthesis of ILs were analytical grade and they are as follows: 1-methylimidazole (Sigma Aldrich, >99%), 1-butylimidazole (Sigma Aldrich, >99%), bis(chloethylether) (Sigma Aldrich, >99%), acetic acid (Sigma Aldrich, >99%), silver nitrate (Merck), ethyl acetate (Merck) and dichloromethane (Merck).

6.2.2 Synthesis of Solvent Media (ILs)

Synthesis procedures the ILs used in this work has been adopted with modifications from our previously published work (Raj et al., 2017) and typical synthesis method for symmetrical dicationic ILs (Zhang et al., 2007). Based on the pathway, the synthesis procedure involves two steps: quaternization and metathesis. Ether containing alkyl spacer was tethered to the imidazolium ring through S_N2 reaction between 1-alkyl-imidazole and bis(chloethylether). Acetate anion was then introduced through anion exchange (metathesis) reaction of the halide precursor with acid of the corresponding anion. The general synthesis method is described as follows:

1-methylimidzole (10 g, 0.122 mol) was dissolved in dichloromethane (50 mL) and the solution was added with bis(chloroethylether) (8.72 g, 0.061 mol). The mixture was then stirred for 6 h at 70°C under N_2 atmosphere. The product obtained was washed with ethyl acetate (4 × 25 mL) and excess ethyl acetate was removed by using a rotary evaporator. The viscous halide salt (($C_1MIM)_2OC_2$ 2Cl) obtained was then dried in a vacuum oven at 40°C for 24 h. All the steps mentioned above were repeated with 1-butylimidazole. As for the anion exchange, acetic acid (5 g, 0.08 mol) was added dropwise to ($C_1MIM)_2OC_4$ 2Cl (24 g, 0.08 mol) and stirred at 50°C for 12 h. The desired ILs were washed with ethyl acetate to remove the unreacted materials

(a)   (b)

(C$_1$MIM)$_2$OC$_4$2Cl   (C$_4$MIM)$_2$OC$_4$2Cl

(C1MIM)$_2$OC$_4$2Ac

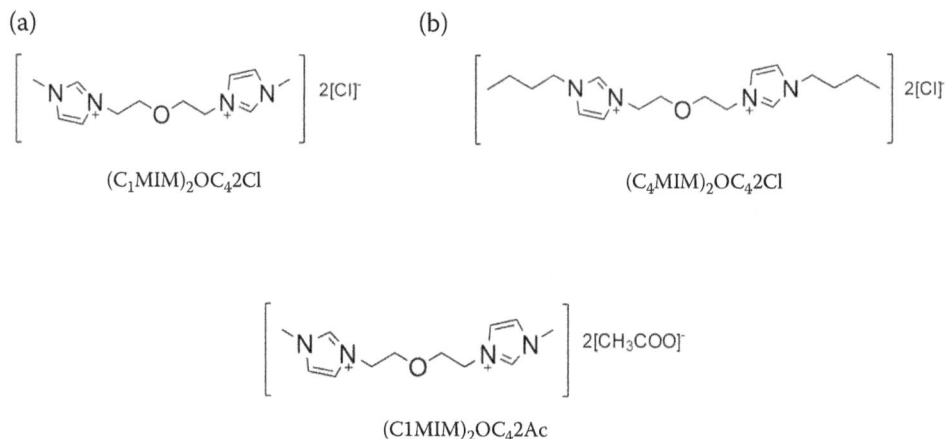

FIGURE 6.1   The molecular structure of ILs used in this study.

and dried in a vacuum oven at 70°C for 24 h. The purity of all the ionic liquids recorded > 99% according to the ^{1}H NMR. The molecular structure of the ILs used in this study are depicted in Figure 6.1. ^{1}H NMR, ^{13}C NMR and water content of the ILs synthesized in this work are reported as below:

**(C$_1$MIM)$_2$OC$_4$2Cl:**^{1}H NMR (500 MHz, [D$_6$]DMSO): δ: 9.44 (s, 2 H), 7.76 (d, 4 H), 4.40 (t, 4 H), 3.92 (s, 6 H), 3.79 (t, 4 H); ^{13}C NMR (125 MHz, [D$_6$] DMSO): δ: 137.46, 123.84, 123.09, 68.47, 49.01, 36.21; Water content (720 ppm)

**(C$_1$MIM)$_2$OC$_4$2Ac:**^{1}H NMR (500 MHz, [D$_6$]DMSO): δ: 11.16 (s, 3 H), 9.49 (s, 2 H), 7.76 (d, 4 H), 4.42 (t, 4 H), 3.92 (s, 6 H), 3.79 (t, 4 H); ^{13}C NMR (125 MHz, [D$_6$] DMSO): δ: 179.44, 137.46, 123.84, 123.09, 68.47, 49.01, 38.91, 36.21; Water content (1,280 ppm)

**(C$_4$MIM)$_2$OC$_4$2Cl:**^{1}H NMR (500 MHz, [D$_6$]DMSO): δ: 9.54 (s, 2 H), 7.83 (d, 4 H), 4.41 (t, 4 H), 4.22 (t, 4 H), 3.79 (t, 4 H), 1.80 (m, 4 H), 1.27 (m, 4 H), 0.90 (t, 6 H); ^{13}C NMR (125 MHz, [D$_6$] DMSO): δ: 137.04, 123.28, 122.55, 68.29, 49.02, 31.86, 19.25, 13.76, 10.72; Water content (590 ppm)

6.2.3   CHARACTERIZATION OF SOLVENT MEDIA (ILs)

The molecular structure of the synthesized solvent media (ILs) were confirmed by ^{1}H and ^{13}C NMR spectroscopy (Bruker Avance 500 MHz spectrometer). Water content in ILs was analyzed by Coulometric Karl Fischer titrator (Mettler Toledo, model DL39). The thermal gravimetric analyzer (Perkin-Elmer, Pyris V-3.81) with heating profile from 50°C to 800°C at a heating rate of 10°C min^{-1} in inert (nitrogen atmosphere). Viscosity of the ILs were measured using Anton Paar viscometer (model SVM3000). The properties of the ILs are tabulated in Table 6.2.

6.2.4   CELLULOSE DISSOLUTION IN ILs AND CHARACTERIZATION

A varying amount of MCC mixture by wt% was prepared in 1 g of ILs. The mixture was then heated up at 20°C in the nitrogen atmosphere alongside vigorous stirring (700–800 rpm). Dissolution time was determined by recording the minimum time required for the complete dissolution. As for the temperature factor, increasing the 5°C interval controlled by the oil bath was used until an optically clear solution was observed. This is only carried out when the dissolution is ineffective. Therefore, the lowest temperature to give a clear solution (observable with naked eyes) was taken as the dissolution temperature. The extent of cellulose dissolution was further confirmed

TABLE 6.2

Molecular Weight (M), Viscosity (η) and Thermal Decomposition Temperature (T_d (10%)) of the ILs

ILs	Appearance	M (g.mol)	η (mPa.s)/20°C	T_d (°C)
$(C_1MIM)_2OC_22Cl$	Viscous Liquid	307.22	586.23	196.59
$(C_4MIM)_2OC_22Cl$	Viscous Liquid	391.38	633.56	270.65
$(C_1MIM)_2OC_22Ac$	Viscous Liquid	354.40	523.37	174.88

Standard uncertainties, u (η) ± 0.35% mPa.s; Standard uncertainties, u ($T_{10\%}$) ± 0.1°C.

using an optical microscope (Axio Scope A1, Carl Zeiss). To regenerate the cellulose, the dissolved cellulose was precipitated by adding ethanol and filtered using filter paper. Finally, the regenerated cellulose was then washed with ethanol to ensure complete removal of ILs and dried at 100°C for 24 h in vacuum.

6.2.5 CRYSTALLINITY AND THERMAL STABILITY OF MCC

MCC cellulose before and after regeneration were subjected to thermogravimetry analysis using thermal gravimetric analyzer (Perkin-Elmer, Pyris V-3.81) with a heating profile from 50°C to 800°C at a heating rate of 10°C min^{-1} in inert (nitrogen atmosphere). XRD D8 ADVANCE (Brucker, USA) equipped with Cu Kα source at room temperature was used to measure the crystallinity of the samples. The sample scans were collected from 2Θ = 2° to 60° with the rate of 0.01°/min and peak height method (Park et al., 2010) was used to calculate the crystallinity index from the XRD patterns.

6.2.6 KAMLET-TAFT PARAMETER

Polarity of ILs is another key index that determines solvation ability, reaction rates and mechanism and product yields (MacFarlane et al., 2006). In this case, polarity of ILs is one of the major components used to investigate ILs-cellulose interaction. Therefore, solvatochromic probes which allows the assessment of the polarity of the ILs can be computed by Kamlet-Taft equation (Kamlet and Taft, 1976):

$$XYZ = (XYZ)_0 + s(\pi^* + d\delta) + a\alpha + b\beta \tag{6.1}$$

where XYZ is the result of a particular solvent-dependent process, $(XYZ)_0$ is the value for the reference system, π^* represents the solvent's dipolarity/polarizability, α is the hydrogen-bond donating ability, β is the hydrogen-bond accepting ability and δ is a correction term. The parameters a, b and s represent the solvent-independent coefficients.

Kamlet taft parameters in this study were measured according to a well-established procedure (Cláudio et al., 2014) where all ILs were dried at 50°C for 48 h under reduced pressure before the measurement. N, N-diethyl-4-nitroaniline and 4-nitroaniline (dyes) were used to determine the hydrogen bond basicity of ILs. Each dye was dissolved in dichloromethane (DCM) separately to make up a solution with the concentration of 1×10^{-5} ppm. Two grams of dried ILs were added to the DCM -dye solution in a conical flask. The DCM was finally removed at 60°C under vacuum for 4 h. UV–vis spectrophotometer (Biochrome, LS60) was used to generate the spectrum of the sample at 25°C.

TABLE 6.3

MCC Dissolution Profile of ILs

ILs	Diss. Capacity	Diss. Temperature (°C)	Diss. Time (h)
$(C_1MIM)_2OC_4$ 2Cl	29.5	60	24
$(C_4MIM)_2OC_4$ 2Cl	17.3	95	36
$(C_1MIM)_2OC_4$ 2Ac	38.2	60	24
$(C_1MIM)_2OC_4$ 2Ac + DMSO	48.4	30	10

∗ wt% of MCC to IL was used to measure the dissolution efficiencies.

6.3   RESULTS AND DISCUSSION

6.3.1   Cellulose Dissolution by ILs

Table 6.3 compares the dissolution capacity of different ILs under various times and temperatures. The dissolution capacity increased in the ILs order of $(C_4MIM)_2OC_4$ 2Cl > $(C_1MIM)_2OC_4$ 2Cl > $(C_1MIM)_2OC_4$ 2Ac. Generally, the trend exhibited on dissolution capacity of the ILs can be discussed from various angles. For instance, imidazolium based monocationic ILs with chloride anion such 1-butyl-3-methylimidazolium chloride, [BMIMCl] has been widely reported in cellulose dissolution in which the dissolution capacity ranging from 15–20 wt% (Wang, Gurau, and Rogers, 2012). Along that line $(C_1MIM)_2OC_4$ 2Cl dicationic ILs used in this experiment recorded the dissolution capacity of 29.5 wt%, higher than the dissolution capacity of monocationic [BMIMCl]. This can be highly associated with the effect of cation and anion of ILs, which is one of the major contributing factors in this case. In $(C_1MIM)_2OC_4$ 2Cl, the size of the cation doubled where two imidazolium group linked by an alkoxy group as compared to monocationic [BMIMCl]. In other words, the increase of dissolution capacity of cellulose from monocationic [BMIMCl] to dicationic $(C_1MIM)_2OC_4$ 2Cl can be strongly associated with the presence of a greater number of aromatic imidazolium group. The nature of the imidazolium group which is highly polarizable (Ferreira et al., 2011) due to the charge delocalization in the aromatic rings leads to a lower relative interaction strength between the cations and anions (Fernandes et al., 2011). Thus, it facilitates anion to form hydrogen bonds with cellulose easily. Another prominent factor would be the presence of ether group as a spacer that links the two-imidazole group in $(C_1MIM)_2OC_4$ 2Cl ILs. Generally, most imidazolium and pyridinium based dicationic ILs with halide anions are in solid form (Anderson et al., 2005), which limits their usage as an extractant due to poor mass transfer. In the case of dicationic $(C_1MIM)_2OC_4$ 2Cl ILs used in this study, it appeared as a viscous liquid rather than in solid form. This behavior can be linked to the presence of alkoxy chain spacer in on the cation $(C_1MIM)_2OC_4$ 2Cl ILs, which increases the chain flexibility due to free axial rotation. This phenomenon reduced the aggregation/interaction of the chains of neighboring cations and resulted in lower viscosity (Raj et al., 2019; Tang, Baker, and Zhao, 2012). In addition, the presence of basic oxygen atoms as alkoxy spacer in $(C_1MIM)_2OC_4$ 2Cl ILs may also have contributed to the efficient hydrogen bonding with cellulose hydroxyl groups (Zhao et al., 2008). On the other hand, the increase in alkyl chain length on the dicationic ILs used in this study had resulted in the reduction of dissolution capacity of cellulose. $(C_4MIM)_2OC_4$ 2Cl ILs, which contains one butyl group on each imidazolium cation recorded only 17.3 wt% compared to $(C_1MIM)_2OC_4$ 2Cl ILs which contains shorter alkyl length (methyl) group on the imidazolium cation (29.5 wt%). Similar observation was also reported using monocationic ILs where dissolution capacity reduced significantly with increased alkyl length in imidazolium phenolates ILs (Lethesh et al., 2020) and imidazolium chloride ILs (Wang, Gurau, and Rogers, 2012). This mainly could be due to the tendency of long alkyl chain in ILs to increase the bulkiness of the cation and therefore affect the solubility of cellulose by reducing the effective concentration of chloride as the hydrogen bond acceptor (Swatloski et al., 2002). The reduction in dissolution capacity of cellulose using $(C_4MIM)_2OC_4$ 2Cl ILs, which contain relatively high alkyl chain length can be also explained in terms of viscosity. Table 6.2 clearly depicts the increase in viscosity from 586. 23 mPa.s to 633.56 mPa.s by $(C_1MIM)_2OC_4$ 2Cl ILs and $(C_4MIM)_2OC_4$ 2Cl ILs respectively. Longer alkyl chain in ILs results in the expansion of van der Waals forces, which leads to the increase in viscosity (Ebrahimi and Moosavi, 2018). Thus, it is quite clear that the low viscosity of ILs facilitates efficient cellulose dissolution in this case, in line with the other reports on cellulose dissolution using ILs (Fukaya, Sugimoto, and Ohno, 2006; Zhang et al., 2005). On another note, higher viscosity of $(C_4MIM)_2OC_4$ 2Cl ILs also affects the processing condition where dissolution time and temperature are increased.

On the perspective of anion, hydrogen bond acceptors such as chloride (Cl^-) and acetate (Ac^-) are deemed to be more effective in dissolving cellulose based on previous reports (Wang, Gurau, and Rogers, 2012). Therefore, anions used in this study were selected based on this motivation. Table 6.3 shows that the dissolution capacity increased in the anion order Ac^- (38.2 wt%) > Cl^- (29.5 wt%) for the same cation, $(C_1MIM)_2OC_4$. This indicates the ability of Ac^- anion, which has high basicity to favor the formation of hydrogen bonding with hydrogen atoms then Cl^- anion in cellulose (Vo et al., 2011). In addition, lower viscosity was also attained with Ac^- anion as compared to Cl^- anion which facilitates effective dissolution process (Table 6.2). The most interesting element appears when the dissolution capacity of acetate-based ILs were compared between monocationic with dicationic imidazolium ILs. For instance, 1-butyl-3-methylimidazolium acetate, BMIMOAc ILs is the closest monocationic structure which can be compared to the acetate-based dicationic ILs used in this study, $(C_1MIM)_2OC_4$ 2Ac. BMIMOAc recorded dissolution capacity of 12 wt% in the work carried out by Zhao et al. (Zhao, Baker, and Cowins, 2010), whereas $(C_1MIM)_2OC_4$ 2Ac ILs used in this has the dissolution capacity of 38.2 wt%. This remarkable increase could be strongly associated with the increase in the concentration of Ac^- anion which doubled in the ability to form hydrogen bonding with hydroxyl groups of cellulose. Therefore, this observation somewhat provided an insight that acetate bases dicationic ILs with alkoxy spacer have an excellent potential in dissolving cellulose.

Based on the anion performance, $(C_1MIM)_2OC_4$ 2Ac ILs was selected as the best ILs for a further optimization step. This time, the effect of co-solvent was investigated. Many reports suggested that adding polar aprotic solvents (DMSO and DMF) to the ILs contributes to a wide range of advantages such as increased cellulose solubility and shorter dissolution time (Andanson et al., 2014; Rinaldi, 2011). This can be also seen from Table 6.3 where the dissolution capacity of $(C_1MIM)_2OC_4$ 2Ac ILs [38.2 wt%] has increased to 48.4 wt% when DMSO was added in the ratio of 1:1. The driving force of this phenomenon is due to the reduction in viscosity of the ILs when DMSO is added as the co-solvent. Therefore, low viscosity facilitates effective mass transfer during the dissolution process and results in increased dissolution capacity (Stoppa et al., 2009). Moreover, it can be also seen from Table 6.3 that the effect of adding DMSO has significantly reduced the dissolution time and temperature to 10 h and 30°C using $(C_1MIM)_2OC_4$ 2Ac ILs +DMSO, which is also in line with other reports (Andanson et al., 2014; Kostag and El Seoud, 2019).

Another prominent factor that affects the dissolution of cellulose is the polarity of ILs (Cláudio et al., 2014; Fukaya et al., 2008), which can be measured from a β value especially. A β value is a key indication of hydrogen bond basicity (HBA) of a solvent, ILs in this case. The β values of ILs used in this study are tabulated in Table 6.4. Some typical monocationic imidazolium ILs such as AMIMCl, BMIMCl and BMIMAc that are widely reported for cellulose dissolution have β values of 0.83 (Fukaya, Sugimoto, and Ohno, 2006), 0.95 (Lungwitz and Spange, 2008) and 1.09 (Ohno and Fukaya, 2009), respectively. It can be observed from Table 6.4 that dicationic imidazolium ILs for both Cl^- and Ac^- anion recorded significantly higher β values to monocationic imidazolium ILs. On the other note, Ac^- anion exhibited higher value of β compared to Cl^- anion for dicationic imidazolium ILs used in this work, further strengthening the reason behind its highest cellulose dissolution capacity.

TABLE 6.4

H-Bond Basicity Values (β) and Polarity (π*) of the Imidazolium Dicationic ILs

ILs	β (H-Bond Basicity)	π* (Dipolarity)
$(C_1MIM)_2OC_4$ 2Cl	1.13 ± 0.04	1.18 ± 0.02
$(C_4MIM)_2OC_4$ 2Cl	1.20 ± 0.04	1.29 ± 0.02
$(C_1MIM)_2OC_4$ 2Ac	1.35 ± 0.04	1.38 ± 0.03

6.3.2 MORPHOLOGICAL STUDIES OF CELLULOSE

Field emission scanning electron microscope (FESEM) was to observe the morphological change between cellulose and regenerated cellulose in this study. Figure 6.2(a&b) depicts the bulk structure of untreated cellulose and regenerated cellulose after dissolution in $(C_1MIM)_2OC_4$ 2Ac/ DMSO solution, respectively. Figure 6.2(a) exhibited irregular and rough surfaces of individual micron size cellulose fibrils that were separated up to a certain extent. However, this observation diminished where the fibrils were seen to fuse together homogeneously in Figure 6.2(b). This could be highly associated with the dissolution effect demonstrated by ILs (Swatloski et al., 2002).

6.3.3 CRYSTALLINITY INDEX AND THERMAL STABILITY

Figure 6.3 displayed the X-ray diffraction (XRD) diffractogram patterns that were generated for both untreated and treated (regenerated) MCC using $(C_1MIM)_2OC_4$ 2Ac/DMSO solution to assess the change the change in the microcrystalline structure of the cellulose. The untreated MCC clearly exhibited cellulose I structural characteristics with three major fingerprints, i (15.3°), ii (22.0°) and iii (34.8°) (Tang et al., 2012; Zhao et al., 2009). When MCC was treated and regenerated from $(C_1MIM)_2OC_4$ 2Ac /DMSO, two of the crystalline composite peaks of MCC (i & iii) vanished, suggesting the distortion in cellulose I lattice. Moreover, the intensity of the peak (ii) reduced notably and shifted to a lower value of 21.1° and a small peak appeared at 12.6°. These transitions are a clear indication of the formation of cellulose II structure. In order to measure the extent of the disruption in the crystalline nature of the cellulose, crystallinity index, (CrI) was measured according to Park and coworkers (Park et al., 2010), which used a peak height method. The reduction

(a)   (b)

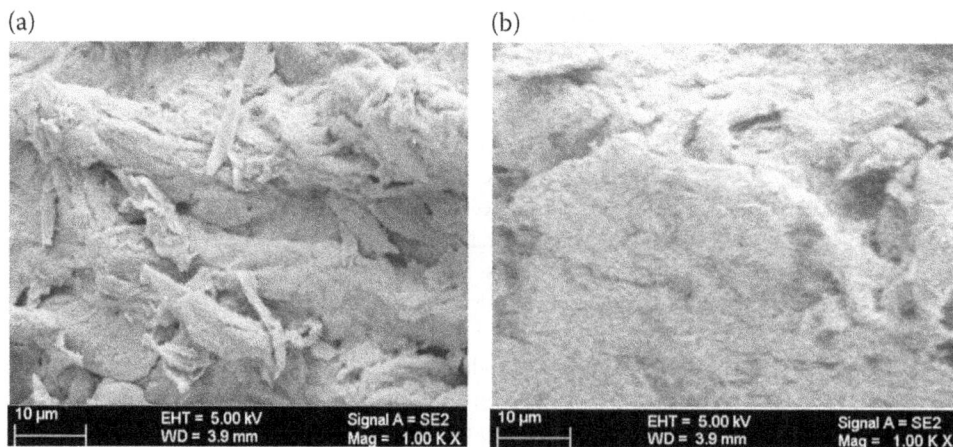

FIGURE 6.2 (a) Untreated MCC, (b) treated and regenerated MCC with ILs.

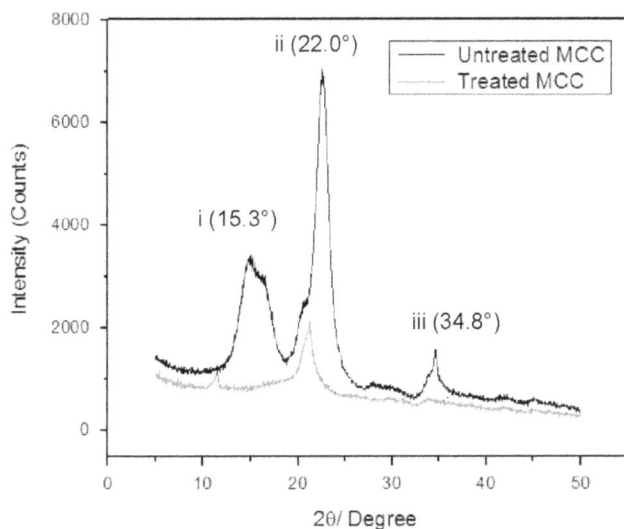

FIGURE 6.3 (XRD) diffractogram untreated and treated (regenerated) MCC using $(C_1MIM)_2OC_4$ 2Ac/DMSO solution.

in CrI value from 83.5 (pure MCC) to 32.4 (regenerated MCC) shows the crystallinity of MCC was significantly disrupted by dissolution using $(C_1MIM)_2OC_4$ 2Ac/DMSO solution. The sign of reduced CrI value of regenerated MCC can be also seen from its thermal behavior.

Thermogravimetric analysis (TGA) was carried out for both untreated and treated MCC cellulose with $(C_1MIM)_2OC_4$ 2Ac/DMSO solution. Figure 6.4 displays the thermal decomposition temperature, T_D, of treated cellulose shifted to a lower value (221°C) compared to untreated cellulose (301°C). Such an observation was also reported widely by few groups (Lethesh et al., 2020; Tang et al., 2012), indicating the reduction in thermal stability of a treated cellulose could be due to the drop in the crystallinity.

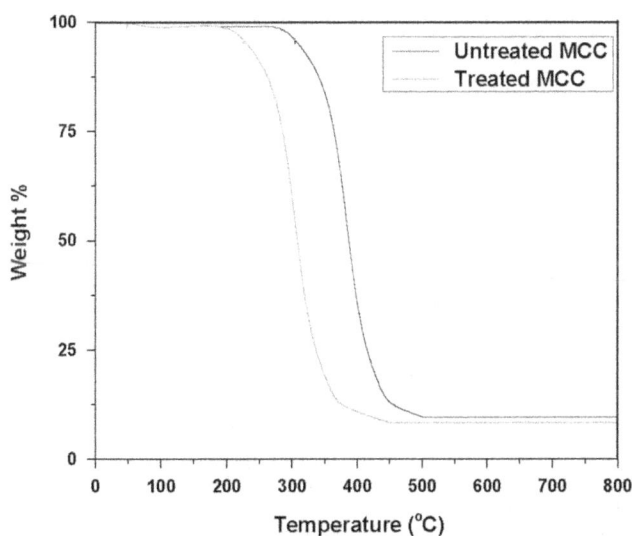

FIGURE 6.4 TGA of untreated and treated (regenerated) MCC using $(C_1MIM)_2OC_4$ 2Ac/DMSO solution.

6.4 CONCLUSION

As a conclusion, dicationic imidazolium ILs linked with alkoxy spacer were developed, carrying chloride and acetate anions. Acetate anions exhibited superior dissolution capacity compared to chloride anions and the dissolution capacity further increased significantly when polar aprotic solvent, DMSO mixed with the ILs. The dissolution capacity of 48.4% was reached at room temperature at a reduced dissolution period. The ability of the $(C_1MIM)_2OC_4$ 2Ac/DMSO solution as an effective solvent with low viscosity can be seen from XRD diffractogram, which exhibited the conversion of cellulose I to cellulose II. The significant reduction in CrI values and degradation temperature of regenerated cellulose also suggests the immense potential of this dicationic ILs in the direction of cellulose dissolution.

DECLARATION OF COMPETING INTEREST

None

ACKNOWLEDGEMENTS

The technical support and fund (PRF 0153AB A-33) produced by the Shale Gas Research Group (SGRG) and Institute of Enhanced Hydrocarbon Recovery (IHR), Universiti Teknologi Petronas.

REFERENCES

Abe, Mitsuru, Kazuki Sugimura, Yoshiharu Nishiyama, Yoshiyuki Nishio. 2017. Rapid benzylation of cellulose in tetra-n-butylphosphonium hydroxide aqueous solution at room temperature. *ACS Sustainable Chemistry, and Engineering*. 5 (6):4505–4510.

Andanson, Jean-Michel, Emilie Bordes, Julien Devémy, Fabrice Leroux, Agilio AH Pádua, and Margarida F Costa Gomes. 2014. Understanding the role of co-solvents in the dissolution of cellulose in ionic liquids. *Green Chemistry*. 16 (5):2528–2538.

Anderson, Jared L, Rongfang Ding, Arkady Ellern, and Daniel W. Armstrong. 2005. Structure and properties of high stability geminal dicationic ionic liquids. *Journal of the American Chemical Society*. 127 (2):593–604.

Brandt, Agnieszka, John Gräsvik, Jason P Hallett, and Tom Welton. 2013. Deconstruction of lignocellulosic biomass with ionic liquids. *Green Chemistry*. 15 (3):550–583.

Cláudio, Ana Filipa M, Lorna Swift, Jason P Hallett, Tom Welton, João AP Coutinho, and Mara G. Freire. 2014. Extended scale for the hydrogen-bond basicity of ionic liquids. *Physical Chemistry Chemical Physics*. 16 (14):6593–6601.

Ebrahimi, Maryam, and Fatemeh Moosavi. 2018. The effects of temperature, alkyl chain length, and anion type on thermophysical properties of the imidazolium based amino acid ionic liquids. *Journal of Molecular Liquids*. 250:121–130.

Fan, Mingming, Jing Yang, Pingping Jiang, Pingbo Zhang, and Sisi Li. 2013. Synthesis of novel dicationic basic ionic liquids and its catalytic activities for biodiesel production. *RSC Advances*. 3 (3):752–756.

Fernandes, Ana M, Marisa AA Rocha, Mara G Freire, Isabel M Marrucho, Joao AP Coutinho, and Luís MNBF Santos. 2011. Evaluation of cation–anion interaction strength in ionic liquids. *The Journal of Physical Chemistry B*. 115 (14):4033–4041.

Ferreira, Ana R, Mara G Freire, Jorge C Ribeiro, et al. 2011. An overview of the liquid– liquid equilibria of (ionic liquid+ hydrocarbon) binary systems and their modeling by the conductor-like screening model for real solvents. *Industrial & Engineering Chemistry Research*. 50 (9):5279–5294.

FitzPatrick, Michael, Pascale Champagne, Michael F Cunningham, and Ralph A. Whitney. 2010. A biorefinery processing perspective: Treatment of lignocellulosic materials for the production of value-added products. *Bioresource Technology*. 101 (23):8915–8922.

Fukaya, Yukinobu, Kensaku Hayashi, Masahisa Wada, and Hiroyuki Ohno. 2008. Cellulose dissolution with polar ionic liquids under mild conditions: Required factors for anions. *Green Chemistry*. 10 (1):44–46.

Fukaya, Yukinobu, Akiko Sugimoto, and Hiroyuki Ohno. 2006. Superior solubility of polysaccharides in low viscosity, polar, and halogen-free 1, 3-dialkylimidazolium formates. *Biomacromolecules*. 7 (12):3295–3297.

Gericke, Martin, Pedro Fardim, and Thomas Heinze. 2012. Ionic liquids—promising but challenging solvents for homogeneous derivatization of cellulose. *Molecules*. 17 (6):7458–7502.

Hammer, RB, and AF Turbak. 1977. *Production of Rayon from Solutions of Cellulose in N2O4-DMF*: ACS Publications.

Isik, Mehmet, Haritz Sardon, and David Mecerreyes. 2014. Ionic liquids and cellulose: Dissolution, chemical modification and preparation of new cellulosic materials. *International Journal of Molecular Sciences*. 15 (7):11922–11940.

Kamlet, Mortimer , and RW Taft. 1976. The solvatochromic comparison method. I. The. beta-scale of solvent hydrogen-bond acceptor (HBA) basicities. *Journal of the American Chemical Society*. 98 (2):377–383.

Kostag, Marc, and Omar A. El Seoud. 2019. Dependence of cellulose dissolution in quaternary ammonium-based ionic liquids/DMSO on the molecular structure of the electrolyte. *Carbohydrate Polymers*. 205:524–532.

Lethesh, Kallidanthiyil Chellappan, Sigvart Evjen, Vishwesh Venkatraman, Syed Nasir Shah, and Anne Fiksdahl. 2020. Highly efficient cellulose dissolution by alkaline ionic liquids. *Carbohydrate Polymers*. 229:115594.

Li, Xin, Haichao Li, Zhe Ling, et al. 2020. Room-Temperature superbase-derived ionic liquids with facile synthesis and low viscosity: Powerful solvents for cellulose dissolution by destroying the cellulose aggregate structure. *Macromolecules*. 53 (9):3284–3295.

Lungwitz, Ralf, and Stefan Spange. 2008. A hydrogen bond accepting (HBA) scale for anions, including room temperature ionic liquids. *New Journal of Chemistry*. 32 (3):392–394.

MacFarlane, Douglas R, Jennifer M Pringle, Katarina M Johansson, Stewart A Forsyth, and Maria Forsyth. 2006. Lewis base ionic liquids. *Chemical Communications*. (18):1905–1917.

Michud, Anne, Marjaana Tanttu, Shirin Asaadi, et al. 2016. Ioncell-F: Ionic liquid-based cellulosic textile fibers as an alternative to viscose and lyocell. *Textile Research Journal*. 86 (5):543–552.

Ohno, Hiroyuki, and Yukinobu Fukaya. 2009. Task specific ionic liquids for cellulose technology. *Chemistry Letters*. 38 (1):2–7.

Park, Sunkyu, John O Baker, Michael E Himmel, Philip A Parilla, and David K. Johnson. 2010. Cellulose crystallinity index: Measurement techniques and their impact on interpreting cellulase performance. *Biotechnology for Biofuels*. 3 (1):1–10.

Pettersen, Roger C. 1984. The chemical composition of wood. *The Chemistry of Solid Wood*. 207:57–126.

Potthast, Antje, Thomas Rosenau, and Paul Kosma. 2006. Analysis of oxidized functionalities in cellulose. *Polysaccharides II*. 1–48.

Raj, Jaganathan Joshua, Sivapragasam Magaret, Matheswaran Pranesh, Kallidanthiyil Chellappan Lethesh, Wilfred Cecilia Devi, and MI Abdul Mutalib. 2019. Dual functionalized imidazolium ionic liquids as a green solvent for extractive desulfurization of fuel oil: Toxicology and mechanistic studies. *Journal of Cleaner Production*. 213:989–998.

Raj, Jaganathan Joshua, Cecilia Devi Wilfred, Syed Nasir Shah, Matheswaran Pranesh, MI Abdul Mutalib, and Kallidanthiyil Chellappan Lethesh. 2017. Physicochemical and thermodynamic properties of imidazolium ionic liquids with nitrile and ether dual functional groups. *Journal of Molecular Liquids*. 225:281–289.

Rinaldi, Roberto. 2011. Instantaneous dissolution of cellulose in organic electrolyte solutions. *Chemical Communications*. 47 (1):511–513.

Sirviö, Juho Antti, and Juha P. Heiskanen. 2020. Room-temperature dissolution and chemical modification of cellulose in aqueous tetraethylammonium hydroxide–carbamide solutions. *Cellulose*. 27 (4):1933–1950.

Stolarska, Olga, Anna Pawlowska-Zygarowicz, Ana Soto, Héctor Rodríguez, and Marcin Smiglak. 2017. Mixtures of ionic liquids as more efficient media for cellulose dissolution. *Carbohydrate Polymers*. 178:277–285.

Stoppa, Alexander, Johannes Hunger, Richard Buchner. 2009. Conductivities of binary mixtures of ionic liquids with polar solvents. *Journal of Chemical, and Engineering Data*. 54 (2):472–479.

Swatloski, Richard P, Scott K Spear, John D Holbrey, and Robin D Rogers. 2002. Dissolution of cellose with ionic liquids. *Journal of the American Chemical Society*. 124 (18):4974–4975.

Tang, Shaokun, Gary A Baker, Sudhir Ravula, John E Jones, and Hua Zhao. 2012. PEG-functionalized ionic liquids for cellulose dissolution and saccharification. *Green Chemistry*. 14 (10):2922–2932.

Tang, Shaokun, Gary A Baker, and Hua Zhao. 2012. Ether-and alcohol-functionalized task-specific ionic liquids: Attractive properties and applications. *Chemical Society Reviews*. 41 (10):4030–4066.

Vo, Huyen Thanh, Chang Soo Kim, Byoung Sung Ahn, Hoon Sik Kim, Hyunjoo. 2011. Study on dissolution and regeneration of poplar wood in imidazolium-based ionic liquids. *Journal of Wood Chemistry Lee, and Technology*. 31 (2):89–102.

Wang, Hui, Gabriela Gurau, and Robin D Rogers. 2012. Ionic liquid processing of cellulose. *Chemical Society Reviews*. 41 (4):1519–1537.

Wasserscheid, Peter, and Thomas Welton. 2008. *Ionic Liquids in Synthesis*: John Wiley & Sons.

Wei-Li, Dai, Jin Bi, Luo Sheng-Lian, Luo Xu-Biao, Tu Xin-Man, and Au Chak-Tong. 2014. Polymer grafted with asymmetrical dication ionic liquid as efficient and reusable catalysts for the synthesis of cyclic carbonates from CO2 and expoxides. *Catalysis Today*. 233:92–99.

Wohlhauser, Sandra, Gwendoline Delepierre, Marianne Labet, et al. 2018. Grafting polymers from cellulose nanocrystals: Synthesis, properties, and applications. *Macromolecules*. 51 (16):6157–6189.

Xu, Airong, Jianji Wang, and Huiyong Wang. 2010. Effects of anionic structure and lithium salts addition on the dissolution of cellulose in 1-butyl-3-methylimidazolium-based ionic liquid solvent systems. *Green Chemistry*. 12 (2):268–275.

Zhang, Hao, Jin Wu, Jun Zhang, and Jiasong He. 2005. 1-Allyl-3-methylimidazolium chloride room temperature ionic liquid: A new and powerful nonderivatizing solvent for cellulose. *Macromolecules*. 38 (20):8272–8277.

Zhang, Zhengxi, Li Yang, Shichun Luo, Miao Tian, Kazuhiro Tachibana, and Kouichi Kamijima. 2007. Ionic liquids based on aliphatic tetraalkylammonium dications and TFSI anion as potential electrolytes. *Journal of Power Sources*. 167 (1):217–222.

Zhao, Hua, Gary A Baker, and Janet V. Cowins. 2010. Fast enzymatic saccharification of switchgrass after pretreatment with ionic liquids. *Biotechnology Progress*. 26 (1):127–133.

Zhao, Hua, Gary A Baker, Zhiyan Song, Olarongbe Olubajo, Tanisha Crittle, and Darkeysha Peters. 2008. Designing enzyme-compatible ionic liquids that can dissolve carbohydrates. *Green Chemistry*. 10 (6):696–705.

Zhao, Hua, Cecil L Jones, Gary A Baker, Shuqian Xia, Olarongbe Olubajo, and Vernecia N. Person. 2009. Regenerating cellulose from ionic liquids for an accelerated enzymatic hydrolysis. *Journal of Biotechnology*. 139 (1):47–54.

7 Swelling Studies on Hydrogel Blend Used in Biomedical Applications

Shreyas Devanathan
University of New South Wales, Sydney, Australia

Sheila Devasahayam
Monash University, Victoria, Australia

Sri Bandyopadhyay
University of New South Wales, Sydney, Australia

CONTENTS

DOI: 10.1201/9781003200710-8

7.1 INTRODUCTION

European Society for Biomaterials defines biomaterials as 'material intended to interface with biological systems to evaluate, treat, augment or replace any tissue, organ or function of the body'. Biomaterials used to make biomedical devices, replace a part or a function of the body in a safe, reliable, economic and physiologically acceptable manner (Ethridge & Hench, 1982). Biomedical devices stay in contact with tissue, blood and biological fluids, and are used for prosthetic, diagnostic, therapeutic and storage applications. They are classified into four groups according to their use in the body: polymers, metals, ceramics and composites. They can be either synthetic or of natural origin and can be made of either hard or flexible materials. The main requirements of biomaterials for biomedical applications include biocompatibility and biofunctionality. They must not react with any tissue in the body, must not be toxic to the body and long-term replacement must not be biodegradable. Biomaterials must not adversely affect the living organism and its components (Bruck, 1980).

Factors that adversely affect the use of biomaterials and the long-term function of the engineered tissue construct include: immunogenicity, host inflammatory responses, fibrous tissue formation, biomaterial degradation and toxicity of degradation products (Norotte et al., 2009). Biomaterials applications face challenges such as replicating complex tissue architecture and arrangement in vitro, understanding extracellular and intracellular modulators of cell function, developing novel materials and processing techniques that are biocompatible with better immune acceptance.

7.1.1 TISSUE ENGINEERING (TE)

Tissue engineering (TE), an important area of biomedical engineering, involves developing bioartificial substitutes for organs and tissues, to generate living tissue ex vivo for replacement or therapeutic applications. This is achieved through materials development, biochemical manipulations, cell culture and genetic engineering. TE induces tissue-specific regeneration processes and overcomes the drawbacks of organ transplantation, e.g., donor shortage or need of immunosuppressive therapy.

TE combines living cells and a natural/synthetic support or scaffold to build a three-dimensional (3D) living construct that has similar or better function, structure and mechanical properties of the tissue that needs replacing. TE depends on biomimetic materials (3D scaffolds), which provide appropriate environment for the new tissue development, structure for cell adhesion, proliferation and extracellular matrix deposition for new tissue restoration (Dolcimascolo et al., 2019). In vitro 3D TE model reduces the use of animals, and offer many benefits over the conventional models, e.g., animal and 2D cell culture models (Caddeo et al., 2017). 3D-printing technologies are applied to biocompatible materials, cells and supporting components and for artificial organ printing and regenerative medicine applications (Murphy & Atala, 2014). 3D-printing technologies for TE include, Powder-based 3D printing, Ink-based 3D printing, Polymerization-based 3D printing and four-dimensional (4D) printing (Liu & Yan, 2018). In 4D printing, additive manufactured structures use smart materials capable of self-transformation into a predefined shape or exert a predefined function depending on the stimuli present in the microenvironment (Goksu et al., 2019).

Success of biomedical implants depend on biomaterial surface design to control the cell–material interactions (Chen, 2016). For a suitable TE/cell and tissue response of biomaterials, surface modification is vital using plasma polymerisation, covalent binding of poly (ethylene glycol) (PEG), heparinisation, peptide functionalisation and calcium phosphate deposition (Williams, 2011).

7.1.2 THIN-FILM COATINGS

Surface coatings can support the biological functions of mammalian cells and subsequent tissue–implant integration, as well as discriminate bacterial colonization at the biological interface

(Chen, 2016). Factors affecting mammalian cell–material interactions and bacterial adhesion include surface biochemistry, surface charge, surface hydrophobicity, surface roughness/topography, surface porosity, surface crystallinity and surface stiffness. Developing suitable materials and surfaces for TE applications relies on cells' response to topography at different length scales (Wang et al., 2019).

Various materials used as coatings for biomedical implants are organic and inorganic coatings. Organic coatings have similar mechanical properties to soft biological tissues, promoting desirable mammalian cell–material interactions. Techniques for organic coatings include dip-coating, spray-coating, and spin-coating (Chen, 2016; Devasahayam & Hussain, 2020). Inorganic coatings promote cell-material interactions and subsequent tissue regeneration. Typical examples of inorganic coatings include, calcium phosphate based bioceramics, titanium oxide layers (or nanotubes) and carbon-based coatings (e.g., diamond-like carbon, carbon nanotubes, graphene coating) (Devasahayam & Hussain, 2020). Titanium oxide coatings provide better corrosion resistance to the implants. Techniques for inorganic coatings include sol gel and dip-coating techniques, spin-coating, layer-by-layer self-assembly, electrophoretic deposition, chemical vapour deposition and the pulsed laser deposition (Mozafari et al., 2016).

7.1.3 SCAFFOLDS

A scaffold is a suitable material for TE. It provides mechanical and porous network structural support, shape, and hierarchy architecture with surface chemistry for cell attachment, cell-cell communication through the porous network, as well as proliferation and differentiation for tissue regeneration.

Porous 3D scaffolds provide the appropriate environment for the regeneration of tissues and organs. Engineered scaffolds (3D biomaterial before cells have been added in vitro or in vivo) induce desirable cellular interactions and facilitate the formation of new functional tissues for medical purposes. Suitability of a scaffold in TE depend on its: biocompatibility, bioactivity, and biodegradability (Dolcimascolo et al., 2019), mechanical properties, scaffold architecture, manufacturing technology and the biomaterial used for scaffold fabrication. Synthetic polymers, ceramics, metals, composites and hydrogels are suitable for scaffolds.

Biocompatibility: A scaffold is biocompatible, when cells: 1) adhere, 2) function normally, 3) migrate onto the surface and eventually through the scaffold and 4) begin to proliferate before laying down new matrix (Fergal, 2011). After implantation, the scaffold or tissue engineered construct (scaffolds which have undergone extensive *in vitro* culture prior to implantation) must cause a negligible immune response to prevent a severe inflammatory response such as reduced healing or trigger rejection.

Biodegradability: TE must allow the body's own cells to eventually replace the implanted scaffold or tissue engineered construct and allow cells to produce their own extracellular matrix. The scaffold or tissue-engineered construct are not meant to be permanent implants, hence must be biodegradable. The by-products of this degradation should not be toxic and must exit the body without interfering with other organs. Degradation and tissue formation should occur simultaneously.

Mechanical Properties: Producing scaffolds with adequate mechanical properties is a big challenge when engineering bone or cartilage. They should exhibit similar mechanical properties to the anatomical site of implantation and be strong enough to allow surgical handling during implantation. Many materials with good mechanical properties and demonstrated potential in vitro fail when implanted in vivo due to insufficient capacity for vascularization.

Scaffold Architecture: The architecture of scaffolds is very important in TE. Scaffold should have an interconnected pore structure and high porosity to guarantee cellular penetration and diffusion of nutrients to cells and allow diffusion of waste products out of the scaffold. The products of scaffold degradation must exit the body without interfering with the other organs and surrounding tissues. The mean pore size of the scaffold need to be large enough to allow cells to migrate into the structure, to bind with the ligands within the scaffold, but small enough with a high specific surface and minimal ligand density to allow efficient binding of a critical number of cells to the scaffold (Fergal, 2011).

7.1.3.1 Manufacturing Technology

Cost effective, scalable manufacturing processes to good manufacturing practice (GMP) standard will ensure successful translation of TE strategies to the clinic, to ensure the product delivery and availability to the clinician.

Porous scaffold production involves solvent casting/particulate leaching method for bone TE (Thomson et al., 1995). Melt moulding/particulate leaching technique, using an unrefined thermoplastic polymer mixed with the porous agent is used to control the pore size and porosity. A variant of melt moulding is an extrusion or injection moulding technique where the porous agent is replaced with a blowing agent based on citric acid to form interconnected and well-shaped pores (Gomes et al., 2001). Other techniques include gas foaming, phase inversion/particulate leaching technique and fibre bonding that allow formation of scaffolds of porous 3D structure of dense frame of synthetic fibres.

Solid freeform fabrication (SSF) allows production of layer-by-layer 3D objects starting from information generated by a CAD system or computer-based medical imaging modalities. Another interesting SFF methodology is fused deposition modelling (FDM). In this case, a filament of thermoplastic material fed and melted inside a heated liquefier head is forced out by an extruder and deposited on a platform to obtain layer by layer 3D objects. Varying the direction of material deposition for each layer allows for changing the pore size and interconnectivity of the scaffold (Hutmacher et al., 2001; Zeltinger et al., 2001).

7.1.4 MEMBRANES

Cell membrane or the plasma membrane located in all cells, separate the core of the cell from the outside environment. Factors determining the use of membranes for various applications include, the molecular weights and physical-chemical properties of the species to be transported or rejected, the interactions between membranes and body fluids, and selecting the materials and processes that causes no harm to the patient. The cell membrane regulates the transport of materials entering and exiting the cell Figure 7.1 (Gahl, n.d.). The rate of solute transportation of a membrane is determined by its intrinsic mass transport properties. Membranes can be both flat sheet and hollow fibre membranes. Biomembranes include floating biomembrane model structure on the air-water interface that mimic the properties and conditions of the cell environment, and a supported

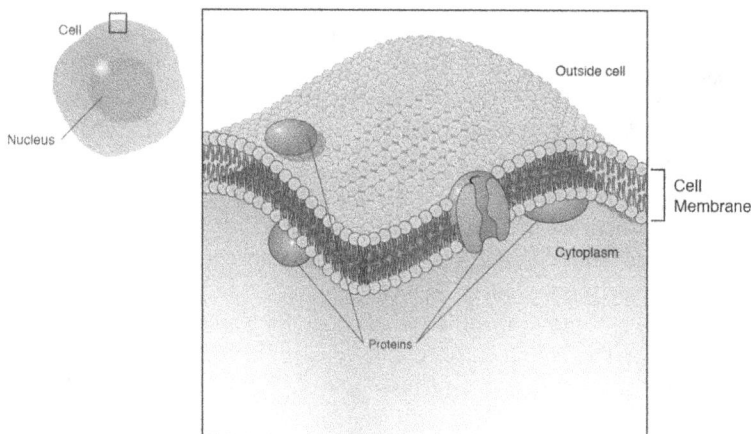

FIGURE 7.1   Cell membrane (Courtesy: National Human Genome Research Institute, https://www.genome. gov/genetics-glossary/Cell-Membrane).

biomembrane or lipid-based structure on a solid substrate, where lipid layers are deposited on surfaces. In nature, most biochemical reactions take place at biomembrane interfaces. Freely floating model membranes allow natural diffusion and migration of molecules.

The cell membrane consisting of a lipid bilayer is semipermeable. Cellulosic membranes generally are less permeable with narrower pores than those prepared with technical polymers. Cellular membranes play crucial roles in biointerfacing, self-identification, signal transduction and compartmentalization. Membranes used in medicine usually are of hollow-fibre configuration, arranged in the "shell-and-tube" configuration or in stacks of cross-wound HF membrane mats. Hollow-fibre membranes are generally prepared by phase inversion by spinning the dissolved or molten polymer in a spinneret with a hollow core, followed by induced polymer precipitation. Supported bilayers of phospholipids with layers of different lipid composition are prepared using Langmuir-Blodgett (LB) and Langmuir–Schaefer (LS) dipping methods (Biology at interfaces: Biomembranes Biolin Scientific, 2018). Medical membranes are classified as either natural or synthetic, depending on whether their backbone is made of cellulose or of a technical polymer.

There has been long-term interest in developing synthetic/artificial membranes that can match the selectivity and high-speed transport offered by natural membranes.

Artificial membranes are used to treat a broad variety of therapeutic purposes, as permselective barriers to permit transport of selected solutes to/from the blood, and to regulate the rate at which solutes are transferred across the membrane to maintain homeostasis (Catapano et al., 2008). Synthetic membranes are solid in nature and can be composed of ceramics, metals, polymers, or combinations thereof (Abetz, Brinkmann & Sözbilir, 2021). It is reported the transport properties of polymer system e.g., proton transfer rates, are similar to biomembranes. Design and scalability of the polymer-based materials pave the way for the design of novel materials that mimic biological materials for a range of other applications (Xu, 2021). Thermoplastic synthetic polymers are used for preparing semipermeable medical membranes owing to their good mechanical resistance and low density. They include acrylic polymers, polyamides, polycarbonates, polyolephines and polysulfones.

7.1.5 HYDROGELS

Hydrogels are attractive biomaterials for many medical applications. Hydrogels resemble living tissue that opens up many opportunities in biomedical areas such as biomaterials, scaffolds for TE, delivery agents and medical devices and implants, absorbents and dentistry (Khan & Tanaka, 2017; Lionetto et al., 2003; Vashuk et al., 2001). Hydrogels are water swellable, yet water insoluble, the cross-linked networks exhibit high water contents and tissue-like elastic properties, making them ideal candidates as scaffolds for growing cells and tissues. The high water content of hydrogels provides an ideal environment for cell survival, and structure which mimics the native tissues. Their unique properties include softness, flexibility, biocompatibility, and their ability to absorb large amounts of water or biological fluids, instant pain killing effect, transparency, good adhesion, oxygen permeability, prevention of loss of fluids present in the body and easy reparability. Hydrogels are serious contenders as wound dressing and wound healing materials. They also have good transport properties and are easily injected into any system and are modified as per practical application requirements.

The main areas of hydrogel applications include contact lenses, wound dressings, drug delivery systems, TE and hygiene products (Enrica & Vitaliy, 2015). Other areas include bone tissue, cartilage tissue, vascular tissues, meniscus, tendon, skin, cornea and soft tissues. Other advantages of hydrogels include timely release of growth factors and nutrients for tissue growth. It is used in targeted drug delivery to kidneys and other (Lionetto et al., 2003) essential areas in the human body (Lee & Mooney, 2001).

Hydrogels have been used as one of the most common TE scaffolds over the past two decades due to their ability to maintain a distinct 3D structure, to provide mechanical support for the cells in the engineered tissues and to simulate the native extracellular matrix (Mantha et al., 2019). A hydrogel is not soluble in water, due to the presence of a 3D polymer network, and swells at equilibrium. The 3D networks in hydrogels are formed either by covalent bonding between polymer chains or physical interactions such as hydrogen boning and ionic bonding. The cross-linked network provides an equilibrium state between dispersing and cohesive forces between polymer chains, resulting in the insolubility of the gel in water. Hydrogel scaffolds are used for fabrication or repair of the lost tissue (Kopeček, 2007) and to tune the physical and mechanical properties according to the desired application. They support cell proliferation, migration and differentiation, to permit oxygen and nutrient transport and to mimic native soft tissue (Spicer, 2020). They have the ability to support cell proliferation, migration and differentiation, to permit oxygen and nutrient transport and to mimic native soft tissue making them attractive material in TE (Spicer, 2020).

Chemical hydrogels can be prepared by '3-D polymerization' where a hydrophilic monomer is polymerized using a polyfunctional cross-linking agent, or by direct cross-linking of water-soluble polymers. 3D printing and smart hydrogels are a potent combination of bioprinting functional 3D tissues. Hydrogels in bioprinting act as a matrix that supports and regulates the cells encapsulated inside the matrix. In 4D printing, a responsive hydrogel fold into a 3D shape upon exposure to a stimulus (Mironov, 2014). Figure 7.2 shows a 4D-printed shape memory polymer gripper working on heat stimulus (Ge et al., 2016).

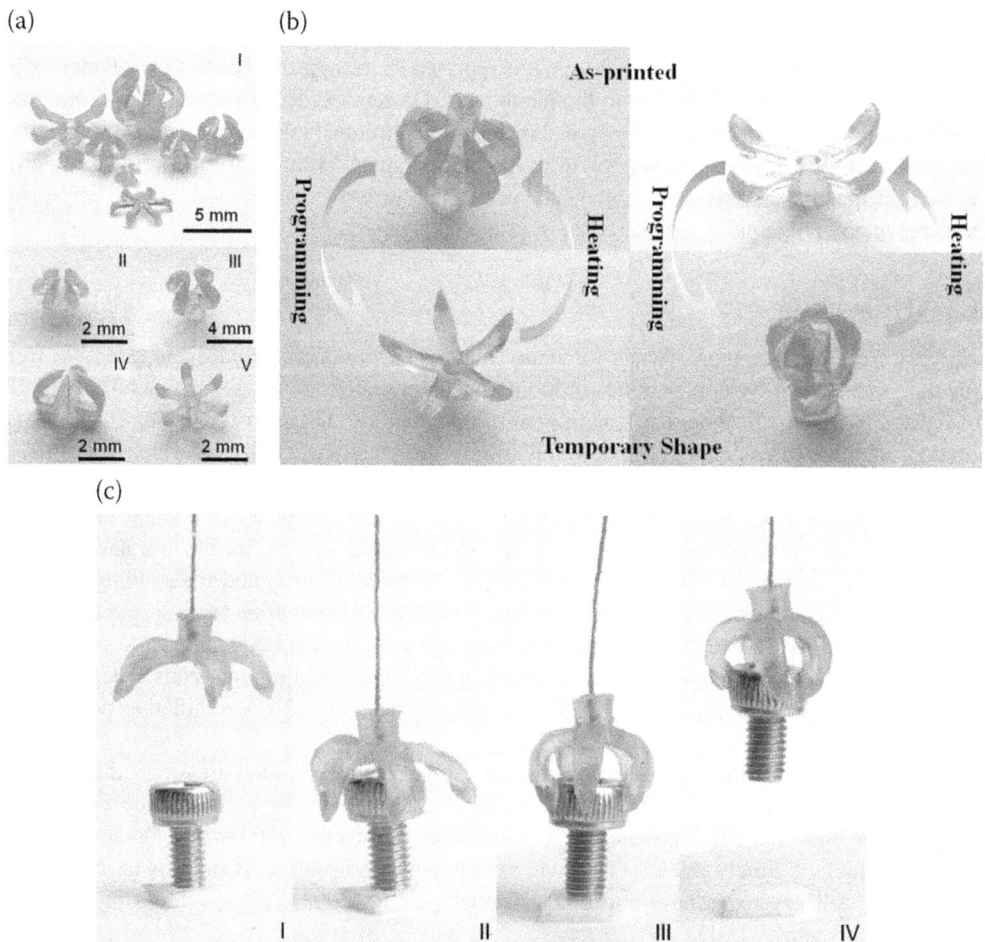

FIGURE 7.2 4D-printed shape memory polymers gripper working on heat stimulus (Ge et al., 2016) (licensed under a Creative Commons Attribution 4.0 International License).

3D polymerization often leaves behind significant amounts of residual monomers that are toxic and can leach out of the hydrogels continuously. Therefore, purification of these materials is necessary. It is possible to avoid the purification of hydrogels by cross-linking ready-made water-soluble polymers such as poly(acrylic acid), poly(vinyl alcohol) (PVA), poly(vinylpyrrolidone) (PVP), poly(ethylene glycol), polyacrylamide and some polysaccharides (Enrica & Vitaliy, 2015). Hydrogels are classified as 'reversible' or 'physical' gels if the network comprises molecular entanglements and/or secondary forces, e.g., ionic, H-bonding or hydrophobic forces. It is possible to dissolve the physical gels by changing the pH, and the ionic strength of solution or temperature (Enrica & Vitaliy, 2015). In 'permanent' or 'chemical' gels, the network comprises of covalent bonds linking different macromolecular chains via cross-linking in dry or wet state (A. Hoffman, 2012). These gels may be polar or non-polar depending on the functional groups present. Swelling behaviour of charged hydrogels changes when pH changes, and shape changes occur when subjected to an electric field (Rosiak & Yoshii, 1999). Their functional groups, and their structure, render them biocompatible and/or biodegradable, as well as impart therapeutic or toxic characteristics. The present work relates to water-soluble polymers like PVP and PVA blends used in biomedical applications that can form film (Jones et al., 2005).

7.1.5.1 Poly(vinyl alcohol) (PVA)

PVA is a linear synthetic polymer produced via partial or full hydrolysis of polyvinyl acetate to remove the acetate groups. PVA is used in controlled drug delivery systems, dialysis membrane, wound dressing, artificial cartilage and tissue engineering scaffold (Zhou et al., 2010), as well as in artificial pancreas, synthetic vitreous body, artificial skin and cardiovascular devices (Abd El-Mohdy & Ghanem, 2009). It is used as a suture material for tight tying, artificial tendons, artificial ligaments and reinforcing fibres for biocomposites in the synthesis of membranes for use in artificial pancreas material (Gough, Scotchford & Downes, 2002). The biodegradability and water solubility of PVA ensure easy degradation and elimination after its use (Park, Park & Ruckenstein, 2001).

The number of hydroxyl groups determine the physical characteristics, chemical properties and mechanical properties of the PVA (Figure 7.3). Even though PVA has higher water absorptivity, its use as a biomedical material is hindered due to its crystallinity and higher hydrophobicity compared to PVP. PVA is highly water soluble and is resistant to most organic solvents (Baker et al., 2012). This enables PVA films to be casted using evaporation technique from aqueous polymer solutions, avoiding the use of organic solvents. The resultant films are clear, homogeneous and resistant to tear. The high content of hydroxyl groups provides PVA and PVA-based materials with other desirable properties e.g., hydrophilicity, nontoxicity, non-carcinogenicity, non-immunogenicity, and inertness in body fluids, making them suitable for biomedical applications.

PVA has good mechanical properties in the dry state, but the high hydrophilicity limits its scope in the wet state, that is, in living environments (Zhou et al., 2010). PVA has poor stability in water because of its high hydrophilic character. Therefore, to render it stable, PVA needs to be treated by copolymerization, grafting, cross-linking and blending.

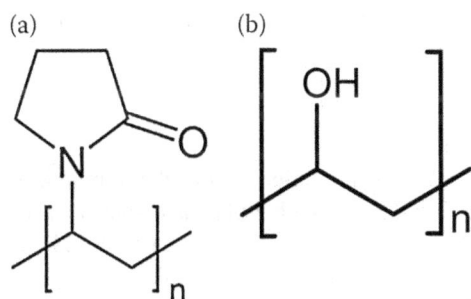

FIGURE 7.3   Molecular structures of (a) PVP and (b) PVA.

7.1.5.2 Poly(vinylpyrrolidone) (PVP)

PVP, a biocompatible engineered polymer, is also hydrophilic, with a higher affinity for water than PVA. PVP is a suitable composite material for enhancing the hydrophilicity of PVA fibre in composites. Under normal conditions, PVP is stable as a solid in solutions. Even in strong acidic conditions, PVP is unusually stable, with no changes in appearance or viscosity, e.g., in 15% HCl at 24°C for two months. PVP has a good stand in pharmacy and medicine due to its very high absorption capacity, complex forming abilities and stability in the biomedical environment (Andrés Bernal, 2012; Robinson et al., 1990; Sionkowska, 2006; Zohuriaan-Mehr & Pourjavadi, 2003).

PVP's biomedical applications include, in controlled drug-release technology, electrochemical devices, as an effective and interesting tissue engineering matrix, as a main component of temporary skin covers, wound dressings, for the preparation of synthetic plasmas (substitute of plasma blood), for creations of hydrogels or thrombo-resistant hydrophilic gels, as a factor giving higher biological activity of bio-artificial polymeric materials and to increase the hydrophilicity of blended polymeric materials.

7.1.5.3 PVA/PVP Blends

The PVP/PVA hydrogels have good water absorptivity and when prepared by low temperature gelation, act as good barriers against the microbes (Azad et al., 2004; Vashuk et al., 2001). The hydroxyl groups of PVAs easily bond with the proton-accepting carbonyl groups in the PVP structure forming hydrogen bonds in the blend. This hydrogen bond influences the solubility and other important properties of the polymer blend (Lai, Du & Li, 2007; Sionkowska, 2006). PVA/PVP blends are stable in physiological environment because of the cross-linking comprising intermolecular hydrogen bonding between the PVA and PVP contributing to their biocompatibility.

The addition of PVP to PVA results in reduction in glass transition temperature, t_g, indicating that the PVP plasticises PVA through PVA/PVP bonding, thus disrupting the crystalline phase of PVA. The crystalline regions of PVA are more accessible to PVP and therefore, the PVA/PVP blends are readily formed (Andrés Bernal, 2012).

Stability or the solubility of the hydrogels in contact with the bodily fluids (Jones et al., 2005; Liu & Yan, 2018; Vashuk et al., 2001) can be controlled by varying the blend composition of the hydrogels and through the use of additives to increase or decrease their solubility in water and to increase or modify their mechanical properties to obtain substantial films for implants in the human body (Bajpai, Bajpai & Shukla, 2001).

This study details the synthesis and the characterisation of hydrogels blends of PVA/PVP. The long-term and short-term swelling attributes of the blend is discussed. The mass loss analysis of the blend when immersed in DI water pertaining to biodegradability; and the chemical morphology and the presence of inter-chain hydrogen bonding in the blends as confirmed using FTIR analysis is presented.

7.2 EXPERIMENTAL

7.2.1 MATERIALS

PVA and PVP are obtained from Sigma Aldrich. Deionized water is used as the solvent for the swelling and mass loss experiments.

7.2.2 SYNTHESIS OF THE BLEND

Different ratios of PVA and PVP are shown in Table 7.1. Polymer solutions (10 wt.%) were placed in beakers and sealed. The solutions were heated at 90°C for 12 h on a hot plate and sonicated for half an hour after stirring to remove the entrapped air bubbles. Each solution was cast into a Petri dish and glass slides and placed at room temperature. The vaporization process took three days to complete. Films with appropriate blend mixtures were collected.

TABLE 7.1

Physical Properties of PVA and PVP Blend

Number	Weight %		Weight Fraction		Density (g/cm³)		Theoretical Density (g/cm³)
	PVP	PVA	PVP	PVA	PVP	PVA	PVP & PVA Mixture
1	100	0	1	0	1.2	1.31	0.83
2	75	25	0.75	0.25	1.2	1.31	0.813
3	50	50	0.5	0.5	1.2	1.31	0.797
4	25	75	0.25	0.75	1.2	1.31	0.78
5	0	100	0	1	1.2	1.31	0.763

7.2.3 SWELLING EXPERIMENTATION

The polymer blends were immersed in DI water for 2 h to obtain a uniform disc-shaped solidified structure (Figure 7.4). The discs were dried in a vacuum oven to remove the impurities at 35°C. The dried discs were weighed and noted as the dry mass (m_{dry}). The discs were immersed in 30 mL DI water. The masses of the swollen gel were measured at intervals of 1 h, 2 h, 4 h, 8 h, 1 day, 2 days going up to 60 days. The solutions were changed after each reading to incorporate accurate readings. After the immersion test, the samples were dried at room temperature and weighed again,

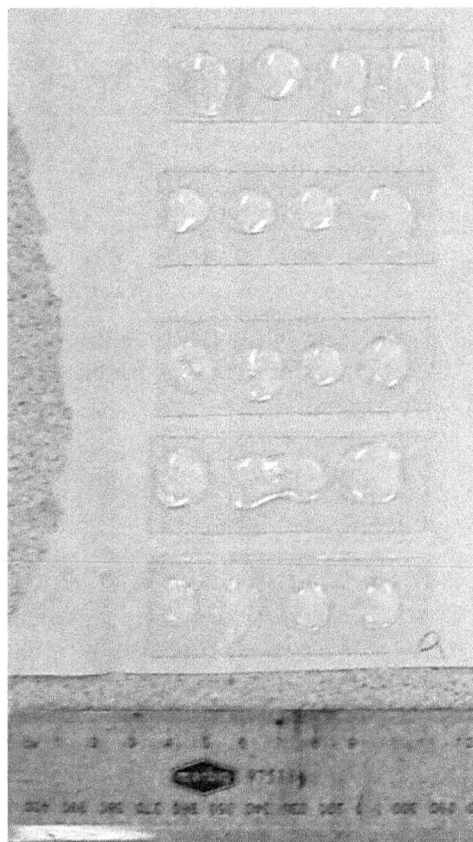

FIGURE 7.4 Samples used for swelling experiments.

which is the mass with respect to immersion time (m). The mass swelling coefficient (q) was calculated by using Equation 7.1:

$$q = \frac{m}{m_{dry}} \qquad (7.1)$$

where m is the mass as a function of time and m_{dry} is the initial dry mass of the polymer.

7.2.4 DEGREE OF SWELLING

For the degree of swelling tests, 2 cm^2 squares were cut and dried at room temperature to constant weight. These samples were immersed in DI water for 30 min to determine the degree of swelling. Readings were taken at the intervals of 1, 5, 10, 20 and 30 min. The degree of swelling of the blend was calculated using Equation 7.2:

$$Degree\ of\ swelling\,(\%) = \frac{(W_s - W_d)}{W_d} \times 100 \qquad (7.2)$$

where W_s and W_d are the weights of the swollen gel and the dried gel, respectively.

7.2.5 MASS LOSS ANALYSIS

The dry mass of the discs was compared against the initial dry masses of the discs. The % polymer mass loss is given by Equation 7.3:

$$Polymer\ Mass\ Loss\ (\%) = \frac{(m_{dry} - m_{60})}{m_{dry}} \times 100 \qquad (7.3)$$

where m_{60} = mass after 60 days of immersion.

7.2.6 FTIR

Attenuated total reflectance-Fourier transform infrared spectroscopy (ATR-FTIR). PerkinElmer, UNSW Australia) over a wavenumber range of 1,000–650 cm^{-1} was conducted on the blends to analyse its chemistry.

7.3 RESULTS

7.3.1 SWELLING COEFFICIENT (Q)

Figure 7.5 shows the comparative swelling characteristics of PVA/PVP blends. Immaculate PVA hydrogels prepared with 25% PVP acquired maximum swelling within two days into immersion. The swelling coefficient increased with increasing PVP content in the blend. The blends prepared with 50% PVP and 75% PVP showed greater swelling compared to the other blends, but were unstable when submerged in DI water. The 75% PVP was the most unstable blend. Due to its amorphous and hydrophilic nature, PVP dominated blends were swollen to a greater extent with less stability (Andres Bernal, Kuritka & Saha, 2000; Peppas & Merrill, 1976).

Water or fluid absorption capacity of the hydrogel is one of the important parameters for wound dressing or other biomedical applications. The favorable property of hydrogels is their ability to swell when they interact with a thermo-dynamical compatible solvent. It is reported swelling is not

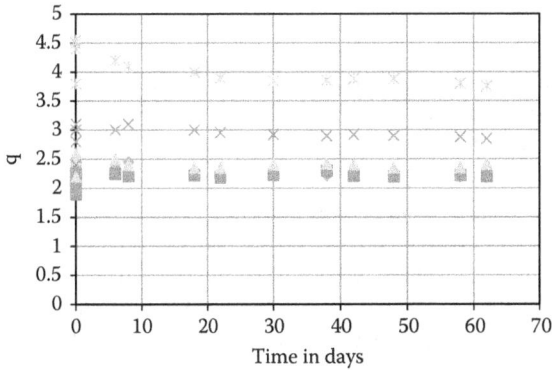

FIGURE 7.5 Comparative swelling analysis of PVA/PVP blends: ◆ 100% PVA; ■ 95% PVA and 5% PVP; ▲ 75% PVA and 25% PVP; ✕ 50% PVA and 50% PVP; ✕ 25% PVA and 75% PVP.

a continual process. The osmotic force is opposed by the elastic force balancing the stretching of the network preventing deformation. In addition, cross-linking also limits the stretching of the network preventing deformation.

The results shown in Figure 7.5 confirm the suitability of these PVA/PVP gels in terms of their swelling characteristics. The PVP ring contains a proton accepting carbonyl group, while PVA has hydroxyl groups and therefore, hydrogen bonds between them, influence the solubility and the mechanical properties of these PVA/PVP blends (Andres Bernal, Kuritka & Saha, 2000; Ping, Nguyen & Néel, 1989).

7.3.2 Degree of Swelling

Figure 7.6 shows the degree of swelling of the hydrogel blends. The degree of swelling increased with an increase in % PVP due to the higher affinity of PVP for water than PVA. This makes the film swell more and absorb higher quantities of water when compared to pure PVA samples. PVP reduces the crystallinity of the PVA in the blend resulting in increased swelling in PVP-dominated blends (Lakouraj, Tajbakhsh & Mokhtary, 2005). There is a sudden decline in water intake or absorption in 100% and 95% PVA after 20 min, whereas the other blends with increased PVP content shows increased water uptake.

7.3.3 Diffusion

The degree of swelling is higher for PVA/PVP blends than just for immaculate PVA. It is because PVP reduces the crystallinity of the PVA in the blend. Water molecules in contact with polymers

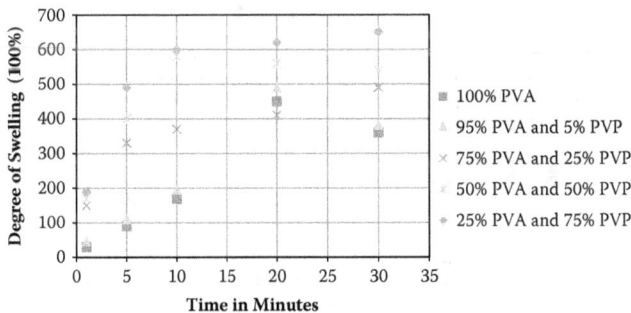

FIGURE 7.6 Degree of swelling of the polymer blends.

can penetrate into the free spaces of the chains and cause the swelling of the film (Chan, Hao & Heng, 1999). In addition, PVPs' higher affinity for water makes the film with PVP swell to a greater extent than pure PVA samples (Lakouraj, Tajbakhsh & Mokhtary, 2005). The swelling is increased during the first 10–20 min. After that the equilibrium is reached and no more water can get into the film in 100 and 95% PVA blends.

The pore size of micro-porous hydrogels is between 100 and 1,000 A° (Perale & Hilborn, 2016). Since the pore size begins to approach the size of the diffusing solutes, the solute transport occurs due to a combination of molecular diffusion and convection in the water-filled pores.

The degree of swelling was analysed according to the film diffusion (diffusion in the absence of product layer), α vs t, where α = degree of swelling/100 and t = time; and ash diffusion models (diffusion in the presence of product layer) for a disc shape, α^2 vs t. Concentrations of PVA and PVP have a big impact on the governing mechanism.

The results (Figures 7.7 and 7.8) show that 95% and 100% PVA follow a film diffusion mechanism as can be seen from the higher R^2 values compared to the ash diffusion model. But as the PVP content increases, the ash diffusion model (diffusion through product layer) is more applicable (with higher R^2 values than the film diffusion), indicating a formation of layer of a blend of amorphous layer progressing from the outer surface into the interior through which the water diffusion occurs.

The addition of PVP to PVA evidenced a reduction of t_g, implying the PVP plasticised PVA as a result of PVA/PVP bonding, disrupting the crystalline phase of PVA. The lower the tg, the faster

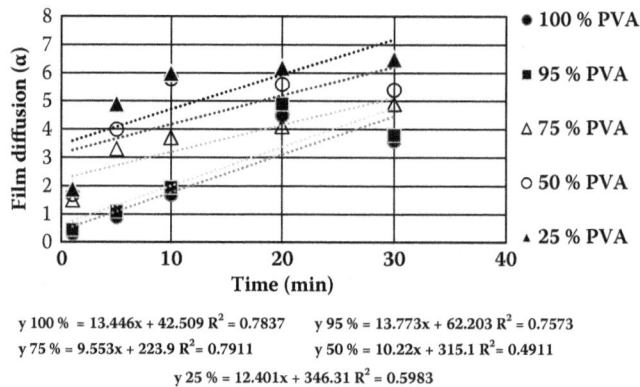

y 100 % = 13.446x + 42.509 R^2 = 0.7837   y 95 % = 13.773x + 62.203 R^2 = 0.7573

y 75 % = 9.553x + 223.9 R^2 = 0.7911   y 50 % = 10.22x + 315.1 R^2 = 0.4911

y 25 % = 12.401x + 346.31 R^2 = 0.5983

FIGURE 7.7   Film diffusion model.

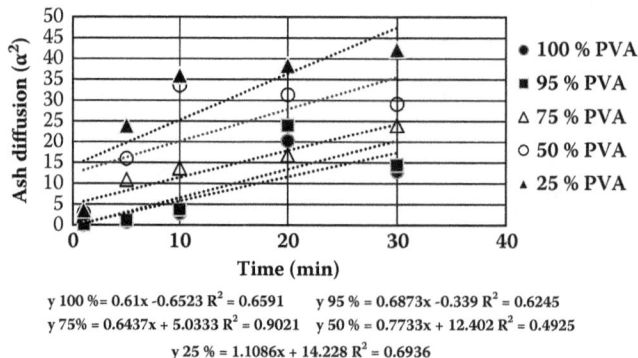

y 100 %= 0.61x -0.6523 R^2 = 0.6591   y 95 % = 0.6873x -0.339 R^2 = 0.6245

y 75%= 0.6437x + 5.0333 R^2 = 0.9021   y 50 % = 0.7733x + 12.402 R^2 = 0.4925

y 25 % = 1.1086x + 14.228 R^2 = 0.6936

FIGURE 7.8   Ash diffusion model.

TABLE 7.2
The Correlation Coefficients and the Rate Constants

% PVA	R^2 α	R^2 α^2	K1 α	K2 α^2
100	0.7837	0.6591	13.44	0.61
95	0.7573	0.6245	13.77	0.69
75	0.7911	0.9021	9.53	0.64
50	0.4911	0.4925	10.22	0.77
25	0.5983	0.6936	12.4	1.11

the diffusion. Water swells the amorphous parts but not the crystalline parts. The diffusivity of water in the amorphous phase is higher than the crystalline parts, because the matrix becomes more mobile (from glassy to rubbery). Water cannot access the crystalline domains at room temperature. The energy barrier for diffusion in a crystalline phase is higher compared to the amorphous phase. The water in the crystalline structure does not move because the energy barrier for creating space involving breaking a lot of H-bonds is too high. Consequently, polymers that can pack well in the glassy or crystalline state have low diffusion constants. The diffusion rates are higher in amorphous material, dominated by ash diffusion (through product layer) (K2, Table 7.2) than in crystalline materials, which is dominated by film diffusion (K1, Table 7.2) Figure 7.9.

7.3.4 Mass Loss Analysis

Mass loss in the hydrogel blend after 60 days of swelling is shown in Figure 7.10. The discs produced from pure PVA had a 3.45% mass loss over a period of 60 days when measured in vitro. As can be seen, PVP increased the solubility of PVA. Increasing the PVP content in the PVA polymer system resulted in increased mass loss. The base rate mass loss is sustained by the blend with 95% PVA and 5% PVP. In 95% PVA and 5% PVP blend, the proximity of the hydrogen bonding in the PVA chain render these segments to solidify. This is the reason why this blend has less mass loss compared to the other blends prepared (Andres Bernal, Kuritka & Saha, 2000).

Interchain hydrogen bonding, as shown in Figure 7.11 between the PVA and PVP, stabilises the polymer network. The hydrogen bonds induce cross-links and constrain the disintegration of the

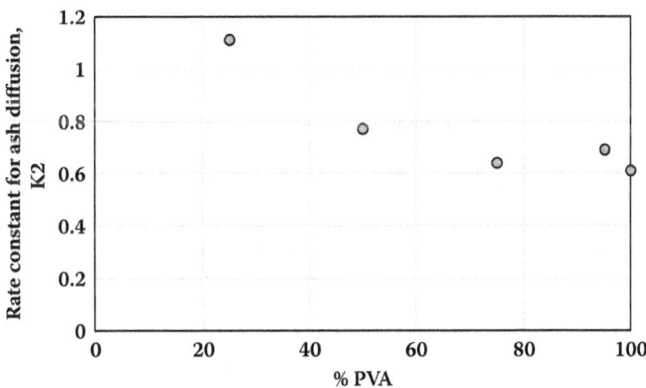

FIGURE 7.9 Rate constant, K2 with respect to % PVA.

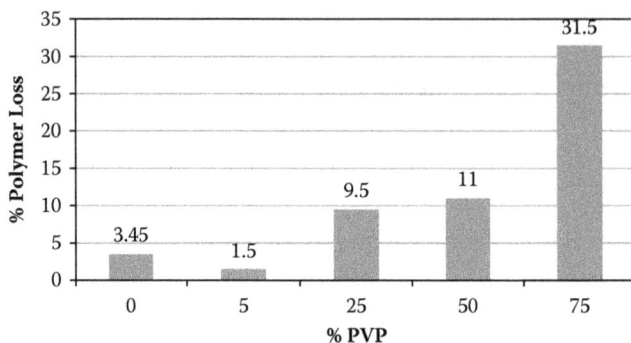

FIGURE 7.10 Polymer mass loss after 60 days of swelling.

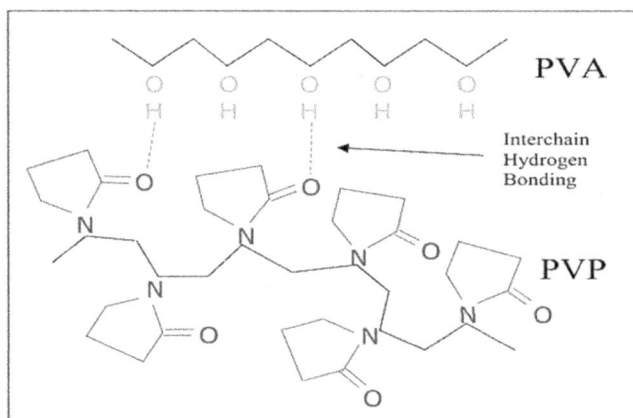

FIGURE 7.11 Inter-chain hydrogen holding inside a PVA/PVP mix between carbonyl bunches on PVP and hydroxyl aggregates on PVA.

polymer from the system into the solution. These cross-linked polymer blends, with the ability to absorb solvent and swell, keep their three-dimensional structure and are promising candidates for long-term implantations (Li, Suo & Deng, 2010). Covalently cross-linked hydrogels are attractive, for their tuneable properties, such as mechanical properties and degradation and find application in cell encapsulation (Nicodemus & Bryant, 2008). There is a fine balance when designing a hydrogel with appropriate properties that exhibit sufficient mechanical integrity without sacrificing cell viability and tissue growth.

The unstable nature of the blends with higher PVP content is attributed to the hindrance to the PVA crystal sections by the bulky pyrrolidone rings as seen in the PVP structure when PVP sterically intrudes on the crystal portions of the PVA polymer. PVP is also more hydrophilic than PVA, as discussed earlier. Due to its amorphous and hydrophilic nature, PVP-dominated blends are swollen to a greater extent and showing less stability. Hence, with increment in % PVP, the polymer network results in a higher % mass loss. For all these reasons, PVP/PVA blends have higher solubility than samples with PVA with the same additives.

7.3.5 FTIR

The FTIR traces of the blends are shown in Figures 7.12a-e. The characteristic vibration bands of PVA are observed at 3,386 cm^{-1} (OH stretching), 2,944 cm^{-1} (–CH stretching), 1,098 cm^{-1}, (C O stretching) and 1,645 cm^{-1} (C O stretching) due to acetate groups which are still present in the partially hydrolysed

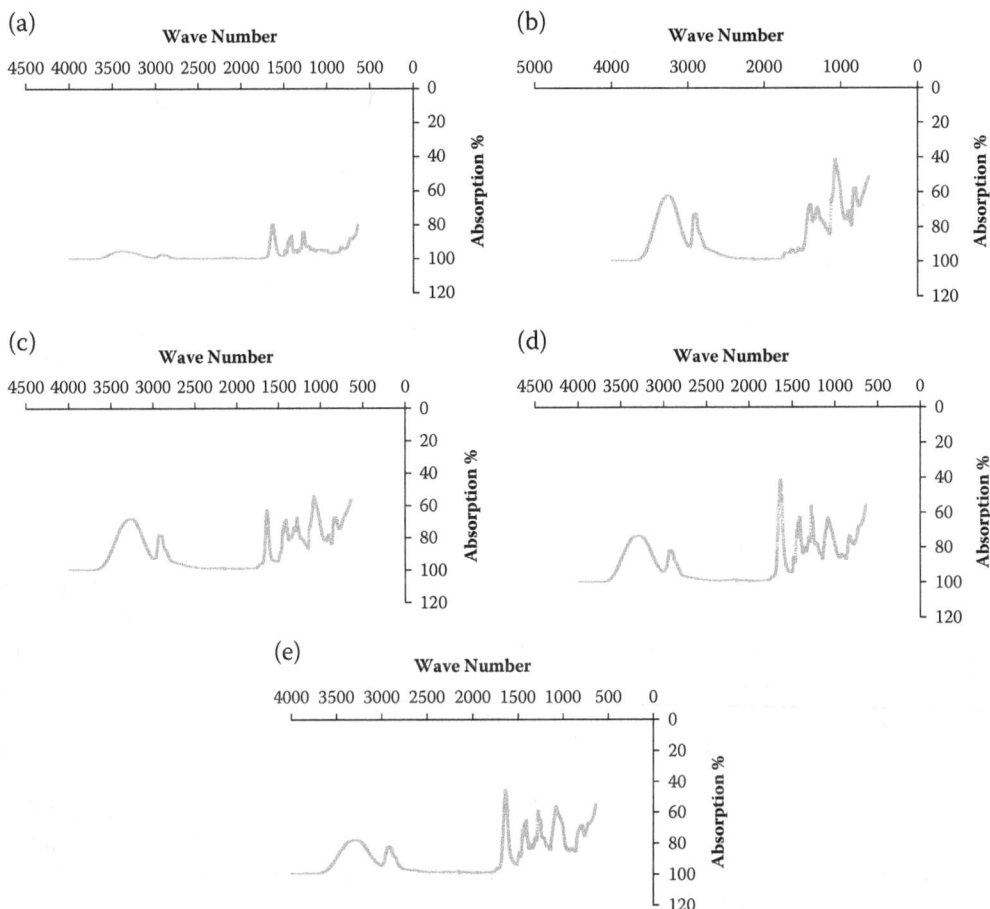

FIGURE 7.12 (a) FTIR of 100% PVA, (b) FTIR of 95% PVA and 5% PVP, (c) FTIR of 75% PVA and 25% PVP, (d) FTIR of 50% PVA and 50% PVP, (e) FTIR of 100% PVP.

form of PVA. The –OH stretching peaks of PVA at 3,385 cm^{-1} and the band of C O of PVA at 1,645 cm^{-1} is shifted to 1,643 cm^{-1}. This shift occurs due to inter-molecular and intra-molecular H-bonding interactions (Mondal et al., 2013).

In the case of PVP, the characteristic vibration bands appeared at 3,423 cm^{-1} (OH stretching), 1,656 cm^{-1} (C O stretching) and 1,292 cm^{-1} (C N stretching). With the addition of PVP in PVA, the shift in OH stretching of PVP from 3,423 cm^{-1} to 3,397 cm^{-1} is due to the strengthening the H-bonds between C...O of PVP and OH bonds of PVA. The conjugation of a lone pair of N with C O increases the C–N single bond strength which causes the slight increase of the C N bond stretching frequency from 1,292 cm^{-1}.

A similar shift of the C...O band around 1,696 to 1,664 cm^{-1} attributed to the CO...HO hydrogen bond is also observed. The broad peak at 1,096 cm^{-1} indicates the C-O stretching vibration. This is a carboxyl stretching band (C–O) and attributed to the crystallinity of the PVA, which is used to assess poly (vinyl alcohol) structure. There is a shift towards a lower frequency (~1,090 to 1,076 cm^{-1}) as the PVP content in the matrix increases. Because of the H-bond interactions, the PVA crystallization is increasingly hindered when the PVP content increases. The cross-linking prevents the PVA fraction in the blends from crystallizing, thus stabilizing the blend properties.

7.4   OTHER APPLICATIONS

In addition to the biomedical applications, the PVA/PVP blends find applications in renewable energy and low-carbon technologies, in membrane technology to separate CO_2, carbon capture and storage. High hydrophilicity of the polymer matrices induces CO_2/HCO_3^- reaction cycle for the selective transport of CO_2. The optimum blend was 50/50 PVA/PVP by weight (Helberg et al., 2020). The PVA/PVP polymeric blend nanocomposite films find applications in optoelectronics and optical devices (Badawi, 2020). PVA/PVP blend doped with sodium fluoride have potential applications in energy storage devices (Mohammed, Alabur & Sankanahally, 2020). Lithium ion conducting polymer electrolytes prepared by enhancing the ionic conductivity of the PVA–PVP polymer blend doping with lithium nitrate find applications in Li ion batteries (Natarajan et al., 2013). A PVA–PVP polymer blend doped with SnO_2 or ammonium acetate finds potential application in sensors and fuel cells (Guruswamy et al., 2018; Shanthi & Muruganand, 2015).

7.5   CONCLUSIONS

Characteristics such as swelling, mechanical properties, degradation and diffusion are associated with the cross-linked structure of the hydrogel and determine the design a hydrogel scaffold. Biodegradable hydrogel design is an important aspect of TE. A fine balance should be maintained when designing a hydrogel with appropriate properties to exhibit sufficient mechanical integrity without sacrificing cell viability and tissue growth. In the present study, the PVA and PVP blends are prepared by casting method and characterised for their suitability as biomaterials. The optimum conditions for their solubility, swelling and the biodegradation are determined. Increasing the PVP component in the blend, increased solubility, swelling and mass loss. The poor mechanical properties of PVP could be overcome by blending with PVA via cross-linking. The cross-linking via H-bonding in the blend is confirmed by FTIR studies. The ash diffusion is dominated in PVP enriched amorphous blends, while film diffusion is dominated in PVA-dominated crystalline blends. The crystallinity of PVA/PVP blends decreases with the increase in the concentration of PVP, as confirmed by FTIR studies.

REFERENCES

Abd El-Mohdy, H. L., and S. Ghanem. 2009. "Biodegradability, antimicrobial activity and properties of PVA/PVP hydrogels prepared by γ-irradiation." *J. Polym. Res.* 16 (1): 1–10.

Abetz, V., T. Brinkmann, and M. Sözbilir. 2021. "Fabrication and function of polymer membranes." *Chemistry Teacher International* 3 (2): 141–154. doi:10.1515/cti-2020-0023.

Azad, A. K., N. Sermsintham, S. Chandrkrachang, and W. F. Stevens. 2004. "Chitosan membrane as a wound-healing dressing: characterization and clinical application." *J. Biomed. Materials Res. Part B-Appplied Materials* 69B: 216–222.

Bernal, A. 2012. "Ballén Bioartificial Polymeric Materials With a Latent Application In Medical Field." Thesis (PhD). Tomas Bata University in Zlin Faculty of Technology, Polymer Centre.

Bernal, A., I. Kuritka, and P. Saha. 2000. "Poly(vinyl alcohol)-poly(vinyl pyrrolidone) blends: Preparation and characterization for a prospective medical application." *Mathematical Methods and Techniques in Engineering and Environmental Science* 1 (1): 430–434.

Badawi, A. 2020. "Engineering the optical properties of PVA/PVP polymeric blend in situ using tin sulfide for optoelectronics." *Appl. Phys. A* 126: 335. doi:10.1007/s00339-020-03514-5.

Bajpai, A. K., J. Bajpai, and S. Shukla. 2001. "Water sorption through a semi interpenetrating polymer network hydrophilic and hydrophobic chains." *React. Funct.Polym* 50: 9–21.

Baker, M. I., S. P. Walsh, Z. Schwartz, and B. D. Boyan. 2012. "A review of polyvinyl alcohol and its uses in cartilage and orthopedic applications." *J Biomed Mater Res.Part B*, 100B, 1451–1457.

Biolin Scientific. 2018."Biology at interfaces: Biomembranes." *Biolin Scientific.* Accessed July 1, 2021. https://www.biolinscientific.com/biology-at-interfaces/biomembranes?utm_term=&utm_campaign=BR-DSA-Biolin-G-EN&utm_source=adwords&utm_medium=ppc&hsa_acc=1094516957&hsa_cam=1334626534&hsa_grp=51529657617&hsa_ad=439661609998&hsa_src=g&hsa_tgt=dsa-19959388920.

Bruck, S. D. 1980. *Properties of Biomaterials in the Physiological Environment.* Boca Raton: CRC Press.

Caddeo, S., M. Boffito, and S. Sartori. 2017. "Tissue engineering approaches in the design of healthy and pathological in vitro tissue models." *Frontiers in Bioengineering and Biotechnology* 5: 40. doi:10.3389/fbioe.2017.00040.

Catapano, G., J. Vienken, W. S. Winston Ho, and T. Matsuura. 2008. "Biomedical applications of membranes." In *Advanced Membrane Technology and Applications*, by Anthony G. Fane and Norman N. Li, 489–517. John Wiley & Sons, Inc. doi:10.1002/9780470276280.

Chan, L., J. Hao, and P. Heng. 1999. "Evaluation of permeability and mechanical properties of composite polyvinyl alcohol films," *Chemical & Pharmaceutical Bulletin* l. 47 (10): 1412–1416.

Chen, J. 2016. "Thin film coatings and the biological interface." In *Thin Film Coatings for Biomaterials and Biomedical Applications*, by Hans J. Griesser, 143–164. Woodhead Publishing. doi:10.1016/B978-1-78242-453-6.01001-4.

Devasahayam, S., and C. M. Hussain. 2020. "Thin-film nanocomposite devices for renewable energy current status and challenges." *Sustainable Materials and Technologies* 26: e00233. doi:10.1016/j.susmat.2020.e00233.

Dolcimascolo, A., G. Calabrese, S. Conoci, and R. Parenti. 2019. "Innovative biomaterials for tissue engineering." In *Biomaterial-supported Tissue Reconstruction or Regeneration*, by Ole Jung, Ralf Smeets and Tadas Koržinska Mike Barbeck. IntechOpen. Accessed June 26, 2021. doi:10.5772/intechopen.83839.

Enrica, C., and V. K. Vitaliy. 2015. "Biomedical applications of hydrogels: A review of patents and commercial products." *European Polymer Journal* 65: 252–267. doi:10.1016/j.eurpolymj.2014.11.024.

Ethridge, E. C. and L. L. Hench 1982. *Biomaterials: An Interfacial Approach.* New York: Academic Press.

Fergal, J. O'Brien. 2011. "Biomaterials & scaffolds for tissue engineering." *Materials Today* 14 (3): 88–95.

Gahl, W. n.d. *Cell Membrane (plasma membrane).* Accessed July 5, 2021. https://www.genome.gov/genetics-glossary/Cell-Membrane.

Ge, Q., A. H. Sakhaei, H. Lee, C. K. Dunn, N. X. Fang, and M. L. Dunn. 2016. "Multimaterial 4D Printing with Tailorable Shape Memory Polymers." *Sci Rep* 6: 31110. doi:10.1038/srep31110.

Goksu, T. D., D. U. Tugba, A. A. Selcen, Y. Deniz, H. Nesrin, and H. Vasif. 2019. "3D and 4D printing of polymers for tissue engineering applications." *Frontiers in Bioengineering and Biotechnology* 7: 164. https://www.frontiersin.org/articles/10.3389/fbioe.2019.00164/full.

Gomes, M. E., A. S. Ribeiro, P. B. Malafaya, R. L. Reis, and A. M. Cunha. 2001. "A new approach based on injection moulding to produce biodegradable starch-based polymeric scaffolds: Morphology, mechanical and degradation behaviour." *Biomaterials* 22 (9): 883–889.

Gough, J., C. Scotchford, and S. Downes. 2002. "Cytotoxicity of glutaraldehyde cross-linked collagen/poly (vinyl alcohol) films is by the mechanism of apoptosis." *J. Biomed. Mater. Res.* 61 (o. 1, p).

Guruswamy, B., V. Ravindrachary, C. Shruthi, Shreedatta, Hegde, R. N. Sagar, and S. D. Praveen. 2018. "Optical, electrical and thermal properties of SnO2 nanoparticles doped poly vinyl alcohol-poly vinyl pyrrolidone blend polymer electrolyte." *Indian Journal of Advances in Chemical Science* 6 (1): 17–20.

Helberg, R. M. L., Z. Dai, L. Ansaloni, and L. Deng. 2020. "PVA/PVP blend polymer matrix for hosting carriers in facilitated transport membranes: Synergistic enhancement of CO2 separation performance." *Green Energy & Environment* 5 (1): 59–68.

Hoffman, A. S. 2012. "Hydrogels for biomedical applications (Review)." *Adv Drug Deliv Rev* 64: 18–23.

Hutmacher, D. W., T. Schantz, I. Zein, K. W. Ng, S. H. Teoh, and K. C. Tan. 2001. "Mechanical properties and cell cultural response of polycaprolactone scaffolds designed and fabricated via fused deposition modeling." *Journal of Biomedical Materials Research* 55 (2): 203–216.

Jones, S. A., G. P. Martin, P. G. Royall, and M. B. Brown. 2005. "Biocompatible polymer blends; effects of physical processing on the molecular interaction of poly (vinyl alcohol) and Poly (vinyl pyrrolidone)." *Journal of Applied Polymer Science* 98 (5): 2290–2229.

Khan, F., and M. Tanaka. 2017. "Designing smart biomaterials for tissue engineering." *International Journal Oof Molecular Sciences* 19 (1): 17. doi:10.3390/ijms19010017.

Kopeček, J. 2007. "Hydrogel biomaterials: A smart future?" *Biomaterials.* 28: 5185–5192. doi:10.1016/j.biomaterials.2007.07.044.

Lai, G., Z. Du, and G. Li. 2007. "The rheological behaviour of collagen dispersion/poly (vinyl alcohol) blends." *Korea-Australia Rheology Journal* 19 (2): 81–88.

Lakouraj, M., M. Tajbakhsh, and M. Mokhtary. 2005. "Synthesis and swelling characterization of cross-linked PVP/PVA hydrogels." *Iranian Polymer Journal* 14 (12): 1022–1030.

Lee, K. Y., and D. J. Mooney. 2001. "Hydrogels for tissue engineering." *Chemical Review* 101: 1869–1880.

Li, J., J. Suo, and R. Deng. 2010. "Structure, mechanical, and swelling behaviors of poly(vinyl alcohol)/SiO2 hybrid membranes." *Journal of Reinforced Plastics and Composites* 29 (4).

Lionetto, F., A. Sannino, G. Mensitieri, and A. Maffezzoli. 2003. "Evaluation of the degree of crosslinking of cellulose-based superabsorbent hydrogels: a comparison between different techniques." *Macromolecular Symposia* 200:199–207.

Liu, J., and C. Yan. 2018. "3D printing of scaffolds for tissue engineering." In *3D Printing*, by Dragan Cvetković. IntechOpen, 137–154. doi:10.5772/intechopen.78145.

Mantha, S., S. Pillai, P. Khayambashi, A. Upadhyay, Y. Zhang, O. Tao, H. M. Pham, and S. D. Tran 2019. "Smart hydrogels in tissue engineering and regenerative medicine." *Materials (Basel)* 12 (20): 3323. doi:10.3390/ma12203323.

Mironov, V. 2014. " 4D bioprinting: Biofabrication of rod-like and tubular tissue engineered constructs using programmable self-folding bioprinted biomaterials." *International Bioprinting Congress. Biopolis (2014)*.

Mohammed, I., M. Alabur, and S. M. Sankanahally. 2020. "Studies on structural characterization and electrical properties of NaF doped PVA-PVP blend electrolyte films." *AIP Conference Proceedings*. 100009. doi:10.1063/5.00.

Mondal, D., M. d. M. R. Mollick, B. Bhowmick, D. Maity, M. K. Bain, D. Rana, A. Mukhopadhyay, K. Dana, and D. Chattopadhyay. 2013. "Effect of poly(vinyl pyrrolidone) on the morphology and physical properties of poly(vinyl alcohol)/sodium montmorillonite nanocomposite films." *Progress in Natural Science: Materials International* 23 (6): 579–587.

Mozafari, M., A. Ramedani, Y. N. Zhang, and D. Mills. 2016. "Thin films for tissue engineering applications." In *Thin Film Coatings for Biomaterials and Biomedical Applications*, by Hans J. Griesser, 167–195. Elsevier. doi:10.1016/B978-1-78242-453-6.00008-0.

Murphy, S. V., and A. Atala 2014. "3D bioprinting of tissues and organs." *Nature Biotechnology* 32: 773–785. doi:10.1038/nbt.2958.

Natarajan, R., S. Subramanian, S. K. M. Prabu, and C. Sanjeeviraja. 2013. "Lithium ion conducting solid polymer blend electrolyte based on bio-degradable polymers." *Bull. Mater. Sci.* 36 (2): 333–333.

Nicodemus, G., and S. Bryant. 2008. "Cell encapsulation in biodegradable hydrogels for tissue engineering applications." *Tissue engineering. Part B, Reviews* 14: 149–165. doi:10.1089/ten.teb.2007.0332.

Norotte, C., Marga, F. S., Niklason, L. E., and Forgacs, G. 2009. "Scaffold-free vascular tissue engineering using bioprinting." *Biomaterials* 30: 5910–5917.

Oka, M. T., P. Noguchi, K. Kumar, T. Ikeuchi, S. H. Yamamuro, Hyon, and Y. Ikada. 1990. "Development of an artificial cartilage."*Clinical Materials* 6: 361–381.

Park, J. S., J. W. Park, and E. A. Ruckenstein. 2001. "Dynamic mechanical and thermal analysis of unplasticized and plasticized poly(vinyl alcohol) /methylcellulose blends." *J. Appl. Polym. Sci.* 80 (10): 1825–1834.

Peppas, N. A., and E. W. Merrill. 1976. "Poly(vinyl alcohol) hydrogels: Reinforcement of radiation-cross-linked networks by crystallization." *J. Polym. Sci. Polym. Chem. Ed.* 14: 441–457. doi:10.1002/pol. 1976.170140215.

Perale, G., and J. Hilborn. 2016. *Bioresorbable Polymers for Biomedical Applications: From Fundamentals to Translational Medicine*. Duxford, UK: Woodhead Publishing.

Ping, Z. H., Q. T. Nguyen, and J. Néel. 1989. *Makromol Chem* 191: 185.

Robinson, B. V., F. M. Sullivan, J. F. Borzelleca, and S. L. Schwartz. 1990. *PVP. A Critical Review of the Kinetics and Toxicology of Polyvinylpyrrolidone (povidone)*. 1st ed. Chelsea: Lewis Publisher Incorporation. doi:ISBN 0-87371-288-9.

Rosiak, J. M., and F. Yoshii. 1999. "Hydrogels and their medical applications(Article)." *Nucl Instrum Meth B,* 151: 56–64.

Shanthi, B., and S. Muruganand. 2015. "Structural, vibrational, thermal, and electrical properties of PVA/PVP biodegradable polymer." *International Journal of Scientific Engineering and Applied Science* 1 (8): 2395–3470.

Sionkowska, A. 2006. "The influence of UV irradiation on surface composition of collagen/PVP blended films." *Applied Surface Science* 253 (4): 1970–1977.

Spicer, C. D. 2020. "Hydrogel scaffolds for tissue engineering: The importance of polymer choice." *Polym. Chem.*, 11: 184–219.

Thomson, R. C., M. J. Yaszemski, J. M. Powers, and A. G. Mikos. 1995. "Fabrication of biodegradable polymer scaffolds to engineer trabecular bone." *Journal of Biomaterials Science, Polymer Edition* 7 (1): 23–38.

Vashuk, E. V., E. V. Vorobieva, I. Basalyga, and N. P. Krutko. 2001. "Water-absorbing properties of hydrogels based on polymeric complexes."*Material Research Innovations* 4: 350–352.

Wang, S., J. Li, Z. Zhou, S. Zhou, and Z. Hu. 2019. "Micro-/nano-scales direct cell behavior on biomaterial surfaces." *Molecules* 24 (1): 75. doi:10.3390/molecules24010075.

Williams, R. 2011. *Surface Modification of Biomaterials, Methods Analysis and Applications.* A volume in Woodhead Publishing Series in Biomaterials. https://www.sciencedirect.com/book/9781845696405/surface-modification-of-biomaterials#book-description.

Xu, T. 2021. "Novel Synthetic Membranes Speed Proton Transport." *Energy.Gov.* March 10. Accessed July 1, 2021. https://www.energy.gov/science/bes/articles/novel-synthetic-membranes-speed-proton-transport.

Zeltinger, J., J. K. Sherwood, D. A. Graham, R. Müeller, and L. G. Griffith. 2001. "Effect of pore size and void fraction on cellular adhesion, proliferation, and matrix deposition." *Tissue Engineering* 7 (5): 557–572. doi:10.1089/107632701753213183.

Zhou, W., B. Gou, M. Liu, R. Liao, A. Rabie, and D. Jia. 2010. "Poly(vinyl alcohol)/halloysite nanotubes bionanocomposite films: properties and in vitro osteoblasts and fibroblasts response." *Journal of Biomedical Materials Research Part A* 93 (4): 1574–1587.

Zohuriaan-Mehr, M. J., and A. Pourjavadi. 2003. "Superabsorbent hydrogels from starch-g-PAN: Effect of some reaction variables on swelling behavior." *Journal of Polymer Materials* 20: 113–120.

Section II

Polymers for Energy and
Specialty Applications

8 Recent Developments of Polymer Electrolytes for Rechargeable Sodium-Ion Batteries

S. Janakiraman
Indian Institute of Technology Indore, Indore, Madhya Pradesh, India

Anandhan Srinivasan
National Institute of Technology Karnataka, Surathkal, Mangalore, Karnataka, India

Venimadhav Adyam
Indian Institute of Technology Kharagpur, Kharagpur, West Bengal, India

CONTENTS

8.1 INTRODUCTION TO SODIUM-ION BATTERIES

Batteries are used in a wide range of applications, including portable gadgets, satellites, computers, medical devices, and electric autos. Simultaneously, nonrenewable resources are fast depleting, and global warming is accelerating due to greenhouse gas emissions. Developing means to efficiently store the generated energy has been an important component of replacing our reliance on renewable resources. The changing nature of sunlight and wind at any given site, in particular, necessitates the incorporation of a storage mechanism into the overall design (Komaba et al. 2011; Omar et al. 2012; Vignarooban et al. 2016). The rechargeable battery has become a key piece of modern technology, enabling a wide range of technologies spanning from consumer electronics and hybrid electric vehicles, and is now a critical part of upcoming technologies such as grid energy storage and all-electric vehicles (Dunn, Kamath, & Tarascon, 2011; O'Heir, 2017). The lithium-ion battery (LIB) in particular has become the most important battery technology in the early 21st century and can be found in all new laptops, cell phones, and electric vehicles as a result of its high efficiency and energy density relative to other electrochemical power sources.

Despite their advantages, LIBs are relatively costly, and the cost of lithium supplies has roughly doubled since 1991 (Komaba et al. 2011). Lithium reserves are concentrated in South America (70% of global deposits) (Pan, Hu, & Chen, 2013), making the lithium battery industry reliant on imported raw lithium resources. Lithium reserves are also in short supply to fulfill the rising demand for a wide range of uses. As a result, new rechargeable battery systems that are made from environmentally benign materials are required. Researchers hope to match the performance of well-known LIBs using sodium-based batteries because sodium is the most abundant, non-toxic, and less expensive metal (Hosaka et al. 2020).

Sodium-ion batteries (SIBs), as an analog with LIB in electrochemistry, have been studied for decades since the 1970s (Whittingham, 1978; Zoń, et al. 2009). However, after the success of the commercial application of LIBs, people pay less attention to SIBs. With the rapidly increasing demand for lithium, it is estimated that the mineable Li will probably exhaust under the average growth of 5% consumption per year. Therefore, in a foreseeable future, the concern of exhausting lithium resources and possible price fluctuation, besides, considering the abundance of sodium, people restart to explore the application of SIBs to partially replace LIBs. It is seen that sodium has a lesser reducing effect than lithium (−2.71 V versus standard hydrogen electrode, contrast to −3.04 V) which means lower operating voltages for SIBs compared with LIBs (Ellis & Nazar, 2012; Kuze, Kageura, & Matsumoto, 2013). Additionally, the capacity of metallic electrodes between Na and Li shows another disadvantage of SIBs. However, when considering the theoretical reversible capacity of the layered oxides as $LiCoO_2$ and $NaCoO_2$ have an identical crystal structure, only a 14% difference exists (Hosaka et al. 2020). Moreover, one of the merits of a larger cation radius is the weaker solvation energy in polar solvents for Na^+ because the solvation energy is an important factor to figure out how alkali-ion insertion happens at the electrolyte contact (Chen & Forsyth, 2016; Hasa, Passerini, & Hassoun, 2016; Pan, Hu, & Chen, 2013). Furthermore, the high ion conductivity of Na^+-based electrolytes has aided the development of sodium-ion batteries significantly, which can enhance the performance and make them as good as a lithium-ion battery concerning conductivity (Freitag et al. 2018).

These batteries work on the principle of the rocking chair concept, where the metal ions (Na^+) shuttle back and forth between the two insertion electrodes during charging and discharging, the rocking chair concept is illustrated schematically in Figure 8.1. Thus, during charging (driven by an electric field), sodium ions are de-intercalated from the cathode and intercalate into the anode. In SIBs, the medium for Na^+ transport in polymer electrolyte (PE), determines the power density. An electrolyte's ionic conductivity is the most significant feature, but it should also have insignificant electronic conductivity to minimize short circuits.

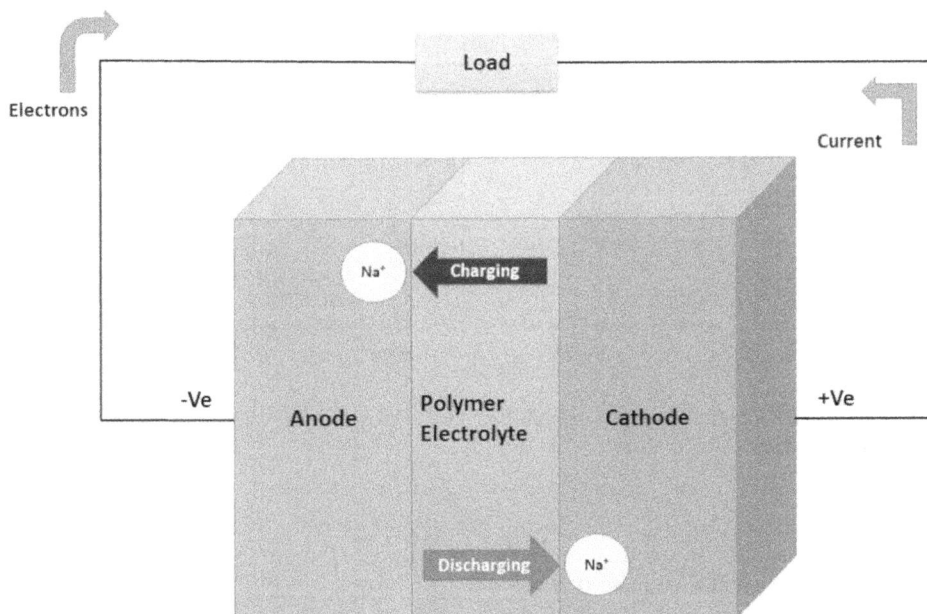

FIGURE 8.1 Charge-discharge mechanism of SIB.

8.2 INTRODUCTION TO POLYMER ELECTROLYTES

The PEs are a relatively modern category of solid electrolytes that has shown enormous promise in the development of many types of electrochemical devices over the past three decades. They have a similar ionic conductivity as liquid electrolytes (LEs) ranging from 10^{-5} to 10^{-2} S cm^{-1}, and solid free-standing consistency. These electrolytes are essentially polymers that incorporate salts, acids, LEs, ionic liquids (ILs), and other ions that supply ions to the insulating polymer matrix. The PEs provide several advantages over other forms of ion-conducting materials, including shape variability, flexibility, and light weight. The major problems that existed in LEs can be overcome in PEs. In SIBs, the PEs are classified into three types like solid polymer electrolyte (SPE), composite solid polymer electrolyte (CSPE), and gel polymer electrolyte (GPE) (Qiao et al. 2020; Yin et al. 2021). The SPE is more safe but has low ionic conductivity at room temperature.

The basic desired qualities of PE materials are listed below:

1. High ionic conductivity at a wide range of temperatures: Ionic conductivity in PEs is closely correlated with crystallinity, glass transition temperature, porosity, and the polymer's ability to absorb solvent from the matrix. Its electrical mobility is proportional to the concentration of charged species.
2. Good chemical and thermal stability: PEs must have good ionic conductivity across a wide temperature range and be structurally stable during manufacturing, cell assembly, storage, and usage to be effective. To enhance the thermal stability of PEs, inorganic nanoparticles such as Al_2O_3, SiO_2, and TiO_2 are used. Another technique is to combine various polymers into a composite to improve the thermal stability of PEs.
3. High electrochemical window: Side reactions at the electrode/electrolyte contact surface should be kept to a minimum.
4. High cation transference number: The transference number of PEs, in addition to their ionic conductivity, is a significant figure of merit when evaluating their efficacy. In single salt PEs, the cationic transference number is the cation mobility relative to the anion. Working species in PEs must have a transference number that is near to unity.

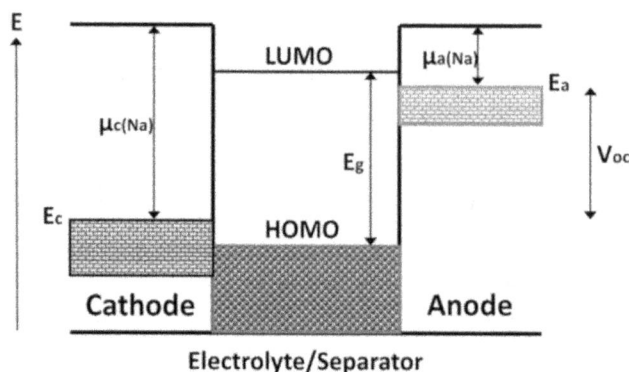

FIGURE 8.2 Thermodynamic stability conditions are shown in the energy diagram of a Na cell (Vignarooban et al. 2016).

5. Good mechanical properties: Because a PE sits between the cathode and the anode, it must be mechanically sound and free of deformation or strain that could jeopardize the battery's stability. They must also be able to withstand changes in volume; however, GPEs sometimes have poor mechanical characteristics as a result of the solvent or plasticizer employed to boost conductivity.
6. Cost-effectiveness: PEs with great characteristics would only be suitable for commercial use if they could be manufactured at a low cost. It necessitates a simple fabrication procedure.

PEs with strong battery voltage stability will be dependent on the electrolyte's lowest unoccupied molecular orbital (LUMO) and highest occupied molecular orbital (HOMO) energy levels, as shown in Figure 8.2 (Vignarooban et al. 2016). For higher thermal stability, the redox energies E_a (anode) and E_c (cathode) should be within the electrolyte's band-gap (E_g).

Aside from morphological differences, PEs and LEs are characterized by their ion conduction mechanism. As part of a solvation shell, dissociated ions in LEs can mainly move freely inside the liquid. Ions travel across the polymer network in SPEs since there are no liquid phases. While GPEs including both liquid and polymer phases can contain both liquid and solid ion conduction channels, the relative contributions will vary depending on the polymer type and amount of liquid phase in the gel. In this chapter, the recent PEs for SIBs, their mechanisms, and applications have been discussed.

8.3 POLYMERS MATRICES, SODIUM SALTS, AND ORGANIC SALTS USED IN SIBS

Before the inclusion of any other component, PEs require at least one host polymer to operate as a base material. Various high molecular weight and dielectric constant polymer hosts like poly(vinyl alcohol) (PVA), poly(vinyl chloride) (PVC), poly(ethylene oxide) (PEO), poly(vinylidene fluoride) (PVDF), polyacrylonitrile (PAN), poly(methyl methacrylate) (PMMA), poly(vinylidene fluoride-hexafluoropropylene)(PVdF), poly(vinylidenefluoride hexafluoropropylene) (PVdF-HFP), etc. have been used. Figure 8.3 compiles a list of the most often used polymers and their chemical structures for producing PEs, which is not exhaustive. Among all PEO-based PEs have been studied extensively in SIBs.

Different solvents are used in organic liquid-based SIBs and LIBs. Because they do not create well-behaving solid electrolyte interphase layers (SEIs), several solvents utilized for LIB electrolytes may not be acceptable for SIBs. In addition, Na^+ is a weaker Lewis acid than Li^+, it will have a less interaction with the salt's anions, affecting the solvents employed, the ideal salt

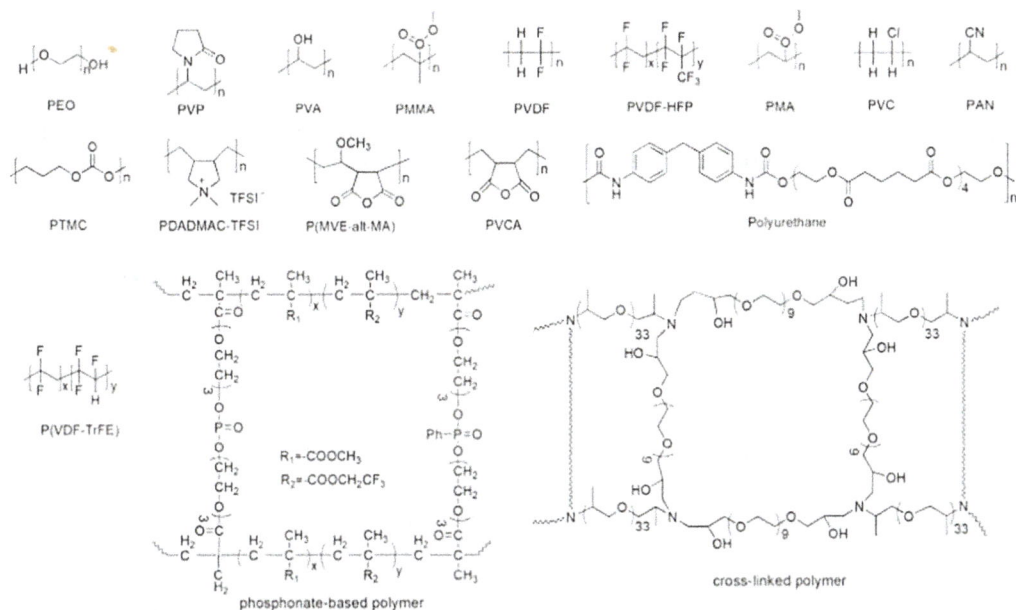

FIGURE 8.3 The common polymers and their chemical structure used for PEs (Qiao et al. 2020).

concentration, and the SEI formation. The most extensively researched LEs for both LIBs and SIBs are linear and cyclic carbonates (Figure 8.4). Unlike LIBs, where PC co-intercalates into graphite and induces exfoliation, PC does not co-intercalate into graphite in LIBs, the most commonly utilized solvent for SIBs is propylene carbonate (PC). To form a stable SEI on HC, however, PC must be combined with additional solvents, such as ethylene carbonate (EC) (Komaba et al. 2011; Ponrouch et al. 2012). EC/diethyl carbonate (DEC) (Komaba et al. 2011), EC/dimethyl carbonate (DMC) (Komaba et al. 2011), and DMC/EC/PC (Ponrouch et al. 2013) are examples of binary solvent-based electrolytes. Organic liquid solvents have dielectric constants and solvents with larger dielectric constants dissolve Na salts more effectively. Because their viscosity at 25°C is 1 mPa s. So to lower viscosity and increase the ionic conductivity of the electrolyte, DMC or DEC are used (Table 8.1). Except for EC, which melts around 35°C, all of the solvents described above are liquids at room temperature (RT).

The most often utilized salt in the preparation of non-aqueous electrolytes for SIBs is sodium hexafluorophosphate (NaPF$_6$). It was selected for its ionic conductivity and SEI formation, as well

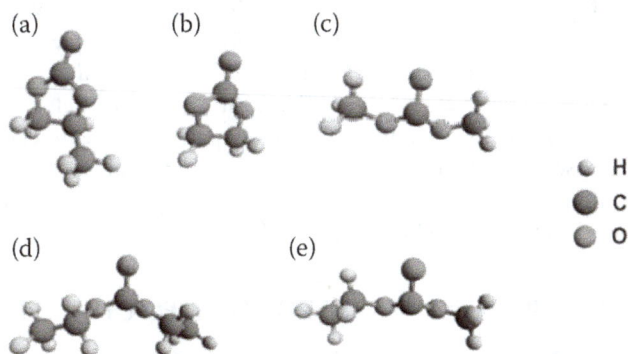

FIGURE 8.4 Chemical structures of (a) PC, (b) EC, (c) DEC, (**d**) DMC, and (e) EMC.

TABLE 8.1

Generally Used Liquid Solvents in Non-Aqueous SIBs Have the Following Properties

Organic Solvent	Dielectric Constant ε	Viscosity at 25°C η (mPa s)	Melting Temperature T_m (°C)	Boiling Temperature T_b (°C)
Propylene carbonate (PC)	66	2.5	−49	242
Ethylene carbonate (EC)	90	1.9 (40°C)	36	248
Diethyl carbonate (DEC)	2.8	0.75	−74	126
Dimethyl carbonate (DMC)	3.1	0.59	4.6	91
Ethyl methyl carbonate (EMC)	3.0	0.65	−53	110

as the fact that it decomposes into F^-, which renders the present aluminium collector inactive. Other salts that have been proposed are mostly sodium counterparts of salts that have been studied in the field of LIBs. sodium bis(trifluoromethanesulfonyl)imide (NaTFSI), sodium perchlorate (NaClO$_4$), sodium tetrafluoroborate (NaBF$_4$), sodium trifluoromethanesulfonate (NaCF$_3$SO$_3$), sodium bis(fluorosulfonyl)imide (NaFSI), sodium bis(trifluoromethanesulfonyl)imide (NaTFSI). Another class of solvents is glymes, which have the general formula $CH_3O(CH_2CH_2O)n\text{-}CH_3$. Non-dendritic sodium plating and stripping have been achieved using LEs containing NaPF$_6$ and glymes (Seh et al. 2015).

8.4 CLASSIFICATION OF POLYMER ELECTROLYTES

SPEs are constructed solely of sodium salts and polymer matrices for sodium batteries. SPEs with inorganic fillers are referred to as CSPEs. PPEs/GPEs are gel/plasticized polymer electrolytes that contain liquid phases in addition to the components found in SPEs or CSPEs. The liquid phase fraction is the difference between PPEs and GPEs, with the former having less than 50% liquid plasticizers, whereas the latter contains more than 50% liquid plasticizers. (Qiao et al. 2020). The chemical structures of various sodium salts are depicted in Figure 8.5. The following sections go over the latest developments in PE-based sodium batteries in depth.

Fenton et al. (Fenton, Parker, & Wright, 1973) reported the first PE in 1973, where the semicrystalline structure of polyethylene oxide (PEO)–alkali salts complexes was discussed. Wright later reported on the electrical characteristics of these polymer-salt complexes, correlating the PEs ionic conductivity with their amorphicity (Wright, 1976). Since then, numerous new ion-conducting PEs with mobile ions such as H^+, Li^+, Na^+, Zn^{2+}, and Mg^{2+}, etc. have been reported in the literature and their potential utility in different all-solid-state electrochemical devices like batteries, fuel cells, sensors, supercapacitors, and dye-sensitized solar cells, etc. have been developed (Adhikari & Majumdar, 2004; Ohno, 1992; Peighambardoust, Rowshanzamir, & Amjadi, 2010; Polymers, 1993; Wang, 2009; Zhou et al. 2003).

8.5 SYNTHESIS AND PROCESSING OF POLYMER ELECTROLYTE MEMBRANES

The PEs for SIBs are in the early stage so all the recent reports have been studied using commercially available polymers like PEO, PVDF, PAN, PMMA. This has substantially advanced the

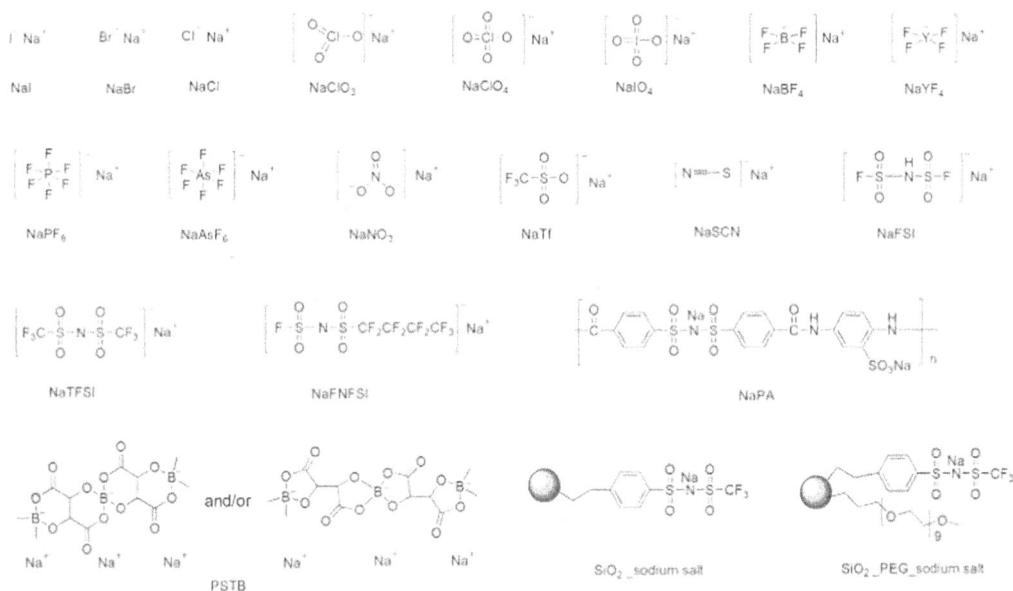

FIGURE 8.5 The chemical structures of salts for SIBs (Qiao et al. 2020).

field. More exploration with new polymers, cross-linking side chains is likely in the future. The fabrication technique used to make polymers is extremely varied. A polymer must still be converted into a PE after it has been manufactured. The charge carriers must be implanted into the structure, either as an SPE or as a GPE, by inflating it with a LE. Furthermore, the bulk polymer must be transformed into a film. The characteristics of the final PE can be influenced by the process of synthesis, allowing for more fine-tuning (Gebert et al. 2021). The following gives an overview of the most prevalent synthesizing techniques.

8.5.1 Solution Casting

This is a standard method for producing PEs films and gels. This method involves adding the required amounts of polymer and complexing salt, which have been dissolved separately in a shared solvent, mixing them, and whirling magnetically for a long enough period to achieve salt complexation with the polymer host. During the stirring process for casting CSPE films, micro/nano-sized filler particles are inserted. The viscous polymer solution is then put into a petri dish to allow the solvent to slowly evaporate and produce the film. Following that, vacuum drying is used to dry the generated SPE films (Chen et al. 2018).

8.5.2 Hot-Pressing

Gray et al. (Gray, MacCallum, & Vincent, 1986) first demonstrated this methodology, which has since been updated by various groups with small tweaks. Compared to typical solution casting, hot-press casting has several advantages, including being a faster, less cost, and solvent-free technique of producing PE films. This method entails mixing the dry polymer and complexing salt powders in the proper ratio (for casting standard SPE membranes) or polymer, complexing salt, and micro-dimension filler particles (for casting CSPE membranes); then heating the homogeneously mixed powder near the T_m of the polymer host for an appropriate amount of time with continued mixing to guarantee complete salt complexation. After that, the slurry is pushed between two cool metal rollers, resulting in an even, stable PE coating of a few mm thickness.

8.5.3 Coating

To make a composite membrane, this approach is used to add a thin layer of polymer films to a porous sublayer. The thin top layer provides ion transport and selectivity, while the sublayer just provides mechanical support. There are a variety of coating processes that can be used. Interfacial polymerization, which involves the reaction of two highly reactive monomers (or a pre-polymer) at the interface of two immiscible solvents, is one type of coating. Another method is dip coating, which involves immersing an asymmetric sustaining layer in a dilute coating solution containing a monomer, or polymer. Due to solvent evaporation, when the membrane is removed from the coating solution, a thin layer forms and adheres (Jeong & Kim, 2004).

8.5.4 Phase Inversion

The phase inversion method includes transitioning a polymer from a liquid to a solid state in a controlled manner. Liquid-liquid demixing is a common way to start the solidification process. One of the liquid phases will solidify during the de-mixing process, resulting in the solid matrix. Controlling the first step of phase transition allows the membrane morphology, such as micro-porous and non-porous structures, to be controlled. Phase inversion can be accomplished using a variety of methods, the most frequent of which is immersion precipitation. In this process, a polymer solution is cast on a suitable substrate and then immersed in a coagulation bath comprising non-solvent (Guillen et al. 2011).

8.5.5 Electrospinning

The initial step in this approach is to spray a polymer solution onto a collector. A voltage is applied between the nozzle and the collector, creating an electric field that charges the solution droplets sprayed into it. The droplets lengthen, as a result, finally forming a continuous cone-shaped stream from the nozzle to the collector. The solvent evaporates as the solution is gathered on the collector surface and long polymer fibres are left behind (Janakiraman et al. 2019b). Electrospinning has a significant benefit over solvent casting in that it allows for considerably more accurate adjustment of the polymer's mechanical properties (Ghosal et al. 2018), PVDF (Janakiraman et al. 2019a), PAN (L. Zhang et al. 2019), and PEO (Freitag et al. 2018) polymers have all been successfully prepared via electrospinning in SIB PEs.

8.5.6 In-Situ Polymerization

Here the battery PEs refer to the polymerization of a polymer in a battery cell utilizing a nonwoven membrane as a scaffold, which is often a commercial polyolefin or glass fibre separator. The scaffold membrane is soaked in a polymer precursor solution, sealed into the battery cell, and the polymerization process begins, which is commonly done by heating the cell. Because of their considerable thickness, this has an impact on energy density, glass fibre separators are inappropriate for commercial application (Kirchhöfer et al. 2014), as well as their reactivity with HF generated during battery operation by the breakdown of hexafluorophosphate salts (Guéguen et al. 2016). Polyethylene has also been used as a scaffold instead of glass fibre in composite separators (Suharto et al. 2018). This method not only improves the separator's safety but however it also permits the use of propylene carbonate-based electrolytes, which are common in research (Ponrouch et al. 2013). However, due to their poor wetting of polyolefin separators, they are inappropriate for use with traditional separators.

8.6 ION CONDUCTION IN POLYMER ELECTROLYTES

Ionic conductivity is the most widely reported physical property in battery electrolyte literature, which is covered in the introduction. This is because it is simple to measure, while also quantifying the most fundamental function of a battery electrolyte: the ability of ions to move through it. This has an immediate impact on battery performance indicators such as capacity, rate capability, coulombic efficiency, and longevity (X. Huang, 2011). A comparable parameter is the sodium ion transference number, which is a measure of the percentage of current carried by Na^+ ions. The transference number has an impact on the battery's cycle stability and safety. The following is a quick rundown of the most prevalent conduction mechanism in PEs.

8.6.1 EARLY PHENOMENOLOGICAL CONCEPTS

The inclusion of dissolved ion-conducting salts in the non-conducting polymer network introduces ionic conductivity in SPEs. In PEO-alkali metal salt complexes, there is an ion transport mechanism. According to the paper, the salt cation is present inside the single or double helices of polyether chains. The cation was expected to hop across the helices in the traditional PEO-salt complexes, while the anions were expected to be almost stationary and put outside the helices (Arya & Sharma, 2017). Later, (Papke, Ratner, and Shriver 1982) proposed that the cation, which is complexed within the helical regions, is transported via the PEs disordered amorphous phase. Berthier et al. (Berthier et al. 1983) suggested through the NMR studies that the cationic and anionic transport through the PE is assisted by local segmental motions of the polymer chains. According to this concept, fast ion transport occurs in polymer-salt complexes via liquid-like ion motion, in which polymer segmental motion promotes ion transport. Recently, Gadjourova et al. (Gadjourova et al. 2001) demonstrated experimentally that, unlike earlier belief, the fast ion conduction in SPEs could be aided by the static and organized crystalline environments in the polymer host.

8.6.2 AMORPHOUS PHASE MODEL

A variety of PEs has been reported in which reduction in crystallinity is observed on the dispersion of ceramic fillers. This crystallinity reduction is closely related to the enhanced conductivity of the CSPEs as compared to the undispersed PEs. Based on this observation, an 'amorphous phase model' has been proposed, which explains the increased conductivity of CSPEs. In crystalline PEO-salt complex systems, for example, the filler particles (e.g., $-Al_2O_3$) operate as nucleation centres and may attach to PEO segments via acid Al surface centres. The crystallization process is aided by the presence of a large number of nucleation centres in the PEs, and as a result, the newly created solid polymeric matrix freezes a larger disorder, which is typical of a liquid state, upon solvent evaporation (Klongkan & Pumchusak, 2015).

8.6.3 THE LEWIS ACID-BASED APPROACH

This approach is based on the idea that the Lewis acid-base reaction equilibrium determines the ionic conductivity of composite electrolytes, which involves the components of the CSPEs i.e., an additive, a polymer, and a salt (with cation acting as Lewis acid and anion as Lewis base). This approach also accounts for the possibility of interactions of the filler with the side-chain OH groups of the polyether (Croce et al. 2001). This approach is very useful especially for designing filler particles that might act as anionic receptors and promote the cationic transport in the electrolyte, which is a preliminary requirement for the electrolyte in a battery system. (Syzek et al. 2007) have used this approach to explain the conductivity and enhancement in the cationic (e.g., Li^+-ion) transport number. The Lewis acid-based technique has also been found to be effective in the development of inorganic fillers with Lewis acid groups. The use of such fillers might lead to an enhancement in conductivity even in electrolytes based on amorphous polymeric materials (Syzdek et al. 2007).

8.6.4 MODELS APPLICABLE TO GEL POLYMER ELECTROLYTES

The GPEs are the LEs-filled polymers, in which conductivity is governed by liquid-like motion (Hueso et al. 2017) and polymer only imparts mechanical stability to membranes. In most of the studied Na$^+$ ion conducting GPEs, the conductivity of the electrolytes has been observed to decrease on the addition of the polymer content, which is due to the increase in the viscosity and the steric hindrance on the addition of polymer structure, resulting in lowering of mobility. The inner resistivity and the molar volume of the polymer also reduce the conductivity of the gel. A few reports are available in which conductivity enhancement is observed on the addition of polymers in the LE (Sekhon, 2003). To explain such kind of conductivity behavior, Chandra et al. (Uma & Nogami, 2008) have proposed a breathing chain model, which states that at low concentration of the polymer in the LE, the electrolyte system behaves as a liquid as a result of dissociation of salt. The viscosity of the LE increases as the polymer content increases, resulting in a reduction in conductivity. The polymer chain "breaths" in which it opens or folds occupying different volumes in the process. This leads to localized pressure change and fluctuation in the volume of the surrounding area. This localized turbulent pressure variation can either assist (i) in the dissociation of the ion pairs resulting in an increase in the free ion concentration or (ii) in enhancing mobility. Saito et al. (Takeda et al. 2020) have proposed a "two-phase model" for ion conduction in porous polyethylene gel membranes. According to this model, the ionic conductivity in the GPEs may occur through two phases, one through the liquid phase and the other through the gel phase. In the liquid phase, ionic mobility is faster through the micro-pores of the polymer matrix and it is dominantly influenced by the amount of LE entrapped in the matrix and the pore size. However, in the gel phase, the ionic conductivity is encouraged by the interaction between LE and the polymer matrix in which ion-ion interaction, ion–polymer interaction, and gelation play a significant role. In both phases, ionic transport increases with the concentration of LEs. Huang and Wunder (H. Huang & Wunder, 2001) hypothesized a method for conductivity in microporous PVDF-HFP/ PS polymer blends, claiming that the larger the amount of LE trapped in the pores, the higher the conductivity of the porous membrane. The liquid trapped in the pores contributed far more to conductivity than the swelling phase.

8.7 APPLICATION OF SOLID POLYMER ELECTROLYTES

SPEs use a macromolecular system capable of dissolving salts/acids. SPE membranes are typically created using a standard solution cast approach; however, a revolutionary hot-press process for casting these films has just been introduced (Gray, MacCallum, & Vincent, 1986). This technique has been adopted worldwide, in which a mixture of constituents, i.e., polymer and salt, is hot pressed to form a dry electrolyte film. SPE, or polymer-salt/acid complexes, are made by complexing appropriate ion donating salts/acids into high molecular weight polymers like PEO, PPO, and PVP. These polymers have a predisposition for building complexes with various ionic salts and acids while remaining in a single-phase and preferably contain polar groups such as -O-, -N-H-, and -C-N- (Di Noto et al. 2011; D. Singh, Bhattacharya, & Virk, 2015). The presence of these inorganic ionic salts or acids is primarily responsible for the ionic conductivity of these PEs. In the formation of polymer–salt/acid complexes, the solvation and lattice energies of polymers and inorganic salts play a role. Polymers with low lattice energy and inorganic salts/acids with low lattice energy encourage enhanced stability in the resulting PE. Among the several polymers investigated for SPE, PEO has been the most studied polymer host for this category of PEs. Due to the presence of ether oxygen in its chain, it can solvate an extremely wide range of ionic salts and acids and provide better ionic conductivity.

Local relaxation and segmental motion of the polymer chains are related to ion transport in PEs, This effect is amplified when the host polymer has a high degree of amorphicity (Wright, 1976). The ionic conductivity of PEs has been found to increase with a salt concentration in general, while

PEO-based solid polymer salt complexes have shown a decrease in conductivity, at greater concentrations of ionic salts (Olsen, Koksbang, & Skou, 1995). The cause of this behaviour has been linked to the obstruction of polymer segmental mobility, which inhibits ion transport and the creation of ion pairs, resulting in a decrease in the number of free ions available for conduction (Petersen, Jacobsson, & Torell, 1992). At higher salt concentrations, the production of positively and/or negatively charged ion triplets has also been seen. Many researchers have confirmed the production of ion pairs at high salt concentrations (Cameron, Ingram, & Harvie, 1989; Jaipal Reddy & Chu, 2002; MacCallum, Tomlin, & Vincent, 1986; Pollock et al. 1994; Schantz, 1991). Thus, the maximum obtainable ionic conductivity in PEO-salt complexes is observed due to an upper permissible limit of the salt concentration in the host polymer. Further, PEO has a melting point of around 70°C. This refers to the transition from semi-crystalline to amorphous phase. PEO exists mostly in an amorphous state at this temperature, allowing for better segmental motion and, as a result, increased ionic mobility in the electrolyte. As a result, a realistic conductivity value (10^{-4} S cm^{-1}) in the PEO-based polymer-salt complex may be easily produced in the temperature range of 70–90°C. Techniques like as cross-linking, block copolymerization, grafting, and others are used to modify PEO chains and have also been found to increase the conductivity of solvent-free SPEs (Dias, Plomp, & Veldhuis, 2000; Duval et al. 1983; Le Nest et al. 1992; Di Noto et al. 2011; Sci & Ed, 1988; D. Singh, Bhattacharya, & Virk, 2015).

A traditional solution casting procedure was used to create the hyper-branched star-like PE film. Pre-prepared CD-(PMMA-b-PPEGMA)$_{21}$ and NaTf with a NaTf mass fraction of 25% were dissolved in THF under magnetic stirring to form a homogeneous solution. A large amount of solvent was evaporated by transferring the solution to a PTFE plate, then dried in a vacuum for 24 h at 50°C. CD-HSPE was assigned to the final hyper-branched star-like PE. A similar process was used to make a PEO/NaTf polymer electrolyte with an EO/Na$^+$ ratio of 20:1. The hyper-branched star-like PE film was made via the traditional solution casting procedure. CD-HSPE was assigned to the final hyper-branched star-like PE. A similar process was used to make a PEO/NaTf polymer electrolyte with an EO/Na$^+$ ratio of 20:1. Figure 8.6a) and b) shows the digital and SEM photograph of an SPE with the constitution of [CD-(PMMA-b-PPEGMA)$_{21}$- 69.3% PEGMA/NaSO$_3$CF$_3$] weight ratio of 3:1, marked as CD-HSPE3. The resulting SPE film is free-standing, flexible, and translucent, as well as having a smooth surface. Figure 8.6c) shows the ionic conductivity of the SPE vs. PEO-SPE at different temperatures. The Arrhenius plot of the CD-HSPE3's conductivity has a curvy appearance. This is consistent with the amorphous nature of the CD-HSPE3 film and reflects the normal behavior of amorphous SPEs. The ionic conductivity of a PEO-based SPE at a [EO]/[Na$^+$] molar ratio of 20 was also displayed as a comparison. CD-HSPE3 exhibits a higher ionic conductivity than PEO-SPE, particularly at temperatures below 65°C when the PEO-based SPE is highly crystalline. At 30 and 60°C, the conductivities of the CD-HSPE3 are 2.83×10^{-5} and 1.31×10^{-4} S cm^{-1}, respectively. which are significantly higher than those of the PEO-based SPE (S. Chen et al. 2018). Because of their high crystallinity and T_g, PEO-based SPEs have low ionic conductivity, which impedes polymeric segment mobility, as noted in the introduction (Z. Zhang et al. 2016). The causes for the CD-HSPE3's increased ionic conductivity can be linked to the three factors listed below (1) the star-like hyper-branched β-CD amorphous structure, which allows more free volume for polymeric segment movement, (2) the –CH$_2$ CH$_2$O– units in PPEGMA segments, which aid sodium salt dissociation, and (3) the steric inside cavity of β-CD core, which provides more free-volume to promote the movement of Na$^+$ ions, is seen graphically in Figure 8.6d). Additionally, the PPEGMA polymer chains threaded along the cavity of the β-CD, forming additional Na$^+$ transport channels.

Zhou et al. (Q. Ma et al. 2017) reported a new SPEs consisting of PEO and sodium (fluorosulfonyl) (n-nona fluorobutanesulfonyl) imide {Na[N(SO$_2$F) (SO$_2$n-C$_4$F$_9$)], NaFNFSI}. At 80°C, the SPE had an ionic conductivity of 3.4×10^{-4} S cm^{-1}. It is observed that the NaFNFSI/PEO had a greater T_{Na}^+ of 0.24 than the NaFSI/PEO, which could be related to the former's bigger anion volume (e.g., FNFSI$^-$ vs FSI$^-$). In addition, with the NaFNFSI-based SPE, the Na/NaCu$_{1/9}$Ni$_{2/9}$Fe$_{1/3}$Mn$_{1/3}$O$_2$ cells demonstrated

(a)

(b)

(c)

(d)

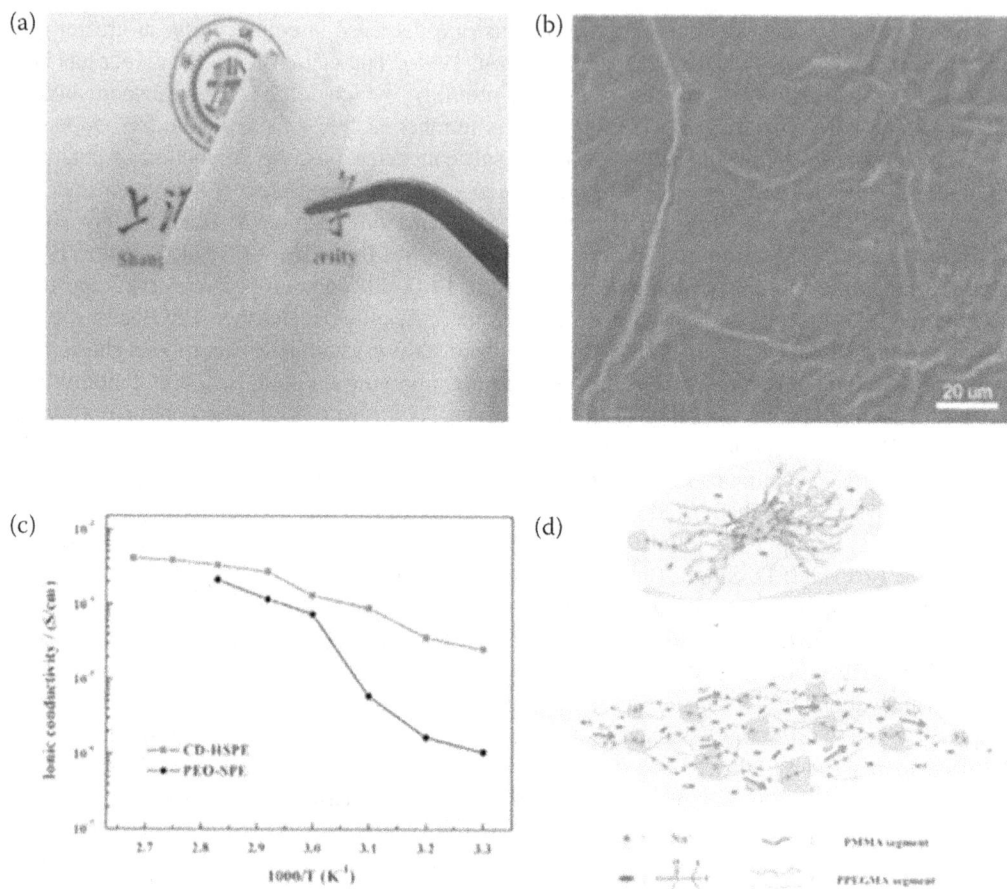

FIGURE 8.6 Photographs showing (a) The SPE for flexibility and transparency; (b) SEM photographs of the SPE membrane; (c) the ionic conductivity of the SPE and the PEO-SPE are compared; (d) the migration of Na$^+$ ions in hyper-branched -CD-based SPE is explained (S. Chen et al. 2018).

a high initial discharge capacity of 122 mAh g^{-1} at 0.1 C and 70% capacity retention after 150 cycles at 1 C (Figure 8.7a) besides outstanding rate performance (Figure 8.7b).

Furthermore, Mindemark et al. (Mindemark et al. 2017) observed that the ionic conductivities of SPEs made of the recently emergent polycarbonate poly (trimethylene carbonate) (PTMC) and various concentrations of NaClO$_4$ or NaTFSI were equivalent to those of Li-based SPEs. At 60°C, PTMC$_3$-NaTFSI had an ionic conductivity of 7.0×10^{-6} S cm^{-1}, allowing Na/Prussian blue cells to operate for multiple cycles with low polarization. The scientists discovered that PTMC$_1$-NaFSI with a high salt concentration (50 mol.%) and a low T_g of −5°C had higher ionic conductivities of 5.7 $\times 10^{-4}$ S cm^{-1} at 60°C due to the formation of salt clusters that enhanced ionic transport via a percolation type ion transport mechanism. On the other hand, the Na/Prussian blue cells using PTMC$_1$ NaFSI in Figure 8.7b) & c). At the same temperature, the cells using PTMC$_5$-NaFSI had a reduced ionic conductivity, had poor cycling performance, and poorer Coulombic efficiency (CE). This inferior performance of the PTMC$_1$-NaFSI-based cell could be attributed to the irreversible reactions between the Na electrode and SPE (Sångeland et al. 2019).

In a nutshell, increasing ionic conductivity in SPEs can be accomplished by adjusting electrolyte salts (e.g., shifting NaTFSI with NaFSI) or polymer matrices (e.g., shifting from PEO to polycarbonates like PTMC5$_4$ and PCL-PTMC) (Sångeland et al. 2019). However, the nature and

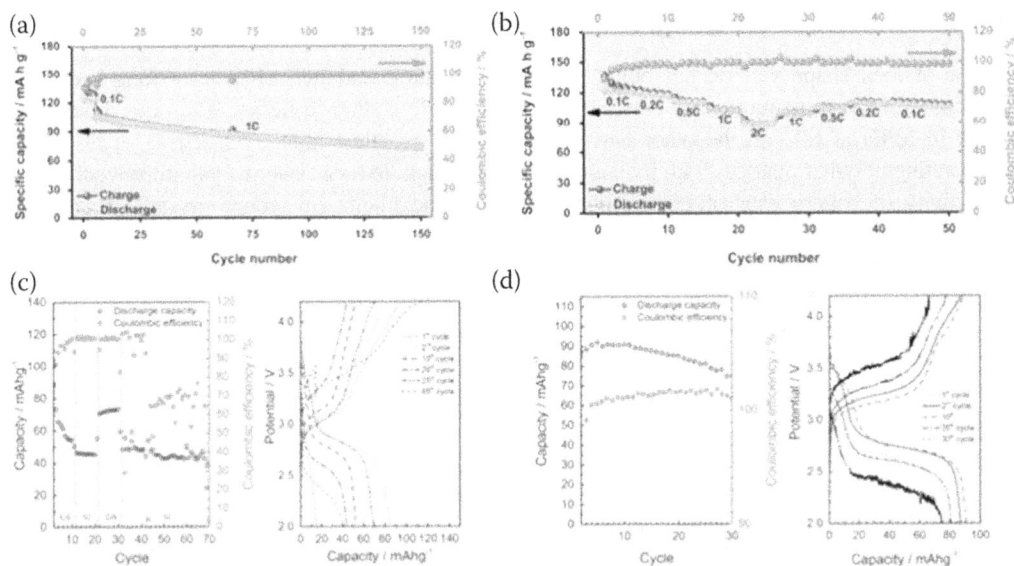

FIGURE 8.7 At 80°C the electrochemical performances of the Na/SPE/Na/NaCu$_{1/9}$Ni$_{2/9}$Fe$_{1/3}$Mn$_{1/3}$O$_2$ cell with NaFNFSI/PEO (EO/Na$^+$ = 15) blended SPE. (a) At 1 C, cycling performance and coulombic efficiencies were measured (after 5 cycles for activation at 0.1 C); (b) Rate capabilities at various C rates (Q. Ma et al. 2017); (c) & (d) discharge capacity and coulombic efficiency per cycle and voltage profiles for a Na/PTMC$_1$-NaFSI/Prussian blue half-cell cycled at 60°C and 40°C (Sångeland et al. 2019).

properties of electrode/electrolyte interphases, particularly the SPE/Na electrode, which govern overall cell performance, cannot be predicted simply from the molecular structure of the electrolyte components; such aspects necessitate in-depth research using both experimental and computational methods. Table 8.2 shows the recent SPEs along with their conductivities and synthesizing techniques. In summary, the inherent features of dry SPEs (e.g., ionic conductivity, T_{Na}^+, interfacial stabilities, etc.) are still inadequate for RT sodium batteries. As a result, new solutions, such as the use of composites, and gelified PEs, have been extensively researched to improve the performance of sodium batteries.

TABLE 8.2

List of Recent SPEs Along with Their Conductivities and Processing Techniques

Electrolyte System	Technique	σ/S cm^{-1}	Ref.
PEO/NaTf (EO/Na = 20)	Ion exchange	3.13×10^{-4} at 85°C	(Doeff et al. 1997)
PVP (70)/ NaClO$_4$ (30)	Solution-casting	2.0×10^{-6} (25°C)	(Subba Reddy et al. 2006)
PEO/ NaAsF$_6$ (EO/Na = 8)	Solvent-casting	1.0×10^{-5} at 45°C	(C. Zhang et al. 2009)
PEO/ NaFSI (EO/Na = 20)	Blending (solvent free)	4.1×10^{-4} (80°C)	(Qi et al. 2016)
PEO/ NaFNFSI (EO/Na = 15)	Solution-casting	3.36×10^{-4} (80°C)	(Q. Ma et al. 2017)
H-β -CD (75)/ NaTf (25)	Solution-casting	1.31×10^{-4} (60°C)	(S. Chen et al. 2018)
POSS-PEG/NaClO$_4$-(EO/Na = 16)	Solution-casting	2.56×10^{-4} (80°C)	(Y. Zheng et al. 2018)
PCL$_{80}$-PTMC$_{20}$ (75)/NaFSI (25)	Solution-casting	3.9×10^{-6} (25°C)	(Sångeland et al. 2019)
PTMC/NaFSI (carbonate/Na = 1or 5)	Solution-casting	5.7×10^{-4} (60°C)	(Sångeland et al. 2019)
PEGDMA-NaFSI (EO/Na = 20:1)	Solvent-free UV-curing	5.6×10^{-4} (60°C)	(Yao et al. 2020)

8.8 APPLICATION OF COMPOSITE SOLID POLYMER ELECTROLYTES

Because of their shape variety, flexibility, and thermodynamic stability, SPEs have long been considered one of the most promising possibilities for high-safe sodium batteries. The poor mobility of cations in SPEs at RT, on the other hand, restricts their practical utility. (H. Zhang et al. 2017). Incorporating polymer matrices with inorganic fillers is thought to be a clever way to improve the ionic conductivity of battery electrolytes. Because: (1) inorganic fillers can reduce crystallinity and T_g, (Capuano, Croce, & Scrosati, 1991); and (2) Polymer chains and salts can interact with surface groups on inorganic surfaces, allowing for faster ion conduction through conductive pathways (Lin et al. 2016; W. Liu et al. 2017; Wieczorek, Florjanczyk, & Stevens, 1995). To fabricate the CSPEs for SIBs, the inactive fillers which are not Na-ion conducting electrolyte such as Al_2O_3, TiO_2 (Ni'Mah et al. 2015), SiO_2 (Praveen, Bhat, & Damle, 2013; Villaluenga et al. 2013), etc. have been used.

Hwang et al (Ni'Mah et al. 2015), produced a PEO-based CSPE with $NaClO_4$ and nanoparticles TiO_2. The ionic conductivity increased to 2.6×10^{-4} S cm^{-1} when 5 wt.% percent TiO_2 was added to the SPE at 60°C, which was twice the pristine SPE (e.g., 1.4×10^{-4} S cm^{-1}). The increased ionic conductivity was attributed to the reduced crystallinity of the PEO phase. Chen et al. (X. Zhang et al. 2018) also found that adding nano-sized fillers to PEs increased their ionic conductivity significantly. The PMA-based CSPE with 3 wt.% Al_2O_3 had high anodic stability of 4.5 V and ionic conductivity of 1.5×10^{-4} S cm^{-1} at 70°C, according to their research. Furthermore, when used in conjunction with the $Na_3V_2(PO_4)_3$ cathode, it has a reversible capacity of 85 mAhg^{-1}, and even after 350 cycles, good capacity retention of 94.1% was achieved.

Liu et al (L. Liu et al. 2019) recently proposed a new approach for fabricating interfacial advantageous CSPEs for SIBs by generating stable SEI layers on Na electrodes (Figure 8.8a) through selective interactions between residual solvent (H_2O), electrolyte salt (NaFSI), and fillers (Al_2O_3). Because of the labile S-F link, it was claimed that FSI$^-$ was relatively unstable, and the anion was hydrolyzed in traces of residual H_2O, resulting in the formation of HF. The Al_2O_3 filler

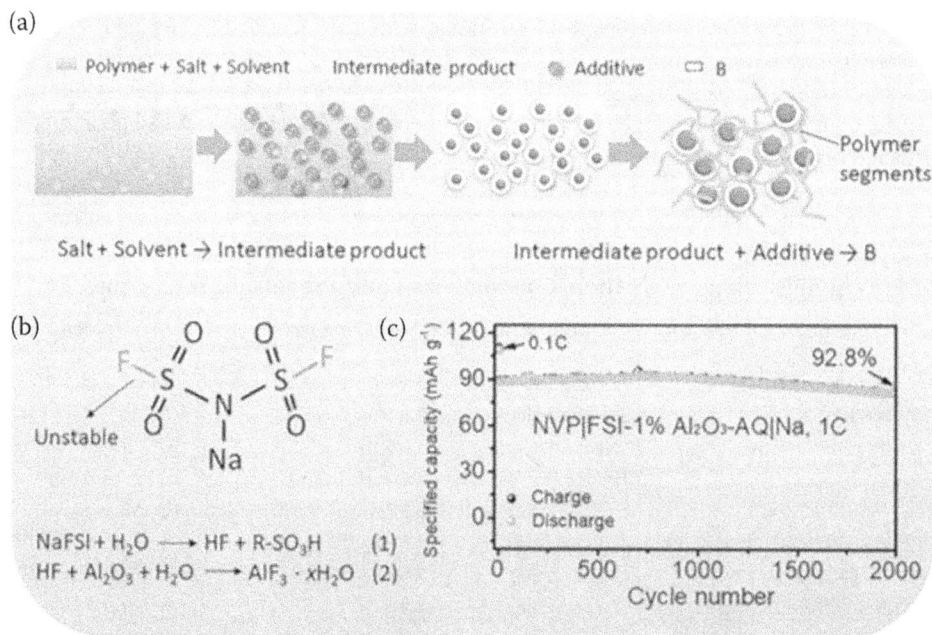

FIGURE 8.8 (a) The additive transforms the intermediate product B (AlF_3. xH_2O) into a stable product B (AlF_3. xH_2O) during the CSPE preparation process; (b) the chemical structure of NaFSI salt; (c) the CSPE-based cell's cycling performance at 1 C (80°C) (L. Liu et al. 2019).

spontaneously interacted with HF to create stable $AlF_3 \cdot xH_2O$ products, leading to the development of SEI layers on the Na electrode (Figure 8.8b). Meanwhile, Al_2O_3 fillers reduce the PEO crystallinity while increasing segmental mobility, resulting in improved ionic conductivity of 1.0×10^{-3} S cm^{-1} at 80°C. As a result, the Na/Na$_3$V$_2$(PO$_4$)$_3$ cell with the NaFSI-based CSPE had a reversible capacity of 110 mAh g^{-1} at 80°C, and excellent long-term cycling performance of 92.8% capacity retention after 2,000 cycles at 1 C (Figure 8.8c).

In addition to inert fillers, inorganic particles with sodium-ion conductivities were created as PE fillers. To develop solvent-free PEO-based CSPEs for sodium batteries, Hu et al. (Z. Zhang et al. 2016) employed two distinct sodium superionic conductors (NASICON) as inorganic fillers (Na$_3$Zr$_2$Si$_2$PO$_{12}$ and Na$_{3.4}$Zr$_{1.8}$Mg$_{0.2}$Si$_2$PO$_{12}$). After the electrolytes have been filled with NASICON-type fillers, the ionic conductivities of CSPEs improved because: (1) the fillers reduced crystallization and improved Na$^+$ migration in the amorphous domain; and (2) the percolation effect, which allowed for the formation of continuous and interconnected Na$^+$ transport channels on the NASICON particle surfaces; and (3) the total conductivity of the as-prepared CSPE was influenced by the movement of Na$^+$ ions in the NASICON bulk. The author postulated a Na$^+$ migration process in the CSPEs, which is seen in Figure 8.9a). When Na$_{3.4}$Zr$_{1.8}$Mg$_{0.2}$Si$_2$PO$_{12}$

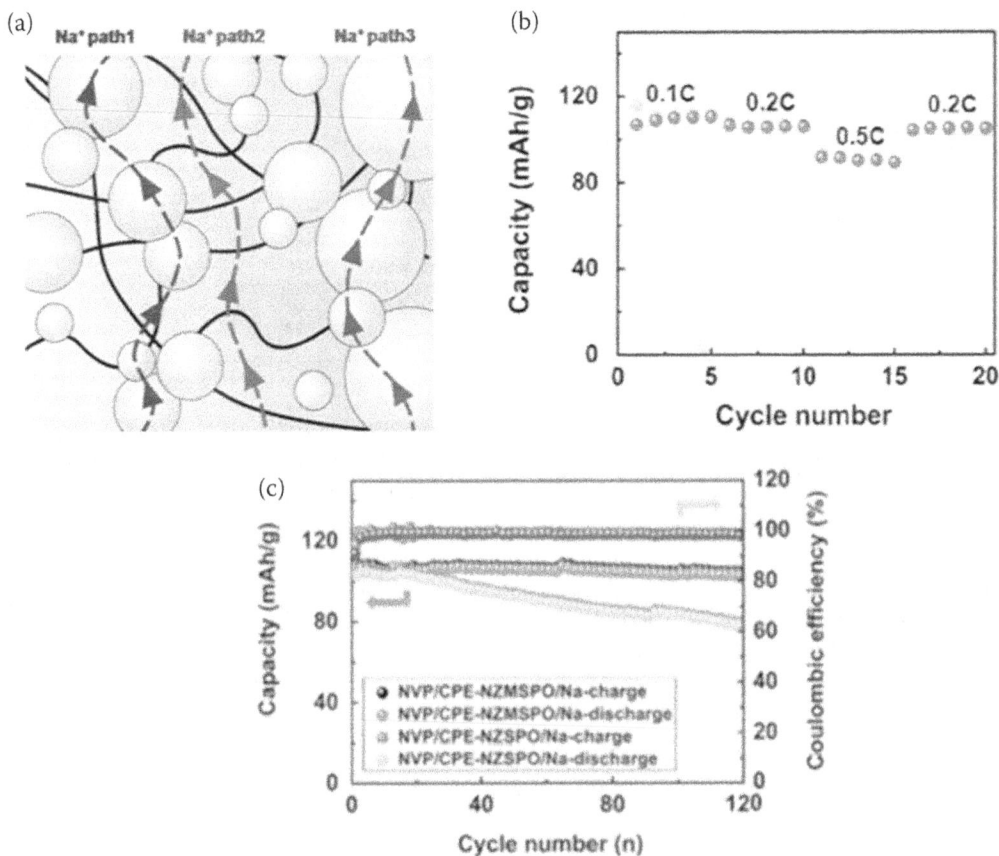

FIGURE 8.9 (a) The CSPEs Na$^+$ transport mechanism is depicted schematically; (b) the rate capability of the Na/NaFSI/PEO/Na$_{3.4}$Zr$_{1.8}$Mg$_{0.2}$Si$_2$PO$_{12}$/Na$_3$V$_2$(PO$_4$)$_3$ cells (Z. Zhang et al. 2017); (c) long cycle performance of the Na/NaFSI/PEO/Na$_{3.4}$Zr$_{1.8}$Mg$_{0.2}$Si$_2$PO$_{12}$/Na$_3$V$_2$(PO$_4$)$_3$ and the Na/NaFSI/PEO/Na$_3$Zr$_2$Si$_2$PO$_{12}$/Na$_3$V$_2$(PO$_4$)$_3$ cells in terms of specific capacity and coulombic efficiency at a current density of 0.1 C (Z. Zhang et al. 2016)

TABLE 8.3

List of Recent CSPEs Along with Their Conductivities and Processing Techniques

Electrolyte System	Technique	σ/S cm^{-1}	Ref.
PEO/NaClO$_4$/TiO$_2$ (EO/Na = 20, 5 wt.% TiO$_2$)	Solution casting	2.62×10^{-4} (60°C)	(Ni'Mah et al. 2015)
NaFSI/PEO/Na$_{3.4}$Zr$_{1.8}$Mg$_{0.2}$Si$_2$PO$_{12}$ (EO/Na = 12, 40 wt.% Na$_{3.4}$Zr$_{1.8}$Mg$_{0.2}$Si$_2$PO$_{12}$)	Solution casting	2.4×10^{-3} (80°C)	(Zhang et al. 2016)
PEO/NaClO$_4$/Al$_2$O$_3$ (EO/Na = 20, 5 wt.% Al$_2$O$_3$)	Solution evaporation	2.0×10^{-4} (80°C)	(Gao et al. 2017)
NaTFSI/PEO/Na$_{3.4}$Zr$_{1.8}$Mg$_{0.2}$Si$_2$PO$_{12}$ (EO/Na = 12, 40 wt.% Na$_{3.4}$Zr$_{1.8}$Mg$_{0.2}$Si$_2$PO$_{12}$)	Solution casting	2.8×10^{-3} (80°C)	(Zhang et al. 2017)
PEO/NaClO$_4$/SiO$_2$ (EO/Na = 13, 3 wt.% SiO$_2$)	Solution casting	6.4×10^{-4} (70°C)	(Wang et al. 2018)
PEO/NaClO$_4$/CQDs (EO/Na = 20, 3 wt.% CQDs)	Solution casting	7.17×10^{-5} (RT)	(Ma et al. 2018)
PMA/PEG/NaClO$_4$/α-Al$_2$O$_3$ (5 wt.% α-Al$_2$O$_3$)	Solution casting	1.46×10^{-4} (70°C)	(X. Zhang et al. 2018)
PVDF (69)/NaClO$_4$ (23)/Na$_3$Zr$_2$Si$_2$PO$_{12}$ (8)	Solution casting	1.1×10^{-4} (RT)	(Lyu et al. 2019)
PEO/NaFSI/Al$_2$O$_3$(EO/Na = 20, 1 wt.% Al$_2$O$_3$)	Blending (solvent free)	1.0×10^{-3} (80°C)	(Liu et al. 2019)
PEO/TiO$_2$/ NaFSI/ (EO/Na = 20, 3 wt.% TiO$_2$)		4.89×10^{-4} (60°C)	(Zhu et al. 2019)
PEO/NaClO$_4$/Na$_3$Zr$_2$Si$_2$PO$_{12}$ (EO/Na = 15, 25 wt.% Na$_3$Zr$_2$Si$_2$PO$_{12}$)	Facile slurry-casting	7.8×10^{-4} (75°C)	(Yu et al. 2019)

(40 wt.%) was added, the ionic conductivity was measured to be 2.4×10^{-3} S cm^{-1}, which is excellent at 80°C. The conductivity of the filler-free electrolyte was significantly lower at 9.0×10^{-4} S cm^{-1} at the same temperature. The CSPE-connected solid-state Na/Na$_3$V$_2$(PO$_4$)$_3$ cell has an initial reversible capacity of 106 mAh g^{-1}, reduced capacity losses after 120 cycles, and better rate performance (Figure 8.9b and c). The CSPE has a high ionic conductivity and good interfacial compatibility with the electrolyte and electrode were responsible for the good performance. Table 8.3 lists some recent CSPEs along with their conductivities. SPEs and CSPEs are potential substitutes to traditional LEs for extremely safe and architecturally flexible sodium batteries; nevertheless, inadequate RT ionic conductivity and poor electrode interfacial compatibility limit their application in practical cells/batteries.

8.9 APPLICATION OF GEL POLYMER ELECTROLYTES

GPEs are an attempt to have an intermediate material having a balance between the high conductivity of organic LEs and the dimensional stability of SPEs. These are solvent swollen polymers, which have a higher ionic conductivity than solvent-free SPEs. GPEs are typically made by mixing a substantial number of liquid solvents and/or plasticizers into a polymer electrolyte matrix (Baskoro, Wong, & Yen, 2019). This occurs in two steps, first, the preparation of a LE by dissolving salt into a polar solvent or ionic liquid, and second, incorporation of the LE into the inert polymeric host. The phase inversion approach, in which a separate porous polymer matrix is created using non-solvent or electrospun and dipped into the LE to absorb into its pores, is another way for preparing GPEs. Solid's cohesive characteristics and liquid's diffusive transport qualities are both influenced by the amount of liquid present in the polymer matrix. However, a substantial volume of liquid in the polymer matrix diminishes the mechanical integrity of GPEs, which can be significantly enhanced by adding cross-linked and/or thermoset components to the GPEs (Mishra, Yadav, & Hashmi, 2020).

The idea of GPEs was first given by Fenullade and Perche in 1975 (Marcoussis, 1975), in which they prepared GPEs having conductivity comparable to the LEs. Since then, GPEs have become an important material for research in the field of rechargeable batteries. Several GPEs have been reported with different host polymers like PEO, PVDF, PAN, PMMA, PVdF-co-HFP, etc. (Yin et al. 2021) and also with some other modified polymer matrices like polymer blends and copolymers (Gao et al. 2016; Gebert et al. 2021). These electrolytes exhibit some unique properties like high room temperature ionic conductivity ($\sim 10^{-4}$–10^{-2} S cm^{-1}), good electrochemical, and interfacial stability.

Kim et al. (2017) reported a free-standing flexible PU and GF/PVdF-HFP based GPE was made by solution casting and non-solvent-induced phase separations, followed by soaking in LE. The ionic conductivity was 8.5×10^{-3} S cm^{-1} and 3.8×10^{-3} S cm^{-1}, respectively. The AC impedance of the cells Na/Na$_3$V$_2$(PO$_4$)$_3$ with LE and PU GPE was tested after cycling. In the cell containing GPE, interfacial resistances were first increased and then stabilized over time. The interfacial resistances in the LE cell, however, were higher. When compared to the pure GF separator, the cell Na/Hard carbon with the GF/PVdF-HFP GPE demonstrated lower interfacial resistance. The total effect of SEI resistance and charge transfer resistance is the interfacial resistance (Cho et al. 2008; Jeong & Kim, 2004; Ryou et al. 2011, 2012). As a result, it's thought that incorporating the GPE boosted sodium ion transport between the electrolyte and electrode. This improved the electrode-electrolyte interfacial compatibility, allowing sodium ion transport to be promoted.

Carbonate solvents, on the other hand, have been restricted because of quick disintegration on the surface of sodium electrodes in batteries (Lee et al. 2016). Before the reductive decomposition of linear carbonates, fluoroethylene carbonate (FEC) was discovered to break down quickly on the anode, helping in the establishment of a preservative surface layer on Na electrodes (Jang et al. 2014, 2015; Lee et al. 2016; Yabuuchi et al. 2014). The FEC additive was used to improve the cycling stability of sodium batteries with gel electrolytes, by preventing the unwanted breakdown of solvents on the Na anode surface.

By raising the C rate from 0.2 C, 0.5 C, 1 C, 2 C, 3 C, 5 C, and 10 C sequentially, the rate capability of the Na/Hard carbon cells using GF and GPEs was investigated (Figure 8.10a). It compares the capacity of cells to rate with various electrolytes. It's worth noting that the cell with GPE had higher discharge capacity than the cell with GF at all current rates tested. It is due to the stripped sodium metal electrode leading to an increase in SEI resistance and reducing cell efficiency. A poly(dimethyl diallyl ammonium) bis(trifluoromethanesulfonyl)imide (PDADMAC-TFSI), NaFSI/C3mpyrFSI, and Al$_2$O$_3$ filler IL-based GPE was recently described by Mecerreyes et al. 106. The iongel membranes had high voltage stability of 5.0 V, whereas the NaFePO$_4$ cells with the composite Na/C anode showed excellent capacity retention with a high CE of > 97% for each C-rate, with a capacity of 114 mAh g^{-1} at C/20 (Figure 8.10b).

In Figure 8.10c, the cell cycling performance and Coulombic efficiency are compared at the 0.5 C charge/discharge condition. The cell with pristine GF had nearly the same initial capacity as other cells with gel polymer electrolytes, but its capacity decline was larger, with an 84.8% retention ratio after the 100th cycle. The retention ratio was enhanced to 88.2%, 90.9%, and 91.6%, respectively, when non-solvent-based gel polymer electrolytes based on methanol, water, and methanol/water were used. Apart from this, Pan et al. developed a single ion conducting polymer electrolyte membrane based on a sodium ion exchanged poly(bis(4-carbonyl benzene sulfonyl)imide-co-2.5-diamino benzesulfonic acid) macromolecule (NaPA). Due to the greater delocalized negative charge in the bis suflonyimide than in sulfonate it had an ionic

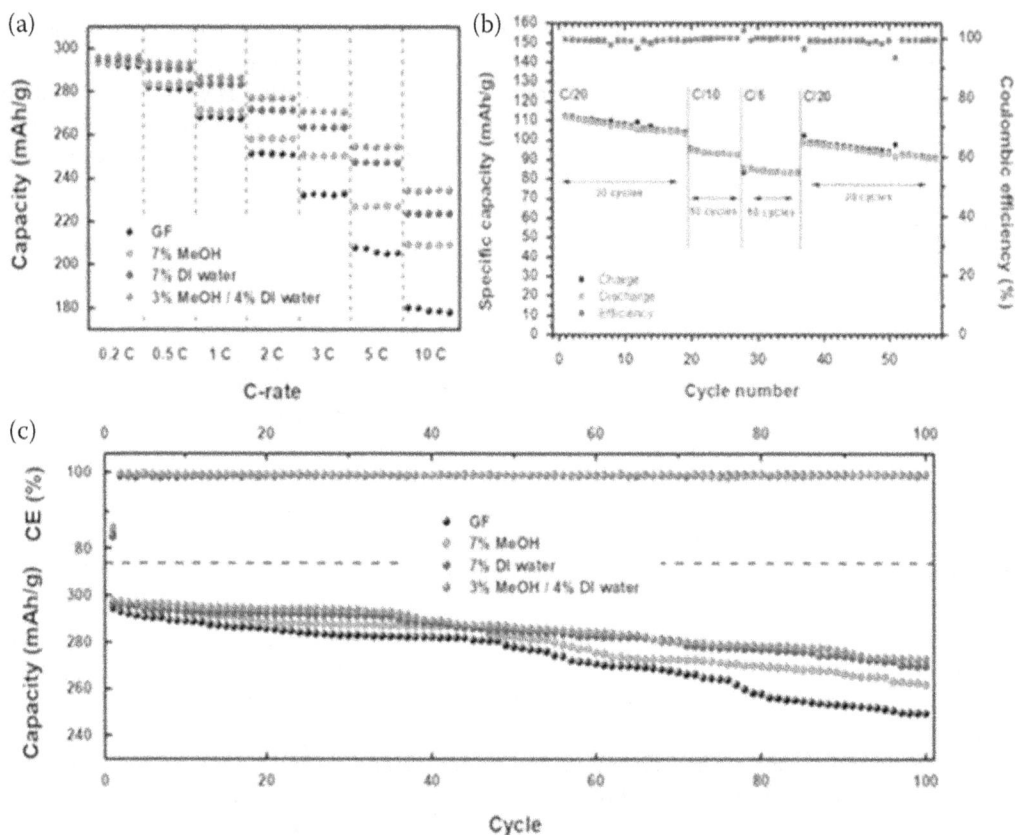

FIGURE 8.10 (a) Discharge capacities of GF and various GPEs at different C-rates (Kim, Chung, & Park, 2018); (b) capacity values and coulombic efficiency of Na-C/NaFePO$_4$ for poly DADMA-TFSI: IL (14 mol. %) 50:50 wt.% + 5 wt.% Al$_2$O$_3$ ion gel at 70°C at different C-rates (C/20, C/10, and C/5) (Fdz De Anastro et al. 2019); (c) discharge capacity and Coulombic efficiency during cycles of GF and various GPEs (Kim, Chung, & Park, 2018).

conductivity of 0.91×10^{-4} S cm^{-1} at 20°C and 4.1×10^{-4} S cm^{-1} at 80°C, respectively. Furthermore, at a 0.2 C rate, the half-cell assembled with Na$_3$V$_2$(PO$_4$)$_3$ cathode delivered 70% of its theoretical specific capacity (Pan et al. 2017). Table 8.4 lists some of the most prevalent GPEs, as well as their conductivities and processing methods. The ionic conductivity of the electrolytes is significantly improved by GPEs, allowing sodium batteries to cycle at ambient temperature. However, the majority of research focuses on volatile and flammable carbonate/ethereal solvents, while employing ILs as plasticizers is excellent for battery safety.

8.10 CONCLUSIONS

SIBs constructed on low-cost materials induced much attention in recent years compared to lithium battery technology. As noticed in the growth of LIBs, liquid electrolytes have a drawback of flammable organic solvents leading to safety concerns. Another approach to use safer batteries is replacing organic solvents with PEs. In this chapter, we have discussed various types of PEs and their ion transport mechanisms for SIBs. For solid/composite electrolytes, the ionic conductivity was found to be low at RT and above 80°C most of the polymers crystallize into liquid. The ionic conductivity can further be improved by adding IL and filler materials to form hybrid polymer

TABLE 8.4

List of Recent GPEs Along with Their Conductivities and Processing Techniques

Electrolyte System	Technique	σ/S cm^{-1}	Ref.
PVDF/NaTf/TEGDME	Solution casting	5.1×10^{-4} (25°C)	(Park et al. 2006)
PVDF-HFP/NaClO$_4$/SiO$_2$ (15 wt. %)/1.0 M in NaTf EC/PC (1:1 Vol.)	Solution casting	4.1×10^{-3} (RT)	(Kumar et al. 2011)
GF/PVDF-HFP/PDA in 1.0 M NaClO$_4$ in PC	Solution casting + hydrophilic coating	5.4×10^{-3} (25°C)	(Gao et al. 2015)
BEMA/PEGMA in 1.0 M NaClO$_4$ in PC	UV cured	5.08×10^{-3} (20°C)	(Bella et al. 2015)
Cellulose/PMMA with 1.0 M NaClO$_4$ in PC/FEC (9:1 Vol.)	Radical polymerization of a precursor + solution drop-cast	6.2×10^{-3} (25°C)	(H. Gao et al. 2016)
PVDF-HFP/SiO$_2$ immersed in 1.0 M NaClO$_4$ in TEGDME	Solution casting	1.0×10^{-3} (25°C)	(Hu et al. 2017)
PVC/NaFSI/EMIFSI	Solution casting	5.6×10^{-3} (45°C)	(Rani et al. 2017)
PVDF-HFP/NaPA immersed in EC/DMC	Ion exchange + Blending	9.1×10^{-5} (20°C)	(Pan et al. 2017)
PSA/PDE with 1.0 M NaClO$_4$ in PC	In situ polymerization	1.2×10^{-3} (RT)	(Zhang et al. 2017)
PEO/NaClO$_4$/PC	Solvent-free photopolymerization	1.0×10^{-3} (25°C)	(Colò et al. 2017)
PEO/NaClO$_4$/SiO$_2$/EMIFSI	Solution casting	1.3×10^{-3} (RT)	(Song et al. 2017)
GF/PVDF-HFP immersed in 1.0 M NaClO$_4$ in PC/EC	Nonsolvent-induced phase separation	3.8×10^{-3} (25°C)	(Kim et al. 2017)
PVDF-HFP immersed in 1.0 M NaPF$_6$ in PC/EC/FEC	phase separation	5.0×10^{-4} (RT)	(Y. Zhang et al. 2018)
MATEPP/MMA/TFMA with 1.0 M NaClO$_4$ in EC/PC/FEC	In situ thermal polymerization	6.29×10^{-3} (RT)	(J. Zheng et al. 2018)
PAN-NFM immersed in 1.0 M NaClO$_4$ in EC/PC/DME	Electrospinning	3.01×10^{-3} (RT)	(Manuel et al. 2018)
PVDF-HFP/EMITf/0.5 M NaTf in EC/PC	Solution casting	5.7×10^{-3} (30°C)	(Kumar 2018)
PVDF-HFP/NaI/Al$_2$O$_3$ immersed in 1.0 M NaClO$_4$ in TEGDME	Not reported	1.2×10^{-3} (25°C)	(Hu et al. 2018)
PVDF-HFP immersed in 1.0 M NaClO$_4$ in PC/FEC	Phase separation	4.2×10^{-3} (RT)	(Guo et al. 2018)
GF/PETEA-THEICTA with 1.0 M NaTFSI in PC/FEC	In situ polymerization	3.85×10^{-3} (25°C)	(Dong Zhou et al. 2018)
PEO/NaTFSI/BMIMTFSI	solution casting	4.1×10^{-4} (30°C)	(Singh et al. 2018)
P(VdF-co-HFP) Immersed in 1.0 M NaClO$_4$ in EC/DEC	Electrospinning	1.13×10^{-3} (RT)	(Janakiraman et al. 2018)
PMMA/PVDF-HFP/Na$_3$Zr$_2$Si$_2$PO$_{-2}$ with 1.0 M NaPF$_6$ in EC/PC	Electrospinning + in situ polymerization	2.78×10^{-3} (30°C)	(Yi et al. 2018)
PAN-NFM immersed in 1.0 M NaTf in PEGDME	Electrospinning	1.0×10^{-3} (30°C)	(Lim et al. 2018)
PVDF-HFP/PVC/NaTf/ TEGDME	Phase transfer + In situ polymerization	1.2×10^{-4} (25°C)	(Chen et al. 2019)
PVDF immersed in 1.0 M NaPF$_6$ in EC/PC	Electrospinning	1.08×10^{-3} (RT)	(Janakiraman et al. 2019b)
PVDF immersed in 1.0 M NaClO$_4$ in EC: DEC	Electrospinning	0.92×10^{-3} (RT)	(Janakiraman et al. 2019a)
PVDF/Celgard/PVDF in 1.0 M NaClO$_4$ in EC: DEC	Electrospinning	1.25×10^{-3} (RT)	(Janakiraman et al. 2020)

electrolytes. Recent research shows that GPEs display high ionic conductivity at RT with excellent electrochemical performance. However, the majority of research focuses on volatile and flammable carbonate/ethereal solvents. While the use of ionic liquids as plasticizers is advantageous in terms of battery safety, it is inefficient in terms of cost. As a result, the future development may be a combination of highly conductive and innately safe PEs for SIBs.

ACKNOWLEDGMENTS

Janakiraman is thankful to the CSIR, India for the grant of research associate fellowship (File: No: 09/079(2859)/2020-EMR-1). He also acknowledges Prof. Prabeer Barpanda, MRC, IISc Bangalore, India for his encouragement and support.

REFERENCES

Adhikari, B., and S. Majumdar. 2004. Polymers in Sensor Applications. *Progress in Polymer Science (Oxford)* 29 (7): 699–766. doi:10.1016/j.progpolymsci.2004.03.002

Arya, A., and A. L. Sharma. 2017. "Polymer Electrolytes for Lithium Ion Batteries: A Critical Study." *Ionics*. 23:497–540. doi:10.1007/s11581-016-1908-6

Baskoro, F., H. Q. Wong, and H. J. Yen. 2019. Strategic Structural Design of a Gel Polymer Electrolyte toward a High Efficiency Lithium-Ion Battery. Review-article. *ACS Applied Energy Materials* 2 (6). American Chemical Society: 3937–3971. doi:10.1021/acsaem.9b00295

Bella, F., F. Colò, J. R. Nair, and C. Gerbaldi. 2015. Photopolymer Electrolytes for Sustainable, Upscalable, Safe, and Ambient-Temperature Sodium-Ion Secondary Batteries. *ChemSusChem* 8 (21): 3668–3676. doi:10.1002/cssc.201500873

Berthier, C., W. Gorecki, M. Minier, M. B. Armand, J. M. Chabagno, and P. Rigaud. 1983. Microscopic Investigation of Ionic Conductivity in Alkali Metal Salts-Poly(Ethylene Oxide) Adducts. *Solid State Ionics* 11 (1): 91–95. doi:10.1016/0167-2738(83)90068-1

Cameron, G. G., M. D. Ingram, and J. L. Harvie. 1989. Ion Transport in Polymer Electrolytes. *Faraday Discussions of the Chemical Society* 88: 55–63. doi:10.1039/DC9898800055

Capuano, F., F. Croce, and B. Scrosati. 1991. Composite Polymer Electrolytes. *Journal of The Electrochemical Society* 138 (7): 1918–1922. doi:10.1149/1.2085900

Chen, F., and M. Forsyth. 2016. Elucidation of Transport Mechanism and Enhanced Alkali Ion Transference Numbers in Mixed Alkali Metal-Organic Ionic Molten Salts.*Physical Chemistry Chemical Physics* 18 (28). Royal Society of Chemistry: 19336–19344. doi:10.1039/c6cp01411a

Chen, S., H. Che, F. Feng, J. Liao, H. Wang, Y. Yin, and Z. F. Ma. 2019. Poly(Vinylene Carbonate)-Based Composite Polymer Electrolyte with Enhanced Interfacial Stability to Realize High-Performance Room-Temperature Solid-State Sodium Batteries. *ACS Applied Materials and Interfaces* 11 (46): 43056–43065. doi:10.1021/acsami.9b11259

Chen, S., F. Feng, Y. Yin, H. Che, X. Z. Liao, and Z. F. Ma. 2018. A Solid Polymer Electrolyte Based on Star-like Hyperbranched B-Cyclodextrin for All-Solid-State Sodium Batteries. *Journal of Power Sources* 399 (August). Elsevier: 363–371. doi:10.1016/j.jpowsour.2018.07.096

Cho, T. H., M. Tanaka, H. Onishi, Y. Kondo, T. Nakamura, H. Yamazaki, S. Tanase, and T. Sakai. 2008. Battery Performances and Thermal Stability of Polyacrylonitrile Nano-Fiber-Based Nonwoven Separators for Li-Ion Battery. *Journal of Power Sources* 181 (1): 155–160. doi:10.1016/j.jpowsour.2008.03.010

Colò, F., F. Bella, J. R. Nair, and C. Gerbaldi. 2017. Light-Cured Polymer Electrolytes for Safe, Low-Cost and Sustainable Sodium-Ion Batteries. *Journal of Power Sources* 365: 293–302. doi:10.1016/j.jpowsour.2017.08.079

Croce, F., L. Persi, B. Scrosati, F. Serraino-Fiory, and E. Plichta. 2001. Role of the Ceramic Fillers in Enhancing the Transport Properties of Composite Polymer Electrolytes. *Electrochimica Acta* 46: 2457–2461.

Dias, F. B., L. Plomp, and J. B. J. Veldhuis. 2000. Trends in Polymer Electrolytes for Secondary Lithium Batteries. *Journal of Power Sources* 88: 169–191. doi:10.1016/S0378-7753(99)00529-7

Doeff, M. M., A. Ferry, Y. Ma, L. Ding, and L. C. De Jonghe. 1997. Effect of Electrolyte Composition on the Performance of Sodium/Polymer Cells. *Journal of The Electrochemical Society* 144 (2): L20–L22. doi:10.1149/1.1837419

Dunn, B., H. Kamath, and J. M. Tarascon. 2011. Electrical Energy Storage for the Grid: A Battery of Choices. *Science* 334 (6058): 928–935. doi:10.1126/science.1212741

Duval, P., H. Le Gac, J. Jones, H. A. Chew, and W. Whltworth. 1994Effect of Grafting Degree and side PEO Chain Length on the Ionic Conductivities of NBR-g-PEO Based Polymer Electrolytes. *Polymer Engineering And Sciences* 34 (17): 1304–1313. doi.org/10.1002/pen.760341702.

Ellis, B. L., and L. F. Nazar. 2012. Sodium and Sodium-Ion Energy Storage Batteries. *Current Opinion in Solid State and Materials Science* 16 (4). Elsevier Ltd: 168–177. doi:10.1016/j.cossms.2012.04.002

Fdz De Anastro, A., N. Lago, C. Berlanga, M. Galcerán, M. Hilder, M. Forsyth, and D. Mecerreyes. 2019. Poly(Ionic Liquid)Iongel Membranes for All Solid-State Rechargeable Sodium Battery. *Journal of Membrane Science* 582 (February). Elsevier B.V.: 435–441. doi:10.1016/j.memsci.2019.02.074

Fenton, D. E., J. M. Parker, and P. V. Wright. 1973. Complexes of Alkali Metal Ions with Poly(Ethylene Oxide). *Polymer* 14 (11): 589. doi:10.1016/0032-3861(73)90146-8

Freitag, K. M., P. Walke, T. Nilges, H. Kirchhain, R. J. Spranger, and L. van Wüllen. 2018. Electrospun-Sodiumtetrafluoroborate-Polyethylene Oxide Membranes for Solvent-Free Sodium Ion Transport in Solid State Sodium Ion Batteries. *Journal of Power Sources* 378 (December 2017). Elsevier: 610–617. doi:10.1016/j.jpowsour.2017.12.083

Gadjourova, Z., Y. G. Andreev, D. P. Tunstall, and P. G. Bruce. 2001. Ionic Conductivity in Crystalline Polymer Electrolytes.*Nature* 412 (August): 520–532.

Gao, H., B. Guo, J. Song, K. Park, and J. B. Goodenough. 2015. A Composite Gel-Polymer/Glass-Fiber Electrolyte for Sodium-Ion Batteries. *Advanced Energy Materials*. doi:10.1002/aenm.201402235

Gao, H., W. Zhou, K. Park, and J. B. Goodenough. 2016. A Sodium-Ion Battery with a Low-Cost Cross-Linked Gel-Polymer Electrolyte. *Advanced Energy Materials* 6 (18). doi:10.1002/aenm.201600467

Gao, R., R. Tan, L. Han, Y. Zhao, Z. Wang, L. Yang, and F. Pan. 2017. Nanofiber Networks of Na3V2(PO4)3 as a Cathode Material for High Performance All-Solid-State Sodium-Ion Batteries. *Journal of Materials Chemistry A* 5 (11): 5273–5277. doi:10.1039/c7ta00314e

Gebert, F., J. Knott, R. Gorkin, S. L. Chou, and S. X. Dou. 2021. Polymer Electrolytes for Sodium-Ion Batteries. *Energy Storage Materials* 36: 10–30. doi:10.1016/j.ensm.2020.11.030

Ghosal, K., A. Chandra, G. Praveen, S. Snigdha, S. Roy, C. Agatemor, S. Thomas, and I. Provaznik. 2018. Electrospinning over Solvent Casting: Tuning of Mechanical Properties of Membranes. *Scientific Reports* 8 (1). Springer US: 1–9. doi:10.1038/s41598-018-23378-3

Gray, F. M., J. R. MacCallum, and C. A. Vincent. 1986. Poly(Ethylene Oxide) "LiCF3SO3" Polystyrene Electrolyte Systems. *Solid State Ionics* 18–19 (PART 1): 282–286. doi:10.1016/0167-2738(86)90127-X

Guéguen, A., D. Streich, M. He, M. Mendez, F. F. Chesneau, P. Novák, and E. J. Berg. 2016. Decomposition of LiPF 6 in High Energy Lithium-Ion Batteries Studied with Online Electrochemical Mass Spectrometry. *Journal of The Electrochemical Society* 163 (6): A1095–A1100. doi:10.1149/2.0981606jes

Guillen, G. R., Y. Pan, M. Li, and E. M. Hoek. 2011. Preparation and Characterization of Membranes Formed by Nonsolvent Induced Phase Separation: A Review. *Industrial and Engineering Chemistry Research* 50 (7): 3798–3817. doi:10.1021/ie101928r

Guo, J. Z., A. B. Yang, Z. Y. Gu, X. L. Wu, W. L. Pang, Q. L. Ning, W. H. Li, J. P. Zhang, and Z. M. Su. 2018. Quasi-Solid-State Sodium-Ion Full Battery with High-Power/Energy Densities. *ACS Applied Materials and Interfaces* 10 (21): 17903–17910. doi:10.1021/acsami.8b02768

Hasa, I., S. Passerini, and J. Hassoun. 2016. Characteristics of an Ionic Liquid Electrolyte for Sodium-Ion Batteries. *Journal of Power Sources* 303: 203–207. doi:10.1016/j.jpowsour.2015.10.100

Hosaka, T., K. Kubota, A. Shahul Hameed, and S. Komaba. 2020. Research Development on K-Ion Batteries. *Chemical Reviews* 120 (14): 6358–6466. doi:10.1021/acs.chemrev.9b00463

Hu, X., G. Dawut, J. Wang, H. Li, and J. Chen. 2018. Room-Temperature Rechargeable Na-SO2 Batteries Containing a Gel-Polymer Electrolyte.*Chemical Communications* 54 (42): 5315–5318. doi:10.1039/c8cc02094a

Hu, X., Z. Li, Y. Zhao, J. Sun, Q. Zhao, J. Wang, Z. Tao, and J. Chen. 2017. Quasi Solid State Rechargeable Na-CO2 Batteries with Reduced Graphene Oxide Na Anodes. *Science Advances* 3 (2): 1–8. doi:10.1126/sciadv.1602396

Huang, H., and S. L. Wunder. 2001. Ionic Conductivity of Microporous PVDF-HFP/PS Polymer Blends. *Journal of The Electrochemical Society* 148 (3): A279. doi:10.1149/1.1351756

Huang, X. 2011. Separator Technologies for Lithium-Ion Batteries.*Journal of Solid State Electrochemistry* 15 (4): 649–662. doi:10.1007/s10008-010-1264-9

Hueso, K. B., V. Palomares, M. Armand, and T. Rojo. 2017. Challenges and Perspectives on High and Intermediate-Temperature Sodium Batteries. *Nano Research* 10 (12): 4082–4114. doi:10.1007/s12274-017-1602-7

Jaipal Reddy, M., and P. P. Chu. 2002. Ion Pair Formation and Its Effect in PEO:Mg Solid Polymer Electrolyte System. *Journal of Power Sources* 109 (2): 340–346. doi: 10.1016/S0378-7753(02)00084-8

Janakiraman, S., M. Khalifa, R. Biswal, S. Ghosh, S. Anandhan, and A. Venimadhav. 2020. High Performance Electrospun Nanofiber Coated Polypropylene Membrane as a Separator for Sodium Ion Batteries. *Journal of Power Sources* 460 (April). Elsevier B.V.: 228060. doi: 10.1016/j.jpowsour.202 0.228060

Janakiraman, S., O. Padmaraj, S. Ghosh, and A. Venimadhav. 2018. A Porous Poly (Vinylidene Fluoride-Co-Hexafluoropropylene) Based Separator-Cum-Gel Polymer Electrolyte for Sodium-Ion Battery. *Journal of Electroanalytical Chemistry* 826. Elsevier B.V: 142–149. doi: 10.1016/j.jelechem.2018.08.032

Janakiraman, S., A. Surendran, R. Biswal, S. Ghosh, S. Anandhan, and A. Venimadhav. 2019a. Electrochemical Characterization of a Polar β-Phase Poly (Vinylidene Fluoride) Gel Electrolyte in Sodium Ion Cell. *Journal of Electroanalytical Chemistry*. doi: 10.1016/j.jelechem.2018.12.011

Janakiraman, S., A. Surendran, R. Biswal, S. Ghosh, S. Anandhan, and A. Venimadhav. 2019b. "Electrospun Electroactive Polyvinylidene Fluoride-Based Fibrous Polymer Electrolyte for Sodium Ion Batteries. *Materials Research Express* 6 (8). IOP Publishing. doi: 10.1088/2053-1591/ab226a

Janakiraman, S., A. Surendran, S. Ghosh, S. Anandhan, and A. Venimadhav. 2018. A New Strategy of PVDF Based Li-Salt Polymer Electrolyte through Electrospinning for Lithium Battery Application A New Strategy of PVDF Based Li-Salt Polymer Electrolyte through Electrospinning for Lithium Battery Application. *Materials Research Express* 6. IOP Publishing. doi: 0000-0002-7765-6484

Jang, J. Y., H. Kim, Y. Lee, K. T. Lee, K. Kang, and N. S. Choi. 2014. Cyclic Carbonate Based-Electrolytes Enhancing the Electrochemical Performance of Na4Fe3(PO4)2(P 2O7) Cathodes for Sodium-Ion Batteries. *Electrochemistry Communications* 44. Elsevier B.V.: 74–77. doi: 10.1016/j.elecom.2014.05.003

Jang, J. Y., Y. Lee, Y. Kim, J. Lee, S. M. Lee, K. T. Lee, and N. S. Choi. 2015. Interfacial Architectures Based on a Binary Additive Combination for High-Performance Sn4P3 Anodes in Sodium-Ion Batteries. *Journal of Materials Chemistry A* 3 (16). Royal Society of Chemistry: 8332–8338. doi: 10.1039/c5ta00724k

Jeong, Y. B., and D. W. Kim. 2004. Effect of Thickness of Coating Layer on Polymer-Coated Separator on Cycling Performance of Lithium-Ion Polymer Cells. *Journal of Power Sources* 128 (2): 256–262. doi: 10.1016/j.jpowsour.2003.09.073

Kim, J. I., Y. Choi, K. Y. Chung, and J. H. Park. 2017. A Structurable Gel-Polymer Electrolyte for Sodium Ion Batteries. *Advanced Functional Materials* 27 (34): 1–7. doi: 10.1002/adfm.201701768

Kim, J. I., K. Y. Chung, and J. H. Park. 2018. Design of a Porous Gel Polymer Electrolyte for Sodium Ion Batteries. *Journal of Membrane Science* 566 (August). Elsevier B.V.: 122–128. doi: 10.1016/j.memsci.2 018.08.066

Kirchhöfer, M., J. Von Zamory, E. Paillard, and S. Passerini. 2014. Separators for Li-Ion and Li-Metal Battery Including Ionic Liquid Based Electrolytes Based on the TFSI- and FSI- Anions. *International Journal of Molecular Sciences* 15 (8): 14868–14890. doi: 10.3390/ijms150814868

Klongkan, S., and J. Pumchusak. 2015. Effects of Nano Alumina and Plasticizers on Morphology, Ionic Conductivity, Thermal and Mechanical Properties of PEO-LiCF3SO3 Solid Polymer Electrolyte. *Electrochimica Acta* 161. Elsevier Ltd.: 171–176. doi: 10.1016/j.electacta.2015.02.074

Komaba, S., W. Murata, T. Ishikawa, N. Yabuuchi, T. Ozeki, T. Nakayama, A. Ogata, K. Gotoh, and K. Fujiwara. 2011. Electrochemical Na Insertion and Solid Electrolyte Interphase for Hard-Carbon Electrodes and Application to Na-Ion Batteries. *Advanced Functional Materials* 21 (20): 3859–3867. doi: 10.1002/adfm.201100854

Kumar, D. 2018. Effect of Organic Solvent Addition on Electrochemical Properties of Ionic Liquid Based Na+ Conducting Gel Electrolytes. *Solid State Ionics* 318 (July). Elsevier: 65–70. doi: 10.1016/j.ssi.2017.09.006

Kumar, D., M. Suleman, and S. A. Hashmi. 2011. Studies on Poly (Vinylidene Fl Uoride-Co-Hexa Fl Uoropropylene) Based Gel Electrolyte Nanocomposite for Sodium Sulfur Batteries. *Solid State Ionics* 202 (1). Elsevier B.V.: 45–53. doi: 10.1016/j.ssi.2011.09.001

Kuze, S., J. Kageura, and S. Matsumoto. 2013. Development of a Sodium Ion Secondary Battery. *Sumitomo* 1–13. http://www.sumitomo-chem.co.jp/english/rd/report/theses/docs/2013E_3.pdf%5Cnpapers3:// publication/uuid/9BDEA394-8ABC-4BC4-B4CB-0E06337D66EB.

Lee, Y., J. Lee, H. Kim, K. Kang, and N. S. Choi. 2016. Highly Stable Linear Carbonate-Containing Electrolytes with Fluoroethylene Carbonate for High-Performance Cathodes in Sodium-Ion Batteries. *Journal of Power Sources* 320. Elsevier B.V: 49–58. doi: 10.1016/j.jpowsour.2016.04.070

Lim, D. H., M. Agostini, J. H. Ahn, and A. Matic. 2018. An Electrospun Nanofiber Membrane as Gel-Based Electrolyte for Room-Temperature Sodium Sulfur Batteries. *Energy Technology* 6 (7): 1214–1219. doi: 10.1002/ente.201800170

Lin, D., W. Liu, Y. Liu, H. R. Lee, P. C. Hsu, K. Liu, and Y. Cui. 2016. High Ionic Conductivity of Composite Solid Polymer Electrolyte via in Situ Synthesis of Monodispersed SiO2 Nanospheres in Poly (Ethylene Oxide). *Nano Letters* 16 (1): 459–465. doi:10.1021/acs.nanolett.5b04117

Liu, L., X. Qi, S. Yin, Q. Zhang, X. Liu, L. Suo, H. Li, L. Chen, and Y. S. Hu. 2019. In Situ Formation of a Stable Interface in Solid-State Batteries. Rapid-communication. *ACS Energy Letters* 4 (7). American Chemical Society: 1650–1657. doi:10.1021/acsenergylett.9b00857

Liu, W., S. W. Lee, D. Lin, F. Shi, S. Wang, A. D. Sendek, and Y. Cui. 2017. Enhancing Ionic Conductivity in Composite Polymer Electrolytes with Well-Aligned Ceramic Nanowires. *Nature Energy* 2 (5): 1–7. doi:10.1038/nenergy.2017.35

Lyu, Y. Q., J. Yu, J. Wu, M. B. Effat, and F. Ciucci. 2019. Stabilizing Na-Metal Batteries with a Manganese Oxide Cathode Using a Solid-State Composite Electrolyte. *Journal of Power Sources* 416 (October 2018): 21–28. doi:10.1016/j.jpowsour.2019.01.082

M. Armand. 1986. *Polymer Electrolytes I* (November). doi:10.1146/annurev.ms.16.080186.001333

Ma, C., K. Dai, H. Hou, X. Ji, L. Chen, D. G. Ivey, and W. Wei. 2018. High Ion-Conducting Solid-State Composite Electrolytes with Carbon Quantum Dot Nanofillers. *Advanced Science* 5 (5): 1–9. doi:10. 1002/advs.201700996

Ma, Q., J. Liu, X. Qi, X. Rong, Y. Shao, W. Feng, J. Nie, et al. 2017. A New Na[(FSO2)(n-C4F9SO2)N]-Based Polymer Electrolyte for Solid-State Sodium Batteriesf. *Journal of Materials Chemistry A* 5 (17). Royal Society of Chemistry: 7738–7743. doi:10.1039/c7ta01820g

MacCallum, J. R., A. S. Tomlin, and C. A. Vincent. 1986. An Investigation of the Conducting Species in Polymer Electrolytes. *European Polymer Journal* 22 (10): 787–791. doi:10.1016/0014-3057(86)90017-0

Manuel, J., X. Zhao, K. K. Cho, J. K. Kim, and J. H. Ahn. 2018. Ultralong Life Organic Sodium Ion Batteries Using a Polyimide/Multiwalled Carbon Nanotubes Nanocomposite and Gel Polymer Electrolyte. *ACS Sustainable Chemistry and Engineering* 6 (7): 8159–8166. doi:10.1021/acssuschemeng.7b04561

Marcoussis, L. D. 1975. Ion-Conductive Macromolecular Gels and Membranes for Solid Lithium Cells.Journal of Applied Electrochemistry 5: 63–69.

Mindemark, J., R. Mogensen, M. J. Smith, M. M. Silva, and D. Brandell. 2017. Polycarbonates as Alternative Electrolyte Host Materials for Solid-State Sodium Batteries. *Electrochemistry Communications* 77. Elsevier B.V.: 58–61. doi:10.1016/j.elecom.2017.02.013

Mishra, K., N. Yadav, and S. A. Hashmi. 2020. "Recent Progress in Electrode and Electrolyte Materials for Flexible Sodium-Ion Batteries." *Journal of Materials Chemistry A* 8 (43): 22507–22543. doi:10.1039/d0ta07188a

Nest, J. F. L., S. Callens, A. Gandini, and M. Armand. 1992. A New Polymer Network for Ionic Conduction. *Electrochimica Acta* 37 (9): 1585–1588. doi:10.1016/0013-4686(92)80116-4

Ni Mah, Y. L., M. Y. Cheng, J. H. Cheng, J. Rick, and B. J. Hwang. 2015. Solid State Polymer Nanocomposite Electrolyte of TiO2/PEO/NaClO4 for Sodium Ion Batteries. *Journal of Power Sources* 278: 375–381. doi:10.1016/j.jpowsour.2014.11.047

Noto, V. D., S. Lavina, G. A. Giffin, E. Negro, and B. Scrosati. 2011. Polymer Electrolytes: Present, Past and Future. *Electrochimica Acta* 57 (1): 4–13. doi:10.1016/j.electacta.2011.08.048

O Heir, J. 2017. Building Better Batteries. *Mechanical Engineering* 139 (1): 10–11.

Ohno, H. 1992. "Applications of Polymer Electrolytes:" 37 (9): 1649–1651.

Olsen, I. I., R. Koksbang, and E. Skou. 1995. Transference Number Measurements on a Hybrid Polymer Electrolyte. *Electrochimica Acta* 40 (11): 1701–1706. doi:10.1016/0013-4686(95)00094-U

Omar, N., M. Daowd, P. van den Bossche, O. Hegazy, J. Smekens, T. Coosemans, and J. van Mierlo. 2012. Rechargeable Energy Storage Systems for Plug-in Hybrid Electric Vehicles-Assessment of Electrical Characteristics. *Energies* 5 (8): 2952–2988. doi:10.3390/en5082952

Papke, B. L., M. A. Ratner, and D. F. Shriver. 1982. Conformation and Ion-Transport Models for the Structure and Ionic Conductivity in Complexes of Polyethers with Alkali Metal Salts. *Journal of The Electrochemical Society* 129(8): 1694–1701. doi:10.1149/1.2124252

Pan, H., Y. S. Hu, and L. Chen. 2013. Room-Temperature Stationary Sodium-Ion Batteries for Large-Scale Electric Energy Storage. *Energy and Environmental Science* 6 (8): 2338–2360. doi:10.1039/c3ee40847g

Pan, Q., Z. Li, W. Zhang, D. Zeng, Y. Sun, and H. Cheng. 2017. Single Ion Conducting Sodium Ion Batteries Enabled by a Sodium Ion Exchanged Poly(Bis(4-Carbonyl Benzene Sulfonyl)Imide-Co-2,5-Diamino Benzesulfonic Acid) Polymer Electrolyte. *Solid State Ionics* 300. Elsevier B.V.: 60–66. doi:10.1016/j.ssi.2016.12.001

Park, C. W., J. H. Ahn, H. S. Ryu, K. W. Kim, and H. J. Ahn. 2006. Room-Temperature Solid-State Sodiumsulfur Battery. *Electrochemical and Solid-State Letters* 9 (3): 123–125. doi:10.1149/1.2164607

Peighambardoust, S. J., S. Rowshanzamir, and M. Amjadi. 2010. Review of the Proton Exchange Membranes for Fuel Cell Applications. *International Journal of Hydrogen Energy*. 35. Elsevier Ltd. doi:10.1016/ j.ijhydene.2010.05.017

Petersen, G., P. Jacobsson, and L. M. Torell. 1992. A Raman Study of Ion-Polymer and Ion-Ion Interactions in Low Molecular Weight Polyether-LiCF3SO3 Complexes. *Electrochimica Acta* 37 (9): 1495–1497. doi:10.1016/0013-4686(92)80097-6

Pollock, D. W., K. J. Williamson, K. S. Weber, L. J. Lyons, and L. R. Sharpe. 1994. Ion Pairing and Ionic Conductivity in Amorphous Polymer Electrolytes: A Structural Investigation Employing EXAFS. *Chemistry of Materials* 6 (11): 1912–1914. doi:10.1021/cm00047a004

Polymers, E. 1993. *Applications of Electroactive Polymers.* doi:10.1007/978-94-011-1568-1

Ponrouch, A., R. Dedryvère, D. Monti, A. E. Demet, J. M. Ateba Mba, L. Croguennec, C. Masquelier, P. Johansson, and M. Rosa Palacín. 2013. Towards High Energy Density Sodium Ion Batteries through Electrolyte Optimization. *Energy and Environmental Science* 6 (8): 2361–2369. doi:10.1039/c3ee41379a

Ponrouch, A., E. Marchante, M. Courty, J. M. Tarascon, and M. Rosa Palacín. 2012. In Search of an Optimized Electrolyte for Na-Ion Batteries.*Energy and Environmental Science* 5 (9): 8572–8583. doi:1 0.1039/c2ee22258b

Praveen, D., S. V. Bhat, and R. Damle. 2013. Role of Silica Nanoparticles in Conductivity Enhancement of Nanocomposite Solid Polymer Electrolytes: (PEGx NaBr): YSiO2. *Ionics* 19 (10): 1375–1379. doi:10.1 007/s11581-013-0871-8

Qi, X., Q. Ma, L. Liu, Y. S. Hu, H. Li, Z. Zhou, X. Huang, and L. Chen. 2016. Sodium Bis(Fluorosulfonyl) Imide/Poly(Ethylene Oxide) Polymer Electrolytes for Sodium-Ion Batteries. *ChemElectroChem* 3 (11): 1741–1745. doi:10.1002/celc.201600221

Qiao, L., X. Judez, T. Rojo, M. Armand, and H. Zhang. 2020. "Review—Polymer Electrolytes for Sodium Batteries." *Journal of The Electrochemical Society* 167 (7): 070534. doi:10.1149/1945-7111/ab7aa0

Rani, M. A. A., J. Hwang, K. Matsumoto, and R. Hagiwara. 2017. Poly(Vinyl Chloride) Ionic Liquid Polymer Electrolyte Based on Bis(Fluorosulfonyl)Amide for Sodium Secondary Batteries. *Journal of The Electrochemical Society* 164 (8): H5031–H5035. doi:10.1149/2.0221708jes

Ryou, M. H., D. J. Lee, J. N. Lee, Y. M. Lee, J. K. Park, and J. W. Choi. 2012. Excellent Cycle Life of Lithium-Metal Anodes in Lithium-Ion Batteries with "Mussel-Inspired Polydopamine-Coated Separators. *Advanced Energy Materials* 2 (6): 645–650. doi:10.1002/aenm.201100687

Ryou, M. H., Y. M. Lee, J. K. Park, and J. W. Choi. 2011. "Mussel-Inspired Polydopamine-Treated Polyethylene Separators for High-Power Li-Ion Batteries." *Advanced Materials* 23 (27): 3066–3070. doi:10.1002/adma.201100303

Sångeland, C., R. Mogensen, et al. 2019. Stable Cycling of Sodium Metal All-Solid-State Batteries with Polycarbonate-Based Polymer Electrolytes. *ACS Applied Polymer Materials* 1 (4): 825–832. doi:10. 1021/acsapm.9b00068

Sångeland, C. R. Younesi, et al. 2019.Towards Room Temperature Operation of All-Solid-State Na-Ion Batteries through Polyester "Polycarbonate-Based Polymer Electrolytes." *Energy Storage Materials* 19 (February). Elsevier Ltd.: 31–38. doi:10.1016/j.ensm.2019.03.022

Schantz, S. 1991. On the Ion Association at Low Salt Concentrations in Polymer Electrolytes; a Raman Study of NaCF3SO3 and LiClO4 Dissolved in Polypropylene Oxide. *The Journal of Chemical Physics* 94 (9): 6296–6306. doi:10.1063/1.460418

Sci, M. G. J. P., and P. P. Ed. 1988. Novel Polymer Electrolytes Based.*Society* 397 (2): 392–397.

Seh, Z. W., J. Sun, Y. Sun, and Y. Cui. 2015. A Highly Reversible Room-Temperature Sodium Metal Anode. *ACS Central Science* 1 (8): 449–455. doi:10.1021/acscentsci.5b00328

Sekhon, S. S. 2003. Conductivity Behaviour of Polymer Gel Electrolytes: Role of Polymer. *Bulletin of Materials Science* 26 (3): 321–328. doi:10.1007/BF02707454

Singh, D., B. Bhattacharya, and H. S. Virk. 2015. "Conductivity Modulation in Polymer Electrolytes and Their Composites Due to Ion-Beam Irradiation." *Solid State Phenomena* 239: 110–148. doi:10.4028/ www.scientific.net/SSP.239.110

Singh, V. K., S. K. Singh, H. Gupta, Shalu, L. Balo, A. K. Tripathi, Y. L. Verma, and R. K. Singh. 2018. Electrochemical Investigations of Na0.7CoO2 Cathode with Peo-Natfsi-Bmimtfsi Electrolyte as Promising Material for Na-Rechargeable Battery. *Journal of Solid State Electrochemistry* 22 (6):1909–1919. doi:10.1007/s10008-018-3891-5

Song, S., M. Kotobuki, F. Zheng, C. Xu, S. V. Savilov, N. Hu, L. Lu, Y. Wang, and W. D. Z. Li. 2017. A Hybrid Polymer/Oxide/Ionic-Liquid Solid Electrolyte for Na-Metal Batteries. *Journal of Materials Chemistry A* 5 (14): 6424–6431. doi:10.1039/C6TA11165C

Subba Reddy, C. V., A. P. Jin, Q. Y. Zhu, L. Q. Mai, and W. Chen. 2006. Preparation and Characterization of (PVP + NaClO4) Electrolytes for Battery Applications. *European Physical Journal E* 19 (4): 471–476. doi:10.1140/epje/i2005-10076-8

Suharto, Y., Y. Lee, J. S. Yu, W. Choi, and K. J. Kim. 2018. Microporous Ceramic Coated Separators with Superior Wettability for Enhancing the Electrochemical Performance of Sodium-Ion Batteries. *Journal of Power Sources* 376 (November 2017). Elsevier: 184–190. doi:10.1016/j.jpowsour.2017.11.083

Syzdek, J., R. Borkowska, K. Perzyna, J. M. Tarascon, and W. Wieczorek. 2007. Novel Composite Polymeric Electrolytes with Surface-Modified Inorganic Fillers. *Journal of Power Sources* 173 (2 SPEC. ISS.): 712–720. doi:10.1016/j.jpowsour.2007.05.061

Takeda, S., Y. Saito, I. Kaneko, and H. Yoshitake. 2020. Effect of Cross-Sectional Shape of Pathway on Ion Migration in Polyethylene Separators for Lithium-Ion Batteries. *Journal of Physical Chemistry C* 124 (3): 1827–1835. doi:10.1021/acs.jpcc.9b09859

Uma, T., and M. Nogami. 2008. Proton-Conducting Glass Electrolyte. *Analytical Chemistry* 80 (2): 506–508. doi:10.1021/ac0706630

Vignarooban, K., R. Kushagra, A. Elango, P. Badami, B. E. Mellander, X. Xu, T. G. Tucker, C. Nam, and A. M. Kannan. 2016. Current Trends and Future Challenges of Electrolytes for Sodium-Ion Batteries. *International Journal of Hydrogen Energy* 41 (4): 2829–2846. doi:10.1016/j.ijhydene.2015.12.090

Villaluenga, I., X. Bogle, S. Greenbaum, I. G. De Muro, T. Rojo, and M. Armand. 2013. Cation Only Conduction in New Polymer-SiO2 Nanohybrids: Na + Electrolytes.*Journal of Materials Chemistry A* 1 (29): 8348–8352. doi:10.1039/c3ta11290j

Wang, X., X. Zhang, Y. Lu, Z. Yan, Z. Tao, D. Jia, and J. Chen. 2018. Flexible and Tailorable Na–CO2 Batteries Based on an All-Solid-State Polymer Electrolyte.*ChemElectroChem* 5 (23): 3628–3632. doi:10.1002/celc.201801018

Wang, Y. 2009. Recent Research Progress on Polymer Electrolytes for Dye-Sensitized Solar Cells. *Solar Energy Materials and Solar Cells* 93 (8). Elsevier: 1167–1175. doi:10.1016/j.solmat.2009.01.009

Whittingham, M. S. 1978. *Chemistry of Intercalation Compounds: Metal Guests in Chalcogenide Hosts.* Vol. 12.

Wieczorek, W., Z. Florjanczyk, and J. R. Stevens. 1995. Composite Based Solid.*Electrochimica Acta* 40 (13–14): 2251–2258.

Wright, P. V. 1976. An Anomalous Transition to a Lower Activation Energy for Dc Electrical Conduction above the Glass-Transition Temperature. *Journal of Polymer Science: Polymer Physics Edition* 14 (5): 955–957. doi:10.1002/pol.1976.180140516

Yabuuchi, N., Y. Matsuura, T. Ishikawa, S. Kuze, J. Y. Son, Y. T. Cui, H. Oji, and S. Komaba. 2014. Phosphorus Electrodes in Sodium Cells: Small Volume Expansion by Sodiation and the Surface-Stabilization Mechanism in Aprotic Solvent. *Chem Electro Chem* 1 (3): 580–589. doi:10.1002/celc.2 01300149

Yao, Y., Z. Wei, H. Wang, H. Huang, Y. Jiang, X. Wu, X. Yao, Z. S. Wu, and Y. Yu. 2020. Toward High Energy Density All Solid-State Sodium Batteries with Excellent Flexibility. *Advanced Energy Materials* 10 (12): 1–9. doi:10.1002/aenm.201903698

Yi, Q., W. Zhang, S. Li, X. Li, and C. Sun. 2018. Durable Sodium Battery with a Flexible Na 3 Zr 2 Si 2 PO 12 -PVDF-HFP Composite Electrolyte and Sodium/Carbon Cloth Anode. *ACS Applied Materials and Interfaces* 10 (41): 35039–35046. doi:10.1021/acsami.8b09991

Yin, H., C. Han, Q. Liu, F. Wu, F. Zhang, and Y. Tang. 2021. Recent Advances and Perspectives on the Polymer Electrolytes for Sodium/Potassium Ion Batteries. *Small*: 2006627. doi:10.1002/smll.202006627

Yu, X., L. Xue, J. B. Goodenough, and A. Manthiram. 2019. A High-Performance All-Solid-State Sodium Battery with a Poly(Ethylene Oxide)-Na3Zr2Si2PO12 Composite Electrolyte. Rapid-communication. *ACS Materials Letters* 1 (1). American Chemical Society: 132–138. doi:10.1021/acsmaterialslett.9b00103

Zhang, C., S. Gamble, D. Ainsworth, A. M. Slawin, Y. G. Andreev, and P. G. Bruce. 2009. Alkali Metal Crystalline Polymer Electrolytes. *Nature Materials* 8 (7). Nature Publishing Group: 580–584. doi:10.1 038/nmat2474

Zhang, H., C. Li, M. Piszcz, E. Coya, T. Rojo, L. M. Rodriguez-Martinez, M. Armand, and Z. Zhou. 2017. Single Lithium-Ion Conducting Solid Polymer Electrolytes: Advances and Perspectives. *Chemical Society Reviews* 46 (3): 797–815. doi:10.1039/c6cs00491a

Zhang, J., H. Wen, L. Yue, J. Chai, J. Ma, P. Hu, G. Ding, et al. 2017. In Situ Formation of Polysulfonamide Supported Poly(Ethylene Glycol) Divinyl Ether Based Polymer Electrolyte toward Monolithic Sodium Ion Batteries.*Small* 13 (2): 1–10. doi:10.1002/smll.201601530

Zhang, L., G. Feng, X. Li, S. Cui, S. Ying, X. Feng, L. Mi, and W. Chen. 2019. Synergism of Surface Group Transfer and In-Situ Growth of Silica-Aerogel Induced High-Performance Modified Polyacrylonitrile Separator for Lithium/Sodium-Ion Batteries. *Journal of Membrane Science* 577 (February). Elsevier B.V.: 137–144. doi:10.1016/j.memsci.2019.02.002

Zhang, X., X. Wang, S. Liu, Z. Tao, and J. Chen. 2018. A Novel PMA/PEG-Based Composite Polymer Electrolyte for All-Solid-State Sodium Ion Batteries. *Nano Research* 11 (12): 6244–6251. doi:10.1007/s12274-018-2144-3

Zhang, Y., Y. An, S. Dong, J. Jiang, H. Dou, and X. Zhang. 2018. Enhanced Cycle Performance of Polyimide Cathode Using a Quasi-Solid-State Electrolyte. Research-article. *Journal of Physical Chemistry C* 122 (39). American Chemical Society: 22294–22300. doi:10.1021/acs.jpcc.8b06513

Zhang, Z., K. Xu, X. Rong, Y. S. Hu, H. Li, X. Huang, and L. Chen. 2017. Na3.4Zr1.8Mg0.2Si2PO12 Filled Poly(Ethylene Oxide)/Na(CF3SO2)2N as Flexible Composite Polymer Electrolyte for Solid-State Sodium Batteries. *Journal of Power Sources* 372 (April). Elsevier: 270–275. doi:10.1016/j.jpowsour.2017.10.083

Zhang, Z., Q. Zhang, C. Ren, F. Luo, Q. Ma, Y. S. Hu, Z. Zhou, H. Li, X. Huang, and L. Chen. 2016. A Ceramic/Polymer Composite Solid Electrolyte for Sodium Batteries. *Journal of Materials Chemistry A* 4 (41). Royal Society of Chemistry: 15823–15828. doi:10.1039/c6ta07590h

Zheng, J., Y. Zhao, X. Feng, W. Chen, and Y. Zhao. 2018. Novel Safer Phosphonate-Based Gel Polymer Electrolytes for Sodium-Ion Batteries with Excellent Cycling Performance. *Journal of Materials Chemistry A* 6 (15). Royal Society of Chemistry: 6559–6564. doi:10.1039/c8ta00530c

Zheng, Y., Q. Pan, M. Clites, B. W. Byles, E. Pomerantseva, and C. Y. Li. 2018. High-Capacity All-Solid-State Sodium Metal Battery with Hybrid Polymer Electrolytes. *Advanced Energy Materials* 8 (27): 1–9. doi:10.1002/aenm.201801885

Zhou, D., G. M. Spinks, G. G. Wallace, C. Tiyapiboonchaiya, D. R. MacFarlane, M. Forsyth, and J. Sun. 2003. Solid State Actuators Based on Polypyrrole and Polymer-in-Ionic Liquid Electrolytes. *Electrochimica Acta* 48 (14-16 SPEC.): 2355–2359. doi:10.1016/S0013-4686(03)00225-1

Zhou, D., Y. Chen, B. Li, H. Fan, F. Cheng, D. Shanmukaraj, T. Rojo, M. Armand, and G. Wang. 2018. A Stable Quasi-Solid-State Sodium–Sulfur Battery. *Angewandte Chemie – International Edition* 57 (32): 10168–10172. doi:10.1002/anie.201805008

Zhu, T., X. Dong, Y. Liu, Y. G. Wang, C. Wang, and Y. Y. Xia. 2019. An All-Solid-State Sodium-Sulfur Battery Using a Sulfur/Carbonized Polyacrylonitrile Composite Cathode. *ACS Applied Energy Materials* 2 (7): 5263–5271. doi:10.1021/acsaem.9b00953

Zoń, J., P. Miziak, N. Amrhein, R. Gancarz, M. Yekini, M. Wazir, N. Bashir, et al. 2009. Showed That Lithium Self-Diffusion Is. *Chemistry – A European Journal* 192 (February): 972–973. http://linkinghub.elsevier.com/retrieve/pii/S0040402008003852%5Cnhttp://doi.wiley.com/10.1002/chem.201503580%5Cnhttp://pid.sagepub.com/lookup/doi/10.1177/0954407013485567%5Cnhttp://pubs.acs.org/doi/abs/10.1021/ja00164a033%5Cnhttp://doi.wiley.com/10.1002/c.

9 Recent Developments in Aromatic Polymer-Based Proton Exchange Membranes

Arijit Ghorai and Susanta Banerjee
Indian Institute of Technology Kharagpur, Kharagpur, India

CONTENTS

9.1 INTRODUCTION

Proton exchange membrane fuel cells (PEMFC) are made massive progress in power generation due to their specific features. A PEMFC cell comprises an anode, cathode, and proton exchange membrane (PEM) and directly transforms the chemical energy of fuels into electrical power. The chemical energy associated with the fuels is directly converted into electricity, water, and heat on reaction with the oxidant in a fuel cell system. Methanol, ethanol, hydrogen are commonly used as fuels in fuel cells [1]. In a fuel cell, the reactions take place through a few steps. In a PEM fuel cell, the hydrogen gas in the anode is converted into hydrogen ions and electrons. The proton is transported through the PEM, reacts with oxygen in the anode, and produces water and heat. The electrons that cannot travel through the PEM are forced to travel through an external circuit to the cathode, generating electricity. Cathodic and anodic reactions in PEMFC are given below:

$$Anode\ electrode: H_2 \rightarrow 2H^+ + 2e^-$$

$$Cathode\ electrode: O_2 + 4H^+ + 4e^- \rightarrow 2H_2O + Heat$$

$$Overall\ PEMFC\ reaction: 2H_2 + O_2 \rightarrow 2H_2O$$

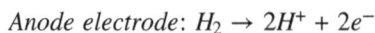

The central part of any PEMFC system is the membrane electrode assembly (MEA), comprised of an electrocatalyst layer and PEM. The job of PEM is the transportation of protons from anode to cathode, a fence to electrons, and stop the fuel crossover [2]. A schematic design of PEMFC is shown in Figure 9.1. The significant advantage of a fuel cell system is its high efficiency, no emission of environmentally fouling gases (CO_2, NO_x, CO, SO_x), and noiseless operation as there is no moving object and with tremendously high consistency. On the contrary, the limitations are the large-scale manufacturing of these power devices and their more significant expense, which can be addressed by employing the new advancements in this area [1].

9.2   GROWTH OF PROTON EXCHANGE MEMBRANES

In the 1970s, DuPont developed a perfluorosulfonic acid-based polymer Nafion® that showed a twofold increment in the membrane's conductivity and expanded the lifetime by four orders in magnitude (h). Shortly, this transformed into a state-of-the-art material in PEMFC. Later, shorter side chains containing advanced perfluorosulfonic acid membranes with a higher proportion of $-SO_3H$ to $>CF_2$ groups had been synthesized by Dow Chemical Company and Asahi Chemical Company [3]. Nafion® is a copolymer with fluoro-3,6-dioxo-4,6-octane sulfonic acid polytetrafluoroethylene (PTFE), where the Teflon unit and sulfonic acid groups have incorporated for providing the hydrophobic and hydrophilic character to the polymer membrane, respectively. These ionic functional groups boost the polymer to absorb plenty of water and consequently affect the hydration of the polymers. Subsequently, the variables influencing the balanced proton exchange level's performance are the degree of hydration and thickness, which function as a vital part in choosing their sustainability for the application on the fuel cell [4]. PEM performance is characterized by its proton conductivity (PC), and the PC is also related to the amount of humidity in the membrane. In order to achieve a highly efficient PEMFC, the PEM must have a set of properties, for example, zero

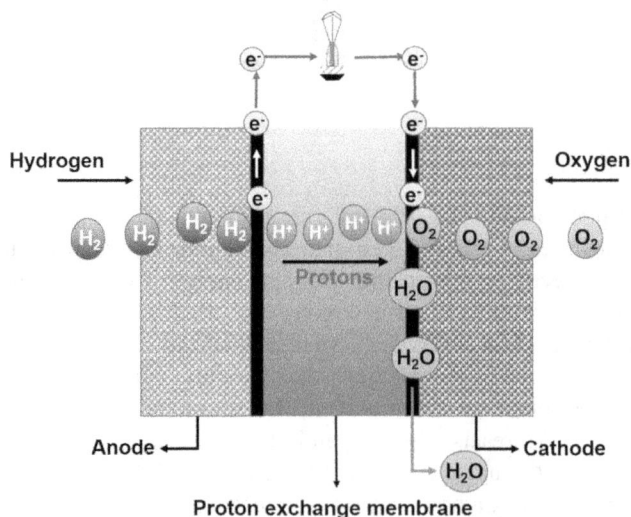

FIGURE 9.1   Schematic diagram of the operation of PEMFC.

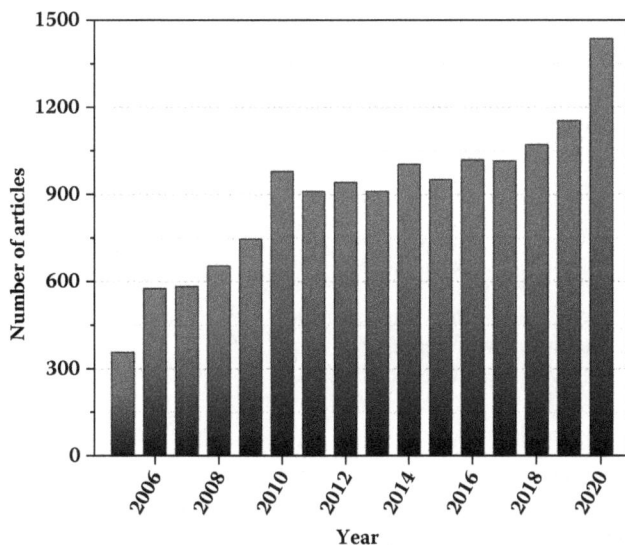

FIGURE 9.2   Total publications per year on PEMFC based on "Web of Science".

electronic conductivity, high proton conductivity, high mechanical strength and stability, electro-chemical durability in working conditions, low fuel crossover, and high oxidative stability [5].

However, the perfluorinated polymers (Nafion® and Flemion®) suffer from limited operating temperature (0°C–80°C), inadequate durability, and high methanol permeability. Extensive efforts have been made to develop alternative aromatic hydrocarbon-based polymers to overcome these hurdles. Many high-performance polymers are exploited in numerous applications because of their structural integrity at high temperatures, a brilliant combination of physical, chemical, mechanical properties, oxidative, and thermal stability. A large number of polymers from different polymeric backbones [poly (arylene ether)s, polyimides, polybenzazoles, polytriazoles, etc.] have been developed and assessed for their suitability for PEM applications [6]. Notably, proton-conducting polymer membranes have mainly consisted of two domains: hydrophobic and hydrophilic systems. The hydrophilic sections containing the acidic groups (sulfonic acid and carboxylic acid groups) are accountable for proton conductivity, while the hydrophobic areas, framed by non-sulfonated polymer sections, furnish the mechanical stability and water management properties of PEM. The microphase segregated structure developed by these two regions controls all the properties of the PEMs [7]. A large number of sulfonated aromatic polymers such as poly(phenylene)s [8], poly(ether ether ketone)s [9,10], poly(ether ether sulfone)s [11–16], poly(arylene ether)s [17], polybenzimidazoles [18], polyamide [19], polyimides [20–23], and polytriazoles [24–31] have been examined as prospective candidate PEM applications. Several research papers appear in this field, as represented in Figure 9.2, indicating the growth in this field.

9.3   PROTON CONDUCTION MECHANISMS IN PROTON EXCHANGE MEMBRANES

The proton conductivity of any PEM materials is considered the most crucial parameter in their potential use in PEMFC. The loss of resistance towards proton conduction in PEM is relative to the membrane's ionic resistance and the high conductivity is especially necessary to obtain high current density. The proton transportation of hydrated polymeric materials at a molecular level is usually described based on the two main mechanisms: (**a**) Grotthuss mechanism or proton hopping (**b**) vehicular mechanism or the diffusion mechanism (water molecules acted as the vehicle in the system) [32–34].

In the Grotthuss process, the hydrogen ions hop from one hydrolyzed ionic site (SO_3^- and H_3O^+) to another through the PEM. The protons, obtained from the oxidation of hydrogen in the

Grotthuss mechanism (Proton hopping)

FIGURE 9.3 Grotthuss type mechanism for proton conduction. Reprinted with permission from Ref. [33]. Copyright (2008), American Chemical Society.

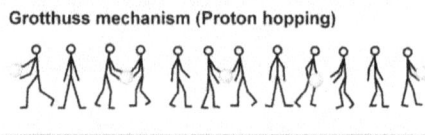

Grotthuss mechanism (Proton hopping)

FIGURE 9.4 Graphical diagram for the vehicular mode of proton conduction. Reprinted with permission from Ref. [33]. Copyright (2008), American Chemical Society.

Vehicle mechanism

anode, are transferred in complex ways to the charge-carrying solvent groups (hydronium ions for water molecules) by forming and breaking hydrogen bonds [34]. In this system, ionic clusters have swelled in water attendance and create a percolation process to transfer protons. A simple sketch of the hopping procedure is displayed in Figure 9.3. The hopping mechanism contributes very little to the conductivity of Nafion-like perfluorinated sulfonic acid membranes.

In vehicular-type proton transportation, the hydrated proton (H_3O^+) diffuses through the water medium because of the electrochemical distinction [33]. Here, the water-associated protons $[H^+(H_2O)_x]$ in consequence of the electro-osmotic drag, convey one or more water molecules through the PEM and moved with them. The accessible volumes inside the polymer chains have an essential role in this type of proton transportation process. It permits the easy movement of the hydrated protons within the PEM. The graphical presentation of the vehicular mode of proton transportation is sketched in Figure 9.4.

9.4 CLASSIFICATION OF AROMATIC-BASED PROTON EXCHANGE MEMBRANES

PEMs alternate to Naffion®. Accordingly, the molecular designs, synthetic methods, fabrications, and characterizations of PEM materials have evolved. Aromatic sulfonated polymers were documented as good contestants in this area. As a result, many polymers were developed, and they have been classified under different groups, e.g., poly(arylene ether)s, polyimides, polybenzimidazoles, and polytriazoles. They are further grouped according to polymer chain structures (linear and non-linear), chemical modifications (cross-linked and composite), and fluorine contents (fluorinated and non-fluorinated). These classifications are given below.

9.4.1 Sulfonated Poly(arylene ether)s

Poly(arylene ether)s are considered high-performance polymers with outstanding properties. Over the years, various sulfonated polymers with common ether linkages were developed for PEM applications. These include sulfonated poly(arylene ether ketone)s, poly(arylene ether sulfone)s, and poly(arylene ether)s, etc. [35]. These polymers were prepared by direct polymerizations of sulfonated aromatic dihalides and bisphenol systems with various monomer molar ratios. With suitable synthetic techniques, both random and block copolymers can be made. A hydrophobic-hydrophilic phase-separated morphology is attained from the random copolymers based on the degree of sulfonations (DS). The multi-block copolymers for PEM applications are synthesized from the end-functionalized oligomeric blocks with various chemical constituents. The combination of supramolecular phase divisions in multi-block copolymers with self-ordering in the molecular scales provides fascinating properties [36].

9.4.1.1 Fluorinated Poly(arylene ether)s

Aromatic fluorinated polymers gained a remarkable level of consideration for their one-of-a-kind properties and high-temperature performances [37,38]. The inclusion of fluorine or fluorinated groups in polymers is the topic of intense field of research since they affects many physical properties of polymers. In this unique state, trifluoromethyl [($-CF_3$), 3 F] hexafluoroisopropylidine [$>C(CF_3)_2$, 6 F] and fluorine [($-$F), F) groups are most widely used for the primary modification of various high-performance polymers, encouraging vital property development. These groups in polymer chains enhance the solubility, ease of membrane formation, increase glass transition temperature, flame retardancy, chemical resistance, and thermomechanical stability [37]. The fluorinated functional groups [3 F, 6 F and F] affect several polymer properties and they are given below:

a. Polymers with C–F bonds show higher thermal and chemical stability because the strength of C–F bonds is higher than that of C–H or C–C bonds [38].
b. The electron-withdrawing effect of 3 F and 6 F groups effectively reduces the electron density of the attached aromatic rings, thereby diminishing the susceptibility to oxidative attack by radicals.
c. The assemblies of fluorinated systems help to improve solubility (usually stated as the "fluorine effect") in comparison to the non-fluorinated polymers, subsequently improve their processing, and make them suitable for a wide range of applications [35,39].

9.4.1.1.1 Random Poly(arylene ether)s

A series of sulfonated poly(aryl ether ketone) copolymers (SPAEK-6F-XX, XX is the mole% of sulfonic acid-containing monomer in the polymer) were reported by Xing et al. The SPAEK-6F-XX (XX = 20, 30, 40, 50, 60, 70, and 80) copolymers were made by the polycondensation of hexafluoroisopropylidene diphenol (6F-BPA) with 4,4'-difluorobenzophenone (DFBP) and 5,5'-carbonylbis(2-fluorobenzenesulfonate) at several molar proportions (Scheme 9.1) [40]. The solvent cast membranes from these polymers were tough and flexible. The sulfonic acid content in the d copolymers was estimated from the 1 H NMR spectral analysis and was in accordance with the monomer feed ratio. The SPAEK-6F polymers (acid form) offered high thermal stabilities (up to ~260°C in air) that are required for PEMFC applications. The SPAEK-6Fs membranes displayed rising trends in the water uptake (WU: 8%–92%) and swelling ratios (SR: 3%–30%) with the increase of sulfonated monomer proportion in the polymer. The WU and SR values were increased with an increase in the temperature, and values at 80°C are given in Table 9.1. Fenton's reagent experiment indicated that SPAEK-6F membranes had average oxidative stability, and the values were 0.8, 1.5, 4.0, 6.5, and >24 h for SPAEK-6F-70, 60, 50, 40, and 30, respectively. The gradual increase in WU for hydrated SPAEK-6F-20, -30, and -40 membranes with retention of mechanical integrity indicated the formation of percolated structures [41]. At a lower degree of sulfonation, a hydrophobic backbone predominates for determining the PEM properties. The proton conductivities (PCs) at 80°C of SPAEK-6F-20, -30, -40, -50, -60 and -70 were 2.8, 11, 30, 80, 110, and 160 mS cm^{-1} (Table 9.1). A combination of PC, water management, and mechanical and oxidative stabilities suggested that SPAEK -6F-40, -50, and -60 membranes in the SPAEK-6F-XX series were the most suitable for PEMFC applications.

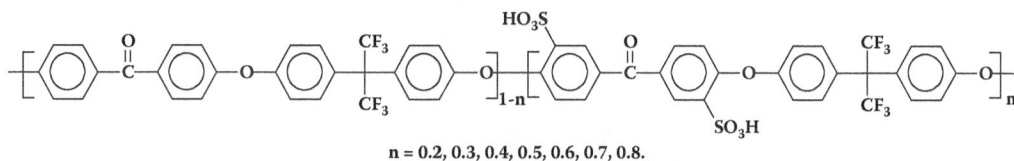

n = 0.2, 0.3, 0.4, 0.5, 0.6, 0.7, 0.8.

SCHEME 9.1 Repeat unit structure of SPAEK-6F-XX polymers [40].

TABLE 9.1
Thermal, Mechanical, and Different PEM-Related Properties of Various Sulfonated Poly(arylene ether)s

Polymer	IEC$_w$	T$_d^{a/b/c}$ (°C)	TS^d (MPa)	YM^e (GPa)	EB^f (%)	WU^g (80°C) (%)	SR^h (80°C) (%)	τ^i (h)	σ^i (30°C) (mS cm^{-1})	σ^i (80°C) (mS cm^{-1})	Ref.
SPAEK-6F-20	0.73	261^a	–	–	–	10	5	–	0.95	2.8	[40]
SPAEK-6F-30	1.07	248^a	–	–	–	14	7	>24	4.5	11	[40]
SPAEK-6F-40	1.38	250^a	–	1.30	39	25	9	6.5	13	30	[40]
SPAEK-6F-50	1.68	275^a	–	0.66	37	68	24	4.0	39	80	[40]
SPAEK-6F-60	1.97	254^a	–	0.76	30	157	52	1.5	46	110	[40]
SPAEK-6F-70	2.23	256^a	–	0.50	15	2400	250	0.8	80	160	[40]
SPAEK-6F-80	2.49	263^a	–	–	–	–	–	–	82	–	[40]
SPEEKK-6F-40	1.17	340^b	52.8	1.61	110	7	5	>12	31	65	[42]
SPEEKK-6F-50	1.43	338^b	60.7	1.2	91	10	8	>12	34	71	[42]
SPEEKK-6F-60	1.68	349^b	47	1.41	79	20	16	2	44	83	[42]
SPEEKK-6F-70	1.92	326^b	36.5	0.9	84	81	60	0.5	50	85	[42]
SPEEKK-6F-80	2.14	320^b	–	–	–	–	–	–	53	96	[42]
SPEEKK-6FP-40	1.2	341^b	52.7	1.21	141	16	6	>12	13	54	[42]
SPEEKK-6FP-50	1.46	358^b	57.7	1.41	117	29	12	>12	32	100	[42]
SPEEKK-6FP-60	1.71	358^b	59.2	0.61	116	54	33	1.5	46	160	[42]
SPEEKK-6FP-70	1.95	324^b	–	–	–	–	–	0.5	63	–	[42]
SPEEKK-6FP-80	2.18	311^b	–	–	–	–	–	–	110	–	[42]
S2-PAES-30	1.23	354^b	37	0.84	21.6	18	–	–	11	34	[53]
S2-PAES-40	1.47	364^b	39	0.93	31	28	–	–	15	51	[53]
S2-PAES-50	1.8	345^b	38	0.75	12.1	37	–	–	29	88	[53]
S2-PAES-60	2.03	338^b	35	0.71	36	60	–	–	85	147	[53]
S4-PAES-30	1.62	366^b	33	0.79	21	27	–	–	18	63	[53]
S4-PAES-40	1.88	342^b	30	0.84	15.2	38	–	–	31	96	[53]
S4-PAES-50	2	337^b	32	0.65	13.1	53	–	–	62	125	[53]
SPAEK-35	1.81	364^b	44		38	23.5	10.9	–	–	68.8	[54]
SPAEK-30	1.63	369^b	54	–	74	19.8	6.8	–	–	35.7	[54]
SPAEK-25	1.36	371^b	58	–	90	16.5	6.3	–	–	32.1	[54]

SPAEK-20	1.28	378[b]	61	–	110	12.4	4.8	–	–	22.7	[54]
SPAEK-15	0.93	390[b]	53	–	110	8.7	3.8	–	–	18.9	[54]
6F-PAEK-SP22	1.79	362[b]	29.5	–	119	46.2	32.8	–	70	136	[56]
6F-PAEK-SP18	1.54	367[b]	33.4	–	110	37.1	19.3	–	56	108	[56]
6F-PAEK-SP14	1.28	390[b]	39.2	–	160	25	14.2	–	32	87	[56]
6F-PAEK-SP10	0.97	421[b]	43.4	–	135	23.1	10.4	–	12	31	[56]
Nafion* 117	0.91	–	33.2	–	205	29.4	23.1	–	53	108	[56]
msPEPOF-100	1.39	392[b]	40.4	1.1	6.2	38	–	6.5	–	74	[67]
dsPEPOF-100	1.41	393[b]	43.3	1.11	10	37	–	26	–	69	[67]
dsPEPOF-110	1.53	389[b]	42.4	1.13	8.2	43	–	8	–	83	[67]
dsPEPOF-120	1.66	384[b]	44.4	0.94	7.4	56	–	5	–	110	[67]
tsPEPOF-100	1.39	398[b]	41.3	1.08	6	28	–	64	–	44	[67]
tsPEPOF-110	1.5	391[b]	43.4	1.13	7	36	–	46	–	75	[67]
tsPEPOF-120	1.62	387[b]	46.5	1.09	6.7	43	–	38	–	92	[67]
tsPEPOF-130	1.78	385[b]	37.8	0.77	9	72	–	30	–	120	[67]
PAQSH-30	0.73	382[b]	67	1.53	16	14	5	25.3	5	11	[11]
PAQSH-40	0.99	355[b]	60	1.33	18	21	7	17.8	6	14	[11]
PAQSH-50	1.25	343[b]	48	1.26	10	32	11	3.6	12	25	[11]
PAQSH-60	1.51	324[b]	44	1.15	6	44	13	2.2	28	50	[11]
HPPQSH-10	0.27	466[c]	61	2.52	6	14	21	>24	–	15[k]	[13]
HPPQSH-20	0.54	373[c]	58	2.03	4	17	24	>24	–	24[k]	[13]
HPPQSH-30	0.81	317[c]	38	1.42	4	21	28	>24	–	27[k]	[13]
HPPQSH-PS	1.14	310[c]	70	1.97	20	23	33	>24	–	32[k]	[13]
HPPQSH-10 PS	1.38	285[c]	69	2.2	9	29	36	22	–	45[k]	[13]
HPPQSH-20 PS	1.62	263[c]	65	2.55	4	37	45	10.2	–	90[k]	[13]
HPPQSH-30 PS	1.86	245[c]	37	1.35	4	42	48	5.5	–	142[k]	[13]
TPQSH-COOH-50	1.67	405[b]	52	1.68	8	32	20	21.2	–	55	[12]
TPQSH-COOH-60	1.96	394[b]	48	1.57	9	154	48	5.5	–	109	[12]
TPQSH-COOH-70	0.91	388[b]	24	1.01	15	643	98	1.2	–	116	[12]
IBQSH-20	0.45	468[b]	80	1.89	14	4	1.6	30.1	1	3	[16]
IBQSH-30	0.68	355[b]	74	1.9	8	8	2.4	25.5	2	6	[16]
IBQSH-40	0.92	344[b]	41	1.8	3	12	4.9	5.3	3	7	[16]

(Continued)

TABLE 9.1 (Continued)

Thermal, Mechanical, and Different PEM-Related Properties of Various Sulfonated Poly(arylene ether)s

Polymer	IEC_W	$T_d^{a/b/c}$ (°C)	TS^d (MPa)	YM^e (GPa)	EB^f (%)	WU^g (80°C) (%)	SR^h (80°C) (%)	τ^i (h)	σ^j (30°C) (mS cm^{-1})	σ^j (80°C) (mS cm^{-1})	Ref.
IBQSH-50	1.16	325^b	57	1.72	6	23	7.5	4.7	7	17	[16]
IBQSH-60	1.39	317^b	34	1.13	4	30	9.7	4.4	21	49	[16]
6FBPAQSH-20	0.16	407^b	52	1.52	7	2	1.3^l	32.4	4	8	[15]
6FBPAQSH-30	0.29	346^b	36	1.39	4	5	2.0^l	27.2	5	11	[15]
6FBPAQSH-40	0.35	312^b	45	1.46	6	8	2.8^l	5.3	19	38	[15]
6FBPAQSH-50	0.44	301^b	35	1.32	3	13	3.6^l	4.8	38	69	[15]
6FBPAQSH-60	0.53	299^b	40	1.17	6	23	5.1^l	4.2	54	98	[15]
BPAQSH-20	0.19	365^b	57	1.55	18	4	2.6	29.6	9	25	[14]
BPAQSH-30	0.29	355^b	49	1.25	21	7	3.3	25.8	13	31	[14]
BPAQSH-40	0.4	347^b	37	1.05	19	12	5.3	4.5	16	40	[14]
BPAQSH-50	0.49	310^b	35	1.02	18	22	6.3	4.3	31	71	[14]
BPAQSH-60	0.56	283^b	27	0.91	4	40	10.2	3.8	40	80	[14]

Notes

a Degradation temperature in TGA
b 5% degradation from TGA
c 10% degradation from TGA
d Tensile strength
e Young's modulus
f Elongation at break
g Water uptake at 80°C
h Swelling ratio (through plane) at 80°C
i Oxidative stability
j Proton conductivity
k Proton conductivity at 90°C
l Swelling at 30°C

Liu et al. prepared two series of trifluoromethyl groups containing sulfonated poly(aryl ether ether ketone ketone)s (SPEEKK- XX, where XX = 40, 50, 60, 70, 80, and 100) from different fluorinated bisphenols [42]. The polymer structures with 6 F or 6 FP unit are presented in Figure 9.2. This study was conducted to compare the effects of different molecular structures and the length effect of hydrophobic parts of the polymers on properties related to PC. The polymers displayed T_g values above 196°C and the first-step weight loss was around 224°C–332°C in TGA, related with the degradation of the sulfonic acid groups of SPEEKK-XX polymers. The salt form of the polymers were thermally stable up to 400°C under N_2 atmosphere. The SPEEKK-6FP polymer series (acid) exhibited higher T_g values (212°C–271°C) because of the presence of bulky 6FP groups than the acid form of SPEEKK-6F polymer series (T_g: 196°C–236°C, degradation temperature: 255°C–358°C), which had an average thermal stability. The Young's modulus (YM) and elongation at break (EB) were 0.61–1.61 GPa and 79%–141%, respectively. The SPEEKK-6FP membranes showed an improvement in WUs and SRs than those of the membranes from SPEEKK-6F. The presence of the bulky pendant 6 FP constituents in SPEEKK-6FP contributed towards attaining large free volumes during close packing of the polymer chains that confined the water molecules. The SRs of 6 FP polymers also raised compare to the more ordered 6F unit of polymer structures. For example, the WU and SR (23°C) of SPEEKK- 6FP-60 and SPEEKK-6F-60 were 15% and 9%; and 6% and 5%, respectively. High IEC value SPEEKK-6FP-70, SPEEKK-6F-70, SPEEKK-6FP-60 and SPEEKK-6F-60 membranes were dissolved into the Fenton reagent after about 0.5, 0.5, 1.5, and 2.0 h, respectively. On the other side, both series of polymers with DS values 50 and 40 could not be dissolved even after 12 h and they sustained dimensional stability after 6 h in water. The PC of the acidified membranes of all polymers increased with increasing temperature. The polymers demonstrated PC at room temperature above 1×10^{-2} S cm^{-1}, which is the least required value as a membrane to act as PEM in fuel cell applications. The PEM-related parameters were also tabulated in Table 9.1. (Scheme 9.2)

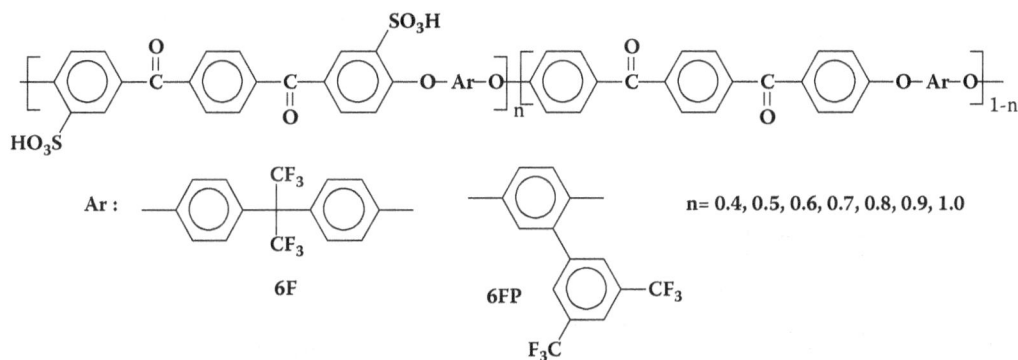

SCHEME 9.2 Repeat unit structure of the SPEEKK-6FP-XX polymers [42].

Liu et al. developed another series of aromatic poly(arylene ether ketone)s, where the post-sulfonation of the polymeric chains were accomplished using concentrated sulfuric acid as sulfonating agent [10]. The sulfonated polymers displayed lower methanol permeability values and sufficient thermal, mechanical, and oxidative stability required for PEM applications. However, the PC values were lower than Nafion®.

In the search for the development of PEM materials through the post-sulfonation route, Jeong et al. used chlorosulfonic acid to prepare several sulfonated poly(arylene ether ketone)s (SPAEK-BPs) with dissimilar DS values (Scheme 9.3) [43]. They demonstrated that only one sulfonic acid group could be placed in each repeating unit by controlling the reaction conditions, such as reaction time and temperature without degrading the polymer main chain. The T_g and T_d values of SPAEK-BPs polymers ranged in 167°C–213°C and 298°C–343°C, respectively. The sulfonated polymers showed high mechanical and dimensional stability. The SPAEK-BP membranes displayed PC values from 17 mS cm^{-1} (SPAEK-BP-20) to 83 mS cm^{-1} (SPAEK-BP-90) at 25°C, and from 21 mS cm^{-1} (SPAEK-BP-20) to 98 mS cm^{-1} (SPAEK-BP-90) at 80°C. The formation of continuous proton transport networks under hydrated conditions lead to the high PC of the SPAEK-BP membranes, which were also similar to Nafion® 117 [44]. The channel development was qualified to the pendant phenyl rings, which increase the water content within the membranes and thus upsurge the PC.

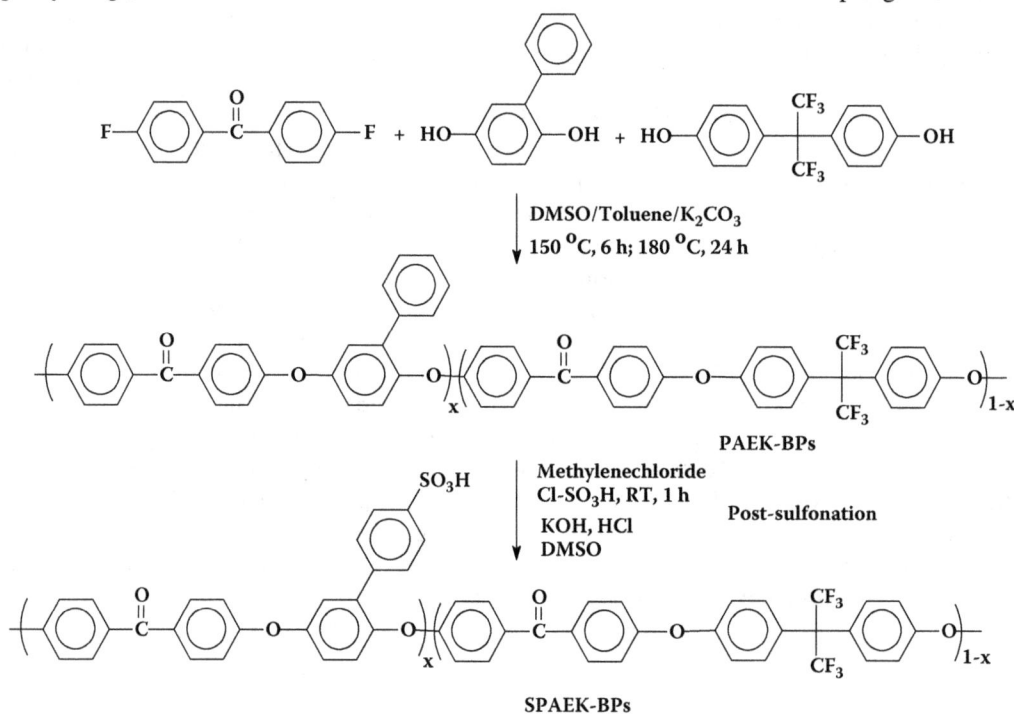

SCHEME 9.3   Scheme of synthesis of SPAEK-BPs copolymers via post-sulfonation [43].

Later, several studies had been done on the development of multiblock copolymers with highly concentrated sulfonic acid groups. Poly(arylene ether)s with concentrated sulfonic acid part or end cap reported by Hay et al. [45–47]. These ionomers had IEC$_w$ of 1.16 meq g^{-1} and comparable PC as Nafion® 117. Comb-shaped copoly(arylene ether)s reported by Guiver et al. [48–50] displayed good phase-separated morphology with well-connected nano ionic channels. The prepared polymers showed an improvement in PC under moderately hydrated states compared to other hydrocarbon-based PEMs. Furthermore, Jannasch et al. developed poly (ether sulfone)s with exceedingly sulfonated units in the polymeric backbones or side chains [51,52]. These sulfonated ionomers were designed to attain increased PC; nevertheless, the methanol permeability (MP) of the membranes was set aside.

Kim et al. reported a comb-shaped poly(arylene ether sulfone) copolymers with side chains included of two to four sulfonic acid groups (Scheme 9.4) [53]. The polymers prepared using sulfonated 4-fluorobenzophenone and 1,1-bis-(4-hydroxyphenyl)-1,4-((4-fluorophenyl) thio) phenyl-2,2,2-trifluoroethane. Sulfonated side chains containing copolymers (S2-PAES-30 to -60 and S4-PAES-30 to -50) demonstrated high thermal stability; T$_{d5\%}$ were ranged from 337°C–366°C (Table 9.1). The polymers exhibited comparable PC and lower MP than Nafion® 1135 (Figure 9.5). As a whole, the balance thermal stability (above 335°C), use of inexpensive monomers, adequate PC and MP, and moderately low WU of the membranes of S2-PAES-50 and S4-PAES-40 polymers made them appealing for PEMFC applications.

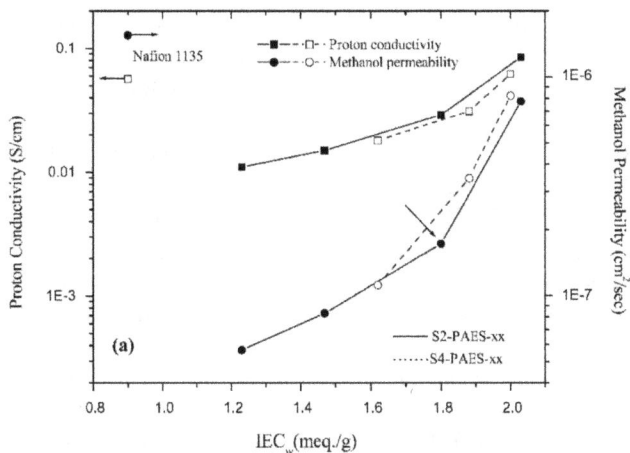

FIGURE 9.5 PC and MP of S2-PAES-XX, S4-PAES-XX and Nafion® 1135 as a function of temperature. Reprinted with permission from Ref. [53]. Copyright (2008), American Chemical Society.

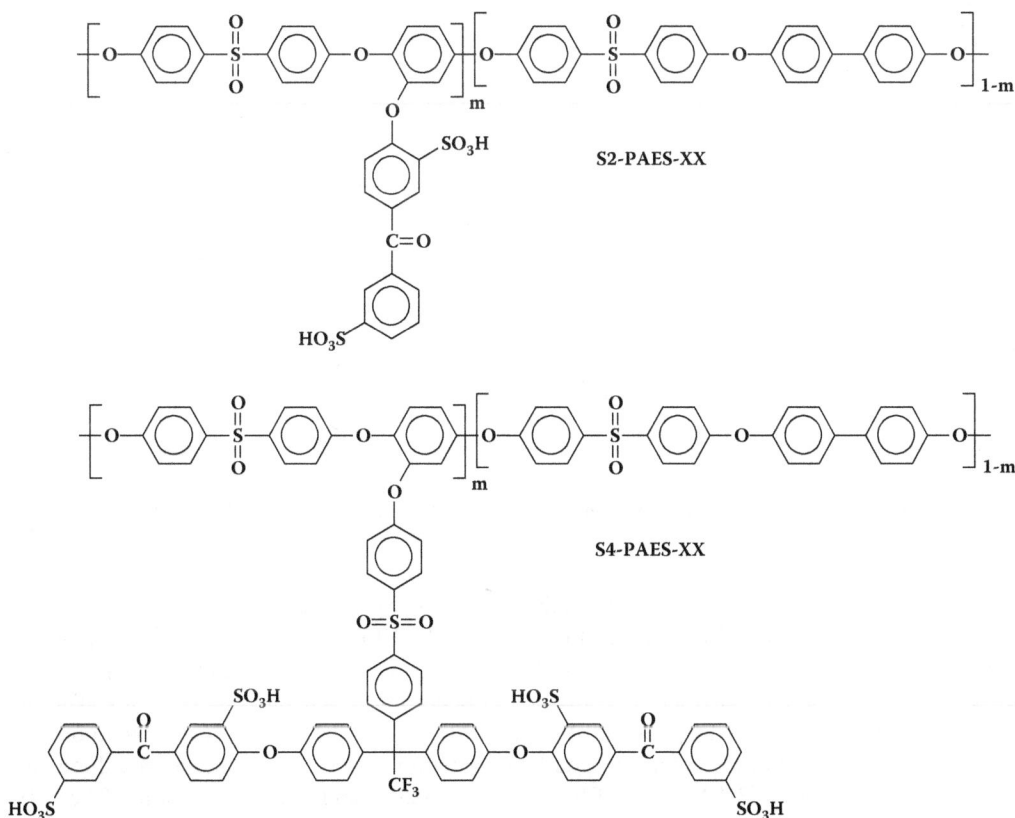

SCHEME 9.4 Comb-shaped sulfonated poly(arylene ether sulfone)s [53].

Pang et al. prepared four sulfonic acid groups containing difluoride monomer via two step synthetic approaches: First, 4,4′-bis(4-fluorobenzoyl)diphenyl was prepared from 4-fluorobenzoyl chloride and diphenyl by Friedel-Crafts reaction followed by sulfonation of 4,4′-bis(4-fluorobenzoyl)diphenyl using fuming sulfuric acid [54]. The monomer was used to prepare tetra-sulfonated poly(arylene ether ketone) copolymers (SPAEK-XX, where XX = 15, 20, 25, 30, and 35) (Scheme 9.5). The densely sulfonated SPAEK-XX copolymers were assessed for PEM applications. The T_g and $T_{d5\%}$ of the polymers were ranged from 193°C to 231°C, and from 364°C–390°C. The TS and EB of SPAEK-15 to -35 are in the range of 44–61 MPa and 38%–110%, respectively. SPAEK-35 (IEC-1.81 meq g^{-1}) had displayed a comparatively low WU (23.5%) and excellent dimensional stability in hot water as the SR (In-plane) at 80°C was 5.1%. WU and SR were both increased with the rise of IEC and temperature (Table 9.1). The highly sulfonated aromatic-based polymer membranes form a continuous and extended ionic network at elevated temperatures, therefore the WU and SR increased abruptly [55]. The polymers did not show any weight loss after 1 h at 80°C in Fenton's reagent, representing their good oxidative stability. As usual, the high IEC SPAEK-35 membrane exhibited the highest proton conductivity compared to other membranes. The PC of SPAEK-35 membrane were 35 and 84 mS cm^{-1} at 40°C and 100°C. It is suggested that combining locally high-density hydrophilic segments with fluorinated hydrophobic units in polymeric chains can well balance dimensional stability and proton transportation.

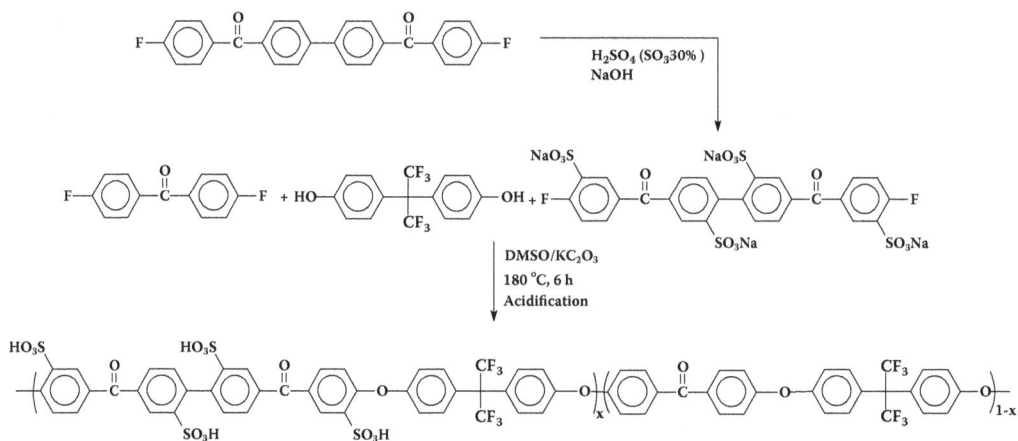

SCHEME 9.5   Ynthetic scheme of tetra-sulfonated SPAEK-XX (XX = 15, 20, 25, 30, and 35) polymers [54].

Wang et al. prepared several fluorinated sulfonated poly(arylene ether ketone)s (6F-PAEK-SPx; x = 18 or 22) by a post sulfonation reaction (Scheme 9.6) [56]. The aim of their work was to enhance the concentration of hydrophilic sulfonic acid groups through pendant groups without affecting the hydrophobic polymer backbone to get better phase separated morphology. The SRs (80°C) of 6F-PAEK-SP18 and 6F-PAEK-SP22 were 11.1 and 19.3%. The hydrophobic fluorinated backbones of these polymers helped in attaining reasonable SRs. The membranes exhibit high mechanical (TS: 29.5 to 43.4 MPa and EB o110.5%–160.3%) and oxidative stability. The IECw values of residual 6F-PAEK-SPx polymers after treated in Fenton's reagent were ranged from 0.69 to 0.85 meq g^{-1}. The high oxidative stability of 6F-PAEK-SP membranes were associated to their localized hydrophilic side chains with moderate sulfonic acid content. The PC of 6F-PAEK-SP22 (1.77 meq g^{-1}) was 148 mS cm^{-1} (100°C) and which is higher than that of Nafion® 117 (Table 9.1). The localized density of sulfonic acid groups through the pendant groups in the polymers facilitate proton transportation in hydrated conditions.

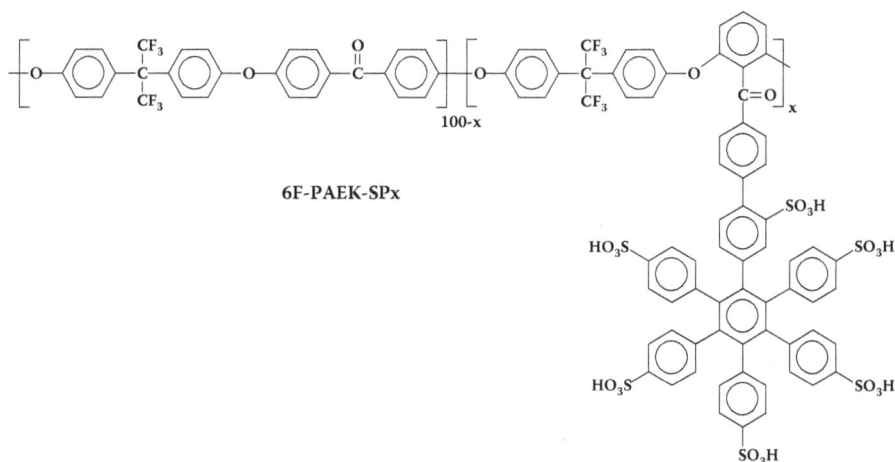

SCHEME 9.6 Sulfonated poly(arylene ether ketone)s with pendant sulfophenyl groups (6F-PAEK-SPx; x = 18 or 22) [56].

Poly(arylene ether phosphine oxide)s are important class of polymers that showed many interesting set of properties suitable for membrane-based applications [57,58]. Thus, several sulfonated poly(arylene ether phosphine oxide) polymers were developed over the years, and their PEM properties have been investigated [59–64]. Membranes from some of these polymers suffered from disproportionate swelling behaviour at high temperatures, which requires to be addressed [60,64,65]. It was noted that the microstructural modification of the membranes could be better governed by putting pendant sulfonic acid groups in the polymer chain. These polymers showed reduced swelling and improved broad membrane properties [66]. Liao et al. methodically explored the significance of the spreading and percentage of pendant sulfonic acid groups in sulfonated poly(arylene ether phosphine oxide)s on general membrane features including the microstructures [67]. They prepared one, two, or three sodium sulfonated groups containing phosphine-oxide-based difluoro monomers and their corresponding poly(arylene ether phosphine oxide)s (Scheme 9.7). The influence of distribution and content of the pendant sulfonic acid side groups were investigated on the PEM properties, which considerably regulated the connectivity of ionic channel networks of the membranes in a hydrated state and significantly controlled the oxidative and dimensional stability. The oxidative stability and other PEM data are presented in Table 9.1. The fabricated membranes exhibited SR similar to that of Nafion® 117 with 1.6 times PC.

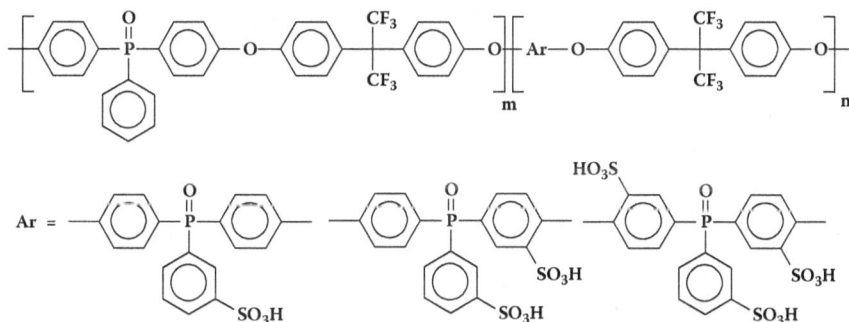

SCHEME 9.7 Sulfonated fluorinated poly(arylene ether phosphine oxide)s with different distributions and proportions of sulfonic acid groups [67].

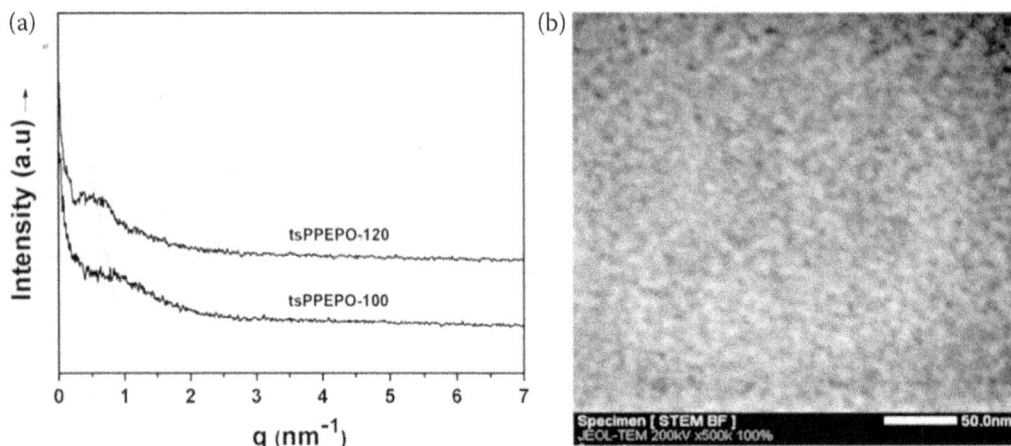

FIGURE 9.6 (a) SAXS profiles of tsPPEPO-120 and tsPPEPO-100 membranes. (b) STEM image of tsPPEPO-120. Reprinted with permission from Ref. [68]. Copyright (2020), American Chemical Society.

Zhang et al. recently prepared a series of trisulfonated poly(phthalazinone ether phosphine oxide)s (tsPPEPO) and investigated their PEM properties (Scheme 9.8) [68]. The tsPPEPO membranes revealed extremely low SR and excellent oxidation stability without forfeiting the PC despite their high DS values. The strong intermolecular interaction through hydrogen bonding in tsBFPPO polymers and distributed hydrophilic-hydrophobic morphology contributed to these properties. The SAXS and STEM (Figure 9.6) of tsPPEPO-100 and tsPPEPO-120 membranes revealed microphase separated morphology and was more prominent with higher DS with the formation of nano-cluster channel networks responsible for better proton conduction. This study indicated that introduction of trisulfonated phosphine oxide and phthalazinone systems together into the polymeric backbone were effective in improving the PEM properties. The tsPPEPO-120 polymer membrane showed a PC of 66 mS cm^{-1} and SR of 11.5% at 80°C, and anticipated as a substitute membrane for its application in PMFC.

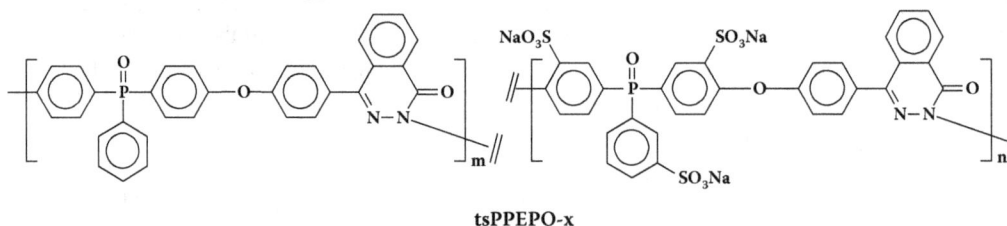

tsPPEPO-x

SCHEME 9.8 Tri-sulfonated Poly(phthalazinone ether phosphine oxide)s (tsPPEPO) [68].

Banerjee et al. prepared numerous fluorinated poly(arylene ether sulfone), BPAQSH-XX co-polymers and investigated their detailed PEM properties (Scheme 9.9) [14]. The copolymers demonstrated excellent oxidative, thermal and mechanical stability subject on their DS values (BPAQSH-20 to -60). The $T_{d5\%}$ and T_g values of BPAQSH-XX copolymer membrane ranged from 283°C–365°C and above 245°C, correspondingly. The copolymers showed promising results towards WU, low SR (Table 9.1) and moderately high oxidative stability. The presence of 3 F and QBF groups were believe to improve the hydrophobicity of the polymers, which provided the membrane stability. The polymers showed nanophase separation with higher DS values that facilitate efficient proton transport within the hydrated membrane.

Sankir et al. had been reported that due to the self-assembled capability of the polymer membranes prepared using 4-4′-hexafluoropropylene diphenol (6 FBPA)-based fluorine. The membranes displayed with more fluorine content in XPS in the polymer air interface [69]. The presence of 6 F groups in the polymers resulted in lowering of WU and efficient water management. Furthermore, several research work directed that the incorporation of bulky pendants and/or bulky rigid units into the aromatic sulfonated polymers increase free volume by inhibiting of polymer chain packing and thereby improve the polymer's solubility and PC [70,71].

Banerjee et al. also investigated the role of fluorine on PEM properties of sulfonated polymers by replacing BPA with 6 FBPA (Scheme 9.9) [15]. The polymers showed high thermal and mechanical stability and exceptional film-forming ability. The values were several times higher than Nafion® 117 and were comparable to their previous BPAQSH-XX series polymers. Significantly, the membranes from 6FBPAQSH copolymers showed very promising results towards their WU, SR, and oxidative stability values and credited to the attendance of additional fluorine in these polymers in the form of 6F groups. The relatively low WU and high PC of the membranes from these polymers, even in their low IEC_W (Table 9.1), representing them with better PEM materials compared to the structurally analogous BPAQSH-XX polymers without 6F groups [14]. The nature of the impedance plots of the 6FBPAQSH-30 copolymer membranes is shown in Figure 9.7 as a representative one. The plot represents that the membranes realize less resistance towards proton conduction as the temperature increased.

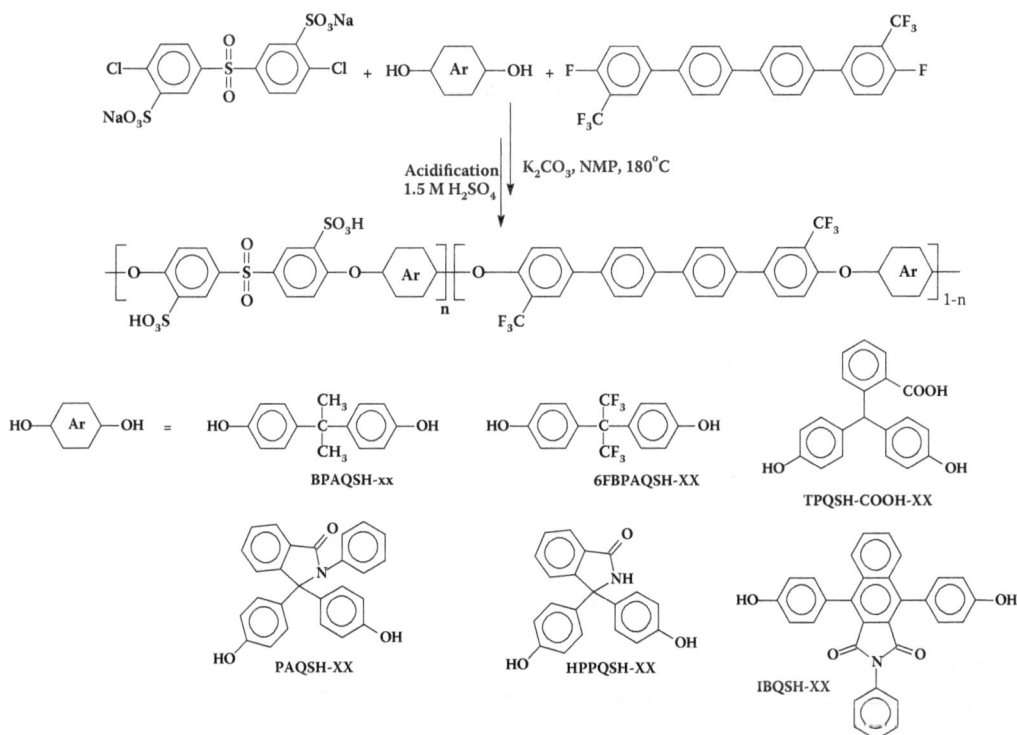

SCHEME 9.9 Different fluorinated sulfonated poly(arylene ether sulfone) copolymers for PEM applications prepared by Banerjee et al. [11–16].

Partially fluorinated poly(imidioarylene ether)s with the sulfonic acid groups was prepared by Banerjee et al. and their mechanical, thermal and chemical stability qualifies them potential candidates in PEM applications [16,72]. The partly fluorinated sulfonated poly(imidioaryl ether sulfone)s (IBQSH-XX) (Scheme 9.9) [16] demonstrated low WU (4%–30% at 80°C, Figure 9.8) and SR

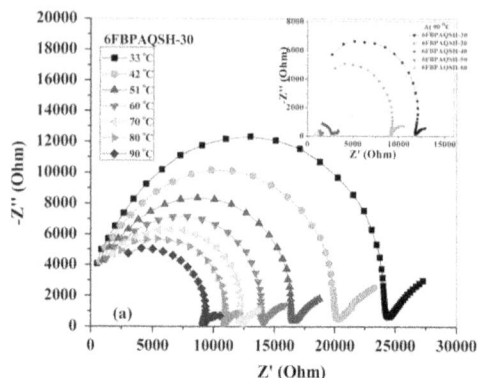

FIGURE 9.7 Temperature dependence impedance plots of 6FBPAQSH-30 membrane. Reprinted with permission from Ref. [15]. Copyright (2013), American Chemical Society.

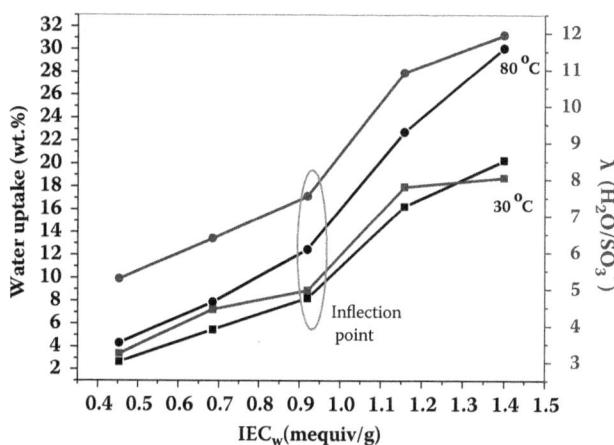

FIGURE 9.8 WU and number of water molecules linked with sulfonic acid group (λ) of IBQSH-XX polymers with respect to their IEC_W. Reprinted from Ref. [16], Copyright (2014), with permission from Elsevier.

(1.6%–9.7% at 80°C), brilliant thermal ($T_{d5\%}$: 468°C–317°C) and mechanical stability (TS: 34–80 MPa), and fairly high oxidation stability (4.4–30.1 h) with a reasonable PC (3–49 mS cm^{-1}) (Table 9.1). However, the PC value was lower related to Nafion® 117, the general properties of IBQSH-XX copolymers made them a fascinating contestant for PEM applications.

Banerjee et al. also prepared phthalimidine-based aromatic fluorinated sulfonated poly(arylene ether sulfone) (PAQSH-XX) copolymers (Scheme 9.9). The polymers displayed high thermal and mechanical stability ($T_{d5\%}$: 324°C–382°C and high TS (43.7–68.8 MPa) and were considerably developed comared to the earlier reported similar BPA based and Nafion® 117 membrane [11]. The attendance of rigid quadriphenyl moiety and bulky N-phenyl-3,3′-bis(4-hydroxyphenyl) phthalimidine and hydrophobic 3 F groups qualified much better dimensional steadiness and water management associated with Nafion® 117 (Table 9.1). The copolymers presented reduced WU than BPA-based copolymers. The strong ionic interactions between sulfonic acid groups and rigid and bulky backbone increase the stiffness of the polymeric network structure restricted the free volume for water absorption and reduced WU of PAQSH-XX copolymer membranes [73]. Though the PAQSH-XX polymers displayed relatively low PC (11–50 mS cm^{-1} at 80°C), this work opens up the scope of further research towards modifying the lactam ring as a possible option for the progress of new PEM materials.

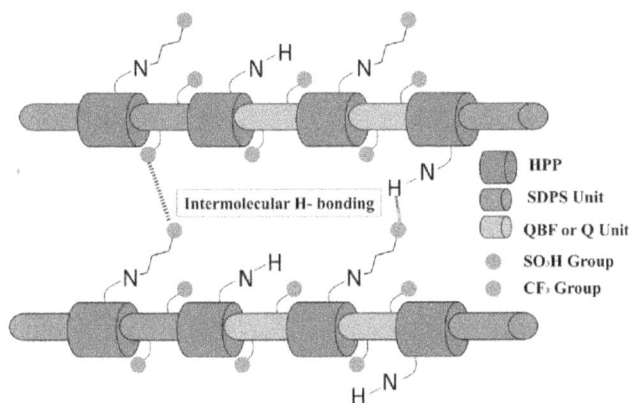

FIGURE 9.9 Schematic depiction of protonic cross linking in acid-form grafted copolymers (HPPQSH-XX PS). Reprinted from ref. [13] with permission from The Royal Society of Chemistry.

Another sequence of sulfonated poly(arylene ether sulfone) (HPPQSH-XX, Scheme 9.9) was developed by Banerjee et al., which were further modified with pendant sulfopropyl groups (HPPQSH-XX PS) via post-sulfonation technique [13]. The morphological study by TEM reveals tiny ionic clusters (5–10 nm) for the HPPQSH-XX membranes, whereas post-sulfonated copolymers displayed large ionic domains (60–100 nm). The cartoon picture of the HPPQSH-XX PS (Figure 9.9) indicated strong intermolecular hydrogen bonding between the polymer chains and both HPPQSH-XX PS and HPPQSH-XX membranes showed high mechanical, oxidative, and dimensional stability with good WU and SR (Table 9.1). The membranes from HPPQSH-30 PS copolymer demonstrated comparable PC [125 mS cm^{-1} (80°C) and 142 mS cm^{-1} (90°C)] like Nafion® 117. Additionally, the simple, easy preparation procedure and exceptional PEM properties of the polymers made them candidate materials for PEMFC applications.

Zhang et al. prepared a sequence of sulfonated poly(arylene ether ketone)s comprising carboxylic acid groups in the side chain. They described that the carboxylic acid group in the agile side-chain increases the PC [74]. Banerjee et al. also prepared carboxylic acid-functionalized TPQSH-COOH-XX copolymers (Scheme 9.9) and investigated their PEM properties (Table 9.1) [12]. The polymers exhibited high thermal and mechanical stability and displayed exceptional high oxidative stability compared to analogous sulfonated copolymers. The role of –COOH groups in stabilizing the trityl radical that is generated by capturing hydroperoxide radicals by these polymers is represented in Figure 9.10. The polymers displayed reasonably high PC, oxidative stability, and SR that qualifies their application as polyelectrolyte membranes in a fuel cell.

9.4.1.1.2 Block Poly(arylene ether)s

Multiblock copolymers are very interesting as they provide unique phase-separated morphologies that significantly help in improving proton conductivity. Several multiblock copolymers were prepared by reacting disulfonated poly(arylene ether sulfone) hydrophilic blocks (BPSH100) and dissimilar hydrophobic blocks [75,76]. Two different multiblock copolymers (BPSHx-BPSy, Scheme 9.10) were prepared by combining phenoxide terminated BPSH100 and hexafluorobenzene or decafluorobiphenyl end-capped unsulfonated poly(arylene ether sulfone) (BPS0) [77]. The high reactivity of the oligomeric blocks allowed reactions at mild conditions (<105°C). The T$_{d5\%}$ temperatures of the polymers varied from 357°C to 367°C that depends on the length of the different oligomeric blocks. The WU and SR were increased with the block length and suggested the developing of hydrophilic domains in an orderly manner within the multiblock copolymers. A similar effect was observed for PC of the polymers with their block lengths. The values were comparable or higher to their corresponding

Highly stable radical mainly due to the conjugation through the phenyl ring.

Resonance stabilization of trityl type radical

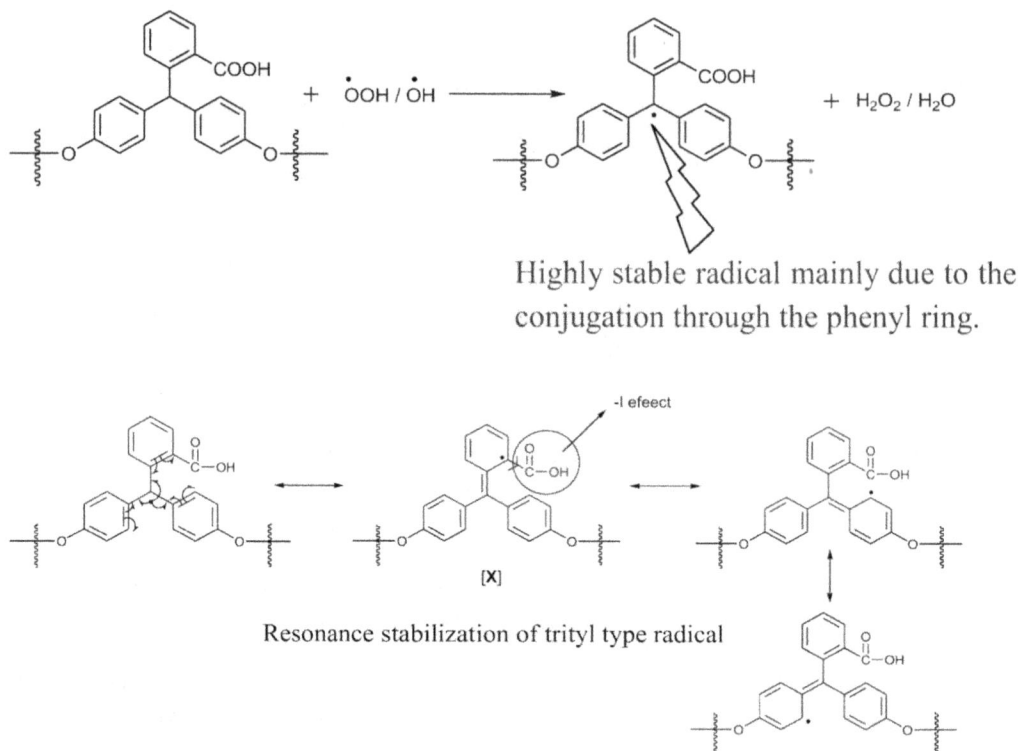

FIGURE 9.10 Possible stabilization pathways of trityl radical by resonance. Reprinted from ref. [12], Copyright (2016), with permission from Elsevier.

random copolymers (BPSH) with the similar IEC, which supported the creation of well-connected ionic networks of hydrophilic domains. The BPSH15-BPS15 (DFBP), BPSH10-BPS5(DFBP), BPSH10-BPS10 (HFB), BPSH20-BPS15 (HFB) block copolymers displayed PC of 120, 160, 100, and 140 mS cm^{-1}, correspondingly. Thus the preparation of sulfonated block copolymers can be an alternative method in manipulating the polymer properties for PEM applications.

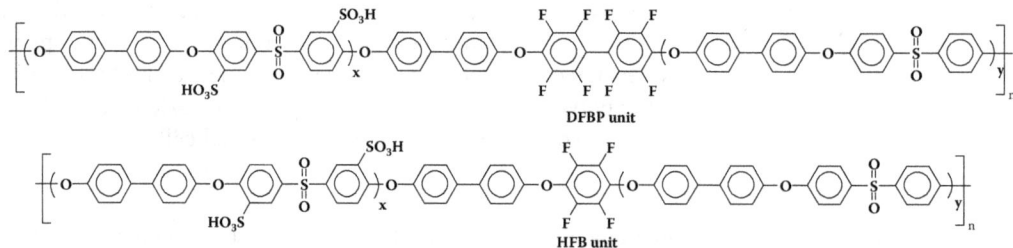

SCHEME 9.10 BPSH–BPS multiblock copolymers for PEM applications [77].

Watanabe et al. reported fluorenyl moiety comprising fluorinated multiblock poly(arylene ether sulfone)s by reacting hydrophobic blocks (naphthalene and biphenyl based) and highly sulfonated hydrophilic block obtained through the oligomeric sulfonation (Figure 9.11 and Scheme 9.11) [78]. The membranes displayed well-phase-separated morphology owing to the sulfonic acid groups' local concentration in the hydrophilic part of the multiblock polymer chains, and longer blocks showed better phase separation. The hydrophobic naphthalene-based multiblock copolymers (IEC: 2.01 meq g^{-1}) unveiled comparable PC to Nafion® NRE-212 membrane (0.91 meq g^{-1}) at >40%

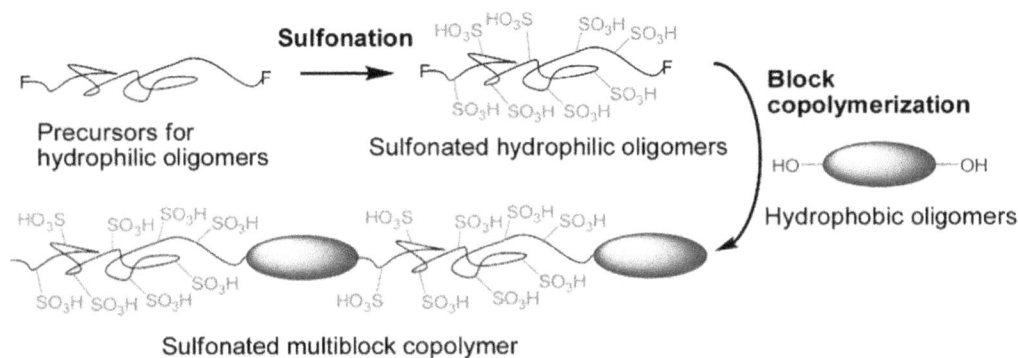

FIGURE 9.11 Graphical representation of sulfonated multiblock poly(arylene ether sulfone)s. Reprinted with permission from ref. [78] Copyright (2011), American Chemical Society.

RH. However, the membranes exhibited moderately low peroxide resistance and high WU and SR. This method allowed the designing of multiblock sulfonated polymeric ionomers to adjustable PEM properties.

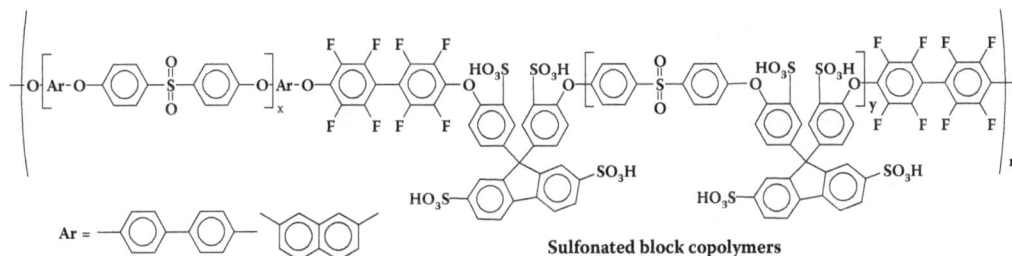

SCHEME 9.11 Multiblock poly(arylene ether sulfone)s with fluorenyl moiety [78].

The PC of PEMs, particularly in low RH is a significant issue, and that needs manipulation in polymer architectures. It is observed that phase-separation among the hydrophobic and hydrophilic sections in polymers, particularly in block copolymers where sulfonic acid groups are concentrated in the hydrophilic block, enhances the PC [77,79–83]. Kreuer et al. developed multiblock copolymers comprising high sulfonated poly(phenylene sulfone) blocks [81]. The polymers formed a well-connected ionic channel for proton transportation, which corresponds to the hydrophilic segments' block length. The membranes exhibited high hydrolytic stability and PC. Especially at higher temperatures, the PC was somewhat higher compared to Nafion® 117 that suggests their use as alternate materials for PEM applications. Jannasch et al. tailored durable PEMs from multiblock poly(arylene ether sulfone)s with varying degrees of sulfonation and block lengths by combining non-sulfonated hydrophobic and extremely sulfonated hydrophilic precursor blocks [82]. Fabricated membranes displayed distinctly phase-separated structures and high PC under low RH. Jung et al. prepared multiblock poly(arylene sulfone) copolymers (sPAS-A/Bs, Scheme 9.12) condensing hydrophilic sulfonated poly(arylene thioether sulfone) and hydrophobic non-sulfonated poly(arylene ether sulfone) oligomers, followed by the oxidation [84]. The membranes from these polymers displayed well-connected nanophase separated ionic domains with improved thermal and oxidative stability, reduced WU and higher PC than the analogous sPTS-A/B membranes. This sPAS-A/Bs membranes also showed enhanced oxidative stability qualified to the occurrence of electron-withdrawing sulfone linkages in the hydrophilic sections. The sPAS-12/4 displayed a about 10% higher PC (191 mS cm^{-1}) compared to sPTS-12/4 (172 mS cm^{-1}) at 80°C in spite of both having almost comparable IEC values and showed similar PC to that of Nafion® 212 at 50% RH.

The single fuel cell of sPAS-A/B polymer membranes displayed a current density of 826 mA cm^{-2} (0.6 V, 70°C, 50% RH). Considering the overall membrane properties of SPAS-A/B copolymers these polymers could be an attractive PEM material for fuel cell applications.

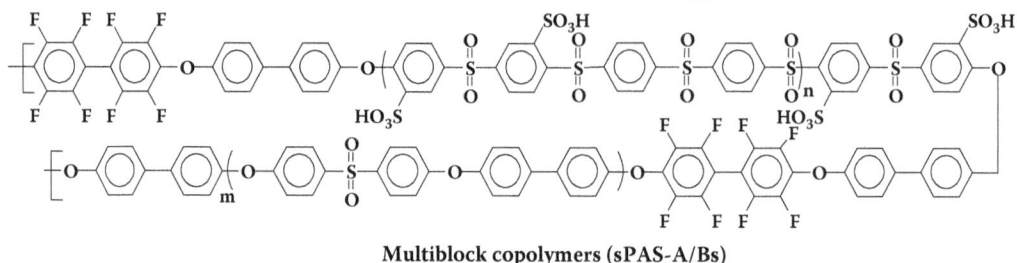

Multiblock copolymers (sPAS-A/Bs)

SCHEME 9.12   Aromatic-based multiblock poly(arylene sulfone) polymers (sPAS-A/Bs) [84].

Guiver et al. prepared a sequence of segmented poly(arylene ether sulfone)s having densely pendant phenyl sulfonic acid groups and fabricated membranes thereof (Scheme 9.13) [49]. The polymers exhibited TS, YM, and EB values between 39.4–68.2 MPa, 0.91–1.31 GPa, and 27.4%–38.2%, respectively, at 25°C and 50% RH. The AFM phase images displayed distinct phase separation among the hydrophobic and hydrophilic sections with inter-connected hydrophilic net-working of ionic clusters originated from the densely populated sulfophenylated groups in the segmented structures [Figure 9.12(a–d)]. The interconnecting network of ionic clusters becomes more prominent with longer hydrophilic blocks (X20Y40 and X20Y30) (X and Y indicate the hydrophilic and hydrophobic blocks). The X20Y20 membrane displayed a PC of 36 mS cm^{-1} (IEC: 1.82 meq g^{-1}) at 80°C under 50% RH, that is similar to Nafion® 112 (40 mS cm^{-1}). Thus, these types of densely sulfonated block polymer membranes were satisfactory for the PEM applications.

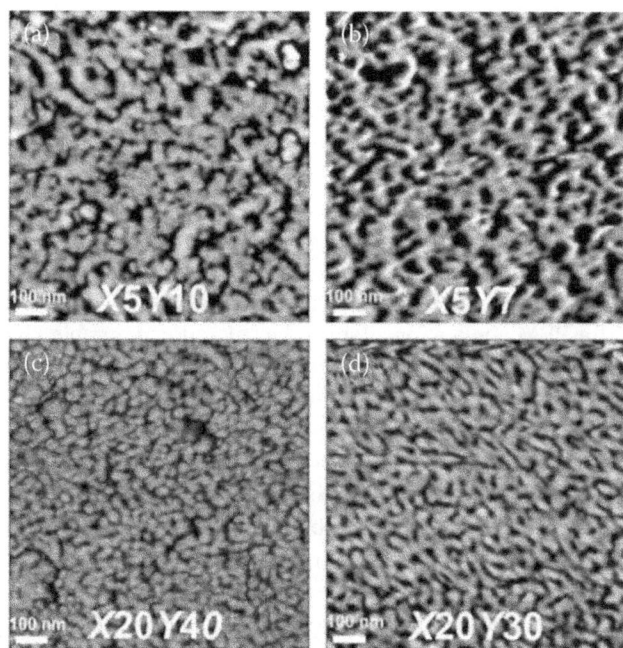

FIGURE 9.12   (a) AFM images of X5Y10, (b) X5Y7, (c) X20Y40, and (d) X20Y30 polymers. Reprinted with permission from ref. [49]. Copyright (2011), American Chemical Society.

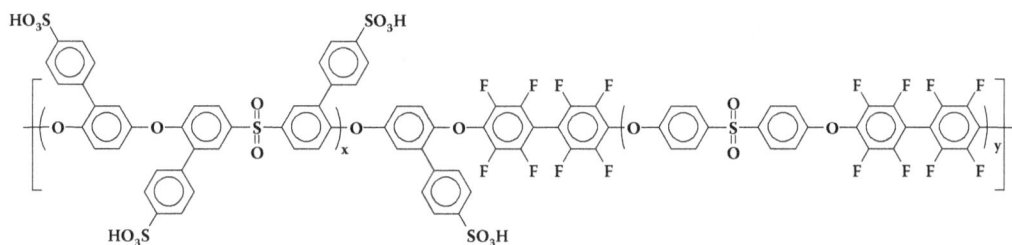

SCHEME 9.13 Highly sulfonated segmented multiblock poly(arylene ether sulfone) [49].

Assumma et al. reported perfluoro-sulfonated multiblock poly(arylene ether sulfone) copolymers (Scheme 9.14). They examined the effects of casting solvent, block lengths, and thermal treatment on different membrane properties including morphology, WU, and PC [85]. It is reported that the solvent selectivity mildly impacted the average size of the ionic domains and the growth upon swelling. At the same time, the supramolecular order of the blocks was strongly affected by the solvent selectivity. The membranes obtained from DMSO exhibited a superior interconnecting network of ionic clusters, higher PC (Figure 9.13), and WU than the membranes cast from DMAc. An improvement in PC (tenfold) was attained after thermal annealing of membranes at 150°C, and the longer block lengths ionomers revealed similar PC to Nafion® 117 at low RH (30%) and 80°C.

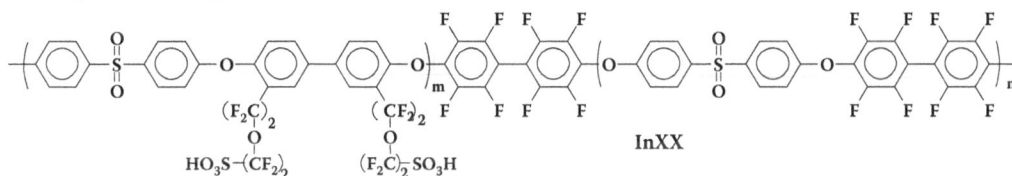

SCHEME 9.14 Perfluoro-sulfonated multiblock poly(arylene ether sulfone) copolymers [85].

9.4.1.2 Non-Fluorinated Poly(arylene ether)s

Several non-fluorinated polymers were developed over the years as a substitute to the perfluorinated ionomers as PEM in fuel cell applications [41,86,87]. Guiver and co-workers reported a sulfonated poly(ether ether ketone) (Scheme 9.15) and studied PEM properties, and discussed the factors the determine the PC [88]. They reported proton conduction in the hydrated state occurred by the transport of H_3O^+ or $H_5O_2^+$ and, the membrane caste from DMF displayed lower PC than that of DMAc.

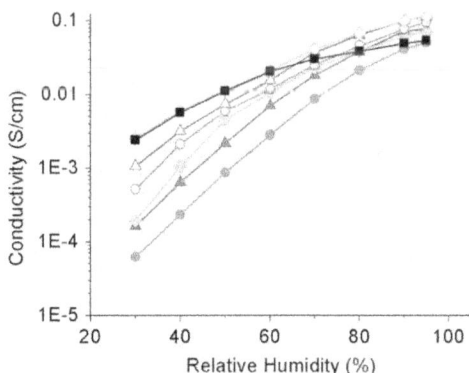

FIGURE 9.13 Proton conductivity at 80°C: Nafion® 117 (■), IN1010$_{DMSO60}$ (▲), IN0505$_{DMSO60}$ (●), IN1515$_{DMSO150}$ (◇), IN0505$_{DMSO150}$ (○), IN1010$_{DMSO150}$ (△), IN1515$_{DMSO60}$ (◆). Reprinted with permission from Ref. [85]. Copyright (2015), American Chemical Society.

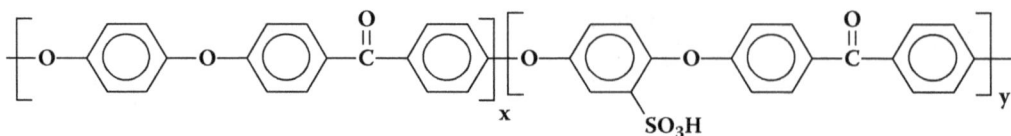

SCHEME 9.15   Non-fluorinated sulfonated poly(ether ether ketone) for PEM applications [88].

A series of sulfonated poly(arylene ether) copolymers with pendant sulfonic acid groups [SC-SPAE, i.e., side-chain-type sulfonated poly(arylene ether)] were synthesized via copolymerization of 4,4'-dichlorodiphenyl sulfone, sodium 4-(4-(2,6-difluorobenzoyl)phenoxy) benzenesufonate, and 4,4'-dihydroxyldiphenylether (Scheme 9.16) [89]. These copolymers displayed brilliant dimensional stability, and their WU and SR values (elevated temperatures) were lower than main polymeric chain-based sulfonated poly(arylene ether) with similar IEC values. For membranes with low sulfonic acid content (IEC < 1.36), the SR and WU increased slightly with increasing temperature (Figure 9.14). Copolymers with high sulfonic acid content (IEC >1.36) displayed a sharp rise in WU and SR at elevated temperature (>80°C) attributed to the development of large and uninterrupted ionic networks [40,90]. PCs of the copolymers were progressively increased with the increasing DS and temperatures and reached 140 mS cm^{-1}.

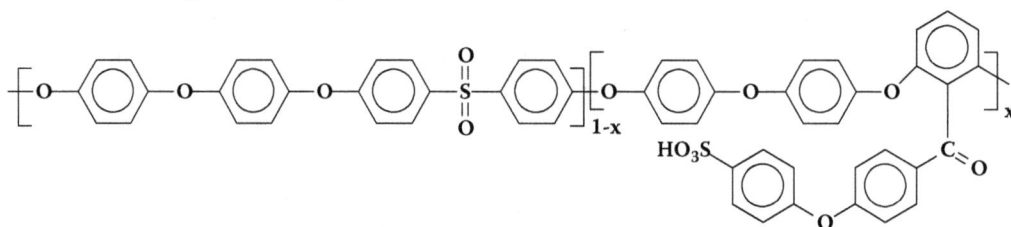

SCHEME 9.16   Sulfonated poly(arylene ether) copolymers with pendant sulfonic acid groups [89].

Polymers containing phosphorus possessed a brilliant property for PEM applications. The polar phosphine oxide group in sulfonated poly(thioether phosphine oxide)s and poly(arylene ether phosphine oxide)s help in attaining better water management and outstanding adhesion performance with different polymers and inorganic substances [91]. Fu et al. prepared several phosphine oxide moieties containing sulfonated poly(arylene ether sulfone)s (sPESPO-XX, Scheme 9.17) by

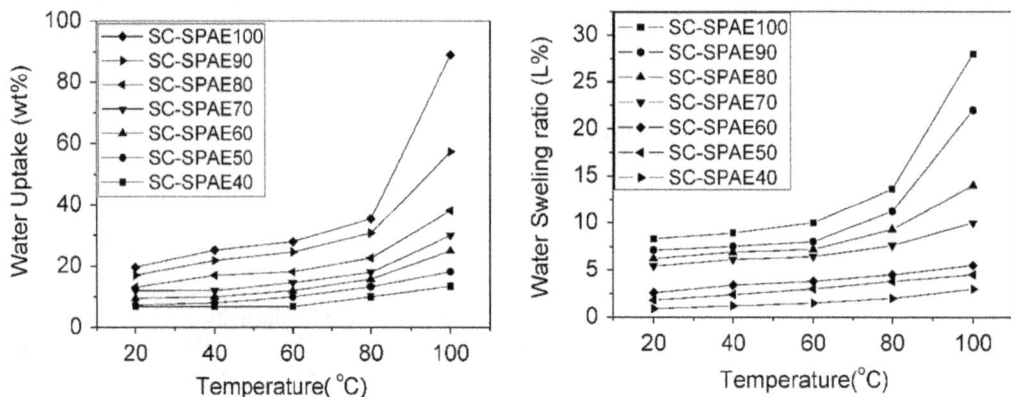

FIGURE 9.14   WU and SR of SC-SPAE membranes as a function of temperature. Reprinted with permission from ref. [89]. Copyright (2007), American Chemical Society.

condensing bis(4-hydroxyphenyl)phenylphosphine oxide, 4-fluorophenyl sulfone, and 3,3′-disulfonate-4,4′-difluorodiphenyl sulfone [62]. The polymers displayed high thermal ($T_{d5\%}$ from 392°C to 407°C), adequate mechanical and oxidative stability with 30.8% WU and 15.8% SR and a PC (80°C) of 87 mS cm^{-1}. The phosphine oxide containing sPESPO-45 polymer membrane showed greater WU than sPES-40 (Scheme 9.17) though they have similar IEC [62]. Additionally, the oxidative stability of the sPESPO-XX membranes was remarkable (54–68 h) and much higher than that of sPES-40 (14 h). This was qualified to the chelation of the phosphine oxide groups with the ferrous ions of Fenton's reagent that unreactive to the peroxide radicals [92,93]. Additionally, the phosphine oxide group can decompose the peroxide radicals and improve the polymers' oxidative stability [94]. Thus, sPESPO-XX polymers might be attractive candidates for PEM applications.

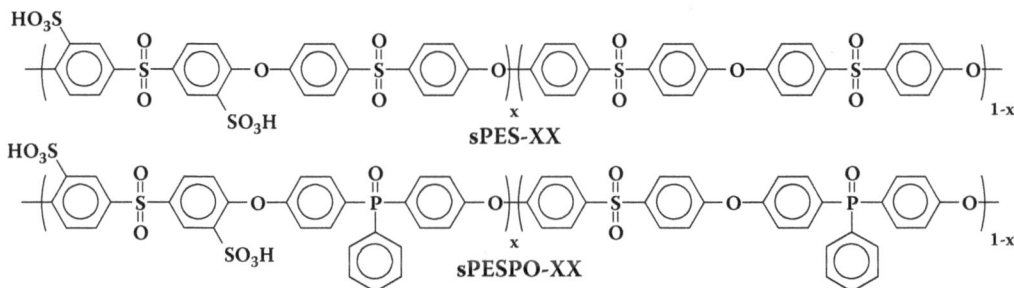

SCHEME 9.17 Structures of different poly(arylene ether phosphine oxide)s [62].

Miyatake et al. reported multiblock sulfonated poly(arylene ether)s by reacting *m*-terphenyl (MTP)-based hydrophobic oligomers with hydrophilic sulfonated blocks (Scheme 9.18) [95]. The membranes of these multiblock polymers were tough and transparent, and TEM images indicated phase segregated morphology. The sequence of the blocks and their compositions contributed towards the morphology as the domain sizes. As usual, the PC and WU exhibited an upsurge trend with the IEC. They reported that the MTP unit contributed towards higher PC [PC: 320 mS cm^{-1}; IEC: 2.13 meq g^{-1}] compared to their earlier polymer that having the same hydrophilic segment but prepared from altered hydrophobic (p-biphenyl) segment [PC: 200 mS cm^{-1}, IEC: 1.69 meq g^{-1}) under similar test conditions. However, this discussion does not talk much about the role of MTP as the polymers under comparison had much different IEC values.

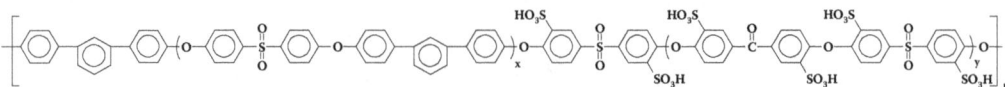

SCHEME 9.18 Structure of multiblock sulfonated poly(arylene ether) with m-terphenyl (MTP) unit [95].

9.4.1.3 Cross-linked and Composite Poly(arylene ether)s

Cross-linking polymer of membranes is an essential approach for overcoming their problems related to long-term durability in PEM applications. Watanabe et al. reported a trifunctional monomer for cross-linking the polymeric chain [96]. McGrath et al. introduced cross-linkable units in polymers that lead to structural changes after cross-linking and increase thermal stability [97,98]. However, cross-linking of the membranes used in PEMFC, particularly sulfonated polymers, without affecting the sulfonic acid group is not studied much.

Lee et al. reported a cross-linkable ethynyl-terminated fluorinated sulfonated poly(arylene ether) ionomer (E-SFQK, Scheme 9.19). After that, the ethynyl end groups were cross-linked at 250°C [Figure 9.15(a)] [99]. The cross-linked membrane (CSFQH) exhibited much less PC, SR, and WU than the non-cross-linked membranes (NSFQH) (PC: 160 to 130 S cm^{-1}; SR: 31% to 13%; WU: 86% to

(a)

(b)

SFQK =

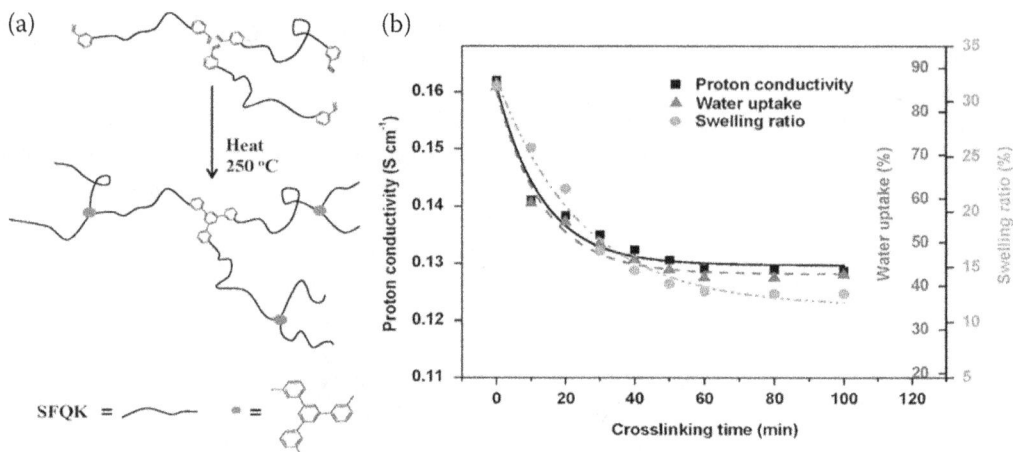

FIGURE 9.15 (B) Possible cross-linking mechanism of polymers. (b) Effect of cross-linking time to various PEM-related properties. Reprinted with permission from ref. [99]. Copyright (2009), American Chemical Society.

42%) and enhanced peroxide stability. The cross-linking time also affects various PEM-related properties Figure 9.15(b). The cross-linked membrane exhibited much less MP (88×10^{-8} cm^2 s^{-1}) than Nafion® 117 (154×10^{-8} cm^2 s^{-1}). The sizes of hydrophilic domains of the CSFQH membrane (5–20 nm) were exceptionally smaller than noncross-linked NSFQH membrane (20–50 nm) suggested a decrease in free volume for the holding of water by the sulfonic acid groups. These results indicated that cross-linking could be an approach for attaining better quality membranes for PEM applications. However, it seems that the cross-linking at 250°C could be detrimental and lead to desulfonation.

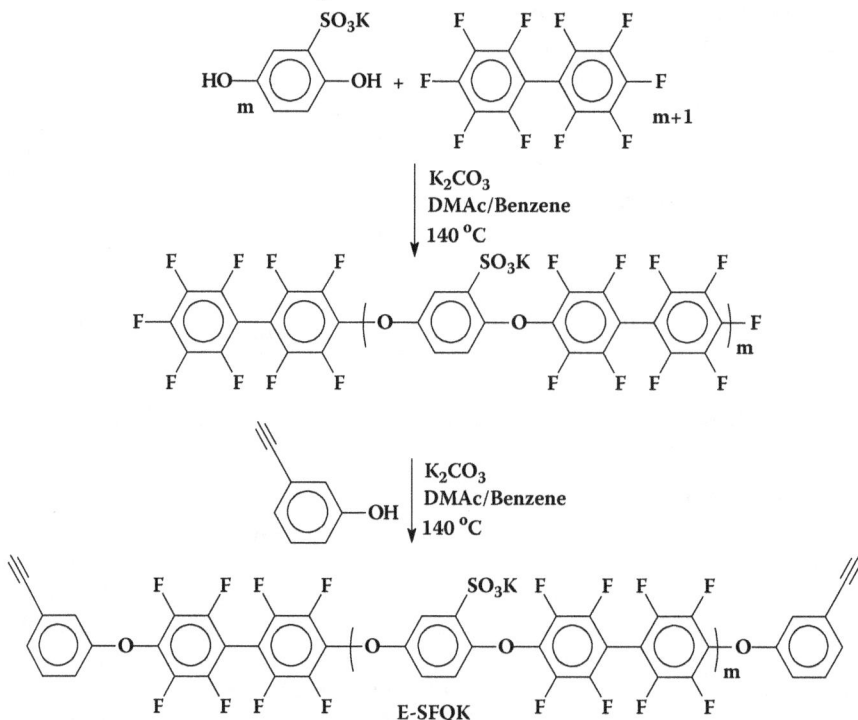

SCHEME 9.19 Synthesis of end-group cross-linkable E-SFQK polymer [99].

Lee and co-workers developed another highly fluorinated ethynyl-terminated sulfonated poly (arylene ether) copolymers (ESFx-6F, Scheme 9.20) [100]. The fluorinated groups (F, 3 F and 6 F) improved thermal and chemical stability with a decreased MP of the membranes [99,101–103]. The cross-linked ESF90-6F membrane displayed the balanced PEM properties with PC of 111 mS cm^{-1} and methanol permeability of 45×10^{-8} cm^2 s^{-1}. These values are better than that of Nafion® 117 (91 mS cm^{-1} and 167×10^{-8} cm^2 s^{-1}).

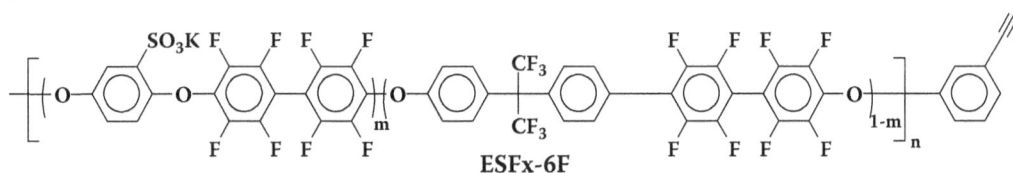

SCHEME 9.20 Chemical structure of ESFx-6Fs polymers [100].

Among the various strategies implemented for improvement the performance of polymeric membranes for their PEM applications, the fabrication of composite membranes using different inorganic fillers is one of them. Usually, hygroscopic compounds [titania (TiO$_2$), silica (SiO$_2$), zirconium phosphate [Zr$_3$(PO$_4$)$_4$, etc.] were incorporated to augment WU and retention of the water by the composite membranes and eventually enhance the PC of PEMs [104,105].

Banerjee et al. developed several composite membranes (SPAES-SS-X, Figure 9.16) from sulfo-propylated polysilsesquioxane (THSPSA) and carboxylic acid functionalized sulfonated poly(arylene ether sulfone) via an *in situ* sol-gel reaction during membrane fabrication step [106]. The sulfonic acid groups of the polymers and the subsequent interaction of the sulfopropylated polysilsesquioxane (SiOPS) formed a network structure for effective proton transport [107]. In addition, the –COOH groups of the copolymers helped for better interaction with the –OH groups of the inorganic phases. The hybrid membranes displayed phase segregated morphology and improved connectivity among the ionic domains than the pristine TPQSH-COOH-50 membrane. The PC of 138 mS cm^{-1} (80°C) was reached for SPAES-SS-10 membrane. However, such high loading of the fillers deteriorates many other PEM properties. Nevertheless, this work demonstrated the preparation of in-situ nanofillers bearing –SO$_3$H groups during the membrane casting step and could be investigated further in obtaining PEMs for fuel cell applications.

9.4.2 Sulfonated Copolytriazoles

Sulfonated polytriazole (SPT) polymers are relatively new candidates for PEM applications than previously discussed polymer classes. However, the click reactions had not adequately employed in synthesizing PEM materials. Ponomarev et al. developed SPT first time by clicking several dialkyne and sulfonated diazide monomers [26,108]. Chang et al. prepared another series of SPTs, fabricated membranes, and investigated their thermal, mechanical, and proton conductivity properties [109]. The high PC of the STP membranes was ascribed to the strong acid-base in-teraction of the sulfonic acid groups and the triazole rings. The triazole ring eventually formed networks between the uniformly dispersed ionic cluster that facilitated the transport of protons. The advantages of using sulfonated polytriazoles for the PEM applications are:

a. The SPT polymers are prepared via efficient and straightforward synthetic routes, and they possessed brilliant membrane forming ability.
b. Polytriazole membranes displayed high mechanical, oxidative, and thermal stabilities.
c. The triazole rings can be represented as efficient proton donors and receptors due to their amphoteric properties and their tautomeric structure. These structures help proton transportation through structural diffusion [110].

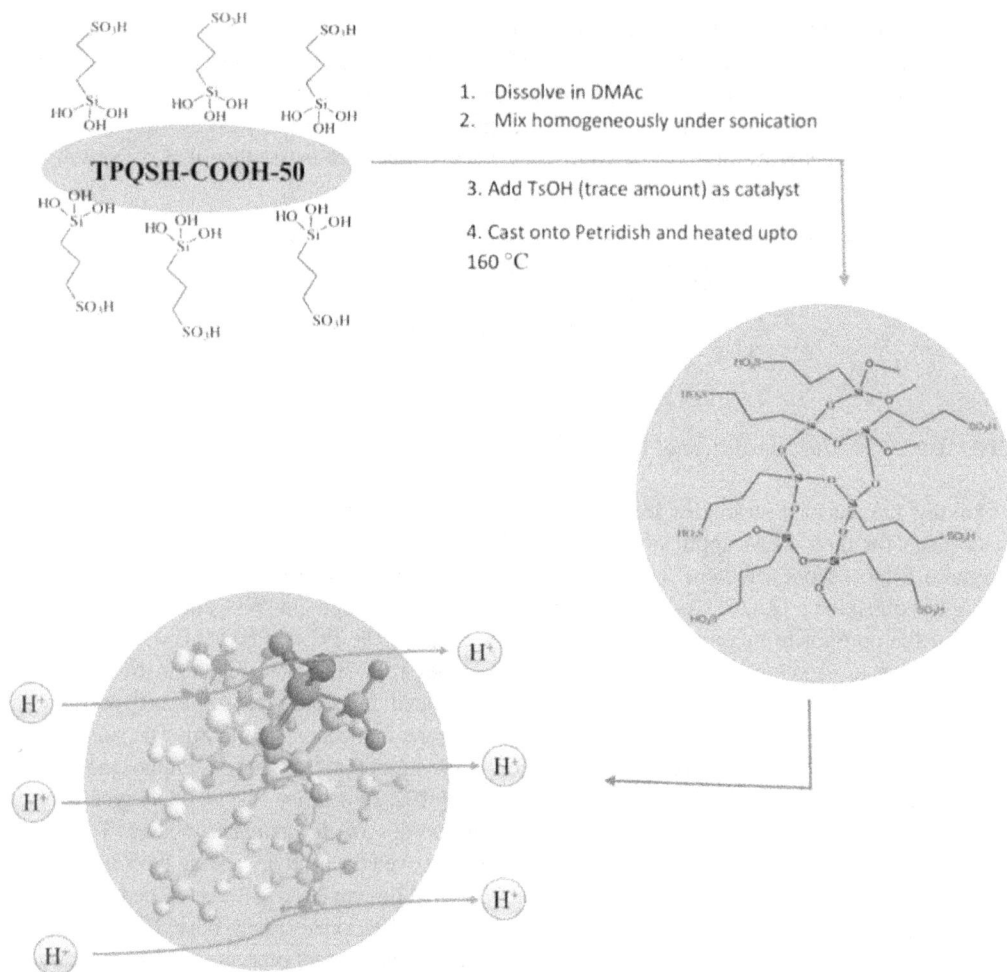

1. Dissolve in DMAc
2. Mix homogeneously under sonication

3. Add TsOH (trace amount) as catalyst

4. Cast onto Petridish and heated upto 160 °C

FIGURE 9.16 Schemetic depiction of the formation of SPAES-SS-X hybrid membranes. Reprinted from ref. [106], Copyright (2020), De Gruyter.

    d. The sulfonic acid group and basic triazole systems created interconnected ionic cluster networks by the strong acid-base interaction, which also favored proton conduction [26].

The SPTs are classified into few groups and discussed below.

9.4.2.1 Fluorinated Copolytriazoles

A set of fluorinated sulfonated copolytriazoles was reported by Banerjee et al. along with the preparation of corresponding fluorinated diazides. The PTAQSH-XX copolymers (Scheme 9.21) were prepared from the copper assisted click polymerization of trifluoromethyl substituted diazide monomer [4,4′-bis[3-trifluoromethyl-4(4-azidophenoxy)phenyl]biphenyl] (QAZ), sulfonated diazide compound [4,4′-diazido-2,2′-stilbenedisulfonic acid disodium salt (DADSDB)] with equimolar amount of a dialkyne monomer [4,4′-(propane-2,2-diyl) di((prop-2-ynyloxy)benzene) (BPEBPA)] [24]. The polymers showed number average molecular weight of about 40,000 g mol^{-1}. The solution cast membranes from these polymers exhibited high thermal (10% decomposition temperature in TGA, T_{d10}: 223°C–308°C) and mechanical stability (TS: 51–65 MPa, EB: 7%–32%, and YM: 1.84–2.24 GPa) (Table 9.2). The membranes from these polymers displayed improved oxidative and hydrolytic stability,

TABLE 9.2

Molecular Weight, Thermal, and Mechanical Properties of Some Fluorinated Polytriazoles

Polymer	M_n^m	$Đ^n$	η_{inh}^o	$T_{d10\%}^p$	TS (MPa)	YM (GPa)	EB (%)	Ref.
PTAQSH-50	43000	3.35	1.67	308	65	2.24	7	[24]
PTAQSH-60	44000	2.59	1.71	273	57	2.10	17	[24]
PTAQSH-70	39000	2.58	1.74	265	53	1.85	32	[24]
PTAQSH-80	35000	2.26	1.78	228	52	1.86	14	[24]
PTAQSH-90	33000	2.52	1.80	223	51	1.84	9	[24]
PTASH-100	19800	1.55	1.84	192	8	1.26	1.5	[24]
PTFQSH-70	24500	2.96	1.52	290	60	1.95	53	[25]
PTFQSH-80	19900	3.55	1.56	280	49	1.93	20	[25]
PTFQSH-90	19300	4.33	1.63	267	42	1.64	16	[25]
PTFSH-100	16300	4.52	1.72	298	58	2.01	34	[25]
PTATSH-60	39000	2.27	1.50	281	62.4	1.85	26	[26]
PTATSH-70	32000	1.90	1.26	261	53.2	1.70	7	[26]
PTATSH-80	37000	2.15	1.42	255	43.6	1.56	8	[26]
PTATSH-90	36,000	2.10	1.39	153	41.3	1.29	14	[26]
PTATSH-100	–	–	–	122	8	1.26	1.5	[26]
PTFOSH-60	25200	2.65	1.35	296	53	2.30	20	[30]
PTFOSH-70	23300	2.95	1.43	288	43	2.07	24	[30]
PTFOSH-80	19800	3.28	1.52	280	41	1.94	28	[30]
PTFOSH-90	18700	3.54	1.61	277	34	1.95	18	[30]
PTEHSH-60	–	–	1.18	251	95	2.18	75	[27]
PTEHSH-70	–	–	1.32	246	92	2.07	70	[27]
PTEHSH-80	–	–	1.39	240	78	1.97	56	[27]
PTEHSH-90	–	–	1.28	232	69	1.55	40	[27]
PTEHSH-100	–	–	1.22	220	64	1.25	34	[27]
PTHQSH-60	994000	2.01	1.95	275	63	1.89	49	[31]
PTHQSH-70	857000	2.28	1.89	268	46	1.82	34	[31]
PTHQSH-80	502000	1.39	1.85	260	42	1.60	36	[31]
PTHQSH-90	445000	1.41	1.80	252	39	1.68	33	[31]
PTSQSH-I	63200	2.78	1.23	254	51	2.02	13	[29]
PTSQSH-II	60700	2.92	1.16	249	36	1.31	11	[29]
PTSQSH-III	67600	2.28	1.12	247	34	1.39	14	[29]
PTSQSH-IV	75300	2.54	1.24	243	30	1.25	19	[29]

Notes

[m] Number average molecular weight

[n] Polydispersity index

[o] Inherent viscosity

[p] 10% degradation from TGA

phase-separated morphology, and PC of 15–90 mS cm^{-1} (Table 9.3). The oxidative stability of the membrane increased with increasing fluorine content, and at the same time, the water holding capacity decreased somewhat due to increased hydrophobicity [6,37]. The efficient synthesis procedure and excellent water management ability make these polymers gifted candidates for PEM applications and open up an area for further investigation.

TABLE 9.3
PEM Properties of Different Sulfonated Polytriazoles

Polymer	IEC$_w$[q] (meq g^{-1})	τ (h)[r]	WU (wt%)[s] 80°C	SR (%)[t] 80°C	σ (mS cm^{-1})[u] 80°C	Ref.
PTAQSH-50	1.15	> 24	10	7.1	12	[24]
PTAQSH-60	1.43	>24	16	7.4	16	[24]
PTAQSH-70	1.72	>24	22	11.8	27	[24]
PTAQSH-80	2.04	19	32	19.2	47	[24]
PTAQSH-90	2.38	12.5	40	28.4	76	[24]
PTASH-100	2.75	3.5	57	–	112	[24]
PTATSH-60	1.47	>24	14	7	15	[26]
PTATSH-70	1.77	>24	18	11	33	[26]
PTATSH-80	2.08	>24	28	17	59	[26]
PTATSH-90	2.40	13	36	25	90	[26]
PTFQSH-70	1.52	>24	12	9	27	[25]
PTFQSH-80	1.79	>24	16	15	46	[25]
PTFQSH-90	2.09	18	30	23	66	[25]
PTFSH-100	2.40	5	42	35	136	[25]
PTFOSH-60	1.57	>24	12	8	14	[30]
PTFOSH-70	1.79	>24	17	10	28	[30]
PTFOSH-80	2.00	>24	28	22	59	[30]
PTFOSH-90	2.20	15	33	29	110	[30]
PTCTSH-60	1.57	>24	45	13	38	[28]
PTCTSH-70	1.88	18	61	31	80	[28]
PTCTSH-80	2.19	10	96	44	118	[28]
PTCTSH-90	2.52	1	200	65	163	[28]
PTCTSH-60/PVA	1.48	>24	9	3	19	[28]
PTCTSH-70/PVA	1.77	>24	16	6	40	[28]
PTCTSH-80/PVA	2.06	15	26	9	78	[28]
PTCTSH-90/PVA	2.38	3	41	20	104	[28]
PTHQSH-60	1.66	>24	12	8.1	20	[31]
PTHQSH-70	2.02	>24	20	11.9	42	[31]
PTHQSH-80	2.40	11.4	34	23.4	81	[31]
PTHQSH-90	2.82	~4.5	65	32.7	142	[31]
PTSQSH-I	2.41	>24	41	15	132	[29]
PTSQSH-II	2.74	>24	60	21	173	[29]
PTSQSH-III	3.10	~8	91	33	248	[29]
PTSQSH-IV	3.49	~2	141	51	304	[29]
PTEHSH-60	1.82	>24	26	15	53	[27]
PTEHSH-70	2.18	>24	37	18	85	[27]
PTEHSH-80	2.56	19	54	23	120	[27]
PTEHSH-90	2.96	10	73	32	163	[27]
PTEHSH-100	3.38	2.5	103	43	215	[27]
PTEOSH-70	2.81	19	67	77	96	[111]
PTEOSH-80	3.11	16	82	106	148	[111]
PTEOSH-90	3.38	12	139	170	196	[111]
PTPBSH-70	1.83	>24	20	12	38	[94]
PTPBSH-80	2.13	>24	29	17	74	[94]
PTPBSH-90	2.44	15	36	25	110	[94]

TABLE 9.3 (Continued)

PEM Properties of Different Sulfonated Polytriazoles

Polymer	IEC$_w$^{q} (meq g^{-1})	τ (h)^r	WU (wt%)^s 80°C	SR (%)^t 80°C	σ (mS cm^{-1})^u 80°C	Ref.
PTPFBSH-60	1.45	39	17	10.4	14	[112]
PTPFBSH-70	1.74	35	25	14.3	45	[112]
PTPFBSH-80	2.06	26	37	19.9	79	[112]
PTPFBSH-90	2.39	18	41	27.5	133	[112]
PTNQSH-50	1.10	>24	14	–	14	[113]
PTNQSH-60	1.37	>24	17	–	31	[113]
PTNQSH-70	1.65	>24	23	–	46	[113]
PTNQSH-80	1.95	>24	27	–	84	[113]
PTPQSH-50	1.17	>24	11	7	26	[114]
PTPQSH-60	1.45	>24	16	9	49	[114]
PTPQSH-70	1.75	>24	33	13	131	[114]
PTPQSH-80	2.07	15	50	20	157	[114]

Notes

^{q} Ion exchange capacity
^{r} Oxidative stability
^{s} Water uptake (wt%)
^{t} Swelling ratios
^{u} Proton conductivity

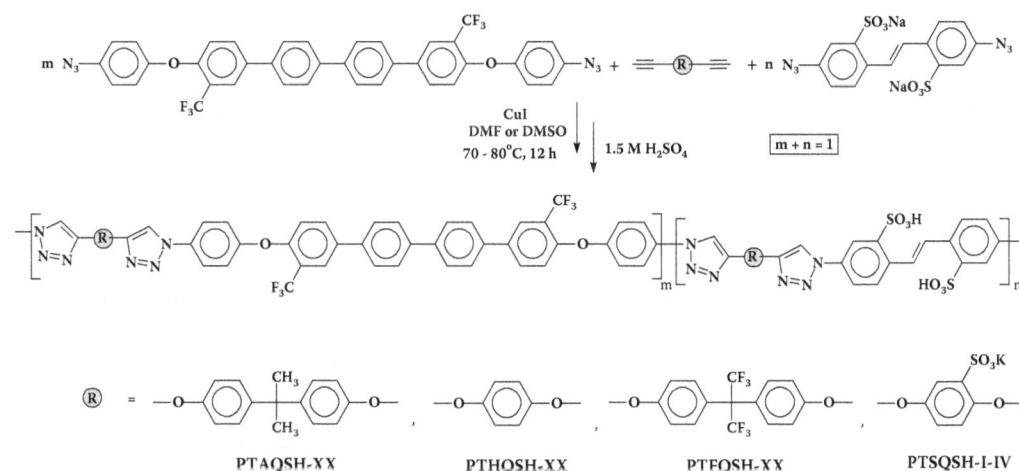

SCHEME 9.21 Synthesis of different fluorinated polytriazoles.

Subsequently, several research articles published by Banerjee et al. increasing the fluorine content in SPT due to its effective contributions towards polymer processability, proton conductivity, and chemical durability. The polymers PTFQSH-XX and PTFOSH-XX (Scheme 9.21) with hexafluoroisopropylidene (6 F) groups were prepared accordingly to increase the fluorine content in the polymers. The molecular mass, thermal and mechanical properties of PTFQSH-XX and PTFOSH-XX are given in Table 9.2. The copolymers displayed lower WU and SR and somewhat higher oxidative stability (18 h for PTFQSH-90 and 15 h for PTFOSH-90) than similar non-fluorinated PTAQSH-XX copolymer membranes (Table 9.3) [25,30]. The attendance of hydrophobic 3 F and 6 F groups in PTFQSH-XX copolymers contributed to a improved phase-segregated morphology and high PC (27–136 mS cm^{-1} at 80°C) [25]. PTFOSH-XX membranes (Scheme 9.22) also showed comparatively high PC (maximum 110 mS cm^{-1} at 80°C for PTFOSH-90) due to higher IEC$_W$ values (Table 9.3) [30].

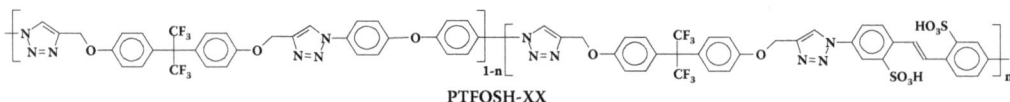

PTFOSH-XX

SCHEME 9.22   Chemical structure of PTFOSH-XX polytriazoles [30].

IEC played a crucial role in determining the PEM properties of any ion-conducting polymers. Hence, any newly developed polymer is scrutinize based on their IEC values for their application in PEM. Several fluorinated sulfonated copolytriazoles, PTHQSH-XX (Scheme 9.21), PTEHSH-XX (Scheme 9.23), and PTSQSH-I to IV (Scheme 9.21) with relatively higher IEC (1.66–2.82 meq g^{-1} for PTHQSH-XX, 1.82–3.38 meq g^{-1} for PTEHSH-XX, and 2.41–3.49 meq g^{-1} for PTSQSH-I to IV) were synthesized by varying the polymer repeat unit structures than the polymers mentioned earlier [27,29,31]. The thermally stable and mechanically robust nature of these polymers (PTHQSH-XX, PTEHSH-XX, and PTSQSH-I to IV) indicated from their molecular weight, T$_{d10\%}$, TS, YM, and EB values, which are provided in Table 9.2. The membranes displayed phase segregated morphology. The ionic channels formed in some cases helped increasing PC as high as 142 mS cm^{-1} (80°C) for PTHQSH-90 copolymer [31]. The transportation of the protons occurred through the channels via the formation of hydrogen bonds (Figure 9.17).

A low molar mass containing dialkyne was used to synthesize PTEHSH-XX copolymers (Scheme 9.23) to increase the IEC$_W$ of the copolymers and investigated its magnitude on proton conductivity [27]. The PTEHSH-XX membranes revealed very high PC in the range 58–235 mS cm^{-1} at 90°C in completely hydrated states. The membranes showed low WU credited to the high fluorine content and hydrophobicity of the copolymers. All other critical PEM-related properties are arranged in Table 9.3.

The PTSQSH-I to IV copolymers (Scheme 9.21) exhibited good solubility, high thermal, mechanical stability and oxidative stability, low WU, and well-interconnected phase-segregated morphology with PC as high as 304 mS cm^{-1} at 80°C (for PTSQSH-IV copolymer) [29]. The PTSQSH-II copolymer with an IEC$_W$ value of 2.74 meq g^{-1} showed in general best membrane performance considering its oxidative stability (more than 24 h), PC (173 mS cm^{-1} at 80°C), and dimensional stability (Table 9.3).

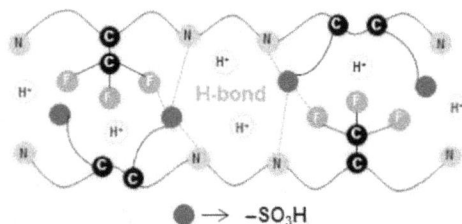

FIGURE 9.17   Proposed model of transportation of the protons through ionic channels and formation of hydrogen bonds in fluorinated sulfonated copolytriazoles. Reprinted with permission from ref. [31]. Copyright © 2017, John Wiley and Sons.

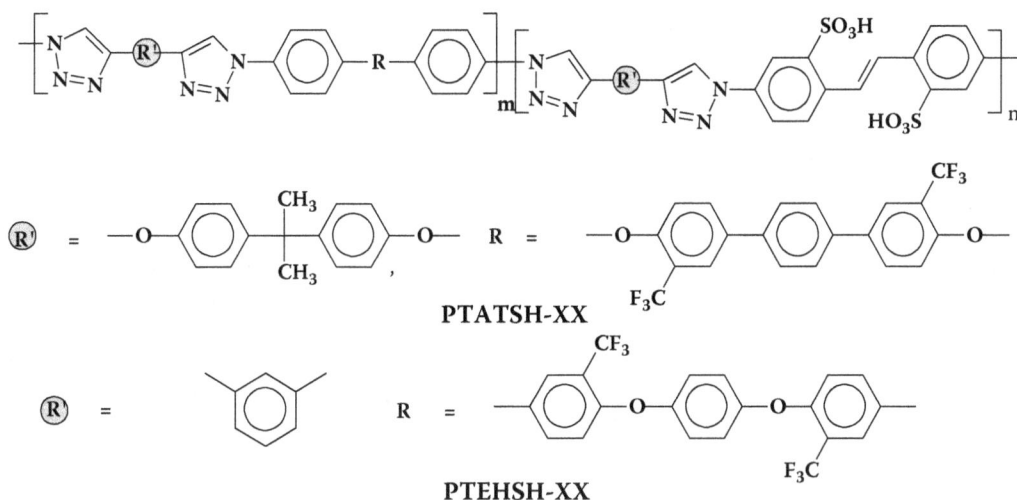

SCHEME 9.23 Reaction scheme of PTATSH-XX and PTEHSH-XX copolymers.

9.4.2.2 Non-Fluorinated Copolytriazoles

Several sulfonic acid groups-containing non-fluorinated copolytriazoles were also reported, and their performance in PEMFC was investigated. Huang et al. reported a sequence of non-fluorinated copolytriazoles by click polymerization (Scheme 9.24) [115]. The polymers displayed thermal stability with $T_{d5\%}$ ranged from 246°C–254°C. The TS, YM, and EB values of the polymers ranged from 21.3–28.2 MPa, 4.08–4.57 GPa, and 5.94%–8.23%, respectively. These X-ray diffraction (XRD) studies indicated the formation of more interconnected channels across the swollen polytriazole membrane in water.

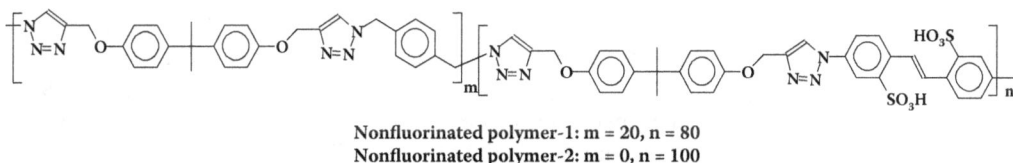

Nonfluorinated polymer-1: m = 20, n = 80
Nonfluorinated polymer-2: m = 0, n = 100

SCHEME 9.24 Reaction scheme of synthesis of non-fluorinated polytriazoles by direct click polymerization reaction [115].

Banerjee et al. prepared another series of non-fluorinated sulfonated polytriazole copolymers (PTEOSH-XX) with various DS (Scheme 9.25) [111]. The $T_{d10\%}$, TS, YM, and EB values of the copolymers PTEOSH-70 to -90 ranged from 232°C–241°C, 41–63 MPa, 1.02–2.18 GPa, and 22%–29%. The oxidative stability values ranged from 12–19 h (Table 9.3). The PC of the copolymer membranes went from 104–215 mS cm^{-1} at 90°C.

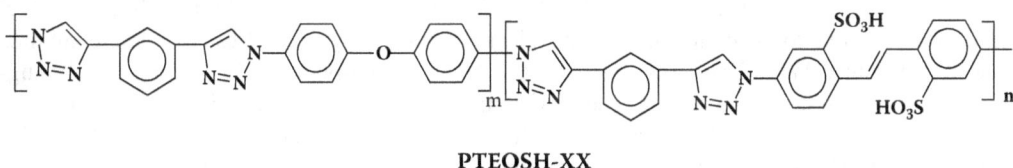

PTEOSH-XX

SCHEME 9.25 Repeat unit structure of PTEOSH-XX polytriazoles [111].

9.4.2.3 Copoly(triazole imide)s

The sulfonated polyimides were synthesized via polycondensation reactions at high temperatures. However, there are numerous complications, including the degradation of the sulfonic acid groups during the high-temperature imidization [22,116]. Recently, Banerjee et al. prepared several new kinds of sulfonic acid-containing copoly(triazole imide)s (PTNQSH-XX) using an imide-based dialkyne monomer (Scheme 9.26) [113]. Membranes were fabricated from these polymers, and their properties were examined. Characteristically, the membranes showed two-step degradation and high mechanical stability. The values of $T_{d10}\%$, TS, YM, and EB were in the range of 280°C–311°C, 49–67 MPa, 1.58–2.06 GPa, and 9%–15%, respectively. The membranes' high oxidative stability (>24 h) was perhaps due to triazole systems in the polymer backbone. The triazole systems formed a chelate with the ferrous ions of Fenton's reagent and helped to inactivate the hydroxyl and hydroperoxyl radicals and thus reduced oxidative attack to the polymer chains [117–119]. The WU and PC of PTNQSH-50 to -80 at 80°C were 14%–27%, 14-84 mS cm^{-1}, respectively (Table 9.3).

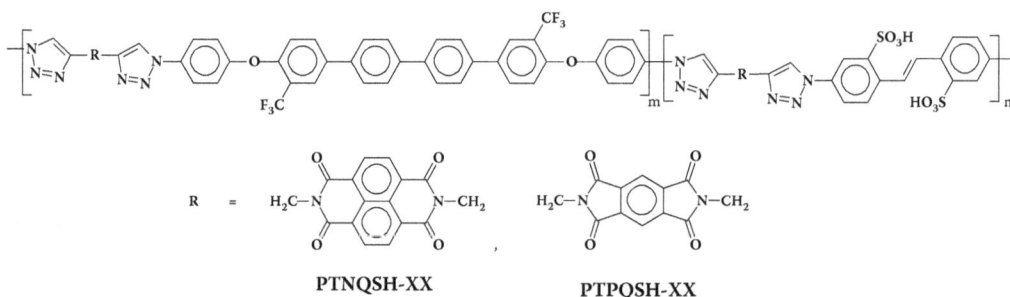

SCHEME 9.26 PTNQSH-XX and PTPQSH-XX copolymers (XX = molar % amount of sulfonated monomer) [113,114].

In continuation of the earlier work, a dialkyne was made from a five-member aromatic dianhydride and used to prepare sulfonated copoly(triazole imide)s (PTPQSH-XX) (Scheme 9.26) [114]. The PTPQSH-XX copolymer membranes revealed high TS (28–45 MPa), YM (0.98–1.34 GPa) and EB (9%–10%) in wet state with $T_{d10\%}$ of 270°C–310°C. The PTPQSH-XX copolymers demonstrated better hydrolytic reliability than the five or six-membered aromatic sulfonated polyimides [37].

9.4.2.4 Phosphorus-Containing Polytriazoles

The positive impact of the phosphine oxide moieties in polymers and their influence on PEM properties inspired researchers to develop phosphorus-containing polymers [61,63]. Banerjee et al. developed several phosphorus-containing diazide monomers and consequently used them to prepare phosphorus-containing sulfonated copolytriazoles (Scheme 9.27) [94]. PTPBSH-XX copolymers appeared to have adequate thermal stability with $T_{d10\%}$ as high as 288°C. The fabricated membranes were both mechanically stable with satisfactorily high EB (29%) in both dry and wet conditions required to form the membrane electrode assembly [15]. Besides, the membrane showed a balanced WU (20%–36% at 80°C) and SR (through-plane: 12%–25% at 80°C), suggesting improved water management. The membranes exhibited increased oxidative stability even for the polymers with high IEC$_w$ values (15 h for PTPBSH-90). This was due to the seizure of •OOH and •OH radicals by the triphenylphosphine oxide present in the polytriazole unit. AFM and SAXS studies designated phase-segregated morphology, upsurges with an increase in the DS of the polymers. The membrane (PTPBSH-90) showed PC up to 119 mS cm^{-1} at 90°C. Improved mechanical, thermal, and peroxide resistance along with the balanced WU, SR, and high PC made these phosphorus-containing polymers as candidate materials for PEM applications [15].

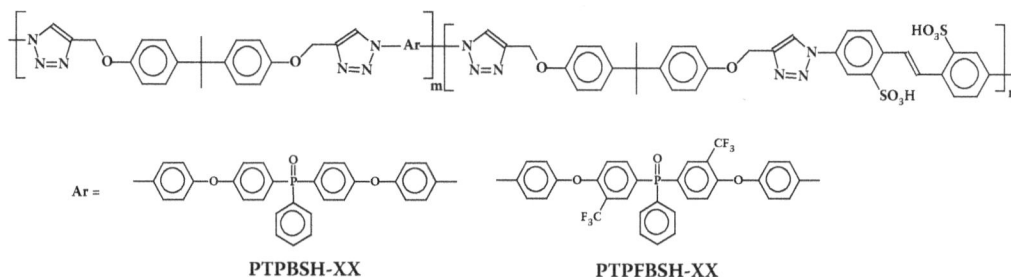

SCHEME 9.27 Repeat unit structures of the phosphorus containing polytriazoles, PTPBSH-XX and PTPFBSH-XX [94,112].

Subsequently, seeing the optimistic effects of phosphorus and fluorine elements in the polymer construction, Banerjee et al. reported a series of sulfonated polytriazoles (PTPFBSH-XX, Scheme 9.27) consisting of trifluoromethyl and phosphine oxide groups [112]. The diazide monomer, bis[4-(4′-azidophenoxy)-3-trifluoromethyl phenyl] phenylphosphine oxide was first synthesized and subsequently used to prepare PTPFBSH-XX copolytriazoles. PTPFBSH-XX polymers were rapidly dissolved numerous polar aprotic solvents, notwithstanding their rigid backbone owing to the combined result of the presence of –O–, –CF$_3$ and polar –P=O units. The solution cast membranes from these polymers were flexible and tough and displayed T$_{d10\%}$, TS, YM, and EB in the range of 284°C–290°C, 49–72 MPa, 0.69–0.90 GPa, and 27%–48%, correspondingly. The improved thermal stability was owing to the interaction among the sulfonic acid groups, triazole rings, and polar phosphine oxide groups that restricts the rotational movements of the polymer chains [59,63,120]. Like the previous PTPBSH-XX polymers, these polymers also showed high peroxide resistance. For instance, PTPFBSH-90 exhibited peroxide radical stability of more than 18 h, and this enhanced oxidative stability is endorsed to the presence of strong electron-withdrawing 3 F groups along with the phosphine oxide moieties. The PC of the PTPFBSH-XX membrane [PTPFBSH-90: (IEC$_W$: 2.39 meq g^{-1}) 142 mS cm^{-1}] under fully hydrated conditions at 90°C was improved by about 20% than PTPBSH-XX polymer membranes. Furthermore, PTPFBSH-XX membranes exhibited microbial fuel cells (MFCs) performance similar to Nafion® 117.

9.4.2.5 Cross-Linked and Composite Polytriazoles

Cross-linking is often familiar in developing PEM, and cross-linking can be attained by covalent, ionic, or other types of interactions. Lee et al. reported cross-linked sulfonated polyamide membranes using dihydroxy functionalized cross-linkers with the carboxylic acid groups of the polyimides [121,122]. Subsequently, Guiver et al. reported cross-linking of sulfonated poly(aryl ether ketone) bearing carboxylic acid groups using poly(vinyl alcohol) (PVA) as cross-linking agent [112].

Banerjee et al. prepared carboxylic acid-containing dialkyne monomer [3,5- bis(prop-2-ynyloxy)benzoic acid (BPBA)] and utilized the same to prepare several SPTs (PTCTSH-XX) [28]. The carboxylic acid (–COOH) groups of PTCTSH-XX functioned as possible cross-linking sites when reacted with PVA as a cross-linker (Figure 9.18).

All PTCTSH-XX and PTCTSH-XX/PVA polymers showed superior thermal stability and higher tensile strength than Nafion® 117 (T$_{d10\%}$: 242°C–267°C, TS: 35–71 MPa). The PTCTSH-XX

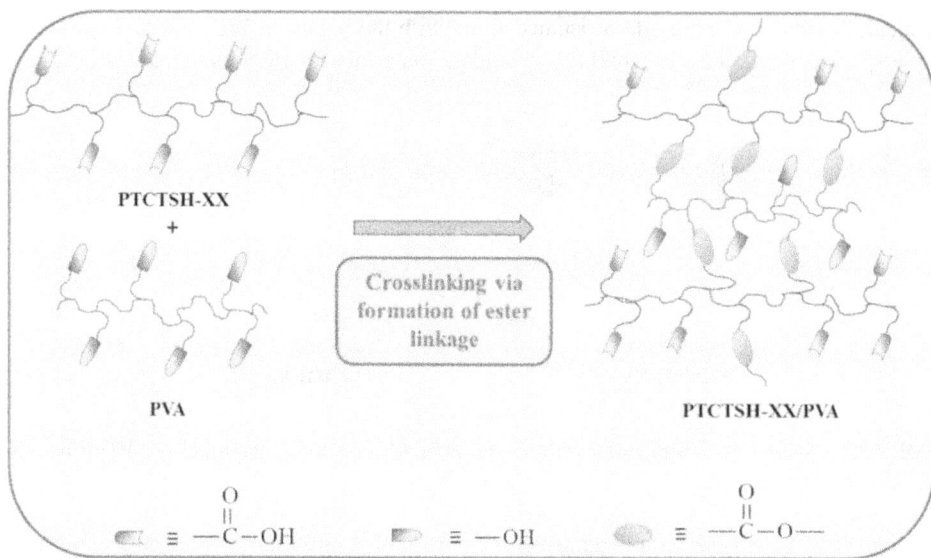

FIGURE 9.18 The schematic representation of cross-linking of PTCTSH-XX copolymers using PVA as a cross-linker. Reprinted from ref. [28], Copyright (2018), with permission from Elsevier.

copolymer membranes presented high WU and low dimensional stability at high DS. However, after cross-linking, PTCTSH-XX/PVA membranes showed five times decrease in WU, and, accordingly, the dimensional stability was also evocatively improved (Table 9.3). The PC of the PTCTSH-XXs membranes ranged from 38–163 mS cm^{-1} (80°C) and 42–176 mS cm^{-1} (90°C) in hydrated conditions. However, the PC values [115 mS cm^{-1} (90°C) for PTCTSH-90/PVA at 2.5 wt% PVA loading) decreased somewhat after cross-linking, though many other PEM-related properties such as WU, dimensional stability (SR), oxidation stability, etc., were improved. The single-cell MFC fabricated from these membranes showed good performance.

The PC of PEM materials is further enhanced by fabricating polymer composites membranes using inorganic fillers with –SO$_3$H groups. A carboxylic acid-containing sulfonated copolytriazoles (PTECHSH-80) reported by Banerjee et al. and used to create composite membranes (Scheme 9.28) [123].

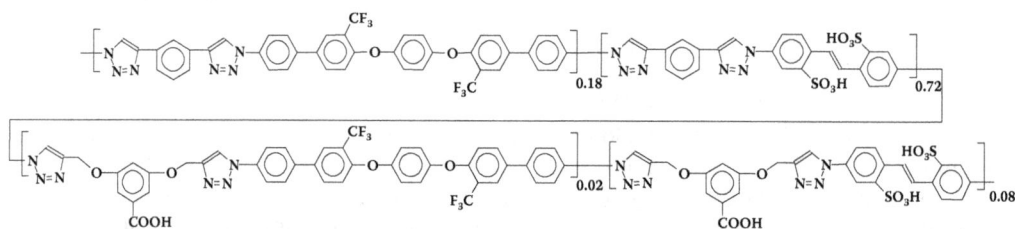

SCHEME 9.28   Semi-fluorinated sulfonated copolytriazoles (PTECHSH-80) copolymer [123].

3-(Trihydroxylsilyl) propane-1-sulfonic acid used as precursor materials and added with PTECHSH-80 copolytriazole to prepare polymer composites PTEHCSH-80/X (X is wt% of the filler, X = 0, 2.5, 5.0, 7.5, 10.0) via sol-gel reaction. PTECHSH-80/X exhibited high thermal stability (T$_{d,10\%}$: 500°C–510°C) and TS, YM, and EB in the range of 67–23 MPa, 0.76–1.35 GPa, and 76%–8%, respectively, depending on the filler loading. The PTEHCSH-80/X membranes showed significantly lower in-plane SRs than the PTEHCSH-80 and Nafion® 117. The sulfopropylated polysilsesquioxane formed interpenetrating cage-like construction in the polymer matrix and restricted movement of water molecules. The PTEHCSH-80/X membranes displayed very high PC [117–182 mS cm^{-1} (80°C) and 131 to 195 mS cm^{-1} (90°C)] depending on filler loading. Hence, fabrication of composite membranes by judiciously selecting inorganic fillers might be an effective approach for developing PEM materials.

9.4.3 SULFONATED POLYIMIDES

Over the years, sulfonated polyamides (SPIs) had accepted as accomplished PEM materials owing to their excellent thermal and mechanical stability, adequate water uptake and chemical stability, and good film-forming ability [124]. Mostly, the polycondensation of the certain diamine and dianhydride monomers is the preferred method for the preparation of SPIs to circumvent the hydrolysis of the imide rings during the post-polymerization acidification method. Furthermore, the direct polycondensation method provides a better opportunity to tune the polymer architecture and thus PEM-related properties by the diamine monomer (non-sulfonated). In general, five-membered polyimides are usually unstable to acids, and thus hydrolysis of the imide rings occurred [20]. However, six-membered naphthalene-based imides exhibited higher chemical, thermal and mechanical stability than similar types of five-membered ring polyamides for their strain-free ring structure and investigated primarily for their PEM applications [22]. McGrath et al. reported six-membered SPIs with flexible ether and sulfone linkages to increase hydrolytic stability and membrane flexibility [125]. Ho et al. developed poly(ethylene oxide) (PEO) soft segment-containing six-membered SPIs to increase membrane robustness and proton conductivity from the polymerization reaction of 1,4,5,8-naphthalenetetracarboxylic dianhydride (NTDA), 4,4′-diaminostilbene-2,2′-disulfonic acid (DSDSA), and PEO-diamine [126]. The recent developments of a few of the six-membered NTDA-based SPIs are categorized, and their PEM properties are deliberated briefly in the consecutive section.

9.4.3.1 Bulky Moiety Containing Polyimides

The incorporation of bulky moieties and rigid structures in the hydrophobic part of the polymeric backbone is a practical approach to tune the PEM properties of the polymeric membranes. Several functionalized bulky moieties and rigid structures (for example; quadriphenyl unit and fluorene system) were introduced by Banerjee et al. to attain good thermal, mechanical and dimensional stability with high PC and well-distributed phase-separated morphology of the PEM materials (Scheme 9.29) [21,22,116,127,128].

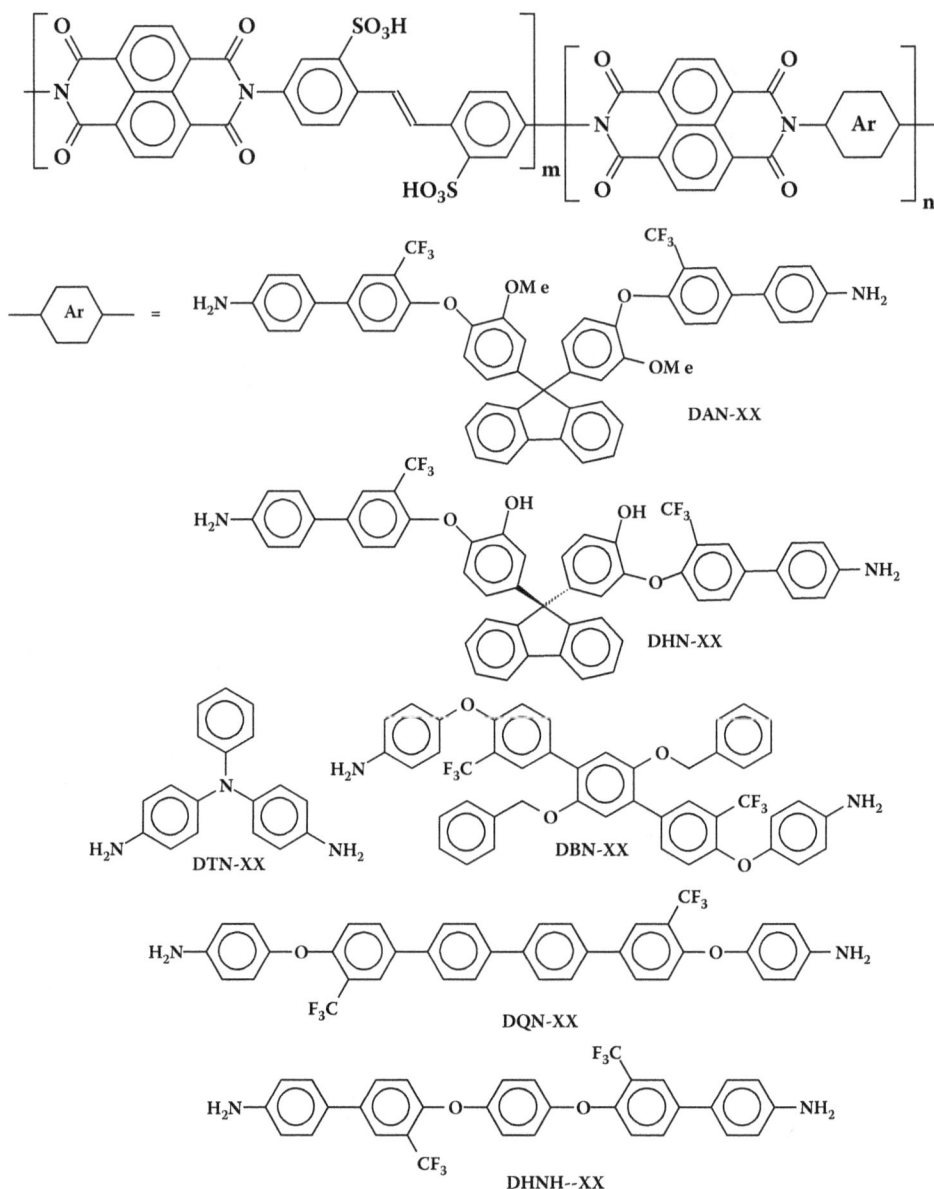

SCHEME 9.29 Different bulky moieties containing fluorinated SPIs [21,22,116,127–129].

The triphenylamine containing SPIs (DTN-XX, Scheme 9.29) were reported by Banerjee et al. exhibited PC of as high 207 mS cm^{-1} (80°C) for DTN-80 (IECw: 2.74 meq g^{-1}) was much higher than Nafion® 117 (135 mS cm^{-1}) under alike test conditions. DTN-XX membranes also displayed much lower O$_2$ gas permeability (DTN-80, P$_{O2}$ = 0.9 barrer) compared to Nafion® 117 (P$_{O2}$ = 3.6 barrer) [129]. Later, several SPIs (DAN-XX, DHN-XX, DBN-XX, DQN-XX, and DHNHXX; Scheme 9.29) were developed by Banerjee et al. and their PEM-related properties (IEC, WU, SR, oxidative stabilities, and PC) are presented in Table 9.4. The fluorenyl moiety containing DAN-80 (Scheme 9.29) (IECw: 2.27 meq g^{-1}) displayed a PC of 149 mS cm^{-1} [116]. The membranes exhibited phase-separat morphology and a steadiness between WU and SR, including reasonably

TABLE 9.4

PEM-Related Properties of Different SPIs Developed by Banerjee et al.

Polymer	IEC$_W$ (meq g^{-1})	WU$_W$ (%) at 80°C	SR (in-plane) (%) at 80°C	Oxidative Stability (h)	σ (80°C) (mS cm^{-1})	Ref.
DHNH40	1.1	15.23	1	20.5	10	[22]
DHNH50	1.41	20.57	2	14.7	28	[22]
DHNH60	1.75	26.29	4	11.2	60	[22]
DHNH70	2.1	33.61	5	6.3	99	[22]
DQN60	1.67	23.84	1.5	20	17	[21]
DQN70	2.03	29.73	2	9	39.6	[21]
DQN80	2.42	35.61	3	4.5	80.6	[21]
TPPO-40	1.18	18	3.0	>24	31	[130]
TPPO-50	1.50	23	4.7	>24	54	[130]
TPPO-60	1.84	41	10.0	20	99	[130]
DFPNH-50	1.36	23	5.3	>24	33	[131]
DFPNH-60	1.70	32	7.5	>24	81	[131]
DFPNH-70	2.05	47	9.0	16	142	[131]
DFPNH-80	2.44	60	11.0	9	181	[131]
DFPNH-90	2.86	70	14.0	2	237	[131]
DPPNH-60	1.88	31	6.0	>24	87	[132]
DPPNH-70	2.23	36	7.0	17	104	[132]
DPPNH-80	2.58	46	9.0	10	202	[132]
DPPNH-90	2.95	64	10.0	7	235	[132]
SPI-COOH-70	2.33	32	10.0	14	114	[133]
SPI/SS-2.5	1.18	42	4.0	12	156	[133]
SPI/SS-5.0	1.50	51	6.0	9	176	[133]
SPI/SS-7.5	1.84	54	7.0	7	194	[133]
SPI/SS-10.0	1.36	62	9.0	6	205	[133]
DAN-50	1.17	17	2.0	>24	20	[116]
DAN-60	1.49	21	3.0	>22	39	[116]
DAN-70	1.85	25	4.0	21	75	[116]
DAN-80	2.27	43	6.0	20	149	[116]
DHN-50	1.18	18	9.0^v	>24	34	[127]
DHN-60	1.51	24	14.0^v	>24	54	[127]
DHN-70	1.87	31	15.0^v	>24	97	[127]
DBN-60	1.56	17	10	>24	85	[128]
DBN-70	1.92	21	11	>24	133	[128]
DBN-80	2.33	28	13	20	171	[128]
DBN-90	2.79	35	18	4	244	[128]
DBN-90-CeO2-2.5	–	31	19	12	208	[128]
DBN-90-CeO2-5.0	–	29	17	18	184	[128]
DBN-90-CeO2-7.5	–	15	16	>24	159	[128]

(Continued)

TABLE 9.4 (Continued)

PEM-Related Properties of Different SPIs Developed by Banerjee et al.

Polymer	IEC$_W$ (meq g^{-1})	WU$_W$ (%) at 80°C	SR (in-plane) (%) at 80°C	Oxidative Stability (h)	σ (80°C) (mS cm^{-1})	Ref.
DFN-70	2.15	33	15.2	>6.5	115	[134]
CDFN-70-I	2.15	29	10.8	>20	99	[134]
CDFN-70-II	2.15	26	8	>24	90	[134]
DFN-75	2.34	44	20.3	>4	128	[134]
DFN-80	2.52	51	24.5	>3.5	139	[134]

Note
^{V} Through plane swelling

oxidative stability required for their PEM application. Another series of SPIs (DHN-XX) copolymers (Scheme 9.29) was developed by the same group, and the fabricated membranes exhibited better oxidative stability than DAN-XX membranes [127]. The higher hydrolytic stability of DHN-XX membranes than DAN-XX membrane was observed because of the development of hydrogen bonds between the −OH groups (phenolic) of DHN polymers and water molecules. The DHN-XX membranes showed higher PC than the analogous DAN-XX (nonhydroxylated equivalents) membranes. The PC (Table 9.4) of DHN-70 copolymer augmented by 30% compared to the DAN-70 copolymer, and this was due to the attendance of −OH groups (phenolic) that generate extra hydrogen bonding with water molecules. The peroxide radical stability of DHN-XX membranes was increased considerably than their nonhydroxylated equivalents. The conjugation effect of phenolic hydroxy groups (Figure 9.19) could be the likely reason for increasing the oxidative stability of these polymers. Later, a series of fluorinated SPIs with pendant benzyl ether groups (DBN-XX, Scheme 9.29) was developed [128], and their PEM-related properties (Table 9.4) were investigated.

9.4.3.2   Phosphorus-Containing Polyimides

The incorporation of phosphorus-containing groups in SPIs is special in developing PEM materials. Banerjee et al. developed several phosphine oxides containing polymers [130–132]. The triphenylphosphine oxide moiety containing co-SPIs (TPPO-XX, Scheme 9.30) [130] created flexible membranes with high thermal and mechanical stability (327°C–363°C). The membranes showed high oxidative and moderate PC (Table 9.4) with low WU and SR. The TPPO-60 membrane (IEC$_W$: 1.84 meq g^{-1}) displayed PC of 99 mS cm^{-1} (80°C) and 107 mS cm^{-1} (90°C) in hydrate state with high oxidative stability (20 h).

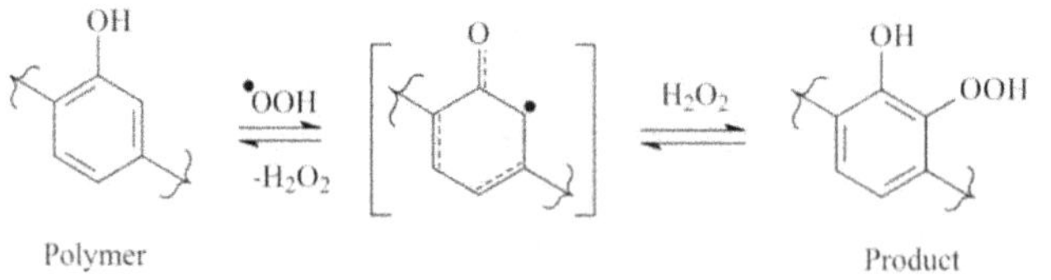

FIGURE 9.19   Stabilization of peroxide radical by −OH groups (phenolic). Reprinted with permission from Ref. [127]. Copyright (2018) American Chemical Society.

Another series of semi-fluorinated SPIs (DFPNH-XX, Scheme 9.30) was developed by his group by means of a trifluoromethyl and phosphine oxide containing diamine monomer [131]. The presence of 3 F and phosphine oxide moieties in the polymers resulted in excellent thermal and mechanical stabilities and formed ionic nanochannels for better proton conduction. Accordingly, the DFPNH-XX copolyimide membranes showed high PC (Table 9.4) and sensibly high oxidative and hydrolytic stability. The DFPNH-90 (IEC$_W$: 2.86 meq g^{-1}) membrane exhibited PC up to 237 mS cm^{-1} (80°C).

The phosphinate -(O=) P-O-C linkage in a cyclic form in DOPO is attractive due to their superior thermal properties, low flammability, organo-solubility. Subsequently, another co-SPIs (DPPNH-XX, Scheme 9.30) with phosphaphenanthrene units were developed by Banerjee et al. [132]. All the synthesized polymers were revealed microphase separated morphology with well-dispersed hydrophilic and hydrophobic phases. The membranes demonstrated high PC and oxidative stability, and low WU and SR (Table 9.4). The DPPNH-80 membrane (IEC$_W$: 2.58 meq g^{-1}) displayed PC as high as 202 mS cm^{-1} (80°C) in a fully hydrated condition.

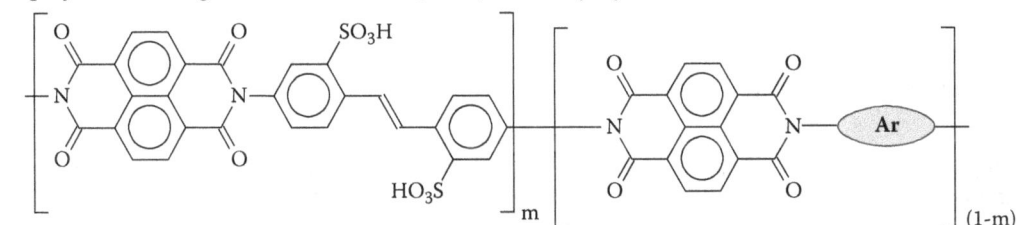

Sulfonated polyimide

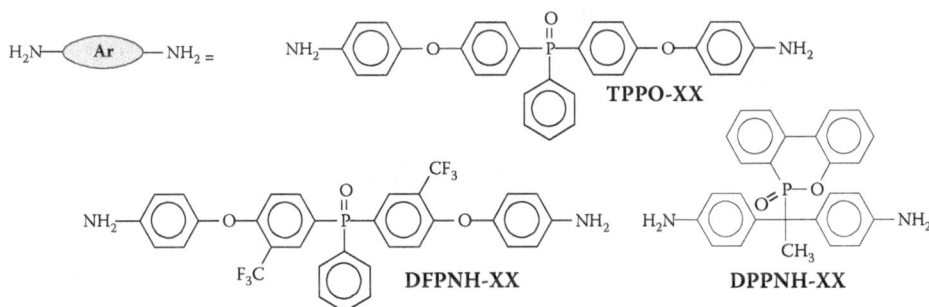

SCHEME 9.30 Synthesis of phosphorus-containing SPIs [130–132].

9.4.3.3 Polyimide-Based Composites

Banerjee et al. reported several cross-linked and composite polyimide membranes for PEM applications [133–135]. A series of hybrid membranes (SPI/S-X, Figure 9.20) were developed from fluorinated copolyimide and colloidal silica particles (~15 nm) [135]. The SPI/S-X membranes exhibited improved mechanical performances (maximum TS: 95 MPa and YM: 2.92 GPa) with a certain wt% of silica loading than pristine SPI membrane. The WU and SR (thickness) values at 80°C of SPI/S-X (X = 0, 2.5, 5.0, 7.5 and 10.0) were ranged in 10.42%– 23.66%, 3.4%–11%, respectively. All the cross-linked membranes showed peroxide stability of >24 h. The maximum PC was achieved from SPI/S-2.5 membrane, and the values were 50.3 mS cm^{-1} at 30°C and 114.8 mS cm^{-1} at 90°C. These results suggested hybrid SPI membranes with cross-linked network structures could be potential candidates for PEM applications.

A –COOH comprising sulfonated copolymer (SPI-COOH-70, Scheme 9.31) was prepared, and a different weight% of a 3-(trihydroxysilyl) propane-1-sulfonic acid was added to it to prepare the

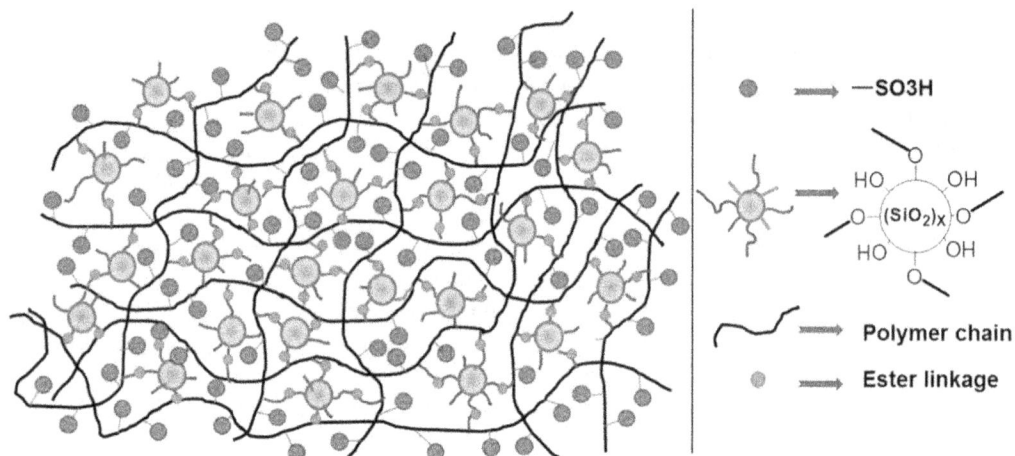

FIGURE 9.20 Cross-linked structure of SPI/silica hybrid membrane (SPI/S-X). Adapted from Ref. [135] with permission from The Royal Society of Chemistry.

composite membranes (SPI/SS-X) via in-situ sol-gel reaction [133]. The PEM properties such as morphology, IEC, WU, and PC of the fabricated composite membranes were investigated with loading of SiOPS filler and temperature. The SPI/SS-X hybrid membranes exhibited excellent thermal and chemical robustness because of the existence of polar phosphinc oxide moieties in the aromatic-based polymer. The SPI/SS-10 membrane demonstrated PC of 205 mS cm^{-1}, greater than (almost two times) pristine SPI-COOH-70 (114 mS cm^{-1}) under comparable test conditions. The progress of PC could be recognised to the pendant sulfonic acid groups in the nano-filler, providing a better proton transportation pathway.

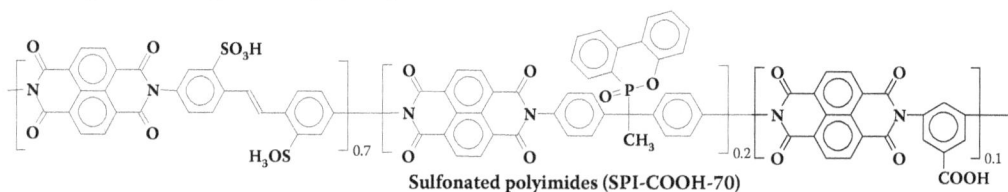

Sulfonated polyimides (SPI-COOH-70)

SCHEME 9.31 Structure of SPI-COOH-70 for the preparation of SPI/SS-X composite membranes [133].

9.5 SUMMARY

Problems related to the energy crisis and environmental pollution have become a matter of considerable concern. PEMFC technology is considered as a possible solution to overcome the mentioned issues and has contemplated as an important technology for future in producing clean energy. The benefits of fuel cells over traditional energy conversion devices are their high efficiency and "green energy" production. The synthesis of sulfonated polymers and their effectiveness as PEMs have been extensively explored by a number of research groups for the development of suitable PEM materials as an alternative of the presently used perfluorosulfonic acid membranes. The purpose of this chapter was to introduce recent advances in aromatic-based PEM materials, including polymer synthesis, properties, and performance as PEM. Some significant points in this chapter are: (a) Combining locally high-density hydrophilic units (mainly –SO$_3$H groups) with fluorinated hydrophobic components in poly(arylene ether) backbone can effectively balance the dimensional stability and proton transportation in hydrated state. (b) The presence of higher fluorene contents [3 F, 6 F and F groups] in the polymeric chain provided well phase-separated morphology, excellent peroxide stability, sufficient WU and SR, and high PC. (c) Incorporation of phosphine oxide systems in polymers [poly(arylene ether)s, polyimides, and polytriazoles] enhanced the overall PEM properties (mainly, water management, oxidative stability and PC). (d) The bulky pendants and/or bulky rigid structural systems into the aromatic sulfonated polymeric backbone triggered the increase of free volume and inhibition of polymer's chain packing, which would further enhance the polymer's solubility and PC. (e) Cross-linking has been used to improve thermal and mechanical properties by creating a well-network construction to decrease WU and MP, but in most cases, it decreases the PC. Considering the PEM properties of different classes of polymers discussed in this chapter, the sulfonated polytriazoles can be considered as an effective alternative PEM material to Nafion®.

ACKNOWLEDGMENT

The funding received from Wacker Metroark Chemicals Pvt. Ltd. (Project No. WACKER-IIT Kgp 2020/03) is gratefully acknowledged.

REFERENCES

[1] Peighambardoust, S. J., Rowshanzamir, S., and Amjadi, M. (2010). Review of the proton exchange membranes for fuel cell applications. *International Journal of Hydrogen Energy*, 35(17), 9349–9384.

[2] Ahmad, M. I., Zaidi, S. M. J., and Rahman, S. U. (2006). Proton conductivity and characterization of novel composite membranes for medium-temperature fuel cells. *Desalination*, 193(1–3), 387–397.

[3] Costamagna, P., and Srinivasan, S. (2001). Quantum jumps in the PEMFC science and technology from the 1960s to the year 2000: Part II. Engineering, technology development and application aspects. *Journal of Power Sources*, 102(1–2), 253–269.

[4] Appleby, A. J., and Foulkes, F.R. (1988). *Fuel cell handbook*. New York:Van Nostrand Reinhold.

[5] Smitha, B., Sridhar, S., and Khan, A. A. (2005). Solid polymer electrolyte membranes for fuel cell applications—a review. *Journal of Membrane Science*, 259(1–2), 10–26.

[6] Banerjee, S., Ghosh, A. (2017). Semifluorinated aromatic polymers and their properties; *Fluorinated polymers*; B. Ameduri, H. Sawada, Eds.; Cambridge, UK: The Royal Society of Chemistry; Vol. 1, pp. 103–188.

[7] Miyatake, K., Oyaizu, K., Tsuchida, E., and Hay, A. S. (2001). Synthesis and properties of novel sulfonated arylene ether/fluorinated alkane copolymers. *Macromolecules*, 34(7), 2065–2071.

[8] Kobayashi, T., Rikukawa, M., Sanui, K., and Ogata, N. (1998). Proton-conducting polymers derived from poly (ether-etherketone) and poly (4-phenoxybenzoyl-1, 4-phenylene). *Solid State Ionics*, 106(3–4), 219–225.

[9] Ghassemi, H., and McGrath, J. E. (2004). Synthesis and properties of new sulfonated poly (p-phenylene) derivatives for proton exchange membranes. I. *Polymer*, 45(17), 5847–5854.

[10] Liu, B., Robertson, G. P., Kim, D. S., Guiver, M. D., Hu, W., and Jiang, Z. (2007). Aromatic poly (ether ketone) s with pendant sulfonic acid phenyl groups prepared by a mild sulfonation method for proton exchange membranes. *Macromolecules*, 40(6), 1934–1944.

[11] Mukherjee, R., Mohanty, A. K., Banerjee, S., Komber, H., and Voit, B. (2013). Phthalimidine based fluorinated sulfonated poly (arylene ether sulfone) s copolymer proton exchange membranes. *Journal of Membrane Science*, 435, 145–154.

[12] Mukherjee, R., Banerjee, S., Komber, H., and Voit, B. (2016). Carboxylic acid functionalized fluorinated sulfonated poly (arylene ether sulfone) copolymers with enhanced oxidative stability. *Journal of Membrane Science*, 510, 497–509.

[13] Mukherjee, R., Banerjee, S., Komber, H., and Voit, B. (2014). Highly proton conducting fluorinated sulfonated poly (arylene ether sulfone) copolymers with side chain grafting. *RSC Advances*, 4(87), 46723–46736.

[14] Mohanty, A. K., Mistri, E. A., Ghosh, A., and Banerjee, S. (2012). Synthesis and characterization of novel fluorinated poly (arylene ether sulfone) s containing pendant sulfonic acid groups for proton exchange membrane materials. *Journal of Membrane Science*, 409, 145–155.

[15] Mohanty, A. K., Mistri, E. A., Banerjee, S., Komber, H., and Voit, B. (2013). Highly fluorinated sulfonated poly (arylene ether sulfone) copolymers: Synthesis and evaluation of proton exchange membrane properties. *Industrial & Engineering Chemistry Research*, *52*(8), 2772–2783.

[16] Mohanty, A. K., Banerjee, S., Komber, H., and Voit, B. (2014). Imidoaryl biphenol based new fluorinated sulfonated poly (arylene ether sulfone) copolymers and their proton exchange membrane properties. *Solid State Ionics*, *254*, 82–91.

[17] Tanaka, M., Fukasawa, K., Nishino, E., Yamaguchi, S., Yamada, K., Tanaka, H., Bae, B., Miyatake, K. and Watanabe, M. (2011). Anion conductive block poly (arylene ether) s: synthesis, properties, and application in alkaline fuel cells. *Journal of the American Chemical Society*, *133*(27), 10646–10654.

[18] Jones, D. J., and Rozière, J. (2001). Recent advances in the functionalisation of polybenzimidazole and polyetherketone for fuel cell applications. *Journal of Membrane Science*, *185*(1), 41–58.

[19] Li, J., Cai, W., Zhang, Y., Chen, Z., Xu, G., and Cheng, H. (2015). Novel polyamide proton exchange membranes with bi-functional sulfonimide bridges for fuel cell applications. *Electrochimica Acta*, *151*, 168–176.

[20] Mistri, E. A., Mohanty, A. K., and Banerjee, S. (2012). Synthesis and characterization of new fluorinated poly (ether imide) copolymers with controlled degree of sulfonation for proton exchange membranes. *Journal of Membrane Science*, *411*, 117–129.

[21] Mistri, E. A., Mohanty, A. K., Banerjee, S., Komber, H., and Voit, B. (2013). Naphthalene dianhydride based semifluorinated sulfonated copoly (ether imide)s: Synthesis, characterization and proton exchange properties. *Journal of Membrane Science*, *441*, 168–177.

[22] Sarkar, P., Mohanty, A. K., Bandyopadhyay, P., Chattopadhyay, S., and Banerjee, S. (2014). Proton exchange properties of flexible diamine-based new fluorinated sulfonated polyimides. *RSC Advances*, *4*(23), 11848–11858.

[23] Mistri, E. A., Banerjee, S., Komber, H., and Voit, B. (2015). Structure–property correlation of semifluorinated 6-membered co-SPIs for proton exchange membrane. *European Polymer Journal*, *73*, 466–479.

[24] Singh, A., Mukherjee, R., Banerjee, S., Komber, H., and Voit, B. (2014). Sulfonated polytriazoles from a new fluorinated diazide monomer and investigation of their proton exchange properties. *Journal of Membrane Science*, *469*, 225–237.

[25] Singh, A., Banerjee, S., Komber, H., and Voit, B. (2016). Synthesis and characterization of highly fluorinated sulfonated polytriazoles for proton exchange membrane application. *RSC Advances*, *6*(16), 13478–13489.

[26] Saha, S., Mukherjee, R., Singh, A., and Banerjee, S. (2017). Synthesis, characterization and investigation of proton exchange properties of sulfonated polytriazoles from a new semifluorinated diazide monomer. *Polymer Engineering & Science*, *57*(3), 312–323.

[27] Saha, S., Banerjee, S., Komber, H., and Voit, B. (2017). Flexible diazide based sulfonated polytriazoles and their proton exchange membrane properties. *Macromolecular Chemistry and Physics*, *218*(14), 1700070.

[28] Saha, S., Kumar, A. G., Noori, M. T., Banerjee, S., Ghangrekar, M. M., Komber, H., and Voit, B. (2018). New cross-linked sulfonated polytriazoles: Proton exchange properties and microbial fuel cell performance. *European Polymer Journal*, *103*, 322–334.

[29] Singh, A., and Banerjee, S. (2018). Synthesis and characterization of highly proton conducting sulfonated polytriazoles. *ACS Omega*, *3*(8), 9620–9629.

[30] Singh, A., Kumar, A. G., Bisoi, S., Saha, S., and Banerjee, S. (2017). Hexafluoroisopropylidene based sulfonated new copolytriazoles: Investigation of proton exchange membrane properties. *E-Polymers*, *17*(2), 107–118.

[31] Singh, A., Bisoi, S., Banerjee, S., Komber, H., and Voit, B. (2017). Hydroquinone based sulfonated copolytriazoles with enhanced proton conductivity. *Macromolecular Materials and Engineering*, *302*(11), 1700208.

[32] Breslau, B. R., and Miller, I. F. (1971). A hydrodynamic model for electroosmosis. *Industrial & Engineering Chemistry Fundamentals*, *10*(4), 554–565.

[33] Ueki, T., and Watanabe, M. (2008) Macromolecules in ionic liquids: Progress, challenges, and opportunities. *Macromolecules*, *41*(11), 3739–3749.

[34] Kreuer, K. D., Paddison, S. J., Spohr, E., and Schuster, M. (2004). Transport in proton conductors for fuel-cell applications: Simulations, elementary reactions, and phenomenology. *Chemical Reviews*, *104*(10), 4637–4678.

[35] Ghosh, A., and Banerjee, S. (2014). Sulfonated fluorinated-aromatic polymers as proton exchange membranes. *e-Polymers*, *14*(4), 227–257.

[36] Noshay, A., and McGrath, J. E. (1977). *Block copolymers: overview and critical survey*. Academic Press, New York.

[37] Banerjee, S. (2015). *Handbook of specialty fluorinated polymers: Preparation, properties, and applications.*, 1st Edition, Elsevier.

[38] Dhara, M. G., and Banerjee, S. (2010). Fluorinated high-performance polymers: Poly (arylene ether) s and aromatic polyimides containing trifluoromethyl groups. *Progress in Polymer Science*, *35*(8), 1022–1077.

[39] Maier, G. (2001). Low dielectric constant polymers for microelectronics. *Progress in Polymer Science*, *26*(1), 3–65.

[40] Xing, P., Robertson, G. P., Guiver, M. D., Mikhailenko, S. D., and Kaliaguine, S. (2004). Sulfonated poly (aryl ether ketone)s containing the hexafluoroisopropylidene diphenyl moiety prepared by direct copolymerization, as proton exchange membranes for fuel cell application. *Macromolecules*, *37*(21), 7960–7967.

[41] Kreuer, K. D. (2001). On the development of proton conducting polymer membranes for hydrogen and methanol fuel cells. *Journal of Membrane Science*, *185*(1), 29–39.

[42] Liu, B., Robertson, G. P., Guiver, M. D., Sun, Y. M., Liu, Y. L., Lai, J. Y., Mikhailenko, S. and Kaliaguine, S. (2006). Sulfonated poly (aryl ether ether ketone ketone)s containing fluorinated moieties as proton exchange membrane materials. *Journal of Polymer Science Part B: Polymer Physics*, *44*(16), 2299–2310.

[43] Jeong, M. H., Lee, K. S., Hong, Y. T., and Lee, J. S. (2008). Selective and quantitative sulfonation of poly (arylene ether ketone)s containing pendant phenyl rings by chlorosulfonic acid. *Journal of Membrane Science*, *314*(1-2), 212–220.

[44] Mauritz, K. A., and Moore, R. B. (2004). State of understanding of Nafion. *Chemical Reviews*, *104*(10), 4535–4586.

[45] Tian, S., Meng, Y., and Hay, A. S. (2009). Membranes from poly (aryl ether)-based ionomers containing multiblock segments of randomly distributed nanoclusters of 18 sulfonic acid groups. *Journal of Polymer Science Part A: Polymer Chemistry*, *47*(18), 4762–4773.

[46] Matsumura, S., Hlil, A. R., Lepiller, C., Gaudet, J., Guay, D., and Hay, A. S. (2008). Ionomers for proton exchange membrane fuel cells with sulfonic acid groups on the end groups: Novel linear aromatic poly (sulfide– ketone) s. *Macromolecules*, *41*(2), 277–280.

[47] Matsumura, S., Hlil, A. R., Lepiller, C., Gaudet, J., Guay, D., Shi, Z., Holdcroft, S. and Hay, A. S. (2008). Ionomers for proton exchange membrane fuel cells with sulfonic acid groups on the end groups: Novel branched poly (ether– ketone) s. *Macromolecules*, *41*(2), 281–284.

[48] Norsten, T. B., Guiver, M. D., Murphy, J., Astill, T., Navessin, T., Holdcroft, S., Frankamp, B. L., Rotello, V. M. and Ding, J. (2006). Highly fluorinated comb-shaped copolymers as proton exchange membranes (PEMs): Improving PEM properties through rational design. *Advanced Functional Materials*, *16*(14), 1814–1822.

[49] Li, N., Hwang, D. S., Lee, S. Y., Liu, Y. L., Lee, Y. M., and Guiver, M. D. (2011). Densely sulfophenylated segmented copoly (arylene ether sulfone) proton exchange membranes. *Macromolecules*, *44*(12), 4901–4910.

[50] Li, N., Wang, C., Lee, S. Y., Park, C. H., Lee, Y. M., and Guiver, M. D. (2011). Enhancement of proton transport by nanochannels in comb-shaped copoly (arylene ether sulfone)s. *Angewandte Chemie*, *123*(39), 9324–9327.

[51] Lafitte, B., and Jannasch, P. (2007). Proton-conducting aromatic polymers carrying hypersulfonated side chains for fuel cell applications. *Advanced Functional Materials*, *17*(15), 2823–2834.

[52] Jutemar, E. P., Takamuku, S., and Jannasch, P. (2011). Sulfonated poly (arylene ether sulfonc) ionomers containing di-and tetrasulfonated arylene sulfone segments. *Polymer Chemistry*, *2*(1), 181–191.

[53] Kim, D. S., Robertson, G. P., and Guiver, M. D. (2008). Comb-shaped poly (arylene ether sulfone)s as proton exchange membranes. *Macromolecules*, *41*(6), 2126–2134.

[54] Pang, J., Shen, K., Ren, D., Feng, S., and Jiang, Z. (2013). Polyelectrolyte based on tetra-sulfonated poly (arylene ether)s for direct methanol fuel cell. *Journal of Power Sources*, *226*, 179–185.

[55] Wang, F., Chen, T., and Xu, J. (1998). Sodium sulfonate-functionalized poly (ether ether ketone)s. *Macromolecular Chemistry and Physics*, *199*(7), 1421–1426.

[56] Pang, J., Jin, X., Wang, Y., Feng, S., Shen, K., and Wang, G. (2015). Fluorinated poly (arylene ether ketone) containing pendent hexasulfophenyl for proton exchange membrane. *Journal of Membrane Science*, *492*, 67–76.

[57] Wang, S., Zhuang, H., Shobha, H. K., Glass, T. E., Sankarapandian, M., Ji, Q., Shultz, A. R., and McGrath, J. E. (2001). Miscibility of poly (arylene phosphine oxide) systems and bisphenol a poly (hydroxy ether). *Macromolecules*, *34*(23), 8051–8063.

[58] Liu, D., Liao, H., Tan, N., Xiao, G., and Yan, D. (2011). Sulfonated poly (arylene thioether phosphine oxide)/sulfonated benzimidazole blends for proton exchange membranes. *Journal of Membrane Science*, *372*(1–2), 125–133.

[59] Fu, L., Xiao, G., and Yan, D. (2012). High performance sulfonated poly (arylene ether phosphine oxide) membranes by self-protected cross-linking for fuel cells. *Journal of Materials Chemistry*, *22*(27), 13714–13722.

[60] Ma, X., Shen, L., Zhang, C., Xiao, G., Yan, D., and Sun, G. (2008). Sulfonated poly (arylene thioether phosphine oxide)s copolymers for proton exchange membrane fuel cells. *Journal of Membrane Science*, *310*(1-2), 303–311.

[61] Ma, X., Zhang, C., Xiao, G., Yan, D., and Sun, G. (2008). Synthesis and characterization of sulfonated poly (phthalazinone ether phosphine oxide)s by direct polycondensation for proton exchange membranes. *Journal of Polymer Science Part A: Polymer Chemistry*, *46*(5), 1758–1769.

[62] Fu, L., Xiao, G., and Yan, D. (2010). Sulfonated poly (arylene ether sulfone) s with phosphine oxide moieties: A promising material for proton exchange membranes. *ACS Applied Materials & Interfaces*, *2*(6), 1601–1607.

[63] Gui, L., Zhang, C., Kang, S., Tan, N., Xiao, G., and Yan, D. (2010). Synthesis and properties of hexafluoroisopropylidene-containing sulfonated poly (arylene thioether phosphine oxide)s for proton exchange membranes. *International Journal of Hydrogen Energy*, *35*(6), 2436–2445.

[64] Fu, L., Liao, H., Xiao, G., and Yan, D. (2012). Sulfonated poly (arylene ether)s with high content of phosphine oxide moieties for proton exchange membranes. *Journal of Membrane Science*, *389*, 407–415.

[65] Ma, X., Zhang, C., Xiao, G., and Yan, D. (2009). Synthesis and properties of sulfonated poly (arylene ether phosphine oxide)s for proton exchange membranes. *Journal of Power Sources*, *188*(1), 57–63.

[66] Liao, H., Xiao, G., and Yan, D. (2013). High performance proton exchange membranes obtained by adjusting the distribution and content of sulfonic acid side groups. *Chemical Communications*, *49*(38), 3979–3981.

[67] Liao, H., Zhang, K., Tong, G., Xiao, G., and Yan, D. (2014). Sulfonated poly (arylene ether phosphine oxide)s with various distributions and contents of pendant sulfonic acid groups synthesized by direct polycondensation. *Polymer Chemistry*, *5*(2), 412–422.

[68] Zhang, S., Liu, H., Tan, Y., Lu, C., Mao, S., Kang, L., and Liao, H. (2020). Novel trisulfonated poly (phthalazinone ether phosphine oxide)s with high dimensional stability for proton exchange membrane. *Energy & Fuels*, *34*(4), 4999–5005.

[69] Sankir, M., Kim, Y. S., Pivovar, B. S., and McGrath, J. E. (2007). Proton exchange membrane for DMFC and H2/air fuel cells: Synthesis and characterization of partially fluorinated disulfonated poly (arylene ether benzonitrile) copolymers. *Journal of Membrane Science*, *299*(1–2), 8–18.

[70] Chikashige, Y., Chikyu, Y., Miyatake, K., and Watanabe, M. (2005). Poly (arylene ether) ionomers containing sulfofluorenyl groups for fuel cell applications. *Macromolecules*, *38*(16), 7121–7126.

[71] Zhong, S., Liu, C., Dou, Z., Li, X., Zhao, C., Fu, T., and Na, H. (2006). Synthesis and properties of sulfonated poly (ether ether ketone ketone) containing tert-butyl groups as proton exchange membrane materials. *Journal of Membrane Science*, *285*(1–2), 404–411.

[72] Kumar Mohanty, A., Kumar Sen, S., and Banerjee, S. (2011). Processable high Tg high strength fluorinated new poly (arylene ether)s containing imido aryl group. *Journal of Applied Polymer Science*, *122*(5), 3038–3047.

[73] Gao, Y., Robertson, G. P., Kim, D. S., Guiver, M. D., Mikhailenko, S. D., Li, X., and Kaliaguine, S. (2007). Comparison of PEM properties of copoly (aryl ether ether nitrile)s containing sulfonic acid bonded to naphthalene in structurally different ways. *Macromolecules*, *40*(5), 1512–1520.

[74] Zhang, Y., Cui, Z., Zhao, C., Shao, K., Li, H., Fu, T., Na, H., and Xing, W. (2009). Synthesis and characterization of novel sulfonated poly (arylene ether ketone) copolymers with pendant carboxylic acid groups for proton exchange membranes. *Journal of Power Sources*, *191*(2), 253–258.

[75] Lee, H. S., Badami, A. S., Roy, A., and McGrath, J. E. (2007). Segmented sulfonated poly (arylene ether sulfone)-b-polyimide copolymers for proton exchange membrane fuel cells. I. Copolymer synthesis and fundamental properties. *Journal of Polymer Science Part A: Polymer Chemistry*, *45*(21), 4879–4890.

[76] Roy, A., Hickner, M. A., Yu, X., Li, Y., Glass, T. E., and McGrath, J. E. (2006). Influence of chemical composition and sequence length on the transport properties of proton exchange membranes. *Journal of Polymer Science Part B: Polymer Physics*, *44*(16), 2226–2239.

[77] Lee, H. S., Roy, A., Lane, O., Dunn, S., and McGrath, J. E. (2008). Hydrophilic–hydrophobic multiblock copolymers based on poly (arylene ether sulfone) via low-temperature coupling reactions for proton exchange membrane fuel cells. *Polymer*, *49*(3), 715–723.

[78] Bae, B., Hoshi, T., Miyatake, K., and Watanabe, M. (2011). Sulfonated block poly (arylene ether sulfone) membranes for fuel cell applications via oligomeric sulfonation. *Macromolecules*, *44*(10), 3884–3892.

[79] Guo, R., Lane, O., VanHouten, D., and McGrath, J. E. (2010). Synthesis and characterization of phenolphthalein-based poly (arylene ether sulfone) hydrophilic– hydrophobic multiblock copolymers for proton exchange membranes. *Industrial & Engineering Chemistry Research*, *49*(23), 12125–12134.

[80] Chen, Y., Lee, C. H., Rowlett, J. R., and McGrath, J. E. (2012). Synthesis and characterization of multiblock semi-crystalline hydrophobic poly (ether ether ketone)–hydrophilic disulfonated poly (arylene ether sulfone) copolymers for proton exchange membranes. *Polymer*, *53*(15), 3143–3153.

[81] Titvinidze, G., Kreuer, K. D., Schuster, M., de Araujo, C. C., Melchior, J. P., and Meyer, W. H. (2012). Proton conducting phase-separated multiblock copolymers with sulfonated poly (phenylene sulfone) blocks for electrochemical applications: Preparation, morphology, hydration behavior, and transport. *Advanced Functional Materials*, *22*(21), 4456–4470.

[82] Takamuku, S., and Jannasch, P. (2012). Multiblock copolymers with highly sulfonated blocks containing di-and tetrasulfonated arylene sulfone segments for proton exchange membrane fuel cell applications. *Advanced Energy Materials*, *2*(1), 129–140.

[83] Takamuku, S., and Jannasch, P. (2012). Multiblock copolymers containing highly sulfonated poly (arylene sulfone) blocks for proton conducting electrolyte membranes. *Macromolecules*, *45*(16), 6538–6546.

[84] Jung, M. S., Kim, T. H., Yoon, Y. J., Kang, C. G., Yu, D. M., Lee, J. Y., Kin, H. J. and Hong, Y. T. (2014). Sulfonated poly (arylene sulfone) multiblock copolymers for proton exchange membrane fuel cells. *Journal of Membrane Science*, *459*, 72–85.

[85] Assumma, L., Nguyen, H. D., Iojoiu, C., Lyonnard, S., Mercier, R., and Espuche, E. (2015). Effects of block length and membrane processing conditions on the morphology and properties of perfluorosulfonated poly (arylene ether sulfone) multiblock copolymer membranes for PEMFC. *ACS Applied Materials & Interfaces*, *7*(25), 13808–13820.

[86] Kerres, J. A. (2001). Development of ionomer membranes for fuel cells. *Journal of Membrane Science*, *185*(1), 3–27.

[87] Rikukawa, M., and Sanui, K. (2000). Proton-conducting polymer electrolyte membranes based on hydrocarbon polymers. *Progress in Polymer Science*, *25*(10), 1463–1502.

[88] Xing, P., Robertson, G. P., Guiver, M. D., Mikhailenko, S. D., Wang, K., and Kaliaguine, S. (2004). Synthesis and characterization of sulfonated poly (ether ether ketone) for proton exchange membranes. *Journal of Membrane Science*, *229*(1–2), 95–106.

[89] Pang, J., Zhang, H., Li, X., and Jiang, Z. (2007). Novel wholly aromatic sulfonated poly (arylene ether) copolymers containing sulfonic acid groups on the pendants for proton exchange membrane materials. *Macromolecules*, *40*(26), 9435–9442.

[90] Wang, F., Hickner, M., Kim, Y. S., Zawodzinski, T. A., and McGrath, J. E. (2002). Direct polymerization of sulfonated poly (arylene ether sulfone) random (statistical) copolymers: candidates for new proton exchange membranes. *Journal of Membrane Science*, *197*(1–2), 231–242.

[91] Shobha, H. K., Smalley, G. R., Sankarapandian, M., and McGrath, J. E. (2000). Synthesis and characterization of sulfonated poly (arylene ether) s based on functionalized triphenyl phosphine oxide for proton exchange membranes. *Polym. Prepr*, *41*, 180–181.

[92] Roustan, J. L., Ansari, N., Lee, F., and Charland, J. P. (1989). Redox reactions involving D7+2NO (Fe)dinitrosyl iron complexes. Oxidation of 2-pyridyldiphenylphosphine into the corresponding phosphine oxide chelating a D6+0NO iron. *Inorg. Chim. Acta*, 155 (1) 11–12.

[93] Valle, G., Casotto, G., Zanonato, P. L., and Zarli, B. (1986). Crystal and molecular structures of two europium (III) nitrate complexes with triphenylphosphine oxide. *Polyhedron*, *5*(12), 2093–2096.

[94] Ghorai, A., Mandal, A. K., and Banerjee, S. (2020). Synthesis and characterization of new phosphorus containing sulfonated polytriazoles for proton exchange membrane application. *Journal of Polymer Science*, *58*(2), 263–279.

[95] Miyake, J., Sakai, M., Sakamoto, M., Watanabe, M., and Miyatake, K. (2015). Synthesis and properties of sulfonated block poly (arylene ether) s containing m-terphenyl groups as proton conductive membranes. *Journal of Membrane Science*, *476*, 156–161.

[96] Chikashige, Y., Chikyu, Y., Miyatake, K., and Watanabe, M. (2006). Branched and cross-linked proton conductive poly (arylene ether sulfone) ionomers: Synthesis and properties. *Macromolecular Chemistry and Physics*, *207*(15), 1334–1343.

[97] Moy, T. M., DePorter, C. D., and McGrath, J. E. (1993). Synthesis of soluble polyimides and functionalized imide oligomers via solution imidization of aromatic diester-diacids and aromatic diamines. *Polymer*, *34*(4), 819–824.

[98] Jayaraman, S., Srinivasan, R., and McGrath, J. E. (1995). Synthesis and characterization of 3-phenylethynyl endcapped matrix resins. *Journal of Polymer Science Part A: Polymer Chemistry*, *33*(10), 1551–1563.

[99] Lee, K. S., Jeong, M. H., Lee, J. P., and Lee, J. S. (2009). End-group cross-linked poly (arylene ether) for proton exchange membranes. *Macromolecules*, *42*(3), 584–590.

[100] Lee, K. S., Jeong, M. H., Lee, J. P., Kim, Y. J., and Lee, J. S. (2010). Synthesis and characterization of highly fluorinated cross-linked aromatic polyethers for polymer electrolytes. *Chemistry of Materials*, *22*(19), 5500–5511.

[101] Miyatake, K., Zhou, H., Matsuo, T., Uchida, H., and Watanabe, M. (2004). Proton conductive polyimide electrolytes containing trifluoromethyl groups: Synthesis, properties, and DMFC performance. *Macromolecules*, *37*(13), 4961–4966.

[102] Lee, K. S., Jeong, M. H., Lee, J. S., Pivovar, B. S., and Kim, Y. S. (2010). Optimizing end-group cross-linkable polymer electrolytes for fuel cell applications. *Journal of Membrane Science*, *352*(1-2), 180–188.

[103] Jeong, M. H., Lee, K. S., and Lee, J. S. (2009). Cross-linking density effect of fluorinated aromatic polyethers on transport properties. *Macromolecules*, *42*(5), 1652–1658.

[104] Adjemian, K. T., Srinivasan, S., Benziger, J., and Bocarsly, A. B. (2002). Investigation of PEMFC operation above 100 C employing perfluorosulfonic acid silicon oxide composite membranes. *Journal of Power Sources*, *109*(2), 356–364.

[105] Zhang, Q., Liu, H., Li, X., Xu, R., Zhong, J., Chen, R., and Gu, X. (2016). Synthesis and characterization of polybenzimidazole/α-zirconium phosphate composites as proton exchange membrane. *Polymer Engineering & Science*, *56*(6), 622–628.

[106] Mukherjee, R., Mandal, A. K., and Banerjee, S. (2020). Sulfonated poly (arylene ether sulfone) functionalized polysilsesquioxane hybrid membranes with enhanced proton conductivity. *E-Polymers*, *20*(1), 430–442.

[107] Xu, K., Chanthad, C., Gadinski, M. R., Hickner, M. A., and Wang, Q. (2009). Acid-functionalized polysilsesquioxane– Nafion composite membranes with high proton conductivity and enhanced selectivity. *ACS Applied Materials & Interfaces*, *1*(11), 2573–2579.

[108] Ponomarev, I. I., Zharinova, M. Y., Petrovskii, P. V., and Klemenkova, Z. S. (2009). New poly (1, 2, 3-triazolesulfonic acids) for proton exchange membranes of fuel cell. *Doklady Chemistry*, *429* (2), 305–310.

[109] Ye, Y. S., Yen, Y. C., Cheng, C. C., Syu, Y. J., Huang, Y. J., and Chang, F. C. (2010). Polytriazole/ clay nanocomposites synthesized using in situ polymerization and click chemistry. *Polymer*, *51*(2), 430–436.

[110] Banerjee, S., Ghorai, A., Roy, S., and Voit, B. (2020). *Preparation of high-performance polymers by click chemistry and their membrane-based application*; Advances in Chemistry Research (ISBN: 978-1-53618-568-3), James C. Taylor, Ed.; New York: Nova Science Publishers; Vol. 64, Chapter 2.

[111] Roy, S., Saha, S., Kumar, A. G., Ghorai, A., and Banerjee, S. (2020). Synthesis and characterization of new sulfonated copolytriazoles and their proton exchange membrane properties. *Journal of Applied Polymer Science*, *137*(13), 48514.

[112] Ghorai, A., Roy, S., Das, S., Komber, H., Ghangrekar, M. M., Voit, B., and Banerjee, S. (2020). Chemically stable sulfonated polytriazoles containing trifluoromethyl and phosphine oxide moieties for proton exchange membranes. *ACS Applied Polymer Materials*, *2*(7), 2967–2979.

[113] Singh, A., Kumar, A. G., Bisoi, S., and Banerjee, S. (2017). New sulfonated copoly (triazole imide)s synthesized by a click chemistry reaction with improved oxidative stability. *New Journal of Chemistry*, *41*(14), 6849–6856.

[114] Singh, A., Kumar, A. G., Saha, S., Mukherjee, R., Bisoi, S., and Banerjee, S. (2019). Synthesis and characterization of chemically stable sulfonated copoly (triazole imide)s with high proton conductivity. *Polymer Engineering and Science*, *59*(11), 2279–2289.

[115] Huang, Y. J., Ye, Y. S., Yen, Y. C., Tsai, L. D., Hwang, B. J., and Chang, F. C. (2011). Synthesis and characterization of new sulfonated polytriazole proton exchange membrane by click reaction for direct methanol fuel cells (DMFCs). *International Journal of Hydrogen Energy*, *36*(23), 15333–15343.

[116] Kumar, A. G., Bera, D., Banerjee, S., Veerubhotla, R., and Das, D. (2016). Sulfonated poly (ether imide) s with fluorenyl and trifluoromethyl groups: Application in microbial fuel cell (MFC). *European Polymer Journal*, *83*, 114–128.

[117] Fujigaya, T., Jiang, D. L., and Aida, T. (2003). Switching of spin states triggered by a phase transition: Spin-crossover properties of self-assembled iron (II) complexes with alkyl-tethered triazole ligands. *Journal of the American Chemical Society*, *125*(48), 14690–14691.

[118] Liao, H., Zhang, K., Xiao, G., and Yan, D. (2013). High performance sulfonated poly (phthalazinone ether phosphine oxide) s for proton exchange membranes. *Journal of Membrane Science*, *447*, 43–49.

[119] Garcia, Y., Kahn, O., Rabardel, L., Chansou, B., Salmon, L., and Tuchagues, J. P. (1999). Two-step spin conversion for the three-dimensional compound tris (4, 4 '-bis-1, 2, 4-triazole) iron (II) di-perchlorate. *Inorganic Chemistry*, *38*(21), 4663–4670.

[120] Tang, J., Wan, L., Zhou, Y., Ye, L., Zhou, X., and Huang, F. (2017). Synthesis and performance study of a novel sulfonated polytriazole proton exchange membrane. *Journal of Solid State Electrochemistry*, *21*(3), 725–734.

[121] Park, H. B., Lee, C. H., Sohn, J. Y., Lee, Y. M., Freeman, B. D., and Kim, H. J. (2006). Effect of cross-linked chain length in sulfonated polyimide membranes on water sorption, proton conduction, and methanol permeation properties. *Journal of Membrane Science*, *285*(1-2), 432–443.

[122] Liu, B., Hu, W., Robertson, G. P., and Guiver, M. D. (2008). Poly (aryl ether ketone)s with carboxylic acid groups: Synthesis, sulfonation and cross-linking. *Journal of Materials Chemistry*, *18*(39), 4675–4682.

[123] Saha, S., and Banerjee, S. (2021). Effect of sulfonic acid-functionalized polysilsesquioxane on new semifluorinated sulfonated polytriazole composites and investigation of proton exchange membrane properties. *Polymer Engineering & Science*, *61*(1), 55–67.

[124] Genies, C., Mercier, R., Sillion, B., Cornet, N., Gebel, G., and Pineri, M. (2001). Soluble sulfonated naphthalenic polyimides as materials for proton exchange membranes. *Polymer*, *42*(2), 359–373.

[125] Einsla, B. R., Hong, Y. T., Seung Kim, Y., Wang, F., Gunduz, N., and McGrath, J. E. (2004). Sulfonated naphthalene dianhydride based polyimide copolymers for proton-exchange-membrane fuel cells. I. Monomer and copolymer synthesis. *Journal of Polymer Science Part A: Polymer Chemistry*, *42*(4), 862–874.

[126] Bai, H., and Ho, W. W. (2008). New poly (ethylene oxide) soft segment-containing sulfonated polyimide copolymers for high temperature proton-exchange membrane fuel cells. *Journal of Membrane Science*, *313*(1-2), 75–85.

[127] Kumar, A. G., Singh, A., Komber, H., Voit, B., Tiwari, B. R., Noori, M. T., Ghangrekar, M. M., and Banerjee, S. (2018). Novel sulfonated co-poly (ether imide) s containing trifluoromethyl, fluorenyl and hydroxyl groups for enhanced proton exchange membrane properties: application in microbial fuel cell. *ACS Applied Materials & Interfaces*, *10*(17), 14803–14817.

[128] Kumar, A. G., Saha, S., Komber, H., Tiwari, B. R., Ghangrekar, M. M., Voit, B., and Banerjee, S. (2019). Trifluoromethyl and benzyl ether side groups containing novel sulfonated co-poly (ether imide)s: Application in microbial fuel cell. *European Polymer Journal*, *118*, 451–464.

[129] Ganeshkumar, A., Bera, D., Mistri, E. A., and Banerjee, S. (2014). Triphenyl amine containing sulfonated aromatic polyimide proton exchange membranes. *European Polymer Journal*, *60*, 235–246.

[130] Mandal, A. K., Bera, D., and Banerjee, S. (2016). Sulfonated polyimides containing triphenylpho sphine oxide for proton exchange membranes. *Materials Chemistry and Physics*, *181*, 265–276.

[131] Mandal, A. K., Bisoi, S., Banerjee, S., Komber, H., and Voit, B. (2017). Sulfonated copolyimides containing trifluoromethyl and phosphine oxide moieties: Synergistic effect towards proton exchange membrane properties. *European Polymer Journal*, *95*, 581–595.

[132] Mandal, A. K., Bisoi, S., and Banerjee, S. (2019). Effect of phosphaphenanthrene skeleton in sulfonated polyimides for proton exchange membrane application. *ACS Applied Polymer Materials*, *1*(4), 893–905.

[133] Mandal, A. K., Ghorai, A., and Banerjee, S. (2020). Sulphonated polysilsesquioxane–polyimide composite membranes: Proton exchange membrane properties. *Bulletin of Materials Science*, *43*(1), 1–11.

[134] Kumar, A. G., Saha, S., Tiwari, B. R., Ghangrekar, M. M., Das, A., Mukherjee, R., and Banerjee, S. (2020). Sulfonated co-poly (ether imide) s with alkyne groups: Fabrication of cross-linked membranes and studies on PEM properties including MFC performance. *Polymer Engineering & Science*, *60*(9), 2097–2110.

[135] Mistri, E. A., and Banerjee, S. (2014). Cross-linked sulfonated poly (ether imide)/silica organic–inorganic hybrid materials: Proton exchange membrane properties. *RSC Advances*, *4*(43), 22398–22410.

ABBREVIATION

3F	Trifluoromethyl [(–CF$_3$)]
6F	Hexafluoroisopropylidine [>C(CF$_3$)$_2$]
AFM	Atomic force microscopy
BPA	4,4′-Isopropylidene diphenol
BPBA	3,5- Bis(prop-2-ynyloxy)benzoic acid
BPEBPA	4,4′-(Propane-2,2-diyl)di((prop-2-ynyloxy)benzene)
DMFCs	Direct methanol fuel cells
DFBP	4,4′-Difluorobenzophenone
DADSDB	4,4′-Diazido-2,2′-stilbenedisulfonic acid disodium salt
DSDSA	4,4′-Diaminostilbene-2,2′-disulfonic acid
DS	Degree of sulfonations
DMF	N, N- dimethylformamide
DMAc	Dimethyl acetamide
DMSO	Dimethyl sulfoxide
D	Polydispersity index
°C	Degree Celsius
EB	Elongation at break
FC	Fuel cell
6F-BPA	Hexafluoroisopropylidene diphenol
HFB	Hexafluorobenzene
η$_{inh}$	Inherent viscosity
IEC$_W$	Weight based ion exchange capacity
IB	3,8-Bis(4-hydroxyphenyl)-N-phenyl-1,2-naphthalimide
MEA	Membrane electrode assembly
M	Molar
Mn	Number average molecular weight
M$_W$	Weight average molecular weight
mA	milli Ampere
MPa	Mega Pascal
meq. g^{-1}	milliequivalent per gram
mS cm^{-1}	milliSiemens per centimeter
MTP	mTerphenyl
MP	Methanol permeability
NTDA	1,4,5,8-naphthalenetetracarboxylic dianhydride
PEMFC	Proton exchange membrane fuel cell
PEM	Proton exchange membrane
PTFE	Polytetrafluoroethylene
PC	Proton conductivity
K$_2$CO$_3$	Potassium carbonate
PVA	Poly(vinyl alcohol)
QBF	4,4'-B(4′-fluoro-3′-trifluoromethyl benzyl) biphenyl

QAZ	4,4'-Bis[3-Trifluoromethyl-4(4-azidophenoxy)phenyl]biphenyl
RH	Relative humidity
THSPSA	Sulfopropylated polysilsesquioxane
SR	Swelling ratios
STP	Sulfonated polytriazole
SDCDPS	3,3'-Disulfonate-4,4'-dichlorodiphenylsulfone
SAXS	Small angle X-ray scattering
SPIs	Sulfonated polyimides
Sigma	Proton conductivity value mS cm^{-1}
TGA	Thermogravimetric analyzer
$T_{d5\%}$	Decomposition temperatures for 5% weight loss in TGA
$T_{d10\%}$	Decomposition temperatures for 10% weight loss in TGA
T_g	Glass transition temperature
TS	Tensile strength
Tau	Time for initiation of breaking of the membrane in Fentons
T	Temperature (K)
TEM	Transmission electron microscopy
WU	Water uptake
XPS	X-ray photoelectron spectroscopy
XRD	X-ray diffraction
YM	Youngs modulus
Z	Real part of impedance
Z	Imaginary part of impedance

10 Recent Advances in Anion Exchange Membranes for Fuel Cell Applications

Vijayalekshmi Vijayakumar and Sang Yong Nam
Gyeongsang National University, Jinju, Republic of Korea

CONTENTS

10.1 INTRODUCTION

The current growing energy and environmental crisis demand viable alternative energy resources. Fossil commodities remain the highest form of energy generation but their harmful effect on the environment proven scientifically causes a sharp decline in the last decade [1]. As a new energy technology, fuel cells have been attracted much attention as one of the most promising clean and efficient power generation technologies for a sustainable future and have become a research hotspot owing to its unique advantages, such as fuel economy and ease of storage and transportation [2]. Among several types of fuel cells, anion exchange membrane fuel cells (AEMFCs) have attracted significant interest in recent years, due to their fundamental advantages of improved electrochemical kinetics especially at oxygen reduction reaction (ORR) thereby providing low activation losses, use of non-precious metal catalysts as cathode catalyst, low cell components' and membrane cost due to less corrosive operating conditions, etc., [3–5]. A typical AEMFC mainly consists of an anode, cathode and membrane electrode assembly (MEA), as shown in Figure 10.1. The MEA is fabricated by assembling an anode gas diffusion layer (GDL), followed by anode catalyst layer, anion exchange membrane (AEM), cathode catalyst layer, and a cathode gas diffusion layer. The GDLs in both electrodes are composed of two layers, a backing layer of carbon paper or carbon cloth and a microporous layer which consist of a mixture of carbon powder and hydrophobic polymer (typically PTFE). The catalyst layers (CLs) are generally made of electrocatalysts and an ionomer, resulting in the formation of triple-phase boundaries for redox reactions. The main role of a GDL is to provide support for both anode and cathode catalyst layers, distributes the reactants uniformly and transports electrons to the current collector [6,7].

Hydrogen is the most commonly used fuel in AEMFCs. Humidified hydrogen charged to the anode flow channels is transported through the anode GDL to the catalyst layer, where it reacts with hydroxide ions to produce water and electrons.

DOI: 10.1201/9781003200710-12

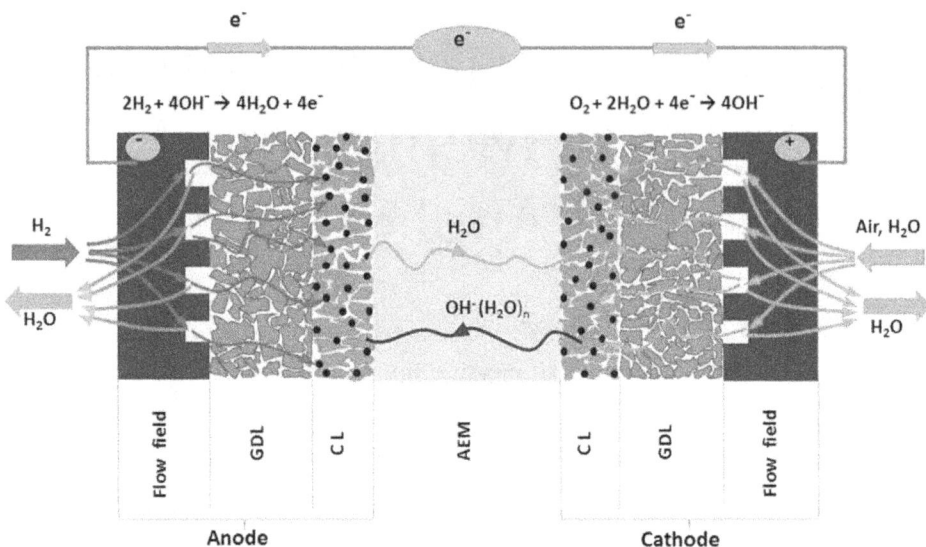

FIGURE 10.1 Schematic representation of a typical H$_2$/AEMFC [Reproduced from reference [8] with permission].

The anode reaction:

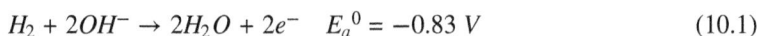

$$H_2 + 2OH^- \rightarrow 2H_2O + 2e^- \quad E_a^0 = -0.83 \ V \qquad (10.1)$$

Humidified oxygen supplied to the cathode flow channel is transported through the cathode GDL to the cathode catalyst layer, where it reduced in the presence of water to produce OH$^-$ ions:

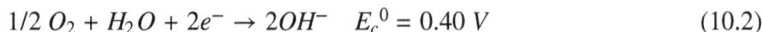

$$1/2 \ O_2 + H_2O + 2e^- \rightarrow 2OH^- \quad E_c^0 = 0.40 \ V \qquad (10.2)$$

Then the generated hydroxide ions are conducted through the AEM for the HOR. The overall cell reaction is expressed as:

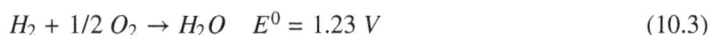

$$H_2 + 1/2 \ O_2 \rightarrow H_2O \quad E^0 = 1.23 \ V \qquad (10.3)$$

The anion exchange membrane (AEM), placed between the anode and cathode is the core component in AEMFCs. The main function of AEM is to prevent physical contact of the electrodes while facilitating free OH$^-$ ion transport (ionic conductor) and barrier for gas (fuel) and electron flow (electronic insulator). The development of AEMFC technology encounters several critical concerns, especially low ionic conductivity and relatively poor alkaline stability of polymeric AEMs and durability of the device. To support a large current with marginal resistive losses, the ion conductivity of AEM should be high enough (100 mS cm^{-1}) [9]. Increasing the number of ionic functional groups can improve ionic conductivity but results in degradation of mechanical stability as a result of excess water uptake. For that reason, a stringent control of the AEM morphology is essential to improve the mechanical stability [10]. Well-designed polymer architecture with tethered organic cations such as block, grafted/comb-shaped or ion-clustered polymer, has been established as the membrane materials with high ionic conductivity. In the aspect of alkaline stability, both polymer backbone and cationic groups present are susceptible to the attack by nucleophilic hydroxide ions under alkaline conditions, bring about the degradation of ionic groups and mechanical properties of the membranes [11]. The detailed mechanism of degradation will be discussed in the forthcoming section.

The principal task of many researchers in this field is to develop anion exchange membranes (AEMs) with high performance and a long lifespan. This book chapter mainly focuses on the efforts of developing such anion exchange membranes (AEMs) in the last five years (2016–2021), including various synthesis approaches together with relations between structure and performance properties related to anion exchange membrane fuel cells (AEMFCs).

10.2 ANION EXCHANGE MEMBRANE MATERIALS FOR FUEL CELLS

10.2.1 AEMs Based on Functionalized Polymers with Different Architecture

The anion exchange membranes with high ionic conductivity and long-term durability remain a challenge in the development of the anion exchange membranes. Conductivity is proportional to the mobility coefficient and concentration of charge carriers. The lower mobility coefficient of OH^- ions (only half) than H^+ ions in infinitely dilute solutions at room temperature leads to low hydroxide ion conductivity. Increasing the concentration of anion exchange group (ion exchange capacity) is the most straightforward method to improve hydroxide ion conductivity however, maintaining dimensional as well as alkaline stability is a challenge [12]. In view of that, a variety of polymeric membranes with advanced architectures have been reported as great potential for fuel cell applications.

In general, cationic functional groups also referred to as head groups such as quaternary ammonium, imidazolium, guanidinium, quaternary phosphonium, benzimidazolium, and metal-organic coordination ions have been employed to the polymer chains, but these groups are easily degradable under alkaline conditions and result in poor durability of AEMs. Quaternary ammonium cations are susceptible to direct nucleophilic substitution (S_N2) as well as β-hydrogen (Hofmann or E2) elimination under strong alkaline conditions at high temperature. Even though imidazolium and guanidinium cations have a stable resonant structure with delocalized charge, the durability of the membrane is still not satisfactory. Low energy barriers in guanidium promote easy degradation under intense alkaline conditions and produce urea derivatives. Ring-opening reaction of imidazole through hydroxyl ion attack at C2 position in high-pH conditions is already reported. Phosphonium salts favour ylide formation upon strong alkaline exposure by Wittig-reaction mechanism. Since quaternary phosphorus atoms are much more electron-withdrawing compared to quaternized nitrogen, which makes α-H in the adjacent alkyl groups more acidic and vulnerable to hydroxyl ion stability of quaternary phosphoniums than that of quaternary ammoniums [13]. To achieve AEMs with superior performance, many researchers focus on designing different molecular structures of polymers. For example, the introduction of spacers between the polymer backbones and head groups, block, graft, cluster type, pendant, and comb-shaped morphology, cross-linked networks, etc., have been proposed and resulted in membranes of a wide range of conductivity levels.

The four most widely studied head groups such as quaternary ammonium (QA), piperidinium (PI), imidazolium (IM) and morpholinium (MO) were functionalized on poly (2,6-dimethyl phenylene oxide) (PPO) by Lim and his coworkers to study its effect on the AEM properties. Owing to the large size of PI moiety and steric hindrance PI-PPO displayed a high ion dissociation rate but at the same time inhibited surface diffusion resulted in low conductivity at 95% RH. At the same time, MO-PPO attributed to the MO hydrogen bond helped to overcome the relative scarcity of water molecules and resulted in higher conductivity at the same condition. PI-PPO and MO-PPO showed high conductivity in water. QA-PPO exhibited high conductivity than the other membranes at 95% RH but the value is lower in water than the PI- and MO-based membranes. The less steric hindrance of ammonium group than PI and MO groups and hence comparatively slow release of bound ions is the reason for the low conductivity. π-π interactions among imidazolium groups led IM-PPO to lower water uptake and conductivity than all the other membranes but its resonance structure resulted in high alkaline stability to the membrane [14]. Chloromethylation of active phenyl group of polyetherimide (PEI), a thermoplastic, chemically stable polymer with high glass transition temperature and successive quaternization with N, N-dimethyl hexylamine, triethylamine and N-methyl morpholine confirmed thermal and mechanical stability as well as consistency under strongly alkaline conditions of 5 M KOH for 48 h at 30°C and high-temperature hydrolytic treatment [15]. Chu et al. observed that anion conductive poly (2,6-dimethyl phenylene oxide) with pendant N-cyclic quaternary ammonium cations (N,N-dimethyl-piperidinium (DMP) and 6-azonia-spiro[5.5]undecan (ASU)) prepared via Cu(I)-catalyzed click chemistry appear to show good alkaline stability. ASU-based membranes showed higher stability than DMP cation based ones. At a high concentration of 10 M NaOH, partial insolubility of membrane due to the cross-linked network formed as a result of chemical degradation through Hoffmann elimination and ring-opening by S_N2 substitution was noticed [11]. Yang et al. developed a series of polystyrene (PS)-grafted poly(2,6-dimethyl-1,4-phenylene oxide) (PPO) with pendant quaternary ammonium groups by the "grafting onto" method via a two-step atom transfer radical polymerization (ATRP) and Cu(I)-catalyzed click chemistry. The results demonstrated a distinct and well-separated ionic domain as well as phase-separated microstructure which possessed high ion exchange capacity (IEC) normalized ion conductivity as well as good dimensional stability to the AEMs. The membranes also exhibited excellent stability in 1 M NaOH at 80°C (10% drop in conductivity after 500 h of alkaline exposure) and power density of 64.4 mW cm^{-1} at 60°C when applied in fuel cells [16].

Tethering functional cation groups covalently on the polymer backbone is healthier to avoid loss of cations and improve hydroxide ion conductivity of the anion exchange membranes. Polyphosphazenes are highly flexible inorganic-organic composite polymers with a linear backbone of alternating phosphorus and nitrogen atoms, each phosphorus bearing two side organic substituents. Polyphosphazene chain carrying tetraphenylphosphonium cation due to its large steric hindrance and absence of labile α-H in benzyl group exhibited significantly enhanced alkaline stability than the polymer with benzyl quaternary phosphonium [13]. To facilitate ionic clustering and conductivity, Weiber and his coworkers designed AEMs based on poly(arylene ether sulfone)s containing randomly distributed biphenyl units tethered with precisely six imidazolium cations. Benzylic bromination followed by Menshutkin reactions formed N-methylimidazolium (NIM) and 1,2,4,5-tetramethylimidazolium (4IM) cations in the copolymer. Remarkably high concentration of aromatic cations with delocalized positive charge facilitated phase separation and ionic dissociation to improve ion conductivity of the membranes. The sterically bulky 1,2,4,5-tetramethylimidazolium (4IM) cation owing to the presence of electron-donating methyl groups was expected to impart high nucleophilic stability to the membrane. However, after 7 h of exposure in 1 M NaOH at 40°C, AEMs had lost their dimensional stability and resulted in a highly swollen insoluble hydrogels [17]. Wan et al. indicated that an orderly arrangement of the cationic group can significantly improve the hydroxyl ion migration efficiency and hence OH⁻ conductivity. They have introduced Gemini surfactant micelles ((maleic alkylene group diethylbis(octyl dimethyl bromide), abbreviated as G$_{16-2-16}$ reversed micelles) into the polymer (polysulfone (PSF)) backbone, opened up a new strategy to design high-performance AEMs. The orderly arranged quaternary ammonium groups in the revered micelles formed facilitated the hydroxyl ion migration resulted in 40–80% improvement in anionic conductivity without increasing ion exchange capacity. Only a 4% reduction in conductivity after 240 h of exposure in 10 M NaOH also supported the achievement of excellent alkaline stability [18].

The electrochemical window of an electrolyte is the voltage range in which the electrolyte is chemically stable and does not undergo any oxidation and reduction when voltage is applied. Quaternary ammonium ions, NR_4^+, are highly inert organic cations toward reduction offer a wide range of electrochemical stability window. The different alkyl substituents (chain length, type, size) have a vital role in the stability of quaternary ammonium cations. The positive charge density of the nitrogen atom in the quaternary ammonium ion can be decreased by electron-donating alkyl groups and improve its stability toward reduction reaction. As the donor-acceptor distance increases, the electron transfer rate decreases and stability increases. The tunnelling distance can be increased by implementing more sterically hindered or bulky alkyl groups, and consequently electrochemical stability of ions toward reduction increases [19]. N-methylpiperidinium linked to poly(biphenyl indole) by means of a long hexyl spacer, enhanced the mobility of N-methylpiperidinium groups and promotes hydrophilic/hydrophobic microphase separation which led to the formation of linked ion transport channels resulted in a conductivity 54 mS cm^{-1} at 20°C and 108 mS cm^{-1} at 80°C [20]. Perfluorinated comb-shaped cationic polymers having long-range ordered -CF$_2$CF$_2$-(CF$_2$CF$_2$)$_n$-CF$_2$-CF$_2$- alkaline stable super hydrophobic backbone with pendant (-CF$_2$CF$_2$SONH-) side chain terminated by long comb hydrophilic cationic groups demonstrated excellent dimensional stability. The large difference in polarity due to super hydrophobic fluorocarbon segment and hydrophilic ionicgroups facilitated well-connected ion clusters formation and exhibited a conductivity of 88.6 mS cm^{-1} at 80°C [21]. Owing to the high electronegativity of fluorine, AEMs containing fluorine could significantly improve the overall performance. Block copolymers of quaternary ammonium poly(arylene ether sulfone) from 9,9′- bis(4-hydroxyphenyl) fluorene, 4,4′-(hexafluoroisopropylidene) diphenol and 4,4′-difluorodiphenyl sulfone afterwards Friedel-Crafts chloromethylation, Menshutkin amination and alkalization exhibited good stability, lower methanol permeability as well as comparable selectivity values with Nafion – 212 suggested its applicability in AEMFCs [22]. It is noteworthy that AEM containing poly(ether sulfone)s with flexible alkyl imidazolium pendants on large planar 6,12-bis(4-hydroxyphenyl)-5,11-dihydroindolo[3,2- b]carbazole (DCP) units exhibited excellent long-term thermal as well as thermochemical stabilities. π–π intermolecular interaction of the planar DCP units stabilized the hydrophobic phase and prevented poly(ether sulfone)s main chains from hydroxyl ion attack [23]. The impact of N3 substitution of different imidazolium cations on the performance of comb-shaped poly(2,6-dimethyl-1,4-phenylene oxide) (PPO) based AEMs were systematically assessed by Wang et al. Two imidazolium with N3-adamantyl substituent as well as four imidazolium with N3-methyl and butyl-substituents were synthesized and used in their work. N3-adamantyl substituted imidazolium showed highest chemical stability than the corresponding N3-methyl and butyl substituted imidazolium and the most stability observed in 1-adamantyl-2-methyl-3-ethylimidazolium, 5 M NaOH aqueous solution at 80°C for 168 h. On the other hand, the AEM functionalized with N3-adamantyl substituted imidazolium was subjected to ring-opening degradation on imidazolium when exposed to 1 M NaOH aqueous solution at 60°C for over 48 h demonstrated a less correlation among modular cation and the corresponding AEM stability [24].

Side-chain-type AEMs are generally reported to exhibit better comprehensive properties such as conductivity, dimensional as well as alkaline stability in comparison with main-chain-type analogues. In comparison to the directly fixed ionic groups on the polymer backbone, side-chain type ionic groups exhibited a strong self-assembly tendency to form large ionic clusters as ion transfer networks. Zhang et al. developed a side-chain-type quaternized aromatic polyelectrolytes via one-step benzylation reaction of commercial poly(ether ether ketone) (PEEK), and the membrane showed high OH$^-$ conductivity of 155 mS cm^{-1}. Furthermore, a peak power density of about 391 mW cm^{-2} was achieved when assembling QBz-PEEK-76.0% AEM in a H$_2$/O$_2$ single cell operated at 70°C [25]. Side-chain-type AEM of poly(arylene ether sulfone)s bearing pendent imidazolium-functionalized polyphenylene oxide (PPO) displayed robust mechanical stability and good thermal properties and ion conductivity of 78.8 mS cm^{-1} at 80°C with reasonable alkaline stability. Open circuit voltage of 0.98 V attained expressed low fuel crossover while operating the single cell [26]. A series of chemically stable and ionically conductive side chain AEMs based on two alkyl-trimethylammonium cations with n-propyl (C$_3$) and n-pentyl (C$_5$) alkyl chains and poly(2,6-dimethyl-1,4-phenylene oxide) (PPO) backbone are reported by Pan et al.. In their work, these two alkyl chains with cations were tethered onto PPO backbones through secondary amine moieties, resulting in two side-chain AEMs, NC3Q-PPO and NC5Q-PPO, respectively, with distinct hydrophilic/hydrophobic microphase separation leading to high ion conductivity and low swelling. Comparatively higher stability was reported in NC5Q-PPO membranes with an n-pentyl amine spacer than NC3Q-PPO with n-propyl amine side-chain materials [27]. The side-chain-type poly (arylene ether sulfone)s block copolymers bearing varied imidazolium functionalized aromatic pendants due to well-connected nano ionic channels also exhibited high ion conductivity and alkaline stability for 144 h at 80°C in 4 M KOH solution (74.8% conductivity retention) [28]. Improvement in alkaline stability and hydroxide ion conductivity through the interactions between strongly polar nitrile groups and side-chain functional cations was reported by Yan et al. They have synthesized poly(ether nitrile) (PEN) via polycondensation of bisphenol A with 2,6-difluorobenzonitrile, followed by grafting with imidazolium and morpholinium functional groups by typical chloromethylation method. Interaction among functional groups was proved by the density functional theory calculations. An increase in the LUMO (lowest unoccupied molecular orbital) energies of the functional groups as well as the reduction in free volume around hydrated cationic groups caused by these interactions contributed to the improvement in alkaline stability of the membranes [29]. Liu et al. in their study paid attention to designing AEM with the unique structural features of both "side-chain type" and "local density type" through densely grafting flexible long chains onto amino groups containing poly (aryl ether ketone) (PAEK). The four long side chains on each repeat unit of main chains and functionalization with three different cycloaminium cations aided micro phase-separated morphologies which limited the absorption of excess water and facilitated the formation of ion-conducting channels inside the membranes. Those membranes exhibited improved hydroxide ion conductivities (74.4 mS cm^{-1}) at 80°C when using N-methyl morpholine as the cation group [30]. Dense flexible side-chain-type copolymers of imidazolium (1-(6-bromohexyl)-3-methylimidazolium bromide) functionalized poly(arylene ether sulfone)s (ImPES)s with distinct nanophase separated morphology also exhibited a conductivity of 112.5 mS cm^{-1} at 80°C [31]. Flexible long-chain spacers bearing various cationic groups such as trimethylammonium (TMA), 1-methylpiperidinium (MPRD), 1-methylpyrrolidinium (MPY), 1-methylimidazolium (Im1) and 1,2-dimethyl imidazolium (Im1,2) were linked to the poly(ether sulfone)s (PES) backbone for preparing AEMs and various electrochemical properties were studied. The stability of AEMs was in the order of PES-MPY>PES-MPRD>PES-TMA>PES-Im1,2 >PES-Im1 and maximum power density of 109.0 mW cm^{-2} and 106.5 mW cm^{-2} was achieved by PES-MPRD and PES-MPY respectively. PES-MPY and PES-MPRD showed the highest stability due to the conformational restrictions imposed by the cyclic structures. The former exhibited even more alkaline tolerance than the latter because PES-MPY has only undergone substitution but no ring-opening reaction, whereas PES-MPRD suffered both methyl substitution and ring-opening reactions [32]. In contrast to a typical grafted AEM, a novel graft-type AEM with fluorocarbon side chains and quaternary ammonium functionalized aromatic backbone having robust and long connective ion transport channels was developed by a controlled atom transfer radical polymerization (ATRP) reaction. 4-fluorostyrene (FPS) monomers was reacted with brominated poly (phenylene oxide) (BPPO) main chain scaffold bearing ATRP initiating sites and then quaternised using trimethylamine. The attained properties demonstrated excellent alkaline stability (100% conductivity retention after exposure in 2 M NaOH for 30 days at 60°C) and good conductivity, but fuel cell performance was not done justice to the high conductivity and stability performances due to the incompatibility among binder and AEM [12]. Lin et al. aimed to reduce the ketone group which is liable to nucleophilic attack and polystyrene-b-poly(ethylene-co-butylene)-b-polystyrene (SEBS) triblock copolymer was grafted with quaternary ammonium groups via long flexible alkyl spacers. Only 7.7% and 13.7% hydroxide ion conductivity reduction was observed after exposure to 1 M aqueous KOH solution at 60°C and 90°C, respectively, for 360 h [33]. Guo and his co-workers reported hexyl spacers with a string of three imidazolium cations grafted fluorine-based poly (arylene ether sulfone) type AEMs having ionic clusters of 20 nm size and conductivity of 120 mS cm^{-1} at 80°C. Those membranes displayed superior dimensional stability than mono cation based AEMs and a peak power density of about 134.4 mW cm^{-2} at 280 mA cm^{-2} during fuel cell operation [34]. Hyper branched polymers, owing to their unique characteristics such as high functionality, good solubility and rigid structure with no entanglements have attracted great attention as good candidates for ion conductive materials. These dendritic macromolecules also have a systematically branched structure. Yang et al. synthesized a series of block polymer AEMs of hyper branched poly(arylene ether ketone) (HBPAEK) with 1-(N′,N′-dimethylamino)-6,12-(N,N,N-trimethylammonium) dodecane bromide (BQA). The dense ionic groups and long flexible spacer between cationic groups and polymer backbone resulted in an enhanced ionic conductivity, mechanical property as well as chemical stability to the membrane. An optimum hydroxide ion conductivity of 122.5 mS cm^{-1} at 80°C and 85% ionic conductivity retention after exposure in 2 M KOH at 80°C for 500 h was reported. H_2/O_2 single fuel cell using the membrane demonstrated a peak power density of 188.6 mW cm^{-2} with an open circuit voltage, 0.96 V [35].

The conventional approach to prepare AEM involves mainly two steps: chloromethylation followed by quaternary amination. Chloromethyl methyl ether (CME) or chloromethyl octyl ether (CMOE) owing to their excellent conversion as well as high yield is the main reagents used for chloromethylation. However, both are carcinogenic and hazardous which directed the researchers to develop another method. Friedel-Crafts alkylation reaction, alkalinization of polymers containing uncharged functional groups via a brominating agent or methyl iodide and bromination using N-bromosuccinimide (NBS) in benzyl or allyl-containing AEM material in the backbone structure, are the various approaches reported to avoid the use of CME and CMOE. Li et al. introduced a flexible pendant functional side chain on polysulfone through Friedel-Crafts alkylation. Friedel-Crafts alkylation, a characteristic reaction of benzene ring means the synthesis of alkyl carbocation by haloalkanes under the action of Lewis acid catalysts ($FeCl_3$, $AlCl_3$, $SnCl_4$, etc.) which act as an electrophile, attacking benzene ring forms carbocation and loses a proton to produce alkylbenzene. In their work, a new kind of alkylation reagent with -$(CH_2)_4$- spacer (1-chlorobutyl-2, 3-dimethylimidazole chloride [CBDMIm]Cl) was introduced as a pendant alkyl side chain on polysulfone resulted in hydrophilic/hydrophobic microphase separation structures to the AEM. The membranes attained a very low swelling and excellent alkaline stability [36]. The use of cost-effective, eco-friendly and highly reactive chloromethyl ethyl ether (CMEE) instead of hazardous and carcinogenic chloromethyl methyl ether for chloromethylation of polyetherimide (PEI) followed by quaternization resulted in conductivity of 0.026–0.032 S cm^{-1} and good chemical stability [15]. Another straightforward method of making high-performance AEMs employs a crown ether structure in polymers. The unique cavity structure of crown ether after initiated in 1967 by Charles Pedersen has attracted the attention of many researchers. The electronegativity due to the presence of oxygen (negatively polarized) inside the molecule mainly applied for complex formation (ion-dipole interaction) with the metal ions and conduction through the metal ions which can help ion migration when used in AEMs [2]. The diameter of metal cation is closely related to the crown ether cavity. The difference in stability constant and selectivity between metal cation and crown ether leads to the difference in ionic conductivity. Figure 10.2 shows the cavity size of several crown ethers and the size of metal ions. Zheng et al. incorporated 4-formyl dibenzo-18-crown-6 ($FDB_{18}C_6$) in polyvinyl alcohol (PVA) followed by KOH treatment form $F_{X\%}$-P-K membranes through complexation of crown ion on K$^+$. Lower than 8.6% conductivity degradation rate after treated with 10 M KOH solution for 240 h at 60°C observed reflected its super alkali resistance [37].

10.2.2 AEMs Based on Cross-Linked Structures

Cross-linked structures are widely developed to reduce swelling and improve the alkaline stability as well as mechanical durability of the membranes. Cross-linking can be divided into two: physical, where cross-links form through physical interaction (ionic, hydrogen bonding, van der Waals interaction, etc.) between molecules as well as chemical crosslinking, chemical bond forms crosslinked network. Du et al. prepared a series of cross-linked phenolphthalein-based poly (arylene

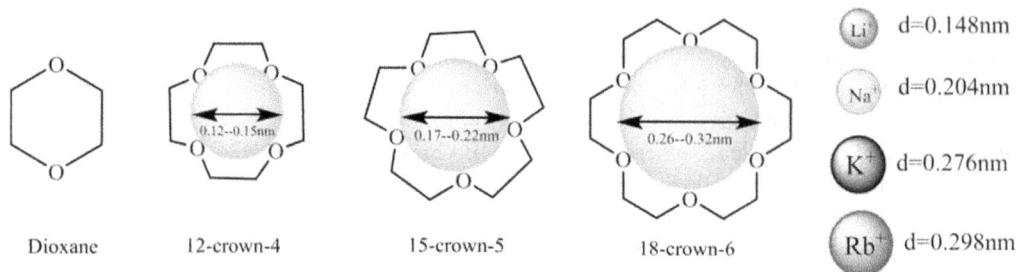

FIGURE 10.2 Size comparison between alkali metal ions and crown ether molecules [Reproduced from reference [37] with permission].

ether ketone) (PEK) copolymer AEMs with a conductivity of 83.6 mS cm^{-1} at 80°C and excellent mechanical, chemical as well as dimensional stability. They have used the method of copoly-condensation between 3,3-bis(4-hydroxy-3,5-dimethylphenyl)–phenolphthalein (DMPPH) and 4, 4'-di-fluorobenzophenone followed by bromination and imidazolium functionalisation using 1-vinyl imidazole. The cross-linked membranes maintained significantly high hydroxide ion conductivity (82.4%) after 400 h of alkaline exposure in 1 M NaOH at 40°C [38]. Crown ethers as a result of the ion-dipole interaction between the positively charged metal ion and the negatively polarized oxygen atoms have attracted great interest in various fields. The application of crown ether as a cross-linker is an effective strategy to enhance the performance of AEMs. Yang et al. prepared brominated poly(2,6-dimethyl-1,4-phenylene oxide) (BPPO) based AEMs and sodium salt of 4,4'(5')-di(hydroxybenzo)-18-crown-6 used as the cross-linker (Figure 10.3). The oxygen atoms in crown ether hold the Lewis base characteristics with strong basicity and the ability to complex with the cation groups to enhance the alkaline resistance. The cation groups that come close to the negatively charged crown ether groups also led to the formation of ionic transport channel, resulted in high hydroxide ion conductivity [39]. Ethylene oxide as a cross-linker helped to form ion clusters and enhance the water retention capacity of the poly(2,6-dimethyl-1,4-phenylene oxide) (PPO), which promoted high conductivity while maintaining high physico-chemical stability. The xBEO-PPO membrane, having bis(ethylene oxide) (BEO) as a cross-linker showed a conductivity of 131.96 mS cm^{-1} in water at 80°C, and 55.21 mS cm^{-1} in 95% RH at

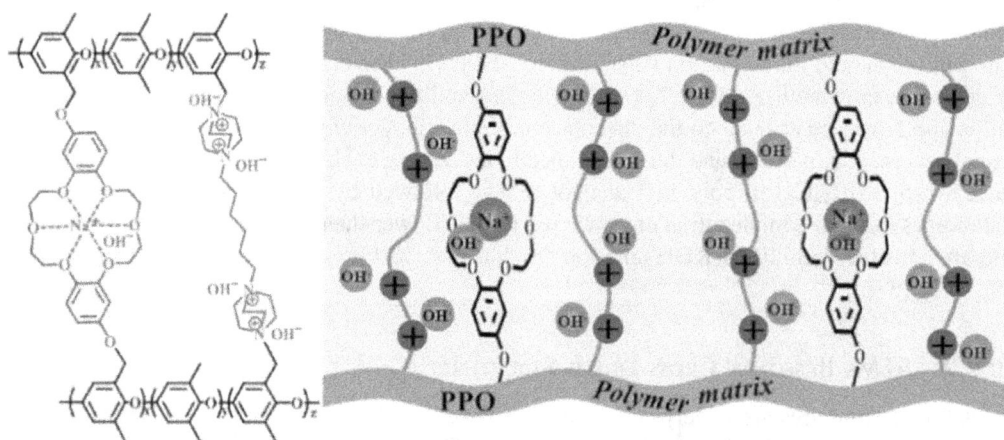

FIGURE 10.3 Chemical structure and sketch of the cross-linked AEMs [Reproduced from reference [39] with permission].

80°C. The single cell using the membrane exhibited a maximum power density of 444 mW cm^{-2} [40]. AEMs with cross-linked ionic regions are a promising strategy to improve the dimensional stability as well as the aggregation of ionic clusters leading to the generation of ion-conducting channels. A series of phenolphthalein containing poly(arylene ether sulfone)s functionalized with 1,2-dimethylimidazole (DIM) and cross-linked with N,N,N',N'-tetramethyl-1,6-diaminohexane (TMHDA) showed high hydroxide ion conductivity in the range of 52.2–143.4 mS cm^{-1} from 30°C to 80°C. The fuel cell assembled with the respective membrane exhibited an OCV of about 1.08 V and a peak power density of 83.6 mW cm^{-2} [41]. Poly (2,6-dimethyl-1,4-phenylene oxide) (PPO) based cross-linked and uncross-linked AEMs with bis-imidazolium cations, 2,3-dimethyl-1-(6-(2-methyl imidazol-1-yl) hexyl)-imidazolium iodide and 1,6-bis (2-methyl imidazol-1-yl) hexane respectively, showed the effect of cross-linking on the membrane performance. Cross-linked membrane exhibited comparatively high conductivity, alkaline stability, oxidation stability and mechanical properties than the uncross-linked one. Moreover, a single cell made with a cross-linked membrane displayed a maximum power density of 212.45 mW cm^{-2}, indicating its potential suitability for fuel cell applications [42].

Among various methods such as olefin metathesis, ionic cross-linking, Fiedel-Crafts alkylation and click chemistry reactions to form the cross-linked network, cross-linking reaction between halide and amine owing to its mild reaction conditions is the most common method. This method generally improves mechanical as well as hydrolytic stability; however, conductivity is not satisfactory. The cross-linking network restricts the polymer chain movement and formation of microphase separation structure, thereby can't construct the efficient ion transport channels in AEMs. Hu et al. demonstrated the effect of cross-linkers' geometric properties like rigidity and length on the structure and performance of AEMs. In their study, cross-linking with a long rigid molecule employed to adjust the polymer chain packing which creates highly efficient ion transport channels in the AEMs [43]. Polystyrene-b-poly(ethylene-co-butylene)-b-polystyrene (SEBS), a commercially available triblock copolymer exhibits a well micro-phase-separated structure where the hard polystyrene phase is separated from the soft phase of poly(ethylene-co-butylene). Various studies have been reported on introducing hydroxide ion conducting head group through chlor-omethylation reaction or Friedel-Crafts acylation reaction to improve ion exchange capacity. However, deterioration in mechanical as well as dimensional stability of the resultant AEMs at a high degree of functionalization and gelation limits its application. Introducing chemical cross-links improves the mechanical properties, but exhibits reduction in ion conductivity. Munsur et al. developed quaternary ammonium functionalized hexyl bis(quaternary ammonium) mediated partially cross-linked SEBSs with different grafting degree of the conducting head groups without gelation up to 75% of total styrene as highly conductive and dimensionally stable AEMs [44]. Norborene and its derivatives owing to their simple controllable polymerization technique have been used in AEMs recently. Cross-linked tetraalkylammonium-functionalized norbornene derivatives prepared through ring-opening metathesis polymerization (ROMP) showed an ion exchange capacity of 2.89 mmol g^{-1}, the ion conductivity of 64.79 mS cm^{-1}, water absorption of 12.5% and tensile strength of 15.18 MPa at 25°C [45].

Various approaches to drive ion conducting channel formation in linear polymer-based AEMs including functional group clustering, regulating the architecture through the block, comb-shaped, side chain, interpenetrating network formation, etc. have been developed. Implementation of the hydrophobic alkyl chains in linear block copolymer form microphase separated morphology resulting in interconnected ionic channels. Moreover, alkyl chains increase the functional groups' transportation, facilitating their aggregation into ionic clusters in side chain type or comb-shaped linear AEMs. The chain length has an important role in the formation of morphology and ion transportation channels. Ge and his coworkers proved the effectiveness of tuning the chain length of diamine cross-linkers in the formation of microphase separated morphology and interconnected ion transportation channels. They have done cross-linking of hyper-branched oligomers poly(4-vinyl benzyl chloride) (HB-PVBC) with diamines of different chain lengths. The optimum length of alkyl chains was six carbons to attain the best overall properties including high hydroxide ion conductivity, excellent alkaline stability, as well as moderate water swelling [46]. Interpenetrating polymer networks are combinations of two or more polymer networks partially interlaced on a molecular scale but not covalently bonded to each other and they cannot be separated unless chemical bonds are broken. AEMs based on an interpenetrating network normally have a hydrophobic cross-linked network and hydrophilic linear main chain bearing cationic functional groups. Good dimensional and chemical stability attributed to these characteristics endorsed much research interest in the development of AEMs based on the interpenetrating network [47]. IPN consisted of quaternized polysulfone (QPSF) and poly(vinyl alcohol) (PVA) exhibited low water uptake and a conductivity of 1.82×10^{-2} S cm^{-1} at 60°C [48]. Semi IPN of quaternized poly(butyl acrylate-co-vinyl benzyl chloride) (QPBV) and poly(vinylidene fluoride-co-hexafluoropropylene) [P(VDF-HFP) followed by cross-linking also resulted in a hydroxide ion conductivity of about 15 mS cm^{-1} at 25°C.

Cross-linking is an effective strategy commercially used to solve the conductivity-swelling dilemma in AEMs, but difficulty to recycling generates solid waste especially when the fuel cell is widely implemented. A reversible disulfide bond cross-linking strategy was developed by Hou and his coworkers in which a PPO-based AEM bearing both quaternary ammonium and thioacetate groups was synthesized by a one-pot reaction at mild condition. The chemistry involved the formation of the insoluble and cross-linked structure after hydrolyzing the thioacetate groups in aqueous NaOH solution followed by air exposure and re-dissolving the mass in dimethylformamide (DMF) after treated with dithiothreitol (DTT). The membranes showed acceptable hydroxide ion conductivity even after two repeated cross-linking and recycling processes [49].

10.2.3   AEMs Based on Blends and Hybrid Composites

It is worth mentioning that various blend membranes have been explored to produce effective alkaline AEM fuel cell membranes. Blending hydrophilic polyethylene glycol (PEG) with poly(2,6 dimethyl-1,4-phenylene oxide) modified by 4,4-diethoxybutan-1-amine and 1-methyl piperidine facilitated the formation of interconnected nano channels in the AEMs for ion conduction. The proposed membrane with 2 wt% PEG 2000 exhibited OH$^-$ conductivity of 97.2 mS cm^{-1} at 80°C and a peak power density of 328 mW cm^{-2} at 60°C when assembled in a single cell fueling with humidified hydrogen and oxygen [50]. Xiao et al. prepared AEMs via blending branched polyethyleneimine (BPEI) and poly(vinyl alcohol) (PVA) by a simple non-toxic method using water as solvent and glutaraldehyde as cross-linking agent. They found that at BPEI: PVA ratio of 1:1, the quaternized BPEI was more easily degraded in an alkaline environment because of phase separation while at 1:1.25, nearly 85% of the original ionic conductivity was maintained within 24 h after the initial rapid degradation, which indicates fairly good alkaline stability. Moreover maximum conductivity reported was 86.0 mS cm^{-1} [51].

Hybridization is a promising solution to regulate the physical and chemical microenvironment of ionic channels and thereby improving the comprehensive properties of AEMs. Hybrid composites combine the features of both organic and inorganic materials. The main role of inorganic materials is to improve water retention within the membrane and thereby conductivity; reinforcing the membrane which improves mechanical as well as thermal stabilities. Tuning the microphase separation structure is a standard method to increase the hydroxide ion conductivity of the membranes. Graphene oxide nanosheets modified with brush-shape poly(vinyl imidazolium-co-ethylene glycol dimethacrylate) chains in imidazolium functionalized bisphenol A-type polysulfone effectively manipulated aggregation of conductive groups at the polymer-filler interface and provided more interconnected ion conduction channels with low resistance for ion transport [52]. Imidazolium polymer brushes-grafted graphene oxides synthesized via ATRP technique embedded into polyvinyl alcohol also guaranteed interconnected ion conduction channels and enhanced ionic conductivity [53].

Fillers with different dimensions (0D, 1D, 2D) functionalized with ionizable groups have been widely used in polymer electrolyte membranes for the last few decades to regulate ion conducting channels. Carbon nanotube (CNT), a typical kind of 1D material due to its large aspect ratio, is attractive in constructing long-range ion transport channels. Qiu and his coworkers investigated the effect of incorporating ionic liquid grafted CNT on conductivity and alkaline stability of imidazolium functionalized poly(ether ether ketone) (ImPEEK). They have selected two different imidazolium-based ionic liquids of different alkaline stability, 1-(3-aminopropyl)-3-methylimidazolium bromide and 1-(3-aminopropyl)-2-methyl-3-butylimidazole bromide to chemically graft CNTs. The additional hydroxide hopping sites and continuous 1D long-range channels imparted the membranes an ion conductivity of 134.52 mS cm^{-1} at 70°C and 100% RH. Butyl imidazole ionic liquid functionalized filler incorporated composites exhibited higher chemical stability than the other owing to the distinctive electron-donating and hyper conjugative effects of the substituents present in it [54]. Titanate nanotubes covalently bonded with 1-methyl-3-(3-trimethoxysilylpropyl) imidazolium chloride ionic liquid also possessed high ion exchange capacity, hydroxyl ion conductivity and power density to quaternized polysulfone membranes [55]. Similarly, cross-linked hybrid composite membranes composed of 1-vinyl-3-butylimidazolium bromide [VBIm][Br] and 1-vinyl-propyl-triethoxysilaneimidazolium chloride [VPSIm][Cl] and tetraethyl orthosilicate (TEOS) exhibited a conductivity of 0.04 S cm^{-1} at 30°C and good alkaline stability, 80% retention of conductivity after being treated with 2 M KOH solution at 80°C for 12 days and 60°C for 20 days [56]. 1-(3-aminopropyl)-3-methylimidazolium salt bromine (IL-NH$_2$) functionalized graphene oxide in poly-benzimidazole (PBI), as well as poly (2, 6-dimethyl-1, 4-phenylene oxide), displayed relatively excellent thermal stability, high ion conductivity as well as an acceptable alkaline stability for AEMs [57,58]. Introduction of 1-methyl-3-methylimidazolium ionic liquid (Me-IL), 1-methyl-3-ethyl imidazolium ionic liquid (Ethyl-IL), 1-methyl-3-hydroxyethyl imidazolium ionic liquid (HOEt-IL) immobilized titanium dioxide in quaternised PPO exhibited enhanced durability and conductive as well as mechanical properties to the membranes [59]. Polyvinyl alcohol functionalized carbon nanotubes using an ozone-mediated method to improve the compatibility with polyvinyl alcohol matrix also exhibited obstructed fuel permeation as well as good cell performance in terms of open-circuit voltage and peak power density [60]. 1-Methyl-3-(3-trimethoxysilylpropyl) imidazolium chloride grafted on to mesoporous silica (SBA-15) without destroying morphology owing to their large porous volume, high surface area, mechanical and thermal stability showed enhanced performance when incorporated into quaternary polysulfone (QPSU) and applied in a fuel cell (IEC: 1.86 meq g^{-1}, conductivity: 1.89×10^{-2} S cm^{-1}) [61].

Polyhedral oligomeric silsesquioxane (POSS) owing to its unique cage-like structure with nanoscale dimension, good thermal and thermo oxidative stability, environmental neutrality, as well as the availability of surface functional cites to structural modification attracted much interest in various fields. Li et al. found that octaphenyl polyhedral oligomeric silsesquioxane (O-POSS) modified with 1-methylimidazole incorporated in imidazolium functionalized poly(ether ether ketone) (ImPEEK) induced hydrophilic clusters via aggregation of imidazolium functional groups from both polymer and nano filler and thus successful formation of efficient ionic channels. Ionic mobility analysis revealed a fast ion transport that resulted in hydroxide conductivities up to 85.3 mS cm^{-1} [62]. Hybrid composite membranes prepared from tetramethylethylenediamine (TMEDA) modified octa(3-chloropropyl) POSS and glycidyl trimethylammonium chloride (GTMAC) functionalized octaammonium POSS respectively incorporated chloromethylated polysulfone (CMPSF) as well as trimethylamine functionalized poly(2,6-dimethyl-1,4-phenylene oxide) are another example of high anion conductive membranes [63,64]. It was also found that the addition of quaternised POSS in quaternary ammonium functionalized polysulfone (QPSU) exhibited high ion exchange capacity (2.03 meq g^{-1}) attributed to the high density of ion-exchange groups and high surface area of the nano filler with low water uptake of about 8.87% [65].

Graphene, a two-dimensional nanomaterial owing to its facile synthesis unique electronic properties and ease of functionalization has potential application in various fields. Graphene oxide and dopamine modified graphene oxide blended with chloromethylated polysulfone followed by quaternisation and alkalization utilized in AEMs resulted in good mechanical properties as well as conductivities of 1.08×10^{-2} S cm^{-1} and 1.07×10^{-2} S cm^{-1} at 60°C, respectively. Besides, conductivity values were maintained at 86.9% and 76% after the alkaline solution stability test at 60°C, implying its ability to meet the stability requirements [66]. Miao and his co-worker also successfully synthesized a series of cross-linked AEMs of quaternised polysulfone and poly-dopamine functionalized graphene oxide based on mussel-inspired chemistry, a combination of hybridization and cross-linking. Michael addition/Schiff base reactions between the amino groups on the polymer side chains and the polydopamine coating on the surface of rGO resulted in cross-linked structure to the membranes with improved ionic conductivity, alkaline stability, oxidative stability, and mechanical properties [67].

The pore-filling method, filling an ionic polymer into the porous support is an important and effective technology to solve the dilemma between conductivity and stability that existed in the development of AEMs. The thickness of the membranes and their uniformity plays an important role in controlling water transport and ionic resistance. Yang et al. conducted a study on the thickness optimization of the composite membranes made of quaternary ammonium functionalized poly(phenylene oxide) filled polyethylene support by varying the viscosity of the filler. Their study illustrated the effect of thickness on various AEM performances [68].

10.3   ALKALINE STABILITY AND FUEL CELL PERFORMANCE

In this section, membranes with comparatively higher alkaline stability as well as fuel cell performance will be discussed. The stability of anion exchange membranes in a strong alkaline environment is a major challenge for large scale implementation of AEMFCs. Polymer backbone and cationic group are prone to be attacked by strongly basic and nucleophilic hydroxide ions under alkaline conditions, resulting in degradation of ionic groups and mechanical strength of anion exchange membranes especially at high temperatures [16]. For example, quaternary ammonium cations, a commonly used one have been confirmed to be liable to the nucleophilic hydroxide attack which involves two principal degradation mechanisms: the second order nucleophilic substitution (S$_N$2) and Hofmann elimination (E2). Nucleophilic substitution (S$_N$2) leads mainly to the formation of alcohol; however, elimination reactions resulted in a double bond. Benzylic derivatives are more susceptible to S$_N$2 substitution and the reaction depends on both strength of the nucleophile as well as the concentration of nucleophile and benzylic ammonium moiety. At the same time, Hofmann elimination (E2) can occur only in the presence of hydrogen atom near the ammonium group (β-hydrogen) [69]. Imidazolium ring is also prone to breakage by nucleophilic attack of OH$^-$ at the C$_2$ position [58]. Therefore, many researchers put their effort into the synthesis of AEMs using a variety of cationic groups and a wide range of alkaline stability was developed. Moreover, the configuration of cationic groups on polymer backbone is also proved to have a great influence on the stability of membranes. Significant degradation was also reported in polymers based on poly(arylene ether sulfone)s and poly (arylene ether ketone)s. Polysulfone and polyketone groups in the backbone are prone to nucleophilic attack by hydroxide. Cleavage of C-O bonds at high pH limits the long-term use of poly(aryl ether) backbone. Overall, the long-term stability of AEMs have relied on the chemical characteristics of the polymer backbone, as well as the position and nature of fixed cation [70].

Compared to other quaternary cations, quaternary ammonium has better stability but, the decomposition via β-hydrogen (Hoffmann elimination) and the possibility of direct nucleophilic substitution under alkaline conditions are the major cause of concerns. Sterically crowded imidazolium site has higher alkaline stability and OH$^-$ ion conductivity. Sana and his coworkers synthesized poly (methylated pyridinium benzimidazolium) iodide (PMPBI) with the expectation of an increase in conductivity as a result of dual conducting sites like pyridinium and imidazolium as well as alkaline stability through steric crowding around the functional groups. The results clearly showed promising conductivity and alkaline stability [71]. Polymers with different aromatic pendants are also proposed to improve the AEM stability. Ye et al. synthesized a copolymer of poly-4-vinyl benzyl chloride-styrene (P(VBC-St)) with different aromatic pendants (1-allyl naphthalene, p-ethyl phenol) and studied their influence on properties after quaternization from the point of view of conjugative effects, steric effect, as well as the electrostatic interactions. All the membranes exhibited excellent alkali stability in 1 M KOH at 80°C for 1,000 h. p-Ethyl phenol with a similar structure to phenol type radical inhibitor as a pendant group could react with reactive intermediates or radicals formed during an alkaline attack and protect the AEM from nucleophilic attack strongly and showed comparatively superior stability than the other [72]. Stability of different amine functionalities, namely, trimethylamine (TMA), 1,4-diazabicyclo[2.2.2]octane (DABCO), 1-azabicyclo[2.2.2]octane (ABCO) and N-methyl piperidine (NMP) were studied after functionalized on vinyl benzyl chloride (VBC) grafted low-density polyethylene (LDPE). The order in increasing AEM stability in terms of IEC percentage loss observed was TMA>ABCO>NMP>DABCO. The difficulty in obtaining 100% mono quaternization and the formation of a less stable piperazine structure from bis-quaternised DABCO resulted in slightly low stability to DABCO. Moreover, peroxide and hydroxide radical attacks on the ternary (benzylic) carbon and formation of vinyl benzyl trimethylammonium hydroxide (VBTMA) was also detected [5].

Alkyl pendant groups on side chains can improve the chemical stabilities of AEMs. Grafting of four long side chains with cycloamonium cations onto amino groups of each repeat unit in poly (aryl ether ketone) and the resultant well solvated hydroxyl ions in micro phase-separated environment favored good alkaline stabilities to the membranes. Poly(aryl ether ketone) with methyl morpholine (PEAK-MmOH-25) showed an optimum power density of 302 mW cm^{-2} at 640 mA cm^{-2} when functioning in a single fuel cell [30]. Becerra-Arciniegas et al. performed a detailed analysis on the stability of poly(2,6-dimethyl-1,4-phenylene oxide) by changing the position of benzyl trimethylammonium groups anchored on the aromatic rings *via* chloromethylation and bromination routes. The backbone degradation occurred mainly *via* hydroxyl ion attack on the ether linkage, and it decreased when the electron density around the oxygens is high. Quaternary ammonium degradation via an S_N2 mechanism was reported in both but, chloromethylated derivatives degrade more than the bromomethylated ones. A detailed picture of different degradation mechanisms is demonstrated in their article [69]. Side-chain-type polymer architectures have been studied widely as AEM for fuel cells but rigid, hydrophobic spacers generally limit self-assembly of ionic side chains and hence ion conductivity. So flexible, the hydrophilic side-chain-type membrane was developed by Zhu et al. via incorporating hydrophilic and rotatable ethylene oxide spacers into the ionic side chains. Improved flexibility facilitated the ether-containing side chain motion to enhance the self-assembly process and extended the ion-conducting region which accelerated the efficiency of both water as well as ion transport during fuel cell operation. This resulted in a maximum power density of 437 mW cm^{-2} when tested in H_2/O_2 AEMFC [73]. Lee and his coworkers synthesized an ion-conducting additive N, N'-didodecyl-hexyl-bis(quaternaryammonium iodide) (DHBQA) with a long alkyl chain (C_{12}) and two quaternary ammonium groups and introduced into a dodecyl-substituted and quaternary ammonium functionalized poly(2,6-dimethyl-1,4-phenylene oxide) to induce strong alkaline stability to the AEM. 17.9% conductivity loss observed after 500 h of exposure in 1 M KOH at 60°C [74].

To get a highly stable anion exchange membrane, Wang et al. designed aryl ether free polystyrene (PS) functionalized with alkaline stable N-cyclic quaternary ammonium (QA), eg. six-membered (DMP) and bis-six-membered N-cyclic QA (ASU) cations via "click chemistry". N-hetero cyclic structure with steric hindrance in quaternary ammonium cations successfully alleviates the substitution and elimination reactions. The ASU-functionalized PS copolymer demonstrated excellent long-term alkaline stability in 1 M NaOH CD$_3$OD/D$_2$O solution at 80°C for 3,000 h without any significant change in chemical structure [3]. Poly(biphenyl indole) without aryl ether synthesized through rapid acid-catalyzed reaction functionalized with N-methyl piperidinium via long alkyl spacer (PBN-Pip) also exhibited outstanding chemical resistance, attributed to the stable poly(biphenyl indole) backbone and hexyl spacers. Steric hindrance exhibited by a long hydrophobic spacer and low cyclic strain as well as the conformation of a specific six-membered alkyl ring structure in N-methyl piperidinium are predominant factors that impart alkali resistance to the membranes. The cell with PBN-Pip exhibited a maximum energy density of 404 mW cm^{-2} (100% RH, 80°C) at a current density of 960 mA cm^{-2} and good cell durability of 120 h [20].

Benzyltrimethylammonium (BTMA) is one of the most commonly used organic cations in anion exchange membranes though, BTMA-based AEMs owing to their low ionic conductivity and insufficient chemical stability limits its real-world fuel cell applications. The ineffective chemical stability of the membranes is limited to the degradation of electron-deficient cationic moieties in presence of nucleophilic and basic OH$^-$ ions, especially at high temperature. Copper-catalyzed azide-alkyne cycloaddition (CuAAC) chemistry, generally known as "click" reaction, is a proficient, reliable as well as forthright method to covalently connect two building blocks comprising different functional groups. Alkyne tethered BTMA isomers reacted with azide functionalized polyphenylene oxide through CuAAC reaction (namely m-QPPO, o-QPPO and p-QPPO) yields membranes with distinct hydrophilic/hydrophobic microphase separation, leading to high conductivity and alkaline stability. m-PPO demonstrated the best cell performance with a peak power density of 333 mW cm^{-2} at a current density of 700 mA cm^{-2} at 60°C [75]. Self-cross linkable PPO via Cu-catalyzed alkyne-azide click chemistry with side chain type architecture having good ion conductivity, chemical and dimensional stability was also obtained by He et al. [76].

Guo et al. reported an interpenetrating network of poly(vinyl benzyl chloride) (PVBC)/poly (vinyl alcohol) (PVA) blend cross-linked with poly(1-vinyl imidazole) (PVIm) with a conductivity of 41.1 mS cm^{-1} at 60°C. Application of bifunctional macromolecular cross-linking by providing both imidazolium cations and cross-linking exhibited excellent alkaline stability in 4 M KOH solution for about 1,320 h. However, maximum power density achieved only 37.1 mWcm^{-2} at a current density of 64 mA cm^{-2} [47]. Tetraalkylammonium functionalized fluoropolyolefin synthesized by direct Ziegler-Natta catalyzed copolymerization and subsequent quaternization owing to the compact entanglement of the polymer chain in solution showed good dimensional stability together with 89.5% of retention in hydroxide conductivity after 700 h of aging in 1 M NaOH at 80°C [4].

The presence of the π conjugated structure in imidazolium cation and steric hindrance make it exhibit high stability. Guanidimidazole, a modified version of imidazole as quaternising as well as cross-linking agent was also reported to improve the alkaline stability of the membranes [77]. Lai et al. reported a copolymer of 1-vinyl imidazole and N-vinyl carbazole quaternised with imidazolium cation exhibited high ion conductivity and desirable methanol permeability for AEMFCs. Furthermore, the membrane maintained ion conductivity and IEC of about 79% and 81%, respectively, without further loss up to 1,400 h of exposure in 1 M KOH solution at 60°C. Carbazole, a nitrogen-containing aromatic heterocyclic compound having a large π-conjugated system with rigid fused rings favored the membranes to facilitate excellent thermal and alkaline stability [78]. Ari and his coworkers showed that the incorporation of electron-withdrawing nitro group in quaternized PPO can weaken the interaction between OH$^-$ group and quaternary ammonium group in the membrane structure and made the quaternized group less prone to degradation through nucleophilic substitution (S$_N$2) [79]. A facile synthetic strategy developed by Mandal et al. to prepare a series of tetra block AEM copolymers based on vinyl addition polymerization of norborene (alternate two blocks each of butyl norborene and bromopropyl norborene) expressed high thermal stability up to 400°C and ionic conductivity of 122.7 mS cm^{-1} at 80°C with an IEC of 1.88 meq g^{-1}. The prepared membrane displayed exceptional long-term chemical stability in 1 M NaOH solution at 80°C with no noticeable degradation (< 1%) over a 1,200 h and an excellent fuel cell performance with an optimum power density of 542.57 mW cm^{-2} at 0.43 V and 1.26 A cm^{-2} [70].

Fabrication of organic-inorganic composite membranes with an appropriate quantity of nano-materials is reported to upgrade the overall performance of the AEMs. Ionic liquid (1-methyl-3-(3-trimethoxysilylpropyl) imidazolium chloride) grafted titanate nanotubes and mesoporous silica in quaternised polysulfone demonstrated a peak power density of 302 mW cm^{-2} and 278 mW cm^{-2} respectively to the fuel cell [55,61]. Quaternized titanate nanotubes in quaternised polysulfone also displayed a peak power density of 285 mW cm^{-2} at 680 mA cm^{-2} [80]. We have published a work, which revealed a promising approach to improve the alkaline stability and cell performance of quaternized poly(2,6-dimethyl-1,4-phenylene oxide) (QPPO)/glycidyl trimethylammonium chloride grafted octaammonium polyhedral oligomeric silsesquioxane (QPOSS) composites. The respective composite membrane with best overall performance exhibited an increment in ionic conductivity up to 200 h of exposure in 1 M KOH solution, after that, it decreased but maintained about 102% of the initial conductivity after prolonged exposure of 1,200 h, indicating its excellent alkaline stability. The steric hindrance offered by the bulky substituents of filler acted as a barrier for nucleophilic attack by excess hydroxide ions. Fuel cells operated using the composite membrane displayed a peak power density of 288 mW cm^{-2} at a current density of 662 mA cm^{-2} [64]. Quaternized POSS/quaternary ammonium functionalized polysulfone composite prepared by Elumalai and Sangeetha also displayed an optimum power density of 321 mW cm^{-2} at a current density of 720 mA cm^{-2} when applied in a single fuel cell at 60°C [65]. Flexible and transparent zwitterionic polymer membranes with Si-O-Si cross-linked network fabricated by sol-gel polycondensation reaction and self-assembly process showed good dimensional, mechanical as well as thermochemical stability. Zheng et al. first synthesized a polymer with cationic quaternary ammonium groups via grafting alkyl siloxane onto cardo poly(aryl ether sulfone ketone)s with pendant tertiary amine groups. The Zwitterionic polymer was then produced through partial hydrolysis of alkyl siloxane in a weak basic medium. Good alkaline stability in 1 M NaOH aqueous solution at 60°C for 600 h was attained in their studies [81].

Han et al. recently designed an efficient cross-linked AEM with improved alkaline stability which is free from the dilemma of reduction in OH$^-$ ion transport and swelling as a result of cross-linking. In their work, novel aggregated and ionic cross-linked membranes of quaternary ammonium polysulfone with microphase morphology (acS$_8$QAPSF, a, and c, indicates aggregation of ion domains and cross-linking, x is the number of non-hydrogen atoms of the secondary aliphatic amine attached side chain) were prepared. The aggregated ionic clusters in the backbone, as well as ionic cross-linked network, hindered both cation and backbone degradations of AEM and the weight loss was only 8% after exposure to 1 M NaOH solution at 80°C for 30 days. The membranes also exhibited 92% and 90.5% IEC and ionic conductivity retention respectively. The fuel cell operated using acS$_8$QAPSF membrane at 60°C showed a peak power density of 0.612 W cm^{-2} [82]. The idea of assembling ionic highways using charge-delocalized cations as cross-linkers is established as high-performance AEMs. Cross-linked polymer membranes comprising ionic highways together with charge-delocalized pyrazolium cations and homoconjugated triptycenes were designed by Kim et al. A series of pyrazolium cross-linked poly(triptycene ether sulfone)s were prepared under the design criteria of (1) constructing a cross-linked network by installing cation on cross-linker to balance IEC as well as mechanical stability, (2) fully substituted cationic rings to avoid reactive C-H groups and improve chemical stability and finally (3) covalent assembly of repeating ionic segments, act as an ionic highway to reduce activation barrier and improve conductivity. The AEMs retain more than 75% of their initial conductivity after 30 days of alkaline stability test in 1 M KOH solution at 80°C. A single cell assembled the membrane at 80°C displayed a maximum power density of 0.73 W cm^{-2} with durability up to 400 h [83]. Addition of triptycene poly(ether sulfones) into 1-methylimidazolium poly(ether sulfone) with good dimensional as well as chemical stability and conductivity up to 0.082 S cm^{-1} was also proved [84].

In the intensive search for highly stable AEMs containing base-stable cation, aliphatic heterocyclic quaternary ammonium cations have been investigated to possess excellent stability than the typical tetraalkylammonium and benzylic cations. Recently Xue and his team incorporated N-spirocyclic quaternary ammonium cations into poly(2,6-dimethyl-1,4-phenylene oxide) (PPO) which resulted in valuable insights toward more effective molecular designs for AEMs. Spirocyclic 3,6-diazaspiro[5,5]undecan-6-ium bromide (DSU) single and double cation polymers were synthesized via nucleophilic substitution reaction between brominated PPO and DSU (Menshutkin reaction). Iodomethane, a quaternisation agent reacting with tertiary piperidine rings in single cation polymer precursors, formed double cation polymers. Double cation polymeric membranes expressed higher stability than the single cation polymer which suggested an important role of quaternary piperidinium in DSU ring in moderating hydroxyl ion attack on the spirocyclic cation centre [85].

Polybenzimidazole (PBI), as an ether-bond free polymer, because of its good thermal as well as chemical stability has attracted much attention but low solubility and conductivity limited its application as AEM material. Recently, a series of ether-bond free polybenzimidazole membranes with pendant imidazolium groups (1-(6-bromohexyl)-2,3-dimethyl imidazolium bromine and 1-(3-bromopropyl)-2,3-dimethyl imidazolium bromine) to solve the low conductivity as well as, alkaline stability issues associated with AEMs, were reported by Lin et al. Membranes also displayed an excellent solubility in dimethylformamide and dimethyl sulfoxide and film-forming properties. The single fuel cell with the corresponding membrane attained a peak power density of 444.5 mW cm^{-2}, demonstrating the potential of ether-bond free functionalized PBI for AEMFC applications [86].

Adequate water maintenance in AEMFC is an important criterion to attain the best cell performance. By maintaining AEM and electrode hydration without flooding or drying out the catalyst or gas diffusion layers, maximum current, as well as power densities of the system, can be attained. The balance between proper membrane hydration and flooded catalyst layers is a challenge facing the development of high-performance AEMFCs these days. A new benchmark for a H$_2$/O$_2$ AEMFC with a peak power density of 1400 mW cm^{-2} at 60°C was achieved through the application of radiation grafted ethylene tetrafluoroethylene (ETFE) based AEM (ETFE-g-poly(vinyl benzyl chloride)) and anion exchange ionomer based powder functionalized with benzyl trimethyl ammonium groups [87].

The Department of Energy's 2015 performance targets required achieving a minimum of 95% alkaline stability for 5,000 h at temperatures up to 120°C, Focusing on the targeted temperature, Nunez and his coworkers provided a keen ^{1}H NMR-based alkaline degradation method using various n-alkyl spacers of different lengths between the polymer backbone and tetraalkylammonium (TAA) cation. Poly(2,6 dimethyl-1,4-phenylene oxide) and polystyrene backbones were compared in their studies. Degradation products were analyzed using ^{1}H NMR and explained the mechanism of deterioration. The major degradation route, substitution at benzyl position directed the authors towards the investigation on the effect of interstitial n-alkyl spacers between the aryl ring and the trimethylammonium cation and the results proved its pronounced effect on the stability of the AEMs [88].

Significant efforts on improving the chemical stability of AEMs focused on the functional groups and polymer backbone have been reported recently. However, systematic study of polymer backbone degradation in an alkaline environment is still lacking. Most of the studies reported were typically conducted by investigating the change in mechanical properties, ion exchange conductivity as well as structural analysis by 2D NMR spectrum after immersing the membrane in alkaline solution at specified conditions. Bae's group has done a systematic investigation on the stability of various aromatic precursor polymer backbones to expand the scope of polymer main chain chemical degradation analysis. They have used gel permeation chromatography (GPC) to detect the location of the repeating unit being degraded in the alkaline exposure. They have preferred GPC because using NMR it is difficult to detect degradation of complex polymeric structures quantitatively as the characterization of end groups of degraded polymer chains is challenging. Nine representative polymer structures including poly(biphenyl alkylene)s, poly(arylene ether)s and polystyrene block copolymers were scrutinized for alkaline stability and the presence of aryl ether bond in the repeating unit was reported as the main cause of stability reduction. Molecule model compound stability studies using NMR and density functional theory (DFT) also revealed that the presence of electron-withdrawing group near aryl ether bond accelerates the degradation [89].

A model to predict the strong impact of operating current density and characteristics of membrane on performance stability and lifetime of AEMFC was developed by Dekel et al. The quantitative evaluation on the effect of membrane thickness reduction and use of more stable functional groups on stability improvement were used in their study. They have predicted a substantial improvement in the long-term cell stability, more than 5,000 h using a theoretical approach, satisfying the DOE 2020s' cell lifetime target for automotive applications. Current researchers in the field of the fuel cell can make use of the material designs and operating parameters provided by the authors for the development of viable AEMFC technology [90].

10.4 SUMMARY AND PERSPECTIVES

Owing to the remarkable advantages such as accelerated reaction kinetics as well as the application of non-precious metals as the catalyst and hence reduction in cost anion exchange membranes has attracted much attention as a polymer electrolyte membrane in fuel cells. Fabrication of highly conductive and stable membrane materials is a current challenge in the development of AEM for fuel cells and this chapter clearly shows great efforts have been made in the last five years (2016–2021) to design and optimize the chemical structure of AEMs for the application of fuel cells.

The degradation mechanism of the traditional quaternary ammonium head groups is well known and intense research efforts on the modification of quaternary ammonium type membranes are reported. Non-ammonium-based cations such as imidazolium, benzimidazolium, quaternary phosphonium, guanidinium and metal-organic coordination ions have been employed to polymer chain and studied the durability of the membranes but the durability was not satisfactory. Introduction of spacers between polymer chains and the head groups and graft, block, cluster-type morphology as well as cross-linked network structures are being studied with the aim of performance improvement. Several promising candidates with satisfactory performances are being reported. Increasing the donor-acceptor distance by implementing more sterically hindered or bulky electron-donating alkyl groups and pendant quaternary ammonium grafting via Cu(I)-catalyzed click chemistry has a positive effect on conductivity as well as alkaline stability of the membranes. Comb-shaped cationic polymers holding long-range ordered super hydrophobic alkaline stable backbone with pendant side chain terminated by long-comb hydrophilic cationic groups also helps to improve the dimensional stability of the membranes. The position for substitution of cationic head groups also has a great effect on the performance of AEMs. Hybridization using functionalized nanomaterials with ionizable groups to ward off chemical attack as well as improve electrochemical performance is also being developed to enhance existing membranes. Pore-filled membranes are futuristic technology that holds the promise to solve the dilemma between conductivity and stability that existed in the development of AEMs. We have also paid attention to the various analysis of degradation mechanisms on functional head groups as well as the main polymer backbone in this chapter.

Encouraging advances achieved so far as reviewed in this chapter anion exchange membranes are still an important topic with great prospect. Regarding future efforts to overcome technological challenges and promote the successful implementation of high-performance AEMs for AEMFCs, we propose the following research directions: tuning of constituents as well as the architecture of both main and side chain structures with improved stability from hydroxide attack; identify the reactive sites of the AEM which is critical for understanding the degradation mechanism and designing new stable AEMs; further fundamental understanding through both computational study and experimental identification/validation of material selection, process and characteristics concerning their cell performance. We strongly believe that with continuous efforts in this area, motivated researchers will make great paces towards success in the commercialization of high-performance AEMFCs.

ACKNOWLEDGMENT

The authors acknowledge the financial assistance received from the National Research Foundation of Korea (NRF), ministry of science and ICT (NRF - 2021 M1A2A2038115). This work was also supported by basic science research program funded by ministry of education (NRF - 2020R1A6A1A03038697).

REFERENCES

[1] Ogungbemi, E.; Ijaodola, O.; Khatib, F.N.; Wilberforce, T.; El Hassan, Z.; Thompson, J; Ramadan, M.; Olabi, A.G. 2019. Fuel cell membranes – Pros and cons. *Energy* 172:155–172.

[2] Zheng, X.; Shang, C.; Yang, J.; Wang, J.; Wang, L. 2017. Preparation and characterization of chitosan-crown ether membranes for alkaline fuel cells. *Syn. Metals* 247:109–115.

[3] Wang, X.; Sheng, W.; Shen, Y.; Liu, L.; Dai, S.; Li, N. 2019. N-cyclic quaternary ammonium-functionalized anion exchange membrane with improved alkaline stability enabled by aryl-ether free polymer backbones for alkaline fuel cells. *J. Membr. Sci.* 587:117135.

[4] Zhang, X.; Chu, X.; Zhang, M.; Zhu, M.; Huang, Y.; Wang, Y.; Liu, L.; Li, N. 2019. Molecularly designed, solvent processable tetraalkylammonium-functionalized fluoropolyolefin for durable anion exchange membrane fuel cells. *J. Membr. Sci.* 574:212–221.

[5] Espiritu, R.; Golding, B.T.; Scott, K.; Mamlouk, M. 2018. Degradation of radiation grafted anion exchange membranes tethered with different amine functional groups via removal of vinylbenzyl trimethylammonium hydroxide. *J. Power Sources* 375:373–386.

[6] Vijayakumar, V.; Nam, S.Y. 2019. Recent advancements in applications of alkaline anion exchange membranes for polymer electrolyte fuel cells. *J. Ind. Eng. Chem.* 70:70–86.

[7] Pan, Z.F.; An, L.; Zhao, T.S.; Tang, Z.K. 2018. Advances and challenges in alkaline anion exchange membrane fuel cells. *Prog. Energy Combust. Sci.* 66:141–175.

[8] Dekel, D.R. 2018. Review of cell performance in anion exchange membrane fuel cells. *J. Power Sources* 375:158–169.

[9] Merle, G.; Wessling, M.; Nijmeijer, K. 2011. Anion exchange membranes for alkaline fuel cells: A review. *J. Membr. Sci.* 377:1–35.

[10] Hao, J.; Gao, X.; Jiang, Y.; Zhang, H.; Luo, J.; Shao, Z,: Yi, B. 2018. Crosslinked high-performance anion exchange membranes based on poly(styrene-b-(ethylene-co-butylene)-b-styrene). *J. Membr. Sci.* 551: 66–75.

[11] Chu, X.; Liu, L.; Huang, Y.; Guiver, M.D.; Li, N. 2019. Practical implementation of bis-six-membered N-cyclic quaternary ammonium cations in advanced anion exchange membranes for fuel cells: Synthesis and durability. *J. Membr. Sci.* 578:239–250.

[12] Ran, J.; Ding, L.; Yu, D.; Zhang, X.; Hu, M.; Wu, L.; Xu, T. 2018. A novel strategy to construct highly conductive and stabilized anionic channels by fluorocarbon grafted polymers. *J. Membr. Sci.* 549:631–637.

[13] Han, H.; Ma, H.; Yu, J.; Zhu, H.; Wang, Z. 2019. Preparation and performance of novel tetraphenylphosphonium-functionalized polyphosphazene membranes for alkaline fuel cells. *Eur. Polym. J.* 114:109–117.

[14] Lim, H.; Lee, B.; Yun, D.; Al Munsur, A.Z.; Chae, J.E.; Lee, S.Y.; Kim, H.J.; Nam, S.Y.; Park, C.H.; Kim, T.H. 2018. Poly(2,6-dimethyl-1,4-phenylene oxide)s with various head groups: Effect of head groups on the properties of anion exchange membranes. *ACS Appl. Mater. Interfaces* 10:41279–41292.

[15] Yadav, V.; Rajput, A.; Sharma, P.P.; Jha, P.K.; Kulshrestha, V. 2020. Polyetherimide based anion exchange membranes for alkaline fuel cell: Better ion transport properties and stability. *Colloids Surfaces A Physicochem. Eng. A.* 588:124348.

[16] Yang, C.; Liu, L.; Huang, Y.; Dong, J.; Li, N. 2019. Anion-conductive poly(2,6-dimethyl-1,4-phenylene oxide) grafted with tailored polystyrene chains for alkaline fuel cells. *J. Membr. Sci.*573:247–256.

[17] Weiber, E.A.; Jannasch, P. 2016. Anion-conducting polysulfone membranes containing hexa imidazolium functionalized biphenyl units. *J. Membr. Sci.* 520:425–433.

[18] Wang, Y.; Wan, H.; Wang, J.; Wang, L.; Feng, R. 2017. Hydroxide ion highway constructed by orderly aligned quaternary ammonium groups in anion exchange membranes. *J. Electrochem. Soc.* 164:F1051–F1062.

[19] Mousavi, M.P.S.; Kashefolgheta, S.; Stein, A.; Buhlmann, P. 2016. Electrochemical stability of quaternary ammonium cations: An experimental and computational study. *J. Electrochem. Soc.* 163: H74–H80.

[20] Wang, K.; Gao, L.; Liu, J.; Su, X.; Yan, X.; Dai, Y.; Jiang, X.; Wu, X.; He, G. 2019. Comb-shaped ether-free poly(biphenyl indole) based alkaline membrane. *J. Membr. Sci.* 588:117216.

[21] Liu, X.; Wu, D.; Liu, X.; Luo, X.; Liu, Y.; Zhao, Q.; Li, J.; Dong, D. 2020. Perfluorinated comb-shaped cationic polymer containing long-range ordered main chain for anion exchange membrane. *Electrochim. Acta* 336:135757.

[22] Hu, Z.; Tang, W.; Ning, D.; Zhang, X.; Bi, H.; Chen, S. 2016. Fluorenyl-containing quaternary ammonium poly(arylene ether sulfone)s for anion exchange membrane applications. *Fuel Cells* 16:557–567.

[23] Zuo, P.; Su, Y.; Li, W. 2016. Comb-like poly (ether-sulfone) membranes derived from planar 6,12-diaryl-5,11-dihydroindolo[3,2-b]carbazole monomer for alkaline fuel cells. *Macromol. Rapid Commun.* 37:1748–1753.

[24] Wang, J.; Wang, X.; Zu, D.; Hua, Y.; Li, Y.; Yang, S.; Wei, H.; Ding, Y. 2018. N3-adamantyl imidazolium cations: Alkaline stability assessment and the corresponding comb-shaped anion exchange membranes. *J. Membr. Sci.* 545:116–125.

[25] Zhang, Z.; Xiao, X.; Yan, X.; Liang, X.; Wu, L. 2019. Highly conductive anion exchange membranes based on one-step benzylation modification of poly(ether ether ketone). *J. Membr. Sci.* 574:205–211.

[26] Lin, C.X.; Zhuo, Y.Z.; Lai, A.N.; Zhang, Q.G.; Zhu, A.M.; Ye, M.L. 2016. Side-chain-type anion exchange membranes bearing pendent imidazolium-functionalized poly (phenylene oxide) for fuel cells. *J. Membr. Sci.* 513:206–216.

[27] Pan, J.; Han, J.; Zhu, L.; Hickner, M.A. 2017. Cationic Side-chain attachment to poly(phenylene oxide) backbones for chemically stable and conductive anion exchange membranes. *Chem. Mater.* 29:5321–5330.

[28] Zhang, X.; Li, S.; Chen, P.; Fang, J.; Shi, Q.; Weng, Q.; Luo, X.; Chen, X.; An, Z. 2018. Imidazolium functionalized block copolymer anion exchange membrane with enhanced hydroxide conductivity and alkaline stability via tailoring side chains. *Int. J. Hydrogen Energy* 43:3716–3730.

[29] Yan, X.; Deng, R.; Pan, Y.; Xu, X.; El Hamouti, I.; Ruan, X.; Wu, X.; Hao, C.; He, G. 2017. Improvement of alkaline stability for hydroxide exchange membranes by the interactions between strongly polar nitrile groups and functional cations. *J. Membr. Sci.* 533:121–129.

[30] Liu, J.; Yan, X.; Gao, L.; Hu, L.; Wu, X.; Dai, Y.; Ruan, X.; He, G. 2019. Long-branched and densely functionalized anion exchange membranes for fuel cells. *J. Membr. Sci.* 581:82–92.

[31] Guo, D.; Lai, A.N.; Lin, C.X.; Zhang, Q.G.; Zhu, A.M.; Liu, Q.L. 2016. Imidazolium-functionalized poly(arylene ether sulfone) anion-exchange membranes densely grafted with flexible side chains for fuel cells. *ACS Appl. Mater. Interfaces* 8:25279–25288.

[32] Liu, F.H.; Lin, C.X.; Hu, E.N.; Yang, Q.; Zhang, Q.G.; Zhu, A.M.; Liu, Q.L. 2018. Anion exchange membranes with well-developed conductive channels: Effect of the functional groups. *J. Membr. Sci.* 564:298–307.

[33] Lin, C.X.; Wang, X.Q.; Hu, E.N.; Yang, Q.; Zhang, Q.G.; Zhu, A.M.; Liu, Q.L. 2017. Quaternized triblock polymer anion exchange membranes with enhanced alkaline stability. *J. Membr. Sci.* 541:358–366.

[34] Guo, D.; Lin, C.X.; Hu, E.N.; Shi, L.; Soyekwo, F.; Zhang, Q.G.; Zhu, A.M.; Liu, Q.L. 2017. Clustered multi-imidazolium side chains functionalized alkaline anion exchange membranes for fuel cells. *J. Membr. Sci.* 541:214–223.

[35] Yang, Q.; Li, L.; Lin, C.X.; Gao, X.L.; Zhao, C.H.; Zhang, Q.G.; Zhu, A.M.; Liu, Q.L. 2018. Hyperbranched poly(arylene ether ketone) anion exchange membranes for fuel cells. *J. Membr. Sci.* 560:77–86.

[36] Li, T.; Yan, X.; Liu, J.; Wu, X.; Gong, X.; Zhen, D.; Sun, S.; Chen, W.; He, G. 2019. Friedel-Crafts alkylation route for preparation of pendent side chain imidazolium-functionalized polysulfone anion exchange membranes for fuel cells. *J. Membr. Sci.* 573:157–166.

[37] Zheng, X.Y.; Song, S.Y.; Yang, J.R.; Wang, J.L.; Wang, L. 2019. 4-Formyl dibenzo-18-crown-6 grafted polyvinyl alcohol as anion exchange membranes for fuel cell. *Eur. Polym. J.* 112:581–590.

[38] Du, X.; Wang, Z.; Liu, W.; Xu, J.; Chen, Z.; Wang, C. 2018. Imidazolium-functionalized poly (arylene ether ketone) cross-linked anion exchange membranes. *J. Membr. Sci.* 566:205–212.

[39] Yang, Q.; Li, L.; Gao, X.L.; Wu, H.Y.; Liu, F.H.; Zhang, Q.G.; Zhu, A.M.; Zhao, C.H.; Liu, Q.L. 2019. Crown ether bridged anion exchange membranes with robust alkaline durability. *J. Membr. Sci.* 578:230–238.

[40] Sung, S.; Mayadevi, T.S.; Min, K.; Lee, J.; Chae, J.E.; Kim, T.H. 2021. Crosslinked PPO-based anion exchange membranes: The effect of crystallinity versus hydrophilicity by oxygen-containing crosslinker chain length. *J. Membr. Sci.* 619:118774.

[41] Lai, A.N.; Guo, D.; Lin, C.X.; Zhang, Q.G.; Zhu, A.M.; Ye, M.L.; Liu, Q.L. 2016. Enhanced performance of anion exchange membranes via crosslinking of ion cluster regions for fuel cells. *J. Power Sources* 327:56–66.

[42] Lin, B.; Xu, F.; Chu, F.; Ren, Y.; Ding, J.; Yan, F. 2019. Bis-imidazolium based poly(phenylene oxide) anion exchange membranes for fuel cells: The effect of cross-linking. *J. Mater. Chem. A* 7:13275–13283.

[43] Hu, C.; Deng, X.; Dong, X.; Hong, Y.; Zhang, Q.; Liu, Q. 2021. Rigid crosslinkers towards constructing highly-efficient ion transport channels in anion exchange membranes. *J. Membr. Sci.* 619:118806.

[44] Al Munsur, A.Z.; Hossain, I.; Nam, S.Y.; Chae, J.E.; Kim, T.H. 2020. Quaternary ammonium-functionalized hexyl bis(quaternary ammonium)-mediated partially crosslinked SEBSs as highly conductive and stable anion exchange membranes. *Int. J. Hydrogen Energy* 45:15658–15671.

[45] Wang, C.; Mo, B.; He, Z.; Shao, Q.; Pan, D.; Wujick, E.; Guo, J.; Xie, X.; Xie, X.; Guo, Z. 2018. Crosslinked norbornene copolymer anion exchange membrane for fuel cells. *J. Membr. Sci.* 556:118–125.

[46] Ge, Q.; Liang, X.; Ding, L.; Hou, J.; Miao, J.; Wu, B.; Yang, Z.; Xu, T. 2019. Guiding the self-assembly of hyperbranched anion exchange membranes utilized in alkaline fuel cells. *J. Membr. Sci.* 573:595–601.

[47] Guo, D.; Zhuo, Y.Z.; Lai, A.N.; Zhang, Q.G.; Zhu, A.M.; Lin, Q. 2016. Interpenetrating anion exchange membranes using poly (1-vinylimidazole) as bifunctional crosslinker for fuel cells. *J. Membr. Sci.* 518:295–304.

[48] Gong, Y.; Liao, X.; Xu, J.; Chen, D.; Zhang, H. 2016. Novel anion-conducting interpenetrating polymer network of quaternized polysulfone and poly (vinyl alcohol) for alkaline fuel cells. *Int. J. Hydrogen Energy* 41:5816–5823.

[49] Hou, J.; Liu, Y.; Ge, Q.; Yang, Z.; Wu, L.; Xu, T. 2018. Recyclable cross-linked anion exchange membrane for alkaline fuel cell application. *J. Power Sources* 375:404–411.

[50] Wan, R.; Xu, S.; Wang, J.; Yang, Y.; Zhang, D.; He, R. 2021. Construction of ion conducting channels by embedding hydrophilic oligomers in piperidine functionalized poly(2, 6-dimethyl-1, 4-phenylene oxide) membranes. *Eur. Polym. J.* 142:110150.

[51] Xiao, Y.; Huang, W.; Xu, K.; Li, M.; Fan, M.; Wang, K. 2018. Preparation of anion exchange membrane with branch polyethyleneimine as main skeleton component. *Mater. Des.* 160:698–707.

[52] Mao, X.; Li, Z.; He, G.; Li, Z: Zhao, J.; Zhang, Y.; Jiang, Z. 2019. Enhancing hydroxide conductivity of anion exchange membrane via incorporating densely imidazolium functionalized graphene oxide. *Solid State Ionics* 333:83–92.

[53] Wang, J.; Chen, H.; Ma, Y.; Bai, H.; Shi, B.; Hou, C.; Wang, J.; Li, Y. 2020. Grafting high content of imidazolium polymer brushes on graphene oxide for nanocomposite membranes with enhanced anion transport. *React. Funct. Polym.* 146:104447.

[54] Qiu, M.; Zhang, B.; Wu, H.; Cao, L.; He, X.; Li, Y.; Li, J.; Xu, M.; Jiang, Z. 2019. Preparation of anion exchange membrane with enhanced conductivity and alkaline stability by incorporating ionic liquid modified carbon nanotubes. *J. Membr. Sci.* 573:1–10.

[55] Elumalai, V.; Sangeetha, D. 2019. Synergic effect of ionic liquid grafted titanate nanotubes on the performance of anion exchange membrane fuel cell. *J. Power Sources* 412:586–596.

[56] Feng, T.; Lin, B.; Zhang, S.; Yuan, N.; Chu, F.; Hickner, M.A.; Wang, C.; Zhu, L.; Ding, J. 2016. Imidazolium - based organic - inorganic hybrid anion exchange membranes for fuel cell applications. *J. Membr. Sci.* 508:7–14.

[57] Wang, C.; Lin, B.; Qiao, G.; Wang, L.; Zhu, L.; Chu, F.; Feng, T.; Yuan, N.; Ding, J. 2016. Polybenzimidazole/ionic liquid functionalized graphene oxide nanocomposite membrane for alkaline anion exchange membrane fuel cells. *Mater. Lett.* 173:219–222.

[58] Yang, Q.; Lin, C.X.; Liu, F.H.; Li, L.; Zhang, Q.G.; Zhu, A.M.; Liu, Q.L. 2018. Poly (2,6-dimethyl-1,4-phenylene oxide)/ionic liquid functionalized graphene oxide anion exchange membranes for fuel cells. *J. Membr. Sci.* 552:367–376.

[59] Chen, Y.; Li, Z.; Chen, N.; Li, R.; Zhang, Y.; Li, K.; Wang, F.; Zhu, H. 2017. A new method for improving the conductivity of alkaline membrane by incorporating TiO_2 - ionic liquid composite particles. *Electrochim. Acta* 255:335–346.

[60] Huang, C.Y.; Lin, J.S.; Pan, W.H.; Shih, C.M.; Liu, Y.L.; Lue, S.J. 2016. Alkaline direct ethanol fuel cell performance using alkali-impregnated polyvinyl alcohol/functionalized carbon nano-tube solid electrolytes. *J. Power Sources* 303:267–277.

[61] Elumalai, V.; Dharmalingam, S. 2016. Synthesis characterization and performance evaluation of ionic liquid immobilized SBA-15/quaternised polysulfone composite membrane for alkaline fuel cell. *Microporous and mesoporous Mater.* 236:260–268.

[62] Li, Z.; Zhang, Y.; Cao, T.; Yang, Y.; Xiong, Y.; Xu, S.; Xu, Z. 2017. Highly conductive alkaline anion exchange membrane containing imidazolium-functionalized octaphenyl polyhedral oligomeric silsesquioxane filler. *J. Membr. Sci.* 541:474–482.

[63] Liang, N.; Tu, Y.; Xu, J.; Chen, D.; Zhang, H. 2017. Hybrid anion exchange membranes with self-assembled ionic channels. *Adv. Polym. Technol.* 37:1732–1736.

[64] Vijayakumar, V.; Son, T.Y.; Kim, H.J.; Nam, S.Y. 2019. A facile approach to fabricate poly(2,6-dimethyl-1,4-phenylene oxide) based anion exchange membranes with extended alkaline stability and ion conductivity for fuel cell applications. *J. Membr. Sci.* 591:117314.

[65] Elumalai, V.; Sangeetha, D. 2018. Anion exchange composite membrane based on octa quaternary ammonium polyhedral oligomeric silsesquioxane for alkaline fuel cells. *J. Power Sources* 375:412–420.

[66] Luo, Z.; Gong, Y.; Liao, X.; Pan, Y.; Zhang, H. 2016. Nanocomposite membranes modified by graphene-based materials for anion exchange membrane fuel cells. *RSC Adv.* 6:13618–13625.

[67] Miao, L.; Bai, Y.; Yuan, Y.; Lu, C. 2018. Mussel-inspired strategy towards functionalized reduced graphene oxide-crosslinked polysulfone-based anion exchange membranes with enhanced properties. *Int. J. Hydrogen Energy* 43:17461–17474.

[68] Son, T.Y.; Ko, T.H.; Vijayakumar, V.; Kim, K.; Nam, S.Y. 2020. Anion exchange composite membranes composed of poly(phenylene oxide) containing quaternary ammonium and polyethylene support for alkaline anion exchange membrane fuel cell applications. *Solid State Ionics* 344:115153.

[69] Becerra-Arciniegas, R.A.; Narducci, R.; Ercolani, G.; Antonaroli, S.; Sgreccia, E.; Pasquini, L.; Knauth, P.; Di Vona, M.L. 2019. Alkaline stability of model anion exchange membranes based on poly(phenylene oxide) (PPO) with grafted quaternary ammonium groups: Influence of the functionalization route. *Polymer* 185:121931.

[70] Mandal, M.; Huang, G.; Kohl, P.A. 2019. Anionic multiblock copolymer membrane based on vinyl addition polymerization of norbornenes: Applications in anion exchange membrane fuel cells. *J. Membr. Sci.* 570–571:394–402.

[71] Sana, B.; Das, A.; Jana, T. 2019. Polybenzimidazole as alkaline anion exchange membrane with twin hydroxide ion conducting sites. *Polymer* 172:213–220.

[72] Ye, N.; Xu, Y.; Zhang, D.; Yang, Y.; Yang, J.; He, R. 2018. High alkaline resistance of benzyl-triethylammonium functionalized anion exchange membranes with different pendants. *Eur. Polym. J.* 101:83–89.

[73] Zhu, Y.; Ding, L.; Liang, X.; Shehzad, M.A.; Wang, L.; Ge, X.; He, Y.; Wu, L.; Varcoe, J.R.; Xu, T. 2018. Beneficial use of rotatable spacer side chains in alkaline anion exchange membranes for fuel cells. *Energy Environ. Sci.* 11:3472–3479.

[74] Lee, B.; Lim, H.; Chae, J.E.; Kim, H.J.; Kim, T.H. 2019. Physically-crosslinked anion exchange membranes by blending ionic additive into alkyl-substituted quaternized PPO. *J. Membr. Sci.* 574:33–43.

[75] Liu, L.; Liu, Z.; Bai, L.; Shao, C.; Chen, R.; Zhao, P.; Chu, X.; Li, N. 2020. Quaternized poly (2, 6-dimethyl-1, 4-phenylene oxide) anion exchange membranes based on isomeric benzyl trimethylammonium cations for alkaline fuel cells. *J. Membr. Sci.* 606:118133.

[76] He, S.; Liu, L.; Wang, X.; Zhang, S.; Guiver, M.D.; Li, N. 2016. Azide-assisted self-crosslinking of highly ion conductive anion exchange membranes. *J. Membr. Sci.* 509:48–56.

[77] Cheng, J.; Yang, G.; Zhang, K.; He, G.; Jia, J.; Yu, H.; Gai, F.; Li, L.; Hao, C.; Zhang, F. 2016. Guanidimidazole-quarternized and cross-linked alkaline polymer electrolyte membrane for fuel cell application. *J. Membr. Sci.* 501:100–108.

[78] Lai, A.N.; Zhou, K.; Zhuo, Y.Z.; Zhang, Q.G.; Zhu, A.M.; Ye, M.L.; Liu, Q.L. 2016. Anion exchange membranes based on carbazole-containing polyolefin for direct methanol fuel cells. *J. Membr. Sci.* 497:99–107.

[79] Ari, G.A.; Iojoiu, C.; Sanchez, J.Y. 2016. Synthesis and characterization of stable anion exchange membranes: the addition of electron-withdrawing group. *J. Natural & Appl. Sci.* 20:442–447.

[80] Elumalai, V.; Sangeetha, D. 2018. Preparation of anion exchangeable titanate nanotubes and their effect on anion exchange membrane fuel cell. *Mater. and design* 154:63–72.

[81] Zheng, J.; Zhang, Q.; Qian, H.; Xue, B.; Li, S.; Zhang, S. 2017. Self-assembly prepared anion exchange membranes with high alkaline stability and organic solvent resistance. *J. Membr. Sci.* 522:159–167.

[82] Han, J.; Lin, B.; Peng, H.; Zhu, Y.; Ren, Z.; Xiao, L.; Zhuang, L. 2020. Aggregated and ionic cross-linked anion exchange membrane with enhanced hydroxide conductivity and stability. *J. Power Sources* 459:227838.

[83] Kim, Y.; Wang, Y.; France-Lanord, A.; Wang, Y.; Wu, Y.C.M.; Lin, S.; Li, Y.; Grossman, J.C.; Swager, T.M. 2019. Ionic highways from covalent assembly in highly conducting and stable anion exchange membrane fuel cells. *J. Am. Chem. Soc.* 141:18152–18159.

[84] Kim, Y.; Moh, L.C.H.; Swager, T.M. 2017. Anion exchange membranes: enhancement by addition of unfunctionalized triptycene poly(ether sulfone)s. *ACS Appl. Mater. Interfaces* 9:42409–42414.

[85] Xue, J.; Liu, X.; Zhang, J.; Yin, Y.; Guiver, M.D. 2020. Poly(phenylene oxide)s incorporating N-spirocyclic quaternary ammonium cation_cation strings for anion exchange membranes. *J. Membr. Sci.*595:117507.

[86] Lin, B.; Xu, F.; Su, Y.; Han, J.; Zhu, Z.; Chu, F.; Ren, Y.; Zhu, L.; Ding, J. 2020. Ether-free polybenzimidazole bearing pendant imidazolium groups for alkaline anion exchange membrane fuel cells application. *ACS Appl. Energy Mater.* 3:1089–1098.

[87] Omasta, T.J.; Wang, L.; Peng, X.; Lewis, C.A.; Varcoe, J.R.; Mustain, W.E. 2018. Importance of balancing membrane and electrode water in anion exchange membrane fuel cells. *J. Power Sources* 375:205–213.

[88] Nunez, S.A.; Capparelli, C.; Hickner, M.Λ. 2016. N-alkyl interstitial spacers and terminal pendants influence the alkaline stability of tetraalkylammonium cations for anion exchange membrane fuel cells. *Chem. Mater.* 28:2589–2598.

[89] Mohanty, A.D.; Tignor, S.E.; Krause, J.A.; Choe, Y.K.; Bae, C. 2016. Systematic alkaline stability study of polymer backbones for anion exchange membrane applications. *Macromolecules* 49:3361–3372.

[90] Dekel, D.R.; Rasin, I.G.; Brandon, S. 2019. Predicting performance stability in anion exchange membrane fuel cells. *J. Power Sources* 420:118–123.

11 Polymeric Materials for Printed Electronics Application

Simran Sharma and Titash Mondal
Indian Institute of Technology, Kharagpur, West Bengal, India

CONTENTS

11.1   INTRODUCTION

In the era of next-generation materials, polymer-based printed electronic devices are a sure-shot future. As suggested by its name, printed electronic devices refer to the electronic devices that use printing as a fabrication technique for their construction (Cruz, Rocha, & Viana, 2018). With recent advancements in technologies and growing requirements for miniaturized electronic

DOI: 10.1201/9781003200710-13

devices, change in form-factors of electronics is rising. So are the researches to cater to the needs of the market. The use of these next-generation materials is not limited to traditional electronics. However, they are catering to non-traditional applications such as healthcare, printed electronics, EMI shielding, tissue engineering, and industrial and structural monitoring. Printing technologies have shown high compatibility and efficiency with polymeric materials, say in the form of inks or substrates. Further advancements in printing technologies have made it practical to print different materials like insulating, conducting, or semiconducting on required substrates. Various electronic devices such as organic field-effect transistors (OFETs), metal-oxide-semiconductors field-effect-transistors (MOSFET), lighting devices, storage devices, and sensors, to name a few, are printed with various printing technologies, including screen printing, inkjet printing, and gravure printing to create functional electronic devices.

The concept of printed electronic devices is not new; in fact, the commercialization of circuit printing started in the 1950s. Charles Ducus made the first significant contribution to this field in 1925 when he filed a patent to print conductors on insulating material with a stencil (Khandpur, 2006). Hitherto, many kinds of research happened, and various technologies for printing circuit boards came into existence, including etching, screen printing, and electrolytic metal deposition, to name a few. Traditional electronics employ the usage of silicon as a semiconducting material; however, functional fillers like graphene, and nanoparticles such as silver,and copper are exploited for preparing printed circuits. Conducting polymers are essential functional materials that can be added with functional filler to achieve the required functional attributes. The existence of electrically conductive polymers came into the limelight only after the 2000s when Hideki Shirakawa, Alan G. MacDiarmid, and Alan J. Heeger were awarded the prestigious Nobel Prize in Chemistry for their work "for the discovery and development of electrically conductive polymers" (Rasmussen, 2011). However, they conducted the study back in 1987. They found a dramatic increase in doped polyacetylene's conductivity, which gave new hope to engineers and scientists to look for alternative technologies that could cater to the need for alternative semiconducting materials. The history of conductive polymers has been an area of discussion for a long time as polyaniline was first reported in the nineteenth century individually by F. Ferdinand Runge, Henry Letheby, John Lightfoot, and Carl Fritzsche by various routes and techniques (Rasmussen, 2017). There is a debate on the discovery of conductive polymers, particularly polyaniline; in the nineteenth century, the synthesis of polyaniline was first reported, but it is only at the end twentieth century that conductive polymers have gained their much-respected place in the domain of the next-generation materials and printed electronic devices as next-generation technologies.

As per Forbes, the world is facing a shortage of semiconducting material in 2021. Due to this, automobile and smartphone companies expect to lose US\$70–80 billion in the second half of 2021. Smartphones and automobile companies face massive losses due to the shortage of semiconducting chips and microcontrollers (Ghosh, 2021; Sharma, 2021). The primary reason for the shortage is the disruption caused in the supply chain, majorly due to the pandemic. The technology to produce semiconductor chips is quite complex and involves many steps, including wafer processing, photolithography, etching, plasma ashing, thermal treatments, and chemical vapor deposition, to name a few, before testing. This complex and lengthy process makes semiconducting chip production a time-consuming process. On the other end, semiconducting polymers are a valuable resource to fabricate electronic devices with a simple printing process that involves only a limited number of steps, including polymer synthesis, and ink formulation followed by printing of the semiconducting polymer onto the substrate and finally curing it. Semiconducting polymers can help overcome the shortage of traditional semiconductors in every aspect once the technology becomes mature, as printing electronic devices is a less time-consuming task with a limited number of steps.

The application of polymers in printed electronic devices involves various sectors, including healthcare, where these materials ensure the degradation of sensors after usage. Due to the large-scale adoption of printed electronics in the healthcare domain, invasive and non-invasive healthcare devices reduce carbon footprints by reducing the volume of electronic waste. The primary application areas of polymeric electronic devices involve organic light-emitting diodes (OLEDs), organic photovoltaics (OPVs), batteries, field-effect transistors, and sensors. Section 11.6.1 will provide a thorough overview of OLEDs.

Further, these devices have found usage in OLED lighting and displays. section 11.6.2 will talk about various sensors in detail that can be used widely in day-to-day life. Batteries are equally essential and high potential application areas that ought to boom in the future and are covered in section 11.6.3. Another central application area of printed electronic devices that has taken the world with it is field-effect transistors. We shall discuss it in section 11.6.4. OPVs are another energy technology based on printed polymers. Section 11.6.5 discusses OPVs in detail (Hosseini et al., 2021). Conductive adhesives perform a crucial role in printed electronics and therefore are discussed in section 11.6.6.

Production of cost-effective and efficient electronic devices for all the discussed applications would not be possible without the right printing technology. Printing technology and ink formulations must be compatible with producing required devices with precision and accurate control that can perform as per the expectations. With the ongoing research, printing techniques are only improving to incline with printed electronic devices' specific requirements. Printing technology has usage in traditional applications such as printing paper, textiles, and plastics. However, research is a central thrust area to make printing a compatible technique for electronics production, and it is understanding the requirements and delivering only the improved results. Discussion on major printing technologies to fabricate polymer-based electronic devices is covered in the next section.

11.2   PRINTING METHODOLOGY

Printing is a widely used mass production technique to locally print colors in the form of images or text on the substrate such as paper, textile, leather, etc. While the advantages of printing couldn't be contained to the printing of colors for conventional applications, printing has found a new application area that is only making the emerging technologies accessible to the masses by providing its benefits to electronic device fabrication. Numerous printing techniques have already been adopted for printing traditional materials, whereas some have proven their worth to fabricate electronic devices. Methods of printing that are keenly being used and researched to apply polymers to construct electronic devices on the requisite substrate are discussed below.

11.2.1   Screen Printing

Screen printing is a contact printing process developed 2,000 years ago to produce repetitive patterns on textiles, and it is the first technique to have found its use to print electronic devices. Screen printing is either a flat-bed machine or a roll-to-roll screen printing machine. Flat-bed screen printing involves using a screen, a flooding blade, a substrate holder, and a squeegee. A squeegee presses against the screen to press the ink through both the flat-bed and rotary printing machines. The working operation of a flat-bed screen printing machine and a rotary printing machine is different. A cylinder contains the ink along with a squeegee in a rotary printing machine. Rotary printing is more expensive than a flat-bed printing machine, but its speed and resolution are better than the former. A mesh is made up of porous fabric or stainless steel with motifs photochemically or manually engraved. Lower viscosity pastes are unsuitable for screen printing as the paste could just run past the mesh without the aid of a squeegee. Therefore, the appropriate viscosity of printing paste is a must. It is suitable for the inks with particle size on the micrometer scale. The resolution of prints is limited in screen printing. It is due to the constraint in the meshing of 30–200 threads per cm. Screen printing inks show true shear thinning characteristics, and the viscosity of ink ranges from 0.5–50 Pa.s (Gregor-Svetec, 2018). Screen printing is advantageous in terms of its simplicity and is one of the widely used printing techniques. It is expected of screen printing to showcase maximum usage in coming years in printed electronic devices, but a significant downside is the lower resolution than other primary printing techniques.

Furthermore, the screen printing technique is helpful to produce encapsulation layers for printed electronic devices effectively. Electrodes and wire printing also employ screen printing technology to serve application areas like solar and fuel cells. Screen printing is the most widely used technique for printing electronic devices in the next decade, as per IDTechEx (Menard et al., 2007; Moonen, Yakimets, & Huskens, 2012; Nisato, Lupo, & Ganz, 2016).

11.2.2 INKJET PRINTING

Inkjet printing is one of the contactless printing techniques that can work with the ink viscosity of 0.001–0.04 Pa.s (Gregor-Svetec, 2018). Inkjet printing employs the usage of the jet that deposits the ink in small amounts. It either works on the principle of drop-on-demand (DOD), which ascertains that the jet dispenses only the required amount of drop on demand, or continuous inkjet (CIJ), which provides a constant stream of drops that are electrically charged and move towards the substrate. The drops act as building blocks in prints due to the higher accuracy and hold on the drops in inkjet printing. Ink formulation is a critical parameter, and various materials, including zero-dimensional (nanoparticles with dimensions less than 100 nm), one-dimensional (outside nanoscale such as nanowires and nanotubes), or two-dimensional material such as graphene, are used as per the requirement (Beedasy & Smith, 2020). A printing machine head cartridge of 1 pL can achieve a print thickness of 20 μm. Inkjet printing is a maskless and contact-less printing technique that ensures the contamination of substrate is minimized. Inkjet printing's success is highly associated with the ink formulation, which depends on the inkjet printing head technology. Right rheology and surface tension should be maintained to develop a stable ink that can develop reliable patterns with the requisite amount of ink. The print resolution is determined by the inkjet nozzle diameter and the spreading behavior of the drop on the substrate. The resolution up to 50–100 μm is achieved with inkjet printing, further improved with hybrid techniques such as inkjet printing employing photolithography to improve the resolution up to 1–20 μm. Inkjet printing has successfully been used to produce full-color OLEDs and circuits and remains the widely used research and development technique (De Gans, Duineveld, & Schubert, 2004; Ko et al., 2007; Gao, Li, & Song, 2017; Al-Halhouli et al., 2019; Kraft et al., 2020).

11.2.3 FLEXOGRAPHY

Flexography is the widely used and mature printing technique in the packaging industry, where the technique is utilized to print polymeric substrates like milk cartons, food and hygiene bags, and disposable cups. It is also becoming a part of intelligent packaging due to its usage in RFID tags' printing. The machine has two units, namely the inking unit and the printing unit. The inking unit consists of an anilox roller and a fluid reservoir, whereas the printing unit comprises a plate cylinder and an impression cylinder. The former unit is responsible for dispensing the right amount of ink onto the anilox roller, whereas the latter unit is responsible for printing the correct printing patterns onto the substrate with the aid of raised ink on the plate cylinder. This technique can provide a layer thickness of tens of nanometers with a resolution of 4,000 PPI, and printed electronic devices constructed with flexography can have a thin structure. Flexography printed organic thin-film transistors are reported to be produced by Thomas Cosnahan et al. at the University of Oxford in 2018, and flexography has been considered viable for producing printed electronic components (Vena et al., 2013; Cosnahan, Watt, & Assender, 2018).

11.2.4 GRAVURE PRINTING

Rotogravure printing is the most commonly used printing technique today and has shown practical usage in printed electronic devices. The printing system consists of an ink pan, and a gravure cylinder, followed by an impression cylinder. Unlike flexography which uses a raised platform for ink transfer onto the substrate, gravure printing involves using a gravure cylinder with patterns engraved on it. The ink is fed continuously from the ink pan to the engraved cells of the gravure cylinder. Cell density on the gravure cylinder ranges from 400 cells per inch, with each cell having a depth of approximately 40 μm and a width of less than 100 μm. The resolution of imprints ranges from several nanometers to tens of micrometers with speeds up to 1 m/s. Inks for gravure printing are supposed to be fluidic with shear viscosity ranging from 0.05–0.2 Pa.s and low surface tension to quickly fill the cells of gravure (Aleeva & Pignataro, 2014). Today, researchers have successfully developed layers of OFETs, OPV, and OLEDs with rotogravure printing. Further research is working to optimize gravure cells' emptying, which causes a thin film on non-printing areas and effective solvent evaporating in the printed patterns remains an area of keen research (Nisato, Lupo, & Ganz, 2016).

11.2.5 REVERSE OFFSET PRINTING

Offset printing is one of the emerging printing technologies that has found its usage in electronic device fabrication. It involves using ink with a viscosity of 1–10 cP applied to a PDMS sheet. After allowing the ink to rest for a few seconds on the sheet until it reaches a semi-dried state, its contact with the cliché removes unnecessary portions of the ink. The sheet is then allowed to contact the substrate where it gets printed (Kusaka, Fukuda, & Ushijima, 2020). As this technology transfers the semi-solid ink to the substrate, the patterns formed are clean; thus, defects such as the halo effect can be avoided. The halo effect is the accumulation of printing ink beyond the edges of the printed areas, thus creating a halo-like outline. A reverse offset printed organic complementary inverter circuit is fabricated by Takeda and coworkers (Takeda et al., 2018). Circuits, RFIDs, and organic thin-film transistors are reported to be printed with this technology due to their plus points, such as the high resolution of producing channel lengths of 1 μm or lesser along with the film thickness of less than 100 nm (M. Kim et al., 2011; Fukuda et al., 2015).

Each of the printing techniques discussed has its own set of pros and cons. While screen printing is ideal for printing multiple thick stacks up to the thickness of 100 μm, inkjet printing is most suitable for research and development activities and can provide print thickness up to 20 μm. However, flexography and gravure printing technologies are the ones that are conquering the market when it comes to mass production.

Further, reverse offset printing is an emerging technology that offers higher resolution prints (Fukuda et al., 2015) and the advantage of having higher operational speeds and accuracy.

It is essential to choose the printing technology as per the final requirement critically. However, the choice of inks that are to be used in printing is also an area of significant concern. Properties of the ink ought to vary with the technology employed to fabricate the device, and therefore it is crucial to monitor and tailor the property of inks. It is beneficial to use polymers to fabricate conductive devices as polymers are compatible materials with printing machines and complement the printing process with their unique characteristics. Ink formulation with polymers is an exciting yet critical area to discover. We shall discuss more on polymeric inks for printed electronic devices in the next section.

11.3 POLYMER INK COMPOSITION FOR PRINTED ELECTRONIC DEVICES

Solution-processable organic semi (conductors)' utility in fabricating low-cost electronic devices with industry-oriented printing techniques makes it crucial to labor on ink formulation. This is important to fabricate devices with precision and accuracy. Even a slight defect in printing can render the device ineffective. Therefore, it is vital to control the printing parameters, and the most critical one is the ink formulation. The utmost interesting aspect of utilizing conducting polymer ink is the ink's tailor ability by playing with the polymer's structure, concentration, molecular weight, and solvent choice accordingly to formulate the desired ink. For example, the polymer can inherently be tailored for improved solubility or electron affinity by adding side chains in the polymer backbone (Aleeva & Pignataro, 2014). Polymers, being macrostructures, can also be modified intrinsically based on their molecular weight to influence ink's rheology. The increasing molecular weight of polymer strongly decreases the printability due to high elongational flow during the nozzle extrusion, as reported by Berend-Jan de Gans et al. in the inkjet printing process (De Gans et al., 2004). Higher elongational flow does not arise with printing techniques such as gravure printing, but rheology, surface tension, and spreading behavior of ink are of utmost concern (Hrehorova, Pekarovicova, & Fleming, 2006). A primary polymeric ink is composed of a functional polymer, solvent, binder, and a rheology modifier. A functional polymer is selected as per the final application. However, only after choosing the suitable functional polymer, does one determines the appropriate solvent, binder, and rheology modifier. The chemistry of the functional polymer is of great interest and is obligatory to deliberate during ink formulation. To print functional polymer onto the substrate, it is essential to solubilize it in an appropriate solvent.

Conversely, the functional polymer will not adhere to the substrate's surface without a binder's aid, an essential component in ink formulation. Even if the formulation of ink consisting of a functional polymer, solvent, and binder can produce the desired electronic device, printing technologies do not support formulated ink. It is the suitable rheology of the ink that makes it a relevant formulation for available printing technologies. The requirement of rheology varies with the adopted printing techniques. For example, inkjet printing ink will be less viscous than the one used in gravure printing. It is essential to tailor the correct rheology of the printing ink. The significant challenges involved in printing conductive inks are attaining uniform and continuous ink deposition (Mendez-Rossal & Wallner, 2019). Therefore, ink formulation is of utmost concern. The conductive polymer's solubility in the respective solvent becomes a vital parameter to consider, depending on many different factors, including molecular weight, crystallinity, branching, degree of cross-linking, and the solvent itself. Generally, as a rule of thumb, non (polar) polymers are dissolved in (non) polar solvents.

Further, the mode of dissolution is swelling of polymer followed by dissolution into smaller polymer coils. While preparing an ink solution, it is worth considering a strong polymer-solvent interaction than a polymer-polymer interaction (Hibon et al., 2020). Quantification of the resulting polymer's solubility can be done with the Hansen solubility parameter, as shown in equation (11.1), in which a single parameter quantifies cohesive energy density;

$$\delta = \sqrt{\frac{E}{V_m}} \tag{11.1}$$

E is the energy of vaporization, and V_m is the molar volume (Schlisske et al., 2020).

Apart from a conductive polymer, binder, solvent, andrheology modifier, ink recipes can constitute other reagents, including surfactants, antimicrobial agents, co-solvents, plasticizers, ionic liquids, etc. depending on the specific requirement. Surfactants are essential in an ink formulation with polymers that require an external component to ensure dispersion in the solvent. Plasticizers like zonyl fluoro surfactants and ionic liquids make the conductive polymer stretchable. In contrast, prints that suffer from defects like the coffee ring effect can be overcome by using co-solvents to create a surface tension gradient in a drop rendering the ink constituents spread evenly inside a drop. Despite a conductivity of 600 S cm^{-1}, PEDOT: PSS ink-based interconnects and circuits suffered due to poor morphology and defects like the coffee ring in the prints. To overcome this problem, dodecylbenzene sodium sulfonate (DBSS) was added to increase the wettability and thereby improving morphology; and a high boiling point and lower surface tension co-solvent is added to the ink that created a surface tension gradient inside the extruded drop, creating a Marangoni effect, which helps to eradicate the coffee ring effect, thereby creating the robust and effective prints (Kraft et al., 2020).

Eom et al. evaluated the morphology of PEDOT: PSS-based thin ink films prepared with or without additives. Glycerol-6% and ethylene glycol butyl ether (EGBE)-0.2% were used as additives to improve the spreading and wetting behavior of printed films on ITO substrate for the preparation of solar cells. The ink with additives had better density and uniform thickness of the thin film and improved morphology than the prints produced from the ink without additives. The improvement may be attributed to ethylene glycol butyl ether (EGBE) surfactant addition that minimized the interfacial tension between the glycerol and PEDOT: PSS (Eom et al., 2009). An ink's conductivity can be increased by the addition of conductive fillers such as graphene, CNTs, fullerenes, etc. They are reported to be added to the formulation along with the basic recipe of ink discussed above (Eom et al., 2010; Dang, Hirsch, & Wantz, 2011; Z. Liu et al., 2015). When the suitable formulation of ink is selected, the constituents' proportions must be adjusted to have precise control over the prints. Therefore, the properties of the respective ink are adjusted to complement the printing technology and substrate. Formulation of the correct ink depends on several factors, as discussed above.

11.4   CONDUCTIVITY IN POLYMERS

For a long time, it has been a common belief that polymers are insulating in nature. With the introduction of conjugated polymers by Shirakawa et al. in 1977 (Derivatives, Louis, & MacDiarmid, 1977), polymers' conductivity became a hot topic of discussion and research. Ways to exploit this nature of conductive polymers in various application areas have been explored, and an era of next-generation materials has started with their emergence. Conductive polymers are reputed candidates to replace traditional materials to produce flexible electronic devices with simple printing techniques. However, the working mechanism of semiconducting polymers is somewhat different from that of traditional semiconductors. Moreover, organic semiconductors' performance is different from traditional semiconductors, limiting their usage in a few application areas.

Traditional electronic devices employ silicon as a semiconducting material, which has electron mobility in 1,500 cm^2 V^{-1} s^{-1} and hole mobility of 450 cm^2 V^{-1} s^{-1}. In contrast, organic semiconductors provide mobility in the range of 1–40 cm^2 V^{-1} s^{-1} for p-type semiconductors and 0.1–5 1–40 cm^2 V^{-1} s^{-1} for n-type semiconductors. Furthermore, the switching speed with printed transistors still lies in the MHz range. In contrast, traditional silicon-based semiconductors provide the switching speed in the GHz range (Swisher, Volkman, & Subramanian, 2015). This considerable difference in mobilities affects the performance of electronic devices. It is the primary reason polymers are still not used in most electronic devices that we use today. Usage of polymer-based electronic devices has shown promising results in low-performance areas where the requirements are not stringent in devices with shorter life spans.

Moreover, semiconducting polymers are effectively exploited in devices such as printed energy harvesting devices, antennas, displays, interconnects, power sources, and power sources. However, their application is still limited in sensors and circuits due to their lower stability and performance (Khan et al., 2020).

Semiconducting polymers are organic and inherently conjugated in nature and possess σ and π bonds at an alternate position. It leads to increased σ bond energy and decreased π bond energy, also known as dimerization. Polymers such as polyacetylene have periodic sp^2 hybridized structures in one dimension. The electronic properties in such polymers arise due to a bandgap's presence due to the splitting of bonding and antibonding orbitals of π bonds occurring due to the overlapping of wave function between two $2p_z$ orbitals of neighboring carbon nuclei. This π-π* gap is known as an energy gap and ranges from 1–3 eV for these polymers. The energy gap is calculated as the difference between the ionization potential (I.P.) and electron affinity (E.A.); where I.P. is the difference between the highest occupied molecular orbital (HOMO) and vacuum level, and E.A. is the difference between the lowest unoccupied molecular orbital (LUMO) and vacuum level. The unique difference in the working mechanism of inorganic semiconductors and organic semiconductors lies in their structure. An inorganic crystal structure remains intact with the presence or absence of electrons in its conduction band. There is a distortion in a lattice structure in an organic semiconductor, leading to electron wave function localization. Electrons having such distortions in lattice structure are termed polarons which polarize the lattice with the extra electron. Two polarons combine to form a bipolaron as well. In the case of holes, hole polarons and bi-polarons are formed. As polarons and bipolarons have a local lattice distortion that moves with the wave function, the kinetic energy is relatively high compared to individual electrons and holes. Furthermore, the higher mass of polarons is responsible for the lower mobility in organic semi-conductors than inorganic semiconductors.

Moreover, polymers do not have perfect periodicity. They are prone to defects, leading to the formation of singlet and triplet states of the exciton (a low-lying excited bound state of electron-hole pair) with discrete states having different binding energies. A singlet exciton decays radiantly with a phonon emission, whereas a triplet exciton decays non-radiatively with multiple phonon emissions. Generally, optical excitation causes the conjugated polymer to excite from a ground state to a singlet exciton state, leading to the phenomena of photoluminescence as exciton tends to come back to the lowest state in 200 femtoseconds, making it an ultrafast process. In electronics, conjugated polymers' significant contribution to the electrical process is based on two significant phenomena: electroluminescence and photoconductivity. Electroluminescence is shown by only those conjugated polymers in which the exciton is in a singlet state with odd parity; when electric bias is applied. Photoconductivity is caused by the carrier charge generation when the polymer is optically excited. However, all of these mechanisms are associated with amorphous conjugated polymers. In contrast, crystalline conjugated polymers such as P3HT do not show luminescence due to their tendency to self-assemble as locally crystalline structures, causing a waveform overlap in adjacent nuclei. It leads to considering a 3D structure of P3HT in which the charge does not move in the form of a single chain polaron, but it has a 3D waveform. Due to the locally crystalline structure of P3HT, it has better hole mobility. P3HT can be p doped due to its lower ionization potential of 5 eV and shows doping and de-doping effects. It tends to dope and de-dope with ambient oxygen, limiting its usage in field-effect transistors for practical usage (Loudon, 1959; Heeger, 2001; Sitch, Halliday, & Monkman, 2001; Meng, 2013).

After a brief understanding of how conductivity works in a polymer, it's essential to look at the processes that can be employed to extract advantages from these polymers. Numerous techniques have been used to produce polymers in general. However, few of them have made a mark in the production of conductive polymers. A brief discussion to produce conductive polymers is discussed in the next section.

11.5 SYNTHESIS OF CONDUCTING POLYMERS

Conductive polymers have found usage in various applications, including OPVs, OLEDs, batteries, etc. Two key routes majorly synthesize synthetic material or organic polymers used to meet these application areas' expectations. The materials are either chemically polymerized or electro-chemically polymerized. Chemical oxidation polymerization is suitable for mass production, whereas electrochemical polymerization is a direct method to produce thin films that can be used in electronic devices. Polymerization of conducting polymers can be achieved with various routes and is discussed in detail below.

11.5.1 SYNTHESIS OF CONDUCTIVE POLYMERS BY CHEMICAL OXIDATION POLYMERIZATION

Chemical oxidative polymerization is a technique used to polymerize conductive monomers. It involves using a solvent as a medium to solubilize the monomers, which are made to react with an oxidizing species such as $CuCl_2$, $AlCl_3$, $FeCl_3$, etc. This process produces HCl as the by-product, thus to avoid harmful by-products like HCl, oxygen is reported to be used as an oxidizing agent (Toshima & Hara, 1995). The oxidizing species produce a cation radical that initiated the polymerization process. The cation-free radical from species like $FeCl_3$ is responsible for creating the conjugated structure with the aid of deprotonation. Polythiophenes like PEDOT can be synthesized from EDOT with the usage of $FeCl_3$ as an oxidizing agent. MWCNT-PEDOT composite is also reported to be synthesized by the polymerization of EDOT monomers with the help of $FeCl_3$ on the MWCNT cables (Reddy et al., 2010). Nanotubes of polyaniline (PANI) have received much attention in the last decade. The chemical oxidation of aniline in a specific condition with organic sulfonic acids has been used to synthesize the same. Chemical oxidative polymerization of aniline with water acting as an oxidizing agent is reported by Ć irić -Marjanovic et al. in 2008 (Ć irić-Marjanovic, Trchová, & Stejskal, 2008).

Polymerization in the vapor phase closely resembles the reaction mechanism in a solution phase but varies in various aspects. There are several reported techniques to employ the vapor phase as a medium for the delivery of monomer and in some cases to the oxidants as well, such as thermal evaporation, vapor phase polymerization (VPP), pulsed laser polymerization, pyrolysis, vapor deposition polymerization, oxidative chemical vapor polymerization (oCVD), and plasma-enhanced chemical vapor polymerization. In most reports, VPP and oCVD are used to synthesize the conductive polymers and are discussed below.

11.5.2 Synthesis of Conductive Polymers by Vapor Phase Polymerization (VPP)

Vapor phase oxidative polymerization is a two-step process and procedures the monomer in vapor form to the polymerization chamber. The need to select a compatible solvent is eliminated in the process, and the polymer can be synthesized directly on the substrate. As polymerization involves using a cation radical as an initiator, the oxidizing agent should be applied to the substrate in the first step. As the reaction proceeds in the vacuum chamber in the second step, vaporized monomers start to polymerize on the substrate at ambient or controlled low pressure. The polymeric film is deposited on the substrate at the end of the polymerization process. Thus, the deposited film is then washed with alcohol to remove adsorbed monomers and unreacted oxidants along with by-products. Polypyrrole was first synthesized using VPP in 1986 on a PVA substrate using $FeCl_3$ as an oxidant. Usage of $CuCl_3$ as an oxidant has also been reported in the preparation of thin-film polypyrrole. However, the conductivity of produced films was lower than commercially available PEDOT:PSS polymer. When PEDOT is VPP synthesized with $FeCl_3$ as an oxidant, initial reports confirm the conductivity yield around 70 S cm^{-1}. Iron (III) p-toluene sulfonate in a mixture with volatile base pyridine is used as an oxidant to prepare PEDOT from VPP with conductivity up to 1,000 S cm^{-1} (Bhattacharyya et al., 2012).

11.5.3 Synthesis of Conductive Polymers by Oxidative Chemical Vapor Deposition (oCVD)

oCVD is a single-step process for the synthesis of conductive polymers. In this process, both the monomer and oxidant are fed to the substrate in the reaction chamber in vapor form. Therefore, the process is even suitable for using volatile oxidants such as bromine. The delivery timing of oxidant and monomer is of utmost concern in this process, and the quality of this synthesized polymer highly depends on it. Anhydrous solutions of Lewis acids such as $FeCl_3$ are used as catalysts and dopants in oxidative chemical polymerization of polypyrroles, polythiophenes, and so on. Organic photovoltaics based on oCVD deposited PEDOT on graphene have been reported to synthesize in a single step.

Both of the discussed techniques based on vapor deposition help prepare co-polymers due to the advantage of controlled feeding of monomers to the reaction chamber. VPP synthesized copolymerized P(EDOT-co-3HT) have shown to have conductivities up to 2.3×10^1 S cm^{-1} (Myers, 1986; Lock, Im, & Gleason, 2006; Bhattacharyya et al., 2012).

11.5.4 SYNTHESIS OF CONDUCTIVE POLYMERS BY PHOTOCHEMICAL POLYMERIZATION

Photochemical polymerization is based on photochemical reactions occurring due to the absorption of light leading to the electrons' excitation in a given molecule. It acts as a photo initiating system for cationic as well as radical-based polymerizations. Photoinitiators including benzoin ether, acetophenones, benzyl oximes, benzyl ketals, acyl phosphine oxide, aminoalkyl phenones, benzophenone, benzyl, thioxanthones, and quinones are used for the photoinitiation in the free radical polymerization. On the contrary, Lewis acids, Bronsted acids, onium salts, and carbonium ions are used in the cationic polymerization photoinitiation. Cationic polymerizations are challenging to control, and therefore living polymerization is difficult to achieve. The splendor of this polymerization is that linear and branched polymers can be synthesized with it (Tasdelen & Yagci, 2011). Photochemical polymerization is an inexpensive and quick method to synthesize conductive polymers. However, only a limited number of polymers are reported to be synthesized by this technique. Preparation of polypyrrole from pyrrole is reported to be done by using visible light for irradiation in the presence of a photosensitizer or an electron acceptor. Similarly, polyaniline has also been synthesized in the presence of H_2O_2 (Awuzie, 2017).

11.5.5 SYNTHESIS OF CONDUCTIVE POLYMERS BY TRANSITION METAL CATALYZED POLYCONDENSATION

Organometallic polycondensation reaction facilitated by transition metal complexes poly condenses the dihaloaromatic (X-Ar-X) compounds to produce polyarylenes $-(Ar)_n-$ by elimination of by-products. Transition metal-catalyzed polycondensations are actually step-growth reactions with changes in the valence of metal. Predominantly, these are two-electron processes such as metal catalysts changing their valence states from Pd(0) to Pd(II) with a reversible attribute except with copper catalysts that change the valence states from Cu(I) to Cu(II). Copper, iron, nickel, palladium, molybdenum, manganese, osmium, and cobalt are common transition metals used in the form of complexes that polycondense the monomers to synthesize polymers. However, palladium and nickel are the transition metals of choice for polycondensation reactions for synthesizing conducting polymers. Polycondensation of polythiophenes such as P3HT in head to tail (H.T.) configuration is of great interest as the compound shows remarkable properties in organic electronic devices. HT-P3HT, also known as regioregular P3HT, can be synthesized by de-brominative polycondensation of 2, 5-dibromo-3-hexylthiophene by using Grignard reagent with transition metal catalyst and is considered one of the successful methods in synthesizing the same (Heitz, 1995; Yamamoto, 2002; Ouchi, Terashima, & Sawamoto, 2009; Tamba et al., 2011).

11.5.6 SYNTHESIS OF CONDUCTIVE POLYMERS BY ELECTROCHEMICAL POLYMERIZATION

An electrochemical cell that converts the electrical energy into chemical energy, called an electrolytic cell, is employed to synthesize conductive polymers. Electrochemical polymerization is carried out in an electrochemical cell either with two electrodes or three electrodes. However, the preferred method is to use three-electrode systems to ensure more control over the synthesized quality. Further, electrochemical polymerization can be categorized into two categories based on the type of radical formation for polymerization initiation, namely, anode electrochemical polymerization or cathode electrochemical polymerization. In the former, radical cations are responsible for starting a reaction with neutral monomers, whereas the latter involves the formation of an anionic radical that initiates the polymerization process. Anode electrochemical polymerization is a general mode of polymerization for conducting polymer synthesis. In the reaction mechanism, the following steps are involved: (i) Soluble monomers are formed as the diffusion layer between the bulk solvent and electrodes; (ii) as the solubility of monomers decreases, thin films are deposited while nucleation and growth of polymer occurring simultaneously; and (iii) the chains grow further to produce a 1D, 2D, or a 3D structure by radical-radical coupling. Aprotic solvents such as DMF, DMSO that have low nucleophilic character are ideal for polymerization as the reaction proceeds with radical cation intermediaries.

Further, the choice of electrolyte also plays a crucial role in shaping the process of polymer synthesis and for this reason, quaternary ammonium salts with structure (R_4NX), where R = alkyl, aryl, X = Br^-, Cl^-, I^-, BF_4^-, PF_6^-, $CH_3C_6H_4SO_3^-$) are helpful in this type of polymerization method. Polythiophene and polyaniline are examples of conducting polymers that are synthesized with this route. Only a few polymers are synthesized with cathode electrochemical polymerization, such as poly(p-phenylenevinylenes). High-quality OLEDs and luminescent films are reported to be synthesized with electrochemical polymerization by Manlin Zhao et al. with N-alkyl carbazole (B. Wang, Tang, & Wang, 1987; Wei et al., 1991; Gurunathan et al., 1999; M. Zhao et al., 2020).

Hitherto, we have discussed the process of developing electronic devices with the aid of polymers. Let us look at the application areas where these devices are serving or ought to serve immensely soon.

11.6 APPLICATIONS

Various thrust areas have been served by printed electronic devices today. They have found applications in sensors for monitoring health care, and industrial and structural monitoring, to name a few. Organic light-emitting diodes (OLEDs) are another central application area that has accepted the usage of organics in electronics and are one of the universally accepted technologies by big giants in the electronics market. OLED-based smartphone and televisions are already doing rounds in the market in the twenty-first century, and technological breakthroughs are only improving their facilitation in device construction. Conductive polymers have also found application in organic photovoltaics (OPVs). However, this particular application is expected to be in its pre-industrial phase and soon is expected to cover a vast market share in photovoltaics. Batteries are being replaced with organics as well; not only are the conductive polymers acting as innovative electrolytes that have substantially changed the design of batteries but they are also being investigated to serve as effective electrode materials that could allow the fabrication of fully printed batteries. Transistor fabrication using polymers is a practical step to replace the circuits in electronic devices. Conductive polymer-based transistors are still lacking in their performance due to lower mobilities than their Si-based counterparts to be used in active electronic devices such as cell phones. However, they effectively serve the low-performance applications such as those involving short-term usage with lower current requirements as in health care applications. Using organics in the high-performance application is still a high thrust area in research, and promising results provide hope to see the replacement of inorganic semiconductors with organic ones shortly.

The primary application areas with ongoing research are reviewed and below are the key findings depicting the relevant research and future scope for growth in the respective applications.

11.6.1 ORGANIC LIGHT-EMITTING DIODES (OLEDS)

According to IUPAC, organic light-emitting diodes are a type of material capable of generating light when an electric current is passed through it. The mainstream usage of OLEDs involves OLED-based lights and displays due to their property of electroluminescence. With the technological breakthrough in fabrication techniques, electronic device construction is changing from a dry process to a wet process; the requirement for solution-processable materials allowed polymers to effectively showcase their potential to be efficiently used in devices like OLEDs (Sekine et al., 2014).

According to IDTechEx, OLED displays and lights combined are expected to share the US $2,000–3,000 million by 2029 in the printed electronics market. J Rogers and Bao reported using organic electrophoretic ink-based p-channel transistors printed on the flexible substrate to drive currents in a paper-like display that can be bent without affecting its operation in 2002 (Rogers & Bao, 2002). Various companies, including Sony, have demonstrated the use of electrophoretic inks to create flexible displays. Recently, LG has become a pioneer in producing OLED TVs by modifying their technology year by year. The most recent commercialization of a nicely printed display is demonstrated by companies like Samsung and Royole Corporation offering foldable displays in their smartphones; Samsung offers a skin phone concept using a foldable display (Huitema, 2012). Zhao et al. prepared a full laser display based on a microlaser array prepared with organic ink.

As reported by Villani et al., piranha solution (H_2SO_4/H_2O_2 4:1 (v/v)) treatment on the ITO coated PET substrate increases the surface energy and polarity, thereby decreasing the contact angle and causing an increase in the wettability to ensure even printing. Piranha solution treated and untreated substrates are analyzed and showed similar transmittance values in UV-Vis-NIR spectrometry in the region 200 to 1,000 nm, as shown in Figure 11.1 (Villani et al., 2009).

Furthermore, 1 H,1 H,2 H,2H-perfluorooctyltriethoxysilane was used as a coating on the substrate to impart a hydrophobic effect to control the printed array morphology. It allowed displaying required colors on the screen effectively. RGB microarrays had well-defined spherical morphology leading to minimized optical scattering and achieving a more substantial microcavity effect (J. Zhao et al., 2019). It is not the only area where organics are limited to use in OLEDs. The traditional use of indium tin oxide as the electrode material in the displays has successfully been challenged by Montanino et al. in their 2019 research to develop gravure printed OLEDs with PEDOT: PSS as the working anode (M. Montanino et al., 2017). Huseynova et al. recently demonstrated the usage of PEDOT: PSS-based solution-processed electrodes in OLEDs (Huseynova et al., 2020). However, the first reported PEDOT: PSS-based OLED with a longer lifetime and higher efficiency was reported by Kim et al. in 2013 (Y. H. Kim et al., 2013). The direction of research in the OLEDs is aiming to fully print OLEDs while fulfilling the technological requirement of the final product and at the same time being compliant with the processing techniques.

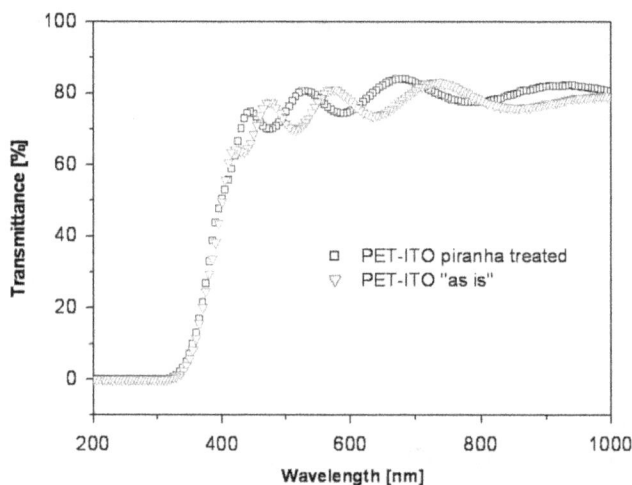

FIGURE 11.1 Transmittance spectra of Piranha solution treated (square dots) and "as is" (downward triangle dots) ITO coated PET substrate when UV-Vis-Spectrometry is performed over a range of 200 nm to 1,000 nm. Transmittance value of 0% at 200 nm for both PET-ITO piranha treated and PET-ITO "as is" starts to increase above 300 nm, it reaches a maximum of approximately 80% at little below 700 nm for PET-ITO piranha treated and with a fringe shift little above than 700 nm for PET-ITO as it is. The fringe shift can be observed throughout the plot. Reprinted (adapted) with permission from Villani, F. et. al. "Inkjet Printed Polymer Layer on Flexible Substrate for OLED Applications." *Journal of Physical Chemistry C* 113 (30): 13398–13402. doi: 10.1021/jp8095538. Copyright (2009) American Chemical Society.

11.6.2 Sensors

Recently, sensing applications involving printed electronic devices have become a sensitive research area, particularly due to their low cost and ease in mass-scale fabrication (Subramanian, Chang, & Liao, 2013). A sensor provides information about the surrounding and physiological environment. A majority of research has showcased the usage of polyethylene terephthalate, polyimide, thermoplastic polyurethane, and polyethylene naphthalate in sensors due to their high flexibility, low Young's modulus, higher elastic limits, and impact resistance; properties ideal for adhering to an external body such as human skin. Usage of paper and dimethylsiloxane (PDMS) has also been confirmed as a substrate in various sensors for their unique characteristics (Cui, 2016; Hoeng, Denneulin, & Bras, 2016; Lessing et al., 2014; Zhou et al., 2018; Maddipatla, Narakathu, & Atashbar, 2020). Moreover, due to polyaniline's high chemical sensitivity, the polymer is exploited in various sensing applications. For the detection of VOCs, polyaniline/MXene-based nanocomposite in which MXene acted as a nanosheet surface and polyaniline nanoparticles effectively-being decorated on it has been efficiently utilized by Zhao et al. The sensor showcased 41.1% ethanol sensitivity and rapid response rate at room temperature. The sensitivity of polyaniline is attributed to the fact that it exists in two emeraldine classes, namely insulating emerald base with conductivity up to 10^{-5} S/cm and emeraldine salt conducting form with conductivity 1,000 S/cm. The former form of polyaniline can be converted to the latter one with the aid of protonic acid. The deprotonating gases cause the depletion regions in the p-n junction of polyaniline-based sensors leading to a decrease in effective conductivity and thereby detecting the gases (Pandey, 2016; L. Zhao et al., 2019). Printed polyaniline nanoparticles-based breath sensors have been developed by Hibbard et al. for the detection of ammonia. The impedance response $[(Z/Z_o) - 1]$ of the sensor against temperature and humidity, volume flow rate, and sample chamber volume has been studied. Results are depicted in Figure 11.2 where affirmative results have been observed depicting its high potential in the applications such as non-invasive breath ammonia testing for health monitoring (Hibbard et al., 2013).

RFID-enabled polyaniline/carbon nanocomposite-based chemi-resistor has been developed for NH_3 and humidity sensing (Quintero Vasquez et al., 2016; Hartwig, Zichner, & Joseph, 2018). Due to electrochemical affinity and high biocompatibility of polypyrrole, polypyrrole-based immunosensor fabrication for biological detection, nucleic acid, and antigen detection are also reported (Taylor, Ramanaviciene, & Ramanavicius, 2013). These breakthroughs in sensing technologies are giving people effective solutions in detection. PEDOT: PSS is a very versatile polymer that has showcased its relevance in a vast area of applications. Some of them are discussed here. PEDOT: PSS-based temperature sensor has been constructed by Wang et al. for wireless health monitoring with high humidity stability. A higher sensitivity of 0.77%°C^{-1} has been achieved in relative humidity ranging from 30% to 80%. The decrease in resistance and increase in conductivity of PEDOT: PSS by limited carrier hopping and tunneling due to an increase in thermal energy leads to the detection of temperature (H. Wang et al., 2020). Moreover, PEDOT: PSS-based inkjet-printed capacitive sensors have also been reported by Salim et. al. (Salim & Lim, 2017). Inkjet-printed gas sensor based on PEDOT: PSS has also been reported by Tseng et al. with a sensitivity of 0.7%/100 ppm for detecting CO_2 with the decrease in resistance due to a decrease in carrier charge concentration (Tseng et al., 2012). Further PEDOT: PSS screen-printed sensors have also been reported to detect ascorbic acid in food due to its high conductivity, commercial availability, good thermal stability, and stability in a wide range of pH. The detection is based on the modulation of channel currents due to electrochemical de-doping and doping phenomena occurring with the electrolyte contact when gate voltages are applied (Contat-Rodrigo et al., 2017). The effective sensing of PEDOT: PSS-based sensors can be utilized for the mass-scale fabrication of low-cost sensing devices. The compatibility of polymers with existing printing technologies is proving them an edge over their non-organic sensing systems. With the ongoing research and their affirmative responses, these advanced materials are set to change sensing technologies' fate.

(a)

(b)

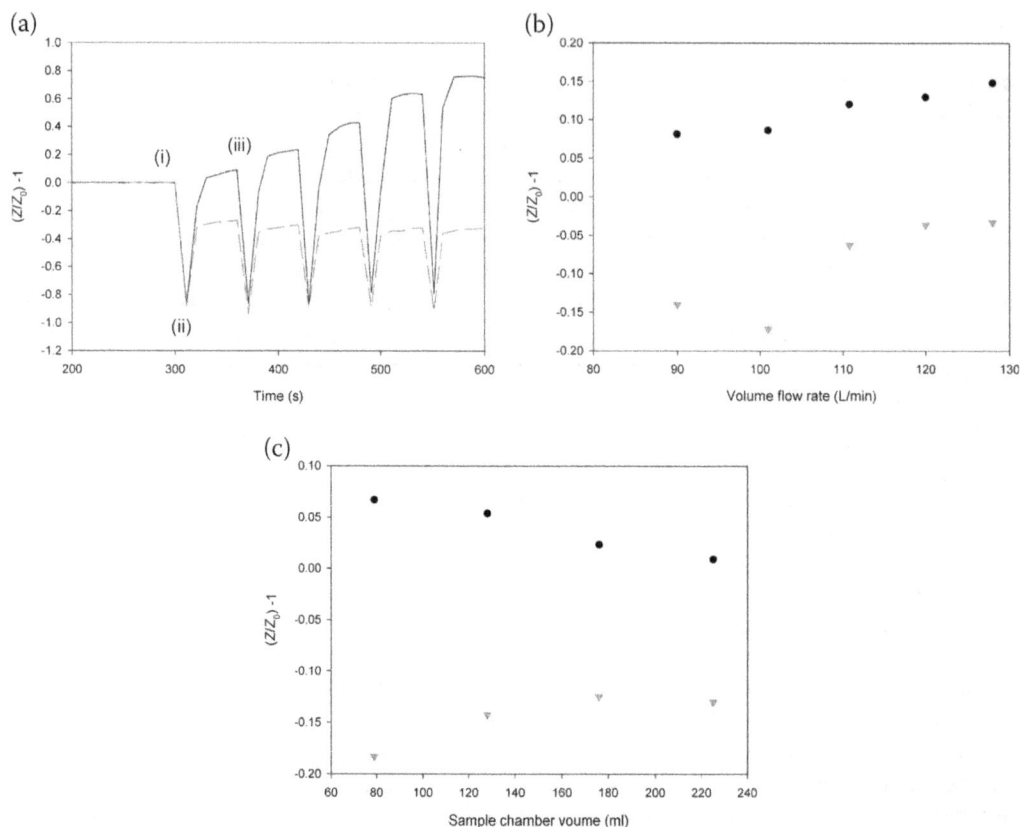

(c)

FIGURE 11.2   Response of sensor impedance with humidity and temperature, volume flow rate, and sample chamber volume. (a) Measurement of five consecutive breaths containing 245 ± 8 ppbv. Air to breath switching (i) leads to the decrease in response caused due to humidity and temperature. (ii) Condensation of water vapors prevents ammonia measurement due to unotimized flow as depicted by dashed red line. (iii) Signal due to humidity is separated and signal due to ammonia is intensified when flow is optimized as depicted by solid black line. (b) Measurement of four cumulative breaths. Impedance response due to humidity (red triangles) and ammonia (black circles) due to sample flow rate is studied. Ammonia response is increased and humidity response is decreased with an increase in flow rate. (c) Measurement of four cumulative breaths. A decrease in ammonia and humidity signal is observed with increase in sample chamber volume. Reprinted (adapted) with permission from Hibbard, Troy. et al. "Point of Care Monitoring of Hemodialysis Patients with a Breath Ammonia Measurement Device Based on Printed Polyaniline Nanoparticle Sensors." *Anal. Chem.* 85 (24): 12158–12165. doi:10.1021/ac403472d. Copyright (2013) American Chemical Society.

11.6.3   BATTERIES

Due to the low cost and fast fabrication requirements to produce thin and flexible batteries, polymers are highly researched materials (Sousa, Costa, & Lanceros-Méndez, 2015). Polymers have successfully found usage in the electrode as well as electrolyte materials found in batteries. However, research is more mature in the polymer-based electrolyte with Li/MnO_2 as electrode material (Maria Montanino et al., 2019; Costa, Gonçalves, & Lanceros-Méndez, 2020). The commercial availability of gel-type or solid electrolyte materials pushes various big companies, including L.G. Chem, Panasonic, DuPont, and Celgard, to use them in their products. Research in switching the electrode material with polymers has come a long way from initial research on polyaniline and polyacetylene-based large area electrodes to carbonyl-based compounds. In the

1960s, William et al. firstly validated that carbonyl based compounds can be used as an active electrode material. Since then, researchers have been improving the quality of carbonyl-based polymers. Polymers such as polyquinones, polyketones, and polyimides have become a promising electrode materials due to their reversible redox stability, high theoretical capacity, flexible structure, fast kinetics, and multi-electron reaction. A major revolution in polymeric electrode material came in 2015 when Song et al. developed anthraquinone-based polymer cathode material where theoretical capacity is decreased by creating static sites by connecting monomers with organic functional groups. Electron transferability and cyclic stability have also been reported to be improved (H. Wang et al., 2020). Fabrication of large-scale fully printed batteries on the flexible substrate requires printable electrodes and electrolyte materials. The reported researches are favorable in pointing out an encouraging trend to have fully printed batteries for critical applications with higher efficiencies.

11.6.4 Field-Effect Transistors (FETs)

Field-effect transistors are helpful in many electronic applications. Be it acting as a switch or a mode of amplification of input signals, FETs have come a long way to create their high reputation in the electronics industry. This component of an electronic circuit is crucial to many applications, and without having a significant breakthrough in designing FETs, the fabrication of fully printable electronics seems out of context. Therefore, many researchers have explored a variety of conjugated polymers that can provide an effective solution.

I.C.s are core electronic components consisting of multiple transistors. Andersson Ersman et al. demonstrated the usage of organic electrochemical transistors based on PEDOT: PSS in fabricating large-scale integrated circuits that rule out to be a significant development in the construction of organic transistors based I.C.s (Andersson Ersman et al., 2019). However, after 2000s, research has been shifted to exploring those materials in fabricating transistors that carry ring structures in polymer backbone with a planer and conjugated solid structure such as acenes; say pentacene, due to their higher carrier mobility than traditional conjugated polymers. The only bottleneck in using acenes in printed electronic devices is their poor solubility in organic solvents. This tailback is successfully removed with the introduction of functional groups in the polymer structure. P3HT is another crucial candidate being studied for application in organic thin-film transistors. Regioregular P3HT has come up as the most attractive material to be used in printable thin-film transistors, and results are providing good insight into the future of this polymer in FETs (Hyun et al., 2015; Chung, Cho, & Lee, 2019). Diketopyrrolopyrrole based semiconductors with optimized side-chain geometry have been reported to show hole mobility of 17.8 cm^2 V^{-1} s^{-1} by Kim et al. It rules out to be an important candidate in manufacturing field-effect transistors (Back et al., 2015). Gravure and flexography printed dielectrics have also been reported to be used in transistors (Cosnahan, Watt, & Assender, 2018; Fattori et al., 2019; Vaklev, Steinke, & Campbell, 2019). Organic nanowire transistor arrays have been prepared by Liu and coworkers with the aid of inkjet printing technique employing PEDOT:PSS PH 500 with conductivity > 1 S cm^{-1}. The inkjet printing process is modified to narrow down the gap between source/drain (S/D) electrodes and mobility as high as 1.26 cm^2. V^{-1}. S^{-1}. A 30 μm jet nozzle has been employed to eject a 40 μm drop by keeping the distance between drops' jetting position (D$_p$) at least 80 μm. The maximum radius of drop spreading (D$_s$) was 35 μm. Error tolerance distance of 10 μm is considered. Due to the higher surface tension of the drops, dried drops showcased the radius (R$_d$) of 10 μm, leaving the effective distance between S/D as 60 μm (refer Figure 11.3(a,b)). As shown in Figure 11.3(c,d), in order to decrease the channel length by keeping Dp the same, three drops are printed at a position, thereby decreasing the channel length to 35 μm with R$_d$ of 25 μm. Further, to decrease the channel length, a substrate is heated to a temperature of 40°C and the first electrode is allowed to dry up before printing the other by delaying the printing time by 2 ms, which resulted in the decrease in channel length to 20 μm, as depicted in Figure 11.3(e,f) (N. Liu et al., 2011).

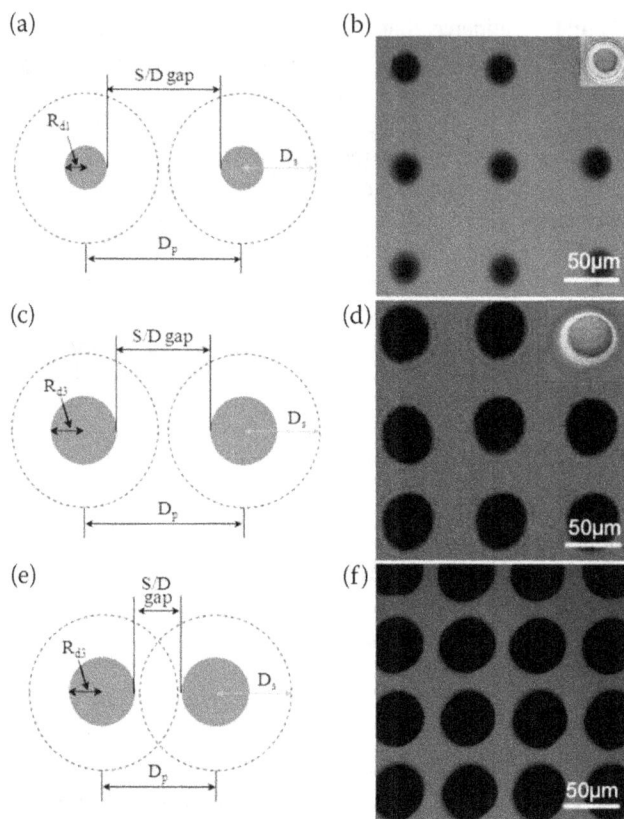

FIGURE 11.3 (a,c,e) Channel lengths of printed S/D electrodes are illustrated. (b,d,f) Showcases actual photographs of the printed S/D electrodes. (a,b) Showcases electrodes printed with one drop. (c,d) Showcases the electrodes printed with three drops. (e,f) Showcases electrodes printed with three drops with substate at 40°C and a delay of 2 ms. Maximum radius of droplet spreading and final radius of droplet spreading and distance between S/D jetting locations is given by D_s, R_d and D_p, respectively. Reprinted (adapted) with permission from Liu, N. et al. "High-Performance, All-Solution-Processed Organic Nanowire Transistor Arrays with Inkjet-Printing Patterned Electrodes," *Langmuir* 27 (24), 14710–14715. doi:10.1021/la2033324. Copyright (2011) American Chemical Society.

With advancements in organic materials, large-scale manufacturing of flexible and high-performance transistors seems evident.

11.6.5   Organic Photovoltaics (OPVs)

OPVs are responsible for generating electricity from sunlight with the aid of the photovoltaic effect. Inorganic P.V.s are already catering to the market's need by acting as active photovoltaic solar cells' materials. After their discovery, organics, mainly conjugated polymers, are actively researched to exploit this application area effectively. The lower bandgap of organic semiconductors makes them a promising material to utilize the sun's light to convert it into electricity effectively. However, as reported by Carle et al. OPVs can be considered in the pre-industrial development stage. Several technological constraints must be dealt with before the production of OPVs on a large scale can be done (Carlé & Krebs, 2013). However, research shows promising results, and OPVs are expected to be a huge market until 2029 as per IDTechEX. P3HT and PEDOT: PSS are the leading organic polymers contributing to the development of next-generation printed OPVs. Inkjet-printed P3HT bulk-hetero junction-based OPVs were reported in 2004 with an efficiency of less than 0.1% (Shah & Wallace, 2004). Quite recently, in 2019, poly(3-hexylthiophene-2,5-diyl):(5Z,5'Z)-5,5'-{[7,7'-(4,4,9,9-tetraoctyl-4,9-dihydro-s-indaceno[1,2-b:5,6-b']dithiophene-2,7-diyl)bis(benzo[c][1,2,5]thiadiazole-7,4-diyl)]bis(methanylylidene)}bis(3-ethyl-2-thioxothiazolidin-4-one) (P3HT: O-IDTBR) based organic solar cells have demonstrated the efficiency of 6% that may be due to the long wavelength inflection point of O-IDTBR leading to efficient charge carrier transport and better performance of solar cells (Corzo et al., 2019; An, Zhong, & Ying, 2020; Fernández Castro et al., 2020). PEDOT: PSS-based solar cells have also demonstrated a good efficiency up to a value of 4.7% (Maisch et al., 2021). Screen printed and inkjet-printed PEDOT: PSS are promising materials to be used in OPVs (Ganesan, Mehta, & Gupta, 2019). Another interesting aspirant in the OPV application is polycarbazole which has demonstrated an efficiency of up to 3.5% (Alem et al., 2018). The encouraging results drive researchers to develop high-efficiency OPVs on a large scale that can replace traditional photovoltaics and can be fabricated easily on various substrates.

11.6.6 CONDUCTIVE ADHESIVES

Conductive polymers are a promising solution for the preparation of fine interconnects at a low cost with low-temperature processing. These are effective alternatives to traditional solders that require high processing temperature for joining interconnects. Conductive adhesives are majorly divided into isotropic conductive adhesives (ICAs) and anisotropic conductive adhesives (ACAs). While ICA can conduct the electricity in all directions, ACAs can conduct the electricity in one dimension only. The advantage of using the latter one is to avoid shorting problems with the former's usage in the electronic circuits. These are generally composites formulated with an insulating polymer resin along with conductive fillers. Depending upon the loading level of filler content, conductive adhesives are divided into subgroups as ICAs and ACAs. They are very crucial in the fabrication of electronic devices for the applications discussed so far. Thin films of ACAs are reported to be used in OLEDs to prepare lightweight, thin-profile, and high-resolution displays (Yim et al., 2008). The utility of intrinsically conductive polymers as adhesives hasn't been researched much, and significant research focuses on producing conductive adhesives with the aid of conductive fillers. However, recently, stencil-printed silicone-based intrinsically conductive adhesives (Silo-ECA) are reported by Zhuo Li et al. with conductivity up to 1.51×10^3 S/cm. Silo-ECAs are a mixture constituting hydride-terminated PDMS (H-PDMS) and vinyl-poly(dimethyl) siloxane (vinyl-PDMS) (Tasdelen & Yagci, 2011; Z. Li et al., 2013, 2015). Further, Mondal et al. reported instant conducting adhesive prepared by grafting silane polymer onto extended graphite. Prepared adhesive showcased increased conductivity as compared to neat silicone-based adhesive due to the formation of a hierarchical array of graphitic plates inside of the polymeric matrix ultimately forming an electrically conductive network (Mondal, Bhowmick, & Krishnamoorti, 2014).

Conductive adhesives are compatible with printing techniques such as screen printing and remain a much-underrated component but crucial in the fabrication of effective electronic devices (Y. Li & Wong, 2006).

The applications mentioned previously are the significant segments where research and development are working as par excellence. It is very much possible to have technology enter day-to-day lives. However, there are numerous other applications where polymers are showcasing their caliber, such as printed antennas, flexible interconnects, printed coils for wireless power transfer, printed power sources, etc.

11.7 CONCLUSION AND FUTURE SCOPE

Printing technologies are an attractive choice for the fabrication of these electronic devices on a large scale at a lower cost while saving a lot of production time compared with the traditional technologies to manufacture electronic devices. Printing techniques such as screen printing, gravure printing, and reverse offset printing are techniques capable of mass production of organic electronic components. Conducting polymers are fabricated on the required substrate to design electronic devices with the aid of printing inks. Therefore, printing ink formulation becomes a crucial step in determining the fabricated device's performance. R&D in ink formulations already helps minimize defects in printing by controlling the critical factors to design high-quality devices with minimized defects. The major component of the ink that is responsible for the conductivity is the conductive polymer. Various routes can synthesize conductive polymer; the most adopted ones are oxidative chemical polymerization, vapor phase polymerizations, and transition metal-catalyzed polycondensation. The primary application areas of printed electronic devices based on polymers are OLEDs, OPVs, batteries, transistors, and sensors at different industrialization levels. Printed electronic devices are becoming a broad and emerging application area for semi (conducting) polymers to showcase their utility. Their application in the core electronic components such as OLEDs with the emergence of OLED-based TVs and smartphones by various companies have already proven their worth and potential to grow further in the domain.

Further, these polymers' utility in non-traditional applications like sensors for healthcare monitoring and structural health monitoring are sensitive areas for research and development that are only pointing towards extraordinary technological breakthroughs to change people's day-to-day lives for the better. Printed batteries and transistors are active research areas that are only improving their fabrication technology and performance to cater to high-performance electronic devices. In applications where the performance requirements are not stringent, the utility of printed batteries and transistors has been explored and exploited. OPVs remains a sensitive area of research that ought to boom in a decade to change energy harvesting's fate by overcoming energy production challenges with renewable resources. There are ample examples and positive results of research that only point to the direction of marvelous growth opportunities and scope for polymers and their application in printed electronic devices.

ACKNOWLEDGMENT

S.S. and T.M. acknowledge the funding support (EEQ/2020/000139) extended by the Science and Engineering Research Board (SERB), India.

REFERENCES

Al-Halhouli, Ala'aldeen, Loiy Al-Ghussain, Saleem El Bouri, Haipeng Liu, and Dingchang Zheng. 2019. "Fabrication and Evaluation of a Novel Non-Invasive Stretchable and Wearable Respiratory Rate Sensor Based on Silver Nanoparticles Using Inkjet Printing Technology." *Polymers* 11 (9): 1518.10.3390/polym11091518

Aleeva, Yana, and Bruno Pignataro. 2014. "Recent Advances in Upscalable Wet Methods and Ink Formulation for Printed Electronics." *Journal of Materials Chemistry C* 49 (32): 6436–6453. 10.1039/C4TC00618F

Alem, Salima, Neil Graddage, Jianping Lu, Terho Kololuoma, Raluca Movileanu, and Ye Tao. 2018. "Flexographic Printing of Polycarbazole-Based Inverted Solar Cells." *Organic Electronics* 52. Elsevier B.V.: 146–152. 10.1016/j.orgel.2017.10.016

An, Kang, Wenkai Zhong, and Lei Ying. 2020. "Enhanced Performance of P3HT-Based Non-Fullerene Polymer Solar Cells by Optimizing Film Morphology Using Non-Halogenated Solvent." *Organic Electronics* 82. Elsevier B.V.: 105701. 10.1016/j.orgel.2020.105701

Andersson Ersman, Peter, Roman Lassnig, Jan Strandberg, Deyu Tu, Vahid Keshmiri, Robert Forchheimer, Simone Fabiano, Göran Gustafsson, and Magnus Berggren. 2019. "All-Printed Large-Scale Integrated Circuits Based on Organic Electrochemical Transistors." *Nature Communications* 10 (1). Springer US: 1–9. 10.1038/s41467-019-13079-4

Awuzie, C. I. 2017. "Conducting Polymers." *Materials Today: Proceedings* 4 (4). Elsevier Ltd: 5721–5726. 10.1016/j.matpr.2017.06.036.

Back, Jang Yeol, Hojeong Yu, Inho Song, Il Kang, Hyungju Ahn, Tae Joo Shin, Soon Ki Kwon, Joon Hak Oh, and Yun Hi Kim. 2015. "Investigation of Structure-Property Relationships in Diketopyrrolopyrrole-Based Polymer Semiconductors via Side-Chain Engineering." *Chemistry of Materials* 27 (5): 1732–1739. 10.1021/cm504545e

Beedasy, Vimanyu, and Patrick J. Smith. 2020. "Printed Electronics as Prepared by Inkjet Printing." *Materials* 13 (3): 1–23. 10.3390/ma13030704

Bhattacharyya, Dhiman, Rachel M. Howden, David C. Borrelli, and Karen K. Gleason. 2012. "Vapor Phase Oxidative Synthesis of Conjugated Polymers and Applications." *Journal of Polymer Science, Part B: Polymer Physics* 50 (19): 1329–1351. 10.1002/polb.23138

Ćirić-Marjanovic, Gordanan, Miroslava Trchova´, and Jaroslav Stejskal. 2008. "The Chemical Oxidative Polymerization of Aniline in Water: Raman Spectroscopy." *Journal of Raman Spectroscopy* 39 (10): 1375–1387.

Carlé, Jon E., and Frederik C. Krebs. 2013. "Technological Status of Organic Photovoltaics (OPV)." *Solar Energy Materials and Solar Cells* 119. Elsevier: 309–310. 10.1016/j.solmat.2013.08.044

Chung, Seungjun, Kyungjune Cho, and Takhee Lee. 2019. "Recent Progress in Inkjet-Printed Thin-Film Transistors." *Advanced Science* 6 (6): 1801445. 10.1002/advs.201801445

Contat-Rodrigo, Laura, Clara Pérez-Fuster, José Vicente Lidón-Roger, Annalisa Bonfiglio, and Eduardo García-Breijo. 2017. "Screen-Printed Organic Electrochemical Transistors for the Detection of Ascorbic Acid in Food." *Organic Electronics* 45. Elsevier B.V.: 89–96. 10.1016/j.orgel.2017.02.037

Corzo, Daniel, Khulud Almasabi, Eloise Bihar, Sky Macphee, Diego Rosas-Villalva, Nicola Gasparini, Sahika Inal, and Derya Baran. 2019. "Digital Inkjet Printing of High-Efficiency Large-Area Nonfullerene Organic Solar Cells." *Advanced Materials Technologies* 4 (7): 1–9. 10.1002/admt.201900040

Cosnahan, Thomas, Andrew A.R. Watt, and Hazel E. Assender. 2018. "Flexography Printing for Organic Thin Film Transistors." *Materials Today: Proceedings* 5 (8). Elsevier Ltd: 16051–16057. 10.1016/j.matpr.2018.05.050

Costa, C. M., R. Gonçalves, and S. Lanceros-Méndez. 2020. "Recent Advances and Future Challenges in Printed Batteries." *Energy Storage Materials* 28. Elsevier B.V.: 216–234. 10.1016/j.ensm.2020.03.012

Cruz, Sílvia Manuela Ferreira, Luís A. Rocha, and Júlio C. Viana. 2018. "Printing Technologies on Flexible Substrates for Printed Electronics." In *Flexible Electronics*, edited by Simas Rackauskas. London: IntechOpen. 10.5772/intechopen.76161

Cui, Zheng. 2016. *Printed Electronics: Materials, Technologies and Applications*. Edited by Zheng Cui. 1st ed. John Wiley & Sons Singapore Pte. Ltd.

Dang, Minh Trung, Lionel Hirsch, and Guillaume Wantz. 2011. "P3HT:PCBM, Best Seller in Polymer Photovoltaic Research." *Advanced Materials* 23 (31): 3597–3602. 10.1002/adma.201100792

Shirakawa, Hideki, Edwin J. Louis, Alan G. Macdiarmid, Chwan K. Chiang , and Alan J. Heeger. 1977. "Synthesis of Electrically Conducting Organic Polymers." Journal of the Chemical Society, Chemical Communications: 578–580.

Eom, Seung Hun, Hanok Park, S. H. Mujawar, Sung Cheol Yoon, Seok Soon Kim, Seok In Na, Seok Ju Kang, Dongyoon Khim, Dong Yu Kim, and Soo Hyoung Lee. 2010. "High Efficiency Polymer Solar Cells via Sequential Inkjet-Printing of PEDOT:PSS and P3HT:PCBM Inks with Additives." *Organic Electronics* 11 (9). Elsevier B.V.: 1516–1522. 10.1016/j.orgel.2010.06.007

Eom, Seung Hun, S. Senthilarasu, Periyayya Uthirakumar, Sung Cheol Yoon, Jongsun Lim, Changjin Lee, Hyun Seok Lim, J. Lee, and Soo Hyoung Lee. 2009. "Polymer Solar Cells Based on Inkjet-Printed PEDOT:PSS Layer." *Organic Electronics* 10 (3). Elsevier B.V.: 536–542. 10.1016/j.orgel.2009.01.015

Fattori, Marco, Joost Fijn, Pieter Harpe, Micael Charbonneau, Stephanie Lombard, Krunoslav Romanjek, Denis Locatelli, Laurent Tournon, Christelle Laugier, and Eugenio Cantatore. 2019. "A Gravure-Printed Organic TFT Technology for Active-Matrix Addressing Applications." *IEEE Electron Device Letters* 40 (10). IEEE: 1682–1685. 10.1109/LED.2019.2938852

Fernández Castro, Marcial, Eva Mazzolini, Roar R. Sondergaard, Moises Espindola-Rodriguez, and Jens Wenzel Andreasen. 2020. "Flexible ITO-Free Roll-Processed Large-Area Nonfullerene Organic Solar Cells Based on P3HT:O-IDTBR." *Physical Review Applied* 14 (3). American Physical Society: 1. 10.1103/PhysRevApplied.14.034067

Fukuda, Kenjiro, Yudai Yoshimura, Tomoko Okamoto, Yasunori Takeda, Daisuke Kumaki, Yoshinori Katayama, and Shizuo Tokito. 2015. "Reverse-Offset Printing Optimized for Scalable Organic Thin-Film Transistors with Submicrometer Channel Lengths." *Advanced Electronic Materials* 1 (8): 1–6. 10.1002/aelm.201500145

Ganesan, S., S. Mehta, and D. Gupta. 2019. "Fully Printed Organic Solar Cells – A Review of Techniques, Challenges and Their Solutions." *Opto-Electronics Review* 27 (3). Association of Polish Electrical Engineers (SEP): 298–320. 10.1016/j.opelre.2019.09.002

De Gans, Berend Jan, Paul C. Duineveld, and Ulrich S. Schubert. 2004. "Inkjet Printing of Polymers: State of the Art and Future Developments." *Advanced Materials* 16 (3): 203–213. 10.1002/adma.200300385.

De Gans, Berend Jan, Emine Kazancioglu, Wilhelm Meyer, and Ulrich S. Schubert. 2004. "Ink-Jet Printing Polymers and Polymer Libraries Using Micropipettes." *Macromolecular Rapid Communications* 25 (1): 292–296. 10.1002/marc.200300148

Gao, Meng, Lihong Li, and Yanlin Song. 2017. "Inkjet Printing Wearable Electronic Devices." *Journal of Materials Chemistry C* 5 (12). Royal Society of Chemistry: 2971–93. 10.1039/c7tc00038c

Ghosh, Palash. n.d. "There's A Worldwide Shortage Of Semiconductors Hurting Car And Smartphone Companies, Here's How Investors Can Benefit." https://www.forbes.com/sites/palashghosh/2021/02/22/theres-a-worldwide-shortage-of-semiconductors-hurting-car-and-smartphone-companies-heres-how-investors-can-benefit/?sh=665a267f795b.

Gregor-Svetec, Diana. 2018. *Intelligent Packaging. Nanomaterials for Food Packaging: Materials, Processing Technologies, and Safety Issues.* Elsevier Inc. 10.1016/B978-0-323-51271-8.00008-5

Gurunathan, K., A. Vadivel Murugan, R. Marimuthu, U. P. Mulik, and D. P. Amalnerkar. 1999. "Electrochemically Synthesized Conducting Polymeric Materials for Applications towards Technology in Electronics, Optoelectronics and Energy Storage Devices." *Materials Chemistry and Physics* 61 (3): 173–191. 10.1016/S0254-0584(99)00081-4

Hartwig, Melinda, Ralf Zichner, and Yvonne Joseph. 2018. "Inkjet-Printed Wireless Chemiresistive Sensors-A Review." *Chemosensors* 6 (4): 1–15. 10.3390/chemosensors6040066

Heeger, Alan J. 2001. "Nobel Lecture: Semiconducting and Metallic Polymers: The Fourth Generation of Polymeric Materials." *Reviews of Modern Physics* 73 (3): 681–700. 10.1103/RevModPhys.73.681

Heitz, W. 1995. "Metal Catalyzed Polycondensation Reactions." *Pure and Applied Chemistry* 67 (12): 1951–1964. 10.1351/pac199567121951

Hibbard, Troy, Karl Crowley, Frank Kelly, Frank Ward, John Holian, Alan Watson, and Anthony J. Killard. 2013. "Point of Care Monitoring of Hemodialysis Patients with a Breath Ammonia Measurement Device Based on Printed Polyaniline Nanoparticle Sensors." *Anal. Chem.* 85 (24): 12158–12165. 10.1021/ac403472d

Hibon, Pauline, Heinz von Seggern, Hsin Rong Tseng, Christoph Leonhard, Manuel Hamburger, and Gaëlle Béalle. 2020. "Improved Thin Film Stability of Differently Formulated, Printed, and Crosslinked Polymer Layers against Successive Solvent Printing." *Journal of Applied Polymer Science* 137 (29): 1–7. 10.1002/app.48895

Hoeng, Fanny, Aurore Denneulin, and Julien Bras. 2016. "Use of Nanocellulose in Printed Electronics: A Review." *Nanoscale* 8 (27): 13131–13154. 10.1039/c6nr03054h

Hosseini, Ensieh S., Saoirse Dervin, Priyanka Ganguly, and Ravinder Dahiya. 2021. "Biodegradable Materials for Sustainable Health Monitoring Devices." *ACS Applied Bio Materials* 4 (1): 163–194. 10.1021/acsabm.0c01139

Hrehorova, Erika, Alexandra Pekarovicova, and Paul D. Fleming. 2006. "Gravure Printability of Conducting Polymer Inks." *Digital Fabrication 2006 –Final Program and Proceedings* 2006: 107–110.

Huitema, Edzer. 2012. "The Future of Displays Is Foldable." *Information Display* 28 (2–3): 6–10. 10.1002/j.2637-496x.2012.tb00470.x

Huseynova, Gunel, Yong Hyun Kim, Jae Hyun Lee, and Jonghee Lee. 2020. "Rising Advancements in the Application of PEDOT:PSS as a Prosperous Transparent and Flexible Electrode Material for Solution-Processed Organic Electronics." *Journal of Information Display* 21 (2). Taylor & Francis: 71–91. 10.1080/15980316.2019.1707311

Hyun, Woo Jin, Ethan B. Secor, Geoffrey A. Rojas, Mark C. Hersam, Lorraine F. Francis, and C. Daniel Frisbie. 2015. "All-Printed, Foldable Organic Thin-Film Transistors on Glassine Paper." *Advanced Materials* 27 (44): 7058–7064. 10.1002/adma.201503478

Khan, Yasser, Arno Thielens, Sifat Muin, Jonathan Ting, Carol Baumbauer, and Ana C. Arias. 2020. "A New Frontier of Printed Electronics: Flexible Hybrid Electronics." *Advanced Materials* 32 (15): 1–29. 10.1002/adma.201905279

Khandpur, R. S. 2006. *Printed Circuit Boards: Design, Fabrication & Assembly.* 1st ed. McGrawHill.

Kim, Minseok, In Kyu You, Hyun Han, Soon Won Jung, Tae Youb Kim, Byeong Kwon Ju, and Jae Bon Koo. 2011. "Organic Thin-Film Transistors with Short Channel Length Fabricated by Reverse Offset Printing." *Electrochemical and Solid-State Letters* 14 (8): 333–336. 10.1149/1.3591435

Kim, Yong Hyun, Jonghee Lee, Simone Hofmann, Malte C. Gather, Lars Müller-Meskamp, and Karl Leo. 2013. "Achieving High Efficiency and Improved Stability in ITO-Free Transparent Organic Light-Emitting Diodes with Conductive Polymer Electrodes." *Advanced Functional Materials* 23 (30): 3763–3769. 10.1002/adfm.201203449

Ko, Seung Hwan, Jaewon Chung, Heng Pan, Costas P. Grigoropoulos, and Dimos Poulikakos. 2007. "Fabrication of Multilayer Passive and Active Electric Components on Polymer Using Inkjet Printing and Low Temperature Laser Processing." *Sensors and Actuators, A: Physical* 134 (1): 161–168. 10.1016/j.sna.2006.04.036

Kraft, Ulrike, Francisco Molina-Lopez, Donghee Son, Zhenan Bao, and Boris Murmann. 2020. "Ink Development and Printing of Conducting Polymers for Intrinsically Stretchable Interconnects and Circuits." *Advanced Electronic Materials* 6 (1): 1–9. 10.1002/aelm.201900681

Kusaka, Yasuyuki, Nobuko Fukuda, and Hirobumi Ushijima. 2020. "Recent Advances in Reverse Offset Printing: An Emerging Process for High-Resolution Printed Electronics." *Japanese Journal of Applied Physics* 59 (SG): SG0802. 10.7567/1347-4065/ab6462

Lessing, Joshua, Ana C. Glavan, S. Brett Walker, Christoph Keplinger, Jennifer A. Lewis, and George M. Whitesides. 2014. "Inkjet Printing of Conductive Inks with High Lateral Resolution on Omniphobic 'RF Paper' for Paper-Based Electronics and MEMS." *Advanced Materials* 26 (27): 4677–4682. 10.1002/adma.201401053

Li, Yi, and C. P. Wong. 2006. "Recent Advances of Conductive Adhesives as a Lead-Free Alternative in Electronic Packaging: Materials, Processing, Reliability and Applications." *Materials Science and Engineering R: Reports* 51 (1–3): 1–35. 10.1016/j.mser.2006.01.001

Li, Zhuo, Kristen Hansen, Yagang Yao, Yanqing Ma, Kyoung Sik Moon, and C. P. Wong. 2013. "The Conduction Development Mechanism of Silicone-Based Electrically Conductive Adhesives." *Journal of Materials Chemistry C* 1 (28): 4368–4374. 10.1039/c3tc30612g

Li, Zhuo, Taoran Le, Zhenkun Wu, Yagang Yao, Liyi Li, Manos Tentzeris, Kyoung Sik Moon, and C. P. Wong. 2015. "Rational Design of a Printable, Highly Conductive Silicone-Based Electrically Conductive Adhesive for Stretchable Radio-Frequency Antennas." *Advanced Functional Materials* 25 (3): 464–470. 10.1002/adfm.201403275

Liu, Nanliu, Yan Zhou, Na Ai, Chan Luo, Junbiao Peng, Jian Wang, Jian Pei, and Yong Cao. 2011. "High-Performance, All-Solution-Processed Organic Nanowire Transistor Arrays with Inkjet-Printing Patterned Electrodes," *Langmuir* 27 (24), 14710–14715. 10.1021/la2033324

Liu, Zhaoyang, Khaled Parvez, Rongjin Li, Renhao Dong, Xinliang Feng, and Klaus Müllen. 2015. "Transparent Conductive Electrodes from Graphene/PEDOT:PSS Hybrid Inks for Ultrathin Organic Photodetectors." *Advanced Materials* 27 (4): 669–675. 10.1002/adma.201403826

Lock, John P., Sung Gap Im, and Karen K. Gleason. 2006. "Oxidative Chemical Vapor Deposition of Electrically Conducting Poly(3,4-Ethylenedioxythiophene) Films." *Macromolecules* 39 (16): 5326–5329. 10.1021/ma060113o

Loudon, Rodney. 1959. "One-Dimensional Hydrogen Atom." *American Journal of Physics* 27 (9): 649–655. 10.1119/1.1934950

Maddipatla, Dinesh, Binu B. Narakathu, and Massood Atashbar. 2020. "Recent Progress in Manufacturing Techniques of Printed and Flexible Sensors: A Review." *Biosensors* 10 (12): 199. 10.3390/bios10120199

Maisch, Philipp, Kai Cheong Tam, DongJu Jang, Marc Steinberger, Fu Yang, Christoph J. Brabec, and Hans-Joachim Egelhaaf. 2021. Inkjet Printed Organic and Perovskite Photovoltaics—Review and Perspectives.*Organic Flexible Electronics*. Edited by Piero Cosseddu & Mario Caironi. Woodhead Publishing. 305–333. 10.1016/b978-0-12-818890-3.00010-2

Menard, Etienne, Matthew A. Meitl, Yugang Sun, Jang Ung Park, Daniel Jay Lee Shir, Yun Suk Nam, Seokwoo Jeon, and John A. Rogers. 2007. "Micro- and Nanopatterning Techniques for Organic Electronic and Optoelectronic Systems." *Chemical Reviews* 107 (4): 1117–1160. 10.1021/cr050139y

Mendez-Rossal, Hector R., and Gernot M. Wallner. 2019. "Printability and Properties of Conductive Inks on Primer-Coated Surfaces." *International Journal of Polymer Science* 2019: 3874181. 10.1155/2019/3874181

Meng, Hsin Fei (Ed.) 2013. *Polymer Electronics*. 1st ed. CRC Press. 10.4032/9789814364041

Mondal, Titash, Anil K. Bhowmick, and Ramanan Krishnamoorti. 2014. "Conducting Instant Adhesives by Grafting of Silane Polymer onto Expanded Graphite." *ACS Applied Materials and Interfaces* 6 (18): 16097–16105. 10.1021/am5040472

Montanino, M., G. Sico, C. T. Prontera, A. De Girolamo Del Mauro, S. Aprano, M. G. Maglione, and C. Minarini. 2017. "Gravure Printed PEDOT:PSS as Anode for Flexible ITO-Free Organic Light Emitting Diodes." *Express Polymer Letters* 11 (6): 518–523. 10.3144/expresspolymlett.2017.48

Montanino, Maria, Giuliano Sico, Anna De Girolamo Del Mauro, and Margherita Moreno. 2019. "LFP-Based Gravure Printed Cathodes for Lithium-Ion Printed Batteries." *Membranes* 9 (6): 5–11. 10.3390/membranes9060071

Moonen, Pieter F., Iryna Yakimets, and Jurriaan Huskens. 2012. "Fabrication of Transistors on Flexible Substrates: From Mass-Printing to High-Resolution Alternative Lithography Strategies." *Advanced Materials* 24 (41): 5526–5541. 10.1002/adma.201202949

Myers, Ronald E. 1986. "Chemical Oxidative Polymerization as a Synthetic Route to Electrically Conducting Polypyrroles." *Journal of Electronic Materials* 15 (2): 61–69. 10.1007/BF02649904

Nisato, Giovanni, Donald Lupo, and Simone Ganz. 2016. *Organic and Printed Electronics*. CRC Press. 10.1201/b20043

Ouchi, Makoto, Takaya Terashima, and Mitsuo Sawamoto. 2009. "Transition Metal-Catalyzed Living Radical Polymerization: Toward Perfection in Catalysis and Precision Polymer Synthesis." *Chemical Reviews* 109 (11): 4963–5050. 10.1021/cr900234b

Pandey, Sadanand. 2016. "Highly Sensitive and Selective Chemiresistor Gas/Vapor Sensors Based on Polyaniline Nanocomposite: A Comprehensive Review." *Journal of Science: Advanced Materials and Devices* 1 (4). Elsevier Ltd: 431–453. 10.1016/j.jsamd.2016.10.005

Quintero Vasquez, A., F. Molina-Lopez, E. C. P. Smits, E. Danesh, J van den Brand, K. Persaud, A. Oprea, et al. 2016. "Smart RFID Label with a Printed Multisensor Platform for Environmental Monitoring." *IOPScience* 1 (2): 025003. 10.1088/2058-8585/1/2/025003

Rasmussen, Seth C. 2017. "The Early History of Polyaniline: Discovery and Origins." *An International Journal of the History of Chemistry Substantia* 1 (12): 99–109. 10.13128/substantia-30

Rasmussen, Seth C. 2011. Revising the history of organic polymers."100+ years of Plastics. Edited by E. Thomas Strom & Seth C. Rasmussen. ACS Symposium Series. 147–163.

Reddy, Kakarla Raghava, Han Mo Jeong, Youngil Lee, and Anjanapura Venkataramanaiah Raghu. 2010. "Synthesis of MWCNTs-Core/Thiophene Polymer-Sheath Composite Nanocables by a Cationic Surfactant-Assisted Chemical Oxidative Polymerization and Their Structural Properties." *Journal of Polymer Chemistry* 48 (7): 1477–1484. 10.1002/pola.23883

Rogers, John A., and Zhenan Bao. 2002. "Printed Plastic Electronics and Paperlike Displays." *Journal of Polymer Science, Part A: Polymer Chemistry* 40 (20): 3327–3334. 10.1002/pola.10405

Salim, Ahmed, and Sungjoon Lim. 2017. "Review of Recent Inkjet-Printed Capacitive Tactile Sensors." *Sensors (Switzerland)* 17 (11): 2593. 10.3390/s17112593

Schlisske, Stefan, Christine Rosenauer, Tobias Rödlmeier, Kai Giringer, Jasper J. Michels, Kurt Kremer, Uli Lemmer, Svenja Morsbach, Kostas Ch Daoulas, and Gerardo Hernandez-Sosa. 2020. "Ink Formulation for Printed Organic Electronics: Investigating Effects of Aggregation on Structure and Rheology of Functional Inks Based on Conjugated Polymers in Mixed Solvents." *Advanced Materials Technologies*, 6(2): 2000335. 10.1002/admt.202000335

Sekine, Chizu, Yoshiaki Tsubata, Takeshi Yamada, Makoto Kitano, and Shuji Doi. 2014. "Recent Progress of High Performance Polymer OLED and OPV Materials for Organic Printed Electronics." *Science and Technology of Advanced Materials* 15 (3): 034203. IOP Publishing. 10.1088/1468-6996/15/3/034203

Shah, Virang G., and David B. Wallace. 2004. "Low-Cost Solar Cell Fabrication by Drop-on-Demand Ink-Jet Printing." *Proceeding of IMAPS 37th Annual International Symposium on Microelectronics*, 1–6. http://www.microfab.com/about/papers/solarcells_IMAPS_2004.pdf.

Sharma, Ashim. n.d. "Not Just Covid-19, Automobile Sector's Other Problem: Microcontroller Shortage." https://www.forbes.com/sites/palashghosh/2021/02/22/theres-a-worldwide-shortage-of-semiconductors-hurting-car-and-smartphone-companies-heres-how-investors-can-benefit/?sh=665a267f795b.

Sitch, C. D., D. P. Halliday, and A. P. Monkman. 2001. "Steady State Photoconductivity in Conjugated Polymers." *Materials Research Society Symposium –Proceedings* 665: 327–331. 10.1557/proc-665-c8.17

Sousa, Ricardo E., Carlos M. Costa, and Senentxu Lanceros-Méndez. 2015. "Advances and Future Challenges in Printed Batteries." *ChemSusChem* 8 (21): 3539–3555. 10.1002/cssc.201500657

Subramanian, Vivek, Josephine Chang, and Frank Liao. 2013. "Printed Organic Chemical Sensors and Sensor Systems."*Applications of Organic and Printed Electronics*. Edited by E. Cantatore. Springer.157–177. 10.1007/978-1-4614-3160-2_8

Swisher, Sarah L., Steven K. Volkman, and Vivek Subramanian. 2015. "Tailoring Indium Oxide Nanocrystal Synthesis Conditions for Air-Stable High-Performance Solution-Processed Thin-Film Transistors." *ACS Applied Materials and Interfaces* 7 (19): 10069–10075. 10.1021/acsami.5b00893

Takeda, Yasunori, Yudai Yoshimura, Rei Shiwaku, Kazuma Hayasaka, Tomohito Sekine, Tomoko Okamoto, Hiroyuki Matsui, Daisuke Kumaki, Yoshinori Katayama, and Shizuo Tokito. 2018. "Organic Complementary Inverter Circuits Fabricated with Reverse Offset Printing." *Advanced Electronic Materials* 4 (1): 5–9. 10.1002/aelm.201700313

Tamba, Shunsuke, Shota Tanaka, Youhei Okubo, Hikaru Meguro, Shuji Okamoto, and Atsunori Mori. 2011. "Nickel-Catalyzed Dehydrobrominative Polycondensation for the Practical Preparation of Regioregular Poly(3-Substituted Thiophene)S." *Chemistry Letters* 40 (4): 398–399. 10.1246/cl.2011.398

Tasdelen, Mehmet Atilla, and Yusuf Yagci. 2011. "Photochemical Methods for the Preparation of Complex Linear and Cross-Linked Macromolecular Structures." *Australian Journal of Chemistry* 64 (8): 982–991. 10.1071/CH11113

Ramanaviciene Almiraand Arunas Ramanavicius. 2013. "Application of Polypyrrole for the Creation of Immunosensors." *Critical Reviews in Analytical Chemistry*, 32(3): 37–41.

Toshima, Naoki, and Susumu Hara. 1995. "Direct Synthesis of Conducting Polymers from Simple Monomers." *Progress in Polymer Science* 20 (1): 155–183. 10.1016/0079-6700(94)00029-2

Tseng, Chun Chieh, Yu Hsien Chou, Ting Wei Hsieh, Min Wen Wang, Youn Yuen Shu, and Ming Der Ger. 2012. "Interdigitated Electrode Fabricated by Integration of Ink-Jet Printing with Electroless Plating and Its Application in Gas Sensor." *Colloids and Surfaces A: Physicochemical and Engineering Aspects* 402. Elsevier B.V.: 45–52. 10.1016/j.colsurfa.2012.03.016

Vaklev, Nikolay L., Joachim H. G. Steinke, and Alasdair J. Campbell. 2019. "Gravure Printed Ultrathin Dielectric for Low Voltage Flexible Organic Field-Effect Transistors." *Advanced Materials Interfaces* 6 (11): 1–6. 10.1002/admi.201900173

Vena, Arnaud, Etienne Perret, Smail Tedjini, Guy Eymin Petot Tourtollet, Anastasia Delattre, Frédéric Garet, and Yann Boutant. 2013. "Design of Chipless RFID Tags Printed on Paper by Flexography." *IEEE Transactions on Antennas and Propagation* 61 (12): 5868–5877. 10.1109/TAP.2013.2281742

Villani, Fulvia, Paolo Vacca, Giuseppe Nenna, Olga Valentino, Gianbattista Burrasca, Tommaso Fasolino, Carla Minarini, and Dario Della Sala. 2009. "Inkjet Printed Polymer Layer on Flexible Substrate for OLED Applications." *Journal of Physical Chemistry C* 113 (30): 13398–13402. 10.1021/jp8095538

Wang, Baochen, Jisong Tang, and Fosong Wang. 1987. "Electrochemical Polymerization of Aniline." *Synthetic Metals* 18 (1–3): 323–328. 10.1016/0379-6779(87)90899-X

Wang, Hao, Hao Wang, Hao Wang, Chang Jiang Yao, Hai Jing Nie, Ke Zhi Wang, Yu Wu Zhong, et al. 2020. "Recent Progress in Carbonyl-Based Organic Polymers as Promising Electrode Materials for Lithium-Ion Batteries (LIBs)." *Journal of Materials Chemistry A* 8 (24): 11906–11922. 10.1039/d0ta03321a

Wei, Yen, Chi-cheung Chan, Jing Tian, Guang-way Jang, and Kesyin F. Hsueh. 1991. "Electrochemical Polymerization of Thiophenes in the Presence of Bithiophene or Terthiophene: Kinetics and Mechanism of the Polymerization." *Chemistry of Materials* 3(5):888–897.

Yamamoto, Takakazu. 2002. "π-Conjugated Polymers with Electronic and Optical Functionalities: Preparation by Organometallic Polycondensation, Properties, and Applications." *Macromolecular Rapid Communications* 23 (10–11): 583–606. 10.1002/1521-3927(20020701)23:10/11<583::AID-MARC583>3.0.CO;2-I

Yim, Myung Jin, Yi Li, Kyoung Sik Moon, Kyung Wook Paik, and C. P. Wong. 2008. "Review of Recent Advances in Electrically Conductive Adhesive Materials and Technologies in Electronic Packaging." *Journal of Adhesion Science and Technology* 22 (14): 1593–1630. 10.1163/156856108X320519

Zhao, Jinyang, Yongli Yan, Zhenhua Gao, Yuxiang Du, Haiyun Dong, Jiannian Yao, and Yong Sheng Zhao. 2019. "Full-Color Laser Displays Based on Organic Printed Microlaser Arrays." *Nature Communications* 10 (1). Springer US: 1–7. 10.1038/s41467-019-08834-6

Zhao, Lianjia, Kang Wang, Wei Wei, Lili Wang, and Wei Han. 2019. "High-performance Flexible Sensing Devices Based on Polyaniline/MXene Nanocomposites." *InfoMat* 1 (3): 407–416. 10.1002/inf2.12032

Zhao, Manlin, Huanhuan Zhang, Cheng Gu, and Yuguang Ma. 2020. "Electrochemical Polymerization: An Emerging Approach for Fabricating High-Quality Luminescent Films and Super-Resolution OLEDs." *Journal of Materials Chemistry C* 8 (16): 5310–5320. 10.1039/c9tc07028a

Zhou, Lu, Mengjie Yu, Xiaolian Chen, Shuhong Nie, Wen Yong Lai, Wenming Su, Zheng Cui, and Wei Huang. 2018. "Screen-Printed Poly(3,4-Ethylenedioxythiophene): Poly(Styrenesulfonate) Grids as ITO-Free Anodes for Flexible Organic Light-Emitting Diodes." *Advanced Functional Materials* 28 (11): 1–7. 10.1002/adfm.201705955

12 Polymer Nanocomposites-Based Wearable Smart Sensors

Mohammed Khalifa
Komepetenzzentrum Holz GmbH, Wood K-plus, Linz, Austria

C. Shamitha
Manipal Institute of Technology Bengaluru, Manipal Academy of Higher Education (MAHE), Bengaluru, India

Sawan Shetty
Manipal Institute of Technology, Manipal Academy of Higher Education (MAHE), Manipal, India

Anandhan Srinivasan
National Institute of Technology Karnataka, Surathkal, Mangalore, Karnataka, India

CONTENTS

DOI: 10.1201/9781003200710-14

12.1   INTRODUCTION

With the advancement of technology and digitalization, a new term, *wearables*, has emerged in smart technology. The word *wearable* is no longer constrained to garments. Instead, it carries metaphors such as health monitoring bands, smart wristwatches, GPS trackers, smart textiles, etc. In the past few years, smart wearables have been considered an essential tool for rendering information specifically for fitness, run tracker, global positioning system (GPS), safety, health care, etc. Wearables are electronic devices that can be worn as gadgets, attached to the helmet, or integrated with the clothing. It is fundamentally a device accomplished to sense, process, and communicate with data storage capabilities. Technological advances have empowered sports people and medical doctors to probe their functional activities, loads, and biomechanics to avoid injuries and maximize their performance. A wearable device can provide precise and uninterrupted physiological measures such as heart rate, sweat rate, skin temperature, skin conductance, eye tracking, etc., consequently allowing physicians to develop the necessary treatment plans (R. T. Li et al., 2016; Pantelopoulos & Bourbakis, 2010; Seshadri et al., 2019; Wanjari & Patil, 2017).

The wearable sensors' global market is expected to see an exponential growth of ~40% compound annual growth rate and is anticipated to reach US$1,083.3 million by 2025. Currently, the United States is among the most popular countries in smart wearable devices, while the Asia Pacific is the fastest-growing region in wearable sensors. With advancements in technology, miniaturization and lightweight have facilitated real-time devices to gain popularity in the market. Sensors such as motion, imaging, position, force, and chemical lead the market, while accelerometers, heart rate monitoring, pressure, and optical sensors are expected to grow rapidly. In addition, wearable devices create key platforms such as the internet of Things (IoT) and machine to machine (M2M), which could upsurge the number of connected devices worldwide. Companies such as Bosch GmBH, ST Microelectronics, Infineon, NXP Semiconductors, Texas Instruments, Knowles Electronics Ltd. are leading the wearable device market (Wearable Sensors Market, 2019; Wearable Sensors Market Size, Share & Trends Analysis Report, 2018; Wearable Sensors Market Size, Trends, Industry Growth Analysis-Global Forecast, 2025).

Wearable devices are becoming more flexible, tiny, and self-powered, which offer new prospects and potential applications. Wearable devices are available in many forms, such as fitness bands, textiles, watches, shoes, bands, patches, necklaces, etc. (Figure 12.1) (Lewis, Pritting, Picazo, & JeanMarie-Tucker, 2020; Lugoda, Hughes-Riley, Morris, & Dias, 2018; Sim, Yoon, & Cho, 2018; Strohmeier, Knibbe, Boring, & Hornbæk, 2018). For example, the human body temperature can provide vital information about a patient's medical condition, including fever, strokes, lung diseases, etc. Biochemical parameters that can be monitored using sensors include saliva, pH content, body

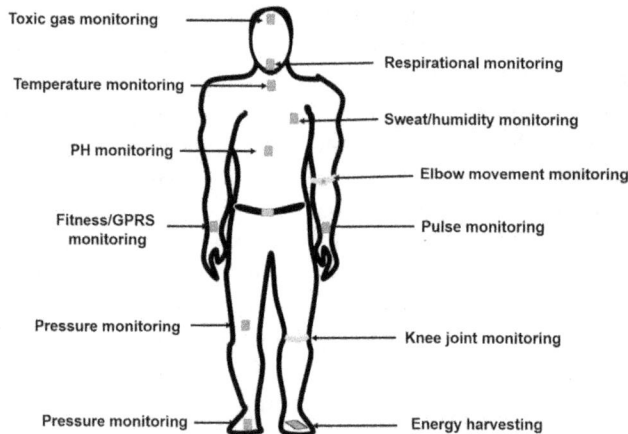

FIGURE 12.1   Wearable sensors for monitoring various parameters.

fluids, fluoride content, oxygen level, and concentration of elements. Several wearable sensors can be used for the measurement of various physical and physiological measurements. Sensors are the heart of wearable technology that is responsible for receiving the desired output. The sensing data can be monitored using an integrated communication network. Wearable sensors are classified into two types, namely physical and chemical sensors. These sensors can be used for drug delivery, health management, robots, thermal wear, etc. Various inputs can be adapted to run these sensors, such as mechanical, electrical, optical, and chemical changes. Mechanical wearable sensors include piezoelectric, piezoresistive, capacitive, and iontronic-based sensors.

12.1.1 Piezoelectric-Based Wearable Sensors

Piezoelectric sensors are based on the piezoelectric effect, wherein the material produces electrical charges upon the application of external load or pressure and vice versa. When the pressure is applied to the sensor, the electric dipoles reorient themselves in one direction. As a result, the surface charge changes and produces voltage (Richardson, 1989). Piezoelectric wearable sensors are used for tactile sensing, motion detection, monitoring pulse pressure, body mobility, and physical activity monitoring (Chung, Lee, Lee, & Kim, 2017; Karan et al., 2018; Khalifa, Mahendran, & Anandhan, 2019; Spanu et al., 2016). Piezoelectric materials such as lead zirconate titanate (PZT) (S. Zhang, Zhang, Wang, Wang, & Pan, 2019), zinc oxide nanowires (Dahiya et al., 2018), barium titanate (BaTiO$_3$) (Alluri et al., 2017), and poly(vinylidene fluoride) (PVDF) (Khalifa & Anandhan, 2019) are amongst the most common materials used for the wearable technology. Piezoelectric sensors are already in the market and are available for various applications. However, the main drawback of these sensors is their charge-leaking characteristic, which could be detrimental for sensing low-frequency signals. Piezoelectric sensors are seen to be potential materials for wearable devices because of their self-powered sensing capabilities.

12.1.2 Piezoresistive Sensors

When a conductive material is subjected to strain, its electrical resistance changes; this phenomenon of electromechanical response is termed *piezoresistance*. Wearable piezoresistance sensors must possess high stretchability and flexibility to yield high sensitivity. Such conditions allow the sensors to mount on the body and be used as a strain/pressure sensor without exhibiting plastic deformation. However, highly stretchable and flexible materials often struggle to achieve high sensitivity and suffer from hysteresis. To improve their sensitivity, structural modifications of material (nano/micro-structuring or addition of nanofillers) are often necessary.

Several publications have reported using piezoresistive wearable devices for pressure, pulse detection, pulsatile blood flow detection, etc. The main drawback of these devices is that they require external power for continuous use for the real-time monitoring of data. Currently, materials such as carbon black (CB), carbon nanotube (CNT), conductive polymers, etc. have been reported to improve the performance of piezoresistive wearable sensors (Ding, Xu, Onyilagha, Fong, & Zhu, 2019; Y. Jung et al., 2019, Rizvi, Cochrane, Biddiss, & Naguib, 2011; Stassi, Cauda, Canavese, & Pirri, 2014). Piezoresistive pressure sensors are devices wherein pressure change converts into electrical resistance change in the sensor material, which have been the most widely used ones in wearable devices. These sensors can be used to detect low-pressure changes such as touch, fluid, or blood flow. Contrary to the strain sensors, these sensors typically consist of two electrodes with minimal electrical resistivity, which the application of pressure can modulate. Piezoresistive pressure sensors generally comprise three components: the active sensor material, substrate, and the electrodes. Various materials, such as graphene, CNT, metal compounds, conductive polymers, nanoparticles, etc. have been used as the active sensor material. However, to augment the sensor performance, modifying the electrodes or incorporating nano-fillers is often necessary. Another limitation of these sensors is that they require an external power source for continuous monitoring. Furthermore, the flexibility, stretchability, sensitivity, robustness, and reliability of these sensors need to be addressed before their real-time applications (Bao, 2000; J. Hu et al., 2020; Yunjing Lu et al., 2019; Tran, Zhang, & Zhu, 2018).

12.1.3 Capacitance-Based Sensors

Capacitive sensors have gained tremendous popularity due to their facile sensing mechanism, which uses a highly compliant dielectric layer between two electrically conductive electrodes. The application strain on the sensor material results in a change of capacitance. These sensors have been widely used in wearable applications for human pressure sensing interfaces (tactile sensing, body pressure monitoring or joint movement detection, etc.). Parallel plate arrangements of the sensors have been widely adopted in the wearable market. The capacitive sensor offers high linearity, stretchability, and low hysteresis. However, these devices often suffer from low sensitivity and undesirable noises from the body or surrounding conditions. Introducing pores, structural modifications, surface modifications, etc. can address these issues, which could help monitor pulse and force tracking (Ding, Xu, Onyilagha, Fong, & Zhu, 2019; Hwang et al., 2019; Jeong, Oh, Woo, & Kim, 2017).

12.1.4 Iontronic Sensors

Recently, elastomers incorporated with ionic liquids have emerged as a susceptible dielectric material. The electron double-layer usually creates the iontronic interface at the interface between the electrode and the electrolyte, and the dielectric/capacitance of the material can be enhanced significantly due to the atomic scale distance between the charges at the interface. Iontronic sensors significantly enhance the piezo-capacitance effect upon the application of pressure. Iontronic sensors have been the potential material for various applications such as pressure monitoring, fluid speed monitoring, robot arm movement detection, e-skins, etc. However, iontronic sensors suffer from low sensitivity and response at high pressure. Also, these sensors are recently developed; incorporating this sensor technology with mass manufacturing is uncertain and challenging. Further, the use of ionic liquids might show some toxicity, which could be risky for wearable devices (Bai et al., 2020; Chhetry, Kim, Yoon, & Park, 2019).

12.1.5 Electrical and Optical Sensors

Electrical sensors measure the change in the electrical resistance, conductivity, or capacitance at the skin surface. In other cases, to monitor the small changes in the charges, high impedance is added to the input value. These sensors should attain good contact with the skin to achieve good sensitivity, which is a challenging task in a dynamic environment. Thus, to improve the signal, the sensor is attached to the skin through an electrolyte gel (wet electrode configuration), which reduces the resistance of the skin. However, prolonged use of wet electrode configuration is restricted due to several factors. Hence, dry electrode configuration has been seen as the alternative approach. To achieve good contact between the skin and the electrode, the air gap between them should be kept at its minimum level, and artifacts between the contacts are essential. It is essential to have ultrathin, low modulus, high stretchable electrodes to achieve good response and prolonged sensor functioning.

Optical sensors have made it big in wearable technology because of their versatility and label-free detection, paving the way in various sensing applications. Optical fibers are distinctive and sensitive to electromagnetic radiations that are capable of detecting nano-scale data. Optical sensors have been implemented for heart rate monitoring, blood pressure monitoring, oxygenation, pressure point determination, and quantification of virus, protein, or ion concentrations. Information acquired from these optical sensors is accurate, which can be assimilated with wearable devices such as wristwatches, fitness bands, etc. These sensors rely on the light source, which is introduced into the body, and the changes in the light signal reveal the information through an optical light detector. Nonetheless, optical sensors for wearable technology are still challenging tasks, and several issues must be addressed (Jin, Jin, & Jian, 2018; Pan, Zhang, Jiang, Zhang, & Tong, 2020; Rehg, Murphy, & Kumar, 2017).

12.1.6 CHEMICAL SENSORS

Chemical sensors have seen a significant upsurge in recent years due to advances in technology and material chemistry. Apart from physical parameters, chemical parameters are considered crucial and need monitoring of individual health performance at a molecular level. Several reports have demonstrated the use of chemical sensors to detect and quantify electrolytes, heavy metals, toxic substances, healing, and sweat rate monitoring. However, chemical sensor technology is still a new concept and poses several challenges because of its complex operation mode. On the other hand, issues related to stability, selectivity, sensitivity, contamination, sensor integration, miniaturization, etc., have hindered its development for practical applications. New materials such as nanostructured materials and multifunctional polymers have emerged to address these issues, which could be a breakthrough in the area of chemical sensor technology (Bandodkar, Jeerapan, & Wang, 2016; Gray, 2019; G. Li & Wen, 2020).

Despite countless advances and laboratory prototypes, only a few wearable sensors have been made into the market. One of the key motive reasons for the significant demand for wearable devices is the rapid emergence of advanced materials with fascinating properties. As a result, wearable devices are becoming cheaper, trustworthy, and smaller. Hence, the role of the sensor materials becomes more relevant for wearable technology. There are various parameters involved in the successful design of sensor material for wearable technology. The key parameters include stability, hysteresis, linearity, sensitivity, flexibility, skin comfort, response time, selectivity, durability, biocompatibility, self-powering, material selection, etc. Optimizing all the parameters is challenging and requires one or two materials combined to form special properties. Among various materials, polymer and polymer nanocomposites have emerged as attractive candidates for wearable sensor fabrication. The polymer-based wearable sensors provide flexibility and offer excellent biocompatibility, skin comfort, and lightweight besides low-cost fabrication. However, polymer-based sensors often suffer from low sensitivity, mechanical integrity, thermal stability, and selectivity. Therefore, polymer nanocomposites have been developed, which overcome the issues mentioned previously. To date, several publications have been reported for smart wearable sensors based on polymer nanocomposites, but the realization for practical implementation is limited.

The smart polymer nanocomposite development always comes from nature's intelligence, which can be used for sensing, actuation, and control functions. Smart polymer nanocomposites are seen as an attractive material in the era of the fourth Industrial Revolution. Smart polymer nanocomposites can be derived from shape memory polymer (Ivens, Urbanus, & De Smet, 2011), electroactive polymer (Khalifa et al., 2019), conductive polymer (Teixeira, Horta-Romarís, Abad, Costa, & Lanceros-Méndez, 2018), stimuli-active polymer, smart electrorheological polymer (Dong, Seo, & Choi, 2019), self-healing and self-cleaning polymer (Reddy et al., 2020; Selim, El-Safty, & Shenashen, 2019), dielectric polymer (R. Li, Zhou, Liu, & Pei, 2017), thermoelectric polymer (Lund, Tian, Darabi, & Müller, 2020), and energy harvesting polymer (Martins, Lopes, & Lanceros-Mendez, 2014). The addition of nanofillers may further increase the smart polymer performance due to their nucleating activity, high surface area, reinforcing influence, and intrinsic functionalities. Smart polymer nanocomposite is widely adopted in applications such as sensors, actuators, drug delivery, wearable devices, stretchable electronics, aerospace, and textile industry (Pielichowski & Pielichowska, 2018).

The collective properties of nanostructures and polymers represent a unique set of functional characteristics and signify an innovative tactic in the technology of materials, which has tremendous potential for smart technology. Nanostructures such as semiconductors, quantum dots, nanotubes, nanosheets, ceramics, nanoshells, metal clusters, etc. have been exploited as the nanoreinforcement materials for polymer nanocomposites. Incorporating nanostructures offers a unique set of physical, chemical, and electronic properties. The performance of smart polymer nanocomposites relies on various features of fillers such as aspect ratio, size, shape, surface area, functionality, compatibility with the matrix, volume fraction, degree of dispersion, etc. Nonetheless, systematic investigations are required to synthesize, process, select, and apply polymer nanocomposite to achieve the desired goal. Smart polymer nanocomposites have been explored for various applications, including smartphones, television, smart homes, GPS, wearable technology sensors, antenna, energy harvesting, and self-powered devices. For example, poly (vinylidene fluoride)-based nanocomposites are well-established materials for piezoelectric and pyroelectric sensors (Bouhfid & Jawaid, 2020; Gray, 2019; Nicolais & Carotenuto, 2013; Ponnamma, Sadasivuni, Cabibihan, & Al-Maadeed, 2017).

In this chapter, with the above motivation, recent advances in polymer nanocomposites as smart sensors for wearable technology have been explored. Various synthetic approaches have also been discussed, which provide an idea on how to fabricate smart polymer nanocomposites for wearable sensor applications.

12.2 POLYMER NANOCOMPOSITE-BASED WEARABLE SENSORS

Polymers contribute as a key component in the sensor unit. This section introduces different types of sensors with their working principles and potential applicability.

12.2.1 STRAIN SENSOR FOR WEARABLE APPLICATION

A piezoresistive strain sensor with excellent flexibility and sensitivity is a vital requirement for wearable application. To encounter these requirements, polymers such as poly(dimethylsiloxane) (PDMS), thermoplastic polyurethane, poly(styrene-butadiene-styrene) (SBS), PVDF, poly(vinyl alcohol), natural rubber, etc., have been used as matrices for the conductive fillers. Polymer nanocomposite based piezoresistive strain sensors are divided into three types: i) filled-type polymer nanocomposite-based sensor, ii) sandwich-type polymer nanocomposite-based sensor, and iii) adsorption-type polymer nanocomposite-based sensor. These strain sensors offer distinct microstructures and sensing performance. Polymer nanocomposite-based piezoresistive strain sensors comprise conductive fillers such as CNT, CB, graphene, polyaniline (PANI) embedded into the polymer matrix by various processing techniques. The repetitive disturbance in the polymer's conductive network is the primary mechanism of the piezoresistive strain sensing. The change in the electrical resistance is governed by the electron tunneling mechanism between the neighboring conductive particles and the electrically conductive pathways formed due to the contact between particles. The formation of a conductive network is dependent on various factors such as state of dispersion, size, and shape of the nanofillers, polymer matrix, morphology of matrix, and processing technique. It is a well-established fact that the state of dispersion of fillers in the polymer matrix defines the strain sensing performance of the nanocomposite (Hu, Karube, Yan, Masuda, & Fukunaga, 2008; Yang Lu, Biswas, Guo, Jeon, & Wujcik, 2019; Xiang, Chen, & Li, 2019; Xie & Zhu, 2018). Furthermore, compatibility and interaction between the filler and polymer greatly influence the strain sensing performance. Strong interaction between them offers effective load transfer from the polymer matrix to conductive particles, making the conductive network easily disturbed under strain loading and relaxing modes.

Moreover, the viscoelastic property of the polymer affects the hysteresis and repeatability of the sensor. Ultra-soft and highly stretchable polymers offer excellent repeatability but possess poor recovery and hysteresis (Amjadi, Yoon, & Park, 2015). Hence, the selection of conductive fillers and polymer matrix and their processing techniques plays an essential role in defining the sensing performance.

CB/thermoplastic polyurethane (TPU) composite was prepared for a flexible strain sensor. The nanocomposite was prepared using compression molding at a pressure of 15 MPa and 200°C. Figure 12.2a shows the digital photograph of CB/TPU nanocomposite depicting its flexibility. The nanocomposite

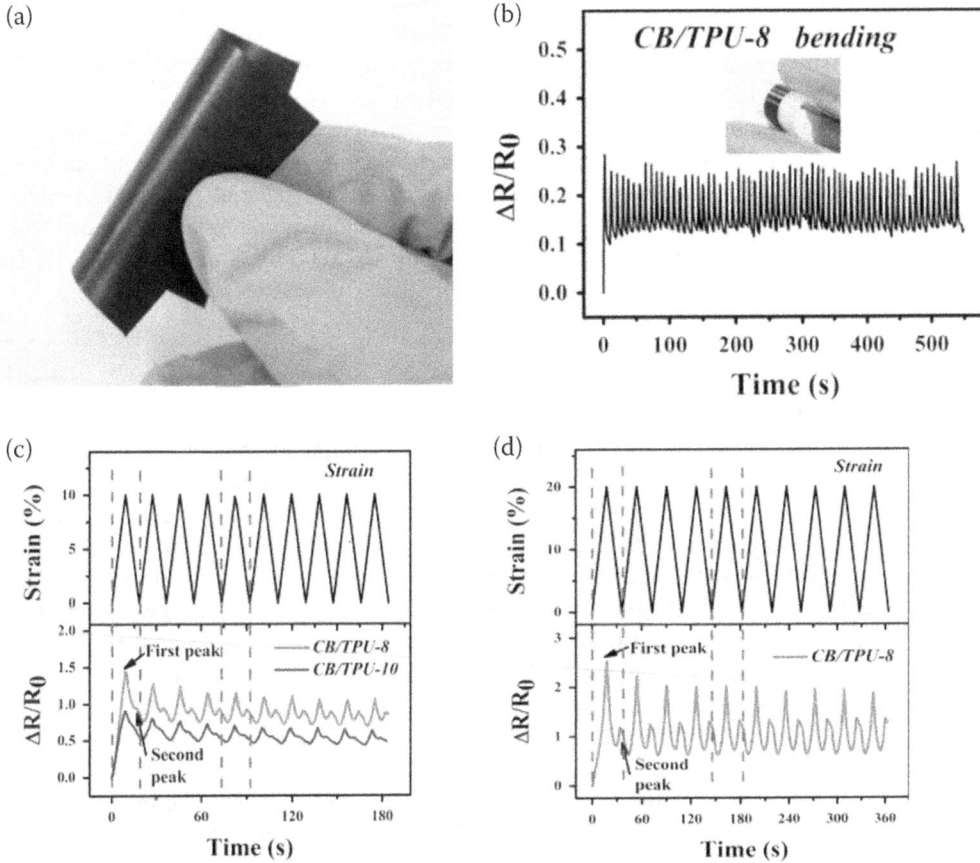

FIGURE 12.2 (a) Digital photograph of CB/TPU nanocomposite film; sensor response under (b) bending cycles, c) stretching and releasing cycles at 10%, and d) stretching and releasing cycles at 20%. [Reproduced with permissions © 2017 Elsevier] (Y. Zheng et al., 2017).

displayed excellent flexibility, and the size of the nanocomposite was $4 \times 0.5 \times 0.04$ cm^3. The nanocomposite's percolation behavior showed a characteristic transition, which is generally observed in carbon-based conductive fillers. The electrical conductivity increased steeply with the increase in CB content due to the high aspect ratio and surface area.

The percolation at low filler content is essential to achieve flexibility of the sensors. Figure 12.2b shows the sensor response upon repeated bending, indicating its adaptability for flexible electronic applications. The response of the sensor in bending mode is dependent on the chord length of the sensor. The sensor response increased with decreasing chord length. Further, the resistance change under repeated stretching-releasing cycles was investigated to determine the sensor's quality and response. The response of the sensor is amplified upon increasing the strain (Figure 12.2c–d). Interestingly, the fluctuation was observed in the form of a shoulder peak during the relaxation. Such fluctuations could be attributed to the deformation of the flexible nanocomposite, which provides the nanocomposite with self-diagnosis ability to identify the conductive network's growth. However, the fluctuations in the response of the sensor are unfavorable for strain sensing applications. CB/TPU nanocomposite-based sensors showed excellent sensitivity with a gauge factor (GF) of 10.8 and exhibited good repeatability (Y. Zheng et al., 2017).

Strain sensors based on the optical platform are seen to be a promising alternative to resistance and capacitance-based sensors, furthering from their unique advantage of high sensitivity, size, electromagnetic resistance, and safety. Despite the great potential, the optical-based strain sensors comprising ceramics and glass are not favorable for wearable technology due to their fragility, stiffness, and non-biocompatible characteristics. Therefore, researchers are keen to develop a highly flexible, stretchable strain sensor based on an optical platform with biocompatible characteristics. To counter the issue, polymer-based optical sensors are seen as the virtuous alternative. PDMS is among the most widely adopted polymers for optical strain sensors due to its high elasticity, chemical inertness, and stability. Another fascinating characteristic is its transparent nature and compatibility as matrix makes it an excellent material for optical sensors. Characteristics of PDMS can be tuned by varying its curing temperature, mixing ratio, and density of the PDMS and curing agent (Gu, Kwon, Ahn, & Park, 2020; Guo, Niu, & Yang, 2017; Guo, Zhou, Yang, Dai, & Kong, 2019; Souri et al., 2020).

12.2.2 Thermoelectric Wearable Sensors

Thermoelectric materials are solid-state devices that convert heat into electricity according to the Seebeck effect. Thermoelectric materials may play an essential role in harvesting energy for wearable devices. A significant advantage of these materials is that they produce electricity without mechanical motions or components, making them more reliable than other harvesting technologies for wearable devices. When a thermoelectric material is attached to human skin, the heat from the body flows through it and the temperature difference between the body and thermoelectric material generates useful energy. These generators potentially allow the wearable devices to function wirelessly and continuously. The efficiency of thermoelectric generators relies on their parameters, such as thermal conductivity, Seebeck coefficient, and electrical conductivity. Optimization of these three parameters to obtain high thermoelectric efficiency is a challenging task. Apart from the thermoelectric efficiency, flexibility of the generator is a vital requirement for wearable devices. Recently, polymer-based thermoelectric materials have been a center of focus due to their unique merits such as lightweight, flexibility, ease of processability, low cost, low thermal conductivity, and facile doping process to achieve high electrical conductivity. Some recently reported studies indicate the use of conducting and non-conducting polymers for the fabrication of thermoelectric generators. Polymers such as PANI, Poly(3-hexylthiophene) (P3HT), PVDF, polyacetylene, polypyrrole, polythiophenes etc have been used for the fabrication of thermoelectric generators (Kroon et al., 2016; Leonov & Vullers, 2009).

Conducting polymers have excellent potential to exhibit high thermoelectric properties and are seen as the next generation thermoelectric material for flexible wearable devices. Their thermoelectric efficiency is strongly reliant on doping, which directly reflects the electrical conductivity and Seebeck coefficient. It is essential to optimize the material parameters and doping level to compromise between the electrical conductivity and Seebeck coefficient to achieve high efficiency. Optimal doping level of the conducting polymer can be achieved using electrochemical synthesis. The process is carried out in a three-electrode cell containing an electrolyte solution with a polymer working electrode, platinum counter electrode, and gold reference electrode. Adoption of electrochemical processes offered high Seebeck coefficient and power factor in the thermoelectric sensors (Bubnova, Berggren, & Crispin, 2012). The Seebeck coefficient of bulk material can be improved by introducing nanofillers or dopants. The energy per carrier transported can be augmented by restricting low-energy charge carriers from participating in the conducting mechanism. This effect is commonly known as the electron energy filtering effect (Berland et al., 2016). PEDOT is among the most suitable candidates for thermoelectric materials, owing to its outstanding properties compared to other conducting polymers. However, its poor solubility and film formation characteristics restrict its implementation. To overcome these issues, poly(styrene sulfonate) (PSS) is generally emulsified, developing the PEDOT: PSS films, which significantly augments the range of applications of PEDOT. Recent reports suggest that the electrical conductivity of PEDOT:PSS film has been increased significantly (~1,500 S.cm^{-1}), but the Seebeck coefficient is low. Several studies have been reported until now, wherein a drastic improvement in the Seebeck coefficient is seen. Thermoelectric materials made of polymer composite or blends have improved the efficiency besides the physical properties. A combination of organic material as filler and polymer with a high Seebeck coefficient is expected to offer high thermoelectric efficiency. Polymer-based nanocomposites with CNT, graphene, and other inorganic nanofillers have been adopted for the same.

12.2.3 PRESSURE SENSORS

Among the wearable sensors, pressure sensors play a crucial role in human health monitoring and medical diagnosis. A signal will be generated in a pressure sensor at a specific pressure range due to the signal transduction mechanism. The low-pressure regime (<10 KPa) and medium-pressure regime (10–100 KPa) are of great interest in wearable healthcare and medical diagnosis. This sensor's main requirement is high sensitivity towards the low-pressure range, fast response, and minimum power usage. The rapid emerging sensor technology leads to sensitive, flexible, and low-cost wearable pressure sensors. The conversion of physical stimuli into electrical stimuli occurs by different transduction mechanisms, including piezoelectric, piezoresistive, and piezocapacitive.

PVDF-based nanocomposites are the promising candidate material in piezoelectric pressure sensors (Khalifa & Anandhan, 2019) developed PVDF/g-C_3N_4 nanocomposite fibers through electrospinning technique. Uniform, bead free, and fine nanofibers of good morphological features were obtained (Figure 12.3a). The composite nanofibers exhibited excellent piezoelectric performance and physicochemical characteristics compared to pristine PVDF nanofibers. This is attributed to the β-phase enhancement due to filler addition and in-situ stretching during electrospinning. The nanogenerator was employed in piezoelectric energy harvesting from various human hand movements (see Figure 12.3b and c). The nanogenerator fabricated from electrospun nanofabric of PVDF/0.75 wt% g-C_3N_4 (PGN-0.75) exhibited the highest piezoelectric response (7.5 V, 2.3 µA). In addition to this, the nanogenerator exhibited good repeatability, stability, and touch sensitivity.

The piezoresistive pressure sensors rely on the mechanism of change in electric resistance due to mechanical stimuli (pressure). These are investigated extensively because of their flexibility, chemical stability, simple device fabrication, and low power consumption. The microstructure associated with the pressure sensor plays an important role in the piezoresistive sensing mechanism. These microstructures assist the reversible conductive pathways when pressure is applied. In order to attain high sensitivity, an increase of contact points or contact area is preferred. So sensors with a 3D conductive network or layered structure are required. Zhang et al., (2018) have developed such multilayer piezoresistive sensors using thiolated graphene (GSH) and PET fabric. The GSH@PET fabric was developed by dipping the PET fabric in GO dispersion followed by the thiolation process (Figure 12.4a). The sensor's sensitivity is directly proportional to the number of fabric layers. The main reason for this is the inherent mechanism associated with the fabric structure. In a smaller pressure regime, the contact area between the GSH@PET fabric increases and the disappearance of spacing between them leads to high sensitivity. At a high-pressure regime, the contact area further increases due to the fabric's elastic deformation, low electric resistance, and high sensitivity. Also, the roughness and high mechanical resilience of PET fabric help in the initial state's fast recovery. This multilayered sensor exhibits an extensive working range of up to 200 KPa (Figure 12.4b). Furthermore, these sensors display high sensitivity and repeatability. The sensor was applied to demonstrate various human body movements, and electronic skin was developed to map the spatial distribution of pressure (Figure 12.4c–e).

12.2.4 HUMIDITY SENSORS

Humidity sensors broaden the potential of wearable devices. The utilization of humidity sensors provides necessary information of the user's conditions such as moisture content, sweat rate, etc. Based on the measuring unit, humidity sensors are classified into two classes: relative humidity

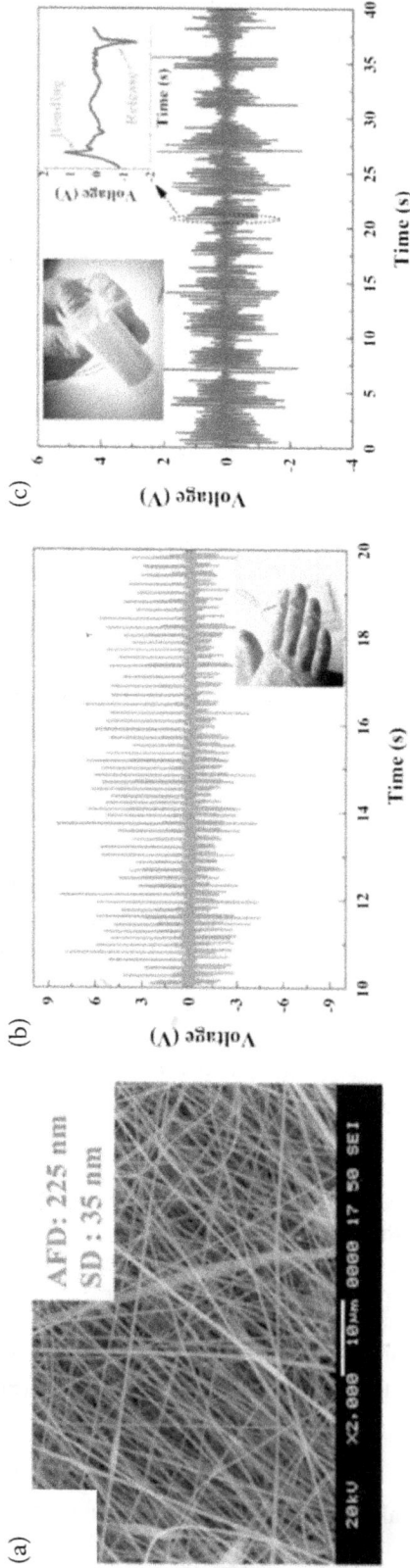

FIGURE 12.3 (a) SEM image of PGN-0.75 electrospun nanofiber (b) and (c) piezoelectric response of PGN-0.75 nanofiber under palm imparting and bending mode, respectively. [Reproduced with permission ©2019 Springer Nature] (Khalifa & Anandhan, 2019).

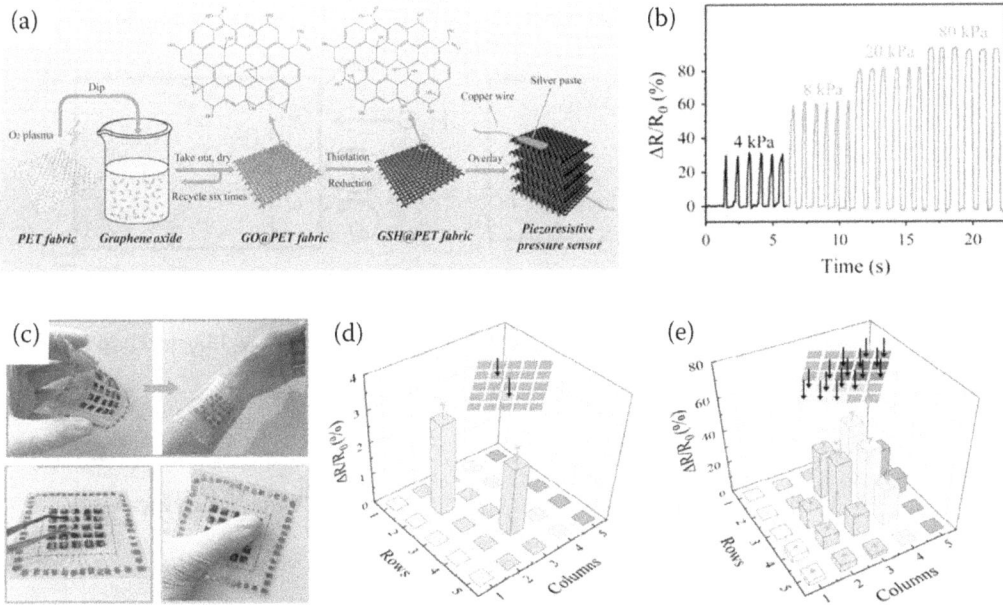

FIGURE 12.4 (a) Schematic representation of fabrication of GSH@PET piezoresistive pressure sensor, (b) response of five-layer sensor, (c), (d), and (e) wearable multilayer piezoresistive pressure sensor as e-skin and their point and plane distribution of pressure. [Reproduced with permission ©2018 American chemical society] (Zhang et al., 2018).

sensor and absolute humidity sensor. In most applications, relative humidity-based sensor is used due to its simpler and low-cost fabrication process. Various techniques have been adopted for humidity sensors, including capacitive, resistive, hygrometer, quartz microbalance, etc. Among them, resistive- and capacitive-based sensors offer several advantages over the other techniques.

Capacitive and resistive humidity sensors are based on the change in the dielectric constant and electrical resistance, respectively. Currently, ceramic, polymer, or a combination of ceramic and polymer-based humidity sensors are the attractive options available in the market. Polymer-based nanostructured materials such as nanowires, nanotubes, nanofibers, p-n heterojunctions, etc., have proven to be reliable and exhibit high performance. Polymer-based sensors are often prepared as thin micrometer-sized films, and their sensing mechanism is analogous to the oxide-based sensors. One vital advantage of polymer-based humidity sensors over other sensors is their room temperature sensing abilities. Recently, polymers containing phosphonium compounds have been developed along with conventional quaternary ammonium and sulfonate compounds. Despite their advantages discussed above, polymer-based humidity sensors exhibit low sensitivity, high hysteresis, and non-linear response. Also, lack of both high-temperature stability and selectivity are other issues to address. Most of the polymer-based humidity sensors are based on the electrical property or dielectric properties, which change upon interaction with water molecules. However, polymer-based humidity sensors possess several limitations, and hence the introduction of inorganic materials in the polymer (to use them as nanocomposites) has gained attention which overcomes those issues. Polymer/ceramic, polymer/metal oxide, polymer/metal particles, acid/polymer, polymer/ionic salts, polymer/conducting material-based nanocomposites have been reported as high-performance sensors, which provide the benefits of both the materials (Farahani, Wagiran, & Hamidon, 2014; Lim, Yap, Chong, & Ahmad, 2014; Lv et al., 2019; Popov, Nikolaev, Timofeev, Smagulova, & Antonova, 2017; S. C. Mukhopadhyay & K. P. Jayasundera, 2013Mukhopadhyay Mukhopadhyay 2013).

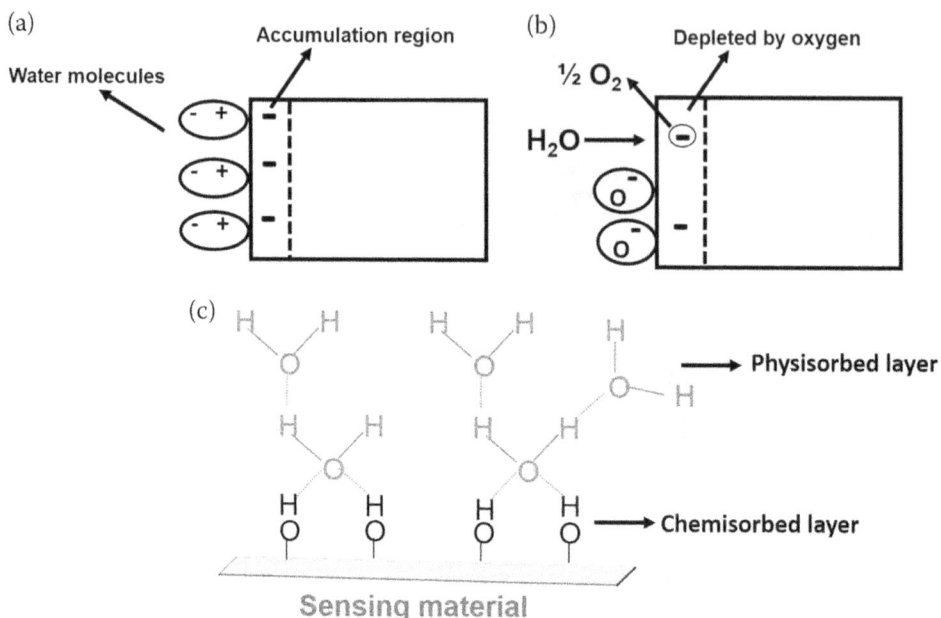

FIGURE 12.5 Plausible mechanism from the donor effect in n-type semiconductors: (a) Electrons attracted by the adsorbed H_2O molecules on the sensor surface, (b) release of electrons by adsorption, and (c) multilayer network depicting Grotthuss mechanism.

There are several reports on graphene/polymer nanocomposite based sensors for humidity sensing. Graphene is one of the most suitable materials for humidity sensing. Water molecules are adsorbed on the sensor surface in the molecular and hydroxyl forms, which results in the change of electrical properties. For p-type semiconductors, a decrease in the electrical conductivity was generally observed, while it increases for n-type active sensing material. Such an effect is related to the electron transportation from the chemically adsorbed water molecules to the sensing material. Another possibility is that H_2O molecules substitute the earlier adsorbed and ionized oxygen molecules, which releases the electrons from the ionized oxygen (Figure 12.5a and b). Electrical properties are sensitive to the electron concentration on the sensor surface; the sensing type is generally called an electronic type. However, most sensors struggled to perform effectively at room temperature; for example, SnO_2, ZnO, and In_2O_3. At low relative humidity conditions, the adsorbed H_2O molecules permit additional conductive pathways for charge transport. At higher relative humidity, the Grotthuss charge transport mechanism arises through a hydrogen bond network (Figure 12.5c). Also, the Grotthuss mechanism is less sensitive to temperature variation due to its low activation barrier for proton transport. However, the transport mechanism is disturbed at a higher temperature due to relatively weak hydrogen bond networks, which is destroyed at a higher temperature (~50°C) (Z. Chen & Lu, 2005; Khalifa, Wuzella, Lammer, & Mahendran, 2020; Lv et al., 2019).

Utilizing such a mechanism, He et al. (J. He et al., 2018) prepared the poly(dopamine) (PDA)/reduced graphene heterogeneous structure for high-performance wearable humidity sensors. A flexible humidity sensor consisting of a PDA/graphene layer structured with a tuned interlayer spacing of 0.7 to 1.4 nm was prepared by drop casting technique on a polyimide substrate. Once the solvent is evaporated, the drop cast PDA/graphene self-assembles to form a thick 40 nm film (GNCP) comprising alternate graphene and PDA molecular layers (Figure 12.6a–c). Such heterogeneous structure could play a critical role in optimizing the charge transport in the polymer. The interlayer arrangement of the sensing junction also affects the sensing performance. Figure

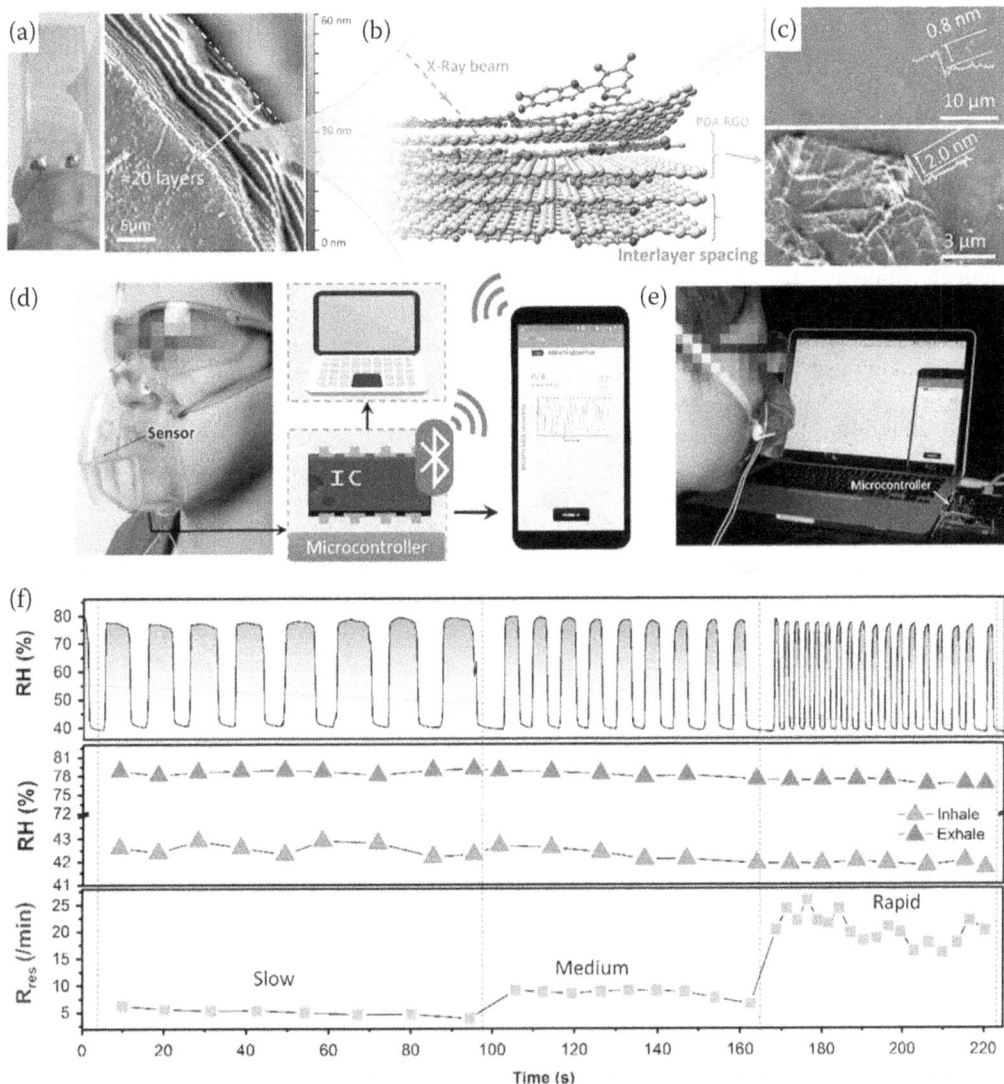

FIGURE 12.6 (a) Digital photograph and AFM image of fabricated GNCP sensor, (b) schematic representation of superlattices composed of graphene/PDA layers, (c) AFM images of graphene and PDA modified graphene, (d) and (e) digital photograph and schematic illustration of GNCP sensor for respiratory monitoring, (f) response of GNCP sensor recorded from respiration. [Reproduced with permissions © 2018 American chemical society] (J. He et al., 2018).

12.6d–e shows the sensor device consisting of a wireless module, breathing mask, and microcontroller to record real-time human respiration by a non-contact method. Monitoring of respiration is vital information to assess the medical condition of the patient. Water vapors are one of the primary modules in respiration. Hence, humidity sensors can help monitor respiration by determining the change in the moisture change. The respiration monitoring was monitored from the nasal breath. When the air is inhaled or exhaled through the mask, the relative humidity varies resulting in a change in the sensor response (Figure 12.6f). The sensor attached to the mask is beneficial since it provides comfort for the user. The sensor monitored the breath at different rates (slow, medium, and rapid breathing state). However, under high wind speed mixed with the exhaled air, the sensor's response, but still the respiratory signal can be distinguished, which allows the device to use it in daily lives as a wearable sensor (J. He et al., 2018).

12.2.5 Temperature Sensor

Temperature is one of the primary and vital physiological parameters of the human body that defines the health condition. The body temperature provides an imperative indicator of hypothermia, hyperthermia, fever, fatigue, metabolism, and depression. Body temperature reflects the health condition and uncharacteristic changes in the body. Hence, body temperature is an essential parameter to monitor. To be wearable, temperature sensors must possess high flexibility, softness, comfort, lightweight, high sensitivity, and skin-friendliness. Several reports attempted to make highly sensitive, soft, and biocompatible sensors. Temperature sensors are required to provide real-time data with high precision. Because the human skin temperature has a small window ranging from 25°C to 40°C with an average body temperature of 37°C, generally, the temperature of the skin can be measured effortlessly by inflexible temperature sensors such as thermometers and thermocouples. The main drawback of these sensors is that they create pressure on the skin for monitoring the temperature. For attaining high performance, the sensor should be well in contact with the body without upsetting human comfort. Textile sensors are one such kind, which attracted considerable attention due to their small and comfortable characteristics. There has been continuous development in the temperature sensors for wearable devices, such as printing technology, nanostructured materials mimicking biological structures (such as octopus feet, mammals, etc.), thermal imaging, etc. (Q. Li, Zhang, Tao, & Ding, 2017; Lima et al., 2011; Sahatiya, Puttapati, Srikanth, & Badhulika, 2016; Soni, Bhattacharjee, Ntagios, & Dahiya, 2020).

There are various kinds of temperature sensors that are generally classified into invasive, semi-invasive, and non-invasive types. Detection of temperature can be classified into two types: contact type and non-contact type. Contact-type temperature detectors include thermistors, integrated circuits, and resistance-type temperature detectors. Non-contact-type devices include thermopiles, pyroelectric sensors, IR radiations, and thermographic imaging. Among them, resistance-based temperature detectors, thermocouples, and thermistors are the most common types of sensors used for temperature detection. The temperature change in the material causes a change in the electrical resistance, which could be due to electron vibrations, which modulate the electron flow. A high degree of accuracy, facile fabrication process, and linear process make the resistance type of temperature sensor more desirable than the other ones. The sensitivity of the resistance-based temperature sensor is directly related to the temperature coefficient of resistance. A higher temperature coefficient suggests higher precision of the temperature sensor. Wheatstone bridge circuit is generally used to avoid errors due to the thermal conductivity of the sensors. The most straightforward configuration of the circuit is ¼ of the bridge. For higher precision and accuracy, three-wire and four-wire circuits can be used. On the other hand, several aspects such as material, size, cost, oxidation resistance, configuration, thermal coefficients etc., should be considered for the temperature sensor's optimum performance. Materials such as copper, silver, gold, platinum, and nickel have been widely used as active materials for resistance-based temperature sensors. Recently, conductive polymers and carbon-based materials are the most attractive materials for temperature-sensing applications (Childs, Greenwood, & Long, 2000; Hurley W. & Schooley F., 1984; S. C. Mukhopadhyay & K. P. Jayasundera, 2013; Mukhopadhyay Mukhopadhyay 2013).

In the last decade, extensive efforts have been made to develop a highly sensitive and flexible temperature sensor for wearable devices. The temperature sensitive material is responsible for the response to thermal signals. The properties of the active material define the performance of the temperature sensor. Sensitivity, thermal coefficient, response and recovery time, stability, hysteresis, resolution, and durability are considered essential for the temperature sensor. A temperature sensor with biocompatibility, skin-friendliness, high selectivity, and flexibility has excellent demand in wearable devices. Currently, carbon-based materials have been used as temperature sensitive active materials. For the flexible temperature sensor, carbon and metal fillers are embedded into a polymer matrix. Carbon-based materials such as CNT, CB, graphene, and graphite are most widely used as conductive fillers due to their excellent electronic, mechanical, and thermal properties. CB incorporated polymer nanocomposites offer a high thermal coefficient of resistance, which defines the temperature sensor's sensitivity. The electrical conductivity of CB and graphite-filled polymer matrix relies on the filler loading. As the filler loading increases, the conductive pathways are formed. At a particular loading, the electrical conductivity upsurges; this is known as the percolation threshold. The percolation threshold for CB and graphite-based polymer composite was observed above 20 wt% (G. H. Chen, Wu, Weng, & Yan, 2001; Yurddaskal, Erol, & Celik, 2017).

Nevertheless, the electrical conductivity increases, but the rigidity of the composite and filler agglomerations increase and thus deteriorate the mechanical properties of the composite, which hinders its application in wearable devices. Compared to CB, graphite is more subtle to temperature, and the ensuing composite requires less filler content. The loose and porous graphite structure known as expanded graphite obtained from graphite intercalation and high-temperature expansion is also useful for improving the electrical properties. On the other hand, graphene has excellent carrier mobility, excellent thermal conductivity, and is also biocompatible. Graphene oxide's low electrical conductivity makes it a poor choice for electronic devices. However, reduced graphene oxide is used widely to provide excellent electronic properties. In addition, CNT has been incorporated in polymers for temperature sensing application. Other than carbon-based conductive fillers, conducting polymers including PEDOT:PSS, polypyrrole, and PANI have great significance due to their suitability for wearable application (Tamura, Huang, & Togawa, 2018; Walczak & Sibiński, 2014).

PEDOT:PSS and PANI were screen printed on cotton. Both of the sensors showed a negative temperature coefficient of resistance. The sensor was reversible but with drift after multiple cycles. The Seebeck coefficient was found to be 15 μV/K and 18 μV/K, respectively, for PANI and PEDOT:PSS-printed cellulose (Seeberg, Royset, Jahren, & Strisland, 2011). A PANI/manganese oxide (Mn_3O_4) nanocomposite-based temperature sensor was prepared using oxidative polymerization process. The resistance of the sensor decreased upon increasing the temperature, suggesting a negative temperature coefficient. The electron mobility enhanced significantly upon the addition of PANI, which improved the response of the sensor (Majid, Awasthi, & Singla, 2007). Hybrid nanocomposite-based on silver nanowires (AgNW), thermally reduced graphene (TRG), and PVDF was fabricated via hot pressing technique. AC conductivity of AgNW/TRG/PVDF hybrid nanocomposite is shown in Figure 12.7a. For AgNW/PVDF and TRG/PVDF

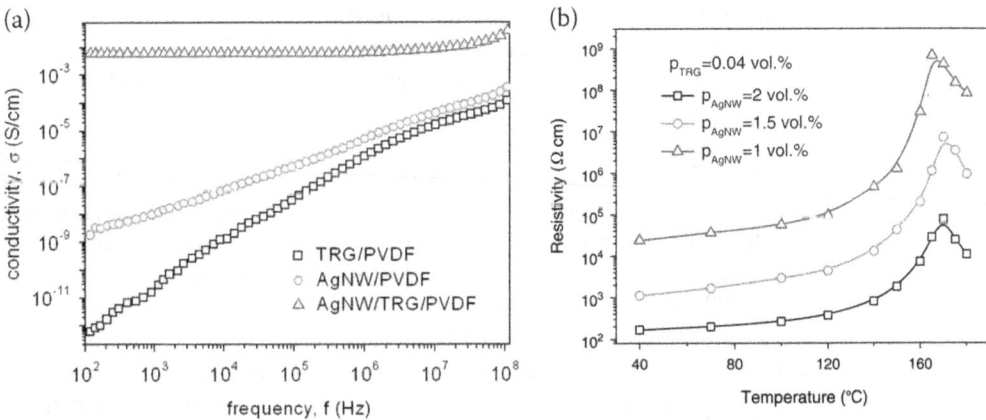

FIGURE 12.7 Ag/TRGPVDF composites: (a) AC conductivity, (b) change in electrical resistivity different temperature. [Reproduced with permissions © 2014 Springer] (L. He & Tjong, 2014).

nanocomposites, conductivity increased linearly with the increase in frequency, indicating the insulator characteristics. For hybrid nanocomposite, the conductivity was frequency independent in the range of 10^2 to 10^7 Hz, indicating the formation of a network within the PVDF matrix. The AgNW effectively bridges TRG particles, resulting in the electron transport between them. Figure 12.7b shows the effect of temperature on the electrical resistivity of AgNW/TRG/PVDF hybrid nanocomposite. The electrical resistivity increased gradually with the increase in the temperature at low temperature (<120°C); after that, there was a steep increase in the resistivity. The increase in resistivity indicates the positive temperature coefficient of resistance. The resistivity increase was more noticeable when the nanocomposite was loaded with 0.04 vol % TRG and 1 vol % AgNW. At temperatures above the melting point of PVDF, the hybrid nanocomposite showed a converse effect, i.e., negative temperature coefficient of resistance. The converse effect is evident due to the filler's re-aggregation in the softened PVDF matrix (L. He & Tjong, 2014).

12.2.6 pH Sensors

pH sensors are essential electrochemical sensors that determine the chemical reactions and variation of body fluids' pH levels giving direct/indirect indication of health conditions of the human body. Various metal oxide-based pH sensors have been reported. However, metal oxide-based sensors lack flexibility, so they are not suitable for flexible electronic applications. Hence, the necessity of polymer nanocomposite-based sensors arises. pH sensors based on polymers exhibit excellent sensing behavior because of their excellent electrochemical properties. Among polymers, polyaniline (PANI) exhibits pH-dependent behavior because of its reversible transformation of emeraldine base and emeraldine salt during acid-base reactions (Alam et al., 2018). Kaempgen and Roth (2006) have developed a polyaniline-coated carbon nanotube (PANI/CNT)-based flexible and transparent pH sensor, which exhibited improved linear response compared to that of pure CNT (Figure 12.8a and b). PANI coating on the CNT network provided excellent stability in all buffer solutions. Better performance was observed for a thin coating of PANI compared to a thicker coating; as the thickness increases, the diffusion of ions will take more time (Shiu, Song, & Lau, 1999). The PANI/CNT sensor exhibited good response time and high selectivity towards H[+] ions. Furthermore, Dang et al. (2018) have reported stretchable wireless sensors based on graphite/ polyurethane (G/PU) composite for sweat pH monitoring. Variation in the sweat pH levels indicates the disease and metabolic activity associated with the human body. The sensing electrode of this potentiometric sensor was fabricated by printing the G/PU composite on the graphene layer followed by curing (see Figure 12.8c and d) and Ag/AgCl was the reference electrode. The significant advantages of the G/PU composite are flexibility, fast response, and low-temperature processing. This sensor's interesting feature is that the pH data can be transmitted wirelessly to a smartphone using a stretchable RFID (radio frequency identification) antenna fabricated shown in Figure 12.8e. This sensor exhibited pH response in the range of 5–9 and sensitivity of 11.13 ± 5.8 mV/pH on human sweat equivalent solution.

Jeon, Kang, and Ha (2020) have reported flexible chemiresistive pH sensors based on single wall nanotube (SWNT) and Nafion® nanocomposites fabricated by screen printing. The chemical structure of the Nafion® and the sensor's optical image are shown in Figure 12.8f–g. This sensor exhibited a wide pH detection limit of 1–12. In acidic conditions, the H[+] ions adsorb at the nanocomposite film's surface, which increases the surface potential of the composite. This leads to the hole transport into SWNTs and reduction of resistance. Whereas in the alkaline condition, the OH- ions interact with the C-H bonds present in the nanocomposite. This leads to the removal of positive ions from the SWNT, and further, the increase of resistance arises (shown in Figure 12.8h). They have developed a real-time detection system to evaluate the water quality using this sensor integrated on a drone with wireless data transmission to a smartphone. Better sensor response was observed for multilayer printed nanocomposites. The pH sensor returns to the original state when the pH buffer solutions are removed (Figure 12.8i).

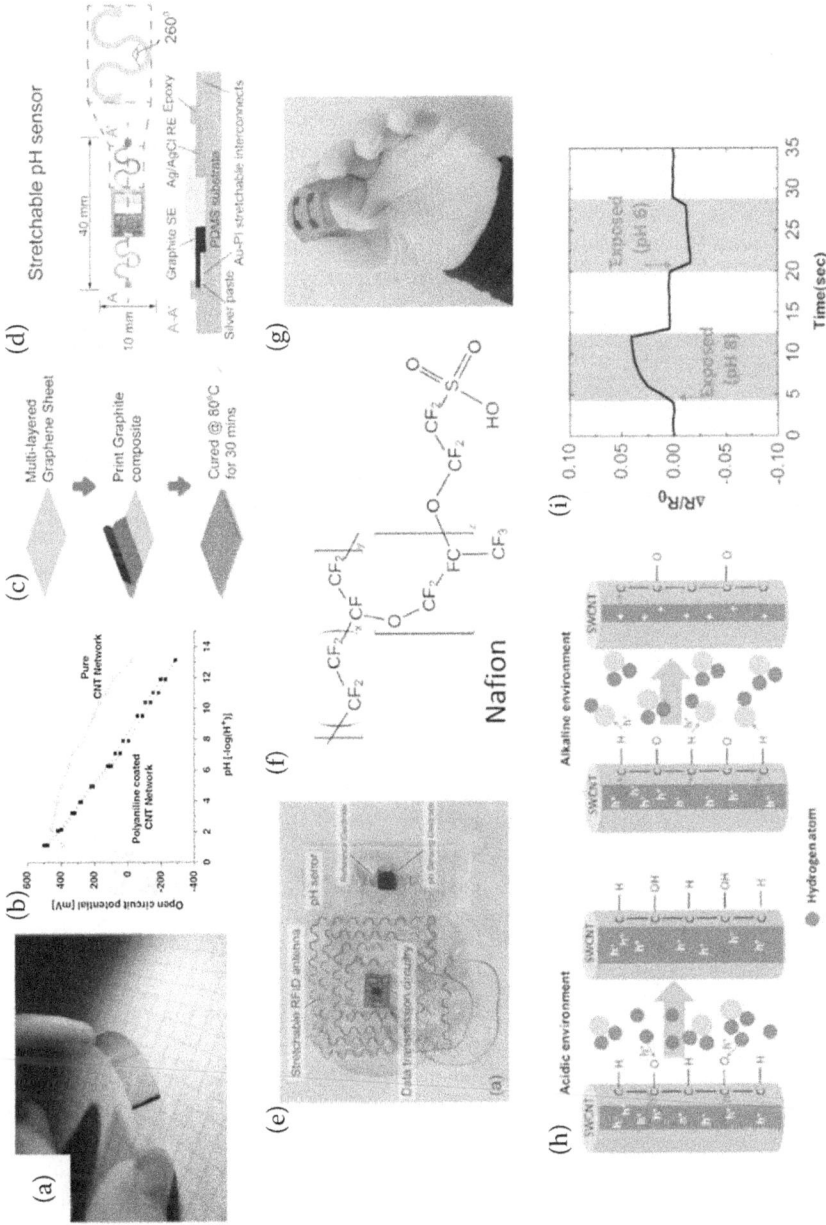

FIGURE 12.8 (a) Flexible and transparent PANI/CNT based pH sensor, (b) open circuit potential versus pH plot of PANI/CNT nanocomposite and pure CNT network [Reproduced with permission © 2006 Elsevier] (Kaempgen & Roth, 2006), (c) schematic of pH sensing electrode fabrication, (d) schematic G/PU nanocomposite based pH sensor, (e) optical image of G/PU nanocomposite based wireless sensor system for sweat pH monitoring [Reproduced with permission ©2018 Elsevier] (Dang et al., 2018), (f) chemical structure of Nafion®, (g) Nafion®/SWNT nanocomposite-based flexible pH sensor, (h) sensing mechanism in chemiresistive pH sensor in acidic/alkane conditions, (i) dynamic response of Nafion®/SWNT nanocomposite-based pH sensor with change in pH. [Reproduced with permission © 2020 Elsevier] (Jeon et al., 2020).

12.2.7 HEART RATE AND PULSE MONITORING

In medical diagnostics, the arterial pulse plays a crucial role in heart rate, blood pressure, and information on probable cardiovascular diseases. Polymer nanocomposites have become an inevitable material in the field of wearable sensors. Jung et al. (2012) reported CNT/PDMS composite materials-based flexible and biocompatible electrodes for ECG monitoring. The CNT/PDMS (polydimethylsiloxane) composites were prepared by the mechanical force dispersion method. The fabrication process of the CNT/PDMS electrode involves two steps: (i) fabrication of PDMS mold and (ii) fabrication of electrode by the replication process. This electrode can be directly connected to conventional ECG equipment. Different compositions of CNT in PDMS were prepared (1, 1.5, 2, and 4.5 wt%), and their performance was compared with the Ag/AgCl reference electrode. The composite containing 4.5 wt% CNT in PDMS exhibited qualitative signals compared to other compositions. The ECG signals obtained for CNT/PDMS composite electrodes under motion were comparable to that of the reference electrode. Moreover, these electrodes exhibited superior cell viability of 95% and long-term wearability.

Furthermore, PDA-rGO/PVA [poly(dopamine) capped reduced graphene oxide/poly(vinyl alcohol)] hydrogel has been used as the biocompatible wearable pulse monitoring sensor (Zhang et al., 2020). The composite material exhibited good mechanical properties owing to the hydrogen bonds present between the different phases (see Figure 12.9a). The high sensitivity of the sensor enabled the detection of pulse signals effectively. This epidermal sensor displayed the artery signal before and after running (see Figure 12.9b). Similarly, Chen et al. (2017) have reported flexible and transparent self-powered sensors based on $BaTiO_3$/PDMS nanocomposite film to provide real-time monitoring of various physiological signals. Spray-coated AgNW (silver nanowire) was used as an electrode, and the spin-coated $BaTiO_3$/PDMS nanocomposite film was used as the sensing material (Figure 12.9c). The fabricated nanogenerator was stretchable, twistable, and foldable (Figure 12.9d).

Moreover, this flexible sensor was connected to the wrist to monitor the real-time radial artery pulse. The pulse signals before and after exercises were measured. Obviously, the pulse signal frequency increased after exercise (see Figure 12.9e). Moghadam, Hasanzadeh, and Simchi (2020) have developed PVDF/microporous zirconium-based metal–organic framework (MOF)-based piezoelectric self-powered wearable sensors for arterial pulse monitoring. The nanocomposite sensing material was fabricated using electrospinning technique. The arterial pulse monitoring was carried out by attaching the sensor to the wrist of a 32-year-old male before and after exercise (Figure 12.9f and g). They were able to distinguish the signals associated with the pulse waveform. The signals consisted of four systolic waves and one diastolic wave, as shown in Figure 12.9h.

12.2.8 PIEZOELECTRIC ENERGY HARVESTING DEVICES

The gathering of low-level kinetic energy from human locomotion by harvesting principles is significant in wearable devices. With the miniaturization of devices, wearable devices require low power for their uninterrupted functioning. Piezoelectric energy harvesting devices can replace batteries, which significantly reduces the wearable device's mass, size, and cost. The most commonly used energy source for a wearable device is the battery. The discharging of the battery is one of the major concerns for the devices' uninterrupted functioning, especially in medical devices. Recharging or swapping of batteries has created new roads to look for better solutions. Also, the toxicity, disposal, and recycling of batteries have a significant environmental issue (Gljuščić, Zelenika, Blažević, & Kamenar, 2019).

There are various techniques adopted to harvest the energy for powering wearable devices. Devices based on electromagnetism, thermoelectricity, triboelectricity, pyroelectricity, piezoelectricity, electrostatic, etc. can efficiently convert different types of energy into electrical signals. Among them, the piezoelectric energy harvesting has attracted significant attention due to its

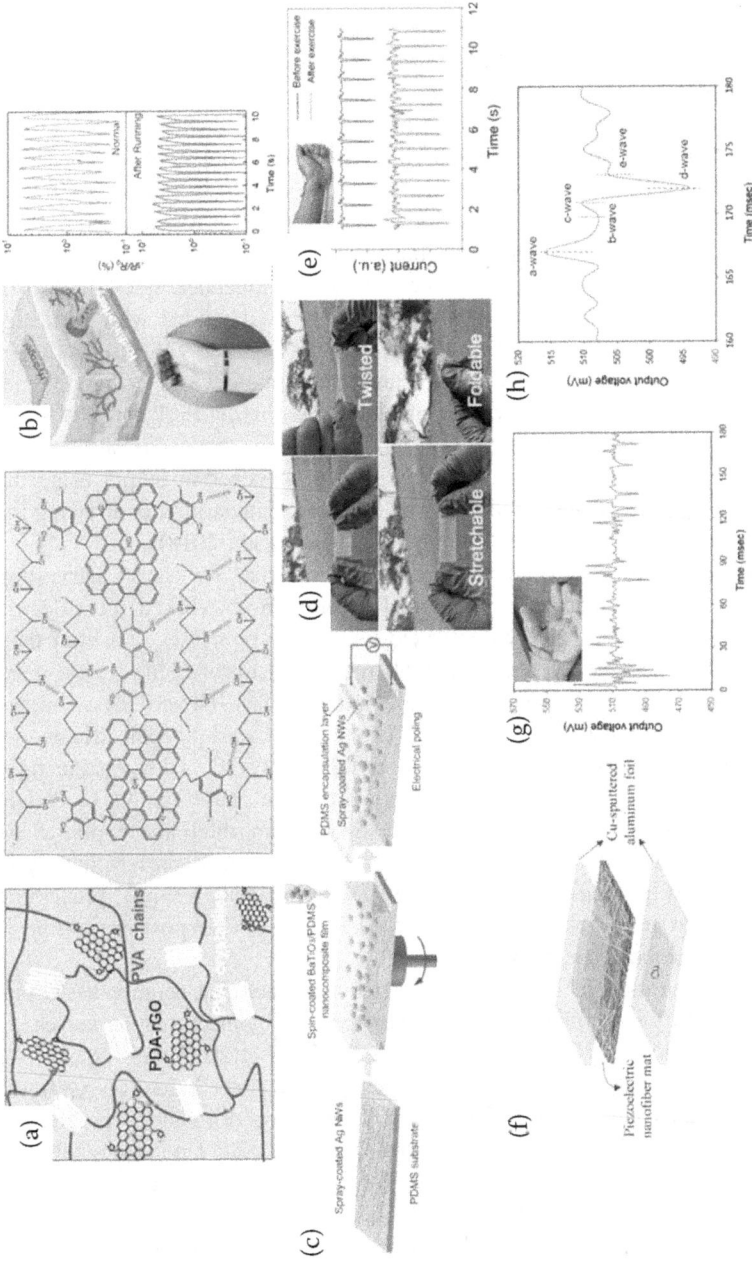

FIGURE 12.9 (a) Schematic representation of crosslinking present in PDA-rGO/PVA hydrogel, (b) schematic representation of PDA-rGO/PVA epidermal sensor connected to the wrist and the corresponding signal obtained [Reproduced with permission ©2020 Royal society of chemistry] (H. Zhang et al., 2020), (c) schematic of fabrication process of BaTiO$_3$/PDMS nanocomposite-based sensor, (d) optical image of BaTiO$_3$/PDMS-based self-powered sensor under various conditions (stretching, folding, and twisting), (e) real-time monitoring of arterial pulse using BaTiO$_3$/PDMS nanocomposite-based sensor connected to wrist [Reproduced with permission © 2017 American chemical society] (X. Chen et al., 2017), (f) schematic representation of fabrication of PVDF/MOF based piezoelectric wearable sensor, (g) and (h) arterial pulse signal obtained using PVDF/MOF based sensor connected to the wrist and its enlarged view. [Reproduced with permission ©2020 American chemical society] (Moghadam et al., 2020).

simple design, low cost, and the availability of a wide range of materials. The mechanical stress on the piezoelectric material converts into useful electrical energy or vice-versa. Energy harvesting utilizing vibrations offers substantial benefits because it is independent of the environmental conditions. The vibrational energy can be gathered from human body locomotion (Xue, Truong, Rantz, & Miah, 2017). Various perovskite materials such as PZT, BaTiO$_3$, potassium-sodium niobate (KNN), lead zirconate (PbZrO$_3$), etc. have been used for energy harvesting, which have high piezoelectric conversion efficiency. However, their fragility, toxicity, and complicated synthesis processes make them impossible to adopt for wearable devices. To achieve high flexibility, these materials have been incorporated in a polymer/elastomer matrix. Cellulose is one of the most desirable piezoelectric materials for wearable energy harvesting devices due to its desirable characteristics, including flexibility, biocompatibility, low cost, and good piezoelectric coefficient. Cellulose nanofibers/PDMS aerogel film are able to consistently generate virtuous piezoelectric voltage up to 60 V and the corresponding power density of 6.3 mW/cm^3 (Sun, Hall, Simoneishvilli, Akdogan, & Safari, 2007; Q. Zheng et al., 2016). BaTiO$_3$ nanofibers prepared by the combination of electrospinning and calcination processes were embedded into the PDMS matrix. The nanogenerator based on PDMS nanocomposite that was embedded with vertically aligned BaTiO$_3$ nanofibers achieved high piezoelectric conversion efficiency with the maximum power output of 0.18 mW, which is sufficient to run wearable electronic devices (Yan & Jeong, 2016). Flexible and durable 3D foam composite was developed by combining PZT with PU foam. The 3D composite structure showed excellent piezoelectric response over thousands of cycles. In addition to the piezoelectric performance, the composite was used to generate output using the pyroelectric effect. Both the effects were superimposable on the composite structure, allowing the harvesting of energy from mechanical and thermal inputs (G. Zhang et al., 2018).

PZT and PVDF are the most widely recognized piezoelectric materials for energy harvesting due to their excellent conversion efficiency. As mentioned before, ceramic-based piezoelectric materials are rigid, fragile, heavy, and involve complicated fabrication processes that limit their use in wearable applications. Besides, piezoelectric materials require post processing such as high voltage poling, mechanical stretching, or annealing to orient the dipoles. Nevertheless, human locomotion consists of high amplitude and low frequency. Thus, PVDF-based piezoelectric energy harvesting devices are suitable for wearable applications. However, PVDF in its pristine form shows poor piezoelectric performance and depends on the electroactive phase content of PVDF. PVDF exhibits five conformations, namely α, β, γ, δ, and ε. Among them, the β and γ -phases are electroactive and possess piezoelectricity. Generally, the non-polar α-phase is present mainly in melt processed PVDF. To convert the non-polar phases of PVDF to polar phases, post-processing is necessary. However, the polar phase content in pristine PVDF even after post-processing is insufficient to generate a high piezoelectric response.

Hence, PVDF is prepared in the form of its nanocomposites by incorporating ceramic-based piezoelectric materials. Another strategy is to enhance the electroactive phase by incorporating nucleating agents such as clay, graphene, oxide, CNT, ionic surfactants etc. Karan et al. (Karan, Mandal, & Khatua, 2015) carried out one such study, wherein iron doped-reduced graphene oxide (Fe-rGO) was introduced into PVDF to enhance the electroactive phase and piezoelectric performance. The Fe-rGO/PVDF nanocomposite was prepared in the form of thin films by solution casting. Upon the addition of Fe-rGO, the γ-phase of PVDF increased to 99%, indicating the filler's nucleating effect.

The formation of the γ-phase could be attributed to the dipole-dipole interaction between the π-electrons in rGO and hydrogen atoms of PVDF. The carbonyl and hydroxyl group of Fe-rGO interacted with the fluorine atoms of PVDF, resulting in the rearrangement of PVDF chains. As a result, the piezoelectric output of the composite increased by at least 12 times than the pristine PVDF films. Nanostructured clay, PANI, ZnO, iron oxide, cellulose, etc. have been used as nucleating agents for the augmentation of the β-phase of PVDF through various processing methods such as spin coating, electrospinning, solution casting, and compression molding (Khalifa, Mahendran, & Anandhan, 2016; Khalifa et al., 2019; Liu, Li, Xu, & Fan, 2010; Sorayani Bafqi,

Bagherzadeh, & Latifi, 2015). Among them, the electrospinning technique is considered to be the most suitable one for enhancing the piezoelectric performance of PVDF. Electrospun PVDF nanofibers do not require post-processing such as poling or mechanical stretching as compared with the other processes. The in-situ poling and stretching of the fibers during the fabrication of PVDF nanofibers enhances the β-phase of PVDF. In another study, $BaTiO_3$/graphene/PVDF hybrid nanocomposite was fabricated by electrospinning process. With the introduction of $BaTiO_3$ and graphene, the β-phase of PVDF increased significantly. $BaTiO_3$ acts as piezoelectric filler, while graphene increases electrical conductivity and acts as a nucleating agent to enhance the β-phase of PVDF. The nanogenerator was flexible and thin, which is comfortable enough to attach to the human body for energy harvesting from human motions. Upon finger pressing, the nanogenerator was able to generate a voltage output of ~6 V. The voltage output of the nanogenerator reached up to 11 V and the power output was 4.1 mW. Such conversion efficiency of the piezoelectric nanogenerator is sufficient to run multiple wearable devices (Shi, Sun, Huang, & Jiang, 2018).

12.3 SUMMARY AND OUTLOOK

The current chapter summarizes the various types of polymer nanocomposite-based sensors with a focus on their application in the field of wearable technology. Polymers contribute significantly to sensor development as they offer flexibility, ease of fabrication, and cost-effectiveness. Furthermore, the performance of the former can be tailored with appropriate additives or chemical modification routes that improve the sensor's responsivity. Recent trends in wearable sensors mandate them to be portable, biocompatible, and multi-tasking; for instance, detecting pressure, temperature, and strain is vital for a precise healthcare monitoring system. To ensure the uninterrupted performance of these sensors, powering via ambient energy harvesting sources has been very attractive. Moreover, the integration of self-powered flexible systems, wireless communication, and the Internet of Things (IoT) in wearable sensors is anticipated to revolutionize human health monitoring system that provides the real-time health status and facilitates the telemedicine system.

Despite the rapid advancement in technology, there are still challenges that hinder the commercialization of these sensors for their practical application. Thus, a viable and indispensable measure comes from the collaborative efforts of scientists, engineers, and clinical practitioners focusing on polymer-based wearable sensors. This necessitates wearable sensors to be portable, intelligent, independent, biocompatible, cost-effective, eco-friendly, biodegradable, and recyclable.

ACKNOWLEDGMENTS

The authors would like to thank Kompetenzzentrum Holz GmbH, Austria and the National Institute of Technology, Karnataka.

REFERENCES

Alam, A. U., Qin, Y., Nambiar, S., Yeow, J. T. W., Howlader, M. M. R., Hu, N. X., & Deen, M. J. (2018). Polymers and organic materials-based pH sensors for healthcare applications. *Progress in Materials Science*, 96, 174–216. 10.1016/j.pmatsci.2018.03.008

Alluri, N. R., Selvarajan, S., Chandrasekhar, A., Saravanakumar, B., Jeong, J. H., & Kim, S. J. (2017). Piezoelectric $BaTiO_3$/alginate spherical composite beads for energy harvesting and self-powered wearable flexion sensor. *Composites Science and Technology*, 142, 65–78. 10.1016/j.compscitech.2017.02.001

Amjadi, M., Yoon, Y. J., & Park, I. (2015). Ultra-stretchable and skin-mountable strain sensors using carbon nanotubes-ecoflex nanocomposites. *Nanotechnology*, 26(37), 375501. 10.1088/0957-4484/26/37/375501

Bai, N., Wang, L., Wang, Q., Deng, J., Wang, Y., Lu, P., & … Guo, C. F. (2020). Graded intrafillable architecture-based iontronic pressure sensor with ultra-broad-range high sensitivity. *Nature Communications*, 11(1), 3–11. 10.1038/s41467-019-14054-9

Bandodkar, A. J., Jeerapan, I., & Wang, J. (2016). Wearable chemical sensors: Present challenges and future prospects. *ACS Sensors*, *1*(5), 464–482. 10.1021/acssensors.6b00250

Bao, M.-H. (2000). Micro mechanical transducers: Pressure sensors, accelerometers, and gyroscopes. In *Handbook of sensors and actuators* (Vol. 8). Retrieved from http://books.google.com/books?hl=en&lr=&id=OwI_xrvrj1kC&pgis=1

Berland, K., Song, X., Carvalho, P. A., Persson, C., Finstad, T. G., & Løvvik, O. M. (2016). Enhancement of thermoelectric properties by energy filtering: Theoretical potential and experimental reality in nanostructured ZnSb. *Journal of Applied Physics*, *119*(12). 10.1063/1.4944716

Bubnova, O., Berggren, M., & Crispin, X. (2012). Tuning the thermoelectric properties of conducting polymers in an electrochemical transistor. *Journal of the American Chemical Society*, *134*(40), 16456–16459. 10.1021/ja305188r

Chen, G. H., Wu, D. J., Weng, W. G., & Yan, W. L. (2001). Dispersion of graphite nanosheets in a polymer matrix and the conducting property of the nanocomposites. *Polymer Engineering and Science*, *41*(12), 2148–2154. 10.1002/pen.10909

Chen, X., Parida, K., Wang, J., Xiong, J., Lin, M. F., Shao, J., & Lee, P. S. (2017). A stretchable and transparent nanocomposite nanogenerator for self-powered physiological monitoring. *ACS Applied Materials and Interfaces*, *9*(48), 42200–42209. 10.1021/acsami.7b13767

Chen, Z., & Lu, C. (2005). Humidity sensors: A review of materials and mechanisms. *Sensor Letters*, *3*(4), 274–295. 10.1166/sl.2005.045

Chhetry, A., Kim, J., Yoon, H., & Park, J. Y. (2019). Ultrasensitive interfacial capacitive pressure sensor based on a randomly distributed microstructured iontronic film for wearable applications. *ACS Applied Materials and Interfaces*, *11*(3), 3438–3449. 10.1021/acsami.8b17765

Childs, P. R. N., Greenwood, J. R., & Long, C. A. (2000). Review of temperature measurement. *Review of Scientific Instruments*, *71*(8), 2959–2978. 10.1063/1.1305516

Chung, S. Y., Lee, H. J., Lee, T. I. l, & Kim, Y. S. (2017). A wearable piezoelectric bending motion sensor for simultaneous detection of bending curvature and speed. *RSC Advances*, *7*(5), 2520–2526. 10.1039/C6RA25797F

Dahiya, A. S., Morini, F., Boubenia, S., Justeau, C., Nadaud, K., Rajeev, K. P., & …Poulin-Vittrant, G. (2018). Zinc oxide nanowire-parylene nanocomposite based stretchable piezoelectric nanogenerators for self-powered wearable electronics. *Journal of Physics: Conference Series*, *1052*(1), 012028–012031. 10.1088/1742-6596/1052/1/012028

Dang, W., Manjakkal, L., Navaraj, W. T., Lorenzelli, L., Vinciguerra, V., & Dahiya, R. (2018). Stretchable wireless system for sweat pH monitoring. *Biosensors and Bioelectronics*, *107*, 192–202. 10.1016/j.bios.2018.02.025

Ding, Y., Xu, T., Onyilagha, O., Fong, H., & Zhu, Z. (2019). Recent advances in flexible and wearable pressure sensors based on piezoresistive 3D monolithic conductive sponges. *ACS Applied Materials and Interfaces*, *11*(7), 6685–6704. 10.1021/acsami.8b20929

Dong, Y. Z., Seo, Y., & Choi, H. J. (2019). Recent development of electro-responsive smart electrorheological fluids. *Soft Matter*, *15*(17), 3473–3486. 10.1039/c9sm00210c

Farahani, H., Wagiran, R., & Hamidon, M. N. (2014). Humidity sensors principle, mechanism, and fabrication technologies: A comprehensive review. *Sensors (Switzerland)*, *14*(5), 7881–7939. 10.3390/s140507881

Gljuščić, P., Zelenika, S., Blažević, D., & Kamenar, E. (2019). Kinetic energy harvesting for wearable medical sensors. *Sensors (Switzerland)*, *19*(22), Article no. 4922. 10.3390/s19224922

Gray, B. L. (2019). Polymer nanocomposites for flexible and wearable fluidic and biomedical microdevices. *2018 IEEE 13th Nanotechnology Materials and Devices Conference (NMDC)*, 1–2. 10.1109/nmdc.2018.8605932

Gu, J., Kwon, D., Ahn, J., & Park, I. (2020). Wearable strain sensors using light transmittance change of carbon nanotube-embedded elastomers with microcracks. *ACS Applied Materials and Interfaces*, *12*(9), 10908–10917. 10.1021/acsami.9b18069

Guo, J., Niu, M., & Yang, C. (2017). Highly flexible and stretchable optical strain sensing for human motion detection. *Optica*, *4*(10), 1285. 10.1364/optica.4.001285

Guo, J., Zhou, B., Yang, C., Dai, Q., & Kong, L. (2019). Stretchable and temperature-sensitive polymer optical fibers for wearable health monitoring. *Advanced Functional Materials*, *29*(33), 1–8. 10.1002/adfm.201902898

He, J., Xiao, P., Shi, J., Liang, Y., Lu, W., Chen, Y., & … Chen, T. (2018). High performance humidity fluctuation sensor for wearable devices via a bioinspired atomic-precise tunable graphene-polymer heterogeneous sensing junction. *Chemistry of Materials*, *30*(13), 4343–4354. 10.1021/acs.chemmater.8b01587

He, L., & Tjong, S.-C. (2014). Electrical behavior and positive temperature coefficient effect of graphene/polyvinylidene fluoride composites containing silver nanowires. *Nanoscale Research Letters*, *9*(1), 375. 10.1186/1556-276X-9-375

Hu, J., Yu, J., Li, Y., Liao, X., Yan, X., & Li, L. (2020). Nano carbon black-based high performance wearable pressure sensors. *Nanomaterials*, *10*(4), Article no. 664. 10.3390/nano10040664

Hu, N., Karube, Y., Yan, C., Masuda, Z., & Fukunaga, H. (2008). Tunneling effect in a polymer/carbon nanotube nanocomposite strain sensor. *Acta Materialia*, *56*(13), 2929–2936. 10.1016/j.actamat.2008.02.030

Hurley, W. and Schooley, F. (1984). Calibration of temperature measurement device thermocouple calibration installed in buildings. Washington D.C., U.S: Government Printing Office.

Hwang, B. U., Zabeeb, A., Trung, T. Q., Wen, L., Lee, J. D., Choi, Y. I., & … Lee, N. E. (2019). A transparent stretchable sensor for distinguishable detection of touch and pressure by capacitive and piezoresistive signal transduction. *NPG Asia Materials*, *11*(1), Article no. 23. 10.1038/s41427-019-0126-x

Ivens, J., Urbanus, M., & De Smet, C. (2011). Shape recovery in a thermoset shape memory polymer and its fabric-reinforced composites. *Express Polymer Letters*, *5*(3), 254–261. 10.3144/expresspolymlett.2011.25

Jeon, J. Y., Kang, B. C., & Ha, T. J. (2020). Flexible pH sensors based on printed nanocomposites of single-wall carbon nanotubes and Nafion. *Applied Surface Science*, *514*(February), 145956. 10.1016/j.apsusc.2020.145956

Jeong, Y. J., Oh, T. I., Woo, E. J., & Kim, K. J. (2017). Integration of piezo-capacitive and piezo-electric nanoweb based pressure sensors for imaging of static and dynamic pressure distribution. *Proceedings of the Annual International Conference of the IEEE Engineering in Medicine and Biology Society, EMBS*, (July), 21–24. 10.1109/EMBC.2017.8036753

Jin, H., Jin, Q., & Jian, J. (2018). Smart materials for wearable healthcare devicesIn. J. H. Ortiz (Ed). *Wearable Technologies*. (1 ed., pp. 109–138).10.5772/intechopen.76604

Jung, H. C., Moon, J. H., Baek, D. H., Lee, J. H., Choi, Y. Y., Hong, J. S., & Lee, S. H. (2012). CNT/PDMS composite flexible dry electrodesfor long-term ECG monitoring. *IEEE Transactions on Biomedical Engineering*, *59*(5), 1472–1479. 10.1109/TBME.2012.2190288

Jung, Y., Jung, K., Park, B., Choi, J., Kim, D., Park, J., & … Cho, H. (2019). Wearable piezoresistive strain sensor based on graphene-coated three-dimensional micro-porous PDMS sponge. *Micro and Nano Systems Letters*, *7*(1), Article no. 20. 10.1186/s40486-019-0097-2

Kaempgen, M., & Roth, S. (2006). Transparent and flexible carbon nanotube/polyaniline pH sensors. *Journal of Electroanalytical Chemistry*, *586*(1), 72–76. 10.1016/j.jelechem.2005.09.009

Karan, S. K., Maiti, S., Kwon, O., Paria, S., Maitra, A., Si, S. K., & … Khatua, B. B. (2018). Nature driven spider silk as high energy conversion efficient bio-piezoelectric nanogenerator. *Nano Energy*, *49*, 655–666. 10.1016/j.nanoen.2018.05.014

Karan, S. K., Mandal, D., & Khatua, B. B. (2015). Self-powered flexible Fe-doped RGO/PVDF nanocomposite: an excellent material for a piezoelectric energy harvester. *Nanoscale*, *7*(24), 10655–10666. 10.1039/C5NR02067K

Khalifa, M., & Anandhan, S. (2019). Synergism of graphitic-carbon nitride and electrospinning on the physico-chemical characteristics and piezoelectric properties of flexible poly(vinylidene fluoride) based nanogenerator. *Journal of Polymer Research*, *26*(3), 1–13. 10.1007/s10965-019-1738-0

Khalifa, M., Mahendran, A., & Anandhan, S. (2016). Probing the synergism of halloysite nanotubes and electrospinning on crystallinity, polymorphism and piezoelectric performance of poly(vinylidene fluoride). *RSC Advances*, *6*(115), 114052–114060. 10.1039/C6RA20599B

Khalifa, M., Mahendran, A., & Anandhan, S. (2019). Durable, efficient, and flexible piezoelectric nanogenerator from electrospun PANi/HNT/PVDF blend nanocomposite. *Polymer Composites*, *40*(4), 1663–1675. 10.1002/pc.24916

Khalifa, M., Wuzella, G., Lammer, H., & Mahendran, A. R. (2020). Smart paper from graphene coated cellulose for high-performance humidity and piezoresistive force sensor. *Synthetic Metals*, *266*(May), 116420–116426. 10.1016/j.synthmet.2020.116420

Kroon, R., Mengistie, D. A., Kiefer, D., Hynynen, J., Ryan, J. D., Yu, L., & Müller, C. (2016). Thermoelectric plastics: From design to synthesis, processing and structure-property relationships. *Chemical Society Reviews*, *45*(22), 6147–6164. 10.1039/c6cs00149a

Leonov, V., & Vullers, R. J. M. (2009). Wearable thermoelectric generators for body-powered devices. *Journal of Electronic Materials*, *38*(7), 1491–1498. 10.1007/s11664-008-0638-6

Lewis, Z. H., Pritting, L., Picazo, A. L., & JeanMarie-Tucker, M. (2020). The utility of wearable fitness trackers and implications for increased engagement: An exploratory, mixed methods observational study. *Digital Health*, *6*, 1–12. 10.1177/2055207619900059

Li, G., & Wen, D. (2020). Wearable biochemical sensors for human health monitoring: Sensing Materials and manufacturing technologies. *Journal of Materials Chemistry B*, *8*(16), 3423–3436. 10.1039/c9tb02474c

Li, Q., Zhang, L. N., Tao, X. M., & Ding, X. (2017). Review of flexible temperature sensing networks for wearable physiological monitoring. *Advanced Healthcare Materials*, *6*(12), 1–23. 10.1002/adhm.201601371

Li, R. T., Kling, S. R., Salata, M. J., Cupp, S. A., Sheehan, J., & Voos, J. E. (2016). Wearable performance devices in sports medicine. *Sports Health*, *8*(1), 74–78. 10.1177/1941738115616917

Li, R., Zhou, J., Liu, H., & Pei, J. (2017). Effect of polymer matrix on the structure and electric properties of piezoelectric lead zirconatetitanate/polymer composites. *Materials*, *10*(8), 945. 10.3390/ma10080945

Lim, W. H., Yap, Y. K., Chong, W. Y., & Ahmad, H. (2014). All-optical graphene oxide humidity sensors. *Sensors (Switzerland)*, *14*(12), 24329–24337. 10.3390/s141224329

Lima, F. P., Burnett, N. P., Helmuth, B., Kish, N., Aveni-Deforge, K., & Wethey, D. S. (2011). Monitoring the intertidal environment with biomimetic devices. *Biomimetic Based Applications*, (May 2014), 422–463. 10.5772/14153

Liu, Y. L., Li, Y., Xu, J. T., & Fan, Z. Q. (2010). Cooperative effect of electrospinning and nanoclay on formation of polar crystalline phases in poly(vinylidene fluoride). *ACS Applied Materials and Interfaces*, *2*(6), 1759–1768. 10.1021/am1002525

Lu, Yang, Biswas, M. C., Guo, Z., Jeon, J. W., & Wujcik, E. K. (2019). Recent developments in bio-monitoring via advanced polymer nanocomposite-based wearable strain sensors. *Biosensors and Bioelectronics*, *123*, 167–177. 10.1016/j.bios.2018.08.037

Lu, Yunjing, Tian, M., Sun, X., Pan, N., Chen, F., Zhu, S., & …Chen, S. (2019). Highly sensitive wearable 3D piezoresistive pressure sensors based on graphene coated isotropic non-woven substrate. *Composites Part A: Applied Science and Manufacturing*, *117*, 202–210. 10.1016/j.compositesa.2018.11.023

Lugoda, P., Hughes-Riley, T., Morris, R., & Dias, T. (2018). A wearable textile thermograph. *Sensors (Switzerland)*, *18*(7), 1–22. 10.3390/s18072369

Lund, A., Tian, Y., Darabi, S., & Müller, C. (2020). A polymer-based textile thermoelectric generator for wearable energy harvesting. *Journal of Power Sources*, *480*, 228836. 10.1016/j.jpowsour.2020.228836

Lv, C., Hu, C., Luo, J., Liu, S., Qiao, Y., Zhang, Z., & … Watanabe, A. (2019). Recent advances in graphene-based humidity sensors. *Nanomaterials*, *9*(3), Article no. 422. 10.3390/nano9030422

Majid, K., Awasthi, S., & Singla, M. L. (2007). Low temperature sensing capability of polyaniline and Mn3O4 composite as NTC material. *Sensors and Actuators, A: Physical*, *135*(1), 113–118. 10.1016/j.sna.2006.06.055

Martins, P., Lopes, A. C., & Lanceros-Mendez, S. (2014). Electroactive phases of poly(vinylidene fluoride): Determination, processing and applications. *Progress in Polymer Science*, *39*(4), 683–706. 10.1016/j.progpolymsci.2013.07.006

Mekhzoum, M. E. M., Qaiss, A. el kacem, and Rachid, B. (Eds.) (2020). Introduction: different types of smart materials and their practical application. In *Polymer Nanocomposite-Based Smart Materials*. (1st ed., pp. 1–19).10.1016/c2018-0-02562-1

Moghadam, B. H., Hasanzadeh, M., & Simchi, A. (2020). Self-powered wearable piezoelectric sensors based on polymer nanofiber-metal-organic framework nanoparticle composites for arterial pulse monitoring. *ACS Applied Nano Materials*, *3*(9), 8742–8752. 10.1021/acsanm.0c01551

Mukhopadhyay, S. C., Jayasundera, K. P., & Fuchs, A. (2013). *Advancement in Sensing Technology New Developments and Practical Applications* (Subhas C. Mukhopadhyay, Krishanthi P. Jayasundera, A. Fuchs, Eds.). Springer-Verlag Berlin Heidelberg.

Nicolais, L., & Carotenuto, G. (2013). *Nanocomposites: In Situ Synthesis of Polymer-Embedded Nanostructures - Luigi Nicolais, Gianfranco Carotenuto*. Retrieved from http://eu.wiley.com/WileyCDA/WileyTitle/productCd-0470109521.html

Pan, J., Zhang, Z., Jiang, C., Zhang, L., & Tong, L. (2020). A multifunctional skin-like wearable optical sensor based on an optical micro-/nanofibre. *Nanoscale*, *12*(33), 17538–17544. 10.1039/d0nr03446k

Pantelopoulos, A., & Bourbakis, N. G. (2010). A survey on wearable sensor-based systems for health monitoring and prognosis. *IEEE Transactions on Systems, Man and Cybernetics Part C: Applications and Reviews*, *40*(1), 1–12. 10.1109/TSMCC.2009.2032660

Pielichowski, K., & Pielichowska, K. (2018). Polymer nanocomposites. In Sergey Vyazovkin , Nobuyoshi Koga , and Christoph Schick. *Handbook of Thermal Analysis and Calorimetry* (Vol. 6). Elsevier, 431–438. 10.1016/B978-0-444-64062-8.00003-6

Ponnamma, D., Sadasivuni, K. K., Cabibhan, J.-J. & Al-Maadeed, M. A. -A. (2017). Smart Polymer Nanocomposites: Energy Harvesting, Self-Healingand Shape Memory ApplicationsIn. D. Ponnamma, K. K. Sadasivuni, J. -J. Cabibhan, & M. A. -A. Al-Maadeed, Eds.). Springer Series on Polymer and Composite Materials. (1st ed.) https://doi.org/10.1007/978-3-319-50424-7

Popov, V. I., Nikolaev, D. V., Timofeev, V. B., Smagulova, S. A., & Antonova, I. V. (2017). Graphene-based humidity sensors: The origin of alternating resistance change. *Nanotechnology*, *28*(35), 1–21. 10.1088/1361-6528/aa7b6e

Reddy, K. R., El-Zein, A., Airey, D. W., Alonso-Marroquin, F., Schubel, P., & Manalo, A. (2020). Self-healing polymers: Synthesis methods and applications. *Nano-Structures and Nano-Objects*, *23*, 100500. 10.1016/j.nanoso.2020.100500

Rehg, J. M., Murphy, S. A., & Kumar, S. (2017). Mobile health: Sensors, analytic methods, and applications. *Mobile Health: Sensors, Analytic Methods, and Applications*, 1–542. 10.1007/978-3-319-51394-2

Richardson, P. D. (1989). Piezoelectric polymers. *IEEE Engineering in Medicine and Biology Magazine*, *8*(2), 14–16. 10.1109/51.31634

Rizvi, R., Cochrane, B., Biddiss, E., & Naguib, H. (2011). Piezoresistance characterization of poly(dimethyl-siloxane) and poly(ethylene) carbon nanotube composites. *Smart Materials and Structures*, *20*(9), 42–47. 10.1088/0964-1726/20/9/094003

Sahatiya, P., Puttapati, S. K., Srikanth, V. V. S. S., & Badhulika, S. (2016). Graphene-based wearable temperature sensor and infrared photodetector on a flexible polyimide substrate. *Flexible and Printed Electronics*, *1*(2). Article no. 025006. 10.1088/2058-8585/1/2/025006

Seeberg, T. M., Royset, A., Jahren, S., & Strisland, F. (2011). Printed organic conductive polymers ther-mocouples in textile and smart clothing applications. *Proceedings of the Annual International Conference of the IEEE Engineering in Medicine and Biology Society, EMBS*, (0314), 3278–3281. 10.1109/IEMBS.2011.6090890

Selim, M. S., El-Safty, S. A., & Shenashen, M. A. (2019). Superhydrophobic foul resistant and self-cleaning polymer coating. In Sushanta K. Samal, Smita Mohanty, and Sanjay Kumar Nayak. *Superhydrophobic Polymer Coatings*. Elsevier. 181–203. 10.1016/b978-0-12-816671-0.00009-6

Seshadri, D. R., Li, R. T., Voos, J. E., Rowbottom, J. R., Alfes, C. M., Zorman, C. A., & Drummond, C. K. (2019). Wearable sensors for monitoring the internal and external workload of the athlete. *Npj Digital Medicine*, *2*(1), Article no. 71. 10.1038/s41746-019-0149-2

Shi, K., Sun, B., Huang, X., & Jiang, P. (2018). Synergistic effect of graphene nanosheet and BaTiO3nanoparticles on performance enhancement of electrospun PVDF nanofiber mat for flexible piezoelectric nanogenerators. *Nano Energy*, *52*(May), 153–162. 10.1016/j.nanoen.2018.07.053

Shiu, K. K., Song, F. Y., & Lau, K. W. (1999). Effects of polymer thickness on the potentiometric pH responses of polypyrrole modified glassy carbon electrodes. *Journal of Electroanalytical Chemistry*, *476*(2), 109–117. 10.1016/S0022-0728(99)00372-1

Sim, J. K., Yoon, S., & Cho, Y. H. (2018). Wearable sweat rate sensors for human thermal comfort mon-itoring. *Scientific Reports*, *8*(1), 1–11. 10.1038/s41598-018-19239-8

Soni, M., Bhattacharjee, M., Ntagios, M., & Dahiya, R. (2020). Printed temperature sensor based on PEDOT: PSS-graphene oxide composite. *IEEE Sensors Journal*, *20*(14), 7525–7531. 10.1109/JSEN.2020.2969667

Sorayani Bafqi, M. S., Bagherzadeh, R., & Latifi, M. (2015). Fabrication of composite PVDF-ZnO nanofiber mats by electrospinning for energy scavenging application with enhanced efficiency. *Journal of Polymer Research*, *22*(7), 130–138. 10.1007/s10965-015-0765-8

Souri, H., Banerjee, H., Jusufi, A., Radacsi, N., Stokes, A. A., Park, I., & …Amjadi, M. (2020). Wearable and stretchable strain sensors: Materials, sensing mechanisms, and applications. *Advanced Intelligent Systems*, *2*(8), 2000039. 10.1002/aisy.202000039

Spanu, A., Pinna, L., Viola, F., Seminara, L., Valle, M., Bonfiglio, A., & Cosseddu, P. (2016). A high-sensitivity tactile sensor based on piezoelectric polymer PVDF coupled to an ultra-low voltage organic transistor. *Organic Electronics: Physics, Materials, Applications*, *36*, 57–60. 10.1016/j.orgel.2016.05.034

Stassi, S., Cauda, V., Canavese, G., & Pirri, C. F. (2014). Flexible tactile sensing based on piezoresistive composites: A review. *Sensors (Switzerland)*, *14*(3), 5296–5332. 10.3390/s140305296

Strohmeier, P., Knibbe, J., Boring, S., & Hornbæk, K. (2018). Hybrid resistive/capacitive etextile input. *TEI 2018 –Proceedings of the 12th International Conference on Tangible, Embedded, and Embodied Interaction*, *2018-January*(March), 188–198. 10.1145/3173225.3173242

Sun, J. J., Hall, A., Simoneishvilli, I., Akdogan, E. K., & Safari, A. (2007). Novel piezoelectric sensors and actuators. In *International Microelectronics and Packaging Society –3rd IMAPS/ACerS International Conference and Exhibition on Ceramic Interconnect and Ceramic Microsystems Technologies, CICMT 2007*.

Tamura, T., Huang, M., & Togawa, T. (2018). Current developments in wearable thermometers. *Advanced Biomedical Engineering*, *7*(April), 88–99. 10.14326/abe.7.88

Teixeira, J., Horta-Romarís, L., Abad, M. J., Costa, P., & Lanceros-Méndez, S. (2018). Piezoresistive response of extruded polyaniline/(styrene-butadiene-styrene) polymer blends for force and deformation sensors. *Materials and Design*, *141*, 1–8. 10.1016/j.matdes.2017.12.011

Tran, A. V., Zhang, X., & Zhu, B. (2018). Mechanical structural design of a piezoresistive pressure sensor for low-pressure measurement: A computational analysis by increases in the sensor sensitivity. *Sensors (Switzerland)*, *18*(7), Article no. 2023. 10.3390/s18072023

Walczak, S., & Sibiński, M. (2014). Flexible, textronic temperature sensors, based on carbon nanostructures. *Bulletin of the Polish Academy of Sciences: Technical Sciences*, *62*(4), 759–763. 10.2478/bpasts-2014-0082

Wanjari, N. D., & Patil, S. C. (2017). Wearable devices. *2016 IEEE International Conference on Advances in Electronics, Communication and Computer Technology, ICAECCT 2016*, (february), 287–290. 10.1109/ICAECCT.2016.7942600

Wearable Sensors Market. (2019). Retrieved from https://www.marketsandmarkets.com/Market-Reports/wearable-sensor-market-158101489.html

Wearable Sensors Market Size, Share & Trends Analysis Report. (2018). Retrieved from https://www.grandviewresearch.com/press-release/wearable-sensors-market

Wearable Sensors Market Size, Trends, Industry Growth Analysis-Global Forecast. (2025). Retrieved from https://www.marketresearchfuture.com/reports/wearable-sensors-market-955

Xiang, D., Chen, Q., & Li, Y. (2019). Strain sensing behavior of conductive polymer/carbon nanotube composites coated fiber. *AIP Conference Proceedings*, *2065*(February), 1–6. 10.1063/1.5088286

Xie, L., & Zhu, Y. (2018). Tune the phase morphology to design conductive polymer composites: A review. *Polymer Composites*, *39*(9), 2985–2996. 10.1002/pc.24345

Xue, T., Truong, B. D., Rantz, R., & Miah, H. (2017). Energy Harvesting for Wearable Devices. In *The University of Utah*. Retrieved from https://iss.mech.utah.edu/energy-harvesting-for-wearable-devices/

Yan, J., & Jeong, Y. G. (2016). High performance flexible piezoelectric nanogenerators based on $batio_3$nanofibers in different alignment modes. *ACS Applied Materials and Interfaces*, *8*(24), 15700–15709. 10.1021/acsami.6b02177

Yurddaskal, M., Erol, M., & Celik, E. (2017). Carbon black and graphite filled conducting nanocomposite films for temperature sensor applications. *Journal of Materials Science: Materials in Electronics*, *28*(13), 9514–9518. 10.1007/s10854-017-6695-y

Zhang, G., Zhao, P., Zhang, X., Han, K., Zhao, T., Zhang, Y., & …Wang, Q. (2018). Flexible three-dimensional interconnected piezoelectric ceramic foam based composites for highly efficient concurrent mechanical and thermal energy harvesting. *Energy and Environmental Science*, *11*(8), 2046–2056. 10.1039/c8ee00595h

Zhang, H., Ren, P., Yang, F., Chen, J., Wang, C., Zhou, Y., & Fu, J. (2020). Biomimetic epidermal sensors assembled from polydopamine-modified reduced graphene oxide/polyvinyl alcohol hydrogels for the real-time monitoring of human motions. *Journal of Materials Chemistry B*, *8*(46), 10549–10558. 10.1039/d0tb02100h

Zhang, L., Li, H., Lai, X., Gao, T., Yang, J., & Zeng, X. (2018). Thiolated graphene@polyester fabric-based multilayer piezoresistive pressure sensors for detecting human motion. *ACS Applied Materials and Interfaces*, *10*(48), 41784–41792. 10.1021/acsami.8b16027

Zhang, S., Zhang, L., Wang, L., Wang, F., & Pan, G. (2019). A Flexible e-skin based on micro-structured PZT thin films prepared: Via a low-temperature PLD method. *Journal of Materials Chemistry C*, *7*(16), 4760–4769. 10.1039/c8tc06350h

Zheng, Q., Zhang, H., Mi, H., Cai, Z., Ma, Z., & Gong, S. (2016). High-performance flexible piezoelectric nanogenerators consisting of porous cellulose nanofibril (CNF)/poly(dimethylsiloxane) (PDMS) aerogel films. *Nano Energy*, *26*, 504–512. 10.1016/j.nanoen.2016.06.009

Zheng, Y., Li, Y., Dai, K., Liu, M., Zhou, K., Zheng, G., & …Shen, C. (2017). Conductive thermoplastic polyurethane composites with tunable piezoresistivity by modulating the filler dimensionality for flexible strain sensors. *Composites Part A: Applied Science and Manufacturing*, *101*, 41–49. 10.1016/j.compositesa.2017.06.003

13 An Insight into the Synthesis and Optoelectronic Properties of Thiophene-2,4,6-Triaryl Pyridine-Based D-A-D Type π-Conjugated Functional Materials

Viprabha K. and Udaya Kumar D.
National Institute of Technology Karnataka (NITK), Surathkal, Mangaluru, India

CONTENTS

DOI: 10.1201/9781003200710-15

13.1 INTRODUCTION

Recently, alongside the rise in awareness of energy and environmental issues, conjugated conducting materials (polymers/molecules) have attracted tremendous attention as a rapidly developing field to conquer these issues. Conjugated conducting or functional materials are organic materials consisting of alternate single and double bonds along the molecular chain, composed mainly of carbon, hydrogen and heteroatoms such as oxygen, nitrogen and sulfur (Figure 13.1). The overlap of π–orbitals in such systems leads to the extended and delocalized conjugation which originates conductivity in the system. These conjugated materials possessing the optical, electronic, magnetic and electrical properties of a metal, have been fascinating numerous researchers all over the world due to their practical application in modern technology. These metallic characteristics of conjugated materials turn them into the category of so-called *synthetic metals*.

Though the initial impetus for the plethora of work on conducting polymers (CPs) or, more precisely, intrinsically conducting polymers (ICPs) roots back to the 1960s, the remarkable breakthrough in the entire field of π–conjugated organic materials was commenced nearly four decades back when the ground-breaking discovery of profound increase in the electrical conductivity in polyacetylene was achieved upon I_2 vapor doping by Alan Heeger, Alan MacDiarmid, and Hideki Shirakawa in 1977 (Chiang, Druy, et al., 1978; Chiang et al., 1977; Chiang, Park, et al., 1978; Shirakawa et al., 1977). In 2000, the work was further reinforced with awarding the Nobel Prize in chemistry to these pioneers "for the discovery and development of electrically conductive polymers". Since then, a variety of π–conjugated materials (polymers, oligomers or small molecules) based on aromatic precursors such as p-phenylenevinylene, thiophene, triphenylamine, carbazole, fluorene and their derivatives (Junkers et al., 2012; McCullough, 1998; Roncali, 1992; Sahin et al., 2011; Skotheim et al., 2007) have been developed (Figure 13.1) and their photophysical and electrochemical properties have also been intensively studied. The semiconducting properties, a broad range of optical absorption, charge mobility and charge storage properties of these conjugated materials enabled the practical usage of these systems in the field of organic electronics and photonics, such as organic solar cells (OSCs) (C.-H. Chen et al., 2019; Duan et al., 2020; Sathiyan et al., 2019), organic light emitting diodes (OLEDs) (Chidirala et al., 2016; C.-K. Liu et al., 2018; Wang et al., 2019), nonlinear optics (NLO) (Divyasree et al., 2018; Ju et al., 2019), sensors (Tan & Baycan, 2020), organic field-effect transistors (OFETs) (Kim et al., 2020), electrochromic devices (X. Cheng et al., 2018), and they are also used as electrical transporters or battery electrodes (Liang et al., 2012).

Unlike the synthesis of inorganic materials, which are mostly brittle and hard, and usually involves high temperatures (T ≥ 500°C) (Pell et al., 2004), the conjugated materials are mechanically flexible in nature and can be easily synthesized through robust, cost-effective organic reactions such as condensation reactions, green synthesis, one-pot multicomponent reactions, palladium-catalyzed cross-coupling reactions, and so on (Y.-J. Cheng et al., 2009; Demeter et al., 2014). The modern organic synthetic methods allow facile functionalization of the conjugated

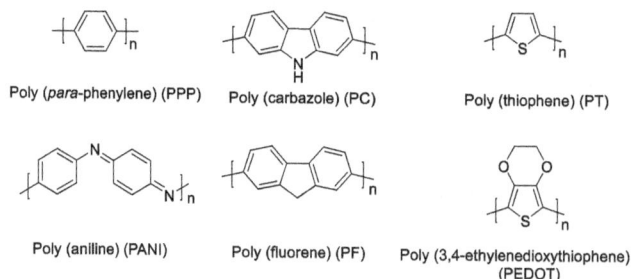

Poly (*para*-phenylene) (PPP)   Poly (carbazole) (PC)   Poly (thiophene) (PT)

Poly (aniline) (PANI)   Poly (fluorene) (PF)   Poly (3,4-ethylenedioxythiophene)
(PEDOT)

FIGURE 13.1   Some important conjugated polymers.

systems which can readily tune their photophysical and electronic properties, making the conjugated materials a mainstay of our technological existence (Forrest, 2004; Gibson et al., 2012).

13.1.1 π–Conjugated Polymers

In this system, π–orbitals are extended throughout the entire conjugated backbone which enable sufficient optical absorption in the ultraviolet–visible (UV–Vis) region and facilitate the intramolecular charge-transfer (ICT) (Farchioni & Grosso, 2013; Nalwa, 1997). These polymers are often functionalized with solubilizing agents, namely, hydrocarbon chains that not only improve the solution processability but also enhances the intermolecular interaction in the solid state (Allard et al., 2008).

The performance of polymer-based materials depends on the molecular weight and distribution of molecular weight, which is called the dispersity (Ð). The inherent drawbacks of conjugated polymer systems are their low solubility in common organic solvents, low crystallinity, and the sample purity which led to the improper charge transport and ultimately diminish the materials performance (Martin & Diederich, 1999). However, the high viscosity of conjugated polymers yields good-quality thin films with uniform and smooth surface, which improves the device performance (Facchetti, 2011).

A number of π–conjugated polymers have been designed and synthesized from electron-rich systems like poly(3-hexylthiophene) (P3HT), electron deficient systems such as poly(benzobisimidazobenzophenanthroline) (PBBL) as well as D–A type polymers like poly(N,N′dialkylperylenedicarboximidedithiophene) (PDIR-T2). Some of them are shown in Figure 13.2.

13.1.2 π–Conjugated Small Molecules

Organic π–conjugated small molecules have greatly been extended in the optoelectronic field for a couple of decades. A wide chemical functionality and simple modifications in synthetic routes of small molecules can tune optical, morphological, electrical, and electrochemical properties. Also, proper selection of alkyl groups improves the solubility of the system (Mishra & Bäuerle, 2012). The purification of organic materials is very essential since the impurities develop charge-carrier traps, which reduces the performance of the device. Small molecule organic materials can be purified by traditional crystallization techniques or column chromatography, that are not applicable to purify high molecular weight polymers. Further, high crystallinity can be achieved in small molecule materials by simple solution-grown crystallization and vacuum deposition techniques which facilitates the intermolecular charge-transport efficiently, and hence, enhances the performance of electronic materials. The small molecules have similar design principles as that of polymers where electron deficient, electron rich and D–A moieties play crucial roles to fine tune their optoelectronic properties. There are numerous reports on π–conjugated small molecules in the literature; some of them are shown in Figure 13.3.

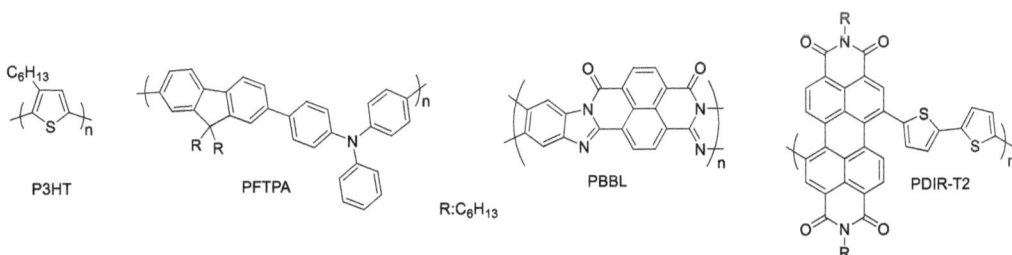

FIGURE 13.2 Chemical structures of P3HT, PFTPA, PBBL and PDIR-T2.

(a)   (b)   (c)   (d)

R: C$_6$H$_{13}$

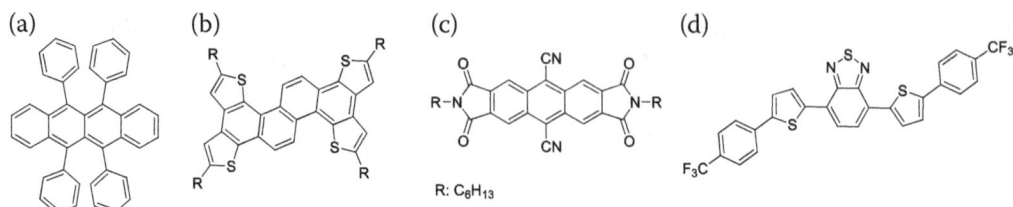

FIGURE 13.3 Molecular structures of (a) rubrene, (b) electron–rich, (c) electron–deficient, and (d) D–A π–conjugated small molecules.

13.2 CONDUCTION IN CONJUGATED SYSTEM

The molecular orbital theory (MOT) becomes more prominent in the case of molecules where two or more π–bonds interact with one another. The π–molecular orbitals are used to describe the energies and locations of π–electrons in conjugated systems. The π–molecular orbitals usually include the frontier molecular orbitals, which are the highest occupied molecular orbital (HOMO), corresponding to fully occupied π–band and the lowest unoccupied molecular orbital (LUMO), corresponding to empty π*–band. The HOMO and the LUMO energy levels are analogous to the conduction band and the valence band, respectively, of inorganic semiconductors. The difference in the energy levels between the HOMO and the LUMO is called the band gap (E_g) (Figure 13.4). The mechanism of conduction is explained by band theory, according to which the half-filled valence band forms a continuous delocalized π–system, which is responsible for the conductivity in the material. In the case of conjugated polymers, the conductivity could be enhanced by the process called *doping*, which involves either oxidation (removal of electrons) or reduction (addition of electrons) of the polymeric system. The mechanism of conductivity in these polymers is

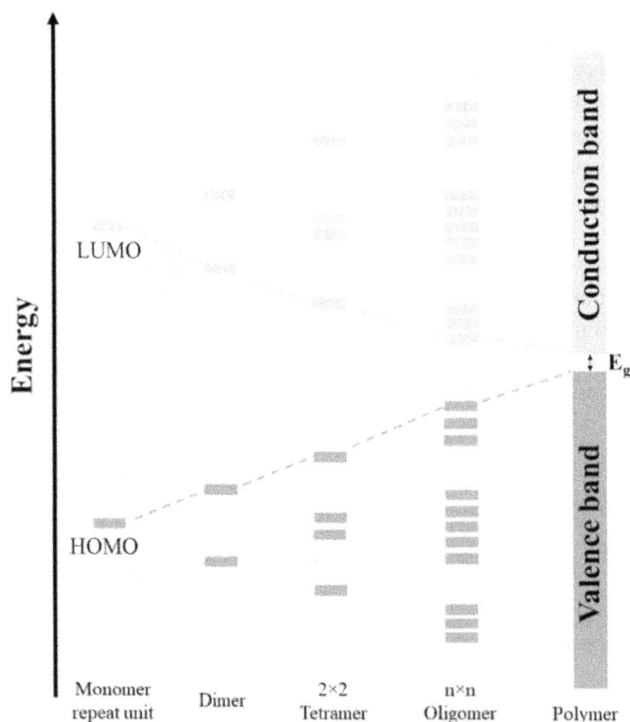

FIGURE 13.4 Schematic of band formation from molecular repeat unit to π–conjugated polymer.

based on the movement of charged defects within the conjugated framework. These charge carriers/defects, either positive (p-type) or negative (n-type), are the products of oxidation or reduction of the polymer, respectively.

13.3 D–A Π–CONJUGATED MATERIALS

The ease of synthesis, structural diversity, tunability, and fascinating optoelectronic properties of carbon and heteroatom-based conjugated materials have attracted particular interest in the field of material science. However, for achieving high-performing devices, the conjugated materials should possess certain physical and chemical properties such as broader solar absorption, low band gap, efficient photoinduced charge transfer and separation, low oxidation potential, and ambipolar charge transport with high mobilities (Nielsen et al., 2013). Therefore, in order to attain desired requirements, modification in the HOMO and the LUMO energy levels is essential. This modification can be facilitated by the D–A approach, wherein the electron rich/donor (D) and the electron deficient/acceptor (A) substituents are arranged alternatively along the conjugated backbone (Havinga et al., 1993). When compared to the conjugated materials consisting of all donor units, D–A conjugated materials with the alternate D and A units absorb solar energy at longer wavelengths. The interaction between a strong electron-donor and a strong electron-acceptor gives rise to an increased double bond character between these units; as a result, there is a significant enhancement of ICT in such D–A systems which extends the conjugation length, leading to a prolonged absorption and a higher absorption coefficient. If the energy of the HOMO level of D and the LUMO level of A moiety is relatively close, there exists a strong intramolecular orbital mixing between D and A units, which considerably lowers the LUMO level and raises the HOMO level, resulting a low-energy gap (Balan et al., 2008; Zotti et al., 1999) as depicted in Figure 13.5. Hence, in a conjugated molecule with an alternating sequence of appropriate D and A units, the hybridization of energy levels of the D and the A moieties can induce an unusually low HOMO–LUMO separation, resulting a reduction in its E_g (Brocks & Tol, 1996). Due to this feature, the D–A conjugated materials have been spotlighted since a couple of decades for their application in many optoelectronic and electronic devices such as OLEDs, OSCs, and OFETs.

In D–A systems, the introduction of strong electron-donor groups raises the HOMO levels, while the introduction of strong electron-withdrawing groups lowers the LUMO levels, which results in the narrow HOMO–LUMO energy gap (Ajayaghosh, 2003). Thus, key design criterion is the selection of proper electron-donor and acceptor units to achieve desired HOMO and LUMO levels beneficial for the development of organic optoelectronic materials. Additionally, the presence of strong electron-donors and acceptors increases the delocalization of π–electrons, making the material highly polarizable which causes remarkable optical nonlinearities in such systems. Since D–A π–conjugated materials possess distinct advantages like lightweight, good film forming property, relatively low manufacturing cost, biocompatibility, moderate to high conductivity and easy tunability of desired properties, they are broadly

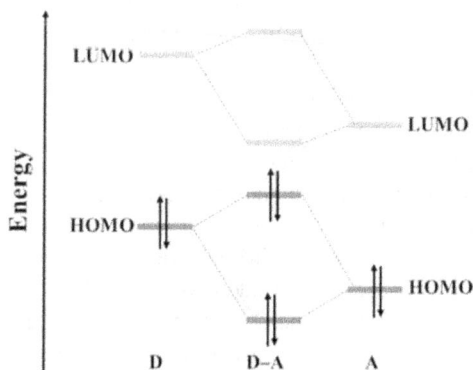

FIGURE 13.5 Imolecular orbital interaction in D-A moieties showing low HOMO–LUMO separation (Mullekom van, 2000).

used as active components in optoelectronic/photonic devices such as OPVs, OLEDs, OFETs, and NLO. Amongst the various applications reported for D–A π–conjugated materials, the NLO attained considerable interest in recent years (Anandan et al., 2018; Q. Xu et al., 2019).

A detailed literature survey reveals that the synthesis of intermediate-sized conjugated molecules or oligomers with extended conjugation can provide desirable qualities of both polymers and small molecules. There has been a dramatic surge in design, synthetic development of conjugated molecules and also, in terms of device efficiency in recent years. Plenty of materials with various D–A combinations such as D–A–D, A–D–A, D–π–A and star-shaped architectures have been reported with excellent optoelectronic performances. Among them, the D–A–D module emerged as one of the most successful and promising modules explored in the optoelectronic and photonic industry (Raynor et al., 2015). Tuning of frontier molecular orbital plays a vital role in achieving desired properties in such π–systems. There are four primary ways to modify the frontier orbital energies: (i) enlarging the π–conjugation, (ii) incorporating planar fused aromatic ring systems so as to increase the planarity of the system, (iii) incorporating strong donor/acceptor functional units alternatively, and (iv) incorporation of polarizable double bonds.

Although, D–A conjugated materials have developed rapidly and became popular in the past couple of decades, they are still the focus of intensive research. There is still a considerable room for the rational design and synthesis of new D–A type materials. For an instance, most of the reported D–A conjugated materials contain simple organic compounds such as thiophene, carbazole, triphenylamine, phenothiazine, fluorene, etc., as powerful electron donors and 1,3,4-oxadiazole, pyridine, cyanovinylene, perylene diimides, 2,1,3-benzothiadiazole, thiadiazole, thiazolo[5,4-d]thiazole, etc., as electron accepting groups which involve various functionalization and different substitution patterns. These substituents play a vital role in "band gap engineering" of organic molecules to fine-tune their optoelectronic properties. However, thiophene-based π–conjugated systems have progressively supplanted other classes of systems in a variety of plastic electronic technologies. The unique combination of efficient electron transfer, structural versatility, environmental stability, a moderate band gap and ease of functionalization of thiophene provides an indispensable role for the thiophene based materials in organic optoelectronic fields such as OSCs, OLEDs, sensors, OFETs, and NLO (Perepichka & Perepichka, 2009; Skabara, 2009). In view of this, we have designed and synthesized two new sets of thiophene-based D–A–D-type conjugated compounds comprising of thiophene as donor and pyridine as acceptor moiety along with π–linkers and other supporting electron donor groups for optoelectronic applications.

13.4   DESIGN CRITERIA AND STRUCTURAL FEATURES OF COMPOUNDS M1, M2, P1, AND P2

Various design strategies like D–A–D, A–D–A and D–π–A have been established by many research groups (Albota et al., 1998; Reinhardt et al., 1998) to enhance the nonlinear properties. In addition, the symmetry of the molecule, proper selection of π–linkers between D and A units, suitable D and A building blocks and their distance play vital roles to fine tune the optical nonlinearity (L. Xu et al., 2017). One of the most widely used high electron withdrawing moiety is pyridine ring. Different substitutions in various positions of pyridine and coordination of different guest units to the nitrogen atom of the pyridine group led to the fine-tuning of optoelectronic properties of pyridine-based materials (Vellis et al., 2008), which impart good electron-transporting abilities, chemical stability, and high thermal stability to the system, so the pyridine has a significant role in the optoelectronic industry (Liaw et al., 2009). Also, the introduction of substituted acrylonitrile groups as side groups/substituents increases the electrochemical stability and mainly lowers the LUMO level to achieve high–performance optoelectronic devices (Mikroyannidis et al., 2009). In addition, molecules with substituted acrylonitrile groups exert immense impact on the ICT from the D to A fragments. On the other hand, the olefinic bond between the aromatic rings improves in planarity, reduces the steric hindrance, extends the π–conjugation (Jebnouni et al., 2018), and modulates the electrochemical properties such as charge transfer ability, light emitting property, etc. (Hwang & Chen, 2002; Qing et al., 2015; Roncali, 1997; Song et al., 1999). Given that, in the present work, a new class of trigonal-shaped molecules (**M1** and **M2**) and polymers (**P1** and **P2**) with D–A–D structural arrangement possessing thiophene acetonitrile (**M1**), phenyl acetonitrile groups (**M2, P2**), and thiophene unit (**P1**) as side groups were designed and synthesized with 2,4,6-trisubstituted pyridine (served as electron acceptor) and thiophene (served as electron donor) backbone to study the effect of conjugation on the fundamental photophysical, electrochemical, and third-order NLO properties. The polymers **P1** and **P2** were synthesized with the expectation that this would bring a drastic improvement in the optical nonlinearity compared to small molecules (**M1** and **M2**).

13.5 DESIGN CRITERIA AND STRUCTURAL FEATURES OF COMPOUNDS M3, M4, M5, AND M6

Similar to thiophene, the triphenylamine moiety, possessing a continuous conjugation between peripheral phenyl groups and the central nitrogen atom is found to be another excellent electron donor (Jana & Ghorai, 2012). In addition, it is an excellent hole transporter and has a high triplet energy of 3.04 eV (Jiang et al., 2009; Sonntag et al., 2005; Tong et al., 2007). The compounds containing carbazole, another electron donating group, also have attracted much attention due to their high thermal stability, solubility, good charge-transport function, extended glassy state, and moderately high oxidation potential (Liaw et al., 2009; Yook & Lee, 2014). Taking this into consideration, two sets of four, butterfly-shaped symmetric molecules (**M3, M4, M5,** and **M6**) were designed. In **M3** and **M4,** the terminal triphenylamine (**M3**)/carbazole (**M4**) moieties (served as electron donors) and the central pyridine core (served as electron acceptor) were connected via thiophene (served as π–linker), respectively, which provided a D–π–A–π–D structural configuration to the system, whereas, in **M5** and **M6,** the peripheral thiophene–pyridine–thiophene unit (D–A–π) was connected to the central triphenylamine (**M5**)/carbazole (**M6**) moiety, which provided D–A–π–D–π–A–D configuration.

13.6 SYNTHESIS OF COMPOUNDS

Two series of eight compounds were synthesized using a proper reaction methodology. The chemical structures of target compounds are shown in Figure 13.6 and the synthetic methodology for the synthesis of target compounds is summarized in Scheme 13.1. In the first step, the intermediate **3** was synthesized by a multicomponent one-pot reaction under microwave irradiation. The intermediate **3** was alkylated using 1-bromodecane, to yield intermediate **4**. Then, formylation of intermediate **4** yielded intermediates **5** and **6**. The intermediates **5** and **6** were reduced to get the intermediates **7** and **8**, respectively, which were further converted into their corresponding Wittig salts **9** and **10**. Finally, the intermediate **5** was subjected to well-known Knoevenagel condensation and Wittig condensation reactions with active methylene compounds *viz.,* thiophene-2-acetonitrile, phenylacetonitrile, 1,4-phenylene diacetonitrile and Wittig salt (**17**) to obtain the target compounds **M1**, **M2**, **P2**, and **P1**, respectively. Similarly, the Wittig reaction of intermediate **10** with 4-formyl triphenylamine (**15**) and intermediate **13** produced final molecules **M3** and **M4**, respectively, while the Wittig reaction of intermediate **9** with bis(4-formylphenyl)phenylamine (**16**) and intermediate **14** produced final molecules **M5** and **M6**, respectively.

13.7 PHOTOPHYSICAL STUDIES

Optical methods are found to be one of the most widely used methods to derive valuable information about the structures of organic materials. As the study of interaction of materials with light is very much useful for their optical characterization, the UV–vis absorption and the fluorescence emission spectroscopic techniques are found to be two well-known tools to study material properties. The UV–vis absorption spectroscopy basically deals with the electronic transitions from

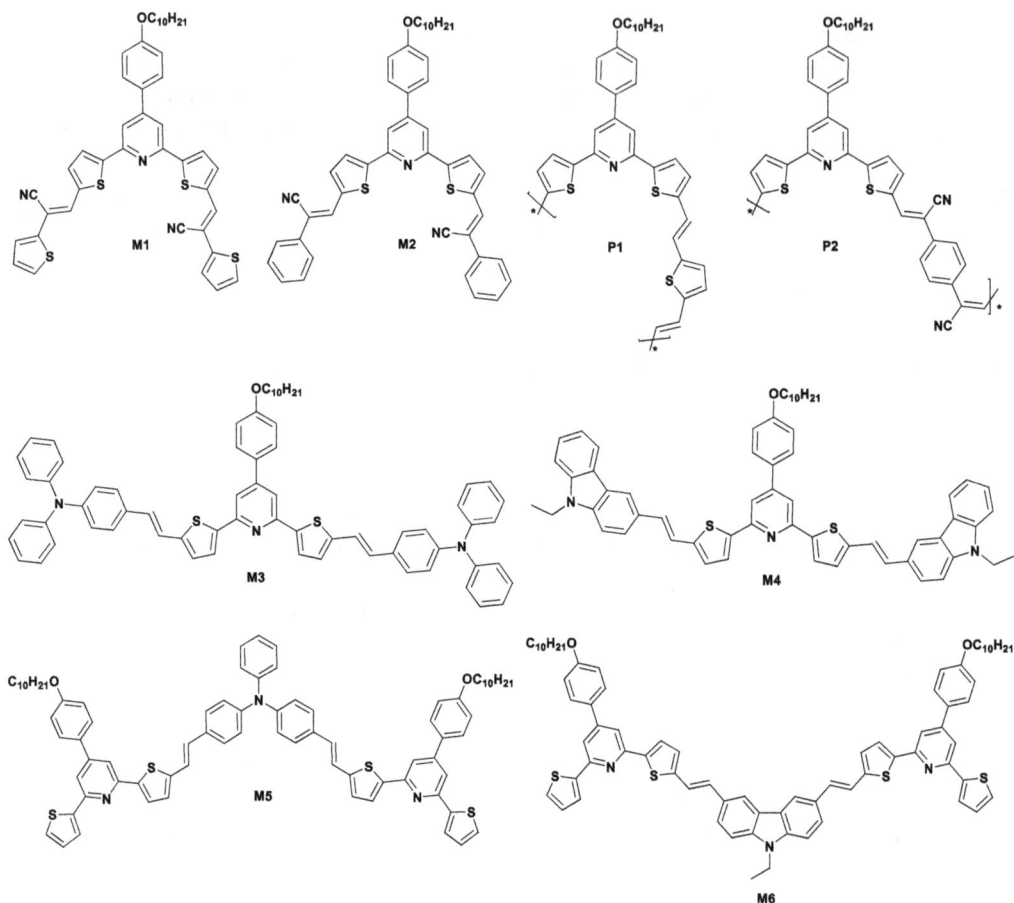

FIGURE 13.6 Chemical structures of **M1, M2, P1, P2, M3, M4, M5** and**M6**.

the ground state or the HOMO to the excited state or the LUMO of the material. The lower the gap between the HOMO and the LUMO, the longer the wavelength of light it can absorb. Therefore, the UV–vis spectroscopy was used to detect the extent of conjugation in the synthesized compounds and also, to investigate the effect of double bonds on the absorption wavelength of the synthesized compounds **M1, M2, P1, P2, M3, M4, M5,** and **M6**. All the spectra were recorded in solution state using CHCl$_3$ as the solvent (1×10^{-5} M) at RT. Fluorescence emission spectroscopy involves the excitation of an analyte by irradiation at a certain wavelength and the emission of radiation of a different wavelength from the excited state to the ground state. Therefore, the fluorescence spectroscopy was used to study the emission properties of the synthesized compounds **M1, M2, P1, P2, M3, M4, M5,** and **M6**. All the spectra were recorded in solution state using CHCl$_3$ as solvent (1×10^{-5} M) at RT. The emission spectra of **M1, M2, P1, P2, M3, M4, M5,** and **M6** were obtained under excitation at their corresponding maximum absorption wavelength (λ_{max}). From the results of optical studies, the important parameters such as absorption and emission region in the electromagnetic spectrum, and the optical band gap (E_g) of **M1, M2, P1, P2, M3, M4, M5,** and **M6** were evaluated, which are key factors for selecting the materials as active materials in photovoltaic/optoelectronic/photonic applications.

The combined normalized UV–vis absorption and PL emission spectra of compounds **M1, M2, P1,** and **P2** are depicted in Figure 13.7 and the data are summarized in Table 13.1. The strong peaks of maximum absorption are observed at wavelengths 395, 377, 312, and 419 nm for **M1, M2, P1,** and **P2**, respectively, as a result of π–π* transition within the conjugated backbone. Similar absorption

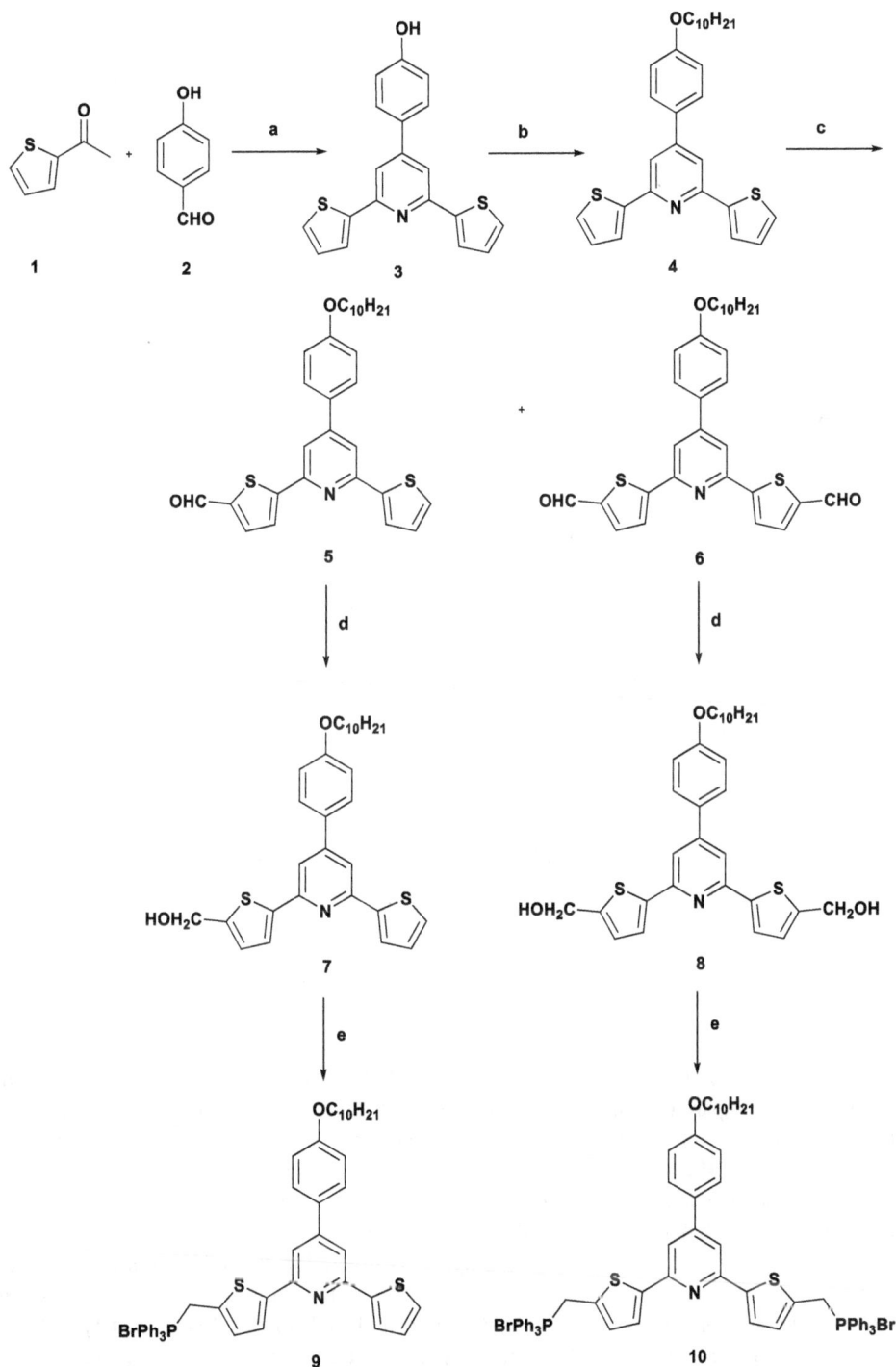

SCHEME 13.1 Synthetic route for **M1, M2, P1, P2, M3, M4, M5,** and **M6**. Reagents and reaction conditions: (a) ammonium acetate, microwave, 400 W, 120°C, 0.5 h, yield: 68%; (b)1-bromodecane, K$_2$CO$_3$, DMF, 85°C–90°C, 8 h, yield: 94%; (c) POCl$_3$, DMF, 90°C–95°C, 48 h, yield: 38%–69%; (d) sodium borohydride, CH$_3$OH, RT, 12 h, yield: 90%–92%; (e) PPh$_3$HBr, CHCl$_3$, 60°C, 3 h, yield: 93%–94%; (f) NaH, C$_2$H$_5$Br, DMF, RT, 12 h, yield: 82%; (g) C$_2$H$_5$ONa, CHCl$_3$, C$_2$H$_5$OH, RT, 12 h, yield: 56%–63%.

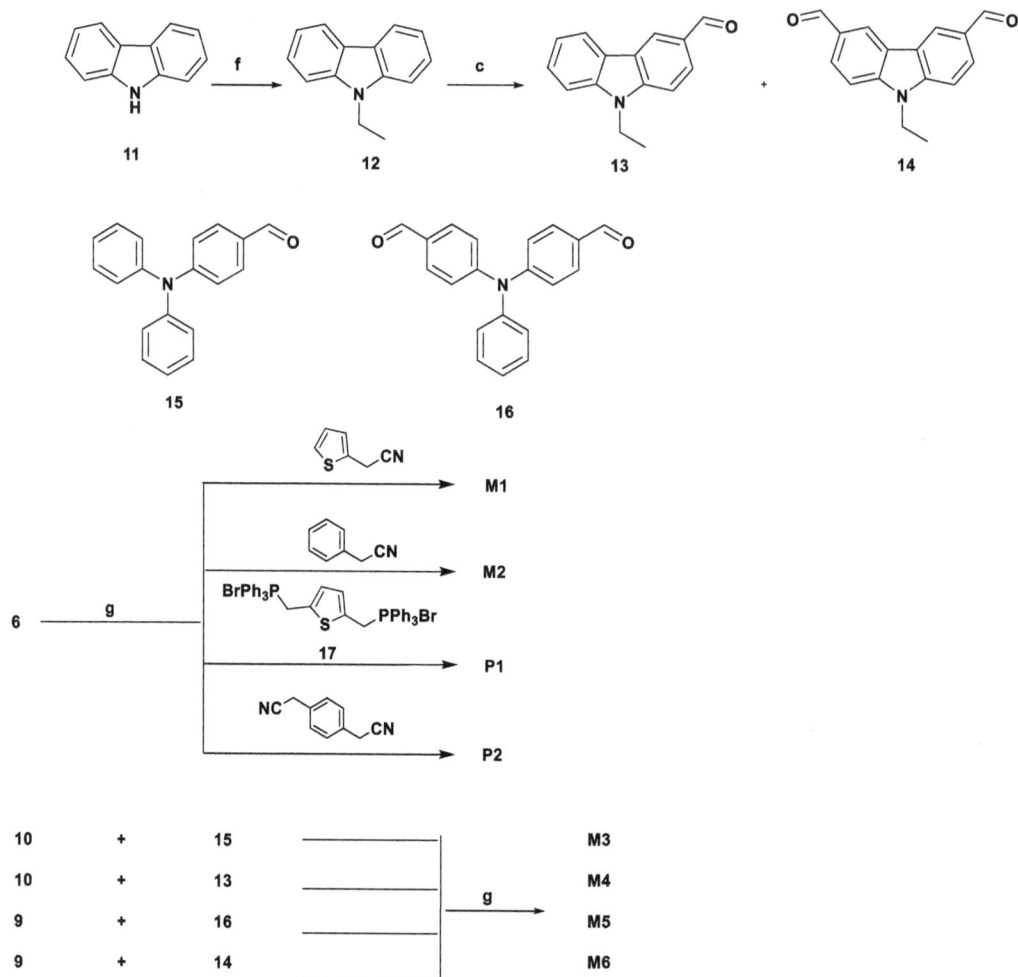

SCHEME 13.1 (*continued*)

profiles were observed in the range of 300 to 500 nm for **M1** and **M2** (Figure 13.7). The redshift in the absorption of **M1** compared to that of **M2** is due to the stronger D–A interaction between the donor thiophene and the acceptor cyanovinylene groups in **M1** compared to that between phenyl and cyanovinylene groups in **M2**. Similarly, due to the presence of cyanovinylene group in **P2**, the λ_{max} of **P2** is bathochromically shifted compared to that of **P1**. As the molecular weights (M_w) of **P1** (M_w = 2,558) and **P2** (M_w = 10,426) are widely different, the blue shift in the absorption spectrum of **P1** could also arise from the short chains of the polymer compared to that of **P2**. Further, all the four compounds were excited at their respective λ_{max} to record the emission profiles. The maximum emission wavelengths are observed at 537, 500, 494, and 562 nm for **M1**, **M2**, **P1**, and **P2**, respectively, a similar trend as those of absorption profiles can be seen. The E_g calculated from the intersection point of absorption and emission spectra are 2.66, 2.74, 2.78, and 2.55 eV for **M1**, **M2**, **P1**, and **P2**, respectively. The presence of strong electron accepting cyanovinylene group results in a considerable reduction of E_g in **M1**, **M2**, and **P2** compared to that of **P1**.

The UV–vis absorption spectra of **M3**, **M4**, **M5**, and **M6** exhibit two absorption bands (Figure 13.8), one located at lower wavelength UV region (280–330 nm), which corresponds to π-π* transition and the other located at higher wavelength visible region (350–460 nm), which attributes to the ICT from the triphenylamine/carbazole moiety (electron donor) to the pyridine (electron acceptor) group.

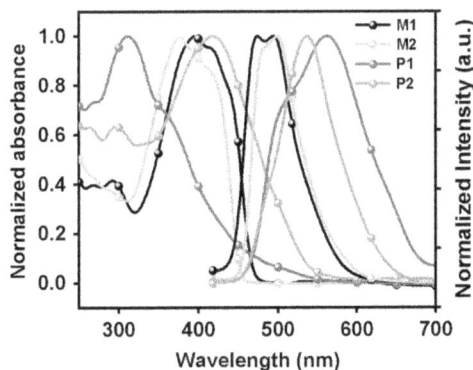

FIGURE 13.7 UV–vis absorption and normalized PL spectra of **M1, M2, P1,** and **P2** (recorded in CHCl$_3$, 10^{-5} M).

TABLE 13.1

Summary of Optical and Electrochemical Data of M1, M2, P1, P2, M3, M4, M5, and M6

Compounds	λ_{max}^{abs} (nm)	λ_{max}^{pl} (nm)	E_g (eV)	E^{ox}_{onset} (V vs SCE)	HOMO (eV)	LUMO (eV)
M1	399	537	2.66	0.82	−5.35[a]	−2.69[b]
M2	377	500	2.74	0.99	−5.52[a]	−2.78[b]
P1	312	494	2.78	0.55	−5.08[a]	−2.30[b]
P2	419	562	2.55	0.91	−5.44[a]	−2.89[b]
M3	409	503	2.64	0.77	−5.31[a]	−2.67[b]
M4	394	493	2.71	0.80	−5.34[a]	−2.63[b]
M5	419	498	2.62	0.71	−5.25[a]	−2.63[b]
M6	404	474	2.76	0.82	−5.36[a]	−2.60[b]

Notes

[a] Experimental values from CV using Equation 13.1 with Fc/Fc$^+$ as internal standard.

[b] Experimental values using Equation 13.2.

E_g Optical band gap calculated from the intersection of normalized absorption and emission spectra.

E^{ox}_{onset} Experimental onset oxidation potential vs SCE.

The observed λ_{max} are 409, 394, 419, and 404 nm, respectively. From the absorption spectra it is clear that the variation of peripheral groups brings about the effectual modification in the conjugation which also affects the absorption of the molecules. For example, the higher electron donating ability of the triphenylamine group in **M3** and **M5** compared to that of carbazole in **M4** and **M6**, bathochromically shifts the absorption of **M3** and **M5** compared to that of **M4** and **M6** (Figure 13.8). The higher λ_{max} of **M5** is also resulted from the extended conjugation in **M5** compared to that of **M3, M4,** and **M6**. The E_g determined from the point of intersection of absorption and emission spectra is 2.64, 2.71, 2.62, and 2.76 eV for **M3, M4, M5,** and **M6**, respectively. Further, from the fluorescence spectra the emission peaks were observed at 503, 450, 488, and 461 nm for **M3, M4, M5,** and **M6**, respectively (Figure 13.8), by exciting at their corresponding λ_{max}, suggesting the green emission from the molecules. The absorption and emission data are listed in Table 13.1.

13.8 ELECTROCHEMICAL STUDIES

As the electrochemistry relates the flow of electrons to the chemical changes, the electrochemical study involves the understanding of basic electronic structures of organic materials. It describes their oxidation and reduction potentials, which are highly useful in determining their HOMO–LUMO energy

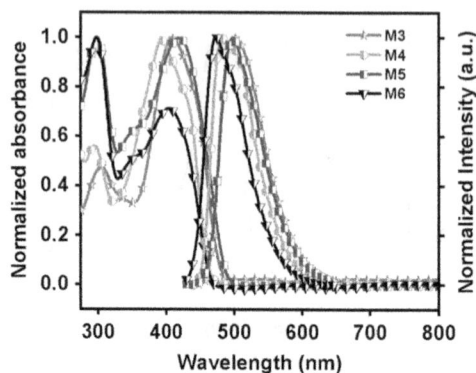

FIGURE 13.8 UV–vis absorption spectra at RT and fluorescence emission spectra at RT of **M3, M4, M5 and M6** (measured in CHCl$_3$, 10^{-5} M).

levels and charge carrying properties. The cyclic voltammetry (CV) is a popular and powerful electrochemical technique, generally employed to investigate the oxidation and the reduction processes of molecular species and is also utilized to estimate the HOMO and LUMO energy levels of the molecules (Y. Li et al., 1999). In the present work, the CV studies were carried out to determine the oxidation potentials and hence, to evaluate the HOMO/LUMO energy levels of the compounds **M1, M2, P1, P2, M3, M4, M5,** and **M6,** which are very much essential in molecular design and controlling the band gap; and also, for the proper selection of electrodes in optoelectronic devices.

The cyclic voltammograms of **M1, M2, P1,** and **P2** are shown in Figure 13.9a–d, respectively. Using the first oxidation potential the HOMO energies were determined using Equation 13.1, whereas, LUMO energies were calculated using Equation 13.2.

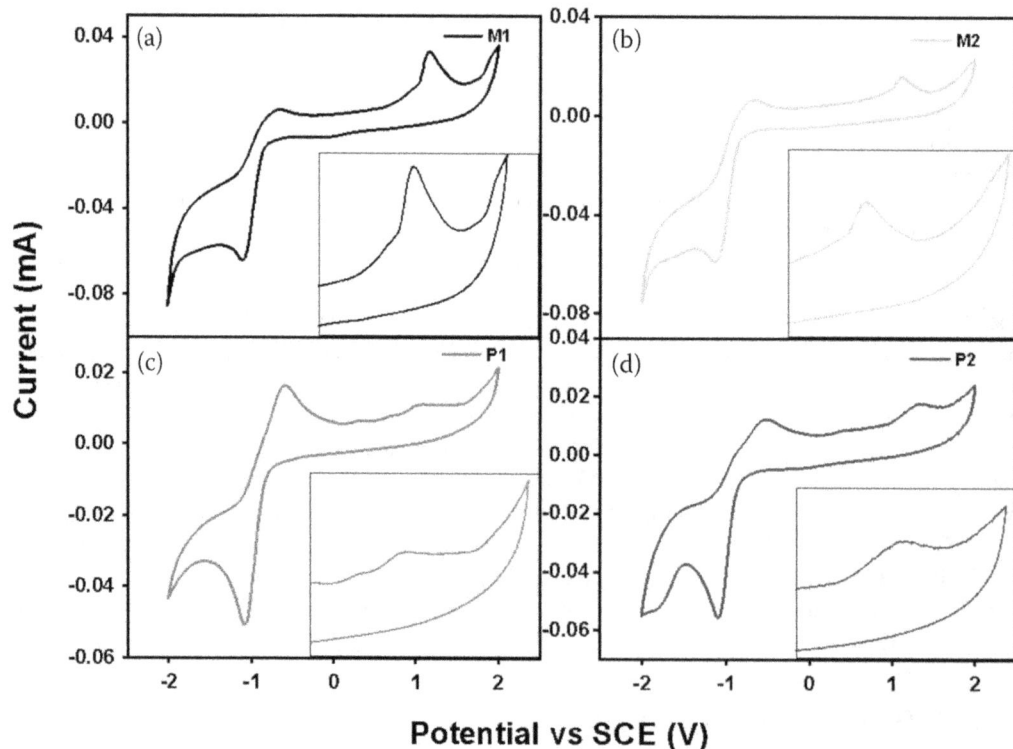

FIGURE 13.9 Cyclic voltammograms of (a) **M1,** (b) **M2,** (c) **P1,** and (d) **P2.** Insets show the enlarged image of oxidation peaks in the anodic region.

$$E_{HOMO} = -[E^{ox}_{onset} + 4.8eV - E_{FOC}] \tag{13.1}$$

where E^{ox}_{onset} and E_{FOC} are the onset oxidation potentials of the compounds and ferrocene, respectively, and −4.8 eV is the HOMO energy level of ferrocene against vacuum.

$$E_{LUMO} = E_{HOMO} + E_g \tag{13.2}$$

The observed onsets of oxidation potentials are at 0.82, 0.99, 0.55 and 0.91 V for **M1, M2, P1,** and **P2**, respectively, using which the energies of HOMO levels were calculated, and are found to be −5.35, −5.52, −5.08, and −5.44 eV and the energies of LUMO levels are −2.69, −2.78, −2.30, and −2.89 eV for **M1, M2, P1,** and **P2**, respectively. The higher HOMO of molecule **M1** compared to M2 is due to the presence of higher electron donating thiophene in it. Similarly, the polymer **P1** possesses comparatively higher HOMO energy level than that of **P2** due to the higher electron donating ability of thiophene and absence of cyano group in **P1**. The electrochemical data are summarized in Table 13.1.

Figure 13.10a–d depict the cyclic voltammograms of **M3, M4, M5,** and **M6,** respectively. The onsets of first oxidation peaks are observed at 0.77, 0.80, 0.71, and 0.82 V vs SCE for **M3, M4, M5,** and **M6,** respectively. Multiple oxidation peaks can be seen in the figure which arise due to the multistep oxidation process resulting from the oxidation of different units present in the molecules. For example, the peak between 0.8 and 1.5 V in **M3** and **M5** could be due to the oxidation of the triphenylamine unit. Similarly, the oxidation peak at ~1.2 V in **M4** and **M6** may be due to the oxidation of N-substituted carbazole units. The oxidation peaks observed at ~1.67–1.7 V could be associated with oxidation of vinylene bonds in all the four molecules. The HOMO energy of **M3, M4, M5,** and **M6** calculated from Equation 13.1 is −5.31, −5.34, −5.25, and −5.36 eV, respectively. Similarly, the LUMO energy of **M3, M4, M5,** and **M6** determined from the Equation 13.2 is −2.67,

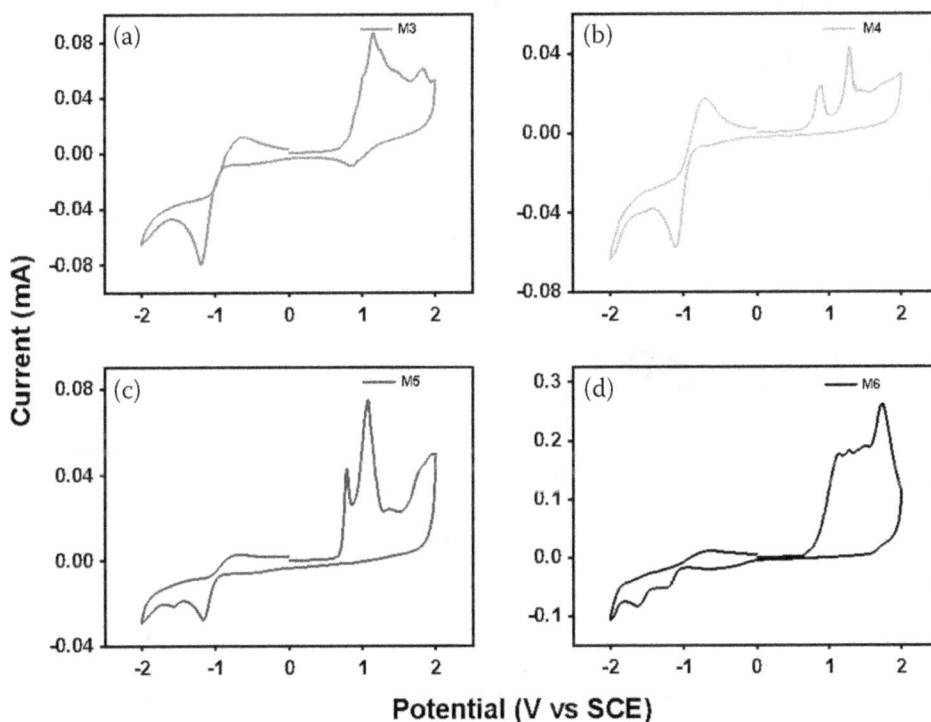

FIGURE 13.10 Cyclic voltammograms of (a) **M3,** (b) **M4,** (c) **M5,** and (d) **M6.**

−2.63, −2.63, and −2.60 eV, respectively. As a result of the presence of stronger electron donating triphenylamine group, molecules **M3** and **M5** possess a lower oxidation potential compared to carbazole containing **M4** and **M6** which shifts the HOMO level to higher energy in **M3** and **M5** compared to that of **M4** and **M6**. The summary of electrochemical data is given in Table 13.1.

13.9 THEORETICAL STUDIES

The density functional theory (DFT) and time-dependent density functional theory (TD-DFT) are two widely used theoretical methods to predict the ground and the excited state properties as well as to understand the spatial arrangement and the distribution of electrons in the molecules. These computational studies help to rationalize the relationship between the molecular structure and device performance (Martsinovich & Troisi, 2011). They are also useful in designing new organic materials for achieving enhanced device performances.

The distribution of electrons and the electronic structure of the synthesized compounds were studied by DFT calculations. The electronic structure of **M1**, **M2**, and two dimers simulating the structure of the polymers **P1** and **P2** were determined using tools of DFT as implemented in Gaussian 09 package using the B3LYP hybrid functional with 6-31 G(d,p) basis set. Figure 13.11 shows the optimized geometry, HOMO and LUMO energy levels of **M1**, **M2**, **P1**, and **P2**. The decyloxy chain was considered as a methoxy group for the easy calculation. As shown in Figure 13.11, the decyloxy substituted phenyl group is twisted by an angle of ~30° from the thiophene–pyridine plane, except this, there is a uniform distribution of HOMO and LUMO orbitals over central pyridine, spacer thiophene, cyanovinylene, and on the peripheral groups in **M1** and **M2** due to the structural planarity. This spatial distribution suggests the $\pi–\pi^*$ type transition in **M1** and **M2**, whereas, in the case of **P1** the HOMO is distributed majorly on thiophene and vinylene linkage present in the right wing but distributed partially on thiophene and vinylene linkage present in the left wing, while, the LUMO is localized majorly on central pyridine, thiophene, and the vinylene linkage present in the right wing. This type of distribution may be

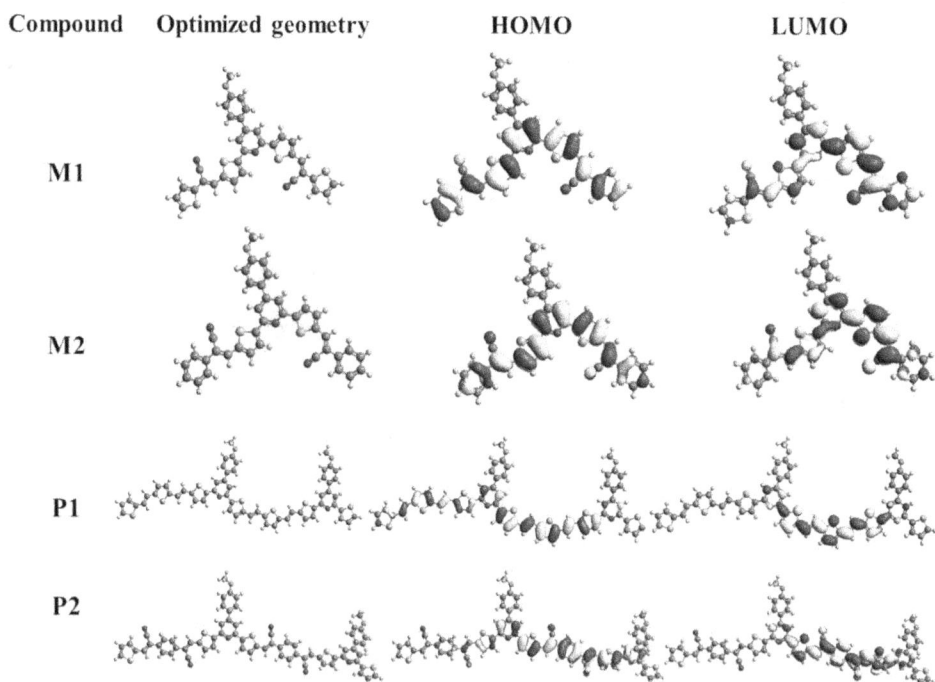

| Compound | Optimized geometry | HOMO | LUMO |

M1

M2

P1

P2

FIGURE 13.11 Optimized ground state geometry and HOMO/LUMO distribution of **M1, M2, P1,** and **P2**.

attributed to the twisting of the polymeric chain from the planarity. Similarly, there can be seen a major distribution of HOMO on thiophene and 1,4-phenyleneacrylonitrile group present in the right wing and a minor distribution on central pyridine and thiophene present in the left wing in **P2**. As a result of twisting of 1,4-phenyleneacrylonitrile group in the left wing, the HOMO distribution is restricted to thiophene unit, while the LUMO in **P2** is majorly localized on central pyridine, thiophene and 1,4-phenyleneacrylonitrile group present in the right wing. Further, the TD-DFT calculations were carried out to study the excited state electronic transitions. Structural optimizations were done on various conformational isomers (dimers) of **P1** and one simple structure (monomer, **P1f**). Of the five conformations **P1a**, **P1b**, **P1c**, **P1d**, and **P1e**, **P1e** is of lowest energy conformation (Figure 13.13). The simple structure (**P1f**) cannot be compared in energy to the others as it has a smaller number of total atoms. Similarly, structural optimizations were performed on **P2** (dimer) and one simple structure (monomer, **P2f**). Simulated absorption spectra were taken for all the conformers along with **M1** and **M2**. Out of all the five conformers of **P1**, **P1e** exhibits a closer agreement with experimental results. However, the experimental absorption spectrum of **P1** can be described better with the electronic structure of the monomer **P1f**. This latter fact suggests that for a more accurate simulation of the absorption spectrum of **P1** it is necessary to add in a larger number of units including several conformations. Similarly, simulated spectra of **P2** and **P2f** showed good results in terms of comparison with experimental data and it seems that extending the number of units shifts the spectrum to a longer wavelength region. Figure 13.12 shows the predicted spectra for **M1**, **M2**, and for the most stable conformations (dimer) and monomer of **P1** and **P2** compared to the experimental solution spectrum.

Figure 13.14 depicts the optimized geometry in the ground state and distribution of electrons in HOMO and LUMO of **M3, M4, M5,** and **M6**. In order to reduce the calculation time the methoxy group was considered instead of decyloxy group. From the figure, it is observed that HOMO is majorly located on spacer thiophene and vinylic linkage with no distribution on decyloxy substituted phenyl group which is situated at angle of ~41° from the thiophene-pyridine plane in **M3** and **M4**. The HOMO is further extended on peripheral triphenylamine i.e., mainly on the nitrogen atom and phenyl group attached to the vinylic bond in **M3** and ethyl carbazole in **M4**. Further, HOMO is partially spread over central pyridine (Figure 13.14). On the other hand, the LUMO is localized to a higher extent on central pyridine and spacer thiophene units; and lesser extent over vinylic linkage in **M3** and **M4**. Further, minor distribution of LUMO can be seen on the substituted phenyl group of triphenylamine and ethyl carbazole in **M3** and **M4**, respectively. Whereas in **M5** and **M6**, the major distribution of HOMO was observed on spacer thiophene, vinylic linkage, central triphenylamine in **M5** and on central carbazole in **M6**, while major distribution of LUMO was observed on electron acceptor pyridine ring, spacer thiophene, and vinylic linkage that is further extended over substituted phenyl rings of triphenylamine and carbazole unit in **M5** and **M6**, respectively. From the DFT study it is analyzed that HOMO is restricted on electron donors while LUMO is mainly localized on electron withdrawing pyridine units, this type of distribution suggests the D-A type charge transfer in the molecules. Also, from the electron density distribution of HOMO and LUMO levels, a significant orbital overlap is observed which suggests the ambipolar property, which lead for the proper charge transport which helps to maintain the active electronic communication between the donor and the acceptor (C. Li et al., 2018; C.-K. Liu et al., 2018).

In order to examine the excited state transition, TD-DFT calculations were performed. The simulated spectra of **M3, M4, M5,** and **M6** are shown in Figure 13.15. The appearance of distinct bands at higher energy and lower energy region indicates the existence of π-π* and charge-transfer transition, respectively.

13.10 THERMAL PROPERTIES

For the material to use as a potent candidate in optoelectronic devices, it should be thermally stable. The thermogravimetric analysis (TGA) was used to determine the thermal stability of the material

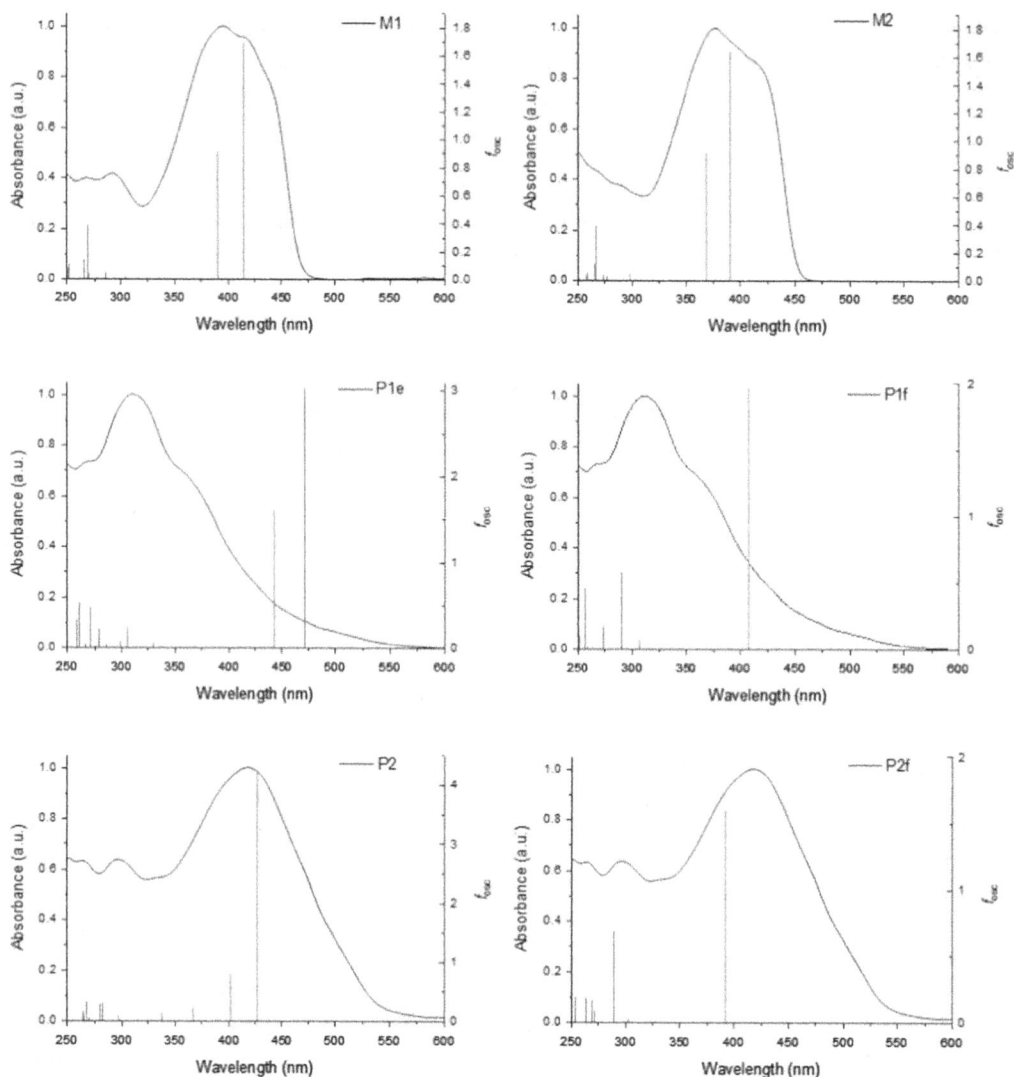

FIGURE 13.12 Predicted spectra (lines) for **M1**, **M2**, and, most stable conformations (dimers) and monomers of **P1** and **P2** compared to the experimental solution spectrum.

which measures the amount of weight change of the material as a function of temperature. It mainly provides information about the phase transitions and the thermal decompositions of the materials.

As it can be seen in Figure 13.16, both **M1** and **M2** are stable up to a temperature of ~350°C without any weight loss, and there can be seen abrupt weight loss with complete decomposition of both the molecules after 350°C (Figure 13.16). The onsets of decomposition temperature (T_d) corresponding to **M1** and **M2** are 347°C and 369°C, respectively. The higher T_d of **M2** compared to that of **M1** is due to the presence of a more stable 2-phenylacrylonitrile unit in **M2**. Similarly, the onset of T_d for **P1** is observed at 260°C. The weight loss at 76°C corresponds to the moisture and that at 260°C is due to the degradation of polymer backbone, whereas **P2** is thermally stable up to 384°C and there observed a complete decomposition of polymer backbone with a sharp weight loss above 384°C. Since 1,4-phenylacrylonitrile is more stable than thiophene, the presence of 1,4-phenylacrylonitrile in **P2** is fetching the higher stability to **P2** compared to that of **P1**, as in the case of **M2**. The presence of cyanovinylene moiety in **M1**, **M2,** and **P2** makes them thermally stable

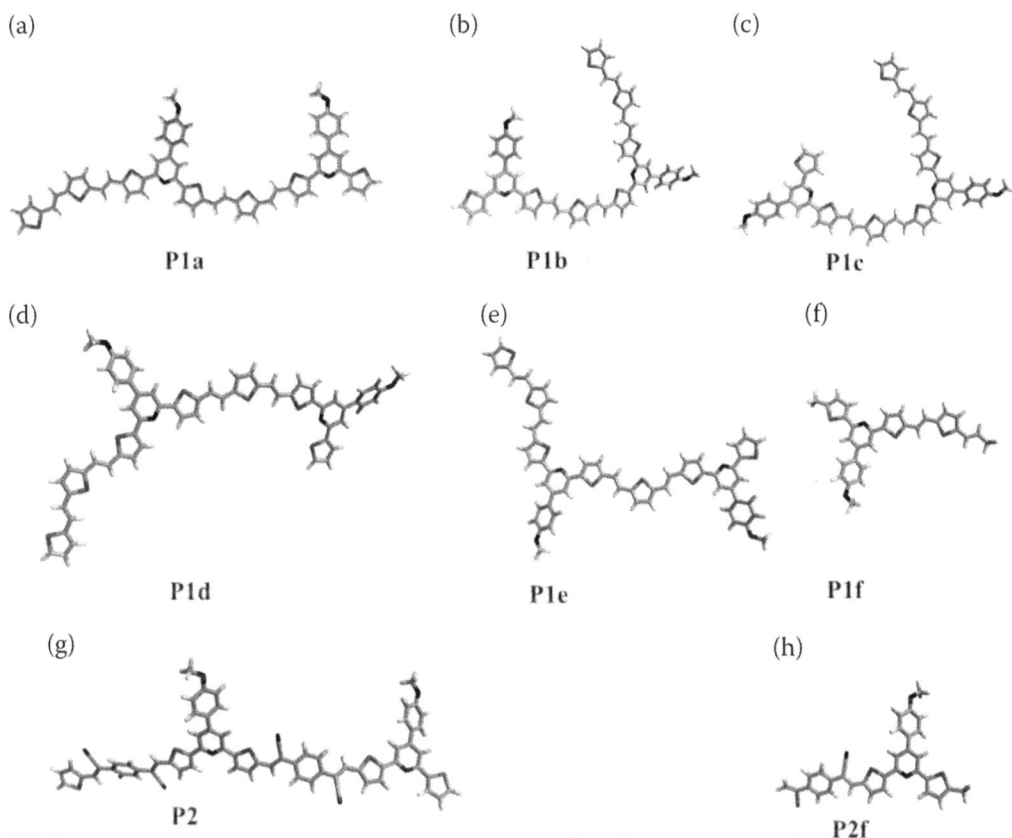

FIGURE 13.13 (a)–(e) The optimized geometry of **P1** dimer in conformations to e, (f) optimized geometry of **P1** monomer, (g) optimized geometry of **P2** dimer, (h) optimized geometry of **P2** monomer.

(Park et al., 2016) and as a result of which they exhibit higher T_d than that of **P1**. The TGA results reveal that the materials are stable enough to be used in photovoltaics/optoelectronics.

The TGA plots of **M3, M4, M5,** and **M6** are depicted in Figure 13.17 which suggests that the molecules possess high degradation temperatures and extremely good thermal stabilities. The T_d observed at 426°C and 432°C (at 5% weight loss) correspond to **M3** and **M4**, respectively, whereas, the T_d observed at 417°C (at 20% weight loss) and 428°C (at 10% weight loss), correspond to **M5** and **M6**, respectively. In addition, minor degradation observed in **M5** and **M6** at 218°C and 240°C (2% weight loss), respectively, is due to the decomposition of peripheral (4-(4-(decyloxy)phenyl)-2,6-di(thiophen-2-yl)pyridine) unit present in them. Further, the higher T_d of **M4** and **M6** compared to that of triphenylamine containing **M3** and **M5** is mainly due to the presence of structurally more rigid carbazole units in them. The excellent thermal stability exhibited by all the four molecules is attributed to their large molecular weights with the bulky D–π–A–π–D/ D–A–π–D–π–A–D structural arrangement (Wang et al., 2019), which can lead to an amorphous and homogeneous film and thus, improve the quality of the film at high temperatures, which is essential for their practical use in laser photonics.

13.11 NONLINEAR OPTICS (NLO)

The contributions of NLO to the entirety of science and especially to our daily affairs, are praiseworthy. So far almost nine Nobel Prizes awarded in physics and chemistry owe credit to NLO

Molecules	Optimized geometry	HOMO	LUMO

M3

M4

M5

M6

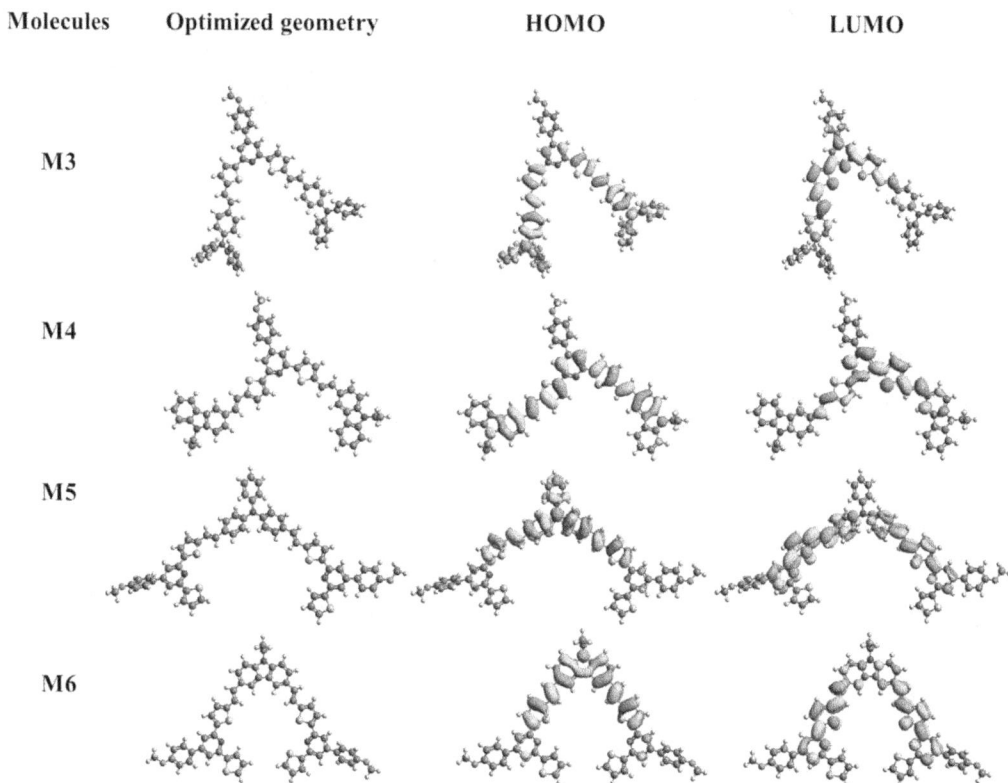

FIGURE 13.14   The optimized molecular structures and HOMO/LUMO distributions of **M3, M4, M5,** and **M6.**

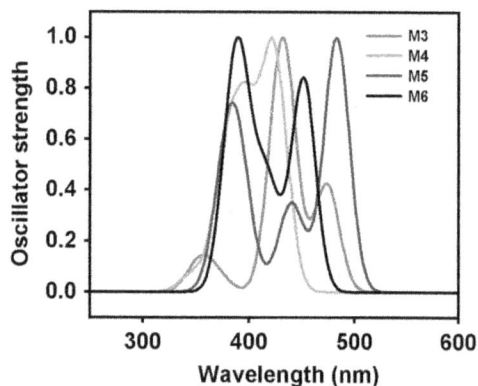

FIGURE 13.15   Simulated absorption spectra of **M3, M4, M5,** and **M6.**

(Garmire, 2013). This field has been at its height since 1961, after the invention of the laser by Theodore Maiman in 1960 (Bloembergen, 1984; Maiman, 1960) followed by the observation of second harmonic generation (SHG) by Franken and coworkers in quartz, a year later. Later on, NLO contributed in diversifying laser, light-material interaction and information technology. Various organic and inorganic systems such as polymers, dyes, semiconductors, liquid crystals, nanoparticles, nanocrystals, nanocomposites, gases, plasma, etc., are being investigated for several NLO applications. Most of the NLO applications are still in their infancy owing to the lack of ideal materials displaying ease in processability and an ability to interface with other materials.

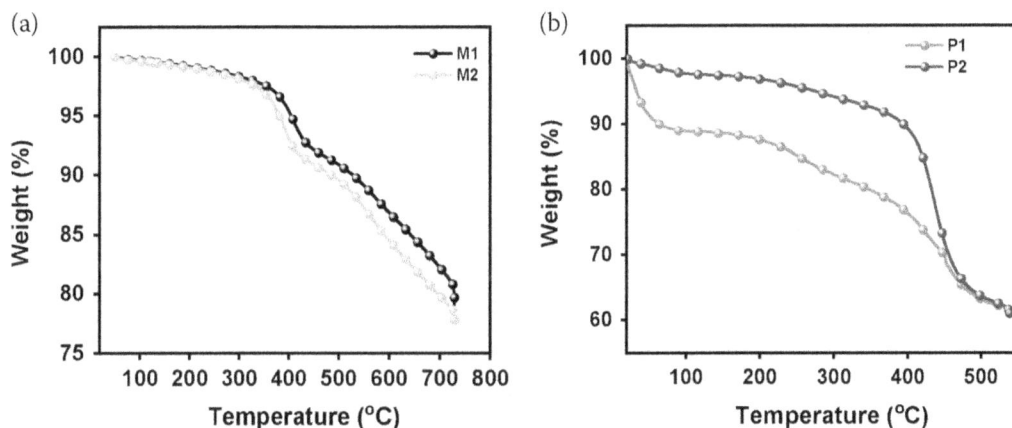

FIGURE 13.16 TGA plots of (a) **M1** and **M2**, (b) **P1** and **P2**.

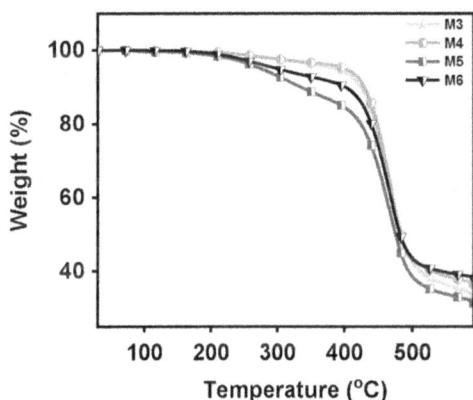

FIGURE 13.17 TGA plots of **M3, M4, M5,** and **M6.**

Researchers in NLO are keenly in search of ideal materials with requisite specifications so as to furnish the needs of the optical industry and research.

13.11.1 Conjugated Materials for NLO

In 1991, Cheng et al. reported the nature of π–conjugated bonds in a series of benzene and stilbene derivatives towards nonlinearity (L. T. Cheng et al., 1991). The three main basic requirements for organic materials to exhibit NLO property are (i) polarizability (electrons need to be greatly perturbed from their equilibrium positions), (ii) acentric crystal packing, and (iii) asymmetric charge distribution (incorporation of D–A molecules). Further, the materials must possess (a) large oscillator strength for electronic transitions from the ground state to the excited state, (b) excited states close in energy to the ground state, and (c) a large difference between the ground and excited state dipole moments. As D–A π–systems allow charge transfer between electron donating and electron withdrawing moieties through π–electron delocalization, which increases the molecular polarizability, making the organic materials superior to inorganics both in the speed of response and in the magnitude of the third-order effect, many of them are studied for NLO properties and device applications. So far, many conjugated organic molecules are recognized as the key components in emerging electronic and photonic technologies like optical computing, optical limiters for sensors and eye protection, telecommunication etc, due to their ease and low cost of synthesis, high laser damage thresholds, large optical nonlinearity, and short response time (Dalton et al., 2009; J. Liu et al., 2015; Rajeshirke et al., 2018; Thakare et al., 2017). Though significant advances have already been established in this field, there is still a scope for improving the third-order nonlinear absorption and hence, the limiting action of the molecules. In this regard, search for efficient NLO materials has been gaining much interest in recent years.

13.11.2 THIRD-ORDER NONLINEARITY

The third-order nonlinear optical susceptibility ($\chi^{(3)}$) is responsible for various third-order nonlinear optical phenomena. Though $\chi^{(3)}$ is a general phenomenon, it strongly depends on the material property and input intensity. The real part of $\chi^{(3)}$ is directly related to nonlinear refraction (NLR) and the imaginary part is related to nonlinear absorption (NLA). Some of the major phenomena associated with $\chi^{(3)}$ are discussed below.

13.11.2.1 Nonlinear Absorption (NLA) and Optical Limiting (OL)

Various phenomena responsible for the NLA property exhibited by the optical materials are: (i) saturable absorption (SA) wherein the transmittance increases with the increase in optical intensity, (ii) reverse saturable absorption (RSA), (iii) two-photon absorption (2PA), and (iv) multiphoton absorption (MPA). In RSA, 2PA and MPA the transmittance decreases with the increase of optical intensity. These properties make the optical materials promising in various applications of science and technology. For example, SA materials are used in lasers as Q-switching elements, and 2PA, 3PA, MPA, and RSA materials are used in two-photon microscopy and optical limiters.

Optical limiting i.e., controlling the intensity of light in a stable and foreseeable way is the most fundamental innovation in photonic technology, which has wide applications in the field of optical communications and optical computing (Chen 2001; Hernandez et al. 2000; Sutherland, 1989). The optical limiting is the phenomenon of decrease in the transmittance of a nonlinear material on increasing the intensity of the incident optical beam. A perfect optical limiter is the one which exhibits constant transmittance at low input fluence and reduced transmittance at high input fluence. The OL data are extracted from the graphs of normalized transmittance obtained from NLA analysis against input fluence. Optical limiters can be exploited in applications such as pulse shaping, pulse smoothing, pulse compression, laser pulse regulation, mode locking, etc. However, their main application is to safeguard the sensors, detectors and human eyes from the high-intensity light (Anand et al., 2011).

The mechanism of number of NLA processes can be represented in an energy level diagram as shown in Figure 13.18. In SA material, intense absorption at the excitation wavelength takes place, the ground state population depletes significantly and finally leads to the saturation of absorption. The absorption cross section for ground state (σ_g) is higher than that of the excited state (σ_e) (Figure 13.18). The 2PA occurs when two photons which are of same or different energy are absorbed simultaneously from the ground state to a higher excited state. And when molecules are excited from an already excited state to a higher excited state, it is denoted as excited state absorption (ESA). For these types of transitions, the σ_{e1} should be significantly greater than σ_g (Figure 13.18).

13.11.2.2 Nonlinear Refraction (NLR)

When a Gaussian profiled beam propagates through the medium, besides NLA it can also undergo phase distortions. The spatial and temporal characteristics of a propagating beam depend largely on the material property and the intensity of light. One of the important properties i.e., the NLR. The third-order NLR plays a crucial role in mode-locking, wave-mixing, self-phase modulation, phase conjugation, photo refractivity, nonlinear waveguides and interfaces, spatial solitons, and optical bistability.

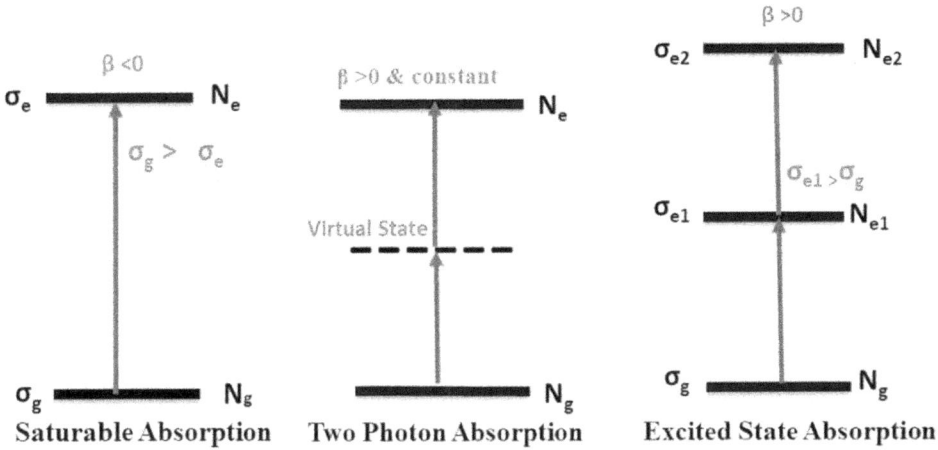

FIGURE 13.18 SA, TPA and RSA in multi-level system.

13.11.3 Z–Scan

Among various experimental techniques, Z-scan, introduced by Sheik Bahae and his coworkers in 1990 (Mansoor Sheik-Bahae et al., 1990), is a simple, sensitive and widely used method for determining the third-order NLO parameters. Both the magnitude and the sign of NLA and NLR coefficients can be figured out from the experiment.

13.11.3.1 Experimental Setup

A frequency doubled Q switched Nd:YAG laser (Quanta–Ray INDI–40) operating at 532 nm wavelength, 7 ns pulse width, and 10 Hz repeating rate excitation source was used for the Z–scan experiment. Using a beam splitter, the laser beam was split into a reference beam and the sample beam, which was passed through the sample taken in a 1 mm thick quartz cuvette through a convex lens of focal length 150 mm. The sample cuvette was placed on a computer controlled translational stage and was moved along the direction of the laser beam in z–direction from +z to −z about 20 mm on either side of the focus of the lens in predetermined steps (1,000 microns). The convex lens was adjusted in such a way that the laser beam provides maximum energy to the sample at the focus which then equally decreases on the either side of the focus. The beam waist at the focus and Rayleigh range of the beam were estimated to be 17.56 μm and 1.82 mm, respectively. During the CA scan, an aperture of 3 mm diameter was placed in front of the detector. Both the reference beam and the transmitted beam from the sample were detected using two identical pyroelectric detectors (RjP–735, laser Probe Inc, USA) and were collected in the energy meter (Rj–7620, Laser Probe, Inc, USA). Since the sample thickness was much smaller than Rayleigh range, the experiment was carried out by adopting a thin sample approximation method (M. Sheik-Bahae et al., 1990). The schematic representation of the Z–scan setup is shown in Figure 13.19.

Using open aperture (OA) Z–scan analysis the NLA and optical limiting parameters were obtained, while, using closed aperture (CA) Z–scan analysis the NLR parameters were recorded. The optical transmittance of the sample in the OA Z–scan analysis was recorded as a function of input intensity. The output transmittance from the sample was plotted against the position of the samples which is known as the z-scan curve. The nature of the nonlinearity: RSA, 2PA, 3PA, SA, and so on will be obtained by the shape of the z-scan curve. Theoretical fitting of the experimental data provides the values of the parameters such as NLA coefficient (β) and saturation intensity (I_s) of the sample. All the experiments were carried out in the "single shot" mode. To avoid thermal effects in the sample, appropriate time intervals were provided between successive pulses. The

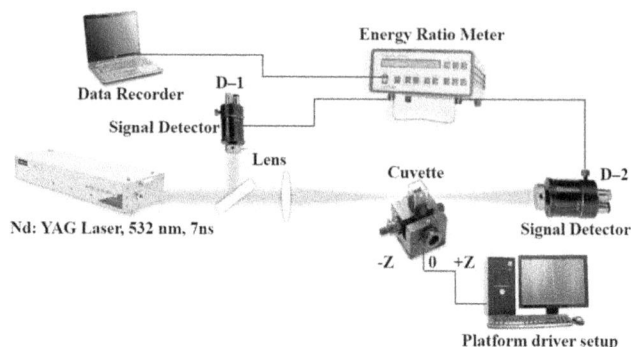

FIGURE 13.19 Schematic representation of Z–scan experimental setup.

sample was taken in 1 mm cuvette wherein the linear transmittance was fixed between 50%–85% at the excitation wavelength for all the molecules.

13.12 THIRD-ORDER NLO STUDIES

13.12.1 NLA STUDIES

The NLA of the compounds under OA condition was determined using ultra-short 50 μJ laser excitation (532 nm, 7 ns pulse width), corresponding to on–axis intensity of 1.386 GW cm^{-2}. The linear transmittance was set at 60%–65% for **M1, M2, P1,** and **P2.** The OA Z–scan curves obtained from **M1, M2, P1,** and **P2** in CHCl$_3$ are shown in Figure 13.20a–d, respectively. When the laser beam interacts with the compounds, there is a gradual decrease in the transmittance towards the focal point which reaches a minimum with the deep transmittance trough at the focus (z = 0), producing a valley pattern i.e., the curves are symmetric at z = 0, which signifies the RSA behavior of the compounds with the positive NLA of the incident light. This RSA is combined with both TPA and ESA in the nanosecond time scale, which is collectively called "the effective TPA" process (Vishnumurthy et al., 2011). The obtained experimental data are fitted to the theoretical model for the ESA-assisted TPA process.

The NLA coefficient (β_{eff}) of OA Z–scan recordings were determined by theoretically fitting to Equation 13.3 and is given by:

$$\alpha(I) = \frac{\alpha_o}{1 + I/I_s} + \beta_{eff} I \qquad (13.3)$$

where α_0 is the linear absorption coefficient, I is the incident laser intensity, I_s is the saturation intensity, and β_{eff} is the NLA coefficient associated to the RSA response (Divyasree et al., 2018).

Pulse propagation equation to calculate β_{eff} is given by:

$$\frac{dI}{dz} = -\left(\frac{\alpha_o}{1 + \frac{I}{I_s}}\right) I - \beta_{eff} I^2 \qquad (13.4)$$

where z is the propagation distance within the sample. The first term in Equation (13.4) expresses the SA and the next term indicates the effective TPA part.

The normalized transmittance (T) is given by:

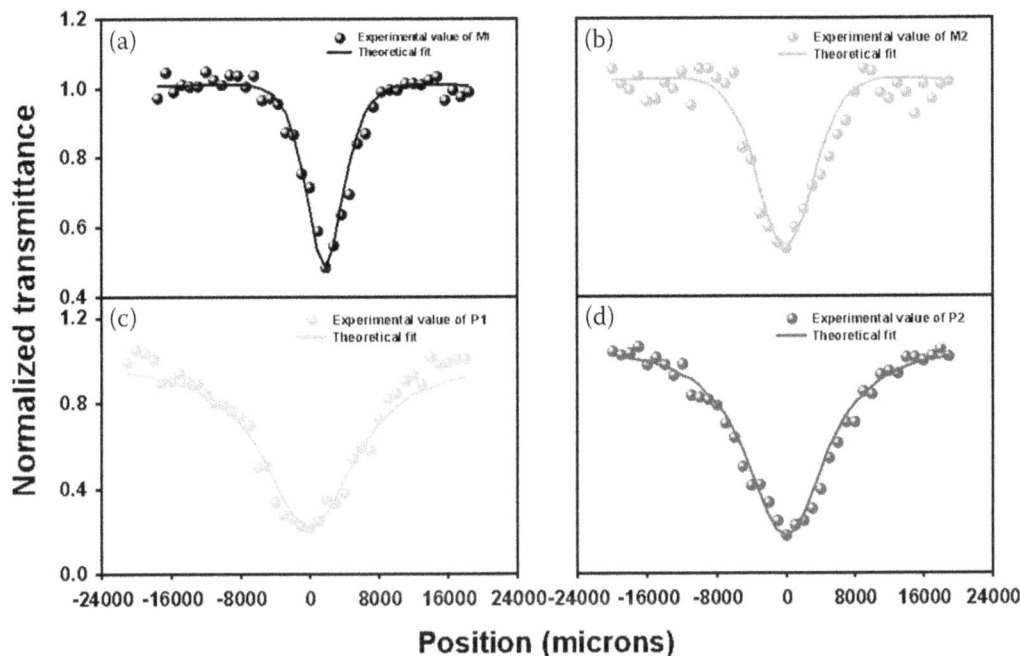

FIGURE 13.20 Z–scan curves of (a) **M1**, (b) **M2**, (c) **P1**, and (d) **P2** under OA configuration.

$$T(z) = \frac{1}{\sqrt{\pi} \, q_0(z)} \int_{-\infty}^{\infty} \ln\left[1 + q_0(z)e^{-\tau^2}\right] d\tau \tag{13.5}$$

where $q_0(z) = \beta_{eff} I_0 L_{eff} / \left(1 + \frac{z^2}{z_0^2}\right)$ and $L_{eff} = (1 - e^{-\alpha_o L})/\alpha_0$

L_{eff} is the effective sample length, L is the sample length, α_0 is the unsaturated linear absorption coefficient, z is the position of the sample, $z_o = \pi \omega_0^2 / \lambda$ is the Raleigh range, ω_0 is the beam waist radius at the focal point, and λ is the wavelength of the laser beam.

The imaginary part of the third-order nonlinear susceptibility $(\chi^{(3)})$ is given by:

$$\operatorname{Im} \chi^{(3)} = n_o^2 \varepsilon_0 c \lambda \beta_{eff} / 2\pi \tag{13.6}$$

where n_0 is the linear refractive index, c is the speed of light, and ε_0 is the permittivity of free space.

The experimental data were fitted to the theoretical model using the Equations 13.3 and 13.5 to obtain NLA parameters (Table 13.2).

The calculated β_{eff} values for **M1, M2, P1,** and **P2** are 1.48×10^{-10}, 2.08×10^{-10}, 6.12×10^{-10}, and 7.02×10^{-10} m W^{-1}, respectively.

The linear transmittance for **M3, M4, M5,** and **M6** was maintained at 70%–72%. The OA Z–scan curves (in CHCl$_3$ solution) of **M3, M4, M5,** and **M6** are shown in Figure 13.21a–d, respectively. The results of OA Z–scan fitted using Equations 13.3–13.6 disclose the improved NLO responses of the synthesized molecules. **M3, M4, M5,** and **M6** possess β_{eff} values of 4.45, 3.65, 3.91, and 2.58×10^{-10} m W^{-1}, respectively. The obtained β_{eff} values of all the synthesized compounds in the present work are close to and predominantly better than some of the reported materials in the literature (Table 13.2).

TABLE 13.2

Comparison of NLA and NLR Parameters of Synthesized Compounds with Similar Results Reported

Sample	β_{eff} (×10^{-10} mW^{-1})	Limiting Threshold (Jcm^{-2})	n_2^I (esu)	$\chi^{(3)}$ (esu) (10^{-12})	References
M1	1.48	8.34	–	4.67596	This work
M2	2.08	4.16	–	6.56524	This work
P1	6.12	1.69	–	19.2738	This work
P2	7.02	1.42	–	22.1203	This work
M3	4.45	1.73	7.06357×10^{-11}	14.0157	This work
M4	3.65	2.72	6.21902×10^{-11}	11.5239	This work
M5	3.91	2.68	7.63576×10^{-11}	12.3421	This work
M6	2.58	4.68	4.78997×10^{-11}	8.1398	This work
	10	2.46	–	–	(Gopi et al., 2020)
	11.6	1.03	–	32.331	(Vintu et al., 2019)
5,10-A$_2$B$_2$ porphyrin–SnCl$_2$_CCTMS complex	6.4	–	–	–	(Zawadzka et al., 2013)
7CB	2.0	–	–	7.42	(Praseetha et al., 2019)
7CB + 0.75% CdSe QD	5.5	3.02	–	20.4	(Praseetha et al., 2019)

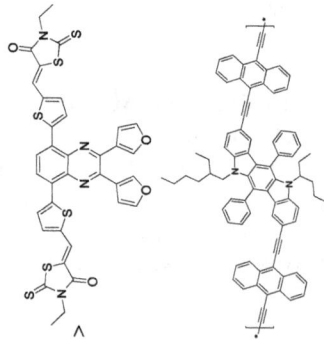

7CB + 1.0% CdSe QD	7.8	2.1	–	28.9	(Praseetha et al., 2019)
Phenothiazine (PTZ)–silver (Ag) organometallic hybrid system	3.60	3.43	8.0	–	(Shiju et al., 2019)
Graphene oxide–Fe_3O_4	2.6	2.82	–	–	(Zhang et al., 2011)
	2.0	1.15	–	–	(Gopi et al., 2020)
Zn-complexed D–π–A type porphyrin derivative, (Por–Zn–N)	4.8	–	–	–	(Mi et al., 2016)

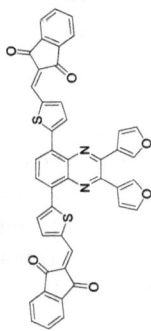

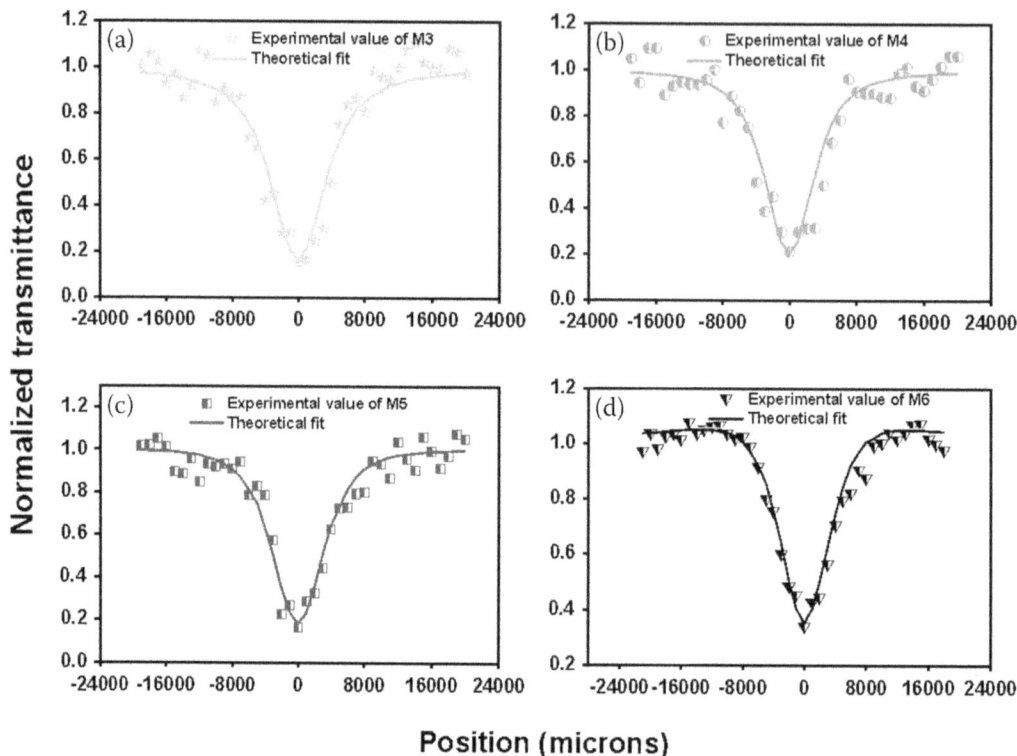

FIGURE 13.21   Z–scan curves of (a) **M3**, (b) **M4**, (c) **M5**, and (d) **M6** under OA configuration.

13.12.2   OPTICAL LIMITING STUDIES

The graphs of normalized transmittance (y-axis) obtained from OA Z–scan analysis plotted against input fluence (x-axis) give the OL data of synthesized compounds. The onsets of limiting action (the value of input fluence at which the intensity of output transmittance starts decreasing) are observed at 1.11, 0.83, 0.22 and 0.16 J cm^{-2} and the limiting threshold (LT) values (the value of input fluence at which the intensity of output pulse becomes 50% of the initial value) are 8.34, 4.16, 1.69, and 1.42 J cm^{-2} for **M1**, **M2**, **P1**, and **P2**, respectively (Figure 13.22a–d). From the results it is observed that compared to **M1** and **M2** the polymers **P1** and **P2** possess remarkably low onset and LT values. The obtained values are closer and better than some of the similar materials (Table 13.2). For an instance, Gopi et al. (Gopi et al., 2020) studied the NLO properties of a new solution processable push–pull molecule based on quinoxaline with the 3-ethyl rhodanine as acceptor unit. The molecule exhibited an effective TPA property with high β_{eff} value of 10.0 × 10^{-10} m W^{-1} and the LT of 2.46 J cm^{-2}. A novel anthracene supported 5,11-dihydroindolo [3,2-*b*] carbazole- based polymer was synthesized by Vintu et al. (Vintu et al., 2019) and studied the NLO properties for the practical application, which showed a very high β_{eff} of 11.6 × 10^{-10} m W^{-1} with the LT of 1.03 J cm^{-2}. The NLO properties of two 5,10-A$_2$B$_2$ porphyrin series supported by different metals in the core were studied by Zawadzka et al. (Zawadzka et al., 2013). The porphyrin-tin (IV) complex, SnCl$_2$_CCTMS, exhibited RSA behavior with the high β_{eff} of 6.4 × 10^{-10} m W^{-1}. Upon doping with CdSe quantum dot (QD) in different concentrations a substantial enhancement in the NLO property of 4′-Heptyl-4-biphenylcarbonitrile (7CB) nematic liquid crystal was observed by Praseethae al. (Praseetha et al., 2019). The pristine 7CB nematic liquid crystal possessed β_{eff} of 2.0 × 10^{-10} m W^{-1} while after doping with 0.75 and 1% of CdSe QD, there

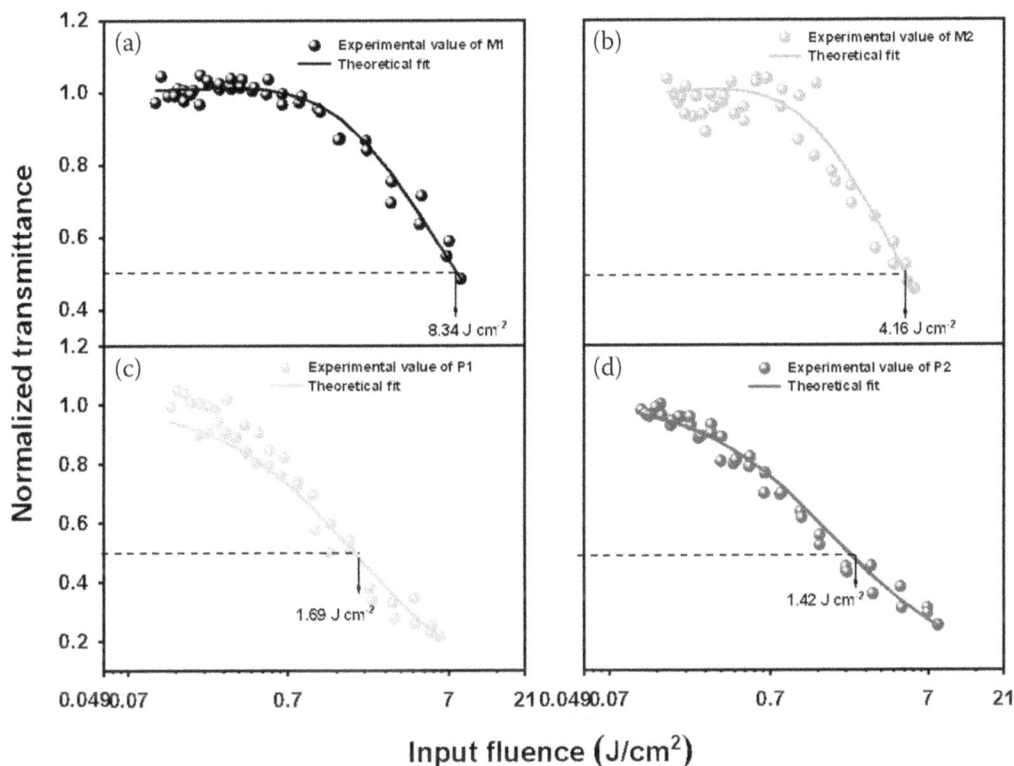

FIGURE 13.22 Optical limiting curves of (a) **M1**, (b) **M2**, (c) **P1**, and (d) **P2** at an input intensity of 1.386 GW cm^{-2}.

observed a raise in β_{eff} to 5.5×10^{-10} m W^{-1} with the LT of 3.02 J cm^{-2} and 7.8×10^{-10} m W^{-1} with the LT of 2.1 J cm^{-2}, respectively.

The enhancement in the optical nonlinearity of **M1, M2, P1,** and **P2** is solely resulting from the conjugation. For example, the higher NLO property of **M2** and **P2** is mainly due to the presence of phenyl acrylonitrile moiety, which extends the conjugation in **M2** and in **P2** compared to that of thiophene acrylonitrile in **M1** and thiophene alone in **P1**. Moreover, the polymers **P1** and **P2** show predominant increment of β_{eff} and substantial reduction in optical limiting values compared to that of molecules **M1** and **M2** that is solely from the higher extension of conjugation, the improved interaction between the electron donor and acceptor units, the increased ICT and the polarizability in **P1** and **P2**. Unfortunately, **M1, M2, P1,** and **P2** did not show NLR properties as the NLA was much stronger than nonlinear refraction of the materials.

The onsets of limiting action of **M3, M4, M5,** and **M6** are at 0.18, 0.49, 0.38, and 0.65 J cm^{-2} and the LT values are 1.73, 2.72, 2.68, and 4.68 J cm^{-2} for, respectively (Figure 13.23). The obtained very low optical limiting values and LT values of **M3, M4, M5,** and **M6** are better than and analogous to recently reported values (Table 13.2). For an instance, **M2**, synthesized in the previous set, showed excellent NLO properties with β_{eff} and LT of 2.08×10^{-10} m W^{-1} and 4.16 J cm^{-2}, respectively, under similar conditions. Shiju et al. (Shiju et al., 2019) observed nearly two orders of higher β_{eff} (3.60×10^{-10} m W^{-1}) and LT (3.43 J cm^{-2}) of phenothiazine (PTZ)–silver (Ag) organometallic hybrid system than pristine PTZ. The graphene oxide–Fe$_3$O$_4$ hybrid material exhibited an enhanced β_{eff} of 2.6×10^{-10} m W^{-1} with the LT of 2.82 J cm^{-2}, which was studied by Zhang et al. (Zhang et al., 2011). Gopi et al. (Gopi et al., 2020) observed an excellent TPA with very high β_{eff} of 2.0×10^{-10} m W^{-1} and low LT of 1.15 J cm^{-2} for quinoxaline based D–A

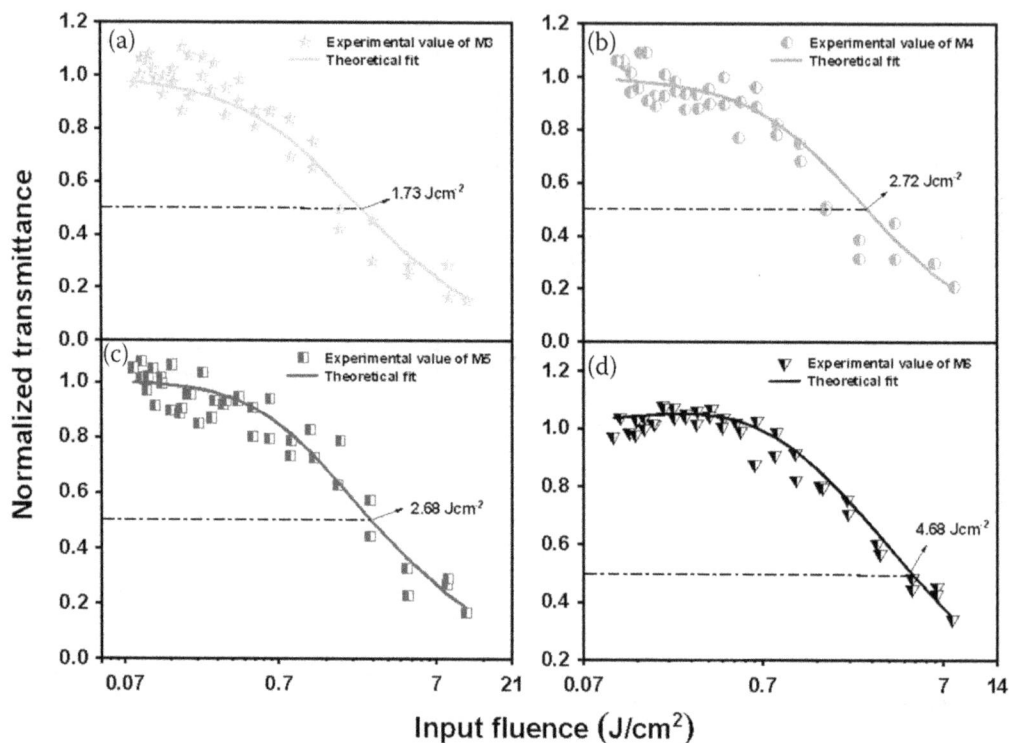

FIGURE 13.23 Optical limiting curves of (a) **M3**, (b) **M4**, (c) **M5**, and (d) **M6** at an input intensity of 1.386 GW cm^{-2}.

molecule using 1,3 indandione as electron acceptor. Further, Mi et al. (Mi et al., 2016) obtained a high β_{eff} of 4.8×10^{-10} m W^{-1} with RSA behavior for Zn-complexed D–π–A type porphyrin derivative, (Por–Zn–N). In the present study, the presence of strong electron donor and acceptor groups provides prolonged π–conjugation via the excellent ICT interactions which improves the structural planarity and also the polarization in the molecules. As a result, all the four molecules exhibit remarkably high β_{eff} value and extremely low LT values. Further, the prominent increment in the β_{eff} value and remarkable reduction in the LT value in **M3** and **M5** compared to that of **M4** and **M6** is mainly due to the presence of higher electron donating triphenylamine in them.

Though the structural arrangement of **M1** and **M2**, (molecules of previous set) as well as **M3** and **M4** of the present series are similar, the bulkier peripheral groups (triphenylamine/carbazole) in **M3** and **M4** leads to the improved ICT, extended conjugation, and increased polarity which leads to the enhanced NLO properties compared to the previous molecules. Moreover, the similar LT value of **M3** (1.73 J cm^{-2}) to that of polymer **P2** (1.42 J cm^{-2}) suggests that the proper modification in the structure and selection of specific donor and acceptor groups in small molecules could also lead to comparable NLO properties as that of polymers and could be a better choice instead of polymeric materials in optical limiting devices. Thus, these materials could act as effectual limiters which would safeguard the system exposed to high intensity sources.

13.12.3 NONLINEAR REFRACTION (NLR) STUDIES

The division of CA data by the corresponding OA Z–scan data yields pure NLR components. Figure 13.24a–d represent the NLR curves of **M3**, **M4**, **M5**, and **M6**, respectively. The negative nonlinearity was observed in **M3**, **M4**, **M5**, and **M6**, which is clearly seen by the signature curve

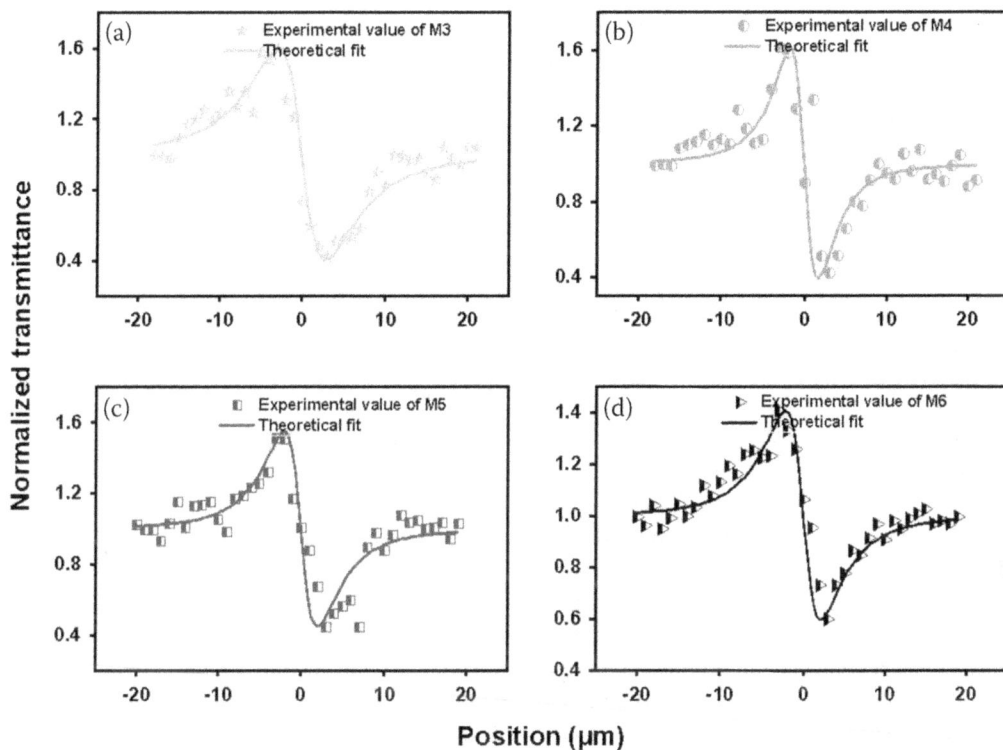

FIGURE 13.24 Z–scan curves of (a) **M3**, (b) **M4**, (c) **M5**, and (d) **M6** under CA configuration.

which consists of a prefocal peak followed by null and postfocal valley, indicating the negative sign of refractive index. Such types of curves represent the self-defocusing nature of the molecules, thus, molecules in the present study act like diverging lenses. By fitting experimental data to the Equation 13.7, the normalized transmittance (T) at CA condition was determined, and is given by:

$$T(z, \Delta\Phi_o) = 1 + \frac{4\left(\frac{z}{z_o}\right)}{\left(\left(\frac{z}{z_o}\right)^2 + 1\right)\left(\left(\frac{z}{z_o}\right)^2 + 9\right)}\Delta\Phi_o \qquad (13.7)$$

where T is the normalized transmittance, $\Delta\Phi_o$ is the on–axis nonlinear phase shift at the focus. Therefore, the nonlinear refractive index (n_2) (the NLR coefficient) is calculated by:

$$n_2(esu) = \frac{cn_0\lambda\Delta\Phi_o}{80\pi^2 I_o L_{eff}} \qquad (13.8)$$

The real $\chi^{(3)}$ is related to n_2 by the relation (in electrostatic unit):

$$\text{Real}\,\chi^{(3)} = 2n_o^2\varepsilon_o cn_2 \qquad (13.9)$$

The NLR parameters of **M3, M4, M5,** and **M6** are listed in Table 13.2.

From the reported literature it is observed that except some pure organic materials which showed improved nonlinearity without any doping, most of the organic compounds are doped with some inorganic materials and the enhancement in the nonlinearity is mostly due to the doping effect. However, the materials in the current work exhibited substantial increment in the β_{eff} of the order 10^{-10} m W^{-1} with prominent nonlinear response and remarkable optical limiting behavior with very low LT in their pristine form. The results are better than some of the well-known materials, which makes them a suitable replacement or better material for optical power limiting devices in laser photonics.

13.13 CONCLUSION

In conclusion, four novel D–A–D type trigonal-shaped nonlinear materials **M1, M2, P1,** and **P2** were designed and synthesized in one set. The opto-electronic properties of polymers were compared with small molecules. The substituted acrylonitrile group in **M1, M2,** and **P2** bath-ochromically shifted the absorption and emission maxima; and enhanced the thermal stability compared to that of **P1**. The lengthened conjugation in **P1** and **P2** raised the HOMO of **P1** and **P2** to higher energy region. The presence of phenylacrylonitrile unit resulted in the enhanced non-linear absorption in **M2** and **P2** compared to **M1** and **P1**. The higher degree of conjugation in **P1** and **P2** improved the electron donor and acceptor interaction, thus, the ICT and the polarizability of the polymers, which resulted in the predominant increment in the β_{eff} (6.12 × 10^{-10} and 7.02 × 10^{-10} m W^{-1} for **P1** and **P2**, respectively) and substantial reduction in optical limiting values (1.69 and 1.42 J cm^{-2} for **P1** and **P2**, respectively) compared to that of **M1** and **M2**.

In the other set, four butterfly-shaped D–π–A–π–D/ D–A–π–D–π–A–D molecules comprised of 2,4,6-trisubstituted pyridine, thiophene, triphenylamine, and carbazole moieties were synthesized and their optoelectronic properties were investigated. As a result of higher donating ability of triphe-nylamine unit and due to the extended conjugation, there is a red shift in the absorption and emission maxima and also there is an increase in the HOMO energy level of **M3** and **M5** compared to **M4** and **M6**. All four molecules exhibit excited state assisted TPA with the β_{eff} as high as 10^{-10} m W^{-1}. The molecules **M3** and **M5** show remarkably high β_{eff} of 4.45 and 3.91 × 10^{-10} m W^{-1}, and extremely low limiting threshold of 1.73 and 2.68 J cm^{-2}, respectively. Out of all synthesized materials, the compounds **P1, P2, M3,** and **M5** exhibit significantly superior NLO properties compared to the other molecules in the present study, and are better than some of the reported NLO materials. Therefore, **P1, P2, M3,** and **M5** could be promising candidates for all-optical limiting devices in laser photonics.

REFERENCES

Ajayaghosh, A. (2003). Donor–acceptor type low band gap polymers: Polysquaraines and related systems. *Chemical Society Reviews, 32*(4), 181–191. 10.1039/B204251G

Albota, M., Beljonne, D., Brédas, J.-L., Ehrlich, J. E., Fu, J.-Y., Heikal, A. A., Hess, S. E., Kogej, T., Levin, M. D., Marder, S. R., McCord-Maughon, D., Perry, J. W., Röckel, H., Rumi, M., Subramaniam, G., Webb, W. W., Wu, X.-L., & Xu, C. (1998). Design of organic molecules with large two-photon absorption cross sections. *Science, 281*(5383), 1653–1656. 10.1126/science.281.5383.1653

Allard, S., Forster, M., Souharce, B., Thiem, H., & Scherf, U. (2008). Organic semiconductors for solution-processable field-effect transistors (OFETs). *Angewandte Chemie International Edition, 47*(22), 4070–4098. 10.1002/anie.200701920

Anand, B., Ntim, S. A., Muthukumar, V. S., Sai, S. S. S., Philip, R., & Mitra, S. (2011). Improved optical limiting in dispersible carbon nanotubes and their metal oxide hybrids. *Carbon, 49*(14), 4767–4773.

Anandan, S., Manoharan, S., Narendran, N. K. S., Girisun, T. C. S., & Asiri, A. M. (2018). Donor-acceptor substituted thiophene dyes for enhanced nonlinear optical limiting. *Optical Materials, 85*, 18–25. 10.1016/j.optmat.2018.08.004

Balan, A., Gunbas, G., Durmus, A., & Toppare, L. (2008). Donor–acceptor polymer with benzotriazole moiety: Enhancing the electrochromic properties of the "donor unit". *Chemistry of Materials, 20*(24), 7510–7513. 10.1021/cm802937x

Bloembergen, N. (1984). Nonlinear optics—Past, present, and future. *IEEE Journal of Quantum Electronics, 20*(6), 556–558. 10.1109/JQE.1984.1072449

Brocks, G., & Tol, A. (1996). Small band gap semiconducting polymers made from dye molecules: Polysquaraines. *The Journal of Physical Chemistry, 100*(5), 1838–1846. 10.1021/jp952276c

Chen, C.-H., Hsu, Y.-T., Wang, B.-C., Chung, C.-L., & Chen, C.-P. (2019). Thienoisoindigo-based dopant-free hole transporting material for efficient p–i–n perovskite solar cells with the grain size in micrometer scale. *The Journal of Physical Chemistry C, 123*(3), 1602–1609. 10.1021/acs.jpcc.8b10070

Chen, J.-J. (2001). CDMA fiber-optic systems with optical hard limiters. *Journal of Lightwave Technology, 19*(7), 950.

Cheng, L. T., Tam, W., Stevenson, S. H., Meredith, G. R., Rikken, G., & Marder, S. R. (1991). Experimental investigations of organic molecular nonlinear optical polarizabilities. 1. Methods and results on benzene and stilbene derivatives. *The Journal of Physical Chemistry, 95*(26), 10631–10643.

Cheng, X., Ju, X., Du, H., Zhang, Y., Zhao, J., & Xie, Y. (2018). Synthesis and characterization of novel donor–acceptor type electrochromic polymers containing diketopyrrolopyrrole as an acceptor and propylenedioxythiophene or indacenodithiophene as a donor. *RSC Advances, 8*(41), 23119–23129. 10.1039/C8RA03570A

Cheng, Y.-J., Yang, S.-H., & Hsu, C.-S. (2009). Synthesis of conjugated polymers for organic solar cell applications. *Chemical Reviews, 109*(11), 5868–5923. 10.1021/cr900182s

Chiang, C. K., Druy, M. A., Gau, S. C., Heeger, A. J., Louis, E. J., MacDiarmid, A. G., Park, Y. W., & Shirakawa, H. (1978). Synthesis of highly conducting films of derivatives of polyacetylene,(CH) x. *Journal of the American Chemical Society, 100*(3), 1013–1015.

Chiang, C. K., Fincher, C. R., Park, Y. W., Heeger, A. J., Shirakawa, H., Louis, E. J., Gau, S. C., & MacDiarmid, A. G. (1977). Electrical conductivity in doped polyacetylene. *Physical Review Letters, 39*(17), 1098–1101. 10.1103/PhysRevLett.39.1098

Chiang, C. K., Park, Y.-W., Heeger, A. J., Shirakawa, H., Louis, E. J., & MacDiarmid, A. G. (1978). Conducting polymers: Halogen doped polyacetylene. *The Journal of Chemical Physics, 69*(11), 5098–5104.

Chidirala, S., Ulla, H., Valaboju, A., Kiran, M. R., Mohanty, M. E., Satyanarayan, M. N., Umesh, G., Bhanuprakash, K., & Rao, V. J. (2016). Pyrene–oxadiazoles for organic light-emitting diodes: Triplet to singlet energy transfer and role of hole-injection/hole-blocking materials. *The Journal of Organic Chemistry, 81*(2), 603–614. 10.1021/acs.joc.5b02423

Dalton, L. R., Sullivan, P. A., & Bale, D. H. (2009). Electric field poled organic electro-optic materials: State of the art and future prospects. *Chemical Reviews, 110*(1), 25–55.

Demeter, D., Mohamed, S., Diac, A., Grosu, I., & Roncali, J. (2014). Small molecular donors for organic solar cells obtained by simple and clean synthesis. *ChemSusChem, 7*(4), 1046–1050. 10.1002/cssc.201301339

Divyasree, M. C., Vasudevan, K., Basith, K. A., Jayakrishnan, P., Ramesan, M. T., & Chandrasekharan, K. (2018). Third-order nonlinear optical properties of phenothiazine-iodine charge transfer complexes in different proportions. *Optics & Laser Technology, 105*, 94–101.

Duan, L., Chen, Y., Jia, J., Zong, X., Sun, Z., Wu, Q., & Xue, S. (2020). Dopant-free hole-transport materials based on 2,4,6-triarylpyridine for inverted planar perovskite solar cells. *ACS Applied Energy Materials, 3*(2), 1672–1683. 10.1021/acsaem.9b02152

Facchetti, A. (2011). π-Conjugated polymers for organic electronics and photovoltaic cell applications. *Chemistry of Materials, 23*(3), 733–758. 10.1021/cm102419z

Farchioni, R., & Grosso, G. (2013). *Organic Electronic Materials: Conjugated Polymers and Low Molecular Weight Organic Solids.* Springer Science & Business Media.

Forrest, S. R. (2004). The path to ubiquitous and low-cost organic electronic appliances on plastic. *Nature, 428*(6986), 911–918. 10.1038/nature02498

Garmire, E. (2013). Nonlinear optics in daily life. *Optics Express, 21*(25), 30532–30544. 10.1364/OE.21.030532

Gibson, G. I., McCormick, T. M., & Seferos, D. S. (2012). Atomistic band gap engineering in donor–acceptor polymers. *Journal of the American Chemical Society, 134*(1), 539–547. 10.1021/ja208917m

Gopi, V., Subbiahraj, S., Chemmanghattu, K., & Ramamurthy, P. C. (2020). 2,3-di(2-furyl) quinoxaline bearing 3 -ethyl rhodanine and 1,3 indandione based heteroaromatic conjugated T-shaped push-pull chromophores: Design, synthesis, photophysical and non-linear optical investigations. *Dyes and Pigments, 173*, 107887. 10.1016/j.dyepig.2019.107887.

Havinga, E. E., ten Hoeve, W., & Wynberg, H. (1993). Alternate donor-acceptor small-band-gap semi-conducting polymers; Polysquaraines and polycroconaines. *Synthetic Metals, 55*(1), 299–306. 10.1016/0379-6779(93)90949-W

Hernández, F. E., Yang, S., Stryland, E. W. V., & Hagan, D. J. (2000). High-dynamic-range cascaded-focus optical limiter. *Optics Letters, 25*(16), 1180–1182. 10.1364/OL.25.001180

Hwang, S.-W., & Chen, Y. (2002). Photoluminescent and electrochemical properties of novel poly (aryl ether) s with isolated hole-transporting carbazole and electron-transporting 1, 3, 4-oxadiazole fluorophores. *Macromolecules, 35*(14), 5438–5443.

Jana, D., & Ghorai, B. K. (2012). Triphenylpyridine-based star-shaped π-conjugated oligomers with triphenylamine core: Synthesis and photophysical properties. *Tetrahedron Letters, 53*(14), 1798–1801. 10.1 016/j.tetlet.2012.01.116

Jebnouni, A., Chemli, M., Lévêque, P., Fall, S., Majdoub, M., & Leclerc, N. (2018). Effects of vinylene and azomethine bridges on optical, theoretical electronic structure and electrical properties of new anthracene and carbazole based π-conjugated molecules. *Organic Electronics, 56*, 96–110.

Jiang, Z., Chen, Y., Yang, C., Cao, Y., Tao, Y., Qin, J., & Ma, D. (2009). A fully diarylmethylene-bridged triphenylamine derivative as novel host for highly efficient green phosphorescent OLEDs. *Organic Letters, 11*(7), 1503–1506. 10.1021/ol9001152

Ju, C., Li, X., Yang, G., Yuan, C., Semin, S., Feng, Y., Rasing, T., & Xu, J. (2019). Polymorph dependent linear and nonlinear optical properties of naphthalenyl functionalized fluorenones. *Dyes and Pigments, 166*, 272–282. 10.1016/j.dyepig.2019.03.030

Junkers, T., Vandenbergh, J., Adriaensens, P., Lutsen, L., & Vanderzande, D. (2012). Synthesis of poly (p-phenylene vinylene) materials via the precursor routes. *Polymer Chemistry, 3*(2), 275–285. 10.1039/ C1PY00345C

Kim, M., Ryu, S. U., Park, S. A., Choi, K., Kim, T., Chung, D., & Park, T. (2020). Donor–acceptor-conjugated polymer for high-performance organic field-effect transistors: A progress report. *Advanced Functional Materials, 30*(20), 1904545. 10.1002/adfm.201904545

Li, C., Li, Z., Liang, J., Luo, H., Liu, Y., Wei, J., & Wang, Y. (2018). A twisted phenanthroimidazole based molecule with high triplet energy as a host material for high efficiency phosphorescent OLEDs. *Journal of Materials Chemistry C, 6*(47), 12888–12895. 10.1039/C8TC04218G

Li, Y., Cao, Y., Gao, J., Wang, D., Yu, G., & Heeger, A. J. (1999). Electrochemical properties of luminescent polymers and polymer light-emitting electrochemical cells. *Synthetic Metals, 99*(3), 243–248. 10.1016/ S0379-6779(99)00007-7

Liang, Y., Tao, Z., & Chen, J. (2012). Organic electrode materials for rechargeable lithium batteries. *Advanced Energy Materials, 2*(7), 742–769. 10.1002/aenm.201100795

Liaw, D.-J., Wang, K.-L., Pujari, S. P., Huang, Y.-C., Tao, B.-C., Chen, M.-H., Lee, K.-R., & Lai, J.-Y. (2009). A novel, conjugated polymer containing fluorene, pyridine and unsymmetric carbazole moieties: Synthesis, protonation and electrochemical properties. *Dyes and Pigments, 82*(2), 109–117. 10.1 016/j.dyepig.2008.12.004

Liu, C.-K., Chen, Y.-H., Long, Y.-J., Sah, P.-T., Chang, W.-C., Chan, L.-H., Wu, J.-L., Jeng, R.-J., Yeh, S.-C., & Chen, C.-T. (2018). Bipolar 9-linked carbazole-π-dimesitylborane fluorophores for nondoped blue OLEDs and red phosphorescent OLEDs. *Dyes and Pigments, 157*, 101–108. 10.1016/j.dyepig.2018.04.035

Liu, J., Yang, Y., Liu, X., & Zhen, Z. (2015). Physical attachment of NLO chromophores to polymers for great improvement of long-term stability. *Materials Letters, 142*, 87–89.

Maiman, T. H. (1960). *Stimulated optical radiation in ruby.Nature* 187, 493–494. 10.1038/187493a0

Martin, R. E., & Diederich, F. (1999). Linear monodisperse π-conjugated oligomers: Model compounds for polymers and more. *Angewandte Chemie International Edition, 38*(10), 1350–1377. https://doi.org/10.1 002/(SICI)1521-3773(19990517)38:10<1350::AID-ANIE1350>3.0.CO;2-6

Martsinovich, N., & Troisi, A. (2011). Theoretical studies of dye-sensitised solar cells: From electronic structure to elementary processes. *Energy & Environmental Science, 4*(11), 4473–4495. 10.1039/C1 EE01906F

McCullough, R. D. (1998). The chemistry of conducting polythiophenes. *Advanced Materials, 10*(2), 93–116. https://doi.org/10.1002/(SICI)1521-4095(199801)10:23.0.CO;2-F

Mi, Y., Liang, P., Yang, Z., Wang, D., Cao, H., He, W., Yang, H., & Yu, L. (2016). Application of near-IR absorption porphyrin dyes derived from click chemistry as third-order nonlinear optical materials. *ChemistryOpen, 5*(1), 71–77. 10.1002/open.201500124

Mikroyannidis, J. A., Stylianakis, M. M., Dong, Q., Zhou, Y., & Tian, W. (2009). New 4, 7-dithienebenzothiadiazole derivatives with cyano-vinylene bonds: Synthesis, photophysics and photovoltaics. *Synthetic Metals, 159*(14), 1471–1477.

Mishra, A., & Bäuerle, P. (2012). Small Molecule Organic Semiconductors on the Move: Promises for Future Solar Energy Technology. *Angewandte Chemie International Edition, 51*(9), 2020–2067. 10.1002/ anie.201102326

Mullekom van, H. A. M. (2000). *The Chemistry Of High And Low Band Gap Π-conjugated Polymers* . Technische Universiteit Eindhoven. 10.6100/IR530045.

Nalwa, H. S. (1997). *Handbook of Organic Conductive Molecules and Polymers, Conductive Polymers: Transport, Photophysics and Applications.* Wiley.

Nielsen, C. B., Turbiez, M., & McCulloch, I. (2013). Recent advances in the development of semiconducting DPP-containing polymers for transistor applications. *Advanced Materials, 25*(13), 1859–1880. 10.1002/adma.201201795

Park, J.-M., Park, S. K., Yoon, W. S., Kim, J. H., Kim, D. W., Choi, T.-L., & Park, S. Y. (2016). Designing thermally stable conjugated polymers with balanced ambipolar field-effect mobilities by incorporating cyanovinylene linker unit. *Macromolecules, 49*(8), 2985–2992. 10.1021/acs.macromol.5b02761

Perepichka, I. F., & Perepichka, D. F. (2009). *Handbook of Thiophene-Based Materials: Applications in Organic Electronics and Photonics, 2 Volume Set.* John Wiley & Sons.

Pell, L. E., Schricker, A. D., Mikulec, F. V., & Korgel, B. A. (2004). Synthesis of amorphous silicon colloids by trisilane thermolysis in high temperature supercritical solvents. *Langmuir, 20*(16), 6546–6548. 10.1021/la048671o

Praseetha, K. P., Divyasree, M .C., Nimmy, John V., Chandrasekharan, K., & Varghese, S. (2019). Enhanced optical nonlinearity in nematic liquid crystal on doping with CdSe quantum dot. *Journal of Molecular Liquids, 273*, 497–503. 10.1016/j.molliq.2018.10.077

Qing, W., Liu, Z., Yang, S., Tan, L., Yang, Y., Zhang, D., & Li, J. (2015). Modulating carrier transfer ability—Linker effect on thieno [3, 4-c] pyrrole-4, 6-dione based conjugated polymers. *RSC Advances, 5*(69), 55619–55624.

Rajeshirke, M., Sreenath, M. C., Chitrambalam, S., Joe, I. H., & Sekar, N. (2018). Enhancement of NLO properties in obo fluorophores derived from carbazole-coumarin chalcones containing carboxylic acid at the N-alky terminal end. *The Journal of Physical Chemistry C. 122*(26),14313–14325.

Raynor, A. M., Gupta, A., Plummer, C. M., Jackson, S. L., Bilic, A., Patil, H., Sonar, P., & Bhosale, S. V. (2015). Significant improvement of optoelectronic and photovoltaic properties by incorporating thiophene in a solution-processable D–A–D modular chromophore. *Molecules, 20*(12), 21787–21801. 10.3390/molecules201219798

Reinhardt, B. A., Brott, L. L., Clarson, S. J., Dillard, A. G., Bhatt, J. C., Kannan, R., Yuan, L., He, G. S., & Prasad, P. N. (1998). Highly active two-photon dyes: Design, synthesis, and characterization toward application. *Chemistry of Materials, 10*(7), 1863–1874. 10.1021/cm980036e

Roncali, J. (1992). Conjugated poly(thiophenes): Synthesis, functionalization, and applications. *Chemical Reviews, 92*(4), 711–738. 10.1021/cr00012a009

Roncali, J. (1997). Synthetic principles for band gap control in linear π-conjugated systems. *Chemical Reviews, 97*(1), 173–206.

Sahin, O., Osken, I., & Ozturk, T. (2011). Investigation of electrochromic properties of poly(3,5-bis(4-methoxyphenyl)dithieno[3,2-b;2′,3′-d]thiophene). *Synthetic Metals, 161*(1), 183–187. 10.1016/j.synthmet.2010.11.020

Sathiyan, G., Ranjan, R., Ranjan, S., Garg, A., Gupta, R. K., & Singh, A. (2019). Dicyanovinylene and thiazolo[5,4-d]thiazole core containing D–A–D type hole-transporting materials for spiro-OMeTAD-free perovskite solar cell applications with superior atmospheric stability. *ACS Applied Energy Materials, 2*(10), 7609–7618. 10.1021/acsaem.9b01598

Sheik-Bahae, M., Said, A. A., Wei, T.-H., Hagan, D. J., & Van Stryland, E. W. (1990). Sensitive measurement of optical nonlinearities using a single beam. *IEEE Journal of Quantum Electronics, 26*(4), 760–769.

Shiju, E., Siji Narendran, N. K., Rao, D. N., & Chandrasekharan, K. (2019). A phenothiazine–silver hybrid system exhibiting switching and photo-induced enhancement in nonlinear optical absorption. *New Journal of Chemistry, 43*(21), 7962–7971. 10.1039/C8NJ06402D.

Shirakawa, H., Louis, E. J., MacDiarmid, A. G., Chiang, C. K., & Heeger, A. J. (1977). Synthesis of electrically conducting organic polymers: Halogen derivatives of polyacetylene, (CH)x. *Journal of the Chemical Society, Chemical Communications, 16*, 578–580. 10.1039/C39770000578

Skabara, P. J. (2009). *Fused oligothiophenes.* Wiley.

Skotheim, T. A., Reynolds, J., & Reynolds, J. (2007). *Handbook of Conducting Polymers, 2 Volume Set.* CRC Press. 10.1201/b12346

Song, S.-Y., Jang, M. S., Shim, H.-K., Hwang, D.-H., & Zyung, T. (1999). Highly efficient light-emitting polymers composed of both hole and electron affinity units in the conjugated main chain. *Macromolecules, 32*(5), 1482–1487.

Sonntag, M., Kreger, K., Hanft, D., Strohriegl, P., Setayesh, S., & de Leeuw, D. (2005). Novel star-shaped triphenylamine-based molecular glasses and their use in OFETs. *Chemistry of Materials, 17*(11), 3031–3039. 10.1021/cm047750i

Sutherland, R. L. (1989). Optical limiters, switches, and filters based on polymer dispersed liquid crystals. *Liquid Crystal Chemistry, Physics, and Applications*, *1080*, 83–90. 10.1117/12.976404

Tan, B., & Baycan, F. (2020). A new donor-acceptor conjugated polymer-gold nanoparticles biocomposite materials for enzymatic determination of glucose. *Polymer*, *210*, 123066. 10.1016/j.polymer.2020.123066

Thakare, S. S., Sreenath, M. C., Chitrambalam, S., Joe, I. H., & Sekar, N. (2017). Non-linear optical study of BODIPY-benzimidazole conjugate by solvatochromic, Z-scan and theoretical methods. *Optical Materials*, *64*, 453–460.

Tong, Q.-X., Lai, S.-L., Chan, M.-Y., Lai, K.-H., Tang, J.-X., Kwong, H.-L., Lee, C.-S., & Lee, S.-T. (2007). High Tg triphenylamine-based starburst hole-transporting material for organic light-emitting devices. *Chemistry of Materials*, *19*(24), 5851–5855. 10.1021/cm0712624

Vellis, P. D., Ye, S., Mikroyannidis, J. A., & Liu, Y. (2008). New divinylene trimers with triphenylpyridine segments: Synthesis, photophysics, electrochemical and electroluminescent properties. *Synthetic Metals*, *158*(21), 854–860. 10.1016/j.synthmet.2008.06.002

Vintu, M., Unnikrishnan, G., Shiju, E., & Chandrasekharan, K. (2019). Indolo[3,2-b]carbazole-based poly (arylene ethynylene)s through Sonogashira coupling for optoelectronic and sensing applications. *Journal of Applied Polymer Science*, *136*(2), 46940. 10.1002/app.46940

Vishnumurthy, K. A., Adhikari, A. V., Sunitha, M. S., Mary, K. A. A., & Philip, R. (2011). Design and synthesis of a new thiophene based donor–acceptor type conjugated polymer with large third order nonlinear optical activity. *Synthetic Metals*, *161*(15), 1699–1706. 10.1016/j.synthmet.2011.06.007

Wang, F., Zhao, Y., Xu, H., Zhang, J., Miao, Y., Guo, K., Shinar, R., Shinar, J., Wang, H., & Xu, B. (2019). Two novel bipolar hosts based on 1,2,4-triazole derivatives for highly efficient red phosphorescent OLEDs showing a small efficiency roll-off. *Organic Electronics*, *70*, 272–278. 10.1016/j.orgel.2019.04.030

Xu, L., Zhang, D., Zhou, Y., Zheng, Y., Cao, L., Jiang, X.-F., & Lu, F. (2017). 4-N, N-bis(4-methoxylphenyl) aniline substituted anthraquinone: X-ray crystal structures, theoretical calculations and third-order nonlinear optical properties. *Optical Materials*, *70*, 131–137. 10.1016/j.optmat.2017.05.025

Xu, Q., Li, Z., Liu, N., Jia, J., Yang, J., & Song, Y. (2019). Third order nonlinear optical properties and transient dynamics of thiophene-contained pyrene derivatives: Effect of peripheral substituent group. *Optics & Laser Technology*, *109*, 666–670. 10.1016/j.optlastec.2018.08.048

Yook, K. S., & Lee, J. Y. (2014). Small molecule host materials for solution processed phosphorescent organic light-emitting diodes. *Advanced Materials*, *26*(25), 4218–4233. 10.1002/adma.201306266

Zawadzka, M., Wang, J., Blau, W. J., & Senge, M. O. (2013). Nonlinear absorption properties of 5,10-A2B2 porphyrins – correlation of molecular structure with the nonlinear responses. *Photochemical & Photobiological Sciences*, *12*(6), 996–1007. 10.1039/C3PP25410K

Zhang, X.-L., Zhao, X., Liu, Z.-B., Shi, S., Zhou, W.-Y., Tian, J.-G., Xu, Y.-F., & Chen, Y.-S. (2011). Nonlinear optical and optical limiting properties of graphene oxide–Fe3O4hybrid material. *Journal of Optics*, *13*(7), 075202. 10.1088/2040-8978/13/7/075202

Zotti, G., Zecchin, S., Schiavon, G., Berlin, A., & Penso, M. (1999). Ionochromic and potentiometric properties of the novel polyconjugated polymer from anodic coupling of 5,5′-Bis(3,4-(ethylenedioxy) thien-2-yl)-2,2'-bipyridine. *Chemistry of Materials*, *11*(11), 3342–3351. 10.1021/cm9904032

14 Recent Applications of Macromolecular Gels for Environmental Remediation

Sayantani Bhattacharya and Raja Shunmugam
Indian Institute of Science Education and Research Kolkata, Mohanpur, Nadia, West Bengal, India

CONTENTS

14.1 INTRODUCTION

Macromolecular gels are defined as three-dimensional cross-linked networks consisting of components like single polymer, different polymer chains or their aggregates capable of retaining large volumes of fluid. In other words, macromolecular gels are soft elastic materials in which the interstitial spaces are filled by a certain solvent [1]. The capacity of the gels to absorb liquid in their network but not dissolving in it, confers the solid-like texture. In fact, Flory has defined that '*a gel has a continuous structure that is permanent on the analytical time scale and is solidlike in its rheological behavior*' [2]. However, due to the unique elasticity, gels are capable of undergoing large deformation, which is in sharp contrast with the solid materials. In principle, any gel is composed of two building blocks: network forming segment and swelling agent[3]. The network structure is generally formulated by small molecules, polymers, particles, crystallites, proteins, etc., and the swelling agent is either water in case of hydrogels or organic medium in case of organogels. However, the most important prerequisite for gel formation using any component is the ability to form a three-dimensional network by physical or chemical cross-linking. There are various ways in which gels can be classified: depending on their origin, swelling medium, constitution and the nature of cross-linking used to fabricate the three-dimensional matrix (Figure 14.1) [4]. In general, the majority of naturally

DOI: 10.1201/9781003200710-16

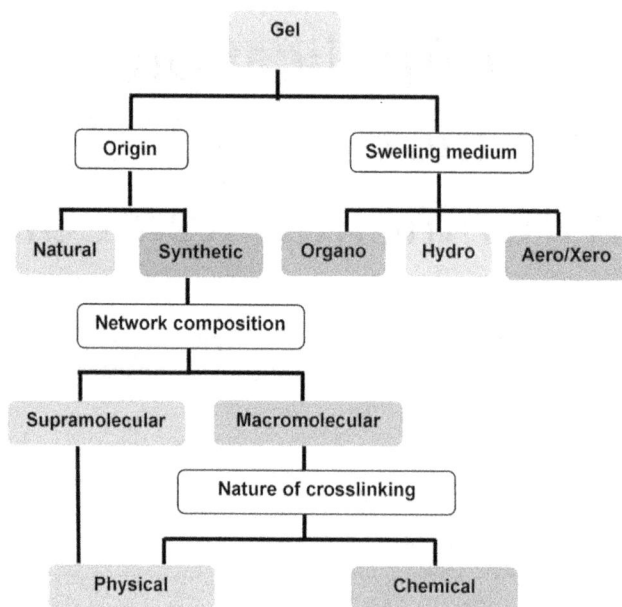

FIGURE 14.1 Typical classification of gels depending on their origin, swelling medium, network composition and type of cross-linking (Chemical Society Reviews 34, 2005, 821–36).

occurring gels are polymeric gels formed via physical cross-linking, especially H-bonding. The examples of such gels include gelatin, agar, starch, pectin, etc. [5]. On the other hand, synthetically developed gels can be categorized as supramolecular gels and macromolecular gels depending upon their constitution. In case of macromolecular gels, the chains are connected either by physical or chemical cross-linking to fabricate the network. Physically cross-linked gels or physical gels are formed by temporary association of polymers chains via hydrogen bonding, ionic interaction, hydrophobic interaction, etc. [6]. Physical gels do not involve any kind of chemical crosslinker during the course of gelation. Hence, they are reversible in nature and they can transform to the 'sol' phase in presence of external stimuli such as temperature, pH, pressure, solvent, etc. Generally, physical gels are weak in nature owing to the restricted mechanical strength derived from weak physical interactions. In contrast, chemically cross-linked gels or chemical gels are permanent in nature as the network is developed by the covalent attachment of suitable functional groups provided by the polymers. These kinds of gels require participation of chemical crosslinkers. The irreversibility of the chemical gels arises from the strong covalent interactions involved to form the cross-linked network. On the contrary, supramolecular gels are derived from low molecular weight compounds that self-assemble to form network junctions via weak non-covalent interactions like van der Waals interaction, metal coordination, donor-acceptor interactions, H-bonding, etc. [7]. Gels are known in literature for centuries considering their various applications but the research on macromolecular gels has received immense attention during the past few decades by the researchers in diverse fields including material science, polymer science, chemical engineering, pharmaceutical chemistry, organic chemistry, etc.

However, the application of polymeric gels for wastewater treatment has been considerably overlooked, until recently when there was a severe outbreak of health hazards about 30 years ago [8]. Since then, this area has witnessed a surge in interest. This is particularly due to the fact that untreated industrial discharges contain various hazardous elements like petrochemical waste, toxic dyes, organic solvents, heavy metals, etc. that upon direct release into the waterways, cause severe water pollution. The contaminated water is a serious threat to mankind and aquatic lives. Thus, efficient removal of pollutants from waterways by economic and sustainable manner is of utmost importance. Further, the practice of traditional decontamination techniques like coagulation-flocculation, ozonation, oxidation, photochemical degradation, etc. are restricted due to their exorbitant price, requirement of specific equipment, possibility of toxic side product generation. In this regard, removal of toxicants by means of adsorption is considered as the most reliable approach for its straightforwardness, economic viability and highest efficiency. Several adsorbent materials such as activated carbon, zeolites, and clays have been exploited for sequestering pollutants [9-11]. But their use in practice is limited due to their poor bio-compatibility, lack of recyclability and environmental toxicity. In this context, polymeric gels serve as optimal sorbents due to the unique combination of engineering adsorption specificity, bio-compatibility and economic viability that cannot be accomplished by any other conventional decon-tamination techniques [12]. Therefore, the following section will focus on the salient properties of polymeric gels that promote removal of pollutants from wastewater.

14.2 KEY FEATURES OF POLYMERIC GELS

14.2.1 Swelling

The most fundamental property of macromolecular gels is the ability to uptake large amounts of solvents into its network and to swell. Polymeric gels are able to swell up to 100-fold compared to their dry state and to hold the solvent in the gel matrix without getting dissolved in it. Swelling of polymeric gels in solvents largely depends on the polymer-solvent interactions. In this context, the theory of solubility parameter is crucial to understand the polymer-solvent interactions. According to Charles M. Hansen, a cross-linked polymer will swell to a maximum in a solvent whose solubility parameter will be in close proximity with its own [13]. Further, the extent of swelling for macromolecular gels is governed by the cross-linking density of the gel. Highly cross-linked gels lead to very tight or compact network structure that restrict the penetration of a given solvent into the gel matrix. Hence, high cross-linking density in a gel results in inadequate swelling. Similarly, low density cross-linked networks will feature enhanced solvent absorption capability. The crosslinker content can be varied to achieve a polymeric gel with desired swelling ability. Owing to this solvent uptake ability, polymeric gels find great application in making diapers, sanitary napkins and agricultural fields [14]. Particularly, sub-stantial water uptake capacity makes the polymeric gels potent candidates for pollutant removal by adsorption. With increased swelling in water, the active adsorption sites present in the gels get exposed to the water-soluble toxicants, leading to enhanced wastewater decontamination.

14.2.2 Rheological Behaviour

The viscoelastic property of a gel is characterized by the rheological measurements. The term "rheology" was introduced by Professor Eugene C. Bingham from the Greek word 'rheos', which means stream or flow [15]. Rheology is the study of flow and deformation of materials. Almdal and coworkers have defined that a gel must show *a flat mechanical spectrum in an oscillatory shear experiment* [16]. This means that a gel must feature a shear storage modulus (G') that shows a distinct plateau extending to times in the order of seconds and a loss modulus (G''), which is considerably smaller in magnitude than that of storage modulus in this region. The storage modulus, which measures the elastic response of a material, is also related to the cross-linking density of polymeric gel. High storage modulus indicates that the material is sufficiently cross-linked that restricts the mobility of polymer chains in the network and leads to a stiff material [17]. Therefore, by varying the cross-linking density, macromolecular gels with required stiffness can be obtained. The stiffness of polymeric gels are crucial when they are used in tissue engineering. However, gels that exhibit rather flow-like behaviour are suitable as injectable scaffolds. Owing to the tunable mechanical strength, flexibility together with the structural properties, polymeric gels serve as efficient adsorbents for removal of toxicants from industrial effluents. Typically, polymeric gels feature enhanced mechanical strength compared to supramolecular gels in which the network is made up of weak non-covalent interactions [18]. Particularly, in the case of polymeric gels, the mechanical strength is not compromised upon the uptake of target pollutant.

14.2.3 CHEMICAL MODIFICATION

One of the unique advantages of synthetic macromolecular gels is the scope of fine-tuning their chemical constituents in order to make them suitable for desired applications. This aspect of polymeric gels make them optimal adsorbents for sequestering pollutants from wastewater. Most of the dyes used in textile industries are ionised to promote water solubility. Further, some of the pharmaceutical wastes, like antibiotics are also charged at ambient water pH. Thus, incorporation of appropriate functional groups in the gel matrix can effectively increase the interactions between the target pollutant and the synthesized gel that subsequently leads to enhanced sequestration of pollutants from wastewater.

14.2.4 MORPHOLOGY

Polymeric gels feature various morphologies, i.e., thin films, monolithic blocks, nanoparticles, microparticles, etc. [19]. Gels with appropriate morphology can be formulated to make it suitable for desired applications. In the case of water remediation, different morphologies offer distinct advantages. For instance, nanoparticles or microparticles exhibit very high pollutant removal efficiency compared to monolithic blocks due to the exposure of large surface area. Tang and his coworkers have demonstrated that the kinetics of Cr (VI) removal by magnetic hydrogel significantly improved (from 2 hours to 5 minutes) just by grinding the bulk hydrogel into power form [20]. However, the micro- or nanoparticles are often difficult to handle especially when it comes to regeneration of the adsorbent. On the other hand, bulk monolithic hydrogels are easy to process, especially post-sequestration. Further, polymeric gels formulated as thin films are particularly interesting for their facile application and recovery [21].

14.2.5 STABILITY AND FACILE REGENERATION

In general, thermal stability together with the resistance to photodegradation, hydrolysis, etc. makes the cross-linked polymeric gels suitable adsorbents that can be used at diverse environmental conditions. For instance, graphene oxide (GO) sheets alone are excellent adsorbents for removal of organic dye pollutants. However, covalent attachment of GO to polymers provides additional cross-linking junctions that improves the stability of the resulting gels. In this context, Peng et al. has utilized thiol modified GO and PEG-diacrylate to fabricate a hydrogel via photomediated thiol-ene click reactions (Figure 14.2) [22]. These hydrogels exhibit excellent dye removal efficiency. Often, cross-linking agents are deliberately added to the gel matrix to increase its stability. Xu et al. has demonstrated fabrication of hydrogel by using epichlorohydrin (ECH) as the potential crosslinker to attach non-functionalized GO with cellulose [23]. The resulting hydrogel exhibited high adsorption capacity for Zn^{2+}, Fe^{3+} and Pb^{2+} with sufficient regeneration ability (up to five cycles without

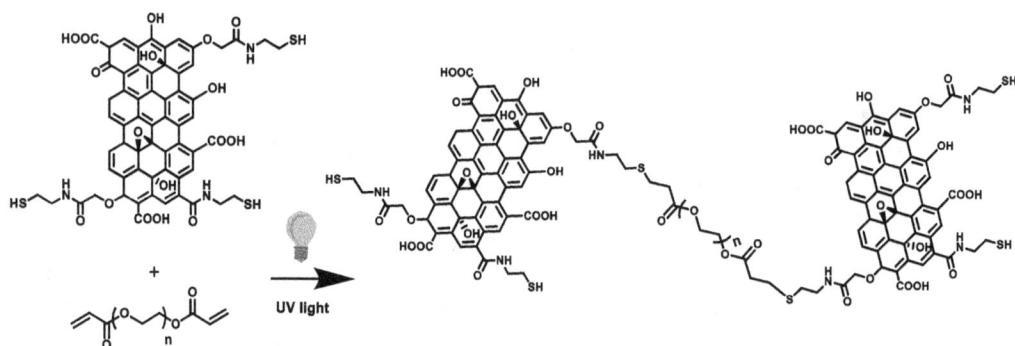

FIGURE 14.2 Formation of gel using thiol-ene click reaction with thiol modified graphene oxide and PEG-diacrylate (Colloids and Surfaces A: Physicochemical and Engineering Aspects 529, 2017, 668–76).

detectable loss in the adsorption capacity). The recyclability of these gels are attributed to the formation of stable ether linkages during the course of gelation. Regeneration of the gel-based adsorbents are practically essential for commercial viability. Further, post-recovery of the adsorbents, the precious metals, dyes can be recycled back to the industries that makes the overall process environmentally friendly as well. Polymeric gels that remove the pollutants via ionic interactions, can be recycled by weakening the electrostatic interaction between the gel and the pollutant. This can be easily achieved by varying the pH of the solution. In addition to that, those polymeric gels can also be recovered via ion-exchange employing large excess of ions that compete with the toxicant.

14.3 CLASSES OF CONTAMINANTS

The surface water which covers up to 70% of the earth, gets polluted by untreated industrial, agricultural or household discharges. Suitably engineered polymeric gel-based adsorbents can effectively remove the pollutants from wastewater. The polymeric gels should bear appropriate functional groups in their framework to establish substantial interactions with the pollutant components that are potentially stronger than the existing interactions with water. Wisely programmed gel-based adsorbents are capable of sequestering target toxicants. Therefore, for the prudent designing of the adsorbent and its subsequent success in decontamination, it is essential to recognize the type and nature of the potential toxicants.

14.3.1 PETROCHEMICALS

An accidental release of liquid petroleum hydrocarbons from tankers, drilling rigs and offshore explorations into ocean or coastal water is disastrous for the environment. Very often, oil spills can extend for hundreds of nautical miles that are detrimental for aquatic animals and plants. The discharged oil harms marine lives either by direct coating or during the cleanup process. Direct coating will lead to the smothering of marine organisms. Highly viscous oils will affect the regular physical activities such as respiration, thermal regulation, feeding, etc. Although contemporary recovery techniques employing biological entities such as microorganisms to degrade the complex hydrocarbons are cost efficient, their use in practice is restricted due to the requirement of long operation time and uncertainty [24]. In this context, polymeric gel-based adsorbents are reliable tools for containing oil spills. Typically, oil spills comprise crude oil, refined petroleum like gasoline, diesel, etc. Hydrocarbons present in oil are immiscible in water. Therefore, for effective remediation, the adsorbent must bear hydrophobic functional groups to establish interactions with the hydrophobic phase. Some other parameters such as porosity or surface area can also be modified to uptake oils of different viscosities [25]. However, after application, the gels should remain capable of floating in the surface water to ensure complete cleanup. For this purpose, organogels are better adsorbents compared to hydrogels as the latter will have a tendency to uptake water that will eventually lead to the sinking.

14.3.2 ORGANIC DYES

Dyes are basically colouring agents that are used to impart tint to materials. The use of dyes has been known for thousands of years. But, the production of synthetic dyes has drastically increased after the discovery of the first synthetic dye back in 1856 [26]. Currently, more than 10,000 types of commercial dyes are manufactured worldwide [27]. Dyes are routinely used in textile, paper, cosmetic, leather and plastics industries. Among these sectors, textile industries discharge the largest amount of effluents. Often, a significant portion of the dyes after their usage, are disseminated into the water without prior treatment. Depending upon the structural difference, dyes exhibit a wide range of colours and they can be classified as cationic, anionic and neutral dyes. Regardless of the structure, dyes are stable in water, resistant to degradation. The dye-infested water does not only lead to undesirable appearance, but is associated with several health risks owing to the severe toxicity and carcinogenicity of the dyes [28,29]. Ingestion of contaminated water leads to irritation of skin, eyes, nausea, headache and other water borne diseases [30,31]. Further, the colored water obstructs the sunlight penetration in water that affects photosynthesis of marine organisms. Therefore, it is essential to remove dissolved organic dyes by suitably programmed polymeric gel-based adsorbents. Adsorbents can be chemically modified to incorporate functional groups that favour the interaction of the gels with the ionic dyes. The driving force for adsorption by synthetically modified macromolecular gels may include non-covalent interactions such as: H-bonding, van der Waals interaction, dipole-dipole interaction or covalent interactions. In conjunction with these, the polymeric gels can be engineered to render biocompatibility and sufficient recyclability that are essential for practical use.

14.3.3 INORGANIC IONS

Besides petrochemicals and organic dyes, inorganic ions and heavy metals are also extremely hazardous for the environment. Toxic cations include ammonium, Pb^{2+}, Hg^{2+}, Cd^{2+}, etc. Anionic species such as phosphates are important for plant growth, but they speed up eutrophication when present in large amounts. Several decontamination techniques such as precipitation, membrane-filtration, ion-exchange, etc. have been explored to separate these metal ions from wastewater. But, the major drawback of these methods is lack of selectivity due to which other benign ions are also removed from water. Polymeric gel-based adsorbents are particularly advantageous in removal of ions as in this case suitable functional groups can be attached to the gel to enable selective ion separation. The presence of ionic recognition sites in the gel matrix promotes the removal. The mechanism of heavy metals or ions separation is associated with the strong interactions of the toxic ions with the oppositely charged entities provided by the gel. For instance, heavy metal ions can be removed by using gels containing Lewis bases like carboxylates and sulfonates, which can establish strong electrostatic interactions. Thus, properly functionalized homo- and co-polymer-based gels can serve as optimal adsorbents for removal of inorganic ions from wastewater.

14.4 MACROMOLECULAR GELS AS ADSORBENTS

14.4.1 OIL-SPILL RECOVERY

Use of polymeric gel-based sorbents for remediation of oil-spill combines the advantage of easy cleanup as well as recovery of the oil that makes the overall process economically feasible. Moreover, oils of a wide range of viscosities can be collected by using prudently designed adsorbents (Table 14.1). For instance, Jeon et al. have developed a unique hydrogel bowl that is capable of removing oil of different viscosities while floating on the water surface [32]. For the fabrication of hydrogel, the authors have carried out polymerisation of acrylamide and N,N'-methylenebisacrylamide with ammonium persulfate (APS) and N,N,N',N'-tetramethylethylenediamine (TEMED) in a bowl-shaped mold. After polymerization, the hydrogel bowl was coated using a long hydrophobic alkyl chain, octadecyltrichlorosilane (OTS). The prepared gel featured macroscopic openings in the side of its walls through which it could effectively uptake oils. This hydrogel bowl was able to effectively remove 40 mL of kerosene in just 24 seconds. It also removed oils with much higher viscosities e.g., high-vacuum pump oil. But, the hydrogel bowl could not be stored in either open air or in water for a longer period of time due to the possibility of quick dehydration and water adsorption respectively. However, it could retain its dimension and efficiency in oil for more than one month. Another advantage of using polymeric hydrogels as adsorbents for crude oil removal is the biocompatibility of the adsorbents that potentially eliminates any scope of secondary pollution. Saruchi and coworkers have developed a biodegradable hydrogel which could remove crude oil from wastewater [33]. For the gel preparation, they have grafted acrylic acid on gum tragacanth by free radical polymerization using glutaraldehyde as the cross-linker. They have varied the acrylic acid concentration and pH of the solution to evaluate the maximum

TABLE 14.1
Comparison of Macromolecular Gels to Combat Oil Spill

Adsorbent	Target Pollutant	Removal Efficiency	Ref.
Acrylamide-based hydrogel bowl coated with octadecyltrichlorosilane (OTS)	Kerosene	40 mL in 24 seconds	Water Research, 145, 2018, 640–49.
Gum tragacanth and poly(acrylic Acid) based hydrogel	Crude oil	2.9 g/g	Petroleum Science and Technology 33, 2015, 278–86.
Wheat bran/divinylbenzene polymer based hybrid gelator	Chloroform	12 g/g	Composites Communications 22 2020, 100471.
Cellulose pulp with sugar-based phase-selective organogelator	Crude oil	16 times increase in weight compared to the mass of gelator	Angewandte Chemie International Edition 56, 2017, 9405–9.
Cross-linked polyolefin terpolymer	Gasoline	41 times increase in weight compared to the polymer weight	Energy and Fuels, 26, 2012, 4896–902.
Electrospun gelatin cross-linked with bis(diaryldiazomethane)	Petrol	5–8 fold increase in weight compared to the adsorbent	ACS Omega 3, 2018, 3928–35.

efficiency of the system. It was observed that the highest removal of crude oil occurred at 40% acrylic acid concentration and at pH = 3 (2.9 g/g). In another work, Cheng-Ma et al. has developed a polymer and wheat bran based oil gelling agent to solidify oil as a gel [34]. To increase lipophilicity and hydrophobicity, wheat barn (WB) was modified with vinyltriethoxysilane followed by grafting on divinylbenzene based porous polymer via free radical polymerization (Figure 14.3). The resulting material demonstrated high oil removal efficiency (12 g/g). Further, the material retained more than 90% oil adsorption efficiency even after ten cycles.

Sureshan et al. have developed an adsorbent material by infusing cellulose pulp with a sugar-based phase-selective organogelator (PSOG), 1,2:5,6-di-O-cyclohexylidene-mannitol [35]. This particular oleogelator could effectively engage in H-bonding with the exposed -OH functionalities of cellulose fibrils and made it temporarily hydrophobic. On coming in contact with oil-water mixture, this material was capable of selectively adsorb the oil by hydrophobic interaction with cellulose fibrils. Thereafter, the gelator was dissolved in oil followed by formation of a gel network with the hydrocarbon via self-assembly through H-bonding. Moreover, the uptake of oil by cellulose fibres within the polymer matrix has substantially improved the gel strength (yield strength increased up to 230-fold) that is very important for practical use as it potentially eliminates the possibility of secondary oil spill. The adsorbed oil could be easily retrieved by distillation.

Despite significant advancement in this field, preparation of adsorbent material to efficiently tackle oil spill with inexpensive starting materials, remained a challenge. To address this concern, Chung et al. have developed an inexpensive polyolefin based oil superabsorbent terpolymer using 1-octene, styrene, divinylbenzene and Ziegler–Natta catalyst followed by thermal cross-linking [36]. This network has similar hydrophobic and oleophilic properties with that of hydrocarbons found in crude oils. These properties together with amorphous morphology and high free volume has promoted the oil adsorption into the network structure. The large amount of oil uptake (around 41 times compared to polymer weight) was associated with low cross-linking density of the network. Due to the high mechanical strength and selective adsorption of only oils makes the material buoyant that is particularly beneficial for practical applications. In addition to this, the

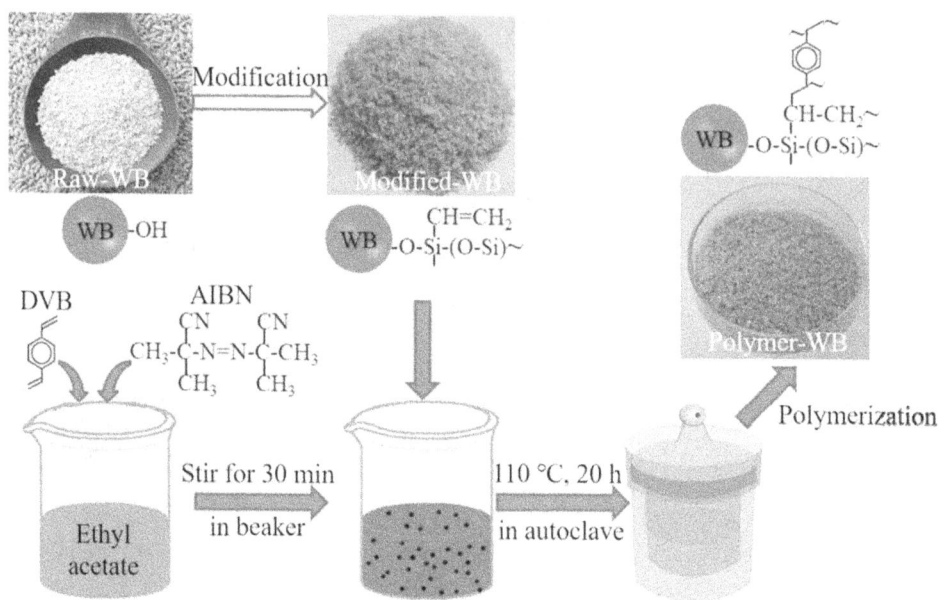

FIGURE 14.3 Schematic representation of the polymer-WB formation. Reproduced with permission from Elsevier, 2020 (Composites Communications 22, 2020, 100471).

polymer network can be thermally degraded below 600°C to get back the low molar mass hydrocarbons during the distillation and the oil can be reused further. This is particularly interesting as it addresses the solid waste disposal problem along with recovery of precious oils.

In another work, Yang et al. have developed a porous organic polymer (POP) to selectively uptake oil from oil-water mixture [37]. To fabricate the material, the surface of the electrospun gelatin membrane was modified and cross-linked by bis(diaryldiazomethane) via carbene insertion method. The fibrous morphology of gelatin along with cross-linking using bis(diaryldiazomethane), provides the required large surface area and hydrophobicity to the network. The material demonstrated high oil uptake capacity (5-16-fold increase in weight compared to the adsorbent weight) that was attributed to the retention of large surface area upon surface modification of gelation fibers.

14.4.2 REMEDIATION OF DYES

Polymeric gel-based adsorbents are the most reliable tool to alleviate dye-based water pollution owing to their selectivity, recyclability and overall robustness. Thus, a great deal of scientific interest is invested in the design and development of befitting hydrogels for targeted dye removal. A comparative study of the recently developed polymeric gel-based adsorbents for remediation of dyes is presented in Table 14.2. Recently, Shunmugam et al. have developed thiol-norbornene-based polymeric gels for removal of cationic and anionic dyes. For removal of cationic dyes, they have synthesised three different PEG containing macromonomers that are end-functionalized with norbornene moieties [38]. Photo-cross-linked network was obtained by reaction of these three macromonomers with a suitable crosslinker and pentaerythritol tetrakis (3-mercaptopropionate) (PETMP). It was very interesting to observe that the dye (rhodamine B, RhB) removal efficiency improved with increase in PEG chain in the network. This is due to the fact that with increase in PEG chain, more amount of water could penetrate into the network resulting in more exposure of the adsorptive sites to the dye molecules leading to amelioration of dye removal. Authors have also intended to remove anionic dyes from wastewater. For that purpose, quaternary ammonium moieties with different alkyl chains were incorporated in the cross-linked network [39]. Here also, the removal capacity of toxic anionic azo dye acid orange 7, improved with increase in alkyl chain in the quaternary ammonium

TABLE 14.2
Comparison of Macromolecular Gels to Remove Dyes from Aqueous Solutions

Adsorbent	Target Pollutant	Removal Efficiency	Ref.
PEG functionalised norbornene-thiol based photo cross-linked gel	Rhodamine B	66.7 mg/g	ACS Omega 6, 2020, 2800–10.
Triphenylphosphonium and quaternary ammonium conjugated thiol-norbornene based photo cross-linked gel	Acid orange 7	105.8 mg/g	New Journal of Chemistry 44, 2020, 14989–99.
Poly(epichlorohydrin)–ethylenediamine hydrogel (PEE-Gel)	Direct red 23	1540.19 mg/g	ACS Sustainable Chemistry and Engineering 5, 2017, 5598–607.
Poly(N-isopropylacrylamide)-based ionic hydrogels cross-linked with imidazolium-based dicationic ionic liquid	Congo red	578.0 mg/g	Polymer Journal 48, 2016, 431–38.
Poly (N-isopropylacrylamide-co-itaconic acid) hydrogels	Brilliant green	228 mg/g	Polymer Bulletin 66, 2011, 551–70.
Acrylic acid–acrylonitrile–N-isopropylacrylamide (AA-AN-NIPAAm) based hydrogel	Methylene blue	2.79 mg/g	Desalination and Water Treatment 57, 2016, 22543–50.
Poly(vinyl alcohol)/cyclodextrin modified poly(acrylic acid)/azobenzene-modified poly(acrylic acid) based hydrogel (PVA/PAA-CD/PAAAzo)	Methylene blue	25.86 mg/g	ACS Omega 5, 2020, 5470–79.
Poly(acrylic acid) based nanocomposite hydrogel (NC gel)	Methylene blue	2100 mg/g	Journal of Materials Chemistry A 6, 2018, 17612–24.
Polyacrylamide-g-chitosan gel	Acid blue 113	255.5 mg/g	Reactive and Functional Polymers 148, 2020, 104491.
Poly(methacryloxyethyltrimethyl ammonium chloride) composite hydrogel	Congo red	1693 mg/g	Journal of Colloid and Interface Science 532, 2018, 680–88.

moiety. Further incorporation of triphenylphosphonium functionality together with quaternary ammonium moiety was found to be conducive for removal of azo dye acid orange 7 [40]. This particular gel was significantly recyclable up to 10 cycles with a loss of only 7% of its adsorption efficiency. Similarly, Li et al. have developed quaternary ammonium-based cationic hydrogel for selective removal of anionic dye [41]. The gel was fabricated by a single step polymerisation using ethylenediamine (EDA) and epichlorohydrin (EPI). This particular gel was capable of removing anionic dye DR23, at pH ranging from 2–12. Owing to the inherent cationic nature of the gel, the ionisation of the quaternary ammonium group was reasonably independent of the pH of solution leading to efficient removal of the anionic dye at both acidic and basic conditions via electrostatic interaction. Further, poly(N-isopropylacrylamide) based hydrogels were also exploited for removal of anionic dyes. For instance, Zhou and coworkers have prepared poly(N-isopropylacrylamide)-based ionic hydrogels via free-radical polymerisation of N-isopropylacrylamide and imidazolium-based dicationic ionic liquid that acted as a crosslinker [42]. 1-vinylimidazole was quaternised with several dibromo compounds with different alkyl chains to prepare the imidazolium-based dicationic ionic liquid. The resulting hydrogels could efficiently remove anionic dyes such as congo red, methyl orange, methyl blue, orange G, bromothymol blue and thymol blue via interfacial interactions. In another work, Emik et al. have developed a temperature and pH-sensitive poly(N-isopropylacrylamide)-based hydrogel via free-radical polymerisation of N-isopropylacrylamide with sodium salt of itaconic acid and *N,N*-methylene bisacrylamide to remove cationic dyes such as Safranine T (ST), Brilliant Green (BG), and Brilliant Cresyl Blue (BCB) from wastewater [43].

Acrylic acid–based polymeric adsorbents are a very popular choice for removal of cationic dyes due to the presence of carboxylic acid functionalities which is readily ionisable. Malana et al. have prepared acrylic acid–acrylonitrile–N-isopropylacrylamide (AA-AN-NIPAAm) based hydrogel by free-radical polymerisation technique [44]. This gel was useful in removing cationic dye methylene blue (MB) with maximum adsorption efficiency 2.79 mg/g. Hou and coworkers have also reported poly(acrylic acid) based adsorbents for cationic dye removal [45]. In this work, hydrogel was fabricated via host-guest interactions and H-bonding, employing poly(vinyl alcohol), azobenzene-modified poly-(acrylic acid), and cyclodextrin-modified poly(acrylic acid) (PVA/PAA-CD/PAA-Azo). Interestingly, this particular gel exhibited a sol-gel transition that was induced by UV light irradiation and temperature. The gel was capable of removing MB dye (efficiency up to 25.86 mg/g) and RhB (maximum efficiency 11.64 mg/g) via electrostatic interactions. However, the dye removal efficiency of acrylic acid based hydrogels can be significantly improved by incorporating suitable crosslinkers that assists in the formation of an optimal three-dimensional network with increased possibility of water penetration inside the gel. Thus, Sun et al has developed poly(acrylic acid)-based nanocomposite hydrogel (NC gel) via free-radical polymerisation using calcium hydroxide (Ca $(OH)_2$) nano-spherulites (CNSs) as potential crosslinkers (Figure 14.4) [46]. The introduction of the CNS crosslinker has strengthened the three-dimensional network that has led to an increase in swelling in water (up to 500 times). Owing to more water penetration in the network, and highly porous morphology, this material exhibited exceptionally high methylene blue dye removal efficiency (2,100 mg/g) at neutral pH conditions. Further, this gel was fairly recyclable up to five cycles that makes it practically useful. In another work, Yadav and coworkers have demonstrated a nanocomposite adsorbent using magnetic iron oxide (Fe_3O_4), activated charcoal, cyclodextrin and sodium alginate polymer by direct mixing with nanofillers [47]. This nanocomposite material showed sufficient cationic dye removal efficiency in both polymeric gel and dry powder phase. However, the dry powder beads featured enhanced dye removal efficiency (10.63 mg/g) compared to polymeric gel (2.079 mg/g). Maximum MB dye removal for this particular adsorbent was almost 99.53% at pH = 6 with recyclability up to five cycles. Further, post-adsorption, the adsorbent can be easily removed with the help of a magnet due to its magnetic response. Feitosa et al. have reported polyacrylamide-g-chitosan–based hydrogels capable of removing toxic azo dye (Acid Blue 113) from wastewater [48]. For the synthesis, polyacrylamide was grafted onto chitosan and cross-linking was carried out with N,N'-methylenebisacrylamide using microwave and conventional techniques. The gel fabricated by microwave assisted synthesis showed enhanced dye removal (255.5 mg/g) compared to its counterpart (151.7 mg/g) prepared by conventional method. The former gel featured a more heterogeneous surface, higher chitosan content that promoted the dye adsorption driven by H-bonding (Figure 14.5). In other work, Ji and coworkers have developed poly(methacryloxyethyl trimethylammonium chloride) composite hydrogel microspheres for anionic dye removal applications [49]. Introduction of polyethersulfone has improved the mechanical properties of the network. This particular gel showed excellent efficiency for removal of anionic dyes (1,693 mg/g for Congo red, 1,491 mg/g for methyl orange and 204.7 mg/g for amaranth, respectively). Further, when the gel was installed in a chromatographic column, more than 95% of methyl orange dye could be separated.

14.4.3 REMOVAL OF INORGANIC IONS

Polymeric gels are more suitable candidates as adsorbents for removal of ions compared to supramolecular gels. Supramolecular gels require certain directional association with the target

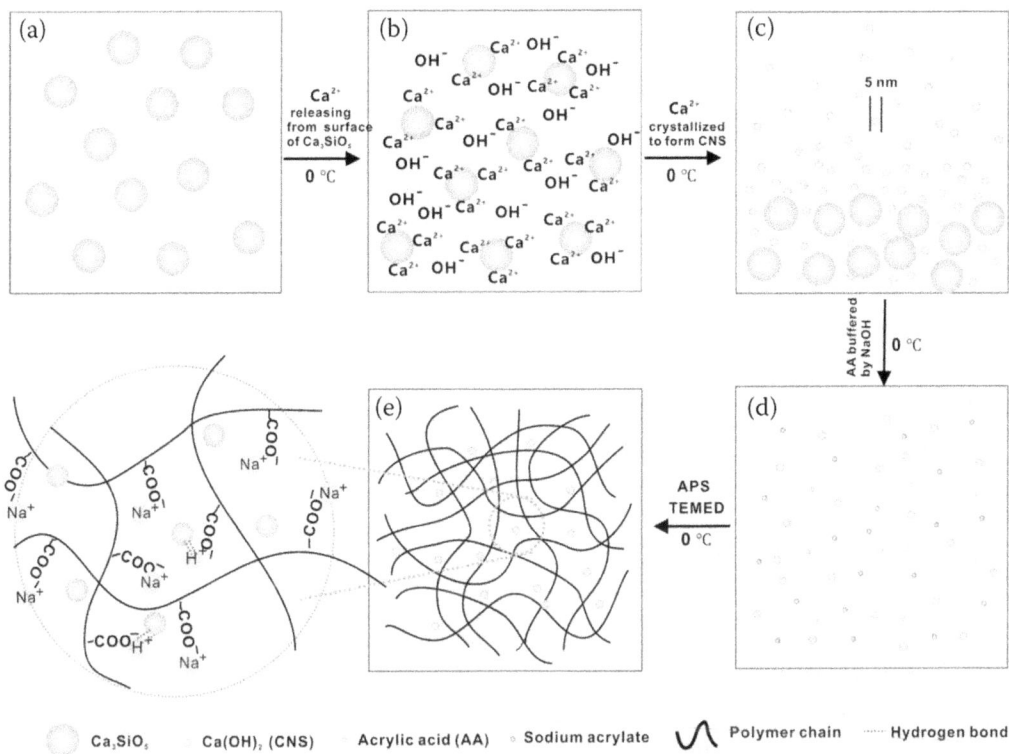

FIGURE 14.4 Schematic representation of the preparation of super-adsorbent NC gel adsorbent. (a) Dispersion of tricalcium silicate (Ca_3SiO_5) in water; (b) release of Ca^{2+} from surface of the Ca_3SiO_5; (c) crystallization of Ca^{2+} to obtain calcium hydroxide ($Ca(OH)_2$) (CNSs); (d) uniform mixture of CNSs with acrylic acid and sodium acrylate; (e) formation of the NC gel. Reproduced with permission from The Royal Society of Chemistry, 2018 (Journal of Materials Chemistry A 6, 2018, 17612–24).

ions for gelation. On the other hand, polymeric gels can be fine-tuned to render selectivity by attaching appropriate functional groups that are able to establish electrostatic or H-bonding interactions with the target toxic ions. Thus, extensive studies were carried out with various polymer backbones (presented in Table 14.3) for remediation of diverse toxic ions. For example, poly (N-isopropylacrylamide) (PNIPAM) is a well-known thermoresponsive polymer comprising both hydrophilic (amide) and hydrophobic (isopropyl) groups. The LCST of this polymer is around 32°C. Thus, this polymer turns hydrophobic from hydrophilic with increase in temperature. This very aspect has been extensively exploited in the field of drug delivery [50]. Besides, this temperature triggered phase change property has been utilized to develop environmentally benign adsorbents. Tokuyama et al. developed a temperature-swing solid-phase extraction (TS-SPE) technique based on N-isopropylacrylamide hydrogel to extract Cu^{2+} [51]. To begin with, Cu^{2+} was complexed with sodium n-dodecylbenzenesulfonate (SDBS) or n-dodecylbenzenesulfonic acid (DBS), both of these surfactants possesses anionic and hydrophobic functionality in their structure. The resulting metal-surfactant complex was selectively adsorbed onto PNIPAM hydrogel above its LCST via hydrophobic interactions. The adsorption of Cu^{2+} ions improved with increase in temperature and maximum adsorption occurred at temperature above LCST. The desorption of the metal-surfactant was easily achieved by cooling the system below LCST. In another work, Zdravkovic´ and coworkers have synthesised temperature and pH-sensitive poly(N-isopropylacrylamide-co-acrylic acid) hydrogels, poly (NIPAM-co-AA), for removal of heavy metals from aqueous solutions [52]. The presence of carboxyl and amide groups in the network structure together with the porous morphology

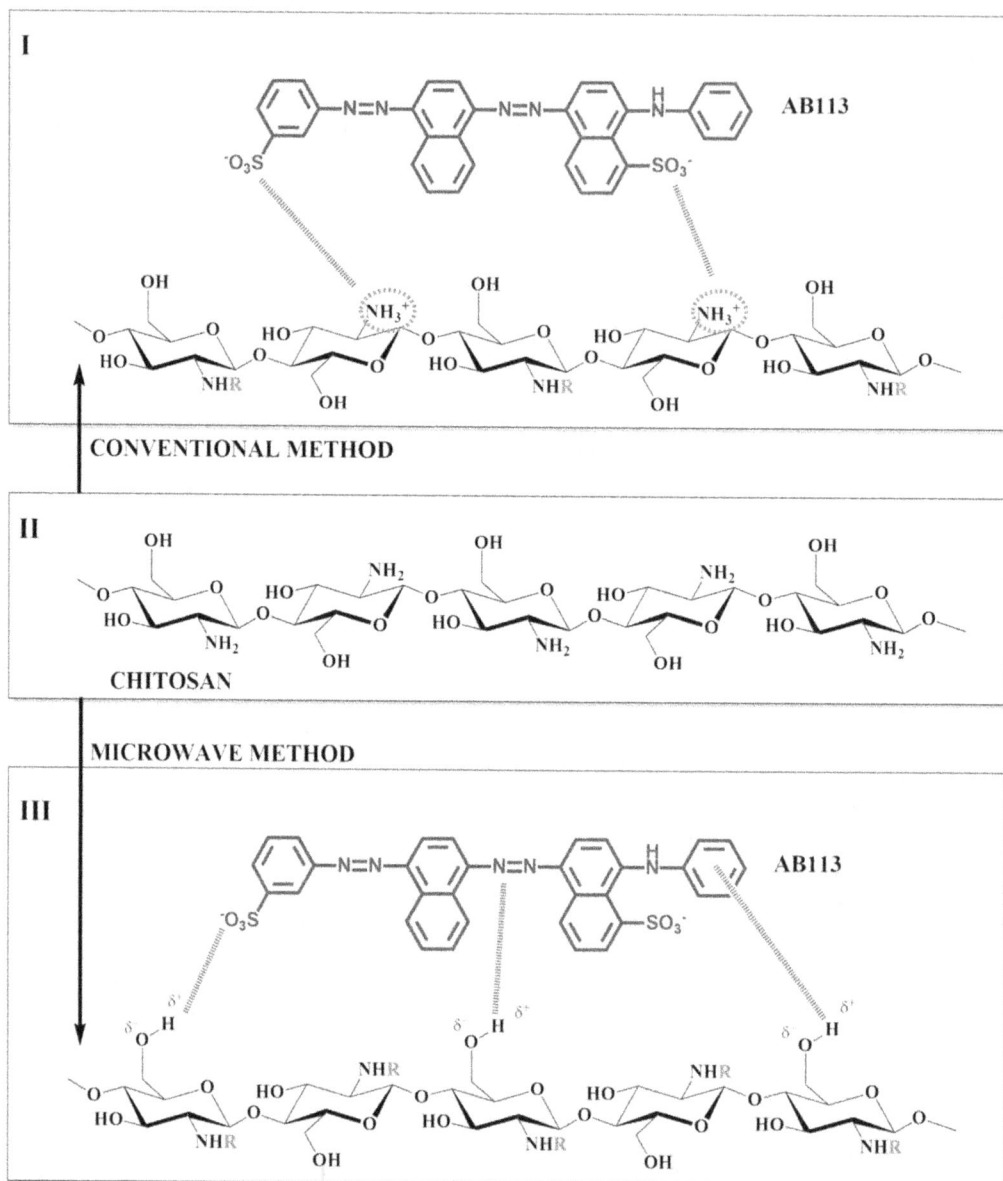

FIGURE 14.5 Possible structure of chitosan before and after the grafting reaction. (I) PAMChC (synthesized by conventional method); (II) chitosan; and (III) PAMChMW-120 (synthesized using microwave with 120 s of exposure under MW) at pH 6 and possible interaction between these hydrogels and dye Acid Blue 113; R = acrylamide (larger possibility) or N,N'-methylenebisacrylamide (smaller possibility). Reproduced with permission from Elsevier, 2020 (Reactive and Functional Polymers 148, 2020, 104491).

promoted removal of Pb(II), Cr(VI) and Mn(II). Chu et al have developed ion-recognition P (NIPAM-co-BCAm) hydrogels for selective removal of Pb(II) [53]. In this work, benzo-18-crown-6-acrylamide (BCAm) was used as the receptor of metal ion and it was polymerized with N-isopropylacrylamide via thermally initiated free-radical polymerisation to obtain the hydrogel. The adsorption of Pb(II) was attributed to the formation of BCAm/Pb^{2+} complexes via host-guest interactions. The adsorption of Pb(II) by this material was temperature dependent which was in line with the thermal responsiveness of poly(NIPAM). Adsorption was favored at

TABLE 14.3

Comparison of Macromolecular Gels for Heavy Metals Removal from Aqueous Solutions

Adsorbent	Target Pollutant	Removal Efficiency	Ref.
Poly(N-isopropylacrylamide) Hydrogel	Cu^{2+}	0.00930 mmol/g	Langmuir 23, 2007, 13104–8.
Poly(N-isopropylacrylamide-co-acrylic acid) hydrogel	$Cr_2O_7^{2-}$	289.85 mg/g	Polymer Bulletin 75, 2018, 4797–821.
	Mn^{2+}	221.73 mg/g	
	Pb^{2+}	588.23 mg/g	
4-vinylpiridine-grafted poly(acrylic acid) hydrogels	Pb^{2+}	117.9 mg/g	Journal of Hazardous Materials 192, 2011, 432–39.
Polyethylene glycol-b-polypropylene glycol (PEG-b-PPG)/ poly(acrylic acid) PAA-based hydrogel	Hg^{2+} Cu^{2+}	222.1 mg/g 283.4 mg/g	Colloids and Surfaces B: Biointerfaces 167, 2018, 176–82.
Carboxymethyl chitosan (CMC)/sodium alginate (SA)/graphene oxide@ Fe_3O_4 gel	Cu^{2+} Cd^{2+} Pb^{2+}	55.96 mg/g 86.28 mg/g 189.04 mg/g	Carbohydrate Polymers 216, 2019, 119–28.
Anion exchanger comprising N,N-diethyl 2-hydroxyethyl ammonium chloride dispersed in chitosan/poly (vinylamine) (PVAm) cyrogel	$Cr_2O_7^{2-}$	317.94 mg/g	Chemical Engineering Journal 330, 2017, 675–91.
Folic acid/polyaniline-based hybrid hydrogel	$Cr_2O_7^{2-}$	148 mg/g	ACS Sustainable Chemistry and Engineering 5, 2017, 9325–37.
Chitosan–poly(vinyl alcohol) hydrogel	Hg^{2+}	585.90 mg/g	Chemical Engineering Journal 228, 2013, 232–42.
La^{3+}/La(OH)₃ loaded acrylamide-based magnetic cationic hydrogel	PO_4^{3-}	90.2 mg/g	Water Research 126, 2017, 433–41.

a temperature below LCST of P(NIPAM-co-BCAm) whereas desorption occurred at temperature higher than LCST. Apart from PNIPAM-based hydrogels, gels based on poly(acrylic acid) are also known as effective sorbents for removal of Pb(II). However, Díaz et al. have demonstrated that the metal ion removal efficiency of PAA hydrogels can be significantly improved by grafting onto 4-vinylpyridine [54]. The Pb(II) ion removal efficiency was greatly influenced by solution pH. Higher adsorption was observed in the pH range 4–6. The maximum Pb(II) removal capacity was 117.9 mg for 1 gram of the net-PAAc-g-4VP hydrogel with 26.74% grafting. Further, Meng and coworkers have developed F127/PAA-based hydrogel for removal of Hg^{2+} and Cu^{2+} from wastewater [55]. In this work, PAA was integrated with biodegradable poly-ethylene glycol-b-polypropylene glycol (PEG-b-PPG) Pluronic F127 copolymers. The resulting gel showed excellent removal efficiency for Hg^{2+} (222.1 mg/g) and Cu^{2+} (283.4 mg/g) ions. In general, acid-based sorbents are not a very popular choice for Hg^{2+} removal. However, in this case the hydrophobic interaction with the polypropylene oxide present in F127 promoted Hg^{2+} removal.

Chitosan-based polymeric gels are also widely studied for removal of metal ions from wastewater due to their inherent biodegradability and possibility of efficient metal coordination using different binding sites. For instance, Luo et al. have developed a chitosan-based unique magnetic composite gel beads for removal of metal ions [56]. The gel was fabricated by mixing carboxymethyl chitosan (CMC), sodium alginate (SA) with graphene oxide@Fe_3O_4 using a calcium chloride cross-linking method. This material was capable of adsorbing metal ions such as Cu^{2+}, Cd^{2+} and Pb^{2+}. The removal was found to be pH dependent and highest adsorption was witnessed at pH 5–6. The maximum dye removal capacity was 55.96, 86.28 and 189.04 mg/g for Cu^{2+}, Cd^{2+} and Pb^{2+}, respectively. The chelation of metal ions with the carboxyl, hydroxyl and N-containing functionalities of the gel beads are the potential driving force for the metal ion sequestration. Moreover, the gel beads were fairly recyclable up to five cycles with retention of 90% efficiency. In another work, Dragan et al. have reported a novel chitosan containing hydrogel-microsphere composite system for removal of hexavalent chromium Cr(VI) [57]. The synthesis of the material was carried out by combining two porous elements as depicted in Figure 14.6. The first component was a strongly basic anion exchanger comprising N,N-diethyl 2-hydroxyethyl ammonium chloride which was evenly dispersed in the second component i.e., chitosan/poly(vinylamine) (PVAm) cryogel prepared by

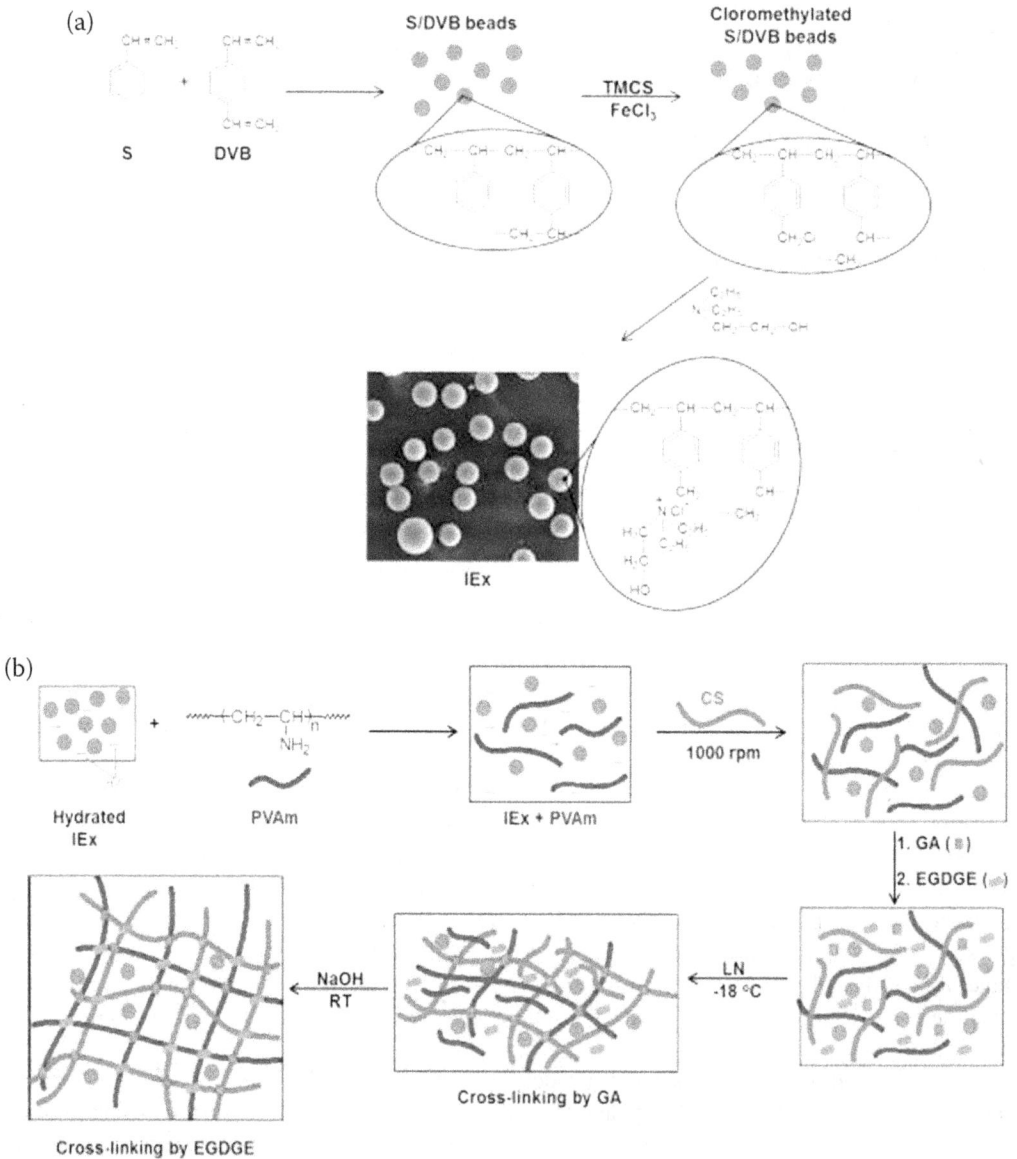

FIGURE 14.6 Schematic of synthesis of CS/PVAm/IEx composite cryogels: (A) Key stages in the preparation of IEx based microspheres; (B) major steps to fabricate CS/PVAm/IEx composite cryogels. Reproduced with permission from Elsevier, 2020 (Chemical Engineering Journal 330, 2017, 675–91).

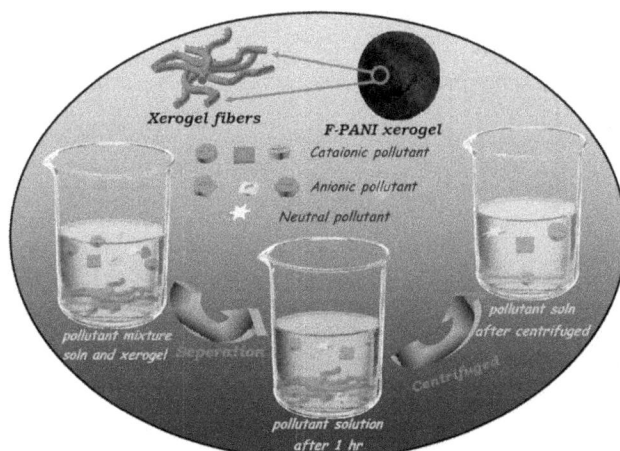

FIGURE 14.7 Schematic diagram of separation mechanism for anionic pollutant removal from mixture using folic acid-polyaniline (PANI)-based porous hybrid hydrogel (F-PANI). Reproduced with permission from American Chemical Society, 2017 (ACS Sustainable Chemistry and Engineering 5, 2017, 9325–37).

cross-linking with glutaraldehyde (GA) and ethylene glycol diglycidyl ether (EGDGE). Adsorption of Cr(VI) was favoured at acidic pH 3–6 and the maximum adsorption was 317.94 mg/g. The removal of Cr(VI) was also investigated by Nandi et al. [58] They have developed a folic acid-polyaniline (PANI) based porous hybrid hydrogel (F-PANI), via in situ polymerization of aniline. In this case folic acid was used as a crosslinker and PANI was present in emeraldine salt (ES) (positively charged) state. The material featured high Cr(VI) removal efficiency due to very high BET surface area and the electrostatic attraction between dichromate anions with positively charged PANI (Figure 14.7). The adsorption was favoured at neutral or acidic pH where Cr(VI) reduced to less toxic Cr (III). The material was recyclable up to five cycles with retention of 73% initial efficacy.

Hg is also recognized as a toxic metal owing to its high tendency of bioaccumulation and volatility. Thus, a lot of polymeric gel-based adsorbents have been explored to remove this ion from wastewater. However, in practical scenarios, Hg (II) coexists in water with other metal ions which interfere in the mercury adsorption by non-selective adsorbents. Therefore, it is imperative to design a selective adsorbent for removal of mercury. Cukrowska et al. have prepared an adsorbent based on cross-linked polyethylenimine and ethylene sulphide [59]. This material demonstrated high Hg (II) removal capability up to 97% in presence of other ions such as: Co(II), Cu (II), Pb(II), Zn(II), Ni(II), Fe(III). In a separate study, Wang and coworkers have developed a chitosan and poly(vinyl alcohol) (CTS-PVA) based hydrogel using glutaraldehyde cross-linking technique and alternate freeze-thaw process [60]. This hydrogel could selectively remove Hg(II) in the presence of Cu (II), Pb (II) and Cd (II) ions and the selectivity coefficient for Hg(II) ion removal was 487.7, 36642.5, 284298.5 times higher compared to Cu(II), Pb(II), and Cd(II) ions, respectively. The presence of pendant $-NH_2$, $-NHCOCH_3$ and $-C=N$ functionalities together with the three-dimensional network have contributed to the selective adsorption of Hg (II). Further, desorption can be achieved in HCl, HNO_3 and NaCl solutions indicating its practical applicability.

Sessler et al. have reported a unique macrocycle based polymeric hydrogel to remove water soluble anions [61]. The network was covalently cross-linked by a tetracationic macrocycle that acted as anion receptor. This material could remove several anions such as NO_3^-, NO_2^-, SO_4^-, HS^-, HSO_4^- and F^- from aqueous solutions via host-guest interaction of the anions with the tetracationic macrocyclic receptor (Figure 14.8). Desorption of the anions was accomplished by treating the material with dilute HCl followed by dialysis. Wang et al. have developed magnetic cationic hydrogel composites comprising La^{3+} and $La(OH)_3$ for selective removal of phosphates

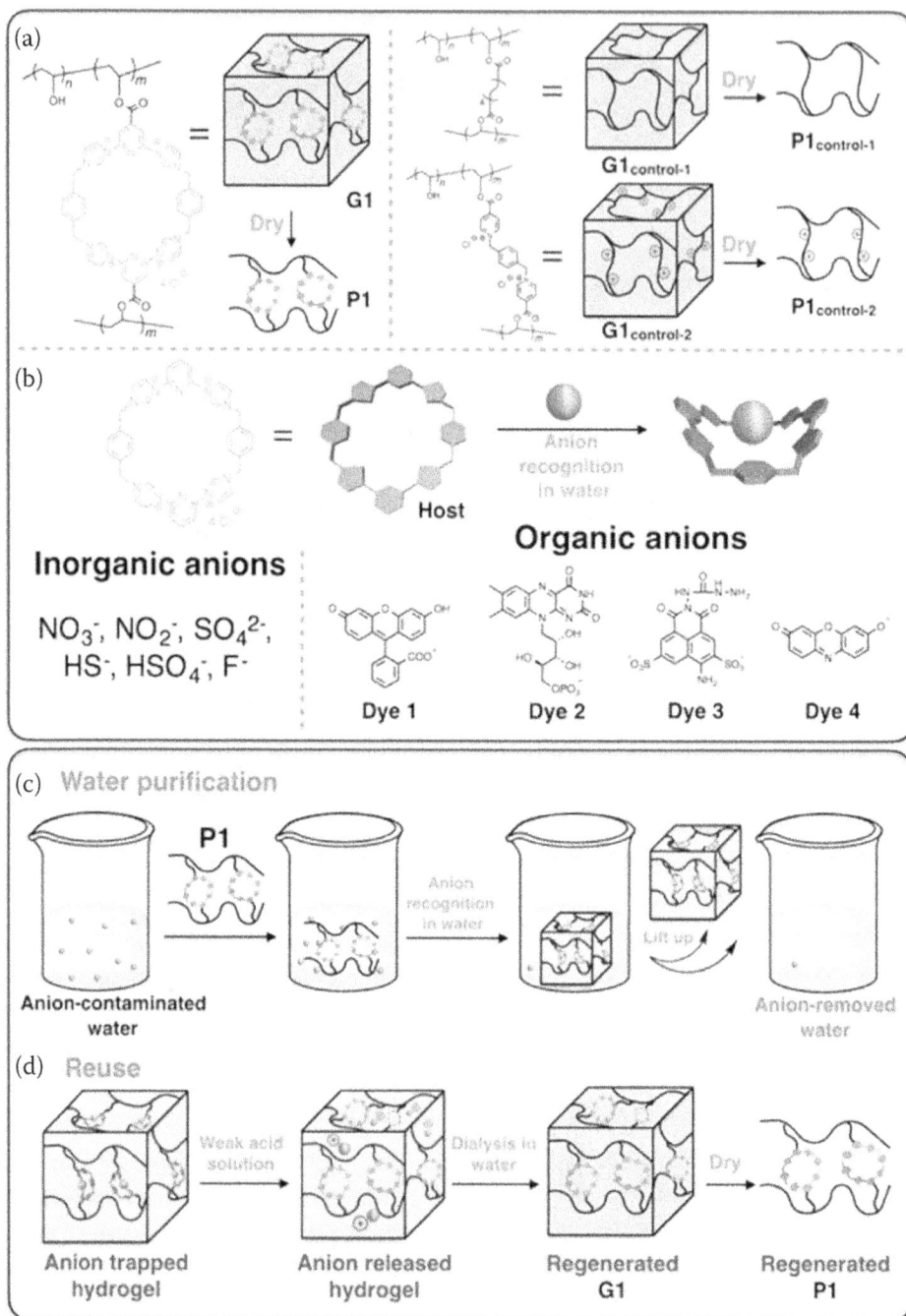

FIGURE 14.8 Macrocycle-based polymeric hydrogel for removal of anions. (**a**) Chemical structures and cartoon representations of tetracationic macrocycle containing polymeric gel; (**b**) proposed interaction of the macrocycle with the anions; (**c**) schematic diagram of anion removal; and (**d**) schematic diagram of regeneration technique. Reproduced with permission from American Chemical Society, 2018 (Journal of the American Chemical Society 140, 2018, 2777–80).

[62]. The adsorption of phosphate was favoured in the pH range 4.5–11 due to the possibility of electrostatic interaction, ligand exchange and Lewis acid-base interactions. Interestingly, the cationic hydrogel was capable of removing phosphate in presence of other interfering anions such as chloride, nitrate, sulfate, silicate, etc. Further, this hydrogel could be regenerated up to five cycles with retention of 72% initial efficiency, using NaOH-HCl as the desorption medium.

14.5 SUMMARY

This chapter depicts the recent developments of polymeric gel-based adsorbents for environmental remediation via sequestering major pollutants like oils, toxic organic dyes and inorganic ions from wastewater. It is apparent from the discussed examples that there has been significant advancement in this field in the last few decades. Considerable attention has been paid to the practical accessibility of the adsorbents. Most of the polymeric gels have been developed by facile synthetic routes without involving stringent conditions or some specific equipment. Most importantly, the environmental toxicity of the polymer components are also considered in many cases. It is vital for the adsorbents to be environmentally friendly in order to eliminate any chance of secondary pollution. Further, the gel-based adsorbents must feature substantial mechanical strength to retain their network structure even after pollutant uptake for easy cleanup post-application. Selectivity towards the target pollutant is another crucial factor and it demands more attention. Most of the adsorbents are tested in controlled laboratory conditions that are far from actual situations where there are multiple competing ions. Thus, binding sites should be prudently chosen and incorporated in the adsorbent in abundance to ensure maximum toxicant removal. Further, the regeneration of adsorbent is an integral part of water treatment. Hence, easy desorption techniques involving readily available reagents adds to the commercial viability of the adsorbent. Thus, this chapter aims to apprise the readers about the prospects of developing eco-friendly and robust polymeric gel-based adsorbents for environmental remediation.

REFERENCES

[1] Almdal, K., J. Dyre, S. Hvidt, and O. Kramer. 1993. Towards a Phenomenological Definition of the Term 'Gel'. *Polymer Gels and Networks* 1, no. 1 (January). 5–17. 10.1016/0966-7822(93)90020-I

[2] Bhattacharya, S., P. Diptendu, and R. Shunmugam. 2020. Triphenylphosphonium Conjugated Quaternary Ammonium Based Gel: Synthesis and Potential Application in the Efficient Removal of Toxic Acid Orange 7 Dye from Aqueous Solution. *New Journal of Chemistry* 44, no. 35 (September). 14989–14999. 10.1039/d0nj02138e

[3] Bhattacharya, S., and R. Shunmugam. 2020a. Quaternary-Ammonium-Based Gels with Varied Alkyl Chains for the Efficient Removal of Toxic Acid Orange 7. *ChemistrySelect* 5, no. 25 (July). 7427–7438. 10.1002/slct.202001527

[4] Bhattacharya, S., and R. Shunmugam. 2020b. Unraveling the Effect of PEG Chain Length on the Physical Properties and Toxicant Removal Capacities of Cross-Linked Network Synthesized by Thiol-Norbornene Photoclick Chemistry. *ACS Omega* 5, no. 6. 2800–2810. 10.1021/acsomega.9b03554

[5] Bläker, C., M. Johanna, P. Christoph, and D. Bathen. 2019. Characterization of Activated Carbon Adsorbents – State of the Art and Novel Approaches. *ChemBioEng Reviews* 6, no. 4 (August). 119–138. 10.1002/cben.201900008

[6] Chen, X., Z. Sukun, Z. Liming, Y. Tingting, and F. Xu. 2016. Adsorption of Heavy Metals by Graphene Oxide/Cellulose Hydrogel Prepared from NaOH/Urea Aqueous Solution. *Materials* 9, no. 7. 1–15. 10.3390/MA9070582

[7] Clark, A. H. 1991. Structural and Mechanical Properties of Biopolymer Gels. *Food Polymers, Gels and Colloids*. 322–338. 10.1533/9781845698331.322

[8] Crini, G. 2006. Non-Conventional Low-Cost Adsorbents for Dye Removal: A Review. *Bioresource Technology* 97, no. 9 (June). 1061–1085. 10.1016/j.biortech.2005.05.001

[9] Das, S., C. Priyadarshi, G. Radhakanta, P. Susmita, M. Sanjoy, P. Aditi, and A. K. Nandi. 2017. Folic Acid-Polyaniline Hybrid Hydrogel for Adsorption/Reduction of Chromium(VI) and Selective Adsorption of Anionic Dye from Water. *ACS Sustainable Chemistry and Engineering* 5, no. 10 (October). 9325–9337. 10.1021/acssuschemeng.7b02342

[10] Dave. 2011. Remediation Technologies for Marine Oil Spills: A Critical Review and Comparative Analysis. *American Journal of Environmental Sciences* 7, no. 5 (May). 423–440. 10.3844/ajessp.2 011.423.440

[11] Dong, S., W. Yili, Z. Yiwen, Z. Xiaohui, and H. Zheng. 2017. La3+/La(OH)3 Loaded Magnetic Cationic Hydrogel Composites for Phosphate Removal: Effect of Lanthanum Species and Mechanistic Study. *Water Research* 126. 433–441. 10.1016/j.watres.2017.09.050

[12] Dragan, E. S., H. Doina, V. D. Maria, and R. I. Olariu. 2017. Kinetics, Equilibrium Modeling, and Thermodynamics on Removal of Cr(VI) Ions from Aqueous Solution Using Novel Composites with Strong Base Anion Exchanger Microspheres Embedded into Chitosan/Poly(Vinyl Amine) Cryogels. *Chemical Engineering Journal* 330 (December). 675–691. 10.1016/j.cej.2017.08.004

[13] Eelkema, R., and A. Pich. 2020. Pros and Cons: Supramolecular or Macromolecular: What Is Best for Functional Hydrogels with Advanced Properties?. *Advanced Materials* 32, no. 20 (May). 1906012. 10.1002/adma.201906012

[14] Flory, P. J. 1974. Introductory Lecture. *Faraday Discussions of the Chemical Society* 57, (January). 7–18. 10.1039/DC9745700007

[15] Gordon, P. F., and P. Gregory. 1987. *Organic Chemistry in Colour*, 10.1007/978-3-642-82959-8

[16] Greene, T., and C. C. Lin. 2015. Modular Cross-Linking of Gelatin-Based Thiol-Norbornene Hydrogels for in Vitro 3D Culture of Hepatocellular Carcinoma Cells. *ACS Biomaterials Science and Engineering* 1, no. 12. 1314–1323. 10.1021/acsbiomaterials.5b00436

[17] Hansen, C. M. 1969. The Universality of the Solubility Parameter. Vol. 14. UTC. https://pubs.acs. org/sharingguidelines.

[18] Hou, N., W. Ran, W. Fan, B. Jiahui, Z. Jingxin, Z. Lexin, J. Hu, S. Liu, and T. Jiao. 2020. Fabrication of Hydrogels via Host-Guest Polymers as Highly Efficient Organic Dye Adsorbents for Wastewater Treatment. *ACS Omega* 5, no. 10. 5470–5479. 10.1021/acsomega.0c00076

[19] Hu, X. S., L. Rui, and G. Sun. 2018. Super-Adsorbent Hydrogel for Removal of Methylene Blue Dye from Aqueous Solution. *Journal of Materials Chemistry A* 6, no. 36 (September). 17612–17624. 10.1039/c8ta04722g

[20] Ji, H., S. Xin, H. Chao, T. Chengqiang, X. Lian, Z. Weifeng, and C. Zhao. 2018. Root-Soil Structure Inspired Hydrogel Microspheres with High Dimensional Stability and Anion-Exchange Capacity. *Journal of Colloid and Interface Science* 532 (December). 680–688. 10.1016/j.jcis.2018.08.036

[21] Ji, X., T. W. Ren, L. Lingliang, G. Chenxing, M. K. Niveen, H. Feihe, and J. L. Sessler. 2018. Physical Removal of Anions from Aqueous Media by Means of a Macrocycle-Containing Polymeric Network. *Journal of the American Chemical Society* 140, no. 8 (February). 2777–2780. 10.1021/ jacs.7b13656

[22] Ju, X. J., B. Z. Shi, Y. Z. Ming, X. Rui, Y. Lihua, and L. Y. Chu. 2009. Novel Heavy-Metal Adsorption Material: Ion-Recognition P(NIPAM-Co-BCAm) Hydrogels for Removal of Lead(II) Ions. *Journal of Hazardous Materials* 167, no. 1–3 (August). 114–118. 10.1016/j.jhazmat.2008.12.089

[23] Kadirvelu, K., M. Kavipriya, C. Karthika, M. Radhika, N. Vennilamani, and S. Pattabhi. 2003. Utilization of Various Agricultural Wastes for Activated Carbon Preparation and Application for the Removal of Dyes and Metal Ions from Aqueous Solutions. *Bioresource Technology* 87, no. 1 (March): 129–132. 10.1016/S0960-8524(02)00201-8

[24] Kato, T., H. Yuki, N. Suguru, and M. Moriyama. 2007. Liquid-Crystalline Physical Gels. *Chemical Society Reviews* 36, no. 12 (November). 1857–1867. 10.1039/b612546h

[25] Kavanagh, G. M., and S. B. Ross-Murphy. 1998. Rheological Characterisation of Polymer Gels. *Progress in Polymer Science (Oxford)*. Elsevier Ltd, January. 10.1016/S0079-6700(97)00047-6

[26] Li, D., L. Qing, B. Ningning, D. Hongzhou, and D. Mao. 2017. One-Step Synthesis of Cationic Hydrogel for Efficient Dye Adsorption and Its Second Use for Emulsified Oil Separation. *ACS Sustainable Chemistry and Engineering* 5, no. 6 (June). 5598–5607. 10.1021/acssuschemeng.7b01083

[27] Lim, J. Y. C., S. G. Shermin, S. S. Liow, K. Xue, and X. J. Loh. 2019. Molecular Gel Sorbent Materials for Environmental Remediation and Wastewater Treatment. *Journal of Materials Chemistry A* 7 (August). 18759–18791. 10.1039/c9ta05782j

[28] Liu, J., Z. Kai, J. Tifeng, X. Ruirui, H. Wei, Z. Lexin, Z. Qingrui, and Q. Peng. 2017. Preparation of Graphene Oxide-Polymer Composite Hydrogels via Thiol-Ene Photopolymerization as Efficient Dye Adsorbents for Wastewater Treatment. *Colloids and Surfaces A: Physicochemical and Engineering Aspects* 529 (September). 668–676. 10.1016/j.colsurfa.2017.06.050

[29] Lv, P., Y. Akram, H. Bin, and P. C. Ma. 2020. Wheat Bran/Polymer Composites as a Solidifier to Gel Oil on Water Surface. *Composites Communications* 22 (December). 100471. 10.1016/j.coco.202 0.100471

[30] Meng, Q., P. Bin, and S. Chong. 2018. Synthesis of F127/PAA Hydrogels for Removal of Heavy Metal Ions from Organic Wastewater. *Colloids and Surfaces B: Biointerfaces* 167 (July). 176–182. 10.1016/j.colsurfb.2018.04.024

[31] Moussavi, G., and M. Mahmoudi. 2009. Removal of Azo and Anthraquinone Reactive Dyes from Industrial Wastewaters Using MgO Nanoparticles. *Journal of Hazardous Materials* 168, no. 2–3 (September). 806–812. 10.1016/j.jhazmat.2009.02.097

[32] Malana, M. A., S. Parveen, and R. B. Qureshi. 2016. Adsorptive Removal of Organic Dyes from Aqueous Solutions Using Acrylic Acid–Acrylonitrile–N-Isopropylacrylamide Polymeric Gels as Adsorbents: Linear and Non Linear Isotherms. *Desalination and Water Treatment* 57, no. 47 (October). 22543–22550. 10.1080/19443994.2015.1132393

[33] Okesola, B. O., and D. K. Smith. 2016. Applying Low-Molecular Weight Supramolecular Gelators in an Environmental Setting-Self-Assembled Gels as Smart Materials for Pollutant Removal. *Chemical Society Reviews* (August). 4226–4251. 10.1039/c6cs00124f

[34] Osada, Y., and J. P. Gong. 1998. Soft and Wet Materials: Polymer Gels. *Advanced Materials* 10, no. 11. 827–837. 10.1002/(SICI)1521-4095(199808)10:11<827::AID-ADMA827>3.0.CO;2-L

[35] Otunola, B. O., and O. O. Ololade. 2020. A Review on the Application of Clay Minerals as Heavy Metal Adsorbents for Remediation Purposes. *Environmental Technology and Innovation* 18 (May). 100692. 10.1016/j.eti.2020.100692

[36] Özkahraman, B., A. Işil, and S. Emik. 2011. Removal of Cationic Dyes from Aqueous Solutions with Poly (N-Isopropylacrylamide-Co-Itaconic Acid) Hydrogels. *Polymer Bulletin* 66, no. 4 (March). 551–570. 10.1007/s00289-010-0371-1

[37] Pan, B., P. Bingcai, Z. Weiming, L. Lu, Z. Quanxing, and S. Zheng. 2009. Development of Polymeric and Polymer-Based Hybrid Adsorbents for Pollutants Removal from Waters. *Chemical Engineering Journal* 151, no. 1–3 (August). 19–29. 10.1016/j.cej.2009.02.036

[38] Prathap, A., and K. M. Sureshan. 2017. Organogelator-Cellulose Composite for Practical and Eco-Friendly Marine Oil-Spill Recovery. *Angewandte Chemie International Edition* 56, no. 32 (August). 9405–9409. 10.1002/anie.201704699

[39] Ramírez, E., S. Guillermina, B. C. Barrera-Díaz, R. Gabriela, and B. Bilyeu. 2011. Use of PH-Sensitive Polymer Hydrogels in Lead Removal from Aqueous Solution. *Journal of Hazardous Materials* 192, no. 2 (August). 432–439. 10.1016/j.jhazmat.2011.04.109

[40] Saad, D. M., E. M. Cukrowska, and H. Tutu. 2013. Selective Removal of Mercury from Aqueous Solutions Using Thiolated Cross-Linked Polyethylenimine. *Applied Water Science* 3, no. 2 (June). 527–534. 10.1007/s13201-013-0100-7

[41] Sangeetha, N. M., and U. Maitra. 2005. Supramolecular Gels: Functions and Uses. *Chemical Society Reviews* 34. 821–836. 10.1039/b417081b

[42] Saruchi, B. S., R. J. Kaith, and V. Kumar. 2015. The Adsorption of Crude Oil from an Aqueous Solution Using a Gum Tragacanth Polyacrylic Acid Based Hydrogel. *Petroleum Science and Technology* 33, no. 3 (February). 278–286. 10.1080/10916466.2014.976310

[43] Shen, D., F. Jianxin, Z. Weizhi, G. Baoyu, Y. Qinyan, and Q. Kang. 2009. Adsorption Kinetics and Isotherm of Anionic Dyes onto Organo-Bentonite from Single and Multisolute Systems. *Journal of Hazardous Materials* 172, no. 1 (December). 99–107. 10.1016/j.jhazmat.2009.06.139

[44] Silva, R. C., B. A. Samile, L. R. C. Pablyana, C. M. P. Regina, and J. P. A. Feitosa. 2020. Effect of Microwave on the Synthesis of Polyacrylamide-g-Chitosan Gel for Azo Dye Removal. *Reactive and Functional Polymers* 148 (March). 104491. 10.1016/j.reactfunctpolym.2020.104491

[45] Štefelová, J., S. Václav, S. Gilberto, T. O. Richard, T. Philippe, Z. Tanja, and H. Sehaqui. 2017. Drying and Pyrolysis of Cellulose Nanofibers from Wood, Bacteria, and Algae for Char Application in Oil Absorption and Dye Adsorption. *ACS Sustainable Chemistry and Engineering* 5, no. 3 (March). 2679–2692. 10.1021/acssuschemeng.6b03027

[46] Tang, S. C. N., W. Peng, Y. Ke, and I. M. C. Lo. 2010. Synthesis and Application of Magnetic Hydrogel for Cr(VI) Removal from Contaminated Water. *Environmental Engineering Science* 27, no. 11 (November). 947–954. 10.1089/ees.2010.0112

[47] Tokuyama, H., and T. Iwama. 2007. Temperature-Swing Solid-Phase Extraction of Heavy Metals on a Poly(N-Isopropylacrylamide) Hydrogel. *Langmuir* 23, no. 26 (December). 13104–13108. 10.1021/la701728n

[48] Tran, V. T., X. T. I. M. Xiubin, C. Jiaxi, J. V. Joost, and I. Jeon. 2018. Hydrogel Bowls for Cleaning Oil Spills on Water. *Water Research* 145 (November). 640–649. 10.1016/j.watres.2018.09.012

[49] Triboni, E. R., B. F. M. Thaisa, and M. J. Politi. 2019. Supramolecular Gels. *Nano Design for Smart Gels*, 35–69. Elsevier. 10.1016/B978-0-12-814825-9.00003-5

[50] Wang, S., and Y. Peng. 2010. Natural Zeolites as Effective Adsorbents in Water and Wastewater Treatment. *Chemical Engineering Journal*. 156, no 1 (January). 11–24. 10.1016/j.cej.2009.10.029

[51] Wang, X., D. Wenye, X. Yuyu, and C. Wang. 2013. Selective Removal of Mercury Ions Using a Chitosan-Poly(Vinyl Alcohol) Hydrogel Adsorbent with Three-Dimensional Network Structure. *Chemical Engineering Journal* 228 (July). 232–242. 10.1016/j.cej.2013.04.104

[52] Warson, H. 2000. Modern Superabsorbent Polymer Technology. *Polymer International* 49, no. 11 (November). 1548–1548. 10.1002/1097-0126(200011)49:11<1548::AID-PI482>3.0.CO;2-D

[53] Wu, Z., D. Weijie, Z. Wei, and J. Luo. 2019. Novel Magnetic Polysaccharide/Graphene Oxide @Fe$_3$O$_4$ Gel Beads for Adsorbing Heavy Metal Ions. *Carbohydrate Polymers* 216 (July). 119–128. 10.1016/j.carbpol.2019.04.020

[54] Xu, X., L. Yang, F. Wenbo, Y. Mingyu, D. Zhen, X. Jiaming, L. Dongxiang, W. Shengjie, X. Yongqing, and M. Cao. 2020. Poly(N-Isopropylacrylamide)-Based Thermoresponsive Composite Hydrogels for Biomedical Applications. *Polymers* 12, no. 3. 580. 10.3390/polym12030580

[55] Yadav, S., A. Anupama, C. Rupa, J. Bhawana, K. S. Ajaya, A. C. C. Sónia, and Md A. B. H. Susan. 2020. Cationic Dye Removal Using Novel Magnetic/Activated Charcoal/β-Cyclodextrin/Alginate Polymer Nanocomposite. *Nanomaterials* 10, no. 1 (January). 1–20. 10.3390/nano10010170

[56] Yu, X., Y. Pengfei, G. M. Mark, W. Liang, X. Jinku, W. Yongqing, L. Lian, and P. Yunlin. 2018. Electrospun Gelatin Membrane Cross-Linked by a Bis(Diarylcarbene) for Oil/Water Separation: A New Strategy to Prepare Porous Organic Polymers. *ACS Omega* 3, no. 4 (April). 3928–3935. 10.1021/acsomega.8b00162

[57] Yuan, X., and T. C. M. Chung. 2012. Novel Solution to Oil Spill Recovery: Using Thermodegradable Polyolefin Oil Superabsorbent Polymer (Oil-SAP). *Energy and Fuels* 26. 4896–4902. 10.1021/ef300388h

[58] Zdravković, A., N. Ljubiša, I. S. Snežana, N. Vesna, N. Stevo, M. Žarko, Ć. Ana, and P. Sanja. 2018. The Removal of Heavy Metal Ions from Aqueous Solutions by Hydrogels Based on N-Isopropylacrylamide and Acrylic Acid. *Polymer Bulletin* 75, no. 10 (October). 4797–4821. 10.1007/s00289-018-2295-0

[59] Zhang, Q., L. Na, W. Yen, and F. Lin. 2018. Facile Fabrication of Hydrogel Coated Membrane for Controllable and Selective Oil-in-Water Emulsion Separation. *Soft Matter* 14, no. 14 (April). 2649–2654. 10.1039/c8sm00139a

[60] Zhou, X., W. Jing, N. Jingjing, and D. Binyang. 2016. Poly(N-Isopropylacrylamide)-Based Ionic Hydrogels: Synthesis, Swelling Properties, Interfacial Adsorption and Release of Dyes. *Polymer Journal* 48, no. 4 (April). 431–438. 10.1038/pj.2015.123

15 GeN-NxT Materials for Tire

Partheban Manoharan and C. Harimohan

Yokohama Off-Highway Tires, Research and Development, Tirunelveli, India

CONTENTS

DOI: 10.1201/9781003200710-17

15.1   INTRODUCTION

The automotive and automobile industries are passing through the season of changeover from internal combustion engines (CEs) to electric vehicles (EVs) across the world. However, the tire is an inevitable product for the auto industry. Today, worldwide production of tires is more than a trillion units per year.[1,2] Tire design and construction has been changed several times over the period as per vehicle requirements and applications maintaining the same torus shape. On the other hand, tire manufacturers around the globe are progressively focusing on eco-friendly and sustainable materials together with Registration, Evaluation, Authorisation, and Restriction of Chemicals (REACH) and SVHC Substance of very high concern (SVHC) compliance to improve the performance of the tire. Consequently, optimizing the tire materials and reducing the burden on our environment is of prime importance. Recent research reveals that regular tires are not suitable for electric vehicles. The reason for this is that the electric vehicles do not have an engine and are operated by strong durable batteries and able to deliver instant high torque with the help of motors. Therefore, it makes zero noise while operating the electrical vehicle (car, bus, commercial vehicles, and so on). The presence of a heavy battery increases the vehicle's weight, which facilitates the fast wear of tires up to 30%.[2] Currently, the tire manufacturers' goal is to support electric vehicles with the quietest tire for the smoothest drive coupled with superior wear resistance.

Tire materials play a vital role in achieving the goals of tire manufacturers. Tires are classified by vehicle type: passenger car, truck, off road, bicycles, motorcycles, industrial, aircraft, military, and race cars. They vary in size, shape, load, and speed rating depending on their application and are designed to perform in a large number of service conditions.[3] By designing each component of tire by choosing appropriate materials and modifying the compound properties, a desired set of performance parameters can be achieved.[4-6] This chapter reviews the next-generation materials used in tires.

15.2   POLYMERS

Polymer is the primary ingredient for tire formulation and plays a significant role in determining their performance. Natural rubber (NR) and synthetic rubbers like styrene butadiene rubber (SBR), polybutadiene rubber (PBD), and butyl rubber (IIR) are mostly used in formulating tread and other components. Their microstructure, glass transition temperature (T_g), and composition have a direct impact on tire performance properties like rolling resistance coefficient (RRC), mileage, traction, permeability, durability, cut and chip, and so on. For example, both RRC and traction are linked with T_g of the polymer, as depicted in Figure 15.1.[7]

15.2.1   NATURAL RUBBER

The excellent properties offered by natural rubber (NR) make it a traditional workhorse for the tire industry. Their stereo-regularity (99.9% cis content) and high molecular weight impart good fatigue life, green strength, low hysteresis, tear resistance, and tack thereby contributing to the overall performance of the tire.[8] Internal components of the tire are predominantly of NR owing to the lower heat generation under flexing and better tack than other synthetic rubbers.[9] Ribbed smoked sheets (RSS) and technically specified block rubber (TSR) are the major forms of NR consumed by the tire industry. A special grade of NR obtained from a blend of latex and cup lump

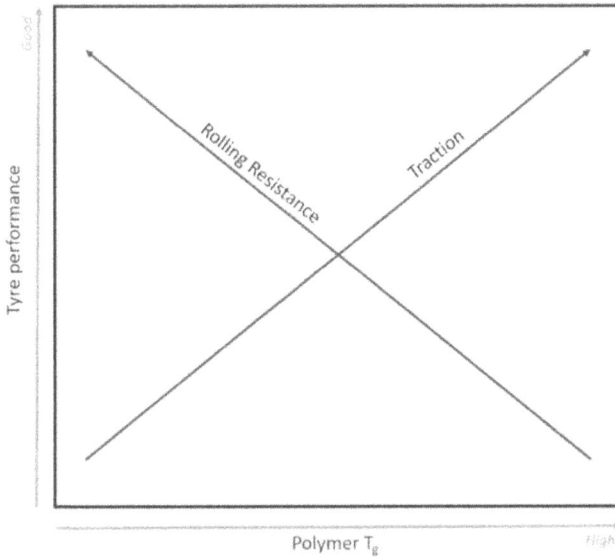

FIGURE 15.1 Dependence of T_g of polymers on rolling resistance and traction. Adapted from.[7]

was identified with high plasticity index. This material has the potential to deliver high physico-mechanical properties. A similar property can be achieved by the direct blending of RSS and TSR. NR is modified chemically or physically to further improve their inherent properties.

15.2.1.1 Epoxidized Natural Rubber

The high level of unsaturation in the main chain leads to undesirable properties such as poor aging retention, low heat, and abrasion resistance. Epoxidation is a chemical reaction that reduces unsaturation by converting the C=C bond into epoxides.[10] Figure 15.2 displays the NR epoxidation process.[11]

The final properties of epoxidized natural rubber (ENR) are decided by the degree of epoxidation. As the epoxidation level increases, the hardness of the rubber increases. It was found that less than 50% epoxidation gives an elastic material with off white colour. An increase in epoxidation increases the gel content i.e., the insoluble part of the polymer. The schematic structure of ENR25 and ENR50 is depicted in Figure 15.3. As epoxidation reduces unsaturation, ENR is better than NR in oxidative stability, imparting improved ageing and heat resistance. Epoxidation also increases the polarity of the rubber so can be reinforced with silica filler without using coupling agents.[12–14] The modification of NR by epoxidation affects the T_g, as shown in Figure 15.4[15].

When ENR was examined as tire tread materials, ENR with 25 mole% epoxidation (ENR25) and with 50 mole% epoxidation (ENR50) displayed good wet grip characteristics.

The ageing character of ENR is dependent on the cure system type. Baker et al. reported that the conventional cure system resulted in vulcanizate with poor ageing characteristics as sulphuric acid from the oxidation of the sulphide ring opening may cause cross-linking which results in poor ageing characteristics.[8] ENR is utilized as a compatibilizer for silica compounds to enhance

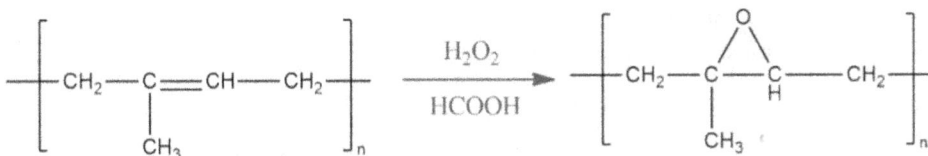

FIGURE 15.2 Epoxidation of NR.[11]

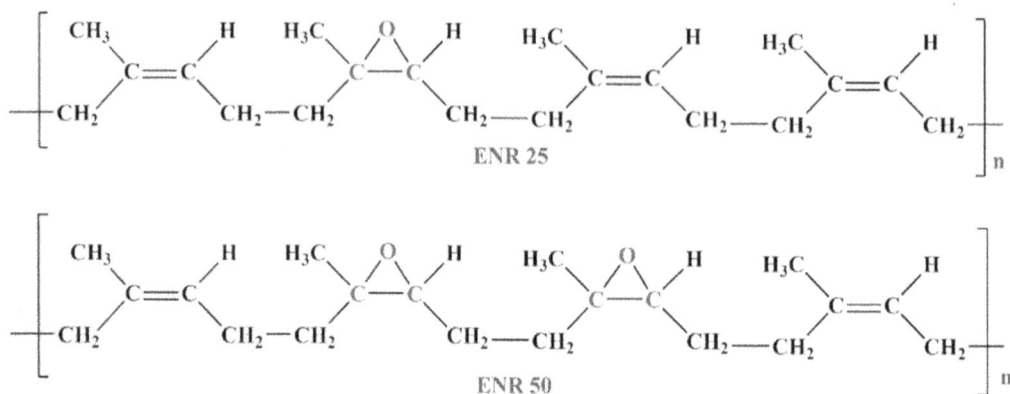

FIGURE 15.3   Structures of ENR25 and ENR50.

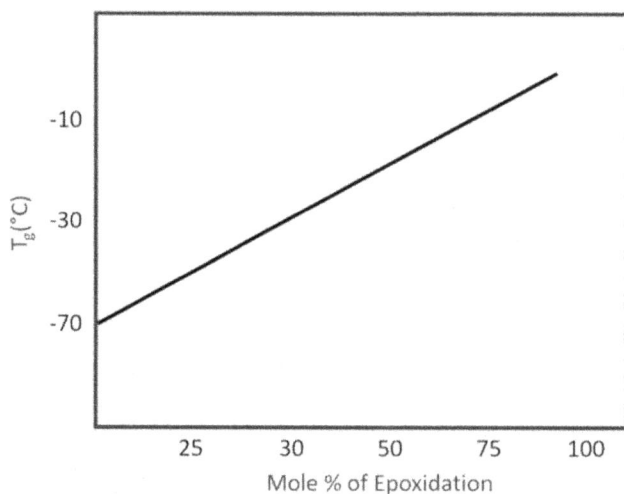

FIGURE 15.4   Variation in Tg with epoxidation.[15]

mechanical properties. Sengloyluan et al. accounted for the use of ENR (2.5–12.5 phr) in NR-silica compound and found improved filler–rubber interaction, silica dispersion, and tensile strength.[16] An increase in ENR content and mole% of epoxide groups increases the interactions between silica and rubber and decreases filler–filler interaction. And this is reflected in the low filler networking factor, Mooney viscosity, Payne effect, flocculation rate constant, and in the high filler–rubber interaction parameter, chemically bound rubber, and improved tensile strength.[16] Figure 15.5 illustrates the plausible interaction mechanism and bonding between ENR and silica particles.[16] Commercial production of epoxidized NR (ENR) has been carried out in Malaysia since 1990, with ENR10 (having 10% mole of epoxide content), ENR25 (having 25% mole of epoxide content), and ENR50 (having 50% mole of epoxide content). However, the properties of ENR10 closely resemble those of NR and it was later withdrawn from the market.[17] Two grades are available under the trade names Epoxyprene 25 and Epoxyprene 50.

15.2.1.2   Oil Extended Natural Rubber

Oil extended natural rubber (OENR) contains 20–25 phr of aromatic or naphthenic oil and the oil is added either in the latex stage or in the dry stage. In general, the increase in oil content reduces tensile strength but possesses better tear and wear resistance when used with butadiene rubber.

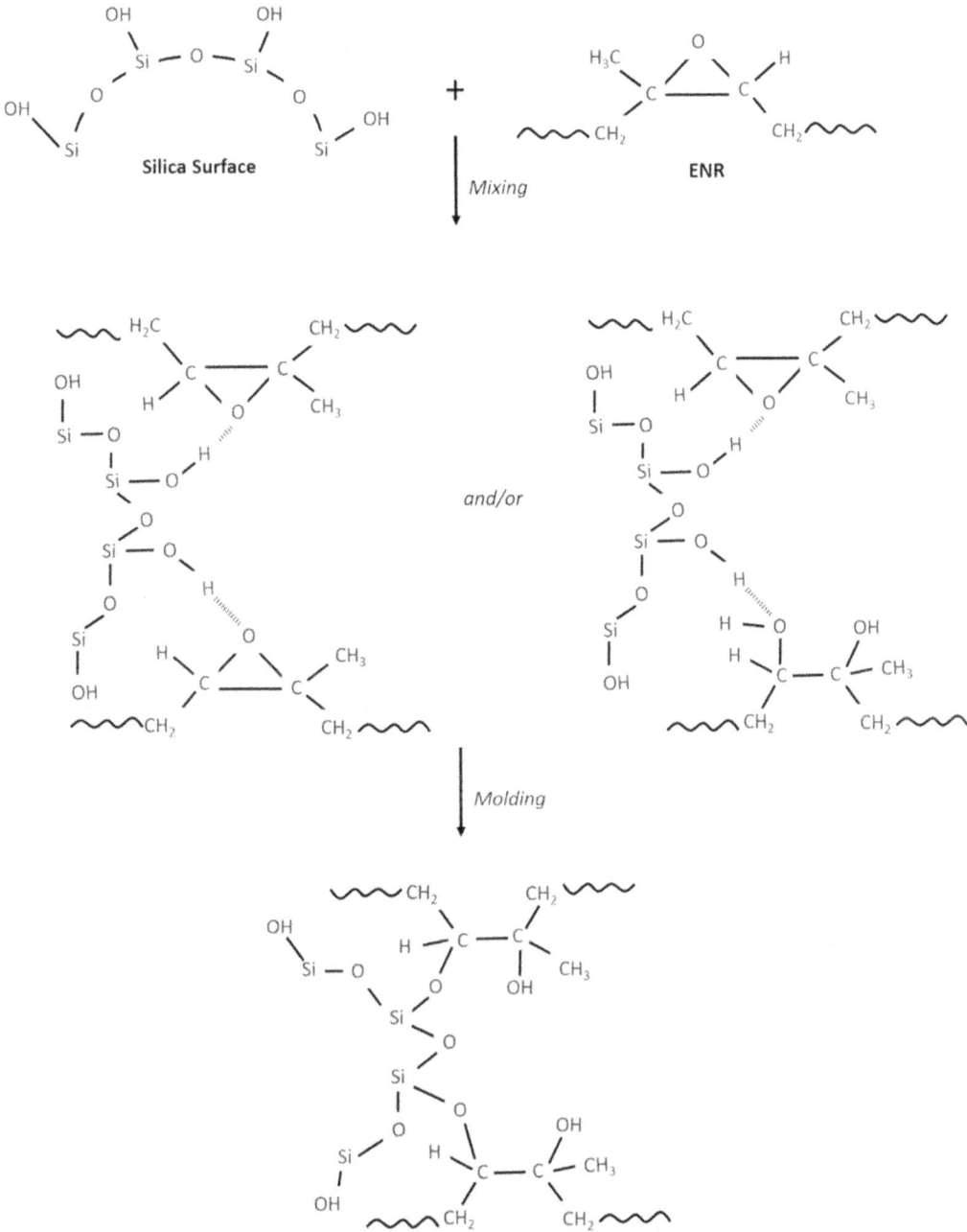

FIGURE 15.5   Plausible mechanism of interaction and bonding between ENR and silica particles.

15.2.2   STYRENE BUTADIENE RUBBER

Styrene butadiene rubber (SBR) is one of the cheapest and highest volume general-purpose synthetic rubbers. It is a random copolymer of butadiene and 10% to 25% styrene monomer. SBR can be produced either by solution polymerization technique (SSBR) or by emulsion polymerization technique (ESBR). Fatigue resistance and low temperature properties of SBR are generally inferior to NR, while their abrasion resistance and heat-aging properties are better. Due to the presence of double bonds in the polydiene backbone, SBR is vulnerable to thermal and oxidative degradation and is generally by cross-linking, leading to an increase in stiffness or embrittlement.

15.2.2.1 Functionalized SBR

SSBR is synthesized by living anionic polymerization, enabling the control over microstructures and the functional groups can be easily grafted to the polymer chain. Functionalization of chain ends with alcohols, amines, amino, and cyano groups have proved to improve polymer-carbon black fillers' interaction.[18,19]

Tin (Sn) coupled chain end modified SSBR with carbon black was reported to have reduced the Payne effect and low hysteresis loss. Improvement in hysteresis loss was noted to be dependent on the Sn content and the type of Sn-C bond, as shown in the Figure 15.6.[20] This improvement is

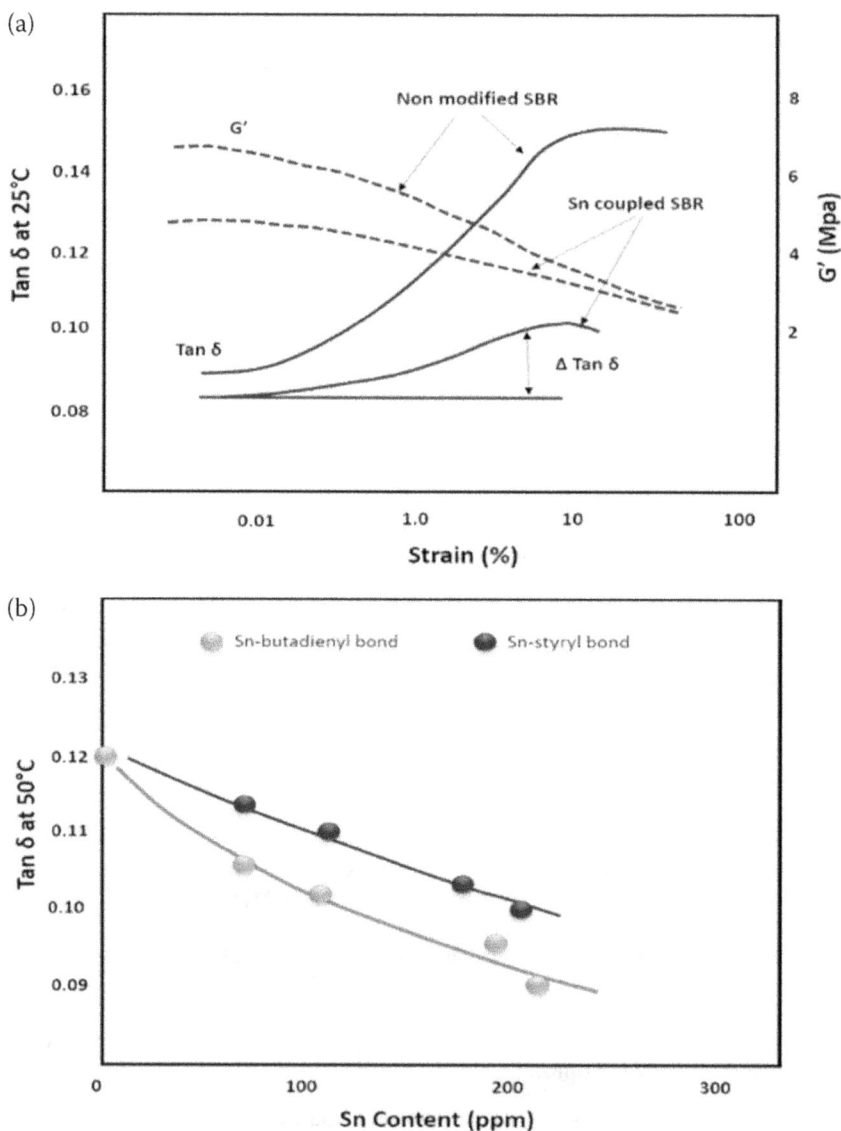

FIGURE 15.6   (a) Impact of tin-coupling on hysteresis loss and payne effect of carbon black–reinforced rubber compounds; (b) effects of types of Sn–C bonds and centration of tin on dynamic hysteresis loss properties of carbon black–reinforced rubber compounds (100 phr SSBR and 50 phr N330 carbon black). Adapted from.[20]

attributed to less formation of carbon black network and lower contribution of free chain ends.[21] Improvement in dynamic properties was reported for SSBR with amine modified chain ends as they strengthen polymer-carbon black interaction and aid better dispersion of carbon black.[22] The carboxylic group forms hydrogen bonds with carbon black particles and thus in-chain modification of SSBR with the carboxylic group gives low hysteresis loss when used with conventional or oxidized carbon black.[21]

The star-shaped polymer, Miktoarm, with Sn centre has attracted more attention recently than the linear integral polymer. Their star-shaped structure and high molecular weight reduces RRC. The weaker Sn-C bond at the centre breaks easily during mixing and thus improves the compatibility with other polymers and the free radical generated can react with the electrons on the carbon black particle.

For silica compounds, a number of functional groups like silanol, alkoxy silane, carboxyl, epoxide, hydroxyl, and polyether are introduced either at the polymer chain ends or along the chains to enhance polymer-filler interaction. The influence on hysteresis loss of silica compounds by different end chain modifications is shown in Figure 15.7.[21] For further improving the dynamic properties, multi-functionalized solution SBRs were developed by copolymerizing functionalized monomers or functionalized initiators. Many mechanisms are involved in the hysteresis loss reduction when a functionalized polymer is used with silica and silane-coupling agents. An improved silica-polymer interaction, better silica dispersion, and less free chain ends are generally believed the primary factors for improved dynamic properties. Synthos rubber introduced silica functionalized high T_g SSBR under the trade names Syntion 2150 × 4 and Syntion 2150 × 1, claiming better wet grip, low RR, and better wear resistance.

Emulsion SBR (ESBR) is synthesized by free radical polymerization and there is little control over microstructure. Several functional groups that can give hydrogen bonds with silica cannot be grafted during living polymerization but can be introduced by copolymerization in emulsion SBR.[23] Hydroxylated methacrylate and acrylates are copolymerized with ESBR to enhance interaction with silica particles. Use of acrylonitrile and hydroxylated methacrylate as a third monomer in ESBR gives compounds with low hysteresis loss.[23]

15.2.3 Isobutylene Isoprene Rubber-IIR (Butyl Rubber)

IIR constitutes 97 to 99.5 mole% isobutylene and 0.5 to 3 mole% isoprene, which provides the double bond for sulfur vulcanization. The higher the mole% of isoprene, the easier the degradation

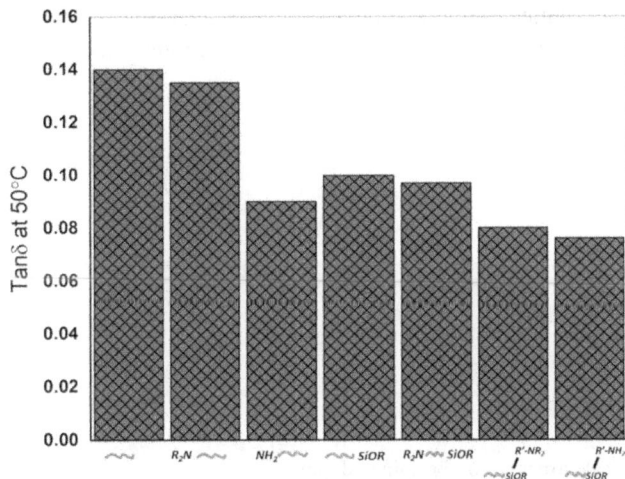

FIGURE 15.7 Effect of different functional groups used for the modification of polymers on the hysteresis loss of rubber. Adapted from.[21]

of the polymer, but speeds up the cure. IIR are produced by the cationic copolymerization of isobutylene and isoprene monomers with aluminium chloride ($AlCl_3$) catalyst in methyl chloride solvent. Because of the saturated backbone, they are resistant to oxygen and ozone, can withstand prolonged exposure to heat, and are usually stable in dilute acids and alkalis. IIR are known for their exceptionally low gas and moisture permeability, making them ideal for inner tubes and high-pressure/vacuum applications.

To enable co-vulcanization with other rubbers such as NR, SBR, and PBD, IIR is often halogenated with bromine or chlorine. The introduction of halogen groups can accelerate curing rate and increase polarity. The enhanced cure properties do not affect impermeability and fatigue properties, thus permitting the development of more durable tubeless tires. Today, halogenated IIR are mostly consumed for tire inner liners and tubes.

15.2.3.1   Brominated Poly (Isobutylene-co-p-methylstyrene)-BIMSM

Brominated poly(isobutylene-co-p-methylstyrene), BIMSM, is a specialty elastomer prepared by carbo-cationic copolymerization of isobutylene and para-methylstyrene. BIMSM copolymers have a saturated backbone with randomly distributed pendant para-methyl substituted aromatic rings. Then, in a bromination process during post-polymerization, a portion of para-methyl groups is converted to bromomethyl groups, imparting the desired amount of benzyl bromide functionality. Figure 15.8 depicts the chemical structure of brominated poly(isobutylene-co-p-methyl styrene).[24] This material possesses a unique combination of low permeability with low T_g and saturated backbone, making the material ideal for applications requiring air barrier properties with ozone, oxidative, and heat-ageing resistance.[6] A new inner liner technology was introduced later by ExxonMobil Chemical Company where BIMSM and nylon were reactive extruded, forming a dynamically vulcanized alloy, Exxcore™ DVA. The material was reported to have ten times the reduction in air permeability compared to conventional halobutyl. This new material has the potential in down-gauging the inner liner and tire light-weighting.[6]

15.2.4   POLYBUTADIENE RUBBER

Polybutadiene (PBD) is a synthetic rubber from the monomer 1,3-butadiene. It is the second-largest in volume of produced synthetic rubber after SBR, of which 70% is consumed by the tire industry. PBD gives excellent elasticity, low hysteresis loss, and a strong resistance to abrasive and fatigue resistance. These unique properties are due to their regularity, linear molecular structure, and low glass transition temperature (T_g), typically lesser than −90°C. However, PBD also gives poor wet traction, so they are usually blended with other elastomers like NR or SBR for tire tread compounds.

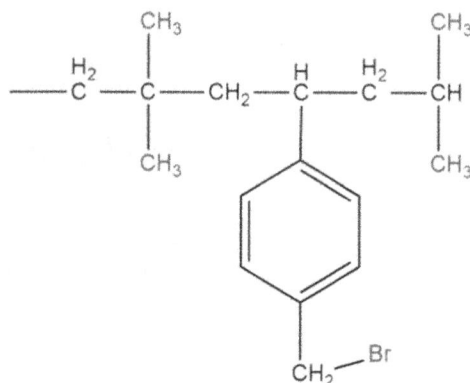

FIGURE 15.8   Chemical structure of brominated poly (isobutylene-co-p-methyl styrene).

TABLE 15.1
Different Catalyst Used in Synthesizing PBD

Catalyst	Microstructure (%)			Mooney Viscosity ML$_{1+4}$ 100°C
	Cis	Trans	Vinyl	
TiX$_4$/AlEt$_3$	90–93	4–2	6–5	45
Co(OCOR)$_2$/AlEt$_2$Cl	96–98	2–1	2–1	30–50
Ni(OCOR)$_2$/AlEt$_3$/ BF$_3$(OEt$_2$)	96–98	2–1	2–1	35–50
Nd(OCOR)$_3$/AlR'$_3$/ Et$_2$Al$_3$Cl$_2$	97–99	2–1	1–0	35–60
Bu-Li	32–36	53–58	8–11	30–70

Cis-1,4 content in PBD plays a vital role in determining elastomer properties and also is beneficial for the strain-induced crystallization. With cis-1,4 content, tensile strength and elongation increases and also improves fatigue and crack resistance. Studies were carried out to narrow the PBD molecular weight distribution ($M_w/M_n > 3.0$), which is related directly to the uniformity of the vulcanized cross-linked network, and was shown to strongly affect the physical properties.[25] Recent research revealed that PBD with high molecular weight, narrow molecular weight distribution (<1.1) and high cis-1,4-regularity (>99%) can be synthesized by using a series of cationic alkyl rare-earth metal catalysts. However, these catalysts are very complicated and expensive, which limited the industrial application of the catalysts.[26]

High cis content polybutadiene rubber (HC PBD) find application in tire, as they are resistant to wear and resistance and flex fatigue. In addition, HC PBD has the lowest T_g of −110°C among the general synthetic elastomers, making it ideal for winter tire tread application. HC PBD is synthesized by anionic coordination polymerization technique by reacting 1,3 butadiene with a Zingler Natta catalyst in organic solvent. Table 15.1 tabulates different catalysts used industrially.[27]

Metal used for catalysts affects both molecular weight and branching. Co, Ni, and Nd are the prime catalysts used to produce HC PBD. HC PBD synthesized with Nd-catalyst exhibited superior tensile strength and low hysteresis loss and wear resistance. Table 15.2 tabulates vulcanizate properties of HC PBD synthesized by different catalysts.[27] Nd allows the insertion of functional groups at the polymer chain ends. End functionalized PBD is reported to have improved resilience and wear resistance.[28] Zeon Corporation has developed terminally modified PBD with a functional group reactive toward carbon black.

15.3 FILLERS

In general, fillers are added to rubber compounds to meet the material properties such as tensile strength, hardness, abrasion resistance, etc.

15.3.1 CARBON BLACK

Carbon black (CB) is the most utilized reinforcing filler in tires. They are elemental carbon and are composed of aggregate particles. CB used in the rubber industry are the ones produced by the oil furnace process and thermal process. Wide varieties of carbon blacks are available and a correct choice is important to achieve performance specifications. Surface area, structure, physical and chemical nature of CB surface, and particle porosity are the important factors that can affect the compound properties.

An increase in CB structure or aggregate size generally gives a better resistance towards fatigue and cuts growth. An increase in surface area results in improved wear resistance and tear strength,

TABLE 15.2

Vulcanizate Properties of HC PBD Synthesized by Different Catalysts

PBD/Catalyst	PBD -Nd	PBD -Co	PBD -Ni
Polymer Properties			
Cis-1,4 (mol%)	96	94	94
Vinyl-1,2 (mol%)	1.3	3.2	2.9
$M_W \times 10^4$ (g/mol)	53	52	63
M_W/M_n	2.8	3.3	4.2
ML_{1+4}, 100°C	43	44	43
Compound Properties			
ML_{1+4} 100°C	72	62	64
Roll processability	Good	Unsatisfied	Good
Mill shrinkage %	29	44	43
Appearance of sheet	Excellent	Poor	Good
Vulcanized Properties			
300% modulus (MPa)	9.1	9.3	7.5
Tensile strength (MPa)	20.0	16.1	18.2
Elongation at break (%)	550	450	580
Wear resistance/index[*]	122	98	100
Tan delta at 50°C[#]	0.144	0.173	0.188

Notes
[*] Lambourn wear (slip ratio 60%, temperature 50°C)
[#] Frequency 15.9 Hz, strain 3%
(Recipe: PBD/CB (N339)/oil/stearic acid/ZnO/IPPD/BBS/S = 100/50/10/2/31/1/1.5 parts)

increase in hysteresis loss, and a drop in resilience. The effect of carbon black loading on compound properties is illustrated in Figure 15.9.[3] The effect of different CB on tire performance was investigated and the following observation were made[29]:

- Decrease in CB loading reduces RRC while an increase in oil content at constant CB loading increases RRC
- Lower particle size CB increases RRC and improves traction
- Broad CB aggregate size distribution with constant surface area and DBP lowers tire RRC

15.3.1.1   Surface-Treated Carbon Black

Improving polymer-filler interaction enhances compound properties. Functionalizing carbon black surfaces with functional groups like carbonyl, phenolic, and quinone provide a chemical bond with the polymer. Ayala et al.[30–32] reported a parameter for polymer-filler interaction from the vulcanizate. They reported the interaction parameter 'σ/n'. 'σ' is the slope of the stress-strain curve associated with polymer filler interaction and 'n', which assesses the filler-filler interaction, is the ratio of dynamic modulus at 1% and 25% strain amplitude. This interaction parameter, 'σ/n', emphasizes the contribution from polymer-filler interaction and reduces the influence of filler networking. Based on the interaction parameter, the following conclusions were made:

- σ/n provides a measurement for CB-polymer interaction for NR-, SBR-, IIR-based compounds
- Higher values of σ/n were found for SBR compounds owing to the aromatic structure influencing carbon black-filler interaction

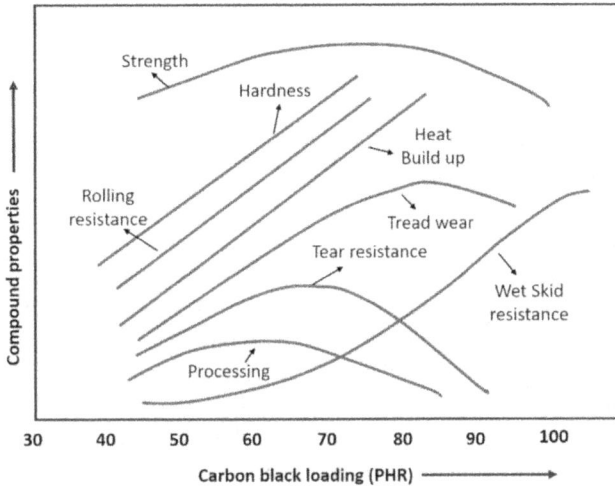

FIGURE 15.9   Effect of carbon black loading on compound properties.[3]

15.3.1.2 Morphologically Modified Carbon Black

Carbon black particles with a broad aggregate size distribution (ASD) give a substantial drop in hysteresis loss compared with carbon black with equivalent surface area. This new technology adjusts carbon black morphology for better wear resistance, maintaining low hysteresis and processability. A broad ASD gives a much reduced Payne effect and hysteresis loss. This new technology develops CB with broad ASD maintaining specific surface area on the level of the reference black.[25,33,34] Cabot Corporation introduced PROPEL E series carbon black with broad ASD.

15.3.2 Silica

Of the non-black filler, silica gives the utmost reinforcement to rubber products. As compared with carbon black, silica can give equivalent tensile and tear strength, but the resilience, set, and abrasion resistance are inferior. Table 15.3 gives the effect of silica and CB on different tire properties.[35] Silica is amorphous in nature and the particle surface is covered with chemically active polar silanol (-Si-OH) groups. The polarity difference between silica and rubber hydrocarbons makes silica less compatible with the general purpose rubber matrix.

The cross-sectional dimension of a primary silica particle is of 5–100 nm and the aggregates are formed by chemical and physical interactions of primary particles and have dimensions of

TABLE 15.3
Carbon Black vs Silica in Tire Properties

Property	Carbon Black	Silica
Rolling resistance	–	+
Ice- or wet grip	~	~/+
Wear resistance	+	–

(+ = better; – = worse; ~ = same/comparable)

100–500 nm. Aggregates condense into agglomerates with a typical dimension of 1–40 μm. Agglomerates of silica more or less disintegrate to the size of aggregates or even primary particles during mixing with rubber. Silica aggregates are similar to those of CB aggregates, but their higher structure gives a better reinforcing effect than the carbon black. Their higher specific component of surface energy results in a stronger tendency to form agglomerates, difficulty in dispersing in the rubber, and even re-agglomeration after mixing.[35]

15.3.2.1 Highly Dispersible Silica

Dispersibility of silica is improved by having a sufficient number of pores on aggregate surface so that rubber can penetrate easily. This also increases the contact area between silica and rubber. Commercially available high dispersible silica affords excellent processability while maintaining important durability properties such as tear strength and cut-growth resistance. There is enormous diversity in silica type based on different production processes and process variations. Rhone Poulenc Chimie (Solvay) reported novel precipitated silica particles that have good dispersion and deagglomeration capacity in rubber matrices compared to conventional silica. A significant improvement in mechanical properties and low Mooney viscosity were also observed with the new silica.[35,36] Table 15.4 tabulates surface properties of different commercially available silica.[3] The greater reinforcement of silica helps to reduce filler content that corresponds to a higher content of elastic rubber in proportion to the damping filler. Thus, using silica as the reinforcing filler is an effective way to reduce RRC.[35]

15.3.3 Carbon Black-Silica Dual Filler

The silica particles are introduced into the CB aggregates via co-fuming process to form the carbon black-silica dual phase filler, CSDPF. As shown in Figure 15.10, the silica particle can be distributed either throughout the aggregates or on the primary carbon black aggregates surface depending on the carbon black manufacturing process and equipment setup.[37,38]. It is reported that in CSDPF, carbon black surfaces have high activity and the silica part has a higher concentration of the silanol group. Therefore, a low level of silane coupling agent is suggested for better performance.

The new generation carbon black-dual phase filler gives an optimized balanced performance from both carbon black and silica filler. The highly active carbon black gives better reinforcement and the high level of silica on the surface imparts improved wet skid resistance. Two different categories of CSDPF were introduced by Cabot –Ecoblack CRX 2XXX and Ecoblack CRX 4XXX series containing silica domain finely distributed throughout the aggregates and relatively higher silica coverage on filler surface, respectively.

15.3.4 Nanoparticles

In the recent past, a substantial effort has been made to develop nanofillers and explore their application in tire components. Nanoclay, organo clay, graphenic nanofiller, and carbon nanotubes

TABLE 15.4
Surface Properties of Silica

Silica	BET Surface Area (m²/g)	CTAB Surface Area (m²/g)
HiSil* EZ HD Silica	167	153
Zeosil* 1165 MP HD Silica	170	155
Ultrasil* VN 338 Silica	180	162
KS* 40438 Silica	170	151

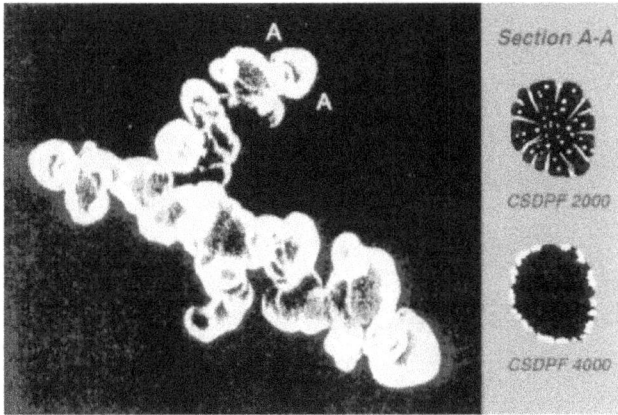

FIGURE 15.10   Schematic representation of 2000 and 4000 series of CSDPF. Adapted from.[21]

have been pulled to enhance the tire performance and finds potential application in treads, inner liners, and other components. When dispersed well in the rubber matrix, nanofillers are expected to impart high reinforcement and improve tire performance owing to their high surface area. To further enhance their efficiency, nanofillers are often functionalized.[39]

15.3.4.1   Nanoclay

Nanoclay is a class of inorganic filler with a layered structure belonging to the phyllosilicate family. They are found to impart mechanical properties as well as other specific properties like impermeability. The crystal structure constitutes a 2-D layer of two fused silicate tetrahedral plates with an edge-shared octahedral sheet of a metal atom such as aluminum (Al) or magnesium (Mg). The thickness of the sheet is around 1 nm and the length may vary from 100 nm to several microns depending upon the specific layered silicate. These silicate platelets stack themselves to a regular van der Waals gap between the layers which is known as interlayer space. The layered silicate is typically characterized by the cation exchange capacity (CEC) due to the surface charge. Montmorillonite (MMT) is the most commonly used one in nanocomposites due to its high CEC capacity. Figure 15.11 shows the chemical structure of MMT that belongs to the 2:1 phyllosilicates family where each crystal layer is composed of a layer of silica tetrahedral sandwiched between two octahedral sheets of aluminium and magnesium hydroxides.[40] Because of its higher aspect ratio (L/D ratio 100–1000), silicates have emerged as potential materials to improve the mechanical properties such as stiffness, strength, and barrier properties without any drastic effect on processability.

The gallery gap between two successive layers is too small; the penetration of polymer molecules will be restricted making their dispersion difficult. Modification of nanoclay with organic molecules is one the techniques that can be employed to enhance the gallery gap between the layers. The pristine-layered silicates usually contain hydrated Na^+ or K^+ ions in the gallery. Organically modified nanoclay can be defined as the replacement of these cations with the organic cations having long organic chains with a positively charged head, as shown in Figure 15.12. They are ionically modified with a quaternary ammonium salt. The corresponding organic cations replace the inorganic ones initially present on the clay surfaces. For this exchangeable adsorption, the amino groups are tethered to the clay surface and the hydrocarbon tail positioned in the gallery space. In this way, the clay becomes compatible with organic molecules; thus termed *organo modified*. Organic modification, in addition to the increment in the interlayer spacing, can also provide functional groups that can interact well with the rubber matrix. However, this technique is not sufficient to obtain uniform dispersion of fully exfoliated clay and thus does not always result in a significant improvement in rubber compound properties.

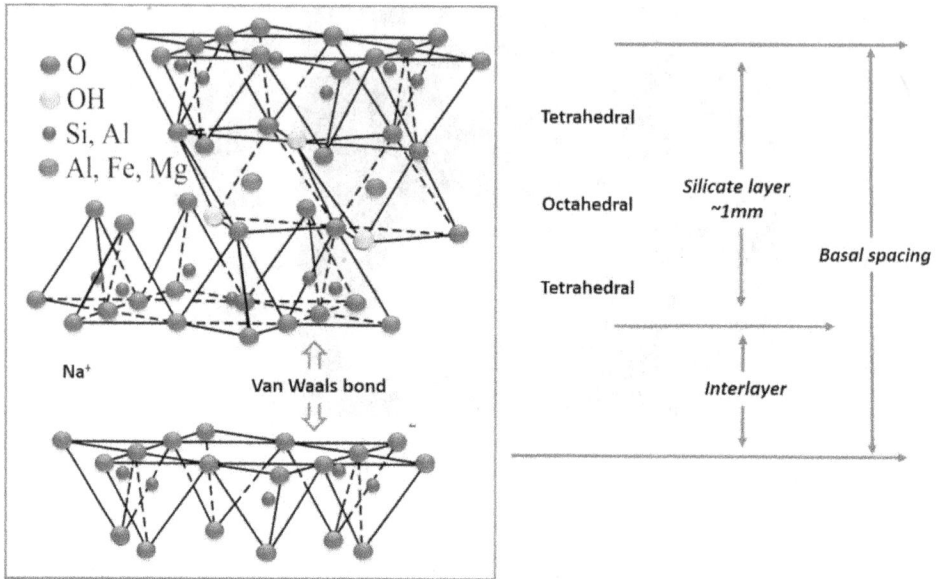

FIGURE 15.11   Representative structure of 2:1 phyllosilicates.

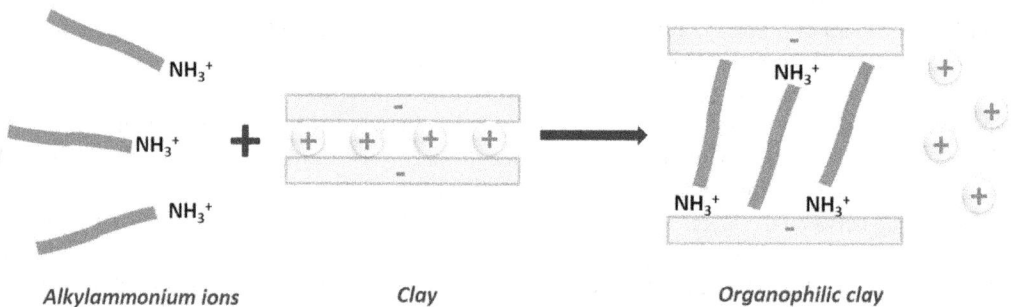

FIGURE 15.12   Increase in the gallery gap due to organo-modification of nanoclay.

A good dispersion of nanoclay in a rubber matrix can be achieved by using a functionalized elastomer in order to promote the good interaction between nanoclay and the polymer chains. Tripathy et al. reported the usage of organoclay in combination with conventional carbon black and found to have low hysteresis loss and maintain good reinforcement.[41] In addition to the ionic modification, nanoclay can also be covalently modified by reacting to the surface –OH groups with alkyl silanes. A large variety of functionalized clay derivatives via silylation reaction of organoclay (cloisite 30B) were reported.

15.3.4.2   Graphene

Graphene is an allotrope of carbon with one atom thick planar sheet (thickness 0.335 nm) of sp^2 carbon atoms arranged in a honeycomb pattern. It is the thinnest known material in the universe to date and possesses extraordinary properties like high surface area, ultra-high mechanical properties, excellent thermal and electrical conductivity and high elasticity.[42] The tire industry has explored graphene in tread, sidewall, and inner liner compounds, making the tires lighter. Use of graphene was found to reduce the Payne effect, RRC, and improved abrasion resistance. Owing to their high conductivity, they can be explored as an antistatic agent in full silica tire tread compounds. Graphene nanoplatelets (GnP) have a plate-like structure with a thickness 5–15 nm and a surface area 2,500 m^2g^{-1}. Their high aspect ratio and two-dimensional plate-like structure enable them to form tortuous pathways in rubber compounds, slowing down the penetration of gas molecules. Recently, studies were done on the barrier properties of GnP in a BIIR inner liner compound and found to have improved properties[43,44].

15.3.4.3 Carbon Nanotubes

Carbon nanotubes, CNTs, are rolled-up graphene sheets into a cylindrical form. Their exceptional mechanical properties and high aspect makes them ideal for rubber reinforcement[45]. Owing to their high surface area and filler-filler interaction, CNTs tend to agglomerate making their dispersion in the rubber matrix a difficult process. It was reported that a combination of graphene and CNT improved the CNT dispersion in rubber matrix[46]. Das et al. reported the formation of graphite-MWCNT filler network in SSBR matrix[47]. Bokobza et al. partially replaced CB with MWCNT in the SBR matrix and an improvement in stress-strain properties was achieved[48]. Minna et al. partially replaced CB with MWCNT in NR/PBD compound and reported an enhanced rubber-filler interaction[49].

15.3.4.4 Graphite Nanofiber

Graphite nanofibers (GNFs) are novel nanofibers containing graphitic planes arranged in either parallel or perpendicular to the axis. These carbonaceous fibers are prepared by heating carbon-containing gases over a catalyst at a temperature ranging from 450°C to 700°C. This new kind of carbon material has high thermal and mechanical stability, low-ohmic resistance, high surface to volume ratio, high electrical conductivity, and large surface active groups. They find potential applications in smart composite material, sensors, electrodes, catalyst support, fuel cells, energy storage devices, etc[50,51]. Their application in tire compounds is less reported. Surya et al. used GNF in NR-based tread compound and found to exhibit a synergetic improvement in mechanical and dynamic mechanical properties with better abrasion resistance[52].

15.3.4.5 Silicon Carbide Nanofiber

Silicon carbide nanofibers (SiCs) have exceptional properties like high mechanical strength, superior thermal and chemical stability, and good thermal conductivity, making them a promising reinforcing material[53]. The use of SiC nanofibers in NR-based tread compounds was experimented. The hybrid compound containing carbon black and SiC nanofibers exhibit enhanced mechanical properties with better traction and low RRC than compounds with only CB[52].

15.3.4.6 Carbon Filament

Carbon filaments also known as CarbonX, is considered as a new phase of sustainable material. It is a carbon material composed of many nano-sized carbon filaments chemically connected to each other forming a three-dimensional micro-sized particle network. Figure 15.13 illustrates the difference of CarbonX from ASTM grade carbon black. CarbonX is the missing link between the performance of nanostructures and the processability of a micron-sized material. With these new fillers, compounds with low RR, better traction and high wear resistance can be obtained. Carbon X has higher aggregate strength, narrower particle size distribution, higher alignment, crystallinity, and bigger crystallite size compared to conventional carbon blacks. Table 15.5 gives different grades of CarbonX available in the market.

15.4 PLASTICIZERS

A plasticizer acts as a process aid as well as a softener. They alter the T_g of the base polymer and lower the viscosity. This will allow further increase in filler content and thus lowers the compound

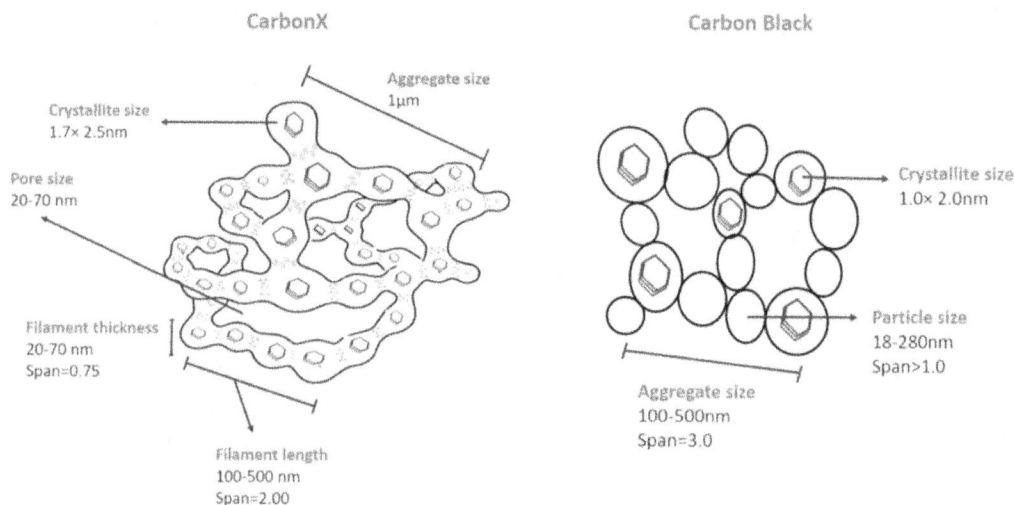

FIGURE 15.13 CarbonX and carbon black properties.

TABLE 15.5
Grades of CarbonX

CarbonX Grade	Surface Area (m²/g)	Filament Thickness (nm)	Filament Length (nm)
XR-1	189	35	78
X1	108	40	74
X3	78	40	102
X5	49	65	200
X7	40	74	240

cost. Plasticizers can modify the final hardness and flexibility of the compound. In the rubber industry, the term *plasticizers is* used for a class of material that include esters, low molecular weight polyethylene, and pine tars. Waxes, oils, and fatty acids are also considered plasticizers. Dibutylphthalate is an example of a frequently used ester plasticizer. Pine tars are often used with NR matrix and found to give good filler dispersion.

15.4.1 Process Oil

Process oils are primarily classified into three categories:

Paraffinic: contain high levels of naphthenic rings and higher levels of alkyl pendant groups, unsaturated hydrocarbon pendant groups, and, most important, fewer naphthenic groups per molecule. Pure paraffins from refined petroleum condense out as wax.

Naphthenic: contain high level of saturated rings.

Aromatic: have high levels of unsaturated rings and pendant alkyl and unsaturated hydrocarbon chains. The predominant structure is aromatic.

Oil tends to migrate or bloom to the compound surface if it is incompatible with the polymer matrix and affects the aesthetic appearance of the product. Compatibility of oil is decided by its composition, viscosity, and molecular weight. Table 15.6 tabulates the properties of paraffinic, naphthenic, and aromatic oil.[3]

TABLE 15.6
Properties of Processing Oils

Physical Property	Aromatic	Naphthenic	Paraffinic
Specific Gravity	0.95–1.0	0.91–0.94	0.85–0.89
Aniline Point (°C)	95.0–150.0	150–210	200–260
Molecular Weight	300–700	300–460	320–650
Aromatic Content (%)	68.0	44.0	15.0

Today, rubber industries started exploring the uses of sustainable biooil in their compound for an eco-friendly approach. They are produced from vegetable oil and its derivatives. Biooils are generally glycerides of organic acids modified to increase the viscosity and introduce oxygenated groups in their chemicals structure for better chemical and physical interaction with rubber. They have unique properties such as better solubility and comparable glass transition temperature (T_g) to aromatic oils. Compared to petroleum-based oils, biooils have a low aniline point, excellent solubility, and low polycyclic aromatic hydrocarbons. WITPROL AES 200 G is a commercially available vegetable oil for the tire industry.

15.5 VULCANIZATION SYSTEM

Vulcanization induces cross-links between the polymer chains to form a cross-link 3D network. A typical vulcanization consists of activators, vulcanizing agents, and accelerators.

15.5.1 CURE ACTIVATORS

Activators (zinc oxide and stearic acid) are used to increase the rate of vulcanization by activating the accelerator. The cure activator system includes an inorganic salt and an organic acid. Zinc oxide is commonly used inorganic salt along with stearic acid to form a rubber soluble soap in the rubber matrix. Solubility of stearic acid in rubber, their low molecular weight, and low melting point make them an ideal organic acid to use along with ZnO to form an efficient activator system for curing reaction.

Zinc oxide, along with fatty acids, is an indispensable activator for sulphur curing. Due to the harmful effect of soluble ZnO compounds to aquatic life and recent over level of Zn in European waterways, the reduction in ZnO level in rubber compounds has become crucial now. Other than ZnO, several other metal oxides like calcium oxide, lead oxide, and magnesium oxide were studied and found to have inadequate cure properties.

Zinc oxide is a dense material, making its dispersion difficult and it tends to form crystals in rubber matrices. The ZnO surface acts as a reactant and catalytic reaction template for curing reaction. Accelerator molecules, sulphur and fatty acids diffuse through the rubber matrix and are adsorbed on the ZnO surface forming an intermediate complex for vulcanization. Thus, the dispersion of inorganic ZnO in an organic polymer matrix is important. Przybyszewska et al. reported that the interface between ZnO and an accelerator plays a significant role in its activation efficiency during a vulcanization reaction.[54] The contact level of such ingredients are mostly determined by the shape, size, and the surface area. Conventional ZnO has particle size in the range 0.3–1.0 μm and surface area 4–6 m^2/g. The next-generation nano-sized ZnO particle with ultra-small nanoscopic granularity and high surface area provide higher activity than the conventional at a much lower concentration. LANXESS have commercialized nano ZnO under the trade name Zinkoxid Aktiv, which has a particle size of 36 nm and surface area 44 m^2/g. The activation efficiency of

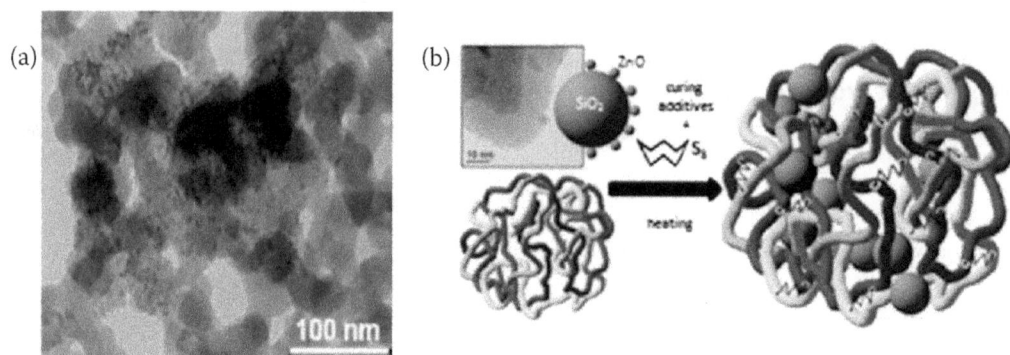

FIGURE 15.14　(a) Transmission electron microscopy (TEM) image of ZnO NPs anchored onto the surface of SiO2 NPs (ZnO/SiO2) (b) Graphical representation of ZnO/SiO2 NPs and their incorporation into rubber matrix after vulcanization. Reproduced from.[58]

nano ZnO was proven in a NR, SBR, PBD matrix.[55,56] The low particle size and increased surface area of nano ZnO helps to form complexes with accelerators, sulphur, and rubber effectively and disperse well in the rubber matrix rather than forming agglomerates.[57] ZnO dispersion in rubber matrices can be further enhanced by anchoring ZnO nanoparticles (NP) on substrates like silica or graphene through chemical or physical bonding. Figure 15.14 shows the TEM image of ZnO NPs anchored onto the surface of SiO2 NPs (ZnO/SiO2) and the graphical representation of ZnO/SiO2 NPs and their incorporation into a rubber matrix after vulcanization.[58] This method depends on the advantage of increased surface area and increased distribution of ZnO, thereby increasing the zinc ions' availability and reactivity towards the curing agents. Susanna et al. reported ZnO NP anchored on a silica surface and defined it as a double function filler, which can behave as a reinforcing filler and as support for Zn NP. ZnO NPs were covalently bonded to the silica surface through Si-O-Zn bond.[58] Lin et al. reported the usage of graphene sheets as the substrate to anchor Zn NP.[59]

15.5.2　Vulcanizing Agent

Sulphur is the widely employed vulcanizing agent in the rubber industry and reacts with most of the unsaturated rubbers. Sulphur is categorized into soluble and insoluble form. Further, treating sulphur with oil to reduce dust and improve wetting of particles ensures a good level of dispersion. Orthorhombic sulfur, S_8, in a ring structure is commonly used. During the thermal process of orthorhombic sulphur at different temperature ranges, sulfur rings undergo ring opening polymerization, followed by a rapid quenching process, which leads to the formation of polymeric sulphur. Unreacted sulphur during the thermal and quenching process is considered as soluble sulphur. The polymeric sulfur, which is a metastable high polymer with molecular weight ranging from 1,00,000 to 3,00,000 is insoluble in rubber matrix, known as insoluble sulphur. Thus, insoluble sulphur is used to prevent blooming in uncured rubber surface and where building tack is required. They give uniform dispersion with less mixing time and improve cross-link uniformity, which results in better compound properties. Above 120°C, the insoluble sulphur reverts to the soluble form. Thus, the mixing temperature plays a vital role. Currently, EASTMAN is offering insoluble sulfur with 10% oil treated grade. This material has enhanced dispersion and improved thermal stability and flow compared to regular insoluble sulphur.

15.5.2.1　Accelerators

Accelerators enable control time and/or temperature required for sulphur vulcanization and thus improving compound properties. Based on their role, accelerators are classified as primary and secondary accelerators. Primary accelerator or single accelerator give sufficient cross-links to produce broad vulcanization plateau and optimum cross-link density. Secondary accelerators are used in combination with a primary accelerator in small quantities (10% to 40% of primary accelerators). They activate the primary accelerator and often give a synergistic effect. Delayed action accelerators give protection against scorching and produce a satisfactory cure.

Most of the accelerators used in the rubber industry fall into one of these groups: aldehydeamines, sulfenamides, thioureas, dithiocarbamates, guanidines, thiurams, thiazoles, and xanthates. Out of these, the sulfinamide class, which includes CBS, TBBS, DCBS, MBS, etc., is the most used in the tire industry due to their delayed action and faster cure rate at vulcanization temperature. They also render a wide range of cross-link density based on their type and dosage.

Activity of sulfenamide accelerators can be:

Scorch safety: Longer (CBS < TBBS < MOR < DCBS)
Cure Rate: Faster (DCBS < MOR < CBS < TBBS)
Crosslink density: Higher at equal dosage (DCBS < MOR < CBS < TBBS)

TMTM, TMTD, DPG, and DOTG are employed as secondary accelerators along with sulphenamide to further increase the cure rate. They improve cross-linking efficiency but also decrease processing safety.

TBBS is modified by the addition of a second benzothiozole group to give TBSI (N-t-butyl-2-benzothiazole-sulfenimide). TBSI is finding a growing market in applications demanding long scorch safety with moderately slow cure rate, providing good properties, and including improved heat buildup. Improved vulcanization properties can be achieved at lower dosages than TBBS or MBS and also add reversion resistance to the compound. The schematic structure of TBBS and TBSI is shown in Figure 15.15[36].

15.6   RETARDERS AND ANTI-REVERSION AGENTS

Retarders are added to improve the induction time or scorch resistance. In general, they are organic acids that can lower the pH of the mixture and thereby retard the vulcanization reaction. Retarders reduce accelerator activity during processing and storage. At curing temperature, they either decompose or do not interfere in accelerator activity. N-Cyclohexylthiophthalimide (CTP) is the most common retarder used in the rubber industry. They will allow processing at elevated temperatures, increasing productivity; 0.1 to 0.3 phr of CTP is used in most of the applications and for butyl compounds, up to 1 phr is used.

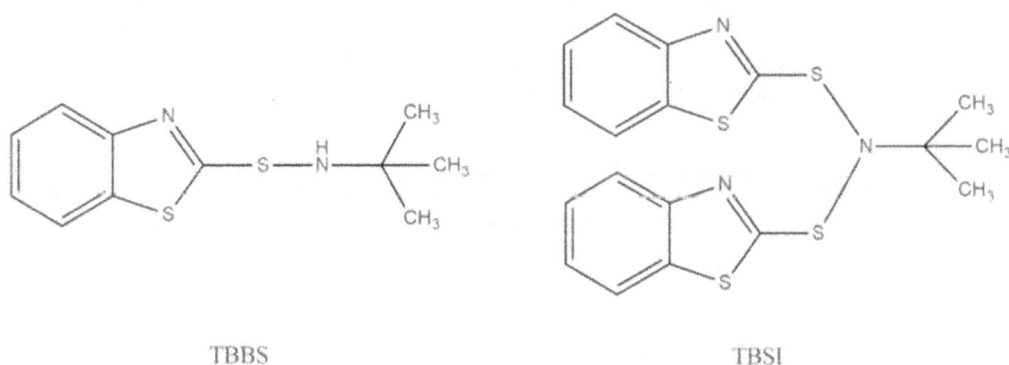

TBBS   TBSI

FIGURE 15.15   Structure of TBBS and TBSI.[36]

Reversion is the loss of network structure in a vulcanizate by non-oxidative thermal ageing and is commonly associated with sulphur-cured isoprene rubber. Anti-reversion agents available commercially are:

Zinc carboxylates: It is a blend of aromatic and aliphatic zinc carboxylic salt. They have higher solubility in the rubber matrix than the zinc salt of stearic acid. They render resistance towards reversion by promoting the formation of sulphidic cross-links of lower rank.

Thiophosphoryl derivative: they provide reversion resistance when used with sulphenamide cure system by forming a high proportion of cross-link network having higher proportion of monosulfidic cross-links. Eg Zinc –O-,O-di-N-butylphosporodithioate (ZBPD)

Bis-(3-triethoxysilylpropyl)-tetrasulfide (TESPT): Apart from a coupling agent, TESPT functions as a sulphur donor, rendering an equilibrium cure system. They promote the formation of cross-links of lower sulphur rank.

N-t-Butyl-2-benzothiazole Sulfenamide (TBSI): It is a commercially available sulphenamide-type accelerator that provides long scorch delay and slow cure rate. It results in cross-links with lower sulphur rank and thus enhancing reversion resistance.

Hexamethylene-1,6-bis thiosulphate disodium dihydrate: Commercially available material forms hybrid cross-links of hexamethylene and sulphidic moieties. This hybrid network increases reversion resistance and is beneficial in maintaining flex/fatigue resistance.

1,3-Bis(citraconimidomethyl)benzene: They impart reversion resistance by cross-link compensation mechanism. It forms cross-links based on carbon-carbon structure compensating the loss of polysulphide cross-link during the reversion process. They react at the onset of reversion and can be added to existing compounds with no further change in formulation.

15.7 ANTIDEGRADANTS

Antidegradants are additives that are added to combat the detrimental changes due to ageing phenomenon and they are categorized as antioxidants and antiozonant. Antidegradants are selected based on their volatility, solubility, staining behaviour, stability, and safety.

15.7.1 TYPES OF ANTIDEGRADANTS

Non-staining antioxidants: This type of antioxidants constitute phosphites, hindered phenol, hydroquinones and hindered bisphenol. Hindered phenols are of low molecular weight material and tend to be volatile. Hindered Bisphenol-4,4'-thiobis(6-t-butyl-m-cresol) is the most persistent antioxidant. Phosphites are used with synthetic rubber formulation and hydroquinones in adhesive formulations.

Staining antioxidants: Polymerized dihydroquinolines and diphenylamine are the two classes of staining antioxidants. Polymerized dihydroquinolines are effective against heavy metal prooxidants and have long durability properties. Their polymeric nature gives low volatility and migratory properties. Diphenylamines-type antioxidants render a direct improvement in fatigue resistance.

Antiozonants: PPDs (para-Phenlenediamines) are antiozonants used in substantial quantities. Figure 15.16 shows the chemical structure of PPD[3]. They give protection against ozone and also enhance resistance against heat, oxygen, metal ions, and fatigue. PPDs are categorized into

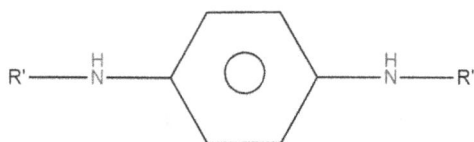

FIGURE 15.16   Chemical structure of PPD.

Dialkyl PPDs: The R group are both alkyl and can range from C3 to C9. They induce high levels of scorching and migrate at much faster rates than other PPDs.

Alkyl-aryl PPDs: One R group is aromatic and the other is an alkyl group. N-1,3-dimethylbutyl-N'-phenyl-p-phenylenediamine is the most used alkyl-aryl PPD. They provide good dynamic and static protection, better processing and scorch safety, and slow rate of migration, making them ideal for long product life.

Diaryl PPDs: Both R groups are aromatic e.g., diphenyl-p-phenylenediamine. They tend to bloom and are less active than alkyl-aryl PPDs.

Waxes: They are used in rubber formulation to improve ozone resistance, particularly under static conditions. They protect by forming a barrier on the rubber surface. Wax migrates from the bulk to the surface maintaining an equilibrium concentration on the surface. Two types of wax are used: microcrystalline wax with melting point 55°C to 100°C extracted from residual heavy lube stock of refined petroleum and paraffinic wax with melting point 35°C to 75°C extracted from light lube distillate of crude oil. Microcrystalline wax migrates at a slower rate than paraffinic wax owing to their high molecular weight and branching. For high temperature, microcrystalline wax performs best while at low temperature, paraffinic wax is best suited. The protective wax film breaks under dynamic conditions after which the antiozonant system in compound formulation serves the purpose. An optimum blend of waxes and PPDs gives a long-term protection against ozone under static and dynamic conditions over a wide temperature range.

15.7.2 BOUND ANTIOXIDANTS

Quinonediimines attached to the elastomer during curing could provide antioxidant activity. This will allow antidegradants being extracted during service thus providing long-term protection. N-1,3-dimethyl butyl-N'-phenyl-p-quinone diimine (QDI) acts both as a antiozonant and bound antidegradants after curing reaction providing long-term protection and also serves as a peptizer, a vulcanizing retarder, and can also enhance polymer-filler interaction.

15.8 PEPTIZING AGENT

A peptizing agent act as a chain-terminating agent and a pro-oxidant during mastication. It will significantly reduce time required to lower the rubber viscosity thereby reducing mixing time and energy. Bis(2-benzamidophenyl)disulphide, zinc-2-benzamidothiophenolate, N-S-Dibenzoyl-2-aminothiophenol, pentachlorothiophenol, zinc pentachlorothiophenolate are some of the commercially available peptizing agents. N-1,3-dimethyl butyl-N´-phenyl-p-quinone diimine (QDI) is also reported to have chemical peptizing activity in NR compounds. Recently, ACMECHEM introduced series of pepitizer under the trade name Peptizol, which includes a mixture of penta chloro thio phenol (PCTP) and Di benzamido diphenyl disulphide (DBD) with activators.

15.9 SPECIAL ADDITIVES

15.9.1 SILICA COUPLING AGENTS

Due to the difference in polarity, hydrophilic silica is less compatible with non-polar rubbers and silica being acidic in nature; they deactivate sulphur-curing reaction, which requires an alkaline condition. The difficulty in silica dispersion and cure retarding effects are much improved with a coupling agent. With better dispersion and improved polymer bonding, the re-agglomeration of silica gets diminished.[60]

A bifunctional organosilane coupling agent has two functions: adhesion to the hydrophilic silica and adhesion or compatibility enhancement to the non-polar polymer, as given in Figure 15.17.[3] It has two classes of functionality: 'X' is a hydrolyzable group like acyloxy, alkoxy, amine, or

$$R \longrightarrow (CH_2)_n \longrightarrow Si \longrightarrow X_3$$

⇑ ⇑ ⇑

Organofunctional group   Linker   Hydrolyzable group

FIGURE 15.17 Chemical structure of an organosilane coupling agent.

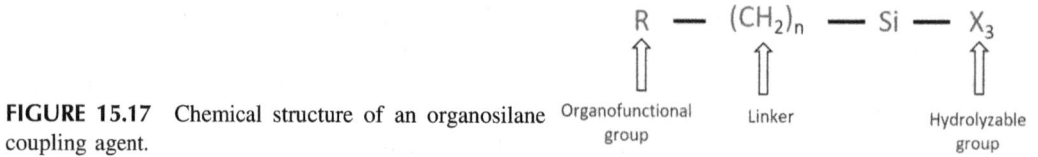

halogen and 'R' is a non-hydrolyzable organic radicle that may possess a functionality with desired characteristics. A bifunctional organosilane coupling agent is generally characterized by its dual functionalities: the silane moiety adheres to the hydrophilic silica surface, and the other part enhances the compatibility with the hydrophobic polymer matrix.[61,62]

Lowering surface energy by reducing specific components of silica, adsorption of hydrocarbon chains on the silica surface can be enhanced and thus the polymer-filler interaction. Coupling agents can be either applied on silica particles as a surface modification or can be added while mixing with other compounding ingredients.

The first coupling agent propagated for rubber application was 3-Mercaptopropyltrimethoxy silane. However, though the addition of it was found to improve mechanical properties, the scorch problem and disagreeable odor during mixing limited their use. Wolf et al. synthesized many silane using trichlorosilane as the starting material and found Bis(triethoxysilylpropyl)tetrasulphide (TESPT) suitable as coupling agents.[63]

TESPT is capable of reacting with the silica surface and the polymer and are commonly used in silica-filled rubber. Utilizing TESPT in tire tread compounds is found to give low RRC with improved physico-mechanical properties. The polysulfide part of TESPT reacts with the polymer and the ethoxysilyl groups on the silicon atom react with the hydroxyl groups present on the silica surface. The average sulfur rank of polysulfide is 3.86. TESPT is unstable at high shear or high temperature. Therefore, during rubber processing, splitting of TESPT molecules takes place, releasing reactive sulfur moieties in silica-filled compounds. An alternative coupling agent, Bis(triethoxysilylpropyl) disulphide (TESPD) was also developed. TESPD is a mixture of polysulfides rather than a pure disulfide. The average sulfur rank is close to 2. Compared to TESPT, it shows higher stability under high shear conditions as well as high thermal stability. Due to its low sulfur content, elemental sulfur is added to gain a reinforcement comparable to that of TESPT. Momentive introduced a new class of coupling agent, NXT silane for superior reinforcement of silica filler in tire tread. NXT silane is chemically 3-Octanoylthio-1-propyltriethoxysilane with Mol. Wt. 364 (Figure 15.18) having low sulfur ranking and found to improve wet traction and reduce RRC.

15.9.2 CARBON BLACK COUPLING AGENTS

Dinitrodiamine compounds were reported to significantly reduce hysteresis loss of CB filled vulcanizate but suffer processability issues.[64]. Based on the studies reported by Howland[65], Walkar[66], and Yamaguchi[64], Sumitomo Chemical Co. proposed carbon black coupling agents, CCA, that can interact with CB and NR molecules. The proposed coupling agents have two functional groups at both CCA chain ends, an amino group that can react with CB and another functional molecule that can interact with double bonds on the NR molecule. Sumitomo commercialized two types of CCA, SUMILINK® 100 and SUMILINK® 200, with two types of functional groups. Table 15.7 summarizes the properties of SUMILINK® 100 and SUMILINK®

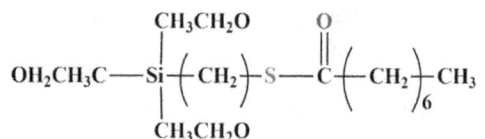

$$OH_2CH_3C \longrightarrow \underset{\underset{CH_3CH_2O}{|}}{\overset{\overset{CH_3CH_2O}{|}}{Si}} \left(CH_2 \right) \! S \longrightarrow \overset{\overset{O}{\|}}{C} \left(CH_2 \right)_{\!6} \! CH_3$$

FIGURE 15.18 Chemical structure of NXT silane.

TABLE 15.7

Properties of Sumilink 100 and Sumilink 200

Product Name	Chemical Name	Chemical Structure	Reaction
SUMILINK®100	S-(3-AminoPropyl) o-Hydrogen Sulfothioate (APHS)		1. mixing stage – CB and – NH₂ coupling 2. vulcanization stage – SSO₃H and rubber molecules coupling
SUMILINK®200	Sodium (Z)-4-((4-AminoPhenyl)amino)-4-Oxobut-2-enoate (SAPO)		1. mixing stage – CB and – NH₂ coupling – double bond in SAPO and rubber molecules coupling 2. vulcanization stage– no coupling reaction

Blue part: interaction with CB Red part: interaction with rubber.

200.[67] Han et al. report the increase in CB-polymer interaction with CCA, which significantly reduces heat generation in NR compounds.

15.9.3 Resin

To modify the viscoelastic properties of rubber compounds, a wide range of resins are used. They can be of natural origin such as rosin and terpene or can be synthetic such as phenol formaldehyde. As per the function, they can be categorized as processing resins, tackifying resins, reinforcing resins, and curing resins (phenolic resin for butyl curing). They are also classified into hydrocarbon resin, petrochemical resins, and phenolic resins in an arbitrary manner.

Hydrocarbon (HC) resins generally have high T_g and at processing temperature they melt and improve compound viscosity. At room temperature, they harden, preserving compound modulus and hardness. Aromatic HC resin serves as a reinforcing resin; aliphatic HC resin improves tack while the intermediate-like coumarone indene resin provides both characteristics. HC resins render:

- Improved physical properties
- Increase fatigue resistance as they improve filler dispersion by wetting filler surface
- Improve cut growth resistance by stress dissipation at the crack tip

Petroleum resins are by-products from oil refining and a wide range of grades are produced. Aliphatic types containing oligomers of isoprene are employed as tackifiers and aromatic resins containing high levels of dicyclopentadiene are more of reinforcing systems.

Phenolic resins are of two types: reactive and non-reactive. Reactive resins have free methylol groups. In the presence of methylene donors, a cross-link network will be formed, enabling them to serve as reinforcing system and adhesion promoters. Non-reactives are mostly oligomers of alkyl phenyl formaldehyde and para alkyl group ranges from C4 to C9. They are used as tackifying resins.

15.9.4 Rubber Metal Adhesion Promoters

The adhesion formed between brass-coated steel cord and rubber compound during vulcanization is an essential characteristic governing the performance and durability of car and truck tires. Brass-plating is a well-known method by which direct bonding can be achieved between rubber compounds and metals such as steel. Interfacial adhesion can depend on many factors, such as the chemical composition and thickness of the brass layer, which may vary quite markedly from almost bare steel to 0.2 um of brass. In addition, certain ingredients in the belt compound formulation greatly influence adhesion and must be optimized to obtain the maximum benefit. By these means, it is very important to achieve a high level of adhesion and sustain this level throughout the service history of the tire. However, the rubber-metal interface is prone to deterioration, particularly at high temperature, high humidity, and high salinity and such conditions will reduce the reinforcement by steel cord to cause a concomitant reduction in the life of the tire. Consequently, various organic cobalt salts are used, alone or in combination with resin systems, to improve and maintain a durable bond at the rubber-metal interface. At present, these organic cobalt salts appear to be the most efficient type of adhesion promoter and could be considered as a benchmark by which the tire industry assesses rubber-metal bond strength.

15.9.4.1 Incorporation of Cobalt

Cobalt can maintain different oxidation states and the ease by which electrons are transferred between these states may influence the migration of ions through the oxide and sulfide layers. The action of cobalt in both the copper sulfide and zinc oxide layers is made possible by the capability of cobalt ions to exist with variable valency (Co2+/Co3+). It is accepted that cobalt is incorporated into the zinc oxide layer early in the vulcanization process, before the onset of sulfidation. This is inferred

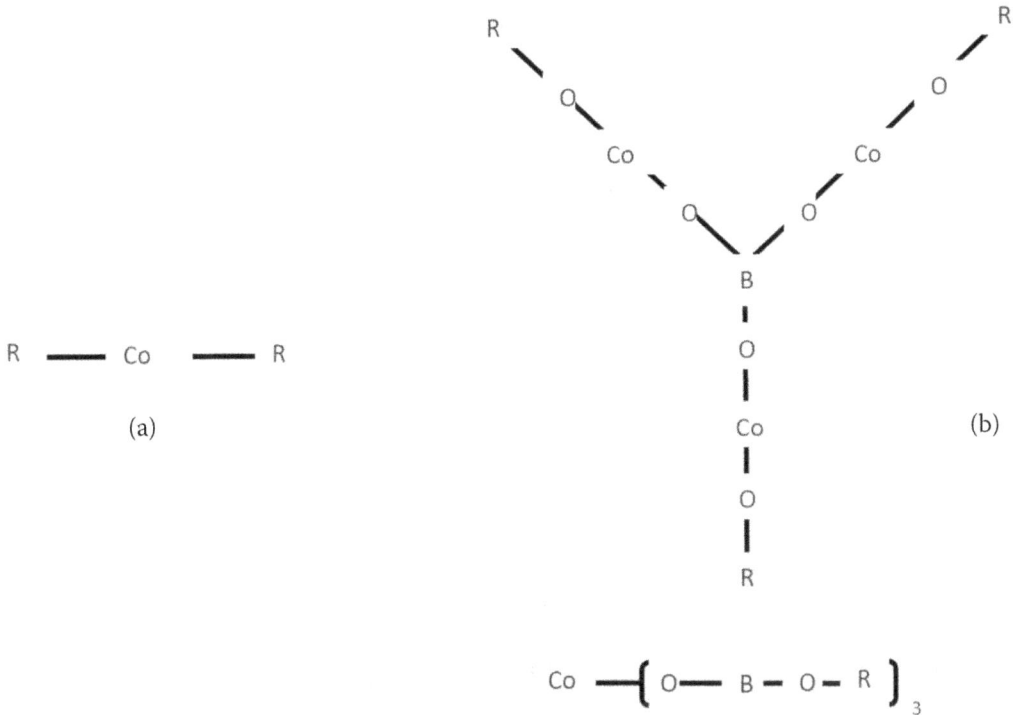

FIGURE 15.19 Chemical structure of (a) cobalt soap, (b) cobalt boroacylate.[68]

from depth profiles that show the distribution of cobalt in the copper sulfide, zinc sulfide, and zinc oxide layers with the peak concentration just into the zinc oxide layer and penetrating to some depth. The maximum in the zinc sulfide concentration occurs in this region and so cobalt is associated with this sulfide phase. During sulfidation, cobalt will be incorporated into the outer copper sulfide layer while it grows, as cobalt sulfide, with the cobalt concentration reducing toward the surface. As it is incorporated, cobalt will lower the defect density of copper sulfide with a concomitant reduction in the diffusion rate of copper and sulfur ions through the copper sulfide layer.

The chemical structure of a more traditional cobalt disoap, such as cobalt stearate, is shown to contain two molecules of acid per cobalt atom (Figure 15.19a) and alongside is a schematic diagram of the second-generation boron-modified cobalt adhesion promoters produced in the early 1980s (Figure 15.19b)[68]. Synthesis of such a type of boroacylate structure produced compounds with an increased cobalt content relative to the carboxylic acid and higher activity than previous adhesion promoters. The purity of these adhesion promoters was also improved by producing compounds with one or more synthetic acids, instead of natural fatty acids used in compounds such as cobalt stearate.[68]

Figure 15.19 Idealized chemical structure of cobalt adhesion promoters. (a) Simple cobalt disoap, such as cobalt stearate, where R = stearic acid. (b) Cobalt boroacylate structure in which R may equal the same or different carboxylic acids. The higher activity of cobalt boroacrylate is associated with relative bond strength on either side of the cobalt atom. DIC Corporation introduced a metal carboxylate containing 22.2% cobalt and neodecanoic acid under the trade name Dicnate NBC-2 for tire applications.

15.10 CONCLUSION

Tire performance requirements are more demanding owing to fuel economy, durability, safety concerns, and quality expectation. A wide range of materials is available to design a tire compound to meet the new electric vehicle tire requirement and other market demands. This chapter discussed the next generation of materials to develop high-performance tires. However, the combination of materials with optimum dosage decides the performance and quality under a given set of service conditions.

ACKNOWLEDGEMENTS

The authors are grateful and thankful for the support from Mr. Sujit Kumar Panda, Mr. Joseph Jose, Mr. Siju, Mr. Venkatraman, Mr. Selvam, Mr. Suryanarayanan, Mr. Sorimuthu, and Mr. Manikandan, Research and Development, Yokohama Off Highway Tires. Special thanks to Dr. Revathy for unconditional support.

REFERENCES

[1] Yoon, B.; Kim, J. Y.; Hong, U.; Oh, M. K.; Kim, M.; Han, S. B.; Nam, J.-D.; Suhr, J., Dynamic viscoelasticity of silica-filled styrene-butadiene rubber/polybutadiene rubber (SBR/BR) elastomer composites. *Composites Part B: Engineering* 2020, *187*, 107865.

[2] Vidhi, R.; Shrivastava, P.; Parikh, A., Social and Technological Impact of Businesses Surrounding Electric Vehicles. *Clean Technologies* 2021, *3*, 81–97.

[3] Rodgers, B.; Waddell, W., The science of rubber compounding. In *Science and technology of rubber.* Elsevier 2005; pp 401–454.

[4] Bhowmick, A., *Rubber products manufacturing technology.* Routledge & CRC Press 2018.

[5] Choi, S. S.; Kwon, H. M.; Kim, Y.; Ko, E.; Lee, K. S., Hybrid factors influencing wet grip and rolling resistance properties of solution styrene-butadiene rubber composites. *Polymer International* 2018, *67*, 340–346.

[6] Ellul, M. D.; Hara, Y., Specialty polymers and dynamically vulcanized alloys for ultra low air permeability tire inner liners. *Rubber Chemistry and Technology* 2018, *91*, 751–756.

[7] Threadingham, D.; Obrecht, W.; Lambert, J.-P.; Happ, M.; Oppenheimer-Stix, C.; Dunn, J.; Krüger, R.; Brandt, H. D.; Nentwig, W.; Rooney, N., Rubber, 3. synthetic. *Ullmann's Encyclopedia of Industrial Chemistry* 2000.

[8] Phinyocheep, P., Chemical modification of natural rubber (NR) for improved performance. In *Chemistry, manufacture and applications of natural rubber.* Elsevier 2014; pp. 68–118.

[9] Hirata, Y.; Kondo, H.; Ozawa, Y., Natural rubber (NR) for the tyre industry. In *Chemistry, manufacture and applications of natural rubber.* Elsevier 2014; pp. 325–352.

[10] Fettes, E. M., *Chemical reactions of polymers.* Interscience Publishers 1964; Vol. 19.

[11] Saengdee, L.; Phinyocheep, P.; Daniel, P., Chemical modification of natural rubber in latex stage for improved thermal, oil, ozone and mechanical properties. *Journal of Polymer Research* 2020, *27*, 1–13.

[12] Cataldo, F., Preparation of silica-based rubber compounds without the use of a silane coupling agent through the use of epoxidized natural rubber. *Macromolecular Materials and Engineering* 2002, *287*, 348–352.

[13] Segatta, T. J.; Sandstrom, P. H.; Verthe, J. J., Sulfur cured rubber composition containing epoxidized natural rubber and silica filler. Google Patents 1995.

[14] Manoharan, P.; Naskar, K., Exploring a highly dispersible silica–elastomer composite for tire applications. *Journal of Applied Polymer Science* 2016, *133*.

[15] Baker, C.; Gelling, I.; Newell, R., Epoxidized natural rubber. *Rubber Chemistry and Technology* 1985, *58*, 67–85.

[16] Sengloyluan, K.; Sahakaro, K.; Dierkes, W. K.; Noordermeer, J. W., Silica-reinforced tire tread compounds compatibilized by using epoxidized natural rubber. *European Polymer Journal* 2014, *51*, 69–79.

[17] Manoharan, P.; Naskar, K., Biologically sustainable rubber resin and rubber-filler promoter: a precursor study. *Polymers for Advanced Technologies* 2017, *28*, 1642–1653.

[18] Jerome, R.; Henrioulle-Granville, M.; Boutevin, B.; Robin, J., Telechelic polymers: Synthesis, characterization and applications. *Progress in Polymer Science* 1991, *16*, 837–906.

[19] Quirk, R. P.; Jang, S. H.; Kim, J., Recent advances in anionic synthesis of functionalized elastomers using functionalized alkyllithium initiators. *Rubber Chemistry and Technology* 1996, *69*, 444–461.

[20] Tsutsumi, F.; Sakakibara, M.; Oshima, N., Structure and dynamic properties of solution SBR coupled with tin compounds. *Rubber Chemistry and Technology* 1990, *63*, 8–22.

[21] Zhang, P.; Morris, M.; Doshi, D., Materials development for lowering rolling resistance of tires. *Rubber Chemistry and Technology* 2016, *89*, 79–116.

[22] Nagata, N.; Kobatake, T.; Watanabe, H.; Ueda, A.; Yoshioka, A., Effect of chemical modification of solution-polymerized rubber on dynamic mechanical properties in carbon-black-filled vulcanizates. *Rubber Chemistry and Technology* 1987, *60*, 837–855.

[23] Thielen, G., Chemically modified emulsion SBR in tire treads. *Rubber Chemistry and Technology* 2008, *81*, 625–637.

[24] Maiti, M.; Sadhu, S.; Bhowmick, A. K., Brominated poly (isobutylene-co-para-methylstyrene) (BIMS)-clay nanocomposites: Synthesis and characterization. *Journal of Polymer Science Part B: Polymer Physics* 2004, *42*, 4489–4502.

[25] McDonald, G.; Hess, W., Carbon black morphology in rubber. *Rubber Chemistry and Technology* 1977, *50*, 842–862.

[26] Liu, J.; Fan, X.; Min, X.; Zhu, X.; Zhao, N.; Wang, Z., Synthesis of high cis-1, 4 polybutadiene with narrow molecular weight distribution via a neodymium-based binary catalyst. *RSC Advances* 2018, *8*, 21926–21932.

[27] Sone, T., Industrial synthetic method of the rubbers. 1. Butadiene rubber. *International Polymer Science and Technology* 2016, *43*, 49–54.

[28] Hattori, I.; Tadaki, T., Improvement of Abrasion Resistance of Polydiene. *Nippon Gomu Kyokaishi (Journal of the Society of Rubber Industry, Japan)* 2007, *80*, 140–146.

[29] Hess, W.; Klamp, W., The effects of carbon black and other compounding variables on tire rolling resistance and traction. *Rubber Chemistry and Technology* 1983, *56*, 390–417.

[30] Ayala, J.; Hess, W.; Dotson, A.; Joyce, G., New studies on the surface properties of carbon blacks. *Rubber Chemistry and Technology* 1990, *63*, 747–778.

[31] Wolff, S.; Görl, U., The influence of modified carbon blacks on viscoelastic compound properties. *Kautschuk und Gummi, Kunststoffe* 1991, *44*, 941–947.

[32] Robertson, C. G.; Hardman, N. J., Nature of Carbon Black Reinforcement of Rubber: Perspective on the Original Polymer Nanocomposite. *Polymers* 2021, *13*, 538.

[33] Stacy, C.; Johnson, P.; Kraus, G., Effect of carbon black structure aggregate size distribution on properties of reinforced rubber. *Rubber Chemistry and Technology* 1975, *48*, 538–547.

[34] Hess, W.; Chirico, V., Elastomer blend properties—Influence of carbon black type and location. *Rubber Chemistry and Technology* 1977, *50*, 301–326.

[35] Ten Brinke, A., *Silica reinforced tyre rubbers*. PhD Thesis Twente University 2002.

[36] De, S. K.; White, J. R., *Rubber technologist's handbook*. Smithers Rapra Publishing 2001; Vol. 1.

[37] Wang, M.; Kutsovsky, Y.; Zhang, P.; Mehos, G.; Murphy, L.; Mahmud, K., Using carbon-silica dual phase filler-Improve global compromise between rolling resistance, wear resistance and wet skid resistance for tires. *Kautsch. Gummi Kunstst.* 2002, *55*, 33–40.

[38] Wang, M.-J.; Kutsovsky, Y.; Zhang, P.; Murphy, L. J.; Laube, S.; Mahmud, K., New generation carbon-silica dual phase filler part I. Characterization and application to passenger tire. *Rubber Chemistry and Technology* 2002, *75*, 247–263.

[39] Lorenz, H.; Fritzsche, J.; Das, A.; Stöckelhuber, K.; Jurk, R.; Heinrich, G.; Klüppel, M., Advanced elastomer nano-composites based on CNT-hybrid filler systems. *Composites Science and Technology* 2009, *69*, 2135–2143.

[40] Golubeva, O. Y.; Korytkova, E.; Gusarov, V., Hydrothermal synthesis of magnesium silicate montmorillonite for polymer-clay nanocomposites. *Russ. J. Appl. Chem.* 2005, *78*, 26–32.

[41] Gopi, J. A.; Patel, S. K.; Chandra, A. K.; Tripathy, D. K., SBR-clay-carbon black hybrid nanocomposites for tire tread application. *Journal of Polymer Research* 2011, *18*, 1625–1634.

[42] Sur, U. K., Graphene: a rising star on the horizon of materials science. *International Journal of Electrochemistry* 2012, *2012*.

[43] Thuruthil Raju, A.; Dash, B.; Dey, P.; Nair, S.; Naskar, K., Evaluation of air permeability characteristics on the hybridization of carbon black with graphene nanoplatelets in bromobutyl rubber/ epoxidized natural rubber composites for inner-liner applications. *Polymers for Advanced Technologies* 2020, *31*, 2390–2402.

[44] Kumar, S.; Chattopadhyay, S.; Padmanabhan, R.; Sreejesh, A.; Nair, S.; Unnikrishnan, G.; Nando, G., Tailoring permeation characteristics of bromobutyl rubber with polyepichlorohydrin and graphene nanoplatelets. *Materials Research Express* 2015, *2*, 105007.

[45] Liu, W.-W.; Chai, S.-P.; Mohamed, A. R.; Hashim, U., Synthesis and characterization of graphene and carbon nanotubes: A review on the past and recent developments. *Journal of Industrial and Engineering Chemistry* 2014, *20*, 1171–1185.

[46] Hu, H.; Zhao, L.; Liu, J.; Liu, Y.; Cheng, J.; Luo, J.; Liang, Y.; Tao, Y.; Wang, X.; Zhao, J., Enhanced dispersion of carbon nanotube in silicone rubber assisted by graphene. *Polymer* 2012, *53*, 3378–3385.

[47] Das, A.; Kasaliwal, G. R.; Jurk, R.; Boldt, R.; Fischer, D.; Stöckelhuber, K. W.; Heinrich, G., Rubber composites based on graphene nanoplatelets, expanded graphite, carbon nanotubes and their combination: A comparative study. *Composites Science and Technology* 2012, *72*, 1961–1967.

[48] Bokobza, L.; Rahmani, M.; Belin, C.; Bruneel, J. L.; Bounia, E.l., Blends, N. E., of carbon blacks and multiwall carbon nanotubes as reinforcing fillers for hydrocarbon rubbers. *Journal of Polymer Science Part B: Polymer Physics* 2008, *46*, 1939–1951.

[49] Poikelispää, M.; Das, A.; Dierkes, W.; Vuorinen, J., The effect of partial replacement of carbon black by carbon nanotubes on the properties of natural rubber/butadiene rubber compound. *Journal of Applied Polymer Science* 2013, *130*, 3153–3160.

[50] Tran, V.-T.; Saint-Martin, J.; Dollfus, P.; Volz, S., High thermoelectric performance of graphite nanofibers. *Nanoscale* 2018, *10*, 3784–3791.

[51] Wu, Y.; Mao, X.; Cui, X.; Zhu, L., Electroanalytical application of graphite nanofibers paste electrode. *Sensors and Actuators B: Chemical* 2010, *145*, 749–755.

[52] Parameswaran, S. K.; Bhattacharya, S.; Mukhopadhyay, R.; Naskar, K.; Bhowmick, A. K., Excavating the unique synergism of nanofibers and carbon black in Natural rubber based tire tread composition. *Journal of Applied Polymer Science* 2021, *138*, 49682.

[53] Vijayan P. P.; Puglia, D.; Dąbrowska, A.; Vijayan P. P.; Huczko, A.; Kenny, J. M.; Thomas, S., Mechanical and thermal properties of epoxy/silicon carbide nanofiber composites. *Polymers for Advanced Technologies* 2015, *26*, 142–146.

[54] Przybyszewska, M.; Zaborski, M., Effect of ionic liquids and surfactants on zinc oxide nanoparticle activity in cross-linking of acrylonitrile butadiene elastomer. *Journal of Applied Polymer Science* 2010, *116*, 155–164.

[55] Akhlaghi, S.; Kalaee, M.; Mazinani, S.; Jowdar, E.; Nouri, A.; Sharif, A.; Sedaghat, N., Effect of zinc oxide nanoparticles on isothermal cure kinetics, morphology and mechanical properties of EPDM rubber. *Thermochim. Acta* 2012, *527*, 91–98.

[56] Pysklo, L.; Pawlowski, P.; Nicinski, K., Study on Reduction of Zinc Oxide Level in Rubber Compounds Part II. *KGK-Kautschuk Gummi Kunstsoffe* 2008.

[57] Sahoo, S.; Maiti, M.; Ganguly, A.; Jacob George, J.; Bhowmick, A. K., Effect of zinc oxide nanoparticles as cure activator on the properties of natural rubber and nitrile rubber. *Journal of Applied Polymer Science* 2007, *105*, 2407–2415.

[58] Susanna, A.; Armelao, L.; Callone, E.; Dirè, S.; D'Arienzo, M.; Di Credico, B.; Giannini, L.; Hanel, T.; Morazzoni, F.; Scotti, R., ZnO nanoparticles anchored to silica filler. A curing accelerator for isoprene rubber composites. *Chem. Eng. J.* 2015, *275*, 245–252.

[59] Lin, Y.; Zeng, Z.; Zhu, J.; Chen, S.; Yuan, X.; Liu, L., Graphene nanosheets decorated with ZnO nanoparticles: facile synthesis and promising application for enhancing the mechanical and gas barrier properties of rubber nanocomposites. *RSC advances* 2015, *5*, 57771–57780.

[60] Dannenberg, E. M., Filler chaices in the rubber industry-the incumbents and some new candidates. 1981.

[61] Reuvekamp, L. A.; Debnath, S.; Ten Brinke, J.; Van Swaaij, P.; Noordermeer, J. W., Effect of zinc oxide on the reaction of TESPT silane coupling agent with silica and rubber. *Rubber Chemistry and Technology* 2004, *77*, 34–49.

[62] Poh, B. a.; Ng, C., Effect of silane coupling agents on the mooney scorch time of silica-filled natural rubber compound. *European Polymer Journal* 1998, *34*, 975–979.

[63] Wolff, S., Chemical aspects of rubber reinforcement by fillers. *Rubber Chemistry and Technology* 1996, *69*, 325–346.

[64] Yamaguchi, T.; Kurimoto, I.; Nagasaki, H.; Okita, T., Coupling agent improves properties. *Rubber World* 1989, *199*, 30–38.

[65] Howland, L. H., Manufacture of rubber. Google Patents 1943.

[66] Walker, L. A.; D'Amico, J. J.; Mullins, D. D., Nitrosoanilines. II. *The Journal of Organic Chemistry* 1962, *27*, 2767–2772.

[67] Ozturk, O., Performance improvement of natural rubber/carbon black composites by novel coupling agents.

[68] Fulton, W. S., Steel tire cord-rubber adhesion, including the contribution of cobalt. *Rubber Chemistry and Technology* 2005, *78*, 426–457.

16 Polymer Composites for Stealth Technology

Deepthi Anna David
Cochin University of Science and Technology (CUSAT), Kerala, India

Vidhukrishnan Naiker
Institute of Chemical Technology, Matunga, Mumbai, India

Jabeen M. J. Fatima
Cochin University of Science and Technology (CUSAT), Kerala, India

Thomas George
Cochin University of Science and Technology (CUSAT), Kerala, India

Pritam V. Dhawale
Institute of Chemical Technology, Matunga, Mumbai, India

Mrudul Vijay Supekar
Cochin University of Science and Technology (CUSAT), Kerala, India

P. M. Sabura Begum
Cochin University of Science and Technology (CUSAT), Kerala, India

Vijay Kumar Thakur
Biorefining and Advanced Materials Research Centre, Dumfries, Edinburgh, United Kingdom

Prasanth Raghavan
Cochin University of Science and Technology (CUSAT), Kerala, India

Institute of Chemical Technology, Matunga, Mumbai, India

Gyeongsang National University, Jinju, Republic of Korea

CONTENTS

DOI: 10.1201/9781003200710-18

16.1   INTRODUCTION

The military finds immense use with stealth technology since it helps with gaining control over strategically important areas and thus destroying main and invasive targets for the safety of the nation. In general, this technology works with the attainment of low notability by reduction of signal detection or reduction in counter measure signals [1,2]. The target emits electromagnetic radiation and acoustic signals which are detected by the detectors like radar. Thus, researchers are trying to prepare composites which do not reflect the electromagnetic waves back to radar [3,4]. To talk about stealth technology, it should be noted that this kind of technology is already present in nature in the form of innumerable creatures who survive by melding with the background which the humans were ignorant about. This is thus used in military research and its equipment which could be hidden from radar [5]. Primarily, as military technology advances, the security issues applied in the earlier stages will be worn out since many advancement studies keep going on which makes the present technology outdated. However, the stealth community, like those linked with technologies for worldwide surveillance and nuclear weapons, opted to keep the sophisticated method under covers. The media also has taken pains to find out the main provenance with enough credentials who are willing to study and explain on how the stealth technology is brought into practice. This makes normal people believe that everything is un-barred now [5]. In this technology, any "observable"-infrared, visible, acoustic is of disquietude. Space and time neglect the actual depth of this. In such a sophisticated era, the radar signature is the basic problem [5].

The culminating concern is that of "blending with the background" as this is essential for military studies and purposes. Examples in nature like a poisonous snake wrapped around a tree that is undetectable at first sight and works like camouflage for hunting its prey and for its safe survival. The contrast is less here and thus can be used for camouflage purposes [5]. Woefully, the way hackneyed radar works causes concern. A beam of radiated energy from an antenna keeps on spreading as it advances and so is when it hits a target. This indicated the deterioration in energy in both directions. It should be noted that for a particular target size, the quantity of energy received back at radar is inversely proportional to the fourth power of target range which is huge [5]. It should be kept in mind that time or space will not allow a full presentation of all of the details on the target that this simple pattern offers. Notwithstanding, to see the value in what is of interest in the decrease of radar reverberation, let us make a couple of focuses about this example. Initially, one sees a reverberation level which changes fiercely. The truth of the matter is that this example was taken for a moderately low radar recurrence, and had a much higher one been utilized, the flap construction would have been so thick as to make the example just about a dark haze. The ordinate is plotted on a logarithmic size (of need), in light of the fact that the real plentifulness fluctuates more than a few significant degrees in parts of a level of pivot [5]. The electromagnetic radiations are transverse waves which radiate through space with different frequencies including radio waves, microwaves, infrared, ultraviolet-visible light, X-rays, gamma, and cosmic rays. Based on the application, the corresponding frequency range can be made use of it and Figure 16.1 includes the electromagnetic spectrum and their potential applications [6].

16.2   RADAR AND ITS WORKING PRINCIPLE

Radio detection and ranging (RADAR), the gadget which is utilized for discovery and going of contacts, free of time and climate conditions, was perhaps the main logical revelations and innovative advancements that rose up out of WWII. Its turn of events, similar to that of most extraordinary developments, was mothered by need. Behind the advancement of radar lay over a hundred years of radio turn of events [7]. Radar is used to recognize and observe the target objects at significant distances by very short eruptions of electromagnetic radiation at the speed of light and the target object will reflect waves which are returned as a reverberation [7]. A radar framework utilizes high-velocity electromagnetic waves to decide the distance, the speed, the heading being voyaged, and the rise height of both fixed and non-fixed objects. These target objects include climate arrangements, engine vehicles, ships, airplanes, shuttles, and even landscapes [8]. Radars can be used for missile guidance (military), navigation, threat detection (military), space exploration/tracking, air traffic control, weather, battlefield, and reconnaissance [8].

The basic working principle behind the radar system is very easy to understand. The electromagnetic radiation transmitted by the radar, at the speed of light (300,000 km/s), is reflected back after hitting on the target object which is shown in Figure 16.2. The incident electromagnetic waves will scatter, transmit, and be reflected by the object which can be an electrically leading surface. Calculating the time taken for reflecting the electromagnetic waves to reach back to the radar antenna gives the distance of the target object (fighter jets, submarines, etc.) from radar. The strength of the electromagnetic radiation reflected also matters. On the other hand, the theory behind the radar working is very complex. Thus, the radar system paved the milestone for a wide range of applications, especially in defense.

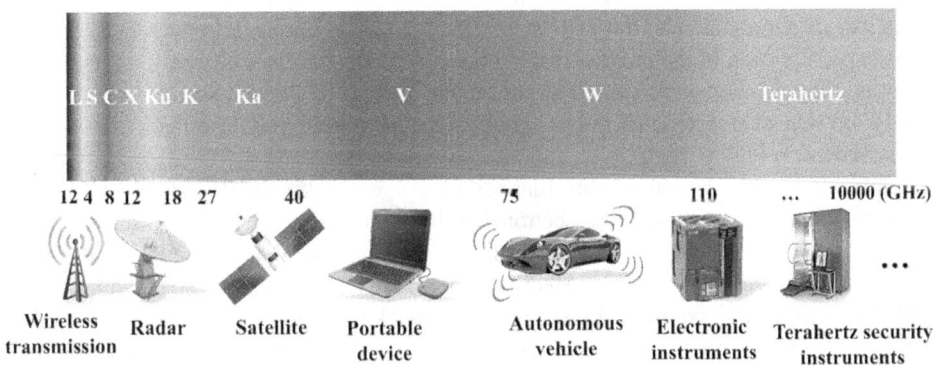

FIGURE 16.1   Electromagnetic spectrum and different microwave and terahertz devices mostly used [6].

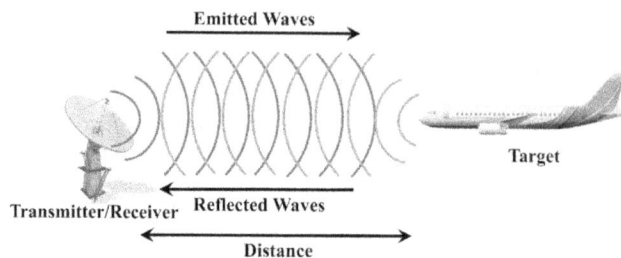

FIGURE 16.2   General working principle of radar [9].

Pulse-modulated radar frameworks differ extraordinarily in detail, the standards are basically something similar for all frameworks. The fundamental radar framework generally incorporates six significant segments, as demonstrated in Figure 16.3 [7].

16.3   STEALTH TECHNOLOGY – A BRIEF HISTORY

Stealth technology which is the low observable technology (LO technology), is a subfield of military tactics as well as the passive and active electronic countermeasures that encompasses a variety of techniques to conceal (invisible or less visible) troops, ships, submarines, planes, missiles, ground vehicles, and satellites from detection by radar, sonar, infrared, and other means. Camouflage, or the ability to blend an object's appearance into the visual background, was the earliest use of this notion. Using the detection technology like radar, sonar, surface-to-air missiles, and others have become more powerful, and thus, the amount to which ground vehicles, aircrafts, and ships have been affected in reaction has risen. Chemicals or high dielectric materials were used on the military uniforms to diminish the impact of their infrared signature. However, the contemporary stealth vehicle is built to have selective spectral signature or treating the surface of the aircraft with materials or construct the aircraft with the right selection of materials which could absorb the radar signals. Predominantly there are two methods for making vehicles invisible to radar.

- The aircraft can be designed in such a way that the electromagnetic radiations it emits are deflected away from the radar systems.
- Materials that absorb radar signals can be used to cover the aircraft, which is simply called radar absorbing materials (RAMs).

As seen in Figure 16.4b, conventional aeroplanes have a round shaped design. This design not only contributes to its aerodynamics, but also makes them an excellent radar reflector. Because of the rounded shape, a portion of the radar signal is bounced back regardless of where it hits the aircraft. The story is different in case of stealth aircraft since it has a totally flat surface with sharp edges (Figure 16.4a). On interception of radar signal by the stealth aircraft, it reflects away at various angles, as shown in Figure 16.4c.

Designing of stealth aircraft in this manner uses a wise combination of RAMs as well as geometry to reduce their radar cross-section has been on track since the military assets were equipped with first radar detection devices in late 1930s. But, the development of RAMs reached its spread during World War II (1939–1945), by Germany and the United States, but the ultimate motive between these two countries have contrast of black and white as Germany was particularly seeking for absorbers which can be used for radar camouflage while the United States directed their research mainly focussing on radar absorbers that can enhance the performance of radar system [10].

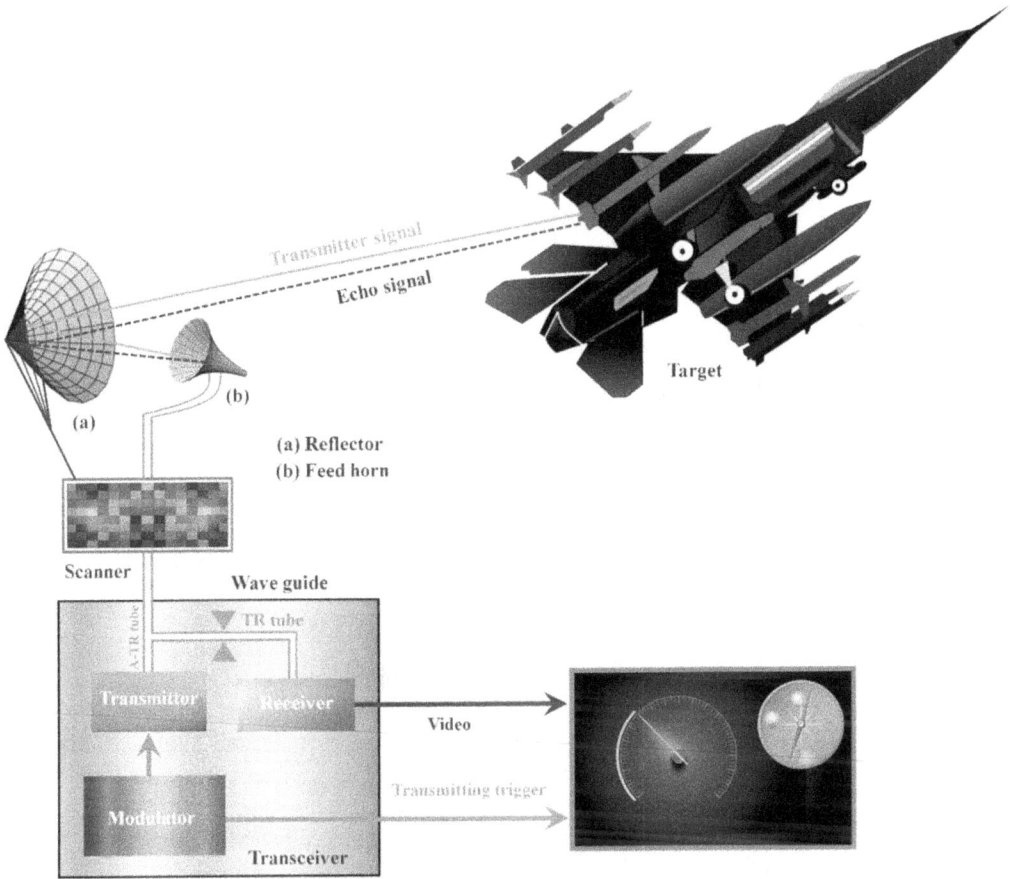

FIGURE 16.3 Basic instrumentation of radar system [7].

Germany successfully developed MacFarlane and Schade, which are two different camouflaging materials, in 1945. The first called "Wesch" and the other as "Jaumann," both prepared with a common material carbon black [11,12]. Both Wesch and Jaumann RAMs were based on polymeric materials. The Wesch material is 0.3 in. thick semi-flexible carbonyl iron powder-loaded rubber-based composite sheet that resonates around at a frequency of 3 GHz. Jaumann RAM belongs to the family of laminar composite, a device with 3 in. thick multiple layers made of rubber and plastic, alternatively. Thus, -20 dB reduction in reflectivity near normal incidence was obtained for 2 to 15 GHz frequency band.

In the United States, during the years 1941–1945, an alternative kind of absorber paint, generally called Halpern Anti-Radiation Paint (HARP) was developed (Halpren and Johnson) in two different variants known as airborne and shipboard versions, in which the thin and light version was of airborne paint. Both variants used rubber-based composite sheets rather than paint and reduced reflection by 15–20 dB at resonance frequency. MX-410, which is an airborne version, has a thickness of 0.025 in. which is merely close to that of paint which uses artificial dielectric materials with high relative permittivity. On the other hand, a rugged shipboard version developed using higher loading of ferrite particles in tough neoprene rubber matrix (binder) which shows absorption broadening near the resonant frequency. The thinnest (0.025 in.) paint coating is considered as an important milestone in the advancement of RAM which makes it possible to use it as an artificial dielectric coating on the core material. The airborne version achieved the dielectric

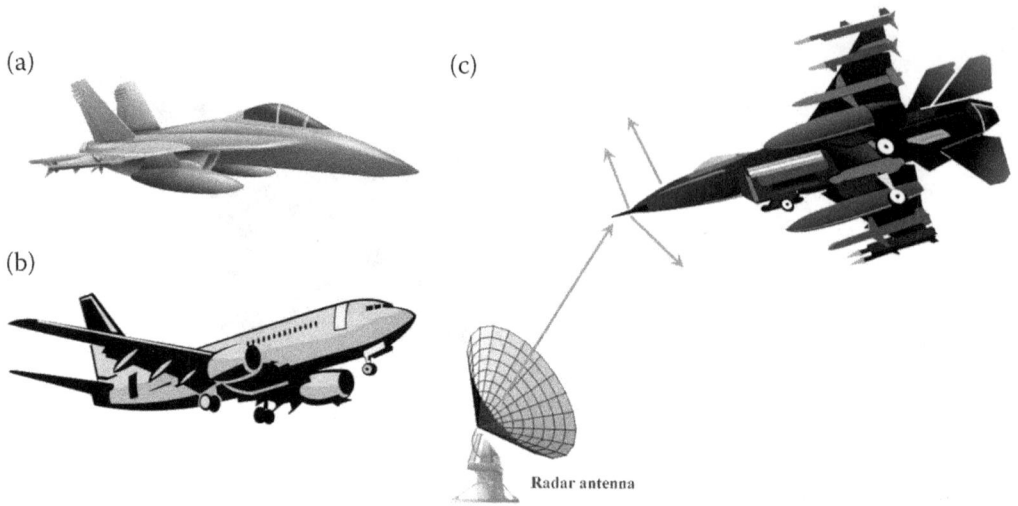

FIGURE 16.4 Pictures of air crafts (a) stealth aircraft, (b) conventional aircraft, and (c) the direction of signal reflects away at different angles, when radar signal hits on a stealth aircraft.

constant of 150 by loading the core rubber matrix with high concentration of largely aligned aluminium disc shaped flakes, while the shipboard variant possessed a significantly lower dielectric constant of 20 that required 0.07 in. thickness to resonate at X-band Similar to MX-410, the shipboard version also could operate in the X-band and its ruggedness was excellent [13].

Around the same time a well-known absorber called the Salisbury screen (SS, Salisbury 1952) RAM, was developed by the Massachusetts Institute of Technology (MIT), US, with a 25% bandwidth at resonance frequency and operated by resonant technique [14]. In SS, a resistive sheet is placed between quarter wavelengths from the scatterer and it was spaced by a less dielectric material and was frequently employed in anechoic chamber design earlier. Internally coating the SS to increase the internal reflection of the waves resulted in the development of a long pyramidal structured high attenuation RAM. Broadband RAM developed after World War II (1945–1950) are employed in anechoic chambers nowadays [13,15].

Another resonant absorber which is widely used is the Dallenbach layer, which consists of a metal plate which is coated with a potential homogeneous lossy layer. The lossy layer's thickness was chosen to match its input impedance to the free space's intrinsic impedance [16,17]. However, all these resonant absorbers were largely narrow band in nature, which necessitated further study into broadband RAMs. Most often the first method to be attempted in this direction was a multilayer Jaumann absorber and hence a majority of the practical broadband RAMs were fabricated using this method. For instance, broadening of Salisbury screens' bandwidth, was achieved by incorporating extra layers of resistive sheets as well as spacers [10].

For certain practical applications, sheets or laminate-type absorbers are not sufficient, where bulk absorbers were employed. In such bulk absorbers, the conductivity is significantly reduced upon carbon infusion in a flexible low-density spongy polymer foam; for example, spongy polyurethane foam. Based on this approach, in 1950s, 2 in, thick "Spongex" was introduced as a commercial broadband radar absorber material by the U.S. Naval Research Laboratory (NRL) based on animal hair with carbon-coating that offered a –20 dB attenuation in reflectance throughout a frequency range of 2.4 to 10 GHz for normal incidence, while thicker materials were utilized for low-frequency applications. In the Spongex, the conductivity was reduced by dipping or spraying carbon over a mat of loosely spun spongy foam similar to animal hair. These bulk absorbers might easily be utilized to control reflections in both indoor (anechoic) and outdoor antenna measurement ranges.

In the late 1950s, the emphasis was turned to larger frames for resolving issues connected with anechoic chamber application and the introduction of RAMS with higher reflection coefficients of –40 to –60 db. Furthermore, decreasing the operational frequency limit and improving measuring techniques for RAM or anechoic chambers were of scientific interest. On the passage of time during the 1970s the U.S. Department of Defence launched project Lockheed Have Blue, in order to build a stealth fighter jet. During the same decade, Meyer and Severin [10,18–20] started research on circuit analog RAM (CA-RAM) devices. They employed circuit theory to depict the absorbing devices' operations, components, and reflectivity model, and so these devices were labelled as circuit analogue devices. These approaches were adapted from the acoustical RAM research program. Resistance-loaded loops, resistive-loaded dipoles, slots in resistive foil, strips of magnetic and resistive materials having various orientations, magnetic loading of resonant materials, and surface shaping were used as key in development of RAM. There are few attempts to modify resonant absorbers in their geometry or structure to reduce the thickness or improve the performance. For an instance circuit analog CA-RAM were fabricated by the deposition of lossy material on a thin lossless film in geometric patterns and the effective resistance of the layer is controlled by the deposition thickness, likewise, effective inductance and capacitance were controlled by the geometry of the pattern. These led to the development of flexible fabrication methods to the tailoring of layers with any value of impedance, inductance and capacitance as well as its characteristic analysis in terms of lumped elements and better performance with reduced-layer thickness [16,18]. These hopeful developments catalyse the research in the field of frequency selective surfaces [21].

The research on CA-RAM devices continued in 1960s and 1970s, and the prospect of magnetic materials as RAM have been extensively researched, which led to the fabrication of thinner RAMs with significantly improved performance that make possible by using magnetic ferrites under layers [22]. Compared to the dielectric RAMs, these thinner magnetic ferrites RAMs have low reflection coefficients and lower operating resonant frequencies [23]. Even though these magnetic RAMs are significantly thinner than dielectric RAMs, they are significantly heavier and more susceptible to disintegration at high resonant frequencies. Also, by employing pyramidal shaped RAMs for the development of anechoic chambers, the reflectivity was reduced by 60 dB at near normal incidence. Later in 1977, Connolly and Luoma used screen printing to control the fabrication of Jaumann RAMs [24] and then RAMs were fabricated from knitted and weaved fabrics, honeycomb or foams, net-like structures, spongy structures, and coated with the paint which is comprised of fibrous or particulate carbon, nickel chromium alloy or evaporated metal [25].

In the 1980s, tremendous improvements happened in the RAM's design process and it was enhanced by optimization techniques [26]. During the same time scientists were interestingly observing that some of the biotech products have ultra-wide-band absorption properties, and succeeding experimental investigations demonstrated that integrating these materials in aircraft design can result in significant RSC reduction. Soon after these observations the U.S. Department of Defense classified these compounds and found that they are retinal compounds that belong to a group of Schiff-base salts that are powdery, black-colored substances, much lighter (1/10 lighter) than ferrite and can be used for the production of RAM paints for stealthy aircrafts [27].

Recently, a new subset of homogeneous materials having chiral properties called chiral materials, have been developed which can be used for RAMs-based applications (Jaggard and Enghetta 1989) and an improved absorption was observed with RAMs fabricated with chiral materials [28]. The performance of the chiral-based RAMs depends on their optical property and circular dichroism as well as the backscatter RCS is not polarization dependent. Unlike dielectric as well as magnetic RAMs, the chiral RAMs are known for their reduced RCS and performance

[29,30]. During the same time, the adoption of various resistive and graded layer profiles enhanced the properties of Jaumann RAM. Various materials such as graphite [26], carbon black [31], ferrites [22], and carbonyl iron [32] and their modified forms are used for the fabrication of RAMs. In addition, the effect of morphology of these materials on RAMs' performance are studied and various morphologies such as wires, rods, sphere, and discs are also micro- and nanosized and used for the preparation of dielectrics [33]. Composite materials and electronically conducting polymers such as polyaniline, polypyrrole, etc., are demonstrated as potential classes of RAMs. And recently lots of developments happened in the field of RAM research as well as advancements in materials and device designs including desired permeability and permittivity calculation using the mixing theory [34] and the optimization methods like numerical optimization techniques and empirical procedures were used to improve the RAM design process [31,35] and are employed in the designing of Jaumann RAMs in 1989 [35]. In recent years, ST was effectively adopted into warfare, when it was utilized by military spies to gather intelligence and invade enemy territory with a surprise attack. ST is known for its air power and worth in contemporary warfare including Gulf Wars (1991), Kosovo conflict (1998), Afghanistan War (2001), Balakot surgical attack by Indian army in Pakisthan (2019), and Isreal-Palastene conflict happened in 2021.

16.4 RADAR CROSS SECTION (RCS) AND FACTORS AFFECTING RCS

The notion of stealth is to operate or hide the target while giving hostile forces no sign of friendly troops' presence. RCS is a crucial detection parameter in stealth technology that determines the virtual area of the target which is visible to the radar [36]. Thus, RCS indicates the ability and magnitude of a target to reflect the electromagnetic radiations from the radar. High RCS implies that a large portion of the surface of the target is vulnerable to reflect back the incident radiations and thus, detected by the radar. Typical radar cross-section (RSC) values of different targets are tabulated in Table 16.1. Likewise, low RCS is nothing but a key to invade the territory of enemies, which is an indication of better stealth property of the missile, ship, aircraft, etc. In the early 1930s itself, studies on the RCS parameters gained the invigorated attention of the researchers countering the development of radar systems [37]. Typical radar cross sections are mentioned below.

Missile = 0.5 sq.m

TABLE 16.1
Typical Radar Section Values of Different Targets

Target	RCS (M^2)
Large commercial airplane	100
Large fighters	5–6
Small fighters	2–3
Man	1
F-117 fighter	0.1
Small bird	0.01
B-2 bomber	0.01
Bug	0.00001

Tactical Jet = 5–100 sq.m

Bomber = 10–1000 sq.m

Ships = 3000–1000000 sq.m

Technically, RCS is a proportion of backscattered power per strong point from the object toward radar to the complete force blocked by the object [38]. The territory characterized by the RCS is not a mathematical cross segment of a target. If there should arise an occurrence of a circle, RCS is maverick of frequency if working at low wavelength (i.e., λ « range and λ « radius) in which the frequency is very high, since the RCS depends on the frequency and is calculated as $4\pi A^2 / \lambda^2$, and here 'A' represents the area of the plate [39].

It can be comparably defined in Equation 16.1 [40]:

$$RCS = \frac{Power\ reflected\ to\ the\ receiver\ per\ solid\ angle}{Incident\ power\ density/4\pi} \tag{16.1}$$

Another relevant arithmetic formula for the RCS of the target object is given in Equation 16.2:

$$\sigma = \frac{4\pi R^2 \times Power\ density\ in\ the\ scattered\ field}{Power\ density\ of\ the\ incident\ plane\ wave} \tag{16.2}$$

RCS can also be explained in terms of decibel square meter (dBsm), which is given in Equation 16.3:

$$\sigma_{(dBsm)} = 10\ \log_{10}(\sigma/\sigma_R) \tag{16.3}$$

where σ_R is the RCS of a reference object with an area of 1 m^2 [40].

RCS is high for the little spheres and fluctuating, and low as there is an increment in the perimeter of the circle. For the roundabout and the circular cylindrical chambers, the pinnacle worth of RCS is at a perspective point of 90. Recurrence shows a direct connection for the chamber, and there is an increment in RCS for enormous chambers [39]. The RCS curve of a stealth aircraft (X-47B) is shown in Figure 16.5 [41].

16.5 FACTORS AFFECTING RCS

The parameters that affect the RCS are size, shape, materials coated on the surface, radar frequency, angle, and orientation of the target object. Flat objects have an RCS of nearly zero, but the aircrafts have a complex structure that can reflect the radar frequencies in which missiles, fuel tank, and even the air intake can increase the RCS value. Perhaps, in order to reduce RCS, the weapons in stealth aircrafts are kept in a bay door, which makes them the stealthiest. Basic variables and techniques used to reduce RCS of a target are explained below.

- *Size and shape:* As the size of the target increases, greater will be the radar reflection and thus, RCS will be high. So, in order to become stealthy, the size of the target should be small. Likewise, the shape of the target enhances the stealth property even up to 99% by scattering the radar radiation in different directions so that only a small portion of radar energy will be reflected back to the radar system. For instance, RCS of F22 is analogous

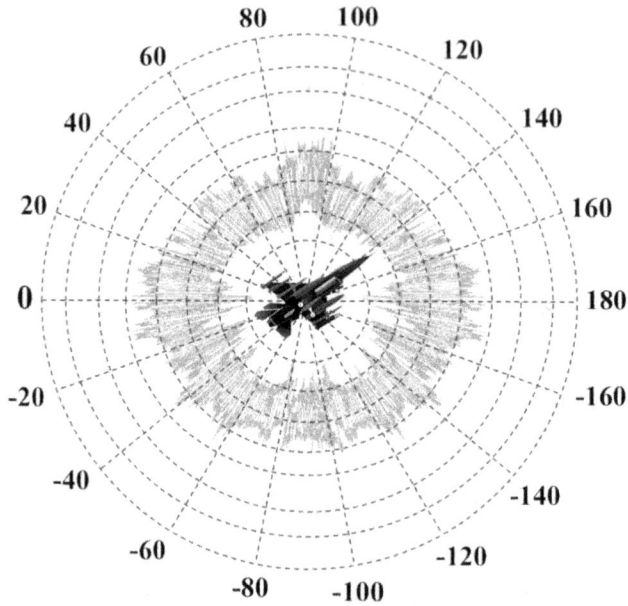

FIGURE 16.5   Radar cross section (RCS) of a stealth aircraft [41].

to a small bird even though it is 62 feet long with wingspan of around 45 feet and it can scatter the radar frequencies on engines as well. Its missile is kept inside the fuselage and its forward cone of angles makes it stealthier. Thus, says that, flat and sharp-edged target objects are found to be stealthier than parabolic shaped targets. Reflections of radar signals from different parts of the stealth aircrafts are shown in Figure 16.6.

• *Materials and radar frequency:* The material used on the surface of the target object can highly influence the RCS value. The materials like metals with high radar reflection properties will result in high RCS. But in stealth technology, materials are wisely chosen and radar absorbing materials which are capable of absorbing the radar frequencies

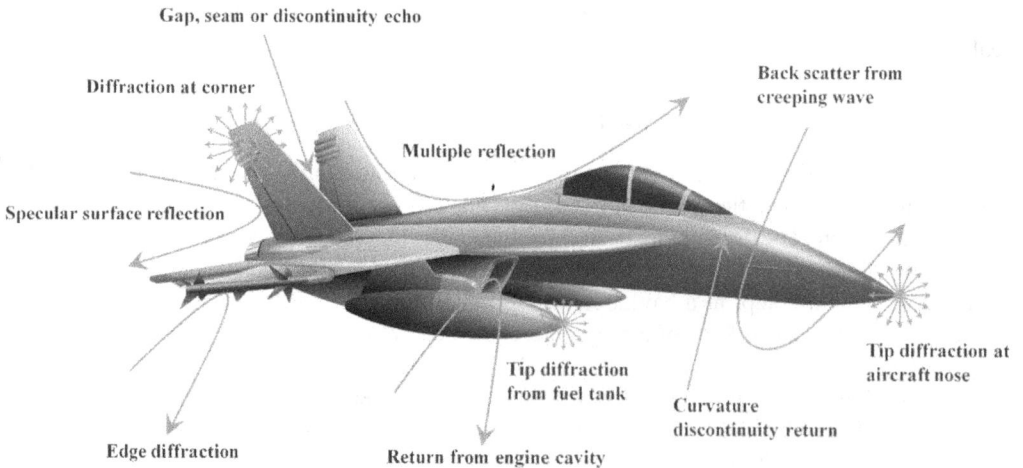

FIGURE 16.6   Some of the reflective corners or parts of the stealth aircrafts, where energy is reflected back toward the target.

without reflecting them, are coated on the surface of the target object. Broadband as well as narrow-band radiation absorbing structures are there, but, the broadband RAS are in the spotlight.

- *Angle and orientation:* The angle of incidence of a radar signal, depends on the geometry of the target object, that can be unpredictable under the aegis of the aerodynamics of the target object. But the reflecting angle depends on the incident angle and so, the geometry is more concerned. Likewise, the orientation of the target also matters in the case of polarization of incoming and outgoing radiations.

16.6 RADAR ABSORBING MATERIALS (RAMS)

Radar absorbing materials (RAMs) are capable of absorbing radar frequencies so that they are less likely to reflect the radar energy back. Dielectric and magnetic RAMs are capable of converting the absorbed radar waves into heat which can be achieved using carbonaceous materials, metals or even conducting polymers. A list of possible materials reported by the researchers for microwave absorbing and shielding (MA&S) applications is tabulated in Table 16.2. Dielectric absorbers function on the basis of ohmic loss of energy which is made possible by the lossy carbon-based materials or metals fillers in a potential polymer matrix. Likewise, magnetic absorbers function on the basis of magnetic hysteresis effect.

RAMs created a jolt in the field of stealth technology in which, the polymer composites are found to be the best. Polymer-based RAMs include a matrix and a reinforcing filler. The filler can be magnetic or conducting one [56]. Embedding matrix can be any potential polymer matrix like nitrile rubber, neoprene rubber, silicone rubber, polyurethane, epoxy resin, etc. On absorbing (partially or completely), a good part of incident radiation from the radar, the signal strength of the reflected waves can be reduced and thus, the damped reflected signal will not reach back to the radar system. RAMs can be used for broadband or narrow bands based on their ability to absorb

TABLE 16.2

List of Possible Materials Reported by the Researchers for Microwave Absorbing and Shielding (MA&S) Applications

Material	Frequency Regime (GHz)	Ref.
Polymer nanocomposites	2–18	[42]
PANI coated magnetite	8.2–12.4	[43]
Flaky cobalt	2–18	[44]
Graphene	1–200	[45]
Fe–Al hybrid nanocomposites	4–14	[46]
Ferrite graphene composites	8.2–12.4	[47]
Flake carbonyl iron	2–18	[48]
TiC/epoxy composites	2–18	[49]
Iron fibers and carbonyl iron	1–6	[50]
Carbon nanotubes (CNTs)	8.2–12.4	[51]
CoFe/Al_2O_3 composite nanoparticles	2–18	[52]
Carbon-based nanocomposites	8.2–12.4	[53]
Iron flake–carbon black composite	2–10	[54]
Barium and strontium hexaferrites	8.2–12.4	[55]

radiation of corresponding frequency. The reflection coefficient of a surface is given in Equation 16.4.

$$r = \frac{\eta_M - \eta_0}{\eta_M + \eta_0} = \frac{Z_M - Z_0}{Z_M + Z_0}; Z = \frac{1}{\eta} \qquad (16.4)$$

where r = reflection coefficient

η_M = admittance of the reflecting substrate

η_0 = admittance of incident medium

Z_M = intrinsic impedance of the reflecting substrate

Z_0 = intrinsic impedance of the incident medium

Thus, based on the conviction of impedance matching, r of a substrate is said to be zero when $\eta_M = \eta_0$. That is to say, when Z_0 matches with Z_M, reflection can be zero, since the intrinsic coefficient is ≈377 Ω for the free space. Thus, an object with intrinsic impedance value of ≈377 Ω, produces zero reflection [25]. For example, metallic airplanes have much different intrinsic impedance from that of free space. So, in order to compensate for the intrinsic impedance, RAMs with magnetic and dielectric materials are coated on the airplane. The overall stealthy performance depends on the thickness of the coated material, geometry which determines the angle of incident radiation and material used [57–59]. The radar cross section and scattering of electromagnetic radiation from the radar antenna, on hitting the target fighter jet (X-47B), is shown in Figure 16.7 [60].

RAMs can dampen the reflected electromagnetic radiation by electromagnetic wave penetration, internal reflection, dielectric relaxation and dielectric loss. The destructive interference can damp the incident electromagnetic waves [61,62]. In the subsequent case, episode waves are counteracted by important 180° stage shift anti-phase reflections that are accomplished by surfaces having reflecting property, that are divided by an ostensible quarter frequency separation. A few RAMs, and furthermore radar-engrossing designs (RAS), use the two standards. In contrast, RAMs are also used for some non-military applications [63,64]. The important requirements for the RAMs include:

- High absorption capacity across a wide range of broad microwave frequencies
- Thin surface coatings
- Low overall density
- Surface coating-layer design that is simple and needs less processing.

16.6.1 CLASSIFICATION OF RADAR ABSORBING MATERIALS (RAMs)

In general, absorbers are divided into two types: impedance matching absorbers as well as resonant absorbers. The major classifications of RAMs are shown in Figure 16.8. The pyramidal absorbers provide gradual transfer of impedance from air to absorber, having a cone-like structure that is perpendicular to the surface. The structure's height and periodicity are approximately one wavelength long. Above a certain operating frequency, this sort of radar absorbing material gives very high attenuation. Because of their small dimensions, tapered loading absorbers are preferable than pyramidal types, but they perform poorly, therefore it is desirable to adjust the impedance gradient over different wavelengths. The material is composed of a mix of less lossy as well as lossy components. The lossy portion is spread parallel to the surface, with a gradient perpendicular to the surface that rises into the material's body. The purpose of matching layer absorbers is to reduce the thickness of above types. This type of absorber uses an absorbing layer in between incident and absorbing media. Between the

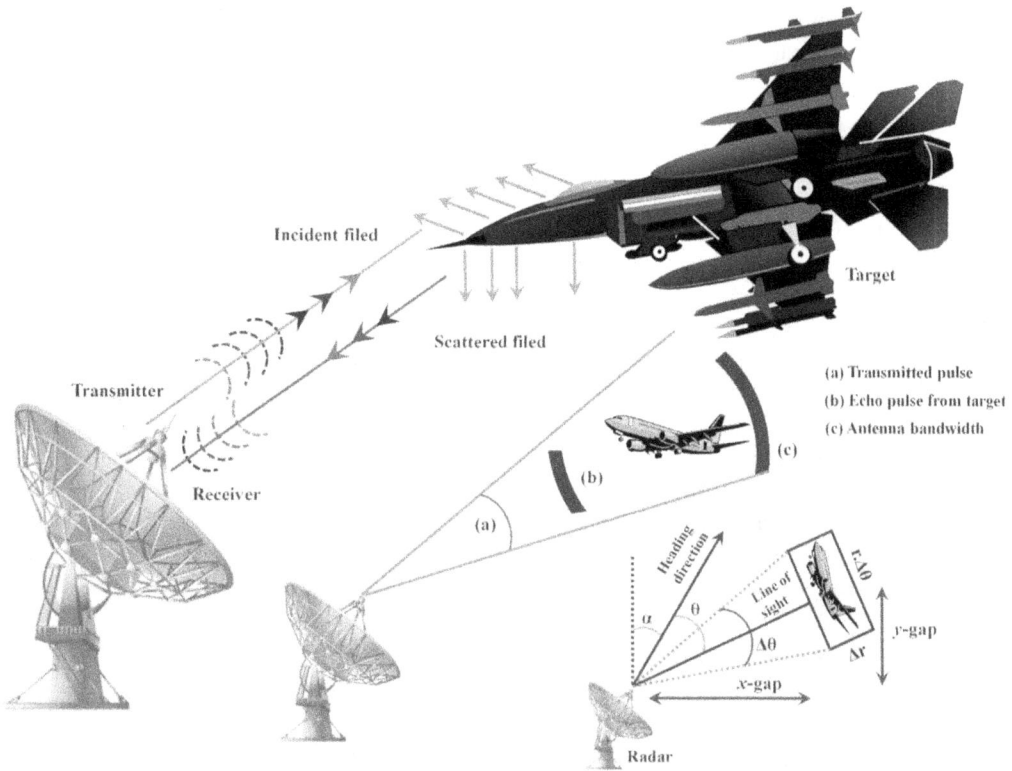

FIGURE 16.7 Schematic illustration of the radar cross section and scattering of radar waves [60].

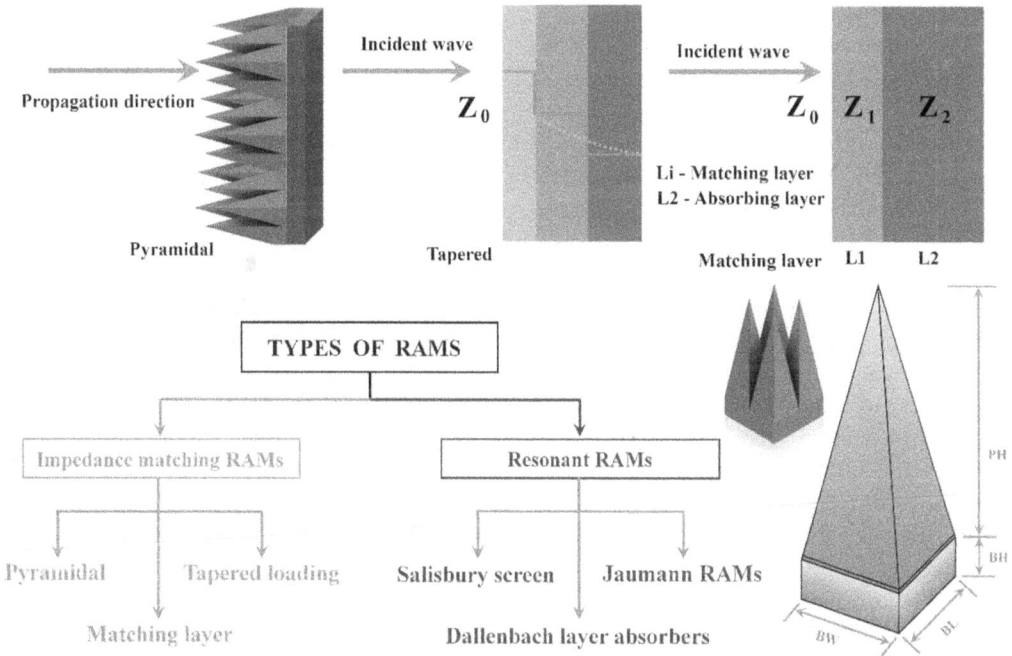

FIGURE 16.8 The general classification of the type of radar absorbing materials and design of pyramidal radar absorbers.

incident and absorbing material, an absorbent layer is used. This transition layer's impedance is between the front and rear layers and their impedance $Z_2 = (Z_1 \times Z_3)^{0.5}$.

16.7 LOSS MECHANISM IN RADAR ABSORBING MATERIALS RAS

16.7.1 LOSS MECHANISM IN MULTI-LAYERED RAS

Broadband RAS often satisfies structural requirements with fibre or fabric reinforced polymer composites, and likewise, RAMs and/or multi-layered grid devices contributing to functional needs. As discussed in the previous section (Section 16.6), pure dielectric and magnetic materials have their own limitations. For example, due to the increased material loading requirements and bulky nature of dielectric materials, absorption at low microwave frequencies is limited, whereas magnetic materials that work on the principle of resonant absorption would provide absorption at low microwave frequencies with low material loading, but with narrow bandwidth. Depending on the patterns, the very easy to process multi-layered grids can offer a wider bandwidth but the thickness and material cost is high. Hence, the material selection and making their combination is very important to the design of broadband RAS. There are very different materials ranging from carbon materials, ceramic materials, metals or metal oxides or magnetic materials and polymeric materials are widely used in RAS, the loss mechanisms vary from materials to materials. Some materials follow the capacitive loss mechanisms while some other materials show the inductive loss mechanisms, hence loss mechanisms involved with dielectric materials explained in detail in the following sections.

16.7.2 LOSS MECHANISM IN COMPOSITES WITH DIELECTRIC FILLERS

Loss mechanisms of different RAMs can be capacitive loss mechanisms, inductive loss mechanism or conductive loss mechanism, etc. Loss mechanisms of dipolar or dielectric and losses owing to conduction exist in nonmagnetic materials. The losses in pure dielectric materials like polymers or ceramics are dipolar in nature while in the magnetic materials, electron spin resonance, hysteresis, conduction, and domain wall resonance influence the loss. In addition, they can exhibit significant conduction losses plus magnetic losses including electron spin resonance, hysteresis, and domain wall resonance.

In the dielectric materials the microwave absorption straightforwardly relies upon permittivity materials ε_r and that can be represented by Equation 16.5:

$$\varepsilon_r = \varepsilon' - j\varepsilon'' \tag{16.5}$$

where ε' and ε'' represent the real and imaginary part of materials' permittivity, respectively.

When microwave radiation falls on the dielectric materials, it leads to the formation of internal fields, which in turn result in the delocalisation of the dipoles as well as the free bound charges like ions and electrons responsible for the dipole displacement current or electronic/ionic conduction. However, the elastic forces as well as the intrinsic friction within the materials will oppose these displacements prompting the losses. The imaginary permittivity of dielectric loss parameter (ε'') can be used to represent this loss collectively.

The loss tangent tan δ, is defined the ratio of ε'' to ε' and can be represented by Equation 16.6:

$$\tan \delta = \varepsilon''/\varepsilon' = \sigma/2\pi f \varepsilon_0 \varepsilon' \tag{16.6}$$

where σ represents the effective conductivity produced by dipole displacement current and electronic/ionic conduction and f denotes the frequency. Similarly, the losses on account of magnetic field can be represented by Equation 16.7:(16.7)

$$\mu_r = \mu' - j\mu'' \tag{16.7}$$

where μ' and μ'' denote the real and imaginary magnetic permeability of the material, respectively.

The magnetic loss tangent, $\tan \delta$ can be expressed as the ratio of μ'' to μ' and can be represented as in Equation 16.8:

$$\text{Tan } \delta = \mu''/\mu' \tag{16.8}$$

It is well known that the vital parameter which dominates the propagation of electromagnetic waves that travels through different mediums depends on its impedance (Z). The correlation between the impedance of the material and permeability or permittivity is can be represented by Equations 16.9 and 16.10:

$$Z = [\mu/\varepsilon]^{1/2} \tag{16.9}$$

$$Z = Z_0 [\mu_r/\varepsilon_r]^{1/2} \tag{16.10}$$

where Z_0 denotes the intrinsic free space impedance.

When electromagnetic waves travel from one material to another, reflections at the interface occur due to the impedance mismatch between the two media. Hence, the reflection coefficient can be represented by Equation 16.11:

$$r = (Z_1 - Z_2)/(Z_1 + Z_0) \tag{16.11}$$

where Z_0 and Z_1 denote the intrinsic free space impedance and boundary surface impedance. In case of metal backed (for electromagnetic measurements) radar absorbing material Z_1 can be represented by Equation 16.12:

$$Z_1 = Z \tan h\gamma d \tag{16.12}$$

where Z and d denote the impedance and thickness of RAM respectively and γ represents the propagation constant of the material, which can be represented by Equation 16.13:

$$\gamma = [j\omega\mu(\sigma + j\omega\varepsilon)^{1/2} \tag{16.13}$$

where σ and ω denote the material conductivity and operating angular frequency, respectively.

The interference of waves reflected at the interfaces of reinforcement layers and at the metal perfectly electrical conducting (PEC) surfaces of reinforcements determines the reflection coefficient (r) of a composite structure with absorbing fillers. The phase of the reflection coefficient is determined by the permittivity, permeability, and thickness of the material, whereas the amplitude is determined by the losses. Figure 16.9 depicts the geometry of a typical multilayer absorbing structure under normal incidence, with incident waves travelling in the positive Z direction and reflected waves travelling in the negative Z direction. Suppose, η_i represents complex intrinsic impedance, l_i represents the thickness and η_i represents the propagation constant, respectively, of last i^{th} layer (where $i = 1, 2, 3, ..., n$) beginning from the side of PEC, ε_0 and μ_0, represents the free space permittivity and permeability values. Owing to the transmission line theory, the impedance came across by the EM waves at layer "I" is defined as:

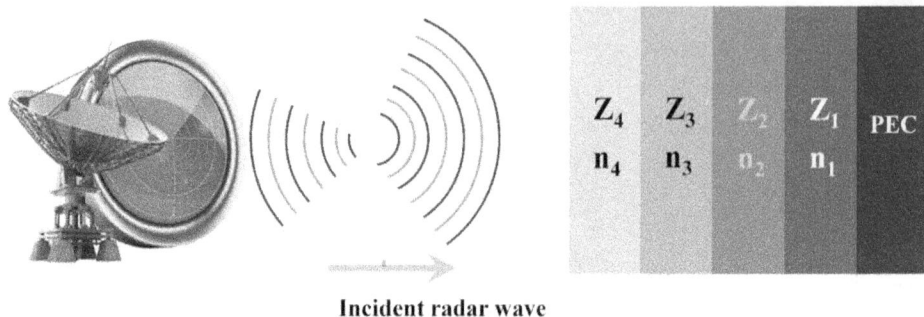

Incident radar wave

FIGURE 16.9   Schematic diagram for multi-layered radar absorbing structures (RAS).

$$\frac{Z_i = \left[Z_{i-1} + \eta_i \tan\!-h(\gamma_i t_i) \right]}{\left[\eta + Z_{i-1} \tan\!-h(\gamma_i t_i) \right]} \tag{16.14}$$

where η_i and γ_i denote:

$$\eta i = Z_0(\mu^*\varepsilon^*)^{1/2} \tag{16.15}$$

$$\gamma_i = j(2\pi/\lambda)[\mu^*\varepsilon^*]^{1/2} \tag{16.16}$$

where Z_0 is 120π, the free space impedance. And ε^* and μ^* corresponds to permittivity and permeability of layer 'I'.

In this case, the last layer having intrinsic impedance η_i and thickness t_1 ($t_4 = t_3 = t_2 = t_1$) has been terminated with PEC, so its impedance can be expressed as a short-circuited transmission line.

$$Z_1 = \eta_1 \tan\!-h(\gamma_1 t_1) \tag{16.17}$$

where η_1 is the intrinsic impedance and r1 is the propagation constant of layer 1.

Using Equations (16.15–16.17), the reflection coefficient for a multi-layered RAS at air interface can be described as

$$r = |Z_n - Z_0|/|Z_n + Z_0| \tag{16.18}$$

And, reflection loss, RL (in dB) of a multilayered RAS can be expressed as:

$$\text{Reflection loss,   RL(dB)} = -20 \log |r| \tag{16.19}$$

The relation between RCS reduction and reflection loss is given in Table 16.3. This clearly indicates that for 10 dB loss the RCS reduction is almost 90%.

16.8   TESTING OF RADAR ABSORBING MATERIALS (RAMS)

When anything shows an indication of its presence, it is detected. The visual, smoke, acoustic, infrared, radar, and contrail characteristics are among the six that an aircraft uses to signal its presence [65]. Various RCS measures are available in accordance with each signature, both outdoors (near field and far field) as well as indoors in an anechoic chamber. The anechoic chambers

TABLE 16.3
RCS Reduction vs. Reflection Loss

RCS Reduction (%)	Reflection Loss (dB)
0	0
90	10
99	20
99.9	30
99.99	40

can assure measurement accuracy by eliminating reflections and disturbances caused by EM waves. The distance between the radar and the target determines which field to use.

It's also feasible to perform lab scale experiments on RAMs to determine EM characteristics including dielectric real and imaginary permittivities, magnetic permeabilities, dielectric loss tangents, loss tangents, and reflection loss (dB). These measurements may be performed in a variety of environments employing transmission line wave guides, resonant cavities, and free space, depending on material properties such as thickness, homogeneity, isotropy, and anisotropy, among others. It is necessary to measure several samples from each type of sample to improve the accuracy of these measures.

For complicated EM characteristics of homogeneous, heterogeneous, isotropic, and thick materials, the transmission line technique with the support of the vector network analyzer (VNA) is commonly employed. In the case of bi-axial samples, however, the waveguide and coaxial methods are preferred. Free space in anechoic chambers may be used to evaluate anisotropic and heterogeneous materials like composites. This enables the measurement of reflection and transmission, as well as the extraction of radar absorption and material impedance properties. These statistical values are used in the development of stealth materials [66]. Figure 16.10 depicts the schematic of several measuring techniques.

16.9 POLYMER MATRICES USED FOR STEALTH TECHNOLOGY

Along with a capable filler, the properties of the embedded matrix also matter on the development of stealth material. Polymer matrices with good thermo-mechanical properties can be considered as a possible contender for stealth technology application. The mechanical property of the embedded matrix like tensile strength, shear resistance, tear strength, Young's modulus, elongation at break, etc. matters. The geometry and composition of radiation absorbing filler particles can affect the overall mechanical property of the polymer composites. The effect of absorbing material on the mechanical properties of the polymer matrix, for some polymer composites, is given in Table 16.4 [67]. The addition of nano-sized fillers positively affects the mechanical and dynamic mechanical properties of the concerned polymer composite along with the electromagnetic absorption property. Polymer matrices like nitrile rubber, natural rubber, neoprene, polyurethane, silicone rubber, epoxy resin, polypropylene, etc. are widely used.

16.10 CLASSIFICATION OF POLYMER COMPOSITES AS RAMS BASED ON FILLER MATERIAL

Polymer composite is a multi-phase material in which reinforcing fillers are uniformly dispersed and distributed in a polymer matrix, offers synergistic mechanical properties that cannot be achieved from the constituent component alone. Polymer composites are designed to possess superior load bearing capacity by transferring the load from the matrix phase to filler phase. Some of the major advantages of polymer composites are their lightweight, high stiffness, strength, cost effectiveness, and corrosion

(a)

Network analyzer

Wave guide (Rx)

Wave guide (Tx)

(b)

Test sample

Horn antenna

(c)

Target

Transmitter signal

Echo signal

Radar antenna
Transmitter

Radar antenna
Receiver

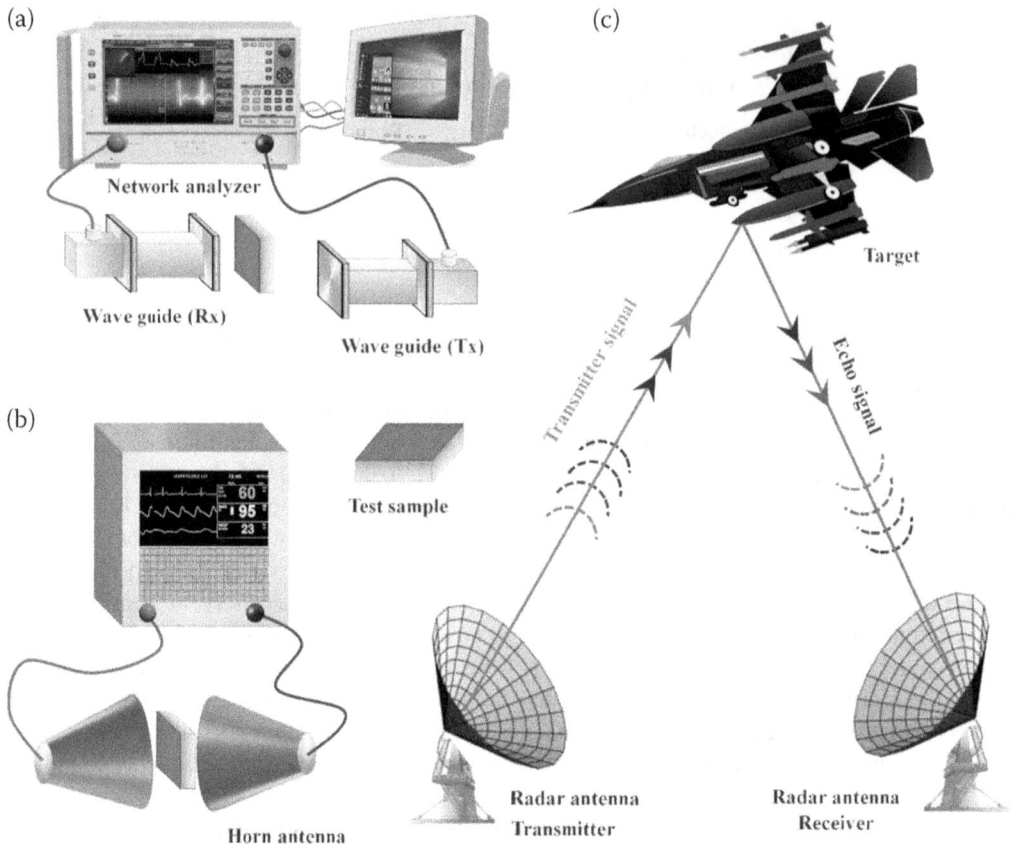

FIGURE 16.10 Test methods for electromagnetic evaluation of radar absorbing materials: (a) wave guide, (b) free space, (c) open range RCS measurements.

resistance. There are different classes of filler materials such as ceramic fillers and carbon nano-materials that are widely used for the preparation of polymer composite based RAMs. Different ceramic fillers such as $BaTiO_3$, ZnO, SiO_2, Al_2O_3, H-BN, TiO_2, ZrO_2, etc. can be employed as ceramic fillers in polymer composite-based RAMs. Figure 16.11 shows the structure of some popular radar absorbing ceramic-type fillers having high dielectric properties.

16.10.1 MAGNETIC ABSORBERS

The magnetic absorbers are known for carbonyl iron, hexaferrites, and spinel ferrites in a polymer matrix, which can absorb the electromagnetic radiation in MHz to GHz range. Tuning the composition and properties of these materials in the core polymer matrix can achieve the absorption frequency in the range of 5–20 GHz based on their molecular size as well as its sintering temperature. A system of two-layer magnetic absorbers can enhance the transfer speed of magnetic materials in which one layer contains the ferrite part and the other layer contains short metal filaments to both the interfaces. The magnetic materials are embedded in a potential matrix which can withstand the temperature and weather conditions. Matrices used can be nitrile rubber, poly-isoprene, silicone rubber, neoprene, etc. [78–80]. Foresee, the thickness as well as the magnetic properties can be tailored based on the requirement and the specific frequency, on demand, which accomplishes the intrinsic impedance around ≈377 Ω and enhanced loss factor. Lossy magnetic materials are found to be the best for better preparation and easy implementation. Examples of

TABLE 16.4

Effect on the Mechanical Properties of Polymer Composites as RAMs [67]

Composite System	Absorbing Filler Material	Mechanical Test Performed	Observation	Frequency (GHz)
Glass epoxy [68]	Multiwalled carbon nanotubes	Tensile, Shear	Tensile strength and modulus increased; Shear properties decreased.	12–18
Glass epoxy [69]	Multiwalled carbon nanotubes	Inter laminar shear strength	Presence of delamination alters RAS performance.	8–12
Glass epoxy [70]	Multiwalled carbon nanotubes and carbon nano fibers (CNFs)	Compression, tension, ILSS	Filler addition increases the viscosity, increase in tensile and compression, ILSS merely affected	12–18
Glass epoxy [71]	Multiwalled carbon nanotubes	Tensile, compression, shear in normal, axial and in-plane direction	Tensile strength, tensile modulus, stiffness: increases in normal mode, not improved in in-plane mode, increases in axial mode. Compressive strength, modulus decreases in normal and in-plane mode but increases in axial mode. Shear properties decreases in all modes	12–18
Polypropylene [72]	Conducting carbon black, carbon fiber	Dynamic mechanical thermal analysis, hardness, impact, tensile	The elastic and storage modulus improved due to modified interfaces in filler-polymer composite; Hardness improved; tensile properties such as modulus and yield stress are also increased; Elongation and toughness reduced.	5–8
Epoxy [73,74]	Multiwalled carbon nanotubes	Tensile	Young's modulus and the yield strength doubled with 1 wt. % and quadrupled with 4 wt.% of filler.	3–18
Glass-epoxy [75]	Carbon black, carbon nanotubes	Tensile	3.8% hike in tensile modulus and 5.7% hike in tensile strength.	8–12
E-glass-carbon-epoxy with ultra-high molecular weight polyethylene fiber [76]	Nano-sized carbon black	Low velocity impact, followed by compression after impact, tension, compression	Tensile properties increased; compression decreased; other properties little affected.	10
Carbon epoxy and PMI foam [77]	CNT	Flexural stiffness by 03-point bending	Flexural strength of RAS designed for 9, 10, and 11 GHz are 89, 90, 87 MPa and specific flexural strength are 0.17, 0.18, 0.18 MPa, respectively	X band

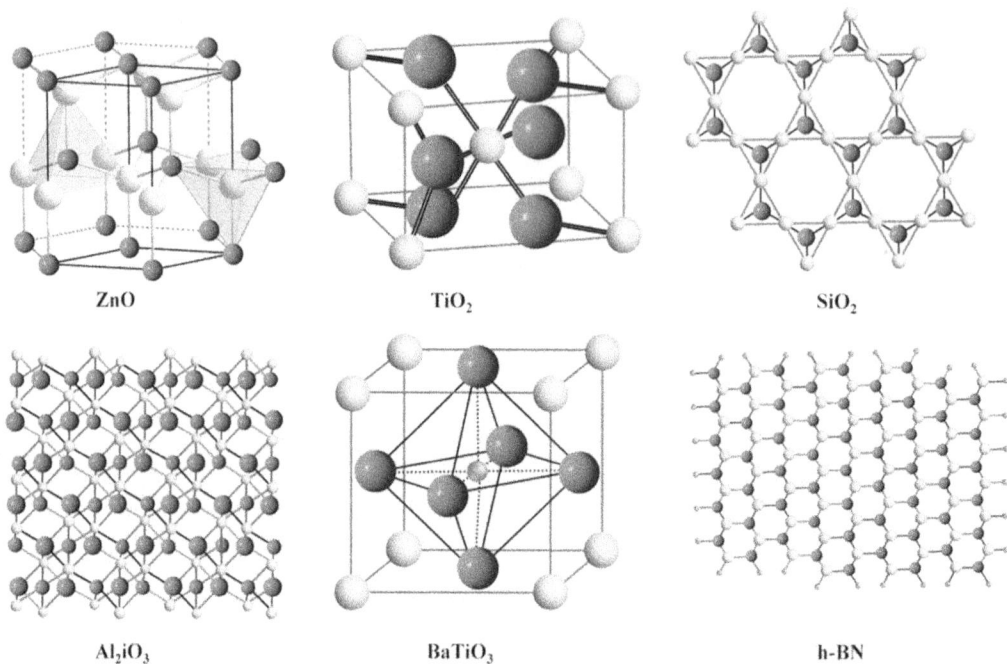

ZnO   TiO$_2$ SiO$_2$

Al$_2$iO$_3$ BaTiO$_3$ h-BN

FIGURE 16.11   Some of the popular radar absorbing ceramic-type fillers based on high dielectric materials employed in polymer composites.

magnetic absorbers that can be used for stealth technology application on aircraft include paints based on spinel ferrites, M-type hexaferrites, and carbonyl iron [81,82].

16.10.2   DIELECTRIC ABSORBERS

Atoms and molecules with perpetual positive and negative charges can be portrayed as dielectric materials in which there is electrostatic attraction in between them. The nuclear and atomic powers make them attracted to each other and are not allowed to dislocate. Be that as it may, on applying electric field to such materials, a few electric dipoles are produced in which they will align themselves along with direction of the applied electric field. Those materials are called dielectric absorbers. The predominant components engaged with microwave ingestion by the dielectric RAMs are the retention of the incident energy and multiple reflections of the incident electromagnetic radiation []. These impacts are because of actual marvels like ohmic losses, and the backscattering of the radiation in various ways by utilizing legitimate geometric shapes of the target object. Dielectric materials are prepared by mixing matrices like polyurethane, epoxy, etc. and incorporation of inorganic materials like carbonaceous materials, nitrides, and so forth or organic dielectric materials [67].

16.10.3   NANOMATERIALS

Nanotechnology complemented with a strong polymer basis is found to be the best possible way to reach the milestone towards the best RAMs. Nanostructured RAMs essentially comprise of the accompanying four sorts: nano-crystal RAMs; centre shell nanocomposite RAMs; nanocomposites of carbon nanomaterials; nanocomposites of multi-walled carbon nanotubes (MWCNT) and inorganic material RAMs [84,85] in potential polymer matrix like thermoplastic as well as thermosetting polymers that can absorb radiation in X-band and even in broadband. Also, nano-materials of the core-shell type influence the development of RAMs with high-absorption dielectric and magnetic characteristics while being light in weight [86,87]. The center shell FeCo(C) amalgam nanoparticles with magnetic FeCo centers and dielectric graphitic shells incorporated RAMs shows acceptable microwave absorption over a wide frequency range [61]. The radiation absorbing property fortified with the power of nanotechnology not only increases the stealthy property but also the thermo-mechanical properties of the polymer composites.

16.10.4 CARBONACEOUS MATERIALS AS FILLERS IN POLYMER COMPOSITES-BASED RAMs

The composites with optimised operating bandwidth are the best along with the weight reduction. The carbonaceous materials as radiation absorbing materials are found to be a promising candidate that can deal with narrow and broadband that can counter with the processes like electromagnetic wave penetration, internal reflection, dielectric relaxation, and dielectric loss. The various carbon entities widely used in polymer composite-based RAMs are shown in Figure 16.12. The properties of the carbonaceous-based composite can be tuned by controlling the composition of carbonaceous materials as well as the embedded matrix, which makes it highly flexible. Nitrile rubber, natural rubber, silicone rubber, polyurethane, epoxy resin, etc. are widely used as the potential matrix with carbon-based conducting fillers [87,89]. Geometry and composition of the RAMs in the matrix and its morphology also matters. The concentration of filler material is important in order to optimize the absorption of the composite and calculate the percolation threshold as well as the thermo-mechanical properties of the concerned polymer composite. If the concentration of the filler is too

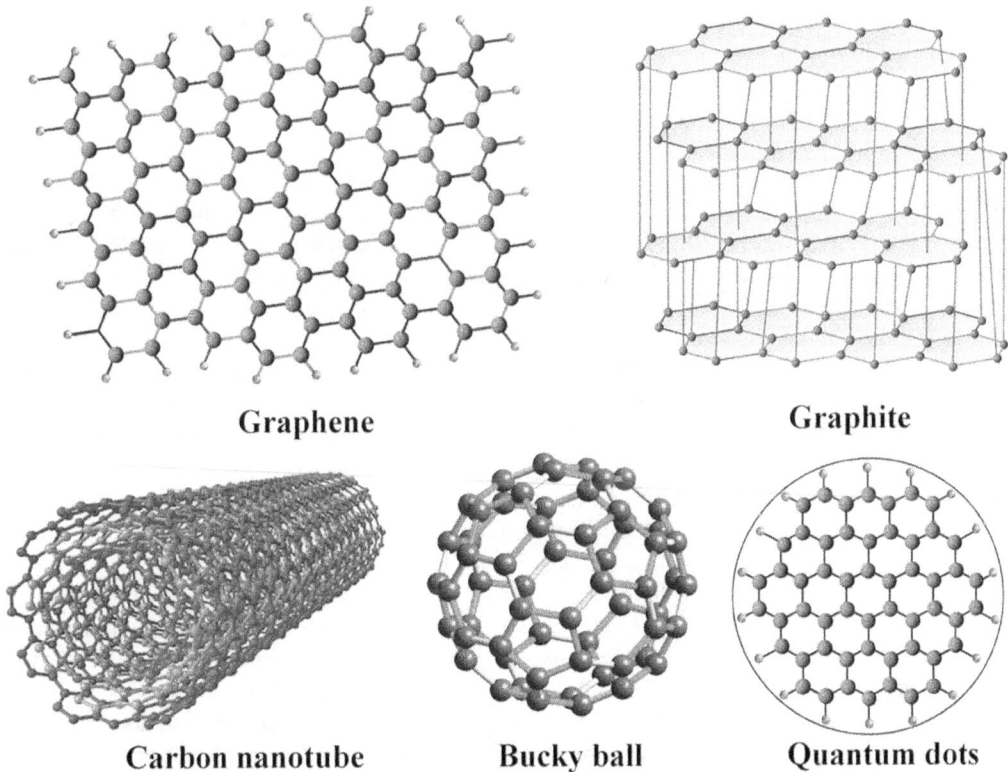

Graphene

Graphite

Carbon nanotube

Bucky ball

Quantum dots

FIGURE 16.12 Some of the popular radar absorbing fillers based on carbon materials employed in polymer composites.

high, the filler particles with very small particle size, get aggregated and result in agglomeration of the filler that negatively affects the thermo-mechanical and absorption properties of the composite. Graphene and multi-walled carbon nanotubes (CNTs) are found to be the best emerging carbonaceous materials being scrutinized for stealth technology applications [90–93].

16.10.4.1 Carbon Black–Based RAMs

Carbon black (CB) is a colloidal form of elemental carbon produced by the incomplete combustion of hydrocarbons that are specifically known for its high surface-area-to-volume-ratio and low density. Inclusion of CB in a potential matrix, can act as the reinforcing matrix as well as the RAMs. The optimized operating broadband frequency for CB can range up to terahertz (THz). CB has been experimented with various polymer matrices like epoxy, polypropylene, polyurethane, etc., and even in hybrid systems like glass/epoxy, etc. CB-reinforced RAMs can be single layered or multi-layered. Single-layered carbon black–reinforced RAMs composed of one-layer carbon black reinforced matrix layer. But, multi-layered carbon black RAMs consist of stacking of more than one layer in which each layer has different electromagnetic wave absorption capacity. So, the net absorption would be the cumulative absorption capability of all the layers. The first layer helps in impedance matching which can also reduce the overall thickness of the RAMs. Further layers help in electromagnetic energy dissipation and reflection of electromagnetic waves towards the internal reflection that results in damping of the electromagnetic wave. Thus, the electromagnetic wave penetrates the first layer of the multi-layered CB reinforced composites and starting from the second layer, there happens internal reflection and thus damping of electromagnetic waves. Proper tuning of the characteristics of each layer is possible and can be considered as an advantage of this type of composites.

CB reinforced polyurethane (PU) resin with PU thickness of 1.3 mm shows 46% absorption in X (8.2–12.4 GHz) and Ku (12–18 GHz) frequency band [94]. High porous CB-low density polypropylene (LDPE) composite shows a maximum absorption of 98.24% for 50 wt.% CB for composite with thickness of 2.4 mm in X-band frequency range [95]. As a potential composite, CB-cement based composite shows an enhanced electromagnetic wave absorption when the carbon black content increases more than 3.0 wt.% in a frequency range of 8–26.5 GHz. This indicates that the CB as reinforcing filler can enhance the dielectric constant of the composite. The minimum reflectivity when the CB content is 2.5 wt.% is found to be 20.30 dB. In between the 14.9 GHz to 26.5 GHz frequency, reflectivity is found to be less than 10 dB. The loss factor also increased as the CB content was increased [96]. Hybrid systems of mixture of CB along with magnetic particles like iron oxides, etc. are also found to be a promising candidate as RAMs [97]. The electromagnetic absorption property of CB-reinforced composite is enhanced as the magnetic particles are added. And the overall absorption would be the conglomerate effect of each filler along with CB. The radar cross-section simulation studies on different shapes of objects on the uncoated and absorber-coated leading edge of stealth aircraft prototype which is a perfect electric conductor (PEC) shell, using the CST microwave studio with bending angle of 30° and length of 45 cm using a plane wave illumination is shown in Figure 16.13. For the main lobe, an RCS reduction of 20 dB was observed. But, in the initial portion (aspect angle = 35°) and terminal portion (aspect angle = 135°), RCS reduction of 14 dB and 13 dB was observed, respectively [98].

16.10.4.2 Graphene-Based RAMs

Graphene is an allotrope of carbon having a single layer of carbon atoms in a two-dimensional hexagonal honeycomb lattice structure that is incredibly strong. Graphene can be synthesized by the chemical, electrochemical, or thermal exfoliation of bulk graphite, CVD method, etc. High mechanical properties and large surface area along with high dielectric properties makes graphene suitable for electromagnetic absorption applications in the range up to broadband terahertz (THz). Varying and thus tuning the composition of graphene in the embedded polymer matrix is the method that can be adopted to optimize the dielectric loss of the graphene-based RAMs. Polymer

(a)

(b)

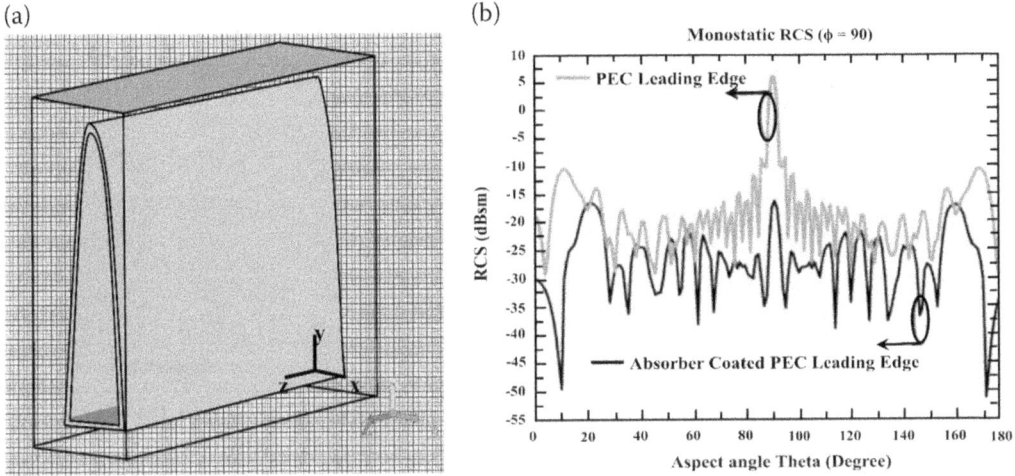

FIGURE 16.13   (a) CST Microwave Studio Model of absorber-coated leading edge of stealth aircraft prototype and (b) comparison of RCS results for uncoated and absorber-coated leading-edge prototype simulation results at 10.3 GHz [98].

matrices like polypropylene [99], epoxy [100], poly(ethylene oxide) [101], etc. were studied. Single-layered graphene-matrix systems are not apt for electromagnetic shielding due to its poor impedance matching property and this can be overcome by adding magnetic particles into the graphene-based composite system. Ultra-light graphene foam and multi-walled carbon nanotubes/multi-walled graphene foam shows better electromagnetic absorption properties due to its enhanced absorption loss and low reflection property in thickness of 3 mm which shows an average terahertz reflection loss up to 23 and 30 dB [102].

Graphene incorporated polypropylene 3D framework shows an absorption of 90% (full-band absorption) in the frequency range of 2–40 GHz and 75–110 GHz which is the highest bandwidth value reported [99]. Green approaches towards the preparation of reduced graphene oxide–based polymer composite using poly(ethylene oxide) as the potential matrix by in-situ reduction of L-ascorbic acid shows good microwave absorption even at 2.6 vol.% with lower reflection loss of –38.8 dB [101]. Ferrite particles decorated graphene-based composites are known for their better electromagnetic absorption properties, and thus can be defined as thin broadband absorbers. An increase in the composition of graphene shows an enhanced absorption bandwidth along with reduction in the thickness and this indicates ferrite decorated or, can say that, magnetic particles decorated graphene based composites are found to be a promising candidate for stealth technology application [47]. Reduced graphene oxides (rGO) within the embedding matrix like epoxy [100] and nitrile rubber [103] are found to be an indelible one to demonstrate as RAMs. The rGO with restoration of π-conjugation can effectively reflect and dissipate the electromagnetic radiation and thus can be used for electromagnetic wave absorbers. Perhaps, doping of graphene with heteroatom can effectively tune the electrical properties of graphene and thus enhance the electromagnetic absorption properties. CVD-grown graphene-based composites are excellent as RAMs due to their ultrahigh conductivity [104]. There are many such recent studies on electromagnetic radiation absorbing materials using graphene as well as in conjunction with metal oxides are well established [103,105,106]. Studies on the hybrid systems of graphene with different magnetic or dielectric materials are also of great interest in which iron containing compounds, conductive polymers, CNTs, oxides of zinc, nickel, cobalt, etc. were also found. The RAM surface based on the structure of a capacitor using single-layered graphene as electrode and its broadband electromagnetic wave absorption spectra at different sheet resistance from visible to microwave region is shown in Figure 16.14 [107].

(a)

(b)

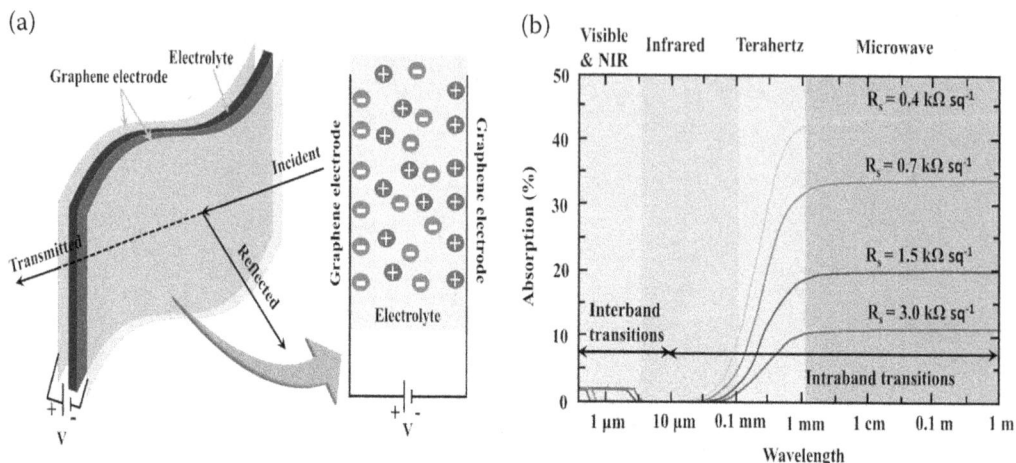

FIGURE 16. 14 (a) Schematic illustration of the graphene-based adaptive microwave surfaces and (b) absorption spectra of graphene with different sheet resistance for the electromagnetic region from the visible to microwave frequencies [107].

16.10.4.3 Carbon Nanotubes (CNTs)–Based RAMs

CNTs are one-dimensional material in nanometer range with tube-like structures having carbon skeletons of sp^2 hybridization. CNTs can be synthesized by a bottom-up approach like thermal chemical vapor decomposition. CNTs-based polymer composites show excellent thermo-mechanical properties and electrical conductivity which makes them good for electromagnetic wave absorption. CNTs in various polymer matrices like varnishes, poly(ethylene terephthalate), poly(trimethylene terephthalate), poly(vinylidene fluoride), etc. were investigated [108,109]. Nano-scale materials embedded in lightweight polymers like epoxy, polyurethane, etc. makes them prominent and even in lower loading of around 0.35%, they show higher conductivity. CNTs-based polymer composites as RAMs are found to be cost-effective. Due to its large surface area, multi-scattering of the electromagnetic wave is possible and thus, the non-magnetic CNTs found its place as RAMs [110,111]. Multi-walled CNTs (MWCNTs) are found to be more prominent than the single-walled CNTs (SWCNTs) due to high permittivity. MWCNTs act as RAMs by the process of dielectric relaxation due to its complex structure along with increased permittivity. The composition of CNTs in the polymer composites plays a key role in tuning the electromagnetic absorption property of the CNTs-based RAMs. The rheological and geometrical percolation threshold depends on each other and the dielectric relaxation.

Epoxy-based CNTs reinforced polymer composites show good absorption property along with an increment in the mechanical property of the polymer composite [73,112]. Hybrid systems of glass-epoxy-MWCNTs absorb in the frequency range 12-18 GHz along with the increase in the tensile and modulus of the polymer composite [68]. The higher concentration of CNTs will get aggregated that leads to non-uniform distribution of CNTs in the embedding polymer matrix which results in the decrement of the thermo-mechanical as well as the absorption properties of the CNTs-based polymer composites. So, the proper tuning of the composition of CNTs are required for CNTs-based polymer composites to act as RAMs. The dispersion process is also vital in determining the electromagnetic absorption property of CNTs-based RAMs. The absorption efficiency of CNTs-filled polymer composites can be simply estimated by measuring the complex permittivity through the calculation of the reflection loss. Figure 16.15 shows the scheme of the reflection of an electromagnetic wave by a perfect conductor after penetrating a polymer based CNT composite layer [113]. As displayed in Figure 16.15, an incident electromagnetic plane wave with two possible polarizations (TE and TM) penetrates a composite layer placed on the top of a perfect conductor, and can be characterized by

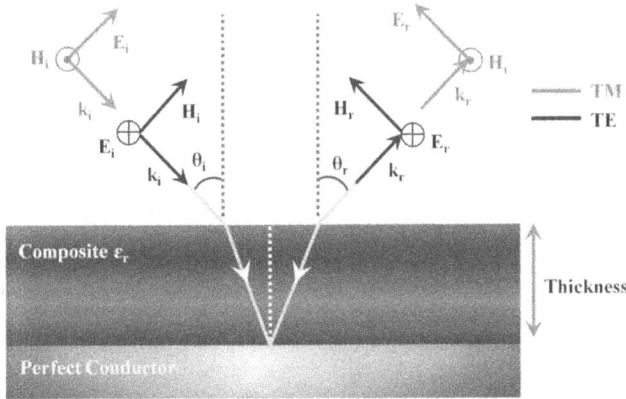

FIGURE 16.15 Reflection loss calculation; scheme of the reflection of an electromagnetic wave by a perfect conductor after penetrating a CNT composite layer.

CNT concentration in the composite and the thickness of the composite layer. This suggests that high density interfacial boundaries or interphases are required for electromagnetic wave absorption. Catalyzing the absorption property of CNTs by the addition of magnetic particles like ferrite shows an enhanced electromagnetic absorption properties due to impedance matching and magnetic loss [114]. Likewise, silver particles modified CNTs-based composites which can absorb broad wavelengths, especially in the range infrared radiation [115]. The addition of silver particles resulted in enhancement up to 673% as well as the infrared emissivity also decreased. The schematic illustration of silver particles modified carbon nanotube paper, and the infrared emissivity with different thickness (67 μm, 130 μm, and 200 μm) and different concentration of silver nitrate solution (0.1 mol/L, 0.3 mol/L and 0.5 mol/L) at 8–14 μm is shown in Figure 16.16 [115]. The silver particles modified CNTs-based composites camouflages the target object by emitting lower energy than the heating plate and thus reduces the temperature to ambient. The infrared emissivity shows a decrease as the concentration of silver nitrate solution is increased and thickness of the composite is increased.

16.10.4.4 Carbon Fibers–Based RAMs

Carbon fibers (CF) are stiff, high tensile strength, thermal resistant, chemical resistant and low-density fibers having diameter ranges from 5 to 10 μm. Due to its low weight-to-strength ratio, CFs

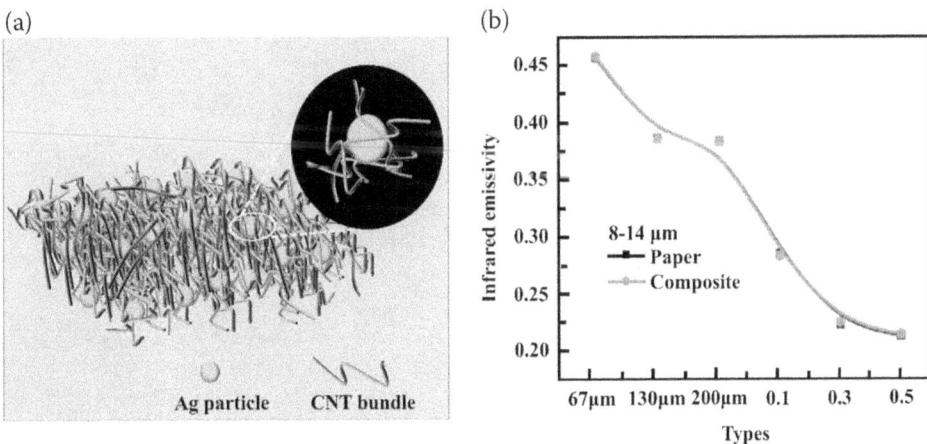

FIGURE 16.16 (a) Silver particles modified carbon nanotube paper and (b) the infrared emissivity of silver particles modified carbon nanotube paper with different thickness and different concentration of silver nitrate solution [115].

are popular as filler in RAMs. CF-reinforced composites are known for its electromagnetic interference shielding rather than microwave absorption due to its quasi-reflectivity at increased concentration. As in the case of CB, CF's concentration and morphology in the composite defines its characteristic property in stealth application. The conductive property of CF contributes to the EMI shielding. Continuous fibers as well as short fibers are known to absorb electromagnetic waves and thus used as RAMs. Polymer matrix like epoxy, polyester, polypropylene, polyacrylonitrile, E-glass-epoxy system, etc. were used. Along with the reduction in the thickness, short fiber-based polymer composites show broadband microwave absorption due to high permittivity, and are prepared by dispersing the short fibers in the embedding matrix. On the other hand, mixtures of short fiber matrices in other potential fillers also show complementary results. Addition of heteroatoms also enhances the electromagnetic absorption of CF-based polymer composites [116]. CF-filled ferrite and M-type ferrite composites prepared by a sol-gel process shows an excellent microwave absorption even in the thickness of 1 mm [117]. A list of different types of RAMs for broadband absorption is summarized in Table 16.5.

TABLE 16.5
List of Different Types of RAMs for Broadband Absorption

RAM Fillers	Type	Frequency (GHz)	Max. Reflection Loss (dB)	Ref.
Carbon black, carbon fiber, multiwalled carbon nanotubes	Semi conducting and conducting	8–18	35	[118]
Carbon black and short carbon fibers	Semiconducting and conducting	8–12	27	[98]
Carbon nanofibers, fullerenes, micro-sized granular graphite, Single walled and multiwalled carbon nanotubes	Semiconducting and conducting	8.2–12.4	10	[53]
Foliated graphite nanocomposites (FGN)	Semiconducting	1–18	37	[119]
Carbon nanotubes	Semiconducting	2–18	24.27	[109]
Multiwalled carbon nanotubes	Semiconducting	8–12	30	[76]
Carbonyl iron, Graphene	Semiconducting	2–18	40.2	[120]
Conducting carbon black	Semiconducting	8.2–12.4	55	[121]
Carbon black	Conducting	8–12	22.2	[122]
Short carbon fiber	Conducting	8.2–12.4	7–11	[123]
Fe_3O_4/TiO_2	Magnetic, dielectric	5.8		[124]
Carbon nanotubes, reduced graphene oxide	Magnetic, dielectric	2–18	55	[125]
NiZnCo ferrites and RGO	Magnetic, dielectric	2–18	51.8	[120]
Fe, Graphene	Magnetic, dielectric	2–18	45	[126]
Ni, Graphene	Magnetic, dielectric	2–18	16	[127]
PAN coated with Ni and CO	Magnetic and dielectric	8–18	32	[128]
Fe_3O_4, RGO	Magnetic and dielectric	2–18	60–70	[120]
Graphene	Dielectric	2–18	37.8	[129]
Reduced graphene oxide	Dielectric	2–18	10–57	[103]
Carbon, reduced graphene oxide	Dielectric	2–18	38.8	[101]
M type Hexaferrite	Magnetic	8.2–12.4	24.5	[130]
MnZn ferrite	Magnetic	2–18	38	[131]
Fe_3O_4	Magnetic	1–18	25–30	[132]
MWCNT/ZnO/ Fe_3O_4	Conducting, magnetic	2–18	40.9	[133]

16.11 CLASSIFICATION OF POLYMER COMPOSITES AS RAMS BASED ON GEOMETRY/WORKING PRINCIPLE

16.11.1 IMPEDANCE MATCHING RAMs

The characteristic impedance of the reflecting target object material which comes in agreement with the free space impedance value, can act as a good RAM. As the impedance contract is higher, the higher will be the reflection, and vice versa. Uniquely, the inherent impedance value of impedance matching RAMs is near the incident medium's intrinsic impedance value (for air its $\approx 377\ \Omega$) [25]. One of the impedance matching RAM is matching layer RAMs. A transition layer is introduced at the interface of the incident surface and the absorbing material which can match the impedance between the both incident and absorbing layer. Thus, the impedance of the incident layer would be compensated with the cumulative impedance of the first and second layer. Another impedance matching RAM is the pyramidal or cone moulded designed pyramidal RAMs which are naturally thick materials which are stretching out vertically to the surface. Pyramidal RAMs were created for the gradual change in the impedance from the air to engrossing material at the interface. Another impedance matching RAM is tapered loading RAMs in which a combination of lossy as well as less lossy material with an ordinary slab-shaped structure.

16.11.2 RESONANT RAMs

In resonant RAMs, the reflection followed by transmittance of the electromagnetic wave at the first interphase results in phase inversion and again the transmitted electromagnetic waves will pass through the RAM itself which is then reflected back by the metal-based layer. As a result, the overall strength of the incident wave is damped. Examples of resonant RAMs are Jaumann layers RAMs, Salisbury RAMs, and Dallenbach RAMs. In Jaumann RAMs, more than one resistive layer was added before the metal backing and the resistive sheets are separated by spacers. Salisbury RAMs are smaller versions of Jaumann RAMs in which a resistive layer is placed in an odd multiple of one-fourth of a wavelength before the metal backing and separated by air. But in the case of Dallenbach RAMs, a dielectric material is backed on the surface of a conducting material like metal.

16.11.3 MISCELLANEOUS RAMs

16.11.3.1 Conducting Polymers

Conducting polymers are organic polymers that have the ability to conduct electricity like metals and semi-conductors. Likewise, they have good mechanical, physical, and chemical properties. Organic monomers like acetylene, aniline, pyrrole, thiophene, and so forth, can be used to synthesize conducting polymers. These organic monomers can act as conjugated polymers like an acceptor or donor pair which can transfer electricity. The structure of the monomer and corresponding conducting polymers are shown in Figure 16.17. Other than as RAMs, these kinds of conducting polymers can be used for a wide range of applications like solar cells, catalysis, etc. The ease in manufacturing of conducting polymers fortified with the power of tunable conductivity found its way as RAMs. Polypyrrole (PPy) and polyaniline (PANI) doped with electron accepting dopants, are most widely used conducting polymers for stealth applications [62]. Partial oxidation of polymer results from the doping makes it electrically conducting by the formation of polarons, bipolarons, and solitons, which will act as the charge carriers. The phonon assisted hopping between random localized states created by the partial oxidation of polymers, becomes the basis of the conductivity of the conducting polymers [134]. But, the insolubility, infusibility, and intractability of the conducting polymers makes it difficult in processing and research shows that few of them are capable of compressing into shapes like PANI or PPy. Hence, primarily blending with thermoplastic materials to form composite is a commonly adopted method for enhancing the

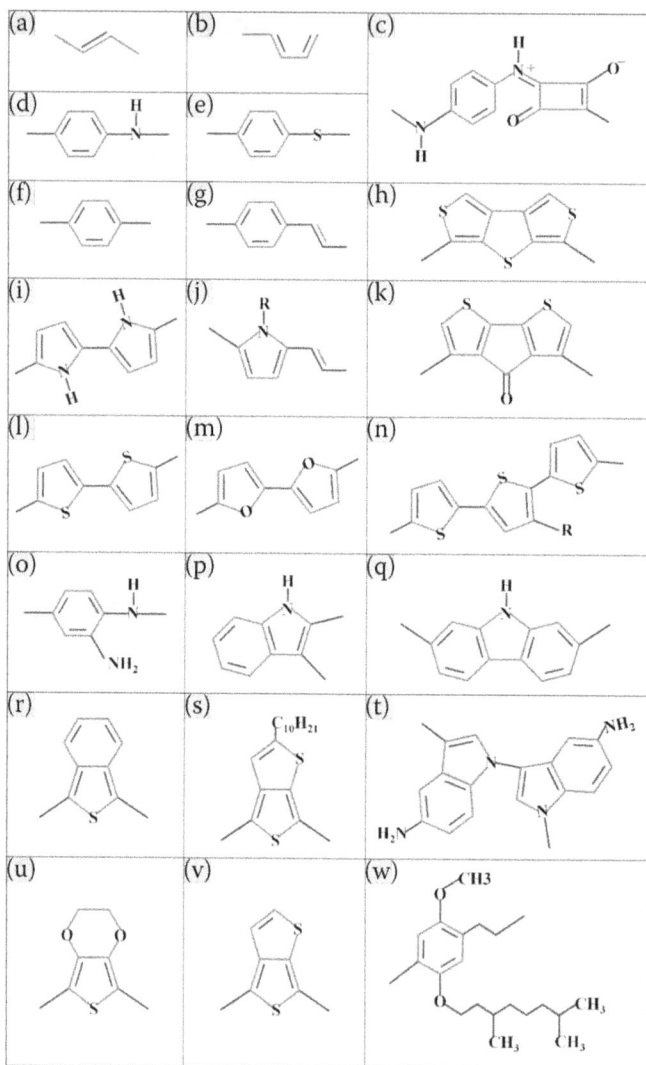

FIGURE 16.17 Structures of different conducting polymers (a) Trans-Poly(acetylene), (b) Cis-Poly (acetylene), (c) Polysquaraies, (d) Poly(aniline), (e) Poly(phenylene sulfide), (f) Poly(para-phenylene), (g) Poly(para-phenylenevinylene, (h) Poly(dithieno-3,4-b: 3', 4'-d)-thiophene, (i) Poly(pyrrole), (j) Poly(1-alkyl-2,5-pyrrlene vinylene), (k) Poy-4H-cyclo-pentadithiophene), (l) Poly(thiophene), (m) Polyfurane, (n) Polyterthiophene, (o) Poly(o-phenylene diamene), (p) 2,3-Polyindole, (q) Poly(carbazole), (r) Poly(iso-thianaphthene), (s) Poly(2-decyl thieno(3,4-thiophene-4,6-dicyl), (t) poly (5-aminoindole), (u) Poly (3,4-ethylenedioxythiophene), (v) Poly(thieno(3,4-b)thiophene) and (w) Poly(2-methoxy-5(3',7'-dimethyl octyloxy)-1,4-phenylene vinylene).

processability of the conducting polymers. PANI or PPy, has been polymerized on the surface of poly(vinyl acetate) (PVAc), poly(vinyl chloride) (PVC) and within the solution or dispersion (in-situ polymerization) to form composites. Chemical oxidation method is the widely used method to prepare conducting polymers or emulsion polymerization has also been used. Electrochemical polymerization is adopted for the preparation of thin films of conducting polymers, for example polythiophene. PANI is soluble in solvents such as dimethylformamide (DMF), or dimethyl sulfoxide (DMSO). Chemical modification has increased the solubility of conductive polymers. Textile materials were utilized as substrates for the fabrication of flexible composites, and they were coated with conducting polymers after being soaked in oxidant and then exposed to monomer [135] or vice versa (soaking in monomer first, followed by oxidant exposure) [136].

In 1989, the first study on the deposition of PANI and PPy onto fabric using an in situ polymerization method was published [137].

16.11.3.1.1 Polypyrrole-Polymer Composites

Polypyrrole (PPy) is a solid organic polymer formed via the oxidative polymerization of pyrrole having the formula $H(C_4H_2NH)_nH$. It is an intrinsically conductive polymer that is utilized in electronics, optics, biology, and medicine. Angeli and Pieroni reported the synthesis of pyrrole blacks from pyrrole magnesium bromide in 1919, which was one of the earliest occurrences of PPy [138]. The physical properties of polypyrrole are poor hence often used as composite with thermoplastic or elastomeric (latex or dry rubber) materials. Polypyrrole shows good electronic conductivity range from 2 to 100 S cm^{-1} depending on the conditions and type of dopant used or degree of doping. They are stable in air up to 150°C, at which point the dopant begins to develop, making them suitable for a variety of applications. The frequency response of polypyrrole- poly(methyl methacrylate) PMMA composites is in the range of 10 kHz to 8 GHz, is also studied [139] and found that the imaginary component of permittivity, and therefore the AC and DC conductivities, were frequency dependent. The composite shows any dependency on the frequency, at low frequencies (<105 Hz). Also, the conductivity of the composite displays percolative behavior and at frequencies up to 109 Hz, subtraction of the DC conductivity indicates a relaxation. It is observed that at frequencies beyond 1 GHz, the electronic conductivity of the composite PPy-PMMA composite is dominated by the AC conductivity resulting from the hopping conduction of charge carriers in the conducting polymer. In case of conducting polymer like PPy composites, the AC conductivity is an intrinsic property however the conductivity of the composite is directly depending on the loading concentration of polypyrrole in the PMMA matrix. Also processing and morphology has a great effect on the permittivity, conductivity and reflectivity of polypyrrole-based products [140]. Polypyrrole/PVC composite sheets fabricated by compression moulding (pressing with a uniaxial pressure of 2,000 bar), and melt injected at a production temperature of 180°C into sheets.

The radioelectric properties studied between 130 MHz and 20 GHz for both a pressed and injected PPy/PVC composite exhibits classical resistive behaviour with a variation of ε'' versus frequency, according to

$$\varepsilon'' \alpha K\omega^{-s} = \sigma(\omega)/(\omega \varepsilon_0)$$

where K denotes a constant, ω denotes angular frequency (rad s^{-1}), and σ denotes conductivity (S cm^{-1}) [141] and thus can say that conductivity depends on the frequency. The compression moulded sample found to be macroscopically conductive while melt-injected material was macroscopically insulating with Maxwell-Wagner-type relaxation. The relaxation frequency is given by $f_r = \sigma_2 / 2\pi\varepsilon_0(2\varepsilon'_1 + \varepsilon'_2)$, where ε_0, is the permittivity in vacuo, σ_2 is the conductivity of the PPy, ε'_1 and ε'_2 correspond to the real permittivities of the matrix (PVC) and the conductive material (PPy), respectively [142]. Due to the insulating nature, tuning the absorption property of compression molded material is difficult, but at the same time, the melt injection moulded material easily forms a Dallenbach resonant layer [140]. However, it has a very narrow absorption and a reflectance of more than −40 dB. Truong et al. fabricated Dallenbach layers from polypyrrole doped with p-toluene sulphonic acid sodium salt, or 5-sulfosalicylic acid dihydrate. The doped PPy conducting fibres were dispersed in a commercial paint via chemical oxidation method, or milled with natural rubber and moulded into flat sheets before being placed to an aluminium backing panel [143,144]. The materials exhibit resonance absorption when the PPy content in the composite is above the percolation threshold. Since the PPy fibers have a high aspect ratio, the percolation threshold for the conducting materials in the composite was found to be 2%–4%. The mill mixing of rubber with fibrous PPy is prone to destroy the original fibrous morphology PPy aggregates, resulting in percolation thresholds up to 16% and leading to low absorbing performances. The natural rubber-conducting fiber (NR-PPy) composites do not produce a macroscopically conductive composite material, however, the rubber composite still shows a reasonably high value of ε'' which makes them convenient in fabricating Dallenbach layers. A 2.5 mm thick paint panel containing 2 wt.% PPy powder has a reflectance of less than -10 dB in the frequency range of 12 to 18 GHz. Carbonyl iron was also used to create a hybrid dielectric/magnetic material, however additional bandwidth was not achieved for this material. From the various studies it was observed that powder processing has a significant influence on the properties of the final material [143,144].

Truong *et al.* also prepared different PPy-based RAM, such as PPy-containing paint/rubber and PPy-coated structural phenolic foams with an impedance gradient [143]. Ppy coated phenolic foam (12–15 mm thick), with a pore size in the range of 1 μm, have been fabricated by soaking the phenolic foam in aqueous ferric chloride followed by exposing to pyrrole vapor from one side. On controlling the exposure time, the gradient of conductivity across the foam thickness was obtained. Also, the uniformly Ppy coated conducting phenolic foam was fabricated by wet chemistry. For that the oxidant-doped foam is submerged in an aqueous pyrrole solution, with the pyrrole solution passing through the foam. The permittivity of the foam was found to be very low values of ε_r', particularly in the range of 1-2 and good values of ε_r'' in the range of 0.5–10. The two major issues in this context of vapour phase polymerization is that the thick coating of the conducting polymer will be formed on one surface of the foam and the poor conductivity of the conducting polymer doped with chloride from the ferric chloride. The wet technique generated a material with superior characteristics, and a gradient could be induced into the material. It was discovered that a 15 mm thick foam produced in this manner [143] outperformed previous reported gradient absorbers [145,146]. PPy can be coated on rigid or flexible open cell foams and the PPy-coated foams are lightweight broadband RAM. The study revealed that the reflection loss is greatly influenced by the thickness and complex permittivity of the materials.

16.11.3.1.2 Polypyrrole-Fabric Composites

Flexible and/or stretchable polypyrrole composite fabrics have been prepared by polymerizing pyrrole in the presence of woven or nonwoven or knitted fabric or fibers. Also mechanically strong rigid or flexible conducting composite papers can be prepared by oxidative polymerization pyrrole with ferric chloride in the presence of polymeric papers or films (cellulose) [135,147]. The electrical conductivity of the composite paper or fabric and morphology of the conducting polymer can be controlled through the manipulation of the chemistry, doping technique and deposition time. Wong *et al.* fabricated tailor-made flexible paper RAM from these materials and studied their absorbing properties [148]. Cotton and polyester fabrics have been treated in the same way [149]. Modelling the phase and amplitude of reflected microwaves from the composite allowed values of the fabric sheets' resistance and capacitance to be calculated as a function of fabric type and PPy loading [148–152]. The studies found that the resistance reduced with Ppy loading while the capacitance improved to maximum at about 2 to 3 mg cm^{-2}. The coating is uniform and smooth which leads to the formation of RAM having lower resistance and higher capacitance, at low loadings of PPy. The coating becomes nonuniform and more particulate which results in short circuits between the fibres (hence a decrease in capacitance), at higher Ppy loadings [150]. When exposed in asymmetrical weaves, the polarization of the microwaves to the weave resulted in large variation in the properties of the PPy fabrics [150]. The polypyrrole coated papers or fabrics (including glass fibres) helps in development of structural RAMs. Wong *et al.* modeled the properties of the fabric-coated RAMs and fabricated Salisbury Screens and Jaumann layers [149,150,152]. The PPy-coated fabrics have been commercialized and marketed by Milliken & Company under the trademark Contex®. These have been utilized as nets for microwave attenuation (trademark Intrigue®) [153] and as Salisbury screens and Jaumann absorbers [151]. Other applications of these coated fabrics and papers also have been realized by patterning the polypyrrole by changing its electrical conductivity [154,155] and employed for "edge-card" materials in low observable aircrafts [156]. The stability of PPy coated PET and PVA fibers have been studied for environmental concerns for use as microwave and millimetre wave obscurants [157].

16.11.3.2 Metallic Thin Films

Intrinsic impedance of metals is found to be zero which is significantly different from intrinsic impedance of free space ($\approx$377 Ω). Thus, metals can reflect electromagnetic radiation since they keep the electric field zero on the outside [158]. Notwithstanding, a few metals can absorb electromagnetic radiation when the thickness is kept to nanometer range. For instance, coating of Kanthal films with thickness changing within the range of 10 to 200 nm can adequately act as RAMs. In contrast with regular RAMs, due to its light weight, the nanometer films can introduce comparative electromagnetic wave attenuation.

16.12 CONCLUSION

Polymer composites-based RAMs adhering to the specific required geometry and electromagnetic absorption capability are the boon for the stealth technology applications. Various researches on the RAMs are in their maturation process and as a military tactics, stealth is found to be an ever ending and pronouncing research area which helms hope on RAMs with better interfacial impedance matching and electromagnetic wave loss ability. Polymer composites-based RAMs are known for their cost effectiveness, high strength-to-weight ratio, and flexible design. So along with radiation absorbing capability, there makes a reduction in the thickness of the RAMs. On the addition of absorption filler material in the polymer matrix, the mechanical properties of the polymer composites are found to be increasing based on the type of filler used. Carbonaceous materials as such or decorated with magnetic or dielectric materials are invigorating stealth technology applications along with potential polymer matrix. Carbon nanotubes (CNTs) and graphene are mostly studied carbon-based nanomaterial RAMs for stealth applications. Impedance matching RAMs and resonant RAMs were used in stealth aircrafts or submarines that used damping of electromagnetic waves or destructive interference that dealt out narrow and broadband width of electromagnetic radiations.

REFERENCES

[1] Kim J, Han K, Hahn JW (2017) Selective dual-band metamaterial perfect absorber for infrared stealth technology. *Sci Rep.* 10.1038/s41598-017-06749-0
[2] Smith FG (1993) The infrared & electro-optical systems handbook: Volume 2 Atmospheric Propagation of Radiation. *Atmos. Propag. Radiat.*
[3] Yoo YJ, Ju S, Park SY, et al. (2015) Metamaterial absorber for electromagnetic waves in periodic water droplets. *Sci Rep* 5. 10.1038/srep14018
[4] Iwaszczuk K, Strikwerda AC, Fan K, et al. (2012) Flexible metamaterial absorbers for stealth applications at terahertz frequencies. *Opt Express* 20:635. 10.1364/oe.20.000635
[5] Bahret WF (1993) The beginnings of stealth technology. *IEEE Trans. Aerospace Electron. Syst.* 29:1377–1385. 10.1109/7.259548
[6] Chen H, Ma W, Huang Z, et al. (2019) Graphene-based materials toward microwave and terahertz absorbing stealth technologies. *Adv Opt Mater*, Volume7, Issue8. 10.1002/adom.201801318
[7] Folger P (2014) Basic radar principles and general characteristics. 1–34. National Geospatial-Intelligence Agency.
[8] Kim Y, Zyl J Van (2000) Basic principles of radar imaging. 1–2.
[9] Ahmad H, Tariq A, Shehzad A, et al. (2019) Stealth technology: Methods and composite materials—A review. *Polym Compos* 40:4457–4472. 10.1002/pc.25311
[10] Emerson WH (1973) Electromagnetic Wave Absorbers and Anechoic Chambers Through the Years. *IEEE Trans Antennas Propag.*, vol. 21, no. 4, pp. 484–490. 10.1109/TAP.1973.1140517

[11] Toit LJ, Du (1994) The design of Jauman absorbers. *IEEE Antennas Propag Mag*, vol. 36, no. 6, pp. 17–25.

[12] MacFarlane GG (1945) Radar camouflage research and development by the Germans. *Unpubl Notes*.

[13] Montgomery CG, Dicke RH, Purcell EM, Purcell EM (1987) *Principles of Microwave Circuits*. IET Electromagnetic Waves Series 25, London, U.K.

[14] Winfield W Salisbury (1952) Absorbent body for electromagnetic waves. US Patent No. 2599944 A.

[15] Ha O, Johnson MJ (2015) Radar: Summary Report of Harp Project.

[16] Ruck GT, Barrick DE, Stuart WD, Krichbaum CK (1970) Radar Cross Section Handbook, Kluwer Academic/Plenum Publishers, New York.

[17] Ruck GT, Barrick DE, Stuart WD, Krichbaum CK, (1970) *Radar Cross Section Handbook*, Volume 2. Kluwer Academic/Plenum Publishers, New York.

[18] Severin H (1956) Nonreflecting absorbers for microwave radiation. *IRE Trans Antennas Propag*, 4(3), 385–392. 10.1109/TAP.1956.1144419

[19] Meyer E, Severin H, Umlauft G (1954) Resonanzabsorber für elektromagnetische Wellen. *Zeitschrift für Phys* 138:465–477.

[20] Meyer E, Severin H (1956) Absorption Anordnungen für elektromagnetische Zentimeterwellen und ihre akustischen Analogien. *Zeitschrift für Angew Phys* 8:105–114.

[21] Munk BA (2000) *Frequency Selective Surfaces: Theory and Design*. John Wiley & Son, Hoboken, USA.

[22] Eugene F. Knott (1993) *Radar Cross Section*. Artech House, Norwood.

[23] Bowman JJ, (1968) Effects of absorbers. In: *Methods of Radar Cross-section Analysis*Elsevier, 210–236.

[24] Connolly TM, Luoma EJ (1977) Microwave absorbers. United States Patent 4,027,384.

[25] Saville P (2005) Review of radar absorbing materials Defence R & D Canada – Atlantic. *Def Res Dev Canada* 62.

[26] Fan Y, Yang H, Liu X, et al. (2008) Preparation and study on radar absorbing materials of nickel-coated carbon fiber and flake graphite. *J Alloys Compd* 461:490–494.

[27] Adam JA (1988) How to design an 'invisible' aircraft. *IEEE Spectr*, 25(4), 26–31.10.1109/6.4529

[28] Van der Plas G, Barel A, Schweicher E (1989) A spectral iteration technique for analyzing scattering from circuit analog absorbers. *IEEE Trans Antennas Propag* 37:1327–1332.

[29] Jaggard DL, Engheta N, Liu J (1990) Chiroshield: A salisbury/dallenbach shield alternative. *Electron Lett.*,26(17), 1332.10.1049/el:19900859

[30] Jaggard DL, Liu JC, Sun X (1991) Spherical chiroshield. *Electron Lett.*, 27(1), 77–79. 10.1049/el:1 9910050

[31] Wu KH, Ting TH, Wang GP, et al. (2008) Effect of carbon black content on electrical and microwave absorbing properties of polyaniline/carbon black nanocomposites. *Polym Degrad Stab.*,93(2), 483–488.10.1016/j.polymdegradstab.2007.11.009

[32] Feng YB, Qiu T, Shen CY (2007) Absorbing properties and structural design of microwave absorbers based on carbonyl iron and barium ferrite. *J Magn Magn Mater*, 318(1–2), 8–13.10.1016/ j.jmmm.2007.04.012

[33] Gaylor K (1989) Radar absorbing materials – Mechanisms and materials. *MRL Tec Rep*.

[34] Saville P, Huber T, Makeiff D (2005) *Fabrication of Organic Radar Absorbing Materials: A Report on the TIF Project*. Defence Research and Development Atlantic Dartmouth, Canada.

[35] Cloete LJ, Du TJH (1989) AP-S: International Symposium Digest: Antennas and Propagation, 1989 (Cat. No.CH2654-2/89). In: *Digest on Antennas and Propagation Society International Symposium*.

[36] Knott EF, Shaeffer JF, Tuley MT (1993) Radar absorbing materials. In: *Radar Cross Section*. Artech House,USA, 297–360.

[37] Vinoy KJ, Jha RM (1995) Trends in radar absorbing materials technology. *Sadhana*, 20(5), 815–850. 10.1007/BF02744411

[38] Knott EF (2008) Radar cross section *Radar Handbook*,. In: 3rd ed. McGraw-Hill Education, New York.

[39] Rajyalakshmi P, Raju GSN (2011) Characteristics of radar cross section with different objects. *Int J Electron Commun Eng* 4:205–216.

[40] Grande O, Cañizo J, Angulo I, et al. (2014) Simplified formulae for the estimation of offshore wind turbines clutter on marine radars. *Sci World J* 2014. 10.1155/2014/982508

[41] Zhao Z, Niu Y, Ma Z, Ji X (2016) A fast stealth trajectory planning algorithm for stealth UAV to fly in multi-radar network. In: 2016 IEEE International Conference on Real-Time Computing and Robotics, RCAR 2016.

[42] Vas JV, Thomas MJ (2018) Electromagnetic shielding effectiveness of layered polymer nano-composites. *IEEE Trans Electromagn Compat.*, 60(2), 376–384. 10.1109/TEMC.2017.2719764

[43] Belaabed B, Lamouri S, Wojkiewicz JL (2018) X-band microwave absorbing properties of epoxy resin composites containing magnetized PANI-coated magnetite. *IEEE Trans Magn,*54(1), 1–8. 10.1109/TMAG.2017.2752147

[44] Gill N, Singh J, Puthucheri S, Singh D (2018) Thin and broadband two-layer microwave absorber in 4–12 GHz with developed flaky cobalt material. *Electron Mater Lett,*14(3), 288–297. 10.1007/s13391-018-0025-2

[45] Baldelli M, Pierontoni L, Belucci S (2016) Learning by using graphene multilayers: An educational app for analyzing the electromagnetic absorption of a graphene multilayer based on a network model. *IEEE Microw Mag*, 17(1), 44–51. 10.1109/MMM.2015.2487918

[46] Najim M, Puthucheri S, Agarwala V, Singh D (2016) ANN-based two-layer absorber design using Fe-Al hybrid nano-composites for broad bandwidth microwave absorption. *IEEE Trans Magn,*52(12), 1–8. 10.1109/TMAG.2016.2598530

[47] Panwar R, Puthucheri S, Singh D, Agarwala V (2015) Design of ferrite-graphene-based thin broadband radar wave absorber for stealth application. *IEEE Trans Magn*, 51(11), 1–4. 10.1109/TMAG.2015.2454431

[48] Qing Y, Zhou W, Huang S, et al. (2014) Evolution of double magnetic resonance behavior and electromagnetic properties of flake carbonyl iron and multi-walled carbon nanotubes filled epoxy-silicone. *J Alloys Compd*, 583, 471–475. 10.1016/j.jallcom.2013.09.002

[49] Wang Y, Luo F, Zhou W, Zhu D (2014) Dielectric and electromagnetic wave absorbing properties of TiC/epoxy composites in the GHz range. *Ceram Int*, 40(7), 10749–10754. 10.1016/j.ceramint.2014.03.064

[50] Ding Q, Zhang M, Zhang C, Qian T (2013) Synthesis and absorbing mechanism of two-layer microwave absorbers containing polycrystalline iron fibers and carbonyl iron. *J Magn Magn Mater,*331, 77–81. 10.1016/j.jmmm.2012.11.005

[51] Micheli D, Apollo C, Pastore R, et al. (2012) Optimization of multilayer shields made of composite nanostructured materials. *IEEE Trans Electromagn Compat*, 54(1), 60–69. 10.1109/TEMC.2011.2171688

[52] Zhen L, Gong YX, Jiang JT, et al. (2011) Synthesis of CoFe/Al2O3 composite nanoparticles as the impedance matching layer of wideband multilayer absorber. *Journal of Applied Physics,*109, 10.1063/1.3564939

[53] Micheli D, Apollo C, Pastore R, Marchetti M (2010) X-Band microwave characterization of carbon-based nanocomposite material, absorption capability comparison and RAS design simulation. *Compos Sci Technol.,*70(2), 400–409. 10.1016/j.compscitech.2009.11.015

[54] Kim JW, Kim SS (2010) Microwave absorbers of two-layer composites laminate for wide oblique incidence angles. *Mater Des,*31(3), 1547–1552. 10.1016/j.matdes.2009.09.054

[55] Parida RC, Singh D, Agarwal NK (2007) Implementation of multilayer ferrite radar absorbing coating with genetic algorithm for radar cross-section reduction at X-band. *Indian J Radio Sp Phys.*36, 145–152.

[56] Wan M (2008) *Conducting Polymers with Micro or Nanometer Structure.* Tsinghua University Press, Beijing & Springer Berlin Heidelberg New York, 1–292. 10.1007/978-3-540-69323-9.

[57] Yuzcelik CK, Jenn D, Adler R (2003) Radar absorbing material designMonterey, California. Naval Postgraduate School, Dudley Knox Library

[58] Säily J, Räisänen A V (2003) Studies on specular and non-specular reflectivities of radar absorbing materials (RAM) at submillimetre wavelengths.

[59] Folgueras LC, Alves MA, Rezende MC (2014) Evaluation of a nanostructured microwave absorbent coating applied to a glass fiber/polyphenylene sulfide laminated composite. *Mater Res* 17:197–202. 10.1590/S1516-14392014005000009

[60] Niu L, Xie Y, Wu P, Zhang C (2020) ARCS: Active radar cross section for multi-radiator problems in complex em environments. *Sensors* (Switzerland). 20(12), 3371. 10.3390/s20123371

[61] Prasad Rambabu N, Eswara Prasad VVK, Wanhill RJH (2017) . *Aerospace Materials and Material Technologies.* Springer Singapore, Volume 1: Aerospace Materials. 1:586. 10.1007/978-981-10-2134-3

[62] Graham Brodie (2018) Energy Transfer from Electromagnetic Fields to Materials. *Electromagnetic Fields and Waves* 53:1689–1699, 10.5772/intechopen.83420.

[63] Padilla WJ, Basov DN, Smith DR (2006) Negative refractive index metamaterials. *Mater Today* 9:28–35. 10.1016/S1369-7021(06)71573-5

[64] Webster RJ (2013) Animal camouflage: Disentangling disruptive coloration from background matching, Carleton University Ottawa, Ontario.

[65] Dranidis DV (2012) Airborne stealth in a nutshell-part I. Mag Comput Harpoon Community http://wwwharpoonhqcom/waypoint/,(Accessed February 2009).

[66] Mazé-Merceur G, Bonnefoy JL, Garat J, Mittra R (1992) Microwave techniques for measurement of radar absorbing materials – A review. In: *IEEE Antennas and Propagation Society, AP-S International Symposium (Digest)*.

[67] Jayalakshmi CG, Inamdar A, Anand A, Kandasubramanian B (2019) Polymer matrix composites as broadband radar absorbing structures for stealth aircrafts. *J. Appl. Polym. Sci*,136, 14. 10.1002/app.47241

[68] Shin JH, Jang HK, Choi WH, et al. (2015) Design and verification of a single slab RAS through mass production of glass/MWNT added epoxy composite prepreg. *J Appl Polym Sci*,132(22). 10.1002/app.42019

[69] Eun SW, Choi WH, Jang HK, et al. (2015) Effect of delamination on the electromagnetic wave absorbing performance of radar absorbing structures. *Compos Sci Technol*,116, 18–25. 10.1016/j.compscitech.2015.04.001

[70] Lee SE, Kang JH, Kim CG (2006) Fabrication and design of multi-layered radar absorbing structures of MWNT-filled glass/epoxy plain-weave composites. *Compos Struct*, 76(4), 397–405. 10.1016/j.compstruct.2005.11.036

[71] Nam YW, Shin JH, Choi JH, et al. (2018) Micro-mechanical failure prediction of radar-absorbing structure dispersed with multi-walled carbon nanotubes considering multi-scale modeling. *J Compos Mater*, 52(12), 1649–1660. 10.1177/0021998317729003

[72] Rezania J, Rahimi H (2017) Investigating the carbon materials' microwave absorption and its effects on the mechanical and physical properties of carbon fiber and carbon black/ polypropylene composites. *J Compos Mater*,51(16), 2263–2276. 10.1177/0021998316669578

[73] Allaoui A, Bai S, Cheng HM, Bai JB (2002) Mechanical and electrical properties of a MWNT/epoxy composite. *Compos Sci Technol*,62(15), 1993–1998. 10.1016/S0266-3538(02)00129-X

[74] Giorcelli M, Savi P, Miscuglio M, et al. (2014) Analysis of MWCNT/epoxy composites at microwave frequency: Reproducibility investigation. *Nanoscale Res Lett.*,9(1), 168. 10.1186/1556-276X-9–168

[75] Choi I, Lee D, Lee DG (2015) Radar absorbing composite structures dispersed with nano-conductive particles. *Compos Struct*,122, 23–30. 10.1016/j.compstruct.2014.11.040

[76] Lee D, Choi I, Lee DG (2015) Development of a damage tolerant structure for nano-composite radar absorbing structures. *Compos Struct*, 119, 107–114. 10.1016/j.compstruct.2014.08.001

[77] Choi I, Kim JG, Seo IS, Lee DG (2012) Radar absorbing sandwich construction composed of CNT, PMI foam and carbon/epoxy composite. *Compos Struct*, 94(9), 3002–3008.10.1016/j.compstruct.2012.04.009

[78] Saini L, Gupta V, Patra MK, et al. (2021) Impedance engineered microwave absorption properties of Fe-Ni/C core-shell enabled rubber composites for X-band stealth applications. *J Alloys Compd*, 869, 159360. 10.1016/j.jallcom.2021.159360

[79] Saini L, Patra MK, Jani RK, et al. (2017) Tunable twin matching frequency (fm1 /fm2) behavior of Ni1-x Znx Fe2 O 4 /NBR composites over 2-12.4 GHz: A strategic material system for stealth applications. *Sci Rep*,7(1). 10.1038/srep44457

[80] Janu Y, Chaudhary D, Singhal N, et al. (2021) Tuning of electromagnetic properties in Ba(MnZn)xCo2(1-x)Fe16O27/NBR flexible composites for wide band microwave absorption in 6–18 GHz. *J Magn Magn Mater*,527, 167666. 10.1016/j.jmmm.2020.167666

[81] Kazantseva NE, Bespyatykh YI, Sapurina I, et al. (2006) Magnetic materials based on manganese-zinc ferrite with surface-organized polyaniline coating. *J Magn Magn Mater* 301:155–165. 10.1016/j.jmmm.2005.06.015

[82] Folgueras LDC, Alves MA, Rezende MC (2009) Electromagnetic radiation absorbing paints based on carbonyl iron and polyaniline. *SBMO/IEEE MTT-S Int Microw Optoelectron Conf Proc*:510–513. 10.1109/IMOC.2009.5427534

[84] Zhao C, Zhang A, Zheng Y, Luan J (2012) Electromagnetic and microwave-absorbing properties of magnetite decorated multiwalled carbon nanotubes prepared with poly(N-vinyl-2-pyrrolidone). *Mater Res Bull* 47:217–221. 10.1016/j.materresbull.2011.11.047

[85] Sharma R, Agarwala RC, Agarwala V (2008) Development of radar absorbing nano crystals by microwave irradiation. *Mater Lett* 62:2233–2236. 10.1016/j.matlet.2007.11.076

[86] Sharma R, Agarwala RC, Agarwala V (2009) Development of electroless (Ni-P)/BaNi0.4Ti0.4Fe11.2O19 nanocomposite powder for enhanced microwave absorption. *J Alloys Compd* 467:357–365. 10.1016/j.jallcom.2007.11.141

[87] Xiao HM, Liu XM, Fu SY (2006) Synthesis, magnetic and microwave absorbing properties of core-shell structured MnFe2O4/TiO2 nanocomposites. *Compos Sci Technol* 66:2003–2008. 10.1016/j.compscitech.2006.01.001

[88] Chakradhary VK, Akhtar MJ (2017) Microwave absorption properties of strontium ferrite and carbon black based nanocomposites for stealth applications. In: Asia-Pacific Microwave Conference Proceedings, APMC.

[89] Faez R, Reis AD, Soto-Oviedo MA, et al. (2005) Microwave absorbing coatings based on a blend of nitrile rubber, EPDM rubber and polyaniline. *Polym Bull*, 55(4), 299–307. 10.1007/s00289-005-0433-y

[90] Lin H, Zhu H, Guo H, Yu L (2008) Microwave-absorbing properties of Co-filled carbon nanotubes. *Mater Res Bull* 43:2697–2702. 10.1016/j.materresbull.2007.10.016

[91] Ibrahim I.R., Matori K.A., Ismail I. et al., (2020) A Study on Microwave Absorption Properties of Carbon Black and Ni0.6Zn0.4Fe2O4 Nanocomposites by Tuning the Matching-Absorbing Layer Structures. *Sci Rep*.,10, 3135 10.1038/s41598-020-60107-1

[92] Sun X-G, Gao M, Li C, Wu Y (2011) Microwave absorption characteristics of carbon nanotubes. *Carbon Nanotub – Synth Charact Appl.* 10.5772/16514

[93] Zhan Y, Zhao R, Lei Y, et al. (2011) A novel carbon nanotubes/Fe3O4 inorganic hybrid material: Synthesis, characterization and microwave electromagnetic properties. *J Magn Magn Mater* 323:1006–1010. 10.1016/j.jmmm.2010.12.005

[94] Gupta KK, Abbas SM, Abhyankar AC (2016) Carbon black/polyurethane nanocomposite-coated fabric for microwave attenuation in X & Ku-band (8–18 GHz) frequency range. *J Ind Text*, 46(2), 510–529. 10.1177/1528083715589752

[95] Ansari A, Akhtar MJ (2018) High porous carbon black based flexible nanocomposite as efficient absorber for X-band applications. *Mater Res Express*. 10.1088/2053-1591/aadb13

[96] Dai Y, Sun M, Liu C, Li Z (2010) Electromagnetic wave absorbing characteristics of carbon black cement-based composites. *Cem Concr Compos*,32(7), 508–513. 10.1016/j.cemconcomp.2010.03.009

[97] Datt G, Kotabage C, Datar S, Abhyankar AC (2018) Correlation between the magnetic-microstructure and microwave mitigation ability of MxCo(1-x)Fe2O4 based ferrite-carbon black/PVA composites. *Phys Chem Chem Phys*. 10.1039/c8cp05235b

[98] Baskey HB, Akhtar MJ, Shami TC (2014) Investigation and performance evaluation of carbon black- and carbon fibers-based wideband dielectric absorbers for X-band stealth applications. *J Electromagn Waves Appl*, 28(14), 1703–1715. 10.1080/09205071.2014.933680

[99] Zhang KL, Zhang JY, Hou ZL, et al. (2019) Multifunctional broadband microwave absorption of flexible graphene composites. *Carbon N Y*. 10.1016/j.carbon.2018.10.024

[100] Zhang X, Wang X, Meng F, et al. (2019) Broadband and strong electromagnetic wave absorption of epoxy composites filled with ultralow content of non-covalently modified reduced graphene oxides. *Carbon N Y*,154, 115–124. 10.1016/j.carbon.2019.07.076

[101] Bai X, Zhai Y, Zhang Y (2011) Green approach to prepare graphene-based composites with high microwave absorption capacity. *J Phys Chem C*, 115(23), 11673–11677. 10.1021/jp202475m

[102] Huang Z, Chen H, Xu S, et al. (2018) Graphene-Based Composites Combining Both Excellent Terahertz Shielding and Stealth Performance. *Adv Opt Mater*, 1801165. 10.1002/adom.201801165

[103] Singh VK, Shukla A, Patra MK, et al. (2012) Microwave absorbing properties of a thermally reduced graphene oxide/nitrile butadiene rubber composite. *Carbon N Y* 50:2202–2208. 10.1016/j.carbon.2012.01.033

[104] Ma L, Lu Z, Tan J, et al. (2017) Transparent Conducting graphene hybrid films to improve electromagnetic interference (EMI) shielding performance of graphene. *ACS Appl Mater Interfaces*, 9(39), 34221–34229. 10.1021/acsami.7b09372

[105] Zhao J, Lin J, Xiao J, Fan H (2015) Synthesis and electromagnetic, microwave absorbing properties of polyaniline/graphene oxide/Fe3O4 nanocomposites. *RSC Adv* 5:19345–19352. 10.1039/c4ra12186d

[106] Wang L, Jia X, Li Y, et al. (2014) Synthesis and microwave absorption property of flexible magnetic film based on graphene oxide/carbon nanotubes and Fe3O4 nanoparticles. *J Mater Chem A* 2:14940–14946. 10.1039/c4ta02815e

[107] Balci O, Polat EO, Kakenov N, Kocabas C (2015) Graphene-enabled electrically switchable radar-absorbing surfaces. *Nat Commun*,6(1). 10.1038/ncomms7628

[108] Gupta A, Choudhary V (2011) Electromagnetic interference shielding behavior of poly(trimethylene terephthalate)/multi-walled carbon nanotube composites. *Compos Sci Technol*, 71(13), 1563–1568. 10.1016/j.compscitech.2011.06.014

[109] Fan Z, Luo G, Zhang Z, et al. (2006) Electromagnetic and microwave absorbing properties of multi-walled carbon nanotubes/polymer composites. *Mater Sci Eng B Solid-State Mater Adv Technol.*, 132(1–2), 85–89. 10.1016/j.mseb.2006.02.045

[110] Gui X, Wang K, Wei J, et al. (2009) Microwave absorbing properties and magnetic properties of different carbon nanotubes. *Sci China, Ser E Technol Sci*, 52(1), 227–231. 10.1007/s11431-009-0020-9

[111] Zhang Y, Li H, Yang X, et al. (2018) Additive manufacturing of carbon nanotube-photopolymer composite radar absorbing materials. *Polym Compos*, 39(S2), E671–E676. 10.1002/pc.24117

[112] Bychanok D, Gorokhov G, Meisak D, et al. (2017) Design of carbon nanotube-based broadband radar absorber for ka-band frequency range. *Prog Electromagn Res M*,53, 9–16. 10.2528/PIERM16090303

[113] Hussein MI, Jehangir SS, Rajmohan IJ, et al. (2020) Microwave Absorbing properties of metal functionalized-CNT-polymer composite for stealth applications. *Sci Rep*, 10(1). 10.1038/s41598-020-72928-1

[114] Kim HM, Kim K, Lee CY, et al. (2004) Electrical conductivity and electromagnetic interference shielding of multiwalled carbon nanotube composites containing Fe catalyst. *Appl Phys Lett*, 84(4), 589–591. 10.1063/1.1641167

[115] Chu H, Zhang Z, Liu Y, Leng J (2016) Silver particles modified carbon nanotube paper/glassfiber reinforced polymer composite material for high temperature infrared stealth camouflage. *Carbon N Y*, 98, 557–566. 10.1016/j.carbon.2015.11.036

[116] Fang J, Chen Z, Wei W, et al. (2015) A carbon fiber based three-phase heterostructure composite CF/Co0.2Fe2.8O4/PANI as an efficient electromagnetic wave absorber in the Ku band. *RSC Adv*, 5(62), 50024–50032. 10.1039/c5ra07192e

[117] Shen G, Xu M, Xu Z (2007) Double-layer microwave absorber based on ferrite and short carbon fiber composites. *Mater Chem Phys*, 105(2–3), 268–272. 10.1016/j.matchemphys.2007.04.056

[118] De Rosa IM, Sarasini F, Sarto MS, Tamburrano A (2008) EMC impact of advanced carbon fiber/carbon nanotube reinforced composites for next-generation aerospace applications. *IEEE Trans Electromagn Compat*, 50(3), 556–563. 10.1109/TEMC.2008.926818

[119] Al-Ghamdi AA, Al-Hartomy OA, Al-Solamy F, et al. (2013) Electromagnetic wave shielding and microwave absorbing properties of hybrid epoxy resin/foliated graphite nanocomposites. *J Appl Polym Sci*, 127(3), 2227–2234. 10.1002/app.37904

[120] Liu P, Yao Z, Zhou J (2015) Controllable synthesis and enhanced microwave absorption properties of silane-modified Ni0.4Zn0.4Co0.2Fe2O4 nanocomposites covered with reduced graphene oxide. *RSC Adv*, 5(114), 93739–93748. 10.1039/c5ra18668d

[121] Chin WS, Lee DG (2007) Development of the composite RAS (radar absorbing structure) for the X-band frequency range. *Compos Struct*, 77(4), 457–465. 10.1016/j.compstruct.2005.07.021

[122] Kwon SK, Ahn JM, Kim GH, et al. (2002) Microwave absorbing properties of carbon black/silicone rubber blend. *Polym Eng Sci*, 42(11), 2165–2171. 10.1002/pen.11106

[123] Cao MS, Song WL, Hou ZL, et al. (2010) The effects of temperature and frequency on the dielectric properties, electromagnetic interference shielding and microwave-absorption of short carbon fiber/silica composites. *Carbon N Y*, 48(3), 788–796. 10.1016/j.carbon.2009.10.028

[124] An YJ, Nishida K, Yamamoto T, et al. (2010) Microwave absorber properties of magnetic and dielectric composite materials. *Electron Commun Japan*, 93(4), 18–26. 10.1002/ecj.10206

[125] Kong L, Yin X, Yuan X, et al. (2014) Electromagnetic wave absorption properties of graphene modified with carbon nanotube/poly(dimethyl siloxane) composites. *Carbon N Y*, 73, 185–193. 10.1016/j.carbon.2014.02.054

[126] Zhao X, Zhang Z, Wang L, et al. (2013) Excellent microwave absorption property of Graphene-coated Fe nanocomposites. *Sci Rep*, 3 (1). 10.1038/srep03421

[127] Fang J, Zha W, Kang M, et al. (2013) Microwave absorption response of nickel/graphene nanocomposites prepared by electrodeposition. *J Mater Sci*, 48(23), 8060–8067. 10.1007/s10853-013-7600-6

[128] Teber A, Unver I, Kavas H, et al. (2016) Knitted radar absorbing materials (RAM) based on nickel-cobalt magnetic materials. *J Magn Magn Mater*, 406, 228–232. 10.1016/j.jmmm.2015.12.056

[129] Chen X, Meng F, Zhou Z, et al. (2014) One-step synthesis of graphene/polyaniline hybrids by in situ intercalation polymerization and their electromagnetic properties. *Nanoscale.* 6(14), 8140–8148. 10.1039/c4nr01738b

[130] Abbas SM, Dixit AK, Chatterjee R, Goel TC (2007) Complex permittivity, complex permeability and microwave absorption properties of ferrite-polymer composites. *J Magn Magn Mater*, 309(1), 20–24. 10.1016/j.jmmm.2006.06.006

[131] Gama AM, Rezende MC, Dantas CC (2011) Dependence of microwave absorption properties on ferrite volume fraction in MnZn ferrite/rubber radar absorbing materials. *J Magn Magn Mater*, 323(22), 2782–2785. 10.1016/j.jmmm.2011.05.052

[132] Yin Y, Zeng M, Liu J, et al. (2016) Enhanced high-frequency absorption of anisotropic Fe3O4/graphene nanocomposites. *Sci Rep*, 6 (1). 10.1038/srep25075

[133] Wang Z, Wu L, Zhou J, et al. (2014) Chemoselectivity-induced multiple interfaces in MWCNT/Fe3O4@ZnO heterotrimers for whole X-band microwave absorption. *Nanoscale.* 6(21), 12298–12302. 10.1039/c4nr03040k

[134] Capaccioli S, Lucchesi M, Rolla PA, Ruggeri G (1998) Dielectric response analysis of a conducting polymer dominated by the hopping charge transport. *J Phys Condens Matter*, 10(25), 5595–5617. 10.1088/0953-8984/10/25/011

[135] Bjorklund RB, Lundström I (1984) Some properties of polypyrrole-paper composites. *J Electron Mater*, 13(1), 211–230. 10.1007/BF02659844

[136] Newman, P., Warren, L., Witucki E (1986) Process for producing electrically conductive composites and composites produced therein. United State Patent: 4,617,228.

[137] Gregory R V, Kimbrell WC, Kuhn HH (1989) Conductive textiles. *Synth Met* 28:823–835.

[138] Angeli A., Pieroni A. (1919) *Gazz Chim Ital* 49 (I):164.

[139] Mohamed ABH, Miane JL, Zangar H (2001) Radiofrequency and microwave (10 kHz–8 GHz) electrical properties of polypyrrole and polypyrrole–poly(methyl methacrylate) composites. *Polym Int* 50:773–777. 10.1002/pi.686

[140] Olmedo L, Hourquebie P, Jousse F (1993) Microwave absorbing materials based on conducting polymers. *Adv Mater*, 5(5), 373–377. 10.1002/adma.19930050509

[141] Olmedo L, Deleuae C, Hourqnebie PFJ (1991) Compos el des Nonvenus MurPriuus I:

[142] Meakins RJ (1961) Mechanisms of dielectric absorption in solids. *Prog Dielectr* 3:151–202.

[143] Truong VV, Turner BD, Muscat RF, Russo MS (1997) Conducting-polymer-based radar-absorbing materials. In: *Smart Materials, Structures, and Integrated Systems*Proceedings of the SPIE, Volume 3241, p. 98–105.

[144] Truong VT, Riddell SZ, Muscat RF (1998) Polypyrrole based microwave absorbers. *J Mater Sci*, 33(20), 4971–4976. 10.1023/A:1004498705776

[145] Ruffoni JM (1992) Foam absorber. US Patent 5151222.

[146] Stubbs HVG, Wickenden BVA, Howell WG, Perry ED (1981) UK Patent GB 2058469 A.

[147] Roberts WP, Scholz LA (1986) Method of forming electrically conductive polymer blends. US Patent 4,604,427.

[148] Wong PTC, Chambers B, Anderson AP, Wright PV (1992) Large area conducting polymer composites and their use in microwave absorbing material. *Electron Lett*, 28(17), 1651. 10.1049/el:19921051

[149] Wong TCP, Chambers B, Anderson AP, Wright PV (1993) Fabrication and evaluation of conducting polymer composites as radar absorbers. In: 1993 Eighth International Conference on Antennas and Propagation, 1993, pp. 934–937 vol.2.

[150] Chambers B., Wong TCP, Anderson AP, Wright PV (1993) Characterisation and modelling of conducting polymer composites and their exploitation in microwave absorbing materials. In: 15th Antenna Measurement Techniques Association Symp.

[151] Wong TCP, Chambers B, Anderson AP, Wright PV (1993) Impedance Characteristics of Conducting Polymer Composites for use in Radar Absorbing Materials. In: Proc 3rd ICEAA Torino.

[152] Wong TCP, Chambers B, Anderson AP, Wright PV (1995) Characterisation of conducting polymer-loaded composite materials at oblique incidence and their application in radar absorbers. In: 1995 Ninth International Conference on Antennas and Propagation, ICAP'95 (Conf. Publ. No. 407). IET, pp 441–444.

[153] Kuhn HH, Child AD, Kimbrell WC (1995) Toward real applications of conductive polymers. *Synth Met* 71:2139–2142.

[154] Adams Jr, Louis W, Gilpatrick MWGRV (1994) Method for generating a conductive fabric and associated product. US Patent 5,292,573.

[155] Pittman, EH, Kuhn HH (1992) Electrically conductive textile fabric having conductivity gradient. US Patent 5,102,727.

[156] Gregory, Richard V, Kimbrell Jr., William C, Cuddihee ME (1992) Electrically conductive polymer material having conductivity gradient. US Patent 5,162,135.

[157] Buckley LJ, Eashoo M (1996) Polypyrrole-coated fibers as microwave and millimeterwave obscurants. *Synth Met*, 78(1), 1–6. 10.1016/0379-6779(95)03561-3

[158] Wang Z, Luo J, Zhao GL (2014) Dielectric and microwave attenuation properties of graphene nanoplatelet-epoxy composites. : Volume 4, Issue 1, AIP Adv 4 10.1063/1.4863687

Index

For Product Safety Concerns and Information please contact our EU
representative GPSR@taylorandfrancis.com
Taylor & Francis Verlag GmbH, Kaufingerstraße 24, 80331 München, Germany

www.ingramcontent.com/pod-product-compliance
Lightning Source LLC
Chambersburg PA
CBHW080137220326
41598CB00032B/5090